# Step-by-Step Programming with
# Base SAS® 9.4
## Second Edition

SAS® Documentation

The correct bibliographic citation for this manual is as follows: SAS Institute Inc. 2016. *Step-by-Step Programming with Base SAS® 9.4, Second Edition*. Cary, NC: SAS Institute Inc.

**Step-by-Step Programming with Base SAS® 9.4, Second Edition**

Copyright © 2016, SAS Institute Inc., Cary, NC, USA

ISBN 978-1-62959-894-9 (Hard copy)
ISBN 978-1-62960-806-8 (PDF)

All Rights Reserved. Produced in the United States of America.

**For a hard-copy book:** No part of this publication may be reproduced, stored in a retrieval system, or transmitted, in any form or by any means, electronic, mechanical, photocopying, or otherwise, without the prior written permission of the publisher, SAS Institute Inc.

**For a web download or e-book:** Your use of this publication shall be governed by the terms established by the vendor at the time you acquire this publication.

The scanning, uploading, and distribution of this book via the Internet or any other means without the permission of the publisher is illegal and punishable by law. Please purchase only authorized electronic editions and do not participate in or encourage electronic piracy of copyrighted materials. Your support of others' rights is appreciated.

**U.S. Government License Rights; Restricted Rights:** The Software and its documentation is commercial computer software developed at private expense and is provided with RESTRICTED RIGHTS to the United States Government. Use, duplication, or disclosure of the Software by the United States Government is subject to the license terms of this Agreement pursuant to, as applicable, FAR 12.212, DFAR 227.7202-1(a), DFAR 227.7202-3(a), and DFAR 227.7202-4, and, to the extent required under U.S. federal law, the minimum restricted rights as set out in FAR 52.227-19 (DEC 2007). If FAR 52.227-19 is applicable, this provision serves as notice under clause (c) thereof and no other notice is required to be affixed to the Software or documentation. The Government's rights in Software and documentation shall be only those set forth in this Agreement.

SAS Institute Inc., SAS Campus Drive, Cary, NC 27513-2414

November 2016

SAS® and all other SAS Institute Inc. product or service names are registered trademarks or trademarks of SAS Institute Inc. in the USA and other countries. ® indicates USA registration.

Other brand and product names are trademarks of their respective companies.

SAS software may be provided with certain third-party software, including but not limited to open-source software, which is licensed under its applicable third-party software license agreement. For license information about third-party software distributed with SAS software, refer to **http://support.sas.com/thirdpartylicenses**.

9.4-P2:basess

# Contents

About This Book . . . . . . . . . . . . . . . . . . . . . . . . . . . . . . . . . . . . . . . . . . . . . . . . . . . . . . *xiii*
What's New in Step-by-Step Programming with Base SAS 9.4 . . . . . . . . . . . . . . . . . . . *xix*
Accessibility Features of Step-by-Step Programming with Base SAS 9.4 . . . . . . . . . . . *xxi*

## PART 1 Introduction to the SAS System 1

### Chapter 1 • What is the SAS System? 3
Introduction to the SAS System . . . . . . . . . . . . . . . . . . . . . . . . . . . . . . . . . . . . . . . . . . . . 3
Components of Base SAS Software . . . . . . . . . . . . . . . . . . . . . . . . . . . . . . . . . . . . . . . . 4
Output Produced by the SAS System . . . . . . . . . . . . . . . . . . . . . . . . . . . . . . . . . . . . . . . 8
Ways to Run SAS Programs . . . . . . . . . . . . . . . . . . . . . . . . . . . . . . . . . . . . . . . . . . . . . 11
Running Programs in the SAS Windowing Environment . . . . . . . . . . . . . . . . . . . . . . . 13
Summary . . . . . . . . . . . . . . . . . . . . . . . . . . . . . . . . . . . . . . . . . . . . . . . . . . . . . . . . . . . . 15
Learning More . . . . . . . . . . . . . . . . . . . . . . . . . . . . . . . . . . . . . . . . . . . . . . . . . . . . . . . . 16

### Chapter 2 • Working with Output Defaults 17
Working with Output Defaults Starting in SAS 9.3 . . . . . . . . . . . . . . . . . . . . . . . . . . . . 17
Learning More . . . . . . . . . . . . . . . . . . . . . . . . . . . . . . . . . . . . . . . . . . . . . . . . . . . . . . . . 23

## PART 2 Getting Your Data into Shape 25

### Chapter 3 • Introduction to DATA Step Processing 27
Introduction to DATA Step Processing . . . . . . . . . . . . . . . . . . . . . . . . . . . . . . . . . . . . . 27
The SAS Data Set: Your Key to the SAS System . . . . . . . . . . . . . . . . . . . . . . . . . . . . . 28
How the DATA Step Works: A Basic Introduction . . . . . . . . . . . . . . . . . . . . . . . . . . . . 33
Supplying Information to Create a SAS Data Set . . . . . . . . . . . . . . . . . . . . . . . . . . . . . 40
Summary . . . . . . . . . . . . . . . . . . . . . . . . . . . . . . . . . . . . . . . . . . . . . . . . . . . . . . . . . . . . 48
Learning More . . . . . . . . . . . . . . . . . . . . . . . . . . . . . . . . . . . . . . . . . . . . . . . . . . . . . . . . 48

### Chapter 4 • Starting with Raw Data: The Basics 51
Introduction to Raw Data . . . . . . . . . . . . . . . . . . . . . . . . . . . . . . . . . . . . . . . . . . . . . . . . 52
Examine the Structure of the Raw Data: Factors to Consider . . . . . . . . . . . . . . . . . . . . 52
Reading Unaligned Data . . . . . . . . . . . . . . . . . . . . . . . . . . . . . . . . . . . . . . . . . . . . . . . . 53
Reading Data That Is Aligned in Columns . . . . . . . . . . . . . . . . . . . . . . . . . . . . . . . . . . 57
Reading Data That Requires Special Instructions . . . . . . . . . . . . . . . . . . . . . . . . . . . . . 60
Reading Unaligned Data with More Flexibility . . . . . . . . . . . . . . . . . . . . . . . . . . . . . . 63
Mixing Styles of Input . . . . . . . . . . . . . . . . . . . . . . . . . . . . . . . . . . . . . . . . . . . . . . . . . . 65
Summary . . . . . . . . . . . . . . . . . . . . . . . . . . . . . . . . . . . . . . . . . . . . . . . . . . . . . . . . . . . . 68
Learning More . . . . . . . . . . . . . . . . . . . . . . . . . . . . . . . . . . . . . . . . . . . . . . . . . . . . . . . . 69

### Chapter 5 • Starting with Raw Data: Beyond the Basics 71
Introduction to Beyond the Basics with Raw Data . . . . . . . . . . . . . . . . . . . . . . . . . . . . 71
Testing a Condition Before Creating an Observation . . . . . . . . . . . . . . . . . . . . . . . . . . 72
Creating Multiple Observations from a Single Record . . . . . . . . . . . . . . . . . . . . . . . . . 73
Reading Multiple Records to Create a Single Observation . . . . . . . . . . . . . . . . . . . . . . 77

Problem Solving: When an Input Record Unexpectedly Does Not
    Have Enough Values ............................................. 84
Summary .................................................................. 87
Learning More ............................................................ 89

## Chapter 6 • Starting with SAS Data Sets ........................................ 91
Introduction to Starting with SAS Data Sets ................................ 91
Understanding the Basics .................................................. 92
Input SAS Data Set for Examples ........................................... 92
Reading Selected Observations ............................................. 95
Reading Selected Variables ................................................ 96
Creating More Than One Data Set in a Single DATA Step ..................... 99
Using the DROP= and KEEP= Data Set Options for Efficiency ................ 101
Summary ................................................................. 102
Learning More ........................................................... 103

## PART 3  Basic Programming  105

## Chapter 7 • Understanding DATA Step Processing ................................ 107
Overview of DATA Step Processing ......................................... 107
Input SAS Data Set for Examples .......................................... 108
Adding Information to a SAS Data Set ..................................... 109
Defining Enough Storage Space for Variables .............................. 114
Conditionally Deleting an Observation .................................... 115
Summary ................................................................. 115
Learning More ........................................................... 116

## Chapter 8 • Working with Numeric Variables ................................... 117
Introduction to Working with Numeric Variables ........................... 117
About Numeric Variables in SAS ........................................... 118
Input SAS Data Set for Examples .......................................... 118
Calculating with Numeric Variables ....................................... 119
Comparing Numeric Variables .............................................. 124
Storing Numeric Variables Efficiently .................................... 126
Summary ................................................................. 127
Learning More ........................................................... 128

## Chapter 9 • Working with Character Variables ................................. 129
Introduction to Working with Character Variables ......................... 129
Input SAS Data Set for Examples .......................................... 130
Identifying Character Variables and Expressing Character Values .......... 131
Setting the Length of Character Variables ................................ 132
Handling Missing Values .................................................. 134
Creating New Character Values ............................................ 137
Saving Storage Space by Treating Numbers as Characters ................... 142
Summary ................................................................. 143
Learning More ........................................................... 144

## Chapter 10 • Acting on Selected Observations ................................. 147
Introduction to Acting on Selected Observations .......................... 147
Input SAS Data Set for Examples .......................................... 148
Selecting Observations ................................................... 149
Constructing Conditions .................................................. 154
Comparing Characters .................................................... 161

## Contents    v

    Summary . . . . . . . . . . . . . . . . . . . . . . . . . . . . . . . . . . . . . . . . . . . . . . . . . . . . . . . . . . . 165
    Learning More . . . . . . . . . . . . . . . . . . . . . . . . . . . . . . . . . . . . . . . . . . . . . . . . . . . 166

### Chapter 11 • Creating Subsets of Observations . . . . . . . . . . . . . . . . . . . . . . . . . . . . . . . . . 169
    Introduction to Creating Subsets of Observations . . . . . . . . . . . . . . . . . . . . . . . 169
    Input SAS Data Set for Examples . . . . . . . . . . . . . . . . . . . . . . . . . . . . . . . . . . . 170
    Selecting Observations for a New SAS Data Set . . . . . . . . . . . . . . . . . . . . . . . 171
    Conditionally Writing Observations to One or More SAS Data Sets . . . . . . . . . 175
    Summary . . . . . . . . . . . . . . . . . . . . . . . . . . . . . . . . . . . . . . . . . . . . . . . . . . . . 180
    Learning More . . . . . . . . . . . . . . . . . . . . . . . . . . . . . . . . . . . . . . . . . . . . . . . . 181

### Chapter 12 • Working with Grouped or Sorted Observations . . . . . . . . . . . . . . . . . . . . . . . 183
    Introduction to Working with Grouped or Sorted Observations . . . . . . . . . . . . . 183
    Input SAS Data Set for Examples . . . . . . . . . . . . . . . . . . . . . . . . . . . . . . . . . . . 184
    Working with Grouped Data . . . . . . . . . . . . . . . . . . . . . . . . . . . . . . . . . . . . . . . 185
    Working with Sorted Data . . . . . . . . . . . . . . . . . . . . . . . . . . . . . . . . . . . . . . . . 192
    Summary . . . . . . . . . . . . . . . . . . . . . . . . . . . . . . . . . . . . . . . . . . . . . . . . . . . . 197
    Learning More . . . . . . . . . . . . . . . . . . . . . . . . . . . . . . . . . . . . . . . . . . . . . . . . 197

### Chapter 13 • Using More Than One Observation in a Calculation . . . . . . . . . . . . . . . . . . . 199
    Introduction to Using More Than One Observation in a Calculation . . . . . . . . . 199
    Input File and SAS Data Set for Examples . . . . . . . . . . . . . . . . . . . . . . . . . . . . 200
    Accumulating a Total for an Entire Data Set . . . . . . . . . . . . . . . . . . . . . . . . . . . 201
    Obtaining a Total for Each BY Group . . . . . . . . . . . . . . . . . . . . . . . . . . . . . . . . 204
    Writing to Separate Data Sets . . . . . . . . . . . . . . . . . . . . . . . . . . . . . . . . . . . . . 206
    Using a Value in a Later Observation . . . . . . . . . . . . . . . . . . . . . . . . . . . . . . . . 209
    Summary . . . . . . . . . . . . . . . . . . . . . . . . . . . . . . . . . . . . . . . . . . . . . . . . . . . . 212
    Learning More . . . . . . . . . . . . . . . . . . . . . . . . . . . . . . . . . . . . . . . . . . . . . . . . 213

### Chapter 14 • Finding Shortcuts in Programming . . . . . . . . . . . . . . . . . . . . . . . . . . . . . . . . . 215
    Introduction to Shortcuts . . . . . . . . . . . . . . . . . . . . . . . . . . . . . . . . . . . . . . . . . 215
    Input File and SAS Data Set . . . . . . . . . . . . . . . . . . . . . . . . . . . . . . . . . . . . . . . 216
    Performing More Than One Action in an IF-THEN Statement . . . . . . . . . . . . . 217
    Performing the Same Action for a Series of Variables . . . . . . . . . . . . . . . . . . . 219
    Summary . . . . . . . . . . . . . . . . . . . . . . . . . . . . . . . . . . . . . . . . . . . . . . . . . . . . 222
    Learning More . . . . . . . . . . . . . . . . . . . . . . . . . . . . . . . . . . . . . . . . . . . . . . . . 223

### Chapter 15 • Working with Dates in the SAS System . . . . . . . . . . . . . . . . . . . . . . . . . . . . . . 225
    Introduction to Working with Dates . . . . . . . . . . . . . . . . . . . . . . . . . . . . . . . . . 226
    Understanding How SAS Handles Dates . . . . . . . . . . . . . . . . . . . . . . . . . . . . . 226
    Input File and SAS Data Set for Examples . . . . . . . . . . . . . . . . . . . . . . . . . . . . 228
    Entering Dates . . . . . . . . . . . . . . . . . . . . . . . . . . . . . . . . . . . . . . . . . . . . . . . . . 228
    Displaying Dates . . . . . . . . . . . . . . . . . . . . . . . . . . . . . . . . . . . . . . . . . . . . . . . 233
    Using Dates in Calculations . . . . . . . . . . . . . . . . . . . . . . . . . . . . . . . . . . . . . . . 237
    Using SAS Date Functions . . . . . . . . . . . . . . . . . . . . . . . . . . . . . . . . . . . . . . . 239
    Comparing Durations and SAS Date Values . . . . . . . . . . . . . . . . . . . . . . . . . . . 241
    Summary . . . . . . . . . . . . . . . . . . . . . . . . . . . . . . . . . . . . . . . . . . . . . . . . . . . . 243
    Learning More . . . . . . . . . . . . . . . . . . . . . . . . . . . . . . . . . . . . . . . . . . . . . . . . 244

## PART 4    Combining SAS Data Sets    247

### Chapter 16 • Methods of Combining SAS Data Sets . . . . . . . . . . . . . . . . . . . . . . . . . . . . . . . 249
    Introduction to Combining SAS Data Sets . . . . . . . . . . . . . . . . . . . . . . . . . . . . 249
    Definition of Concatenating . . . . . . . . . . . . . . . . . . . . . . . . . . . . . . . . . . . . . . . 250

Definition of Interleaving . . . . . . . . . . . . . . . . . . . . . . . . . . . . . . . . . . . . . . . . . . . . . 250
Definition of Merging . . . . . . . . . . . . . . . . . . . . . . . . . . . . . . . . . . . . . . . . . . . . . . . 251
Definition of Updating . . . . . . . . . . . . . . . . . . . . . . . . . . . . . . . . . . . . . . . . . . . . . . 252
Definition of Modifying . . . . . . . . . . . . . . . . . . . . . . . . . . . . . . . . . . . . . . . . . . . . . 253
Comparing Modifying, Merging, and Updating Data Sets . . . . . . . . . . . . . . . . . . . . 254
Learning More . . . . . . . . . . . . . . . . . . . . . . . . . . . . . . . . . . . . . . . . . . . . . . . . . . . . 255

## Chapter 17 • Concatenating SAS Data Sets . . . . . . . . . . . . . . . . . . . . . . . . . . . . . . . . . . . . . 257
Introduction to Concatenating SAS Data Sets . . . . . . . . . . . . . . . . . . . . . . . . . . . . . 257
Concatenating Data Sets with the SET Statement . . . . . . . . . . . . . . . . . . . . . . . . . . 258
Concatenating Data Sets By Using the APPEND Procedure . . . . . . . . . . . . . . . . . . 274
Choosing between the SET Statement and the APPEND Procedure . . . . . . . . . . . . 278
Summary . . . . . . . . . . . . . . . . . . . . . . . . . . . . . . . . . . . . . . . . . . . . . . . . . . . . . . . . . 279
Learning More . . . . . . . . . . . . . . . . . . . . . . . . . . . . . . . . . . . . . . . . . . . . . . . . . . . . 280

## Chapter 18 • Interleaving SAS Data Sets . . . . . . . . . . . . . . . . . . . . . . . . . . . . . . . . . . . . . . . 281
Introduction to Interleaving SAS Data Sets . . . . . . . . . . . . . . . . . . . . . . . . . . . . . . . 281
Understanding BY-Group Processing Concepts . . . . . . . . . . . . . . . . . . . . . . . . . . . 282
Interleaving Data Sets . . . . . . . . . . . . . . . . . . . . . . . . . . . . . . . . . . . . . . . . . . . . . . . 282
Summary . . . . . . . . . . . . . . . . . . . . . . . . . . . . . . . . . . . . . . . . . . . . . . . . . . . . . . . . . 286
Learning More . . . . . . . . . . . . . . . . . . . . . . . . . . . . . . . . . . . . . . . . . . . . . . . . . . . . 287

## Chapter 19 • Merging SAS Data Sets . . . . . . . . . . . . . . . . . . . . . . . . . . . . . . . . . . . . . . . . . . 289
Introduction to Merging SAS Data Sets . . . . . . . . . . . . . . . . . . . . . . . . . . . . . . . . . 289
Understanding the MERGE Statement . . . . . . . . . . . . . . . . . . . . . . . . . . . . . . . . . . 290
One-to-One Merging . . . . . . . . . . . . . . . . . . . . . . . . . . . . . . . . . . . . . . . . . . . . . . . . 290
Match-Merging . . . . . . . . . . . . . . . . . . . . . . . . . . . . . . . . . . . . . . . . . . . . . . . . . . . . 296
Choosing between One-to-One Merging and Match-Merging . . . . . . . . . . . . . . . . . 310
Summary . . . . . . . . . . . . . . . . . . . . . . . . . . . . . . . . . . . . . . . . . . . . . . . . . . . . . . . . . 315
Learning More . . . . . . . . . . . . . . . . . . . . . . . . . . . . . . . . . . . . . . . . . . . . . . . . . . . . 315

## Chapter 20 • Updating SAS Data Sets . . . . . . . . . . . . . . . . . . . . . . . . . . . . . . . . . . . . . . . . . 317
Introduction to Updating SAS Data Sets . . . . . . . . . . . . . . . . . . . . . . . . . . . . . . . . . 317
Understanding the UPDATE Statement . . . . . . . . . . . . . . . . . . . . . . . . . . . . . . . . . . 318
Understanding How to Select BY Variables . . . . . . . . . . . . . . . . . . . . . . . . . . . . . . 318
Updating a Data Set . . . . . . . . . . . . . . . . . . . . . . . . . . . . . . . . . . . . . . . . . . . . . . . . . 319
Updating with Incremental Values . . . . . . . . . . . . . . . . . . . . . . . . . . . . . . . . . . . . . 324
Understanding the Differences between Updating and Merging . . . . . . . . . . . . . . . 326
Handling Missing Values . . . . . . . . . . . . . . . . . . . . . . . . . . . . . . . . . . . . . . . . . . . . 329
Summary . . . . . . . . . . . . . . . . . . . . . . . . . . . . . . . . . . . . . . . . . . . . . . . . . . . . . . . . . 333
Learning More . . . . . . . . . . . . . . . . . . . . . . . . . . . . . . . . . . . . . . . . . . . . . . . . . . . . 333

## Chapter 21 • Modifying SAS Data Sets . . . . . . . . . . . . . . . . . . . . . . . . . . . . . . . . . . . . . . . . 335
Introduction to Modifying SAS Data Sets . . . . . . . . . . . . . . . . . . . . . . . . . . . . . . . . 335
Input SAS Data Set for Examples . . . . . . . . . . . . . . . . . . . . . . . . . . . . . . . . . . . . . . 336
Modifying a SAS Data Set: The Simplest Case . . . . . . . . . . . . . . . . . . . . . . . . . . . . 337
Modifying a Master Data Set with Observations from a Transaction Data Set . . . . . 338
Understanding How Duplicate BY Variables Affect File Update . . . . . . . . . . . . . . 343
Handling Missing Values . . . . . . . . . . . . . . . . . . . . . . . . . . . . . . . . . . . . . . . . . . . . 345
Summary . . . . . . . . . . . . . . . . . . . . . . . . . . . . . . . . . . . . . . . . . . . . . . . . . . . . . . . . . 347
Learning More . . . . . . . . . . . . . . . . . . . . . . . . . . . . . . . . . . . . . . . . . . . . . . . . . . . . 348

## Chapter 22 • Conditionally Processing Observations from Multiple SAS Data Sets . . . . . . 349
Introduction to Conditional Processing from Multiple SAS Data Sets . . . . . . . . . . . 349
Input SAS Data Sets for Examples . . . . . . . . . . . . . . . . . . . . . . . . . . . . . . . . . . . . . 350
Determining Which Data Set Contributed the Observation . . . . . . . . . . . . . . . . . . . 353

Combining Selected Observations from Multiple Data Sets . . . . . . . . . . . . . . . . . . . 358
Performing a Calculation Based on the Last Observation . . . . . . . . . . . . . . . . . . . . 359
Summary . . . . . . . . . . . . . . . . . . . . . . . . . . . . . . . . . . . . . . . . . . . . . . . . . . . . . . . . . . 361
Learning More . . . . . . . . . . . . . . . . . . . . . . . . . . . . . . . . . . . . . . . . . . . . . . . . . . . . . 362

## PART 5  Debugging SAS Programs  363

### Chapter 23 • Analyzing Your SAS Session with the SAS Log . . . . . . . . . . . . . . . . . . . . . . . . . . 365
Introduction to Analyzing Your SAS Session with the SAS Log . . . . . . . . . . . . . . . . 366
Understanding the SAS Log . . . . . . . . . . . . . . . . . . . . . . . . . . . . . . . . . . . . . . . . . . . 366
Locating the SAS Log . . . . . . . . . . . . . . . . . . . . . . . . . . . . . . . . . . . . . . . . . . . . . . . 369
Understanding the Log Structure . . . . . . . . . . . . . . . . . . . . . . . . . . . . . . . . . . . . . . . 369
Writing to the SAS Log . . . . . . . . . . . . . . . . . . . . . . . . . . . . . . . . . . . . . . . . . . . . . . 375
Suppressing Information in the SAS Log . . . . . . . . . . . . . . . . . . . . . . . . . . . . . . . . . 382
Changing the Appearance of the Log . . . . . . . . . . . . . . . . . . . . . . . . . . . . . . . . . . . . 387
Summary . . . . . . . . . . . . . . . . . . . . . . . . . . . . . . . . . . . . . . . . . . . . . . . . . . . . . . . . . . 387
Learning More . . . . . . . . . . . . . . . . . . . . . . . . . . . . . . . . . . . . . . . . . . . . . . . . . . . . . 388

### Chapter 24 • Directing SAS Output and the SAS Log . . . . . . . . . . . . . . . . . . . . . . . . . . . . . . . . 391
Introduction to Directing SAS Output and the SAS Log . . . . . . . . . . . . . . . . . . . . . 391
Input File and SAS Data Set for Examples . . . . . . . . . . . . . . . . . . . . . . . . . . . . . . . 392
Routing the Output and the SAS Log with PROC PRINTTO . . . . . . . . . . . . . . . . . 393
Storing the Output and the SAS Log in the SAS Windowing Environment . . . . . . . 395
Redefining the Default Destination in a Batch or Noninteractive Environment . . . . 396
Summary . . . . . . . . . . . . . . . . . . . . . . . . . . . . . . . . . . . . . . . . . . . . . . . . . . . . . . . . . . 397
Learning More . . . . . . . . . . . . . . . . . . . . . . . . . . . . . . . . . . . . . . . . . . . . . . . . . . . . . 398

### Chapter 25 • Diagnosing and Avoiding Errors . . . . . . . . . . . . . . . . . . . . . . . . . . . . . . . . . . . . . . 399
Introduction to Diagnosing and Avoiding Errors . . . . . . . . . . . . . . . . . . . . . . . . . . . 399
Understanding How the SAS Supervisor Checks a Job . . . . . . . . . . . . . . . . . . . . . . 400
Understanding How SAS Processes Errors . . . . . . . . . . . . . . . . . . . . . . . . . . . . . . . 400
Distinguishing Types of Errors . . . . . . . . . . . . . . . . . . . . . . . . . . . . . . . . . . . . . . . . . 401
Diagnosing Errors . . . . . . . . . . . . . . . . . . . . . . . . . . . . . . . . . . . . . . . . . . . . . . . . . . . 402
Using a Quality Control Checklist . . . . . . . . . . . . . . . . . . . . . . . . . . . . . . . . . . . . . . 412
Learning More . . . . . . . . . . . . . . . . . . . . . . . . . . . . . . . . . . . . . . . . . . . . . . . . . . . . . 412

### Chapter 26 • Finding Logic Errors in Your Program . . . . . . . . . . . . . . . . . . . . . . . . . . . . . . . . . . 415
Finding Logic Errors in Your Program . . . . . . . . . . . . . . . . . . . . . . . . . . . . . . . . . . . 415
Using the DATA Step Debugger . . . . . . . . . . . . . . . . . . . . . . . . . . . . . . . . . . . . . . . . 416
Basic Usage . . . . . . . . . . . . . . . . . . . . . . . . . . . . . . . . . . . . . . . . . . . . . . . . . . . . . . . . 416
Using the Macro Facility with the Debugger . . . . . . . . . . . . . . . . . . . . . . . . . . . . . . 418
Examples . . . . . . . . . . . . . . . . . . . . . . . . . . . . . . . . . . . . . . . . . . . . . . . . . . . . . . . . . . 419

## PART 6  Producing Reports  431

### Chapter 27 • Producing Detail Reports with the PRINT Procedure . . . . . . . . . . . . . . . . . . . . . 433
Introduction to Producing Reports with the PRINT Procedure . . . . . . . . . . . . . . . . 434
Input File and SAS Data Sets for Examples . . . . . . . . . . . . . . . . . . . . . . . . . . . . . . . 434
Creating Simple Reports . . . . . . . . . . . . . . . . . . . . . . . . . . . . . . . . . . . . . . . . . . . . . . 436
Creating Enhanced Reports . . . . . . . . . . . . . . . . . . . . . . . . . . . . . . . . . . . . . . . . . . . . 446
Creating Customized Reports . . . . . . . . . . . . . . . . . . . . . . . . . . . . . . . . . . . . . . . . . . 458

## viii Contents

Making Your Reports Easy to Change . . . . . . . . . . . . . . . . . . . . . . . . . . . . . . . . . . 465
Summary . . . . . . . . . . . . . . . . . . . . . . . . . . . . . . . . . . . . . . . . . . . . . . . . . . . . . . . . 469
Learning More . . . . . . . . . . . . . . . . . . . . . . . . . . . . . . . . . . . . . . . . . . . . . . . . . . . 472

### Chapter 28 • Creating Summary Tables with the TABULATE Procedure . . . . . . . . . . . . . . . . . 473
Introduction to Creating Summary Tables with the TABULATE Procedure . . . . . . . 474
Understanding Summary Table Design . . . . . . . . . . . . . . . . . . . . . . . . . . . . . . . . 474
Understanding the Basics of the TABULATE Procedure . . . . . . . . . . . . . . . . . . . . 476
Input File and SAS Data Set for Examples . . . . . . . . . . . . . . . . . . . . . . . . . . . . . . 479
Creating Simple Summary Tables . . . . . . . . . . . . . . . . . . . . . . . . . . . . . . . . . . . . 480
Creating More Sophisticated Summary Tables . . . . . . . . . . . . . . . . . . . . . . . . . . 485
Summary . . . . . . . . . . . . . . . . . . . . . . . . . . . . . . . . . . . . . . . . . . . . . . . . . . . . . . . . 496
Learning More . . . . . . . . . . . . . . . . . . . . . . . . . . . . . . . . . . . . . . . . . . . . . . . . . . . 499

### Chapter 29 • Creating Detail and Summary Reports with the REPORT Procedure . . . . . . . . . 501
Introduction to Creating Detail and Summary Reports with the REPORT Procedure . 501
Understanding How to Construct a Report . . . . . . . . . . . . . . . . . . . . . . . . . . . . . . 502
Input File and SAS Data Set for Examples . . . . . . . . . . . . . . . . . . . . . . . . . . . . . . 504
Creating Simple Reports . . . . . . . . . . . . . . . . . . . . . . . . . . . . . . . . . . . . . . . . . . . . 505
Creating More Sophisticated Reports . . . . . . . . . . . . . . . . . . . . . . . . . . . . . . . . . . 514
Summary . . . . . . . . . . . . . . . . . . . . . . . . . . . . . . . . . . . . . . . . . . . . . . . . . . . . . . . . 523
Learning More . . . . . . . . . . . . . . . . . . . . . . . . . . . . . . . . . . . . . . . . . . . . . . . . . . . 527

## PART 7  Producing Plots and Charts  529

### Chapter 30 • Plotting the Relationship between Variables . . . . . . . . . . . . . . . . . . . . . . . . . . . . . 531
Introduction to Plotting the Relationship between Variables . . . . . . . . . . . . . . . . . 531
Input File and SAS Data Set for Examples . . . . . . . . . . . . . . . . . . . . . . . . . . . . . . 532
Plotting One Set of Variables . . . . . . . . . . . . . . . . . . . . . . . . . . . . . . . . . . . . . . . . 535
Enhancing the Plot . . . . . . . . . . . . . . . . . . . . . . . . . . . . . . . . . . . . . . . . . . . . . . . . 537
Plotting Multiple Sets of Variables . . . . . . . . . . . . . . . . . . . . . . . . . . . . . . . . . . . . 541
Summary . . . . . . . . . . . . . . . . . . . . . . . . . . . . . . . . . . . . . . . . . . . . . . . . . . . . . . . . 547
Learning More . . . . . . . . . . . . . . . . . . . . . . . . . . . . . . . . . . . . . . . . . . . . . . . . . . . 549

### Chapter 31 • Producing Charts to Summarize Variables . . . . . . . . . . . . . . . . . . . . . . . . . . . . . . . 551
Introduction to Producing Charts to Summarize Variables . . . . . . . . . . . . . . . . . . 552
Understanding the Charting Tools . . . . . . . . . . . . . . . . . . . . . . . . . . . . . . . . . . . . 552
Input File and SAS Data Set for Examples . . . . . . . . . . . . . . . . . . . . . . . . . . . . . . 553
Charting Frequencies with the CHART Procedure . . . . . . . . . . . . . . . . . . . . . . . . 555
Customizing Frequency Charts . . . . . . . . . . . . . . . . . . . . . . . . . . . . . . . . . . . . . . . 563
Creating High-Resolution Histograms . . . . . . . . . . . . . . . . . . . . . . . . . . . . . . . . . 574
Summary . . . . . . . . . . . . . . . . . . . . . . . . . . . . . . . . . . . . . . . . . . . . . . . . . . . . . . . . 586
Learning More . . . . . . . . . . . . . . . . . . . . . . . . . . . . . . . . . . . . . . . . . . . . . . . . . . . 590

## PART 8  Designing Your Own Output  593

### Chapter 32 • Writing Lines to the SAS Log or to an Output File . . . . . . . . . . . . . . . . . . . . . . . . . 595
Introduction to Writing Lines to the SAS Log or to an Output File . . . . . . . . . . . . 595
Understanding the PUT Statement . . . . . . . . . . . . . . . . . . . . . . . . . . . . . . . . . . . . 596
Writing Output without Creating a Data Set . . . . . . . . . . . . . . . . . . . . . . . . . . . . . 596
Writing Simple Text . . . . . . . . . . . . . . . . . . . . . . . . . . . . . . . . . . . . . . . . . . . . . . . 597

Writing a Report .................................................. 603
Summary ........................................................ 610
Learning More ................................................... 611

## Chapter 33 • Understanding and Customizing SAS Output: The Basics ............... 613
Introduction to the Basics of Understanding and Customizing SAS Output ........ 614
Understanding Output .............................................. 614
Input SAS Data Set for Examples .................................... 616
Locating Procedure Output .......................................... 617
Making Output Informative ......................................... 618
Controlling Output Appearance of Listing Output ........................ 624
Controlling the Appearance of Pages .................................. 627
Representing Missing Values ........................................ 638
Summary ........................................................ 641
Learning More ................................................... 642

## Chapter 34 • Understanding and Customizing SAS Output: The Output Delivery System (ODS) ................................................ 643
Introduction to Customizing SAS Output By Using the Output Delivery System .... 644
Input Data Set for Examples ......................................... 644
Understanding ODS Output Formats and Destinations .................... 645
Selecting an Output Format .......................................... 647
Creating Formatted Output .......................................... 648
Selecting the Output That You Want to Format .......................... 661
Customizing ODS Output ........................................... 667
Storing Links to ODS Output ........................................ 678
Summary ........................................................ 680
Learning More ................................................... 683

# PART 9  Storing and Managing Data in SAS Files  685

## Chapter 35 • Understanding SAS Libraries ............................... 687
Introduction to Understanding SAS Libraries ........................... 687
What Is a SAS Library? ............................................. 688
Accessing a SAS Library ............................................ 688
Storing Files in a SAS Library ........................................ 690
Referencing SAS Data Sets in a SAS Library ............................ 691
Summary ........................................................ 693
Learning More ................................................... 694

## Chapter 36 • Managing SAS Libraries .................................. 695
Introduction to Managing SAS Libraries ................................ 695
Choosing Your Tools ............................................... 695
Understanding the DATASETS Procedure .............................. 696
Looking at a PROC DATASETS Session ................................ 697
Summary ........................................................ 698
Learning More ................................................... 698

## Chapter 37 • Getting Information about Your SAS Data Sets .................. 701
Introduction to Getting Information about Your SAS Data Sets ............... 701
Input Data Library for Examples ..................................... 702
Requesting a Directory Listing for a SAS Library ........................ 702
Requesting Contents Information about SAS Data Sets ................... 704
Requesting Contents Information in Different Formats .................... 708

## Chapter 38 • Modifying SAS Data Set Names and Variable Attributes ... 713

Summary ... 710
Learning More ... 710

Introduction to Modifying SAS Data Set Names and Variable Attributes ... 713
Input Data Library for Examples ... 714
Renaming SAS Data Sets ... 714
Modifying Variable Attributes ... 716
Summary ... 723
Learning More ... 724

## Chapter 39 • Copying, Moving, and Deleting SAS Data Sets ... 725

Introduction to Copying, Moving, and Deleting SAS Data Sets ... 725
Input Data Libraries for Examples ... 726
Copying SAS Data Sets ... 727
Copying Specific SAS Data Sets ... 730
Moving SAS Libraries and SAS Data Sets ... 731
Deleting SAS Data Sets ... 734
Deleting All Files in a SAS Library ... 735
Summary ... 736
Learning More ... 737

# PART 10 Understanding Your SAS Environment 739

## Chapter 40 • Introducing the SAS Environment ... 741

Introduction to the SAS Environment ... 741
Starting a SAS Session ... 742
Selecting a SAS Processing Mode ... 743
Summary ... 749
Learning More ... 751

## Chapter 41 • Using the SAS Windowing Environment ... 753

Introduction to Using the SAS Windowing Environment ... 754
Getting Organized ... 755
Finding Online Help ... 758
Using SAS Windowing Environment Command Types ... 758
Working with SAS Windows ... 761
Working with Text ... 766
Working with Files ... 770
Working with SAS Programs ... 775
Working with Output ... 781
Summary ... 790
Learning More ... 792

## Chapter 42 • Customizing the SAS Environment ... 793

Introduction to Customizing the SAS Environment ... 794
Customizing Your Current Session ... 795
Customizing Session-to-Session Settings ... 798
Customizing the SAS Windowing Environment ... 802
Summary ... 807
Learning More ... 809

## PART 11  Appendix  811

### Appendix 1 • Complete DATA Steps for Selected Examples .......................... 813
Complete DATA Steps for Selected Examples ............................. 813
The CITY Data Set ................................................. 814
The UNIVERSITY_TEST_SCORES Data Set ............................ 815
The YEAR_SALES Data Set .......................................... 816
The HIGHLOW Data Set ............................................. 817
The GRADES Data Set .............................................. 818
The USCLIM Data Sets ............................................. 819
The CLIMATE, PRECIP, and STORM Data Sets ........................ 820

### Appendix 2 • DATA Step Debugger Commands ..................................... 823
Dictionary ........................................................ 823

### *Glossary* .................................................. 839
### *Index* ...................................................... 859

# About This Book

## Syntax Conventions for the SAS Language

### Overview of Syntax Conventions for the SAS Language

SAS uses standard conventions in the documentation of syntax for SAS language elements. These conventions enable you to easily identify the components of SAS syntax. The conventions can be divided into these parts:

- syntax components
- style conventions
- special characters
- references to SAS libraries and external files

### Syntax Components

The components of the syntax for most language elements include a keyword and arguments. For some language elements, only a keyword is necessary. For other language elements, the keyword is followed by an equal sign (=). The syntax for arguments has multiple forms in order to demonstrate the syntax of multiple arguments, with and without punctuation.

keyword
> specifies the name of the SAS language element that you use when you write your program. Keyword is a literal that is usually the first word in the syntax. In a CALL routine, the first two words are keywords.

In these examples of SAS syntax, the keywords are bold:

**CHAR** (*string, position*)
**CALL RANBIN** (*seed, n, p, x*);
**ALTER** (*alter-password*)
**BEST** *w*.
**REMOVE** <*data-set-name*>

In this example, the first two words of the CALL routine are the keywords:

**CALL RANBIN**(*seed, n, p, x*)

The syntax of some SAS statements consists of a single keyword without arguments:

**DO**;

> *... SAS code ...*
> **END;**

Some system options require that one of two keyword values be specified:

**DUPLEX | NODUPLEX**

Some procedure statements have multiple keywords throughout the statement syntax:

> **CREATE** <UNIQUE> **INDEX** *index-name* **ON** *table-name* (*column-1* <, *column-2*, ...>)

*argument*
> specifies a numeric or character constant, variable, or expression. Arguments follow the keyword or an equal sign after the keyword. The arguments are used by SAS to process the language element. Arguments can be required or optional. In the syntax, optional arguments are enclosed in angle brackets ( < > ).
>
> In this example, *string* and *position* follow the keyword CHAR. These arguments are required arguments for the CHAR function:
>
> **CHAR** (*string, position*)
>
> Each argument has a value. In this example of SAS code, the argument *string* has a value of 'summer', and the argument *position* has a value of 4:
>
> ```
> x=char('summer', 4);
> ```
>
> In this example, *string* and *substring* are required arguments, whereas *modifiers* and *startpos* are optional.
>
> **FIND**(*string, substring* <, *modifiers*> <, *startpos*>

*argument(s)*
> specifies that one argument is required and that multiple arguments are allowed. Separate arguments with a space. Punctuation, such as a comma ( , ) is not required between arguments.
>
> The MISSING statement is an example of this form of multiple arguments:
>
> **MISSING** *character(s)*;

<LITERAL_ARGUMENT>*argument-1*<<LITERAL_ARGUMENT>*argument-2* ... >
> specifies that one argument is required and that a literal argument can be associated with the argument. You can specify multiple literals and argument pairs. No punctuation is required between the literal and argument pairs. The ellipsis (...) indicates that additional literals and arguments are allowed.
>
> The BY statement is an example of this argument:
>
> **BY** <DESCENDING> *variable-1* <<DESCENDING> *variable-2* ...>;

*argument-1* <*option(s)*> <*argument-2* <*option(s)*> ...>
> specifies that one argument is required and that one or more options can be associated with the argument. You can specify multiple arguments and associated options. No punctuation is required between the argument and the option. The ellipsis (...) indicates that additional arguments with an associated option are allowed.
>
> The FORMAT procedure PICTURE statement is an example of this form of multiple arguments:
>
> **PICTURE** *name* <(*format-option(s)*)>
> <*value-range-set-1* <(*picture-1-option(s)*)>
> <*value-range-set-2* <(*picture-2-option(s)*)> ...>>;

*argument-1=value-1 <argument-2=value-2 ...>*
> specifies that the argument must be assigned a value and that you can specify multiple arguments. The ellipsis (...) indicates that additional arguments are allowed. No punctuation is required between arguments.
>
> The LABEL statement is an example of this form of multiple arguments:
>
> **LABEL** *variable-1=label-1 <variable-2=label-2 ...>*;

*argument-1 <, argument-2, ...>*
> specifies that one argument is required and that you can specify multiple arguments that are separated by a comma or other punctuation. The ellipsis (...) indicates a continuation of the arguments, separated by a comma. Both forms are used in the SAS documentation.
>
> Here are examples of this form of multiple arguments:
>
> **AUTHPROVIDERDOMAIN** (*provider-1:domain-1 <, provider-2:domain-2, ...>*
>
> **INTO** :*macro-variable-specification-1 <, :macro-variable-specification-2, ...>*

*Note:* In most cases, example code in SAS documentation is written in lowercase with a monospace font. You can use uppercase, lowercase, or mixed case in the code that you write.

## *Style Conventions*

The style conventions that are used in documenting SAS syntax include uppercase bold, uppercase, and italic:

**UPPERCASE BOLD**
> identifies SAS keywords such as the names of functions or statements. In this example, the keyword ERROR is written in uppercase bold:
>
> **ERROR** *<message>*;

UPPERCASE
> identifies arguments that are literals.
>
> In this example of the CMPMODEL= system option, the literals include BOTH, CATALOG, and XML:
>
> **CMPMODEL**=BOTH | CATALOG | XML |

*italic*
> identifies arguments or values that you supply. Items in italic represent user-supplied values that are either one of the following:
>
> - nonliteral arguments. In this example of the LINK statement, the argument *label* is a user-supplied value and therefore appears in italic:
>
>   **LINK** *label*;
>
> - nonliteral values that are assigned to an argument.
>
>   In this example of the FORMAT statement, the argument DEFAULT is assigned the variable *default-format*:
>
>   **FORMAT** *variable(s)* *<format>* *<*DEFAULT = *default-format>*;

## *Special Characters*

The syntax of SAS language elements can contain the following special characters:

= 
: an equal sign identifies a value for a literal in some language elements such as system options.

In this example of the MAPS system option, the equal sign sets the value of MAPS:

**MAPS**=*location-of-maps*

< >
: angle brackets identify optional arguments. A required argument is not enclosed in angle brackets.

In this example of the CAT function, at least one item is required:

**CAT** (*item-1* <, *item-2*, ...>)

|
: a vertical bar indicates that you can choose one value from a group of values. Values that are separated by the vertical bar are mutually exclusive.

In this example of the CMPMODEL= system option, you can choose only one of the arguments:

**CMPMODEL**=BOTH | CATALOG | XML

...
: an ellipsis indicates that the argument can be repeated. If an argument and the ellipsis are enclosed in angle brackets, then the argument is optional. The repeated argument must contain punctuation if it appears before or after the argument.

In this example of the CAT function, multiple *item* arguments are allowed, and they must be separated by a comma:

**CAT** (*item-1* <, *item-2*, ...>)

'*value*' or "*value*"
: indicates that an argument that is enclosed in single or double quotation marks must have a value that is also enclosed in single or double quotation marks.

In this example of the FOOTNOTE statement, the argument *text* is enclosed in quotation marks:

**FOOTNOTE** <*n*> <*ods-format-options* '*text*' | "*text*">;

;
: a semicolon indicates the end of a statement or CALL routine.

In this example, each statement ends with a semicolon:

```
data namegame;
   length color name $8;
   color = 'black';
   name = 'jack';
   game = trim(color) || name;
run;
```

## References to SAS Libraries and External Files

Many SAS statements and other language elements refer to SAS libraries and external files. You can choose whether to make the reference through a logical name (a libref or fileref) or use the physical filename enclosed in quotation marks. If you use a logical name, you typically have a choice of using a SAS statement (LIBNAME or FILENAME) or the operating environment's control language to make the reference.

Several methods of referring to SAS libraries and external files are available, and some of these methods depend on your operating environment.

In the examples that use external files, SAS documentation uses the italicized phrase *file-specification*. In the examples that use SAS libraries, SAS documentation uses the italicized phrase *SAS-library* enclosed in quotation marks:

```
infile file-specification obs = 100;
libname libref 'SAS-library';
```

# What's New in Step-by-Step Programming with Base SAS 9.4

## Overview

Step-by-Step Programming with Base SAS 9.4 shows you how to create SAS programs step by step. You are provided with conceptual information and examples that illustrate the SAS concepts. You can execute the programs in this document and view the results. This document contains the basic information that you need to begin writing and debugging your SAS code.

The following enhancements have been made to the documentation:

- additional information about debugging SAS programs
- new method of concatenating SAS variables
- updated sections on Output Delivery System (ODS)

In the third maintenance release for SAS 9.4, the following enhancements have been made to the documentation:

- discussion of the DSD option was added to the documentation about list input
- directions for viewing ODS style templates were updated (see "Customizing ODS Output at the Level of a SAS Job" on page 667)
- discussion of the IN= data set option was added to the documentation about merging data sets

## Debugging SAS Programs

Additional information and examples of SAS log output have been added. Items in the SAS log are explained so that you can more easily debug your own SAS programs.

Documentation for the DATA step debugger has been added. The DATA step debugger is a tool that enables you to find logic errors in your program. A description of the tool and examples are provided. A list of commands that you use with the debugger is also provided.

## Concatenating SAS Variables

A preferred method of concatenating SAS variables has been introduced. You use the CAT function to return a concatenated character string.

## Output Delivery System (ODS)

The sections about the Output Delivery System (ODS) have been updated, and new information has been added. ODS gives you greater flexibility in generating, storing, and reproducing SAS procedure and DATA step output along with a wide range of formatting options. ODS provides formatting functionality that is not available when using individual procedures or the DATA step without ODS.

Beginning with SAS 9.3, the default destination in the SAS windowing environment is HTML, and ODS Graphics is enabled by default. These new defaults have several advantages. Graphs are integrated with tables, and all output is displayed in the same HTML file using a new style. This new style, HTMLBlue, is an all-color style that is designed to integrate tables and modern statistical graphics. The examples in this document now show HTML output.

# Accessibility Features of Step-by-Step Programming with Base SAS 9.4

## Overview

For information about the accessibility of Base SAS, see the *SAS 9.4 Companion for Windows*.

## Part 1

# Introduction to the SAS System

*Chapter 1*
 **What is the SAS System?** . . . . . . . . . . . . . . . . . . . . . . . . . . . . . . . . . . . . . . . *3*

*Chapter 2*
 **Working with Output Defaults** . . . . . . . . . . . . . . . . . . . . . . . . . . . . . . . . . . *17*

# Chapter 1
# What is the SAS System?

| | |
|---|---|
| **Introduction to the SAS System** | 3 |
| **Components of Base SAS Software** | 4 |
| Overview of Base SAS Software | 4 |
| Data Management Facility | 4 |
| Programming Language | 5 |
| Data Analysis and Reporting Utilities | 6 |
| **Output Produced by the SAS System** | 8 |
| Traditional Output | 8 |
| Output from the Output Delivery System (ODS) | 9 |
| **Ways to Run SAS Programs** | 11 |
| Selecting an Approach | 11 |
| SAS Windowing Environment | 11 |
| SAS/ASSIST Software | 12 |
| Noninteractive Mode | 12 |
| Batch Mode | 12 |
| Interactive Line Mode | 13 |
| **Running Programs in the SAS Windowing Environment** | 13 |
| **Summary** | 15 |
| Statements | 15 |
| Procedures | 15 |
| **Learning More** | 16 |

## Introduction to the SAS System

SAS is an integrated system of software solutions that enables you to perform the following tasks:

- data entry, retrieval, and management
- report writing and graphics design
- statistical and mathematical analysis
- business forecasting and decision support
- operations research and project management
- applications development

How you use SAS depends on what you want to accomplish. Some people use many of the capabilities of the SAS System, and others use only a few.

At the core of the SAS System is Base SAS software, which is the software product that you will learn to use in this documentation. This section presents an overview of Base SAS. It introduces the capabilities of Base SAS, addresses methods of running SAS, and outlines various types of output.

# Components of Base SAS Software

## *Overview of Base SAS Software*

Base SAS software contains the following:

- a data management facility
- a programming language
- data analysis and reporting utilities

Learning to use Base SAS enables you to work with these features of SAS. It also prepares you to learn other SAS products, because all SAS products follow the same basic rules.

## *Data Management Facility*

SAS organizes data into a rectangular form or table that is called a SAS data set. The following figure shows a SAS data set. The data describes participants in a 16-week weight program at a health and fitness club. The data for each participant includes an identification number, name, team name, and weight (in U.S. pounds) at the beginning and end of the program.

*Figure 1.1* Rectangular Form of a SAS Data Set

|   | IdNumber | Name | Team | StartWeight | EndWeight |
|---|---|---|---|---|---|
| 1 | 1023 | David Shaw | red | 189 | 165 |
| 2 | 1049 | Amelia Serrano | yellow | 145 | 124 |
| 3 | 1219 | Alan Nance | red | 210 | 192 |
| 4 | 1246 | Ravi Sinha | yellow | 194 | 177 |
| 5 | 1078 | Ashley McKnight | red | 127 | 118 |

In a SAS data set, each row represents information about an individual entity and is called an observation. Each column represents the same type of information and is called a variable. Each separate piece of information is a data value. In a SAS data set, an observation contains all the data values for an entity; a variable contains the same type of data value for all entities.

To build a SAS data set with Base SAS, you write a program that uses statements in the SAS programming language. A SAS program that begins with a DATA statement and typically creates a SAS data set or a report is called a DATA step.

The following SAS program creates a SAS data set named WEIGHT_CLUB from the health club data:

```
data weight_club;   1
   input IdNumber 1-4 Name $ 6-24 Team $ StartWeight EndWeight;   2
   Loss=StartWeight-EndWeight;   3
   datalines;   4
1023 David Shaw          red    189 165   5
1049 Amelia Serrano      yellow 145 124   5
1219 Alan Nance          red    210 192   5
1246 Ravi Sinha          yellow 194 177   5
1078 Ashley McKnight     red    127 118   5
;   6
```

The following list corresponds to the numbered items in the preceding program:

1. The DATA statement tells SAS to begin building a SAS data set named WEIGHT_CLUB.

2. The INPUT statement identifies the fields to be read from the input data and names the SAS variables to be created from them (IdNumber, Name, Team, StartWeight, and EndWeight).

3. The third statement is an assignment statement. It calculates the weight each person lost and assigns the result to a new variable, Loss.

4. The DATALINES statement indicates that data lines follow.

5. The data lines follow the DATALINES statement. This approach to processing raw data is useful when you have only a few lines of data. (Later sections show ways to access larger amounts of data that are stored in files.)

6. The semicolon signals the end of the raw data, and is a step boundary. It tells SAS that the preceding statements are ready for execution.

*Note:* By default, the data set WEIGHT_CLUB is temporary. It exists only for the current job or session. For information about how to create a permanent SAS data set, see "Introduction to DATA Step Processing" on page 27.

## *Programming Language*

### *Elements of the SAS Language*

The statements that created the data set WEIGHT_CLUB are part of the SAS programming language. The SAS language contains statements, expressions, functions and CALL routines, options, formats, and informats – elements that many programming languages share. However, the way you use the elements of the SAS language depends on certain programming rules. The most important rules are listed in the next two sections.

### *Rules for SAS Statements*

The conventions that are shown in the programs in this documentation, such as indenting of subordinate statements, extra spacing, and blank lines, are for the purpose of clarity and ease of use. They are not required by SAS. There are only a few rules for writing SAS statements:

- SAS statements end with a semicolon.
- You can enter SAS statements in lowercase, uppercase, or a mixture of the two.
- You can begin SAS statements in any column of a line and write several statements on the same line.
- You can begin a statement on one line and continue it on another line, but you cannot split a word between two lines.
- Words in SAS statements are separated by blanks or by special characters (such as the equal sign and the minus sign in the calculation of the Loss variable in the WEIGHT_CLUB example).

### Rules for Most SAS Names

SAS names are used for SAS data set names, variable names, and other items. The following rules apply:

- A SAS name can contain from one to 32 characters.
- The first character must be a letter or an underscore (_).
- Subsequent characters must be letters, numbers, or underscores.
- Blank spaces cannot appear in SAS names.

### Special Rules for Variable Names

For variable names only, SAS remembers the combination of uppercase and lowercase letters that you use when you create the variable name. Internally, the case of letters does not matter. "CAT," "cat," and "Cat" all represent the same variable. But for presentation purposes, SAS remembers the initial case of each letter and uses it to represent the variable name when printing it.

## Data Analysis and Reporting Utilities

The SAS programming language is both powerful and flexible. You can program any number of analyses and reports with it. SAS can also simplify programming for you with its library of built-in programs known as SAS procedures. SAS procedures use data values from SAS data sets to produce preprogrammed reports, requiring minimal effort from you.

For example, the following SAS program produces a report that displays the values of the variables in the SAS data set WEIGHT_CLUB. Weight values are presented in U.S. pounds.

```
proc print data=weight_club;
   title 'Health Club Data';
run;
```

This procedure, known as the PRINT procedure, displays the variables in a simple, organized form. The following output displays the results:

*Figure 1.2* Displaying the Values in a SAS Data Set

Health Club Data

| Obs | IdNumber | Name | Team | StartWeight | EndWeight | Loss |
|---|---|---|---|---|---|---|
| 1 | 1023 | David Shaw | red | 189 | 165 | 24 |
| 2 | 1049 | Amelia Serrano | yellow | 145 | 124 | 21 |
| 3 | 1219 | Alan Nance | red | 210 | 192 | 18 |
| 4 | 1246 | Ravi Sinha | yellow | 194 | 177 | 17 |
| 5 | 1078 | Ashley McKnight | red | 127 | 118 | 9 |

To produce a table showing mean starting weight, ending weight, and weight loss for each team, use the TABULATE procedure.

```
proc tabulate data=weight_club;
   class team;
   var StartWeight EndWeight Loss;
   table team, mean*(StartWeight EndWeight Loss);
   title 'Mean Starting Weight, Ending Weight,';
   title2 'and Weight Loss';
run;
```

The following output displays the results:

*Figure 1.3* Table of Mean Values for Each Team

Mean Starting Weight, Ending Weight, and Weight Loss

|  | Mean | | |
|---|---|---|---|
| | StartWeight | EndWeight | Loss |
| Team | | | |
| red | 175.33 | 158.33 | 17.00 |
| yellow | 169.50 | 150.50 | 19.00 |

A portion of a SAS program that begins with a PROC (procedure) statement and ends with a RUN statement (or is ended by another PROC or DATA statement) is called a PROC step. Both of the PROC steps that create the previous two outputs comprise the following elements:

- a PROC statement, which includes the word PROC, the name of the procedure that you want to use, and the name of the SAS data set that contains the values. (If you omit the DATA= option and data set name, the procedure uses the SAS data set that was most recently created in the program.)

- additional statements that give SAS more information about what you want to do, for example, the CLASS, VAR, TABLE, and TITLE statements.

- a RUN statement, which indicates that the preceding group of statements is ready to be executed.

## Output Produced by the SAS System

### *Traditional Output*

A SAS program can produce some or all of the following types of output:

a SAS data set
: contains data values that are stored as a table of observations and variables. It also stores descriptive information about the data set, such as the names and arrangement of variables, the number of observations, and the creation date of the data set. A SAS data set can be temporary or permanent. The examples in this section create the temporary data set WEIGHT_CLUB.

the SAS log
: is a record of the SAS statements that you entered and of messages from SAS about the execution of your program. It can appear as a file on disk, a display on your monitor, or a hard-copy listing. The exact appearance of the SAS log varies according to your operating environment and your site. The output in Log 1.1 on page 9 displays a typical SAS log for the program in this section.

a report or simple listing
: ranges from a simple listing of data values to a subset of a large data set or a complex summary report that groups and summarizes data and displays statistics. The appearance of procedure output varies according to your site and the options that you specify in the program, but the output in Figure 1.2 on page 7 and Figure 1.3 on page 7 illustrate typical procedure output. You can also use a DATA step to produce a completely customized report. For more information, see "Creating Customized Reports" on page 458.

other SAS files such as catalogs
: contain information that cannot be represented as tables of data values. Examples of items that can be stored in SAS catalogs include function key settings, letters that are produced by SAS/FSP software, and displays that are produced by SAS/GRAPH software.

external files or entries in other databases
: can be created and updated by SAS programs. SAS/ACCESS software enables you to create and update files that are stored in databases such as Oracle.

**Log 1.1** *Traditional Output: A SAS Log*

```
NOTE: Additional host information:

 W32_7PRO DNTHOST 6.1.7601 Service Pack 1 Workstation

NOTE: SAS initialization used:
      real time           1.15 seconds
      cpu time            0.87 seconds

1    data weight_club;
2       input IdNumber 1-4 Name $ 6-24 Team $ StartWeight EndWeight;
3       Loss=StartWeight-EndWeight;
4       datalines;

NOTE: The data set WORK.WEIGHT_CLUB has 5 observations and 6 variables.
NOTE: DATA statement used (Total process time):
      real time           0.01 seconds
      cpu time            0.01 seconds


10   ;
11   proc tabulate data=weight_club;
NOTE: Writing HTML Body file: sashtml.htm
12      class team;
13      var StartWeight EndWeight Loss;
14      table team, mean*(StartWeight EndWeight Loss);
15      title 'Mean Starting Weight, Ending Weight,';
16      title2 'and Weight Loss';
17   run;

NOTE: There were 5 observations read from the data set WORK.WEIGHT_CLUB.
NOTE: PROCEDURE TABULATE used (Total process time):
      real time           0.93 seconds
      cpu time            0.64 seconds
```

## Output from the Output Delivery System (ODS)

The Output Delivery System (ODS) enables you to produce output in a variety of formats, such as the following:

- an HTML file
- a traditional SAS Listing (monospace)
- a PostScript file
- an RTF file (for use with Microsoft Word)
- an output data set

The following figure illustrates the concept of output for SAS Version 8.

**Figure 1.4** *Model of the Production of ODS Output*

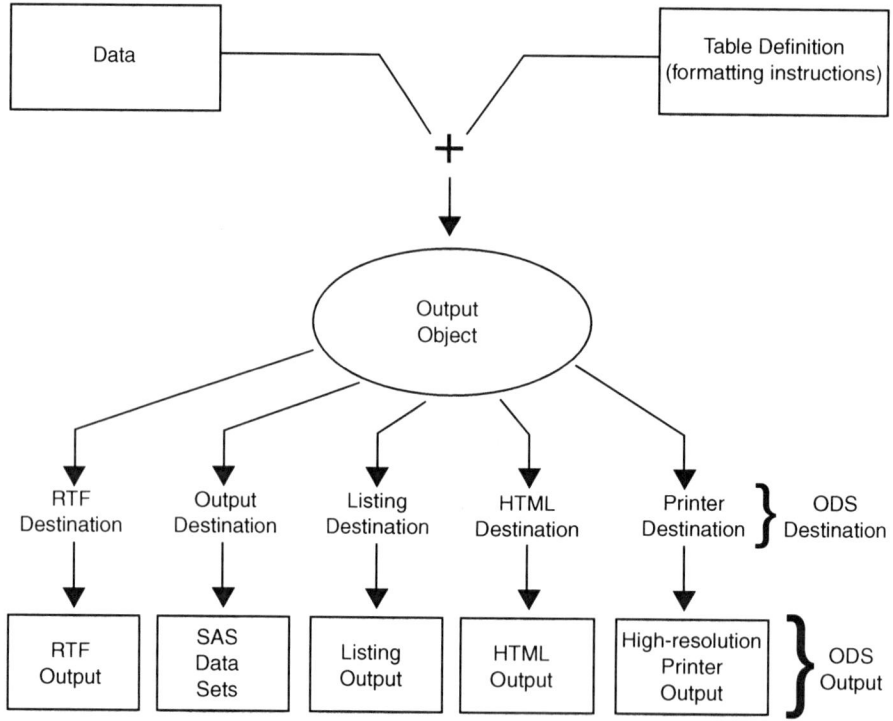

The following definitions describe the terms in the preceding figure:

data
   Each procedure that supports ODS and each DATA step produces data, which contains the results (numbers and characters) of the step in a form similar to a SAS data set.

table definition
   The table definition is a set of instructions that describes how to format the data. This description includes but is not limited to:

   - the order of the columns
   - text and order of column headings
   - formats for data
   - font sizes and font faces

output object
   ODS combines formatting instructions with the data to produce an output object. The output object, therefore, contains both the results of the procedure or DATA step and information about how to format the results. An output object has a name, a label, and a path.

   *Note:* Although many output objects include formatting instructions, not all do. In some cases the output object consists of only the data.

ODS destinations
   An ODS destination specifies a specific type of output. ODS supports a number of destinations, which include the following:

RTF
: produces output that is formatted for use with Microsoft Word.

Output
: produces a SAS data set.

Listing
: produces traditional SAS output (monospace format).

HTML
: produces output that is formatted in Hyper Text Markup Language (HTML). You can access the output on the web with your browser.

Printer
: produces output that is formatted for a high-resolution printer. An example of this type of output is a PostScript file.

ODS output
: ODS output consists of formatted output from any of the ODS destinations.

For more information about ODS output, see Chapter 24, "Directing SAS Output and the SAS Log," on page 391 and Chapter 34, "Understanding and Customizing SAS Output: The Output Delivery System (ODS)," on page 643.

For complete information about ODS, see *SAS Output Delivery System: User's Guide.*

# Ways to Run SAS Programs

## *Selecting an Approach*

There are several ways to run SAS programs. They differ in the speed with which they run, the amount of computer resources that are required, and the amount of interaction that you have with the program (that is, the types of changes that you can make while the program is running).

The examples in this documentation produce the same results, regardless of how you run the programs. However, in a few cases, the way that you run a program determines the appearance of output. The following sections briefly introduce different ways to run SAS programs.

## *SAS Windowing Environment*

The SAS windowing environment enables you to interact with SAS directly through a series of windows. You can use these windows to perform common tasks, such as locating and organizing files, entering and editing programs, reviewing log information, viewing procedure output, setting options, and more. If needed, you can issue operating system commands from within this environment. Or, you can suspend the current SAS windowing environment session, enter operating system commands, and then resume the SAS windowing environment session at a later time.

Using the SAS windowing environment is a quick and convenient way to program in SAS. It is especially useful for learning SAS and developing programs on small test files. Although it uses more computer resources than other techniques, using the SAS windowing environment can save a lot of program development time.

For more information about the SAS windowing environment, see Chapter 41, "Using the SAS Windowing Environment," on page 753.

### SAS/ASSIST Software

One important feature of SAS is the availability of SAS/ASSIST software. SAS/ASSIST provides a point-and-click interface that enables you to select the tasks that you want to perform. SAS then submits the SAS statements to accomplish those tasks. You do not need to know how to program in the SAS language in order to use SAS/ASSIST.

SAS/ASSIST works by submitting SAS statements just like the ones shown earlier in this section. In that way, it provides a number of features, but it does not represent the total functionality of SAS software. If you want to perform tasks other than those that are available in SAS/ASSIST, you need to learn to program in SAS as described in this documentation.

### Noninteractive Mode

In noninteractive mode, you prepare a file that contains SAS statements and any system statements that are required by your operating environment, and submit the program. The program runs immediately and occupies your current workstation session. You cannot continue to work in that session while the program is running,[1] and you usually cannot interact with the program.[2] The log and procedure output go to prespecified destinations, and you usually do not see them until the program ends. To modify the program or correct errors, you must edit and resubmit the program.

Noninteractive execution can be faster than batch execution because the computer system runs the program immediately rather than waiting to schedule your program among other programs.

### Batch Mode

To run a program in batch mode, you prepare a file that contains SAS statements and any system statements that are required by your operating environment, and then you submit the program.

You can then work on another task at your workstation. While you are working, the operating environment schedules your job for execution (along with jobs submitted by other people) and runs it. When execution is complete, you can look at the log and the procedure output.

The central feature of batch execution is that it is completely separate from other activities at your workstation. You do not see the program while it is running, and you cannot correct errors when they occur. The log and procedure output go to prespecified destinations; you can look at them only after the program has finished running. To modify the SAS program, you edit the program with the editor that is supported by your operating environment and submit a new batch job.

When sites charge for computer resources, batch processing is a relatively inexpensive way to execute programs. It is particularly useful for large programs or when you need to use your workstation for other tasks while the program is executing. However, for learning SAS or developing and testing new programs, using batch mode might not be efficient.

---

[1] In a workstation environment, you can switch to another window and continue working.

[2] Limited ways of interaction are available. You can, for example, use the asterisk (*) option in a %INCLUDE statement in your program.

### Interactive Line Mode

In an interactive line-mode session, you enter one line of a SAS program at a time, and SAS executes each DATA or PROC step automatically as soon as it recognizes the end of the step. You usually see procedure output immediately on your display monitor. Depending on your site's computer system and on your workstation, you might be able to scroll backward and forward to see different parts of your log and procedure output, or you might lose them when they scroll off the top of your screen. There are limited facilities for modifying programs and correcting errors.

Interactive line-mode sessions use fewer computer resources than a windowing environment. If you use line mode, you should familiarize yourself with the %INCLUDE, %LIST, and RUN statements in *SAS Statements: Reference*.

## Running Programs in the SAS Windowing Environment

You can run most programs in this documentation by using any of the methods that are described in the previous sections. This documentation uses the SAS windowing environment (as it appears on Windows and UNIX operating environments) when it is necessary to show programming within a SAS session. The SAS windowing environment appears differently depending on the operating environment that you use. For more information about the SAS windowing environment, see Chapter 41, "Using the SAS Windowing Environment," on page 753.

The following example gives a brief overview of a SAS session that uses the SAS windowing environment. When you invoke SAS, the following windows appear.

*Figure 1.5* SAS Windowing Environment

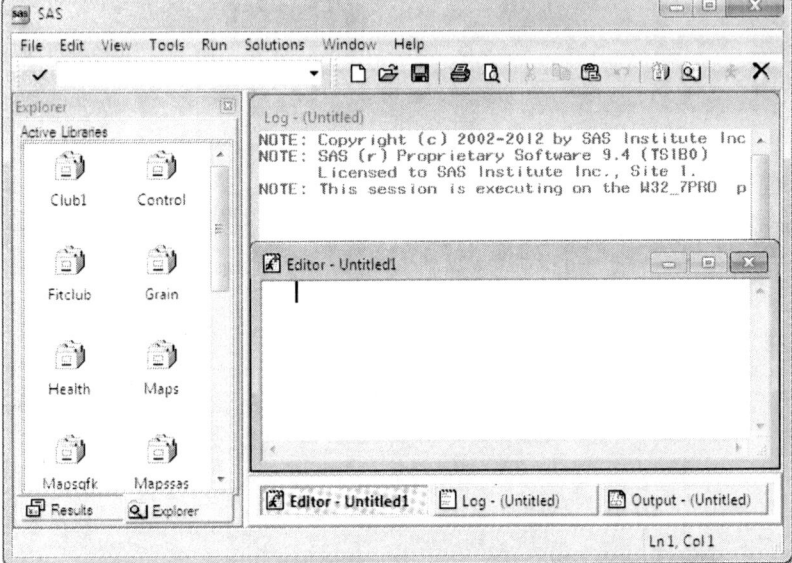

The specific window placement, display colors, messages, and some other details vary according to your site, your monitor, and your operating environment. The window on the left side of the display is the SAS Explorer window, which you can use to assign and locate SAS libraries, files, and other items. The window at the top right is the Log window; it contains the SAS log for the session. The window at the bottom right is the Program Editor window. This window provides an editor in which you edit your SAS programs.

To create the program for the health and fitness club, enter the statements in the Program Editor window. You can turn line numbers on or off to facilitate program creation.

The following display shows the beginning of the program.

*Figure 1.6* Editing a Program in the Program Editor Window

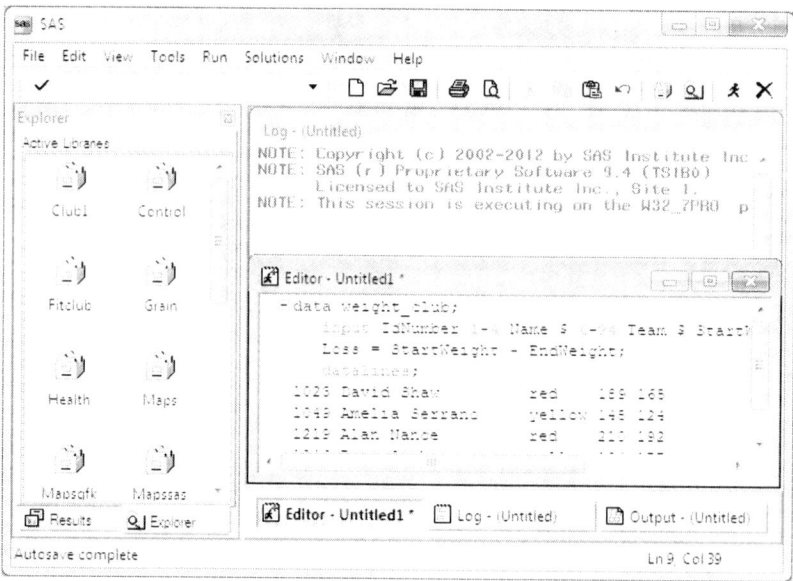

When you fill the Program Editor window, scroll down to continue entering the program. When you finish editing the program, submit it to SAS and view the output. (If SAS does not create output, check the SAS log for error messages.)

The following displays show the first and second pages of the Results Viewer window.

*Figure 1.7* The First Page of Output in the Results Viewer Window

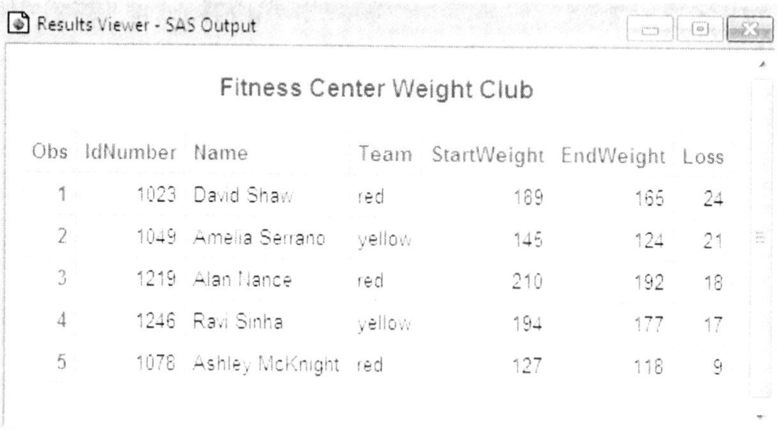

| Obs | IdNumber | Name | Team | StartWeight | EndWeight | Loss |
|---|---|---|---|---|---|---|
| 1 | 1023 | David Shaw | red | 189 | 165 | 24 |
| 2 | 1049 | Amelia Serrano | yellow | 145 | 124 | 21 |
| 3 | 1219 | Alan Nance | red | 210 | 192 | 18 |
| 4 | 1246 | Ravi Sinha | yellow | 194 | 177 | 17 |
| 5 | 1078 | Ashley McKnight | red | 127 | 118 | 9 |

Fitness Center Weight Club

*Figure 1.8  The Second Page of Output in the Results Viewer Window*

[Results Viewer - SAS Output]

**Mean Starting Weight, Ending Weight, and Weight Loss**

| Team | Mean StartWeight | Mean EndWeight | Mean Loss |
|---|---|---|---|
| red | 175.33 | 158.33 | 17.00 |
| yellow | 169.50 | 150.50 | 19.00 |

After you finish viewing the output, you can return to the Program Editor window to begin creating a new program.

By default, the output from all submissions remains in the Output window or the Results Viewer window, and all statements that you submit remain in memory until the end of your session. You can view the output at any time, and you can recall previously submitted statements for editing and resubmitting. You can also clear a window of its contents.

All the commands that you use to move through the SAS windowing environment can be executed as words or as function keys. You can also customize the SAS windowing environment by determining which windows appear, as well as by assigning commands to function keys. For more information about customizing the SAS windowing environment, see Chapter 42, "Customizing the SAS Environment," on page 793.

# Summary

### Statements

DATA *SAS-data-set*;
  begins a DATA step and tells SAS to begin creating a SAS data set. *SAS-data-set* names the data set that is being created.

%INCLUDE *source(s)* </<SOURCE2> <S2=*length*> <*host-options*>>;
  brings SAS programming statements, data lines, or both into a current SAS program.

RUN;
  tells SAS to begin executing the preceding group of SAS statements.

For more information, see *SAS Statements: Reference*.

### Procedures

PROC *procedure* <DATA=*SAS-data-set*>;
  begins a PROC step and tells SAS to invoke a particular SAS procedure to process the SAS data set that is specified in the DATA= option. If you omit the DATA= option, then the procedure processes the most recently created SAS data set in the program.

For more information about using procedures, see the *Base SAS Procedures Guide*.

## Learning More

Basic SAS usage
> For an entry-level introduction to basic SAS programming language, see *The Little SAS® Book: A Primer, Fifth Edition*.

DATA step
> For more information about how to create SAS data sets, see Chapter 3, "Introduction to DATA Step Processing," on page 27.

DATA step processing
> For more information about DATA step processing, see Chapter 7, "Understanding DATA Step Processing," on page 107.

For information about using the SAS environment, see *Getting Started with SAS*.

# Chapter 2
# Working with Output Defaults

**Working with Output Defaults Starting in SAS 9.3** .................... 17
    Overview of Working with Output Defaults ........................ 17
    The Default Destination ........................................ 17
    HTML Output in the SAS Windowing Environment ................... 18
    LISTING Output in the SAS Windowing Environment ................. 19
    ODS Graphics ................................................ 20
    How to Restore 9.2 Behavior ................................... 20

**Learning More** ................................................... 23

## Working with Output Defaults Starting in SAS 9.3

### Overview of Working with Output Defaults

Beginning with SAS 9.3, output in the SAS windowing environment is created by default in HTML. In addition, ODS Graphics is enabled by default. The following sections explain the advantages of these new defaults and how to change the defaults to match those of previous releases:

- ODS Graphics is enabled at SAS start-up.
- The LISTING destination is closed, and the HTML destination is open.
- The default style for the HTML destination is HTMLBlue.
- The default HTML style for z/OS has changed from Default to Htmlblue.
- The default style for the PRINTER destination is Pearl.

LISTING output is the default in previous releases and when you run SAS in batch mode in SAS 9.3. HTML output in the SAS windowing environment is the default in SAS 9.3 for Microsoft Windows and UNIX, but not for other operating systems. Your actual defaults might be different because of your registry or configuration file settings.

### The Default Destination

Beginning with 9.3, by default, in the Windowing environment with the Windows and UNIX operating systems, the LISTING destination is closed and the HTML destination is open. You do not have to submit an ODS HTML statement to generate HTML output, and you do not have to use the ODS HTML CLOSE statement to be able to view your output. However, to create LISTING output, you must either submit the ODS LISTING statement or enable the LISTING destination by other means. For more information, see "How to Restore 9.2 Behavior" on page 20.

The HTML destination does the following:

- generates HTML 4.0 embedded style sheets
- writes output files to the Work directory
- does not require you to specify an ODS HTML CLOSE statement to view your output

These behaviors persist until you explicitly close the ODS HTML destination by specifying the ODS HTML CLOSE statement, and then reopen the HTML destination. After you have closed the HTML statement and issued a new ODS HTML statement, the HTML destination does the following:

- writes output files to the current directory
- does require you to specify an ODS HTML CLOSE statement to view your output

These behaviors persist until you close your SAS session and open a new one.

*CAUTION:*
   **In SAS 9.3, HTML output in the SAS windowing environment is the default for Microsoft Windows and UNIX, but not for other operating systems and not in batch mode.** When you run SAS in batch mode or on other operating systems, the LISTING destination is open and is the default, ODS Graphics is not enabled by default, and the default style for HTML output is Styles.Default. Your actual defaults might be different because of your registry or configuration file settings.

## HTML Output in the SAS Windowing Environment

Beginning with SAS 9.3, the default destination in the SAS windowing environment is HTML, and ODS Graphics is enabled by default. These new defaults have several advantages. Graphs are integrated with tables, and all output is displayed in the same HTML file using a new style. This new style, HTMLBlue, is an all-color style that is designed to integrate tables and modern statistical graphics.

You can view and modify the default settings by selecting **Tools** ⇨ **Options** ⇨ **Preferences** from the menu at the top of the main SAS window. Then open the Results tab. You can remember this sequence using the mnemonic TOPR (pronounced "topper"). The following display shows the SAS **Results** tab with the new default settings specified:

*Figure 2.1  SAS Results Tab with the New Default Settings*

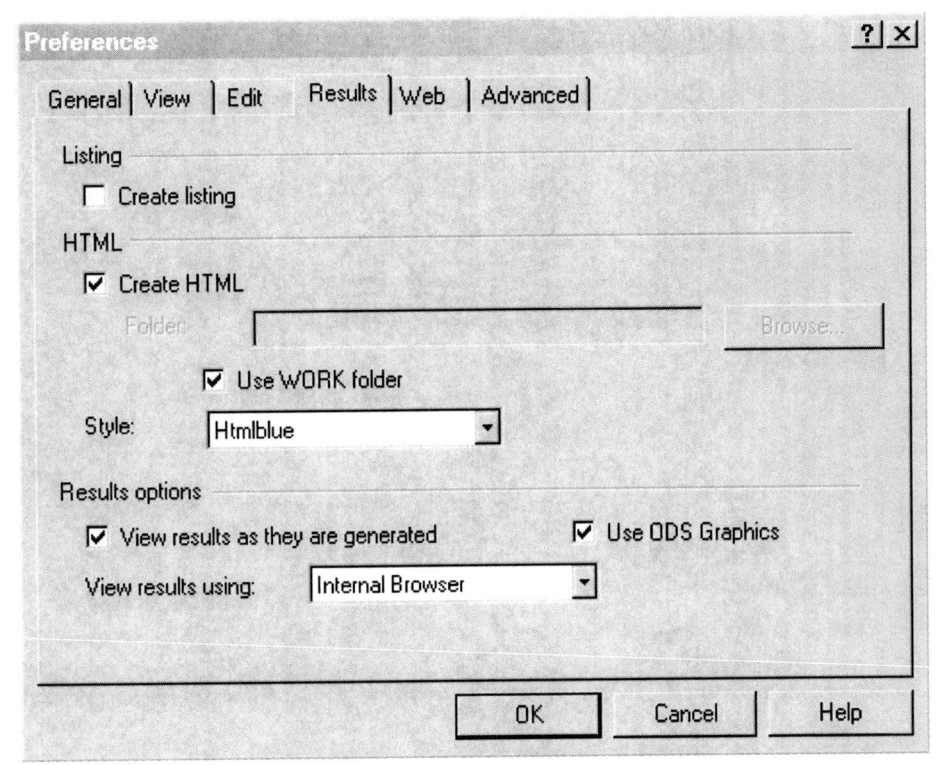

The default settings in the **Results** tab are as follows:

- The **Create listing** check box is not selected, so LISTING output is not created.
- The **Create HTML** check box is selected, so HTML output is created.
- The **Use WORK folder** check box is selected, so both HTML and graph image files are saved in the WORK folder (and not your current directory).
- The default style, HTMLBlue, is selected from the **Style** menu.
- The **Use ODS Graphics** check box is selected, so ODS Graphics is enabled.
- **Internal Browser** is selected from the **View results using:** menu, so results are viewed in an internal SAS browser.

In many cases, graphs are an integral part of a data analysis. However, when you run large computational programs (such as when you use procedures with many BY groups) you might not want to create graphs. In those cases, disabling ODS Graphics improves the performance of your program. You can disable and re-enable ODS Graphics in your SAS programs with the ODS GRAPHICS OFF and ODS GRAPHICS ON statements. You can also change the ODS Graphics default in the **Results** tab.

## *LISTING Output in the SAS Windowing Environment*

Prior to SAS 9.3, SAS output in the SAS windowing environment was created by default in the LISTING destination. In the LISTING destination, tables are displayed in monospace, and graphs are not integrated with tables.

You can create LISTING output by selecting **Tools** ⇨ **Options** ⇨ **Preferences** from the menu at the top of the main SAS window. Then open the **Results** tab. Select the **Create listing** check box, and do not select the **Create HTML** check box.

Prior to SAS 9.3, ODS Graphics was disabled by default. You can enable or disable ODS Graphics by default by using the check box, and you can use the ODS GRAPHICS ON and ODS GRAPHICS OFF statements to enable and disable ODS Graphics in your SAS programs. The following display shows the SAS **Results** tab with the old default settings specified:

*Figure 2.2* SAS Results Tab with the Old Default Settings

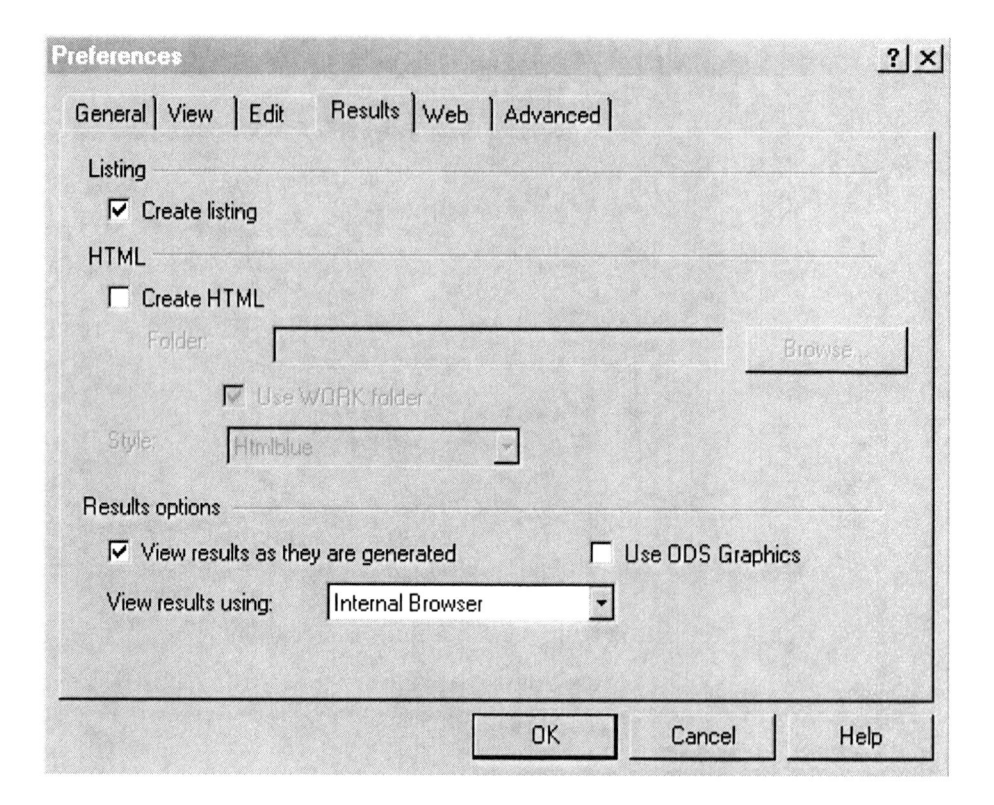

### ODS Graphics

Beginning with SAS 9.3, template-based graphics (frequently referred to as ODS Graphics) are created by default. ODS Graphics includes all graphical output where a compiled ODS template of type STATGRAPH is used to produce graphical output. Supplied templates are stored in Sashelp.Tmplmst. For ODS Graphics, you must use the ODS GRAPHICS statement to control the graphical environment. You do not have to specify the ODS GRAPHICS ON statement to enable ODS Graphics in the SAS Windowing environment in the Windows and UNIX operating systems.

*Note:* The SGSCATTER, SGRENDER, SGPLOT, and SGPANEL procedures always generate graphs, even when ODS Graphics is not enabled.

## How to Restore 9.2 Behavior

### Overview

You can change your output defaults back to 9.2 behavior in one of the following three ways:

- Use the **Results** tab in the Preferences window. This changes the behavior until you change it back.

- Use ODS statements. This change lasts only during your current SAS session.
- Use the ODSSTYLE, ODSDEST, and ODSGRAPHICS system options.

### Using the Preferences Window

Starting in SAS 9.3, the default destination in the SAS windowing environment is HTML, and ODS Graphics is enabled by default. These new defaults have several advantages. Graphs are integrated with tables, and all output is displayed in the same HTML file using a new style. This new style, HTMLBlue, is an all-color style that is designed to integrate tables and modern statistical graphics.

*Note:* The default HTML is HTML4. Changes applied to HTML in the Preferences Window only affect the default HTML version, not HTML5, or other HTML versions.

You can view and modify the default settings by selecting **Tools** ⇨ **Options** ⇨ **Preferences** from the menu at the top of the main SAS window. Then open the Results tab. The settings in your Preferences window persist until you explicitly change them. The following display shows the SAS **Results** tab with the new default settings specified.

To create LISTING output only by default, select **Create listing** and deselect **Create HTML**. To disable ODS Graphics, deselect "Use ODS Graphics".

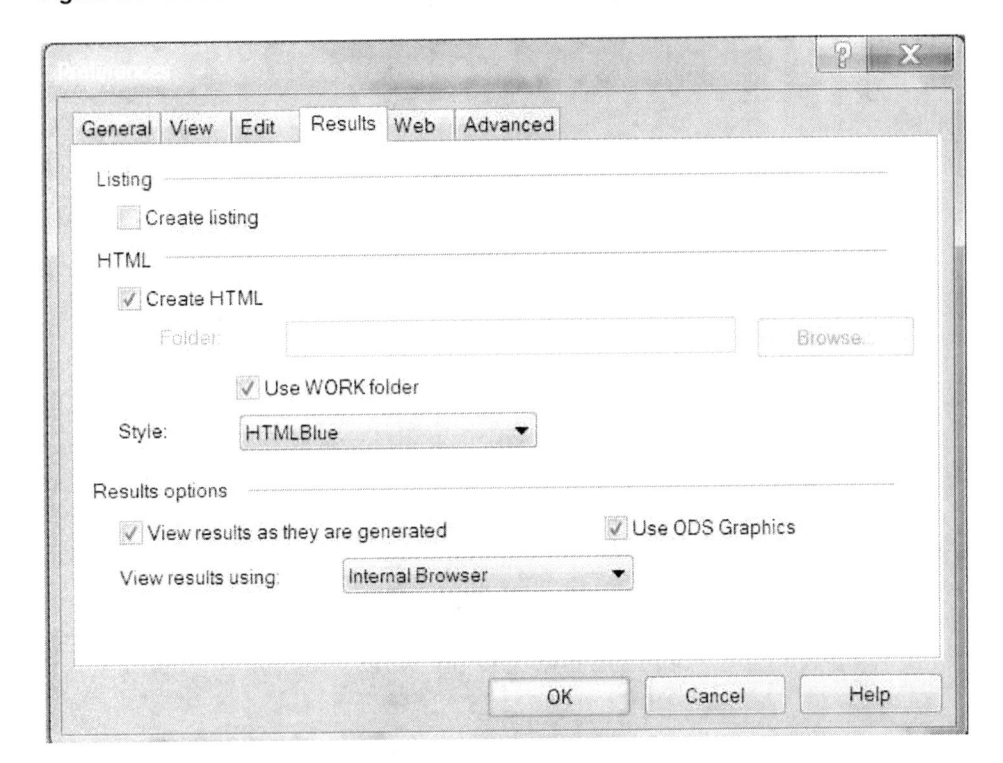

*Figure 2.3* Default Preferences Window in the Windowing Environment

When you set the following selections, your default output destination is LISTING until you explicitly change it using ODS statements, the ODSDEST system option, or the Preferences window. When you have the following selections set, by default, ODS Graphics will always be disabled unless you enable ODS Graphics by specifying the

ODS GRAPHICS ON statement, use the ODSGRAPHICS system option, or change the setting in the Preferences window.

*Figure 2.4* Results Tab Set to Pre-9.3 Defaults

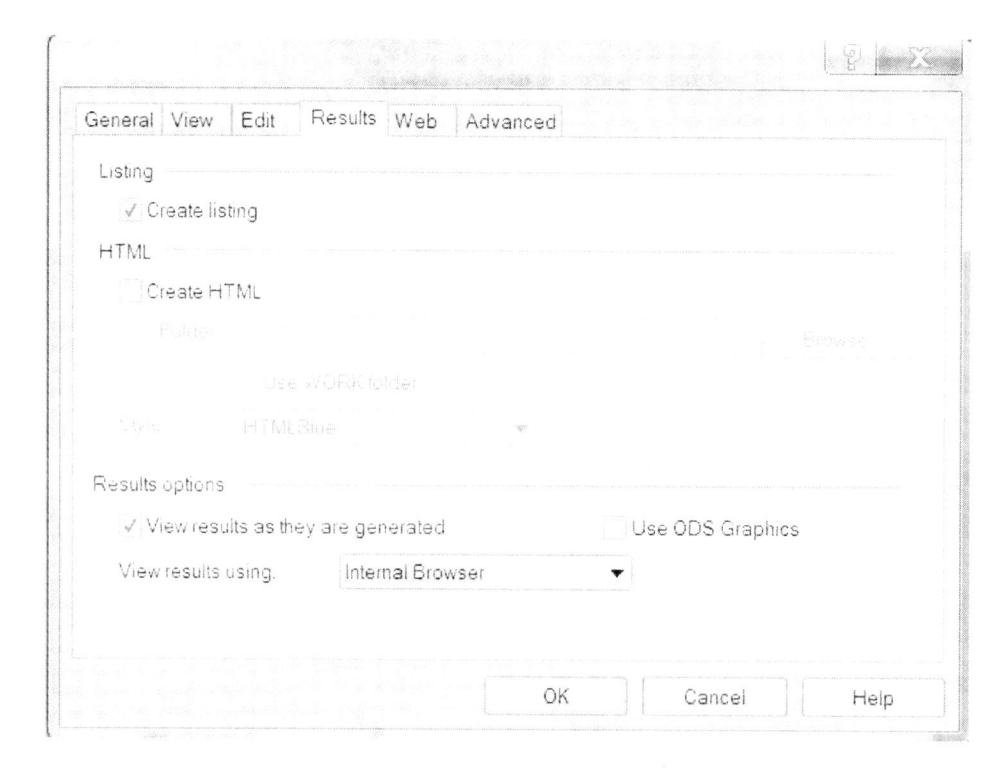

### Using ODS Statements

To change the default destination from HTML to LISTING and to disable ODS Graphics, you can use the following ODS statements:

```
ods graphics off;
ods html close;
ods listing;
```

These statements change the behavior of your current SAS session. When you start a new SAS session, the defaults return to SAS 9.3 behavior.

### Using System Options

There are three new system options for SAS 9.3 that control default output.

ODSSTYLE= System Option
   specifies the default style. To change the default style to Styles.Default, specify **ODSSTYLE=styles.default**. "ODSSTYLE= System Option" in *SAS Output Delivery System: User's Guide*.

ODSGRAPHICS= System Option
   specifies whether ODS Graphics is enabled by default. To disable ODS Graphics by default, specify **ODSGRAPHICS=OFF**. For information about the ODSGRAPHICS= system option, see "ODSGRAPHICS= System Option" in *SAS Output Delivery System: User's Guide*.

ODSDEST= System Option
   specifies the default output destination in the SAS windowing environment. To change the default destination to LISTING, specify **ODSDEST=LISTING**. For

information about the ODSDEST= system option, see "ODSDEST= System Option" in *SAS Output Delivery System: User's Guide*.

## Learning More

- For information about the ODS GRAPHICS statement, see "ODS GRAPHICS Statement" in *SAS Output Delivery System: Procedures Guide*.
- For information about creating an ODS Graphics template, see "TEMPLATE Procedure: Creating ODS Graphics" in *SAS Output Delivery System: Procedures Guide*.
- *SAS ODS Graphics: Procedures Guide*
- *SAS Graph Template Language: User's Guide*
- *SAS Graph Template Language: Reference*
- *SAS ODS Graphics Designer: User's Guide*
- *SAS ODS Graphics Editor: User's Guide*

# Part 2

# Getting Your Data into Shape

*Chapter 3*
**Introduction to DATA Step Processing** . . . . . . . . . . . . . . . . . . . . . . . . . . . *27*

*Chapter 4*
**Starting with Raw Data: The Basics** . . . . . . . . . . . . . . . . . . . . . . . . . . . . *51*

*Chapter 5*
**Starting with Raw Data: Beyond the Basics** . . . . . . . . . . . . . . . . . . . . . *71*

*Chapter 6*
**Starting with SAS Data Sets** . . . . . . . . . . . . . . . . . . . . . . . . . . . . . . . . . . *91*

# Chapter 3
# Introduction to DATA Step Processing

**Introduction to DATA Step Processing** .................................................. **27**
  Purpose .................................................................................................. 27
  Prerequisites ......................................................................................... 28
**The SAS Data Set: Your Key to the SAS System** ..................................... **28**
  Understanding the Function of the SAS Data Set ................................. 28
  Understanding the Structure of the SAS Data Set ................................. 29
  Using FILENAME and LIBNAME Statements ..................................... 31
  Temporary versus Permanent SAS Data Sets ........................................ 32
**How the DATA Step Works: A Basic Introduction** ................................. **33**
  Overview of the DATA Step ................................................................. 33
  During the Compilation Phase .............................................................. 35
  During the Execution Phase .................................................................. 35
  Example of a DATA Step ...................................................................... 36
**Supplying Information to Create a SAS Data Set** ................................... **40**
  Overview of Creating a SAS Data Set ................................................... 40
  Telling SAS How to Read the Data: Styles of Input .............................. 41
  Reading Dates with Two-Digit and Four-Digit Year Values ................. 42
  Defining Variables in SAS ..................................................................... 42
  Indicating the Location of Your Data .................................................... 44
  Using External Files in Your SAS Job ................................................... 45
  Identifying an External File Directly ..................................................... 45
  Referencing an External File with a Fileref ........................................... 46
**Summary** ................................................................................................... **48**
  Statements ............................................................................................. 48
**Learning More** ........................................................................................... **48**

## Introduction to DATA Step Processing

### Purpose

The DATA step is one of the basic building blocks of SAS programming. It creates the data sets that are used in a SAS program's analysis and reporting procedures. Understanding the basic structure, functioning, and components of the DATA step is fundamental to learning how to create your own SAS data sets. In this section, you will learn the following:

- what a SAS data set is and why it is needed
- how the DATA step works
- what information you have to supply to SAS so that it can construct a SAS data set for you

## Prerequisites

You should understand the concepts introduced in Chapter 1, "What is the SAS System?," on page 3 before continuing.

# The SAS Data Set: Your Key to the SAS System

## Understanding the Function of the SAS Data Set

SAS enables you to solve problems by providing methods to analyze or to process your data in some way. You need to first get the data into a form that SAS can recognize and process. After the data is in that form, you can analyze it and generate reports.

The following figure shows this process in the simplest case.

**Figure 3.1** *From Raw Data to Final Analysis*

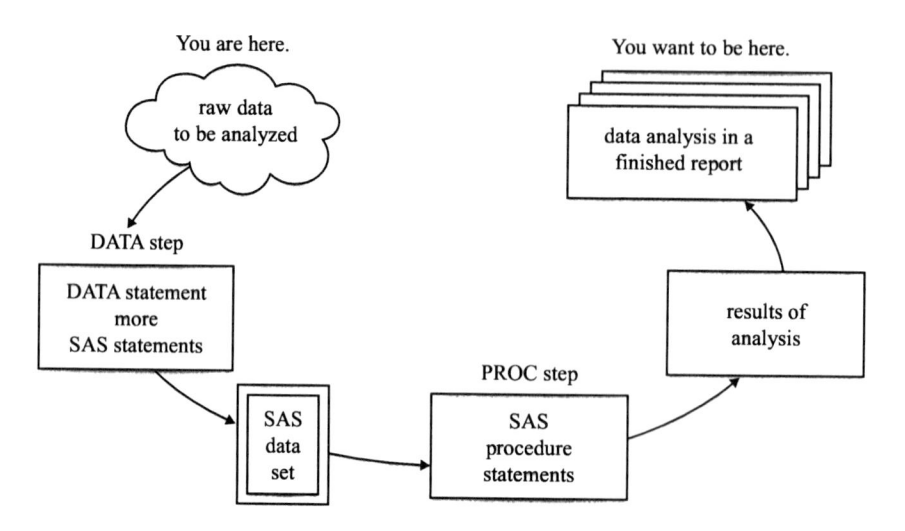

You begin with raw data, that is, a collection of data that has not yet been processed by SAS. You use a set of statements known as a DATA step to get your data into a SAS data set. Then you can further process your data with additional DATA step programming or with SAS procedures.

In its simplest form, the DATA step can be represented by the three components that are shown in the following figure.

*Figure 3.2  From Raw Data to a SAS Data Set*

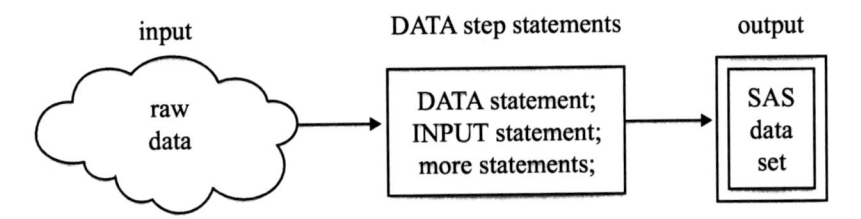

```
    input              DATA step statements            output

    raw       ──▶   DATA statement;         ──▶     SAS
    data            INPUT statement;                data
                    more statements;                set
```

SAS processes input in the form of raw data and creates a SAS data set. When you have a SAS data set, you can use it as input to other DATA steps.

The following figure shows the SAS statements that you can use to create a new SAS data set.

*Figure 3.3  Using One SAS Data Set to Create Another*

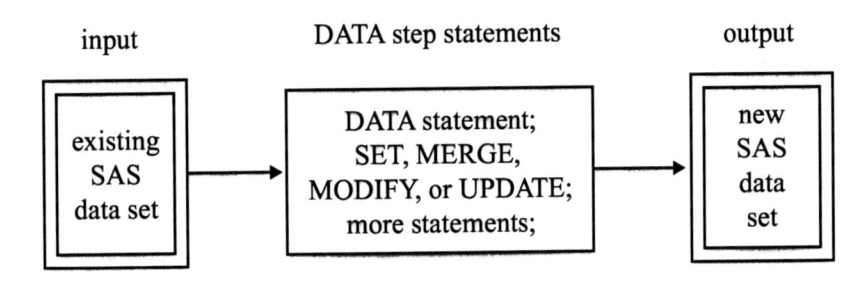

```
    input              DATA step statements            output

   existing           DATA statement;                  new
    SAS      ──▶      SET, MERGE,            ──▶      SAS
   data set        MODIFY, or UPDATE;                 data
                    more statements;                  set
```

## *Understanding the Structure of the SAS Data Set*

Think of a SAS data set as a rectangular structure that identifies and stores data. When your data is in a SAS data set, you can use additional DATA steps for further processing, or perform many types of analyses with SAS procedures.

The rectangular structure of a SAS data set consists of rows and columns in which data values are stored. The rows in a SAS data set are called observations, and the columns are called variables. In a raw data file, the rows are called records and the columns are called fields. Variables contain the data values for all of the items in an observation.

For example, the following figure shows a collection of raw data about participants in a health and fitness club. Each record contains information about one participant.

*Figure 3.4* Raw Data from the Health and Fitness Club

| Id | Name | Team | Starting Weight | Ending Weight |
|---|---|---|---|---|
| 1023 | David Shaw | red | 189 | 165 |
| 1049 | Amelia Serrano | yellow | 145 | 124 |
| 1219 | Alan Nance | red | 210 | 192 |
| 1246 | Ravi Sinha | yellow | 194 | 177 |
| 1078 | Ashley McKnight | red | 127 | 118 |
| 1221 | Jim Brown | yellow | 220 | — |

The following figure shows how easily the health club records can be translated into parts of a SAS data set. Each record becomes an observation. In this case, each observation represents a participant in the program. Each field in the record becomes a variable. The variables represent each participant's identification number, name, team name, and weight at the beginning and end of a 16-week program.

*Figure 3.5* How Data Fits into a SAS Data Set

| | IdNumber | Name | Team | StartWeight | EndWeight |
|---|---|---|---|---|---|
| 1 | 1023 | David Shaw | red | 189 | 165 |
| 2 | 1049 | Amelia Serrano | yellow | 145 | 124 |
| 3 | 1219 | Alan Nance | red | 210 | 192 |
| 4 | 1246 | Ravi Sinha | yellow | 194 | 177 |
| 5 | 1078 | Ashley McKnight | red | 127 | 118 |
| 6 | 1221 | Jim Brown | yellow | 220 | . |

In a SAS data set, every variable exists for every observation. What if you do not have all the data for each observation? If the raw data is incomplete because a value for the numeric variable EndWeight was not recorded for one observation, then this missing value is represented by a period that serves as a placeholder, as shown in observation 6 in the previous figure. (Missing values for character variables are represented by blanks. Character and numeric variables are discussed later in this section.) By coding a value as missing, you can add an observation to the data set for which the data is incomplete and still retain the rectangular shape necessary for a SAS data set.

Along with data values, each SAS data set contains a descriptor portion, as illustrated in the following figure:

**Figure 3.6** Parts of a SAS Data Set

```
                  ┌──────────────┐
                  │  descriptor  │
                  │    portion   │
SAS data set      ├──────────────┤
                  │              │
                  │     data     │
                  │    values    │
                  │              │
                  └──────────────┘
```

The descriptor portion consists of details that SAS records about a data set, such as the names and attributes of all the variables, the number of observations in the data set, and the date and time that the data set was created and updated.

*Operating Environment Information*
Depending on your operating environment and the engine used to write the SAS data set, SAS can store additional information about a SAS data set in its descriptor portion. For more information, refer to the SAS documentation for your operating environment.

## Using FILENAME and LIBNAME Statements

### The FILENAME Statement

The FILENAME statement assigns a fileref to an external file. The association between a fileref and an external file lasts only for the duration of a SAS session or until you change the fileref by using another FILENAME statement.

A FILENAME statement uses the following format:

**FILENAME** *fileref* '*your-input-or-output-file*';

The following FILENAME example assigns the fileref MyWeight to the file named myweight.dat:

```
FILENAME MyWeight 'C:\sasuser\MyWeight\myweight.dat';
```

For more information, see "FILENAME Statement" in *SAS Statements: Reference*.

### The LIBNAME Statement

The LIBNAME statement assigns a libref to a SAS data set or to a DBMS file that can be accessed like a SAS data set. The association between a libref and a SAS library lasts only for the duration of the SAS session or until you change the libref by using another LIBNAME statement.

A LIBNAME statement uses the following format:

**LIBNAME** *libref* '*your-SAS-library*';

The following LIBNAME example assigns the libref Weight to the SAS Library named MyWeight:

```
LIBNAME Weight 'C:\sasuser\MyWeight';
```

For more information, see "LIBNAME Statement" in *SAS Statements: Reference*.

## Temporary versus Permanent SAS Data Sets

### Creating and Using Temporary SAS Data Sets

When you use a DATA step to create a SAS data set with a one-level name, you normally create a temporary SAS data set, one that exists only for the duration of your current session. SAS places this data set in a SAS library referred to as WORK. In most operating environments, all files that SAS stores in the WORK library are deleted at the end of a session.

The following is an example of a DATA step that creates the temporary data set WEIGHT_CLUB.

```
data weight_club;
   input IdNumber Name $ 6-20 Team $ 22-27 StartWeight EndWeight;
   datalines;
1023 David Shaw      red      189 165
1049 Amelia Serrano  yellow   145 124
1219 Alan Nance      red      210 192
1246 Ravi Sinha      yellow   194 177
1078 Ashley McKnight red      127 118
1221 Jim Brown       yellow   220 .
;
```

The preceding program code refers to the temporary data set as WEIGHT_CLUB. SAS. However, it assigns the first-level name WORK to all temporary data sets, and refers to the WEIGHT_CLUB data set with its two-level name, WORK.WEIGHT_CLUB.

The following output from the SAS log displays the name of the temporary data set.

**Log 3.1** *SAS Log: The WORK.WEIGHT_CLUB Temporary Data Set*

```
1   data weight_club;
2      input IdNumber 1-4 Name $ 6-24 Team $ StartWeight EndWeight;
3      Loss=StartWeight-EndWeight;
4      datalines;
```

Because SAS assigns the first-level name WORK to all SAS data sets that have only a one-level name, you do not need to use WORK. You can refer to these temporary data sets with a one-level name, such as WEIGHT_CLUB.

To reference this SAS data set in a later DATA step or in a PROC step, you can use a one-level name:

```
proc print data = weight_club;
run;
```

### Creating and Using Permanent SAS Data Sets

To create a permanent SAS data set, you must indicate a SAS library other than WORK. (WORK is a reserved libref that SAS automatically assigns to a temporary SAS library.) Use a LIBNAME statement to assign a libref to a SAS library on your operating environment's file system. The libref functions as a shorthand way of referring to a SAS library. Here is the form of the LIBNAME statement:

**LIBNAME** *libref 'your-data-library'*;

*libref*
: is a shortcut name to where your SAS files are stored. *libref* must be a valid SAS name. It must begin with a letter or an underscore, and it can contain uppercase and lowercase letters, numbers, or underscores. A libref has a maximum length of 8 characters.

*'your-data-library'*
: must be the physical name for your SAS library. The physical name is the name that is recognized by the operating environment.

Additional restrictions can apply to librefs and physical filenames under some operating environments. For more information, refer to the SAS documentation for your operating environment.

The following is an example of the LIBNAME statement that is used with a DATA step:

```
libname saveit 'your-data-library';  1
data saveit.weight_club;  2
    ...more SAS statements...
;

proc print data = saveit.weight_club;  3
run;
```

The following list corresponds to the numbered items:

1. The LIBNAME statement associates the libref SAVEIT with *your-data-library*, where *your-data-library* is your operating environment's name for a SAS library.

2. To create a new permanent SAS data set and store it in this SAS library, you must use the two-level name SAVEIT.WEIGHT_CLUB in the DATA statement.

3. To reference this SAS data set in a later DATA step or in a PROC step, you must use the two-level name SAVEIT.WEIGHT_CLUB in the PROC step.

For more information, see Chapter 35, "Understanding SAS Libraries," on page 687.

### Conventions That Are Used in This Documentation

Data sets that are used in examples are usually shown as temporary data sets specified with a one-level name:

```
data fitness;
```

In rare cases in this documentation, data sets are created as permanent SAS data sets. These data sets are specified with a two-level name, and a LIBNAME statement precedes each DATA step in which a permanent SAS data set is created:

```
libname saveit 'your-data-library';
data saveit.weight_club;
```

# How the DATA Step Works: A Basic Introduction

### Overview of the DATA Step

The DATA step consists of a group of SAS statements that begins with a DATA statement. The DATA statement begins the process of building a SAS data set and names the data set. The statements that make up the DATA step are compiled, and the syntax is checked. If the syntax is correct, then the statements are executed. In its simplest form, the DATA step is a loop with an automatic output and return action.

The following figure illustrates the flow of action in a simple DATA step.

**Figure 3.7** *Flow of Action in a Typical DATA Step*

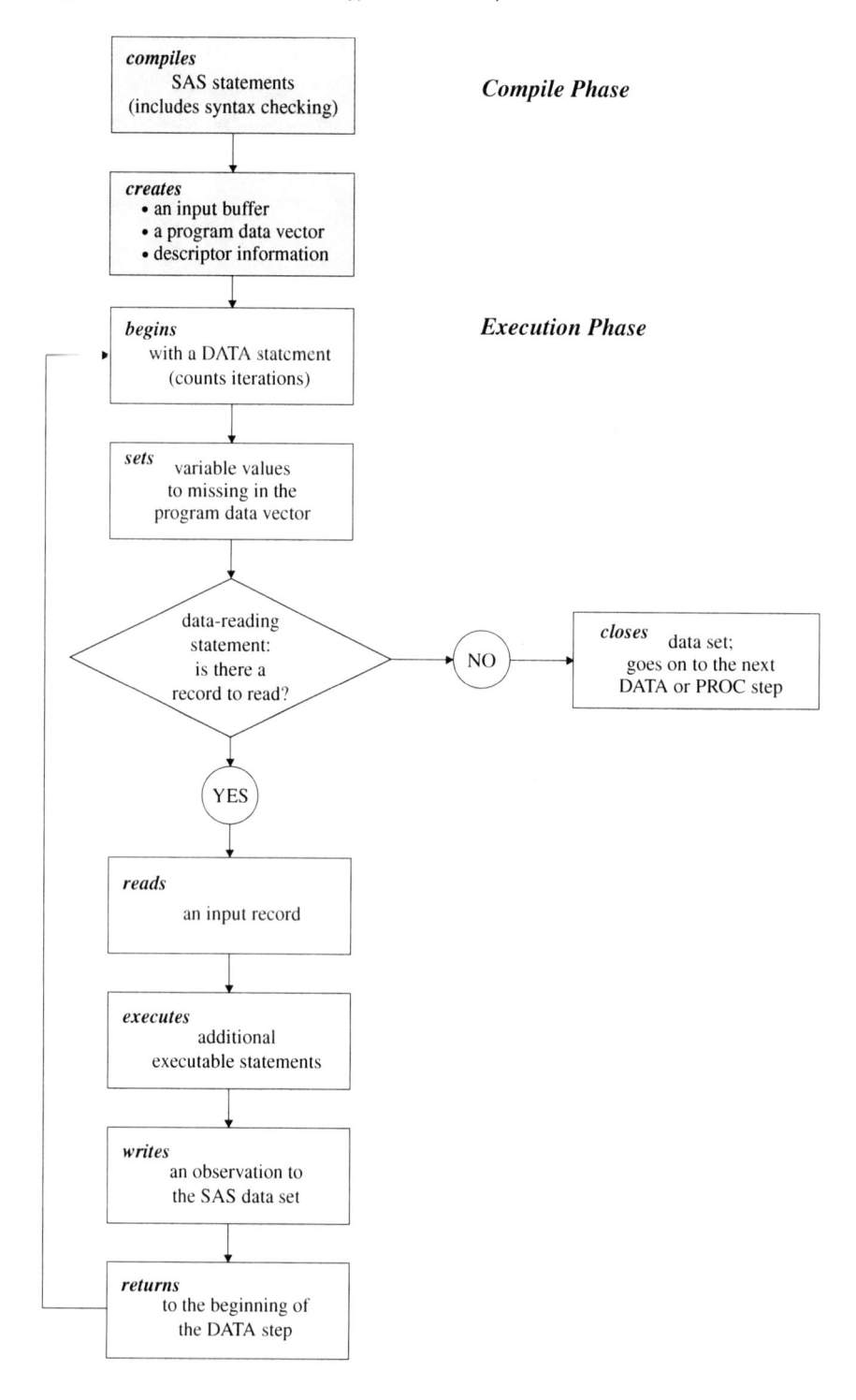

```
         compiles                                   Compile Phase
         SAS statements
      (includes syntax checking)
                │
                ▼
         creates
          • an input buffer
          • a program data vector
          • descriptor information
                │
                ▼
         begins                                    Execution Phase
    ┌──▶ with a DATA statement
    │    (counts iterations)
    │           │
    │           ▼
    │    sets  variable values
    │          to missing in the
    │          program data vector
    │           │
    │           ▼
    │       data-reading                     closes  data set;
    │        statement:       ──NO──▶        goes on to the next
    │       is there a                       DATA or PROC step
    │      record to read?
    │           │
    │          YES
    │           ▼
    │    reads
    │       an input record
    │           │
    │           ▼
    │    executes
    │       additional
    │       executable statements
    │           │
    │           ▼
    │    writes
    │       an observation to
    │       the SAS data set
    │           │
    │           ▼
    │    returns
    └──  to the beginning of
         the DATA step
```

## During the Compilation Phase

When you submit a DATA step for execution, SAS checks the syntax of the SAS statements and compiles them, that is, automatically translates the statements into machine code. SAS further processes the code, and creates the following three items:

input buffer
: is a logical area in memory into which SAS reads each record of data from a raw data file when the program executes. (When SAS reads from a SAS data set, however, the data is written directly to the program data vector.)

program data vector
: is a logical area of memory where SAS builds a data set, one observation at a time. When a program executes, SAS reads data values from the input buffer or creates them by executing SAS language statements. SAS assigns the values to the appropriate variables in the program data vector. From here, SAS writes the values to a SAS data set as a single observation.

  The program data vector also contains two automatic variables, _N_ and _ERROR_. The _N_ variable counts the number of times the DATA step begins to iterate. The _ERROR_ variable signals the occurrence of an error caused by the data during execution. These automatic variables are not written to the output data set.

descriptor information
: is information about each SAS data set, including data set attributes and variable attributes. SAS creates and maintains the descriptor information.

## During the Execution Phase

All executable statements in the DATA step are executed once for each iteration. If your input file contains raw data, then SAS reads a record into the input buffer. SAS then reads the values in the input buffer and assigns the values to the appropriate variables in the program data vector. SAS also calculates values for variables created by program statements, and writes these values to the program data vector. When the program reaches the end of the DATA step, three actions occur by default that make using the SAS language different from using most other programming languages:

1. SAS writes the current observation from the program data vector to the data set.
2. The program loops back to the top of the DATA step.
3. Variables in the program data vector are reset to missing values.

   *Note:* The following exceptions apply:

   - Variables that you specify in a RETAIN statement are not reset to missing values.
   - The automatic variables _N_ and _ERROR_ are not reset to missing.

   For information about the RETAIN statement, see "Using a Value in a Later Observation" on page 209.

If there is another record to read, then the program executes again. SAS builds the second observation, and continues until there are no more records to read. The data set is then closed, and SAS goes on to the next DATA or PROC step.

## Example of a DATA Step

### The DATA Step

The following simple DATA step produces a SAS data set from the data collected for a health and fitness club. As discussed earlier, the input data contains each participant's identification number, name, team name, and weight at the beginning and end of a 16-week weight program:

```
data weight_club; 1
   input IdNumber 1-4 Name $ 6-24 Team $ StartWeight EndWeight; 2
   Loss = StartWeight - EndWeight; 3

   datalines; 4
1023 David Shaw          red     189 165
1049 Amelia Serrano      yellow  145 124
1219 Alan Nance          red     210 192
1246 Ravi Sinha          yellow  194 177
1078 Ashley McKnight     red     127 118
1221 Jim Brown           yellow  220   .
1095 Susan Stewart       blue    135 127
1157 Rosa Gomez          green   155 141
1331 Jason Schock        blue    187 172
1067 Kanoko Nagasaka     green   135 122
1251 Richard Rose        blue    181 166
1333 Li-Hwa Lee          green   141 129
1192 Charlene Armstrong yellow  152 139
1352 Bette Long          green   156 137
1262 Yao Chen            blue    196 180
1087 Kim Sikorski        red     148 135
1124 Adrienne Fink       green   156 142
1197 Lynne Overby        red     138 125
1133 John VanMeter       blue    180 167
1036 Becky Redding       green   135 123
1057 Margie Vanhoy       yellow  146 132
1328 Hisashi Ito         red     155 142
1243 Deanna Hicks        blue    134 122
1177 Holly Choate        red     141 130
1259 Raoul Sanchez       green   189 172
1017 Jennifer Brooks     blue    138 127
1099 Asha Garg           yellow  148 132
1329 Larry Goss          yellow  188 174
; 4
```

The following list corresponds to the numbered items in the preceding program:

1 The DATA statement begins the DATA step and names the data set that is being created.

2 The INPUT statement creates five variables, indicates how SAS reads the values from the input buffer, and assigns the values to variables in the program data vector.

3 The assignment statement creates an additional variable called Loss, calculates the value of Loss during each iteration of the DATA step, and writes the value to the program data vector.

4   The DATALINES statement marks the beginning of the input data. The single semicolon marks the end of the input data and the DATA step.

*Note:* A DATA step that does not contain a DATALINES statement must end with a RUN statement.

## The Process

When you submit a DATA step for execution, SAS automatically compiles the DATA step and then executes it. At compile time, SAS creates the input buffer, program data vector, and descriptor information for the data set WEIGHT_CLUB. As the following figure shows, the program data vector contains the variables that are named in the INPUT statement, as well as the variable Loss. The values of the _N_ and the _ERROR_ variables are automatically generated for every DATA step. The _N_ automatic variable represents the number of times that the DATA step has iterated. The _ERROR_ automatic variable acts like a binary switch whose value is 0 if no errors exist in the DATA step, or 1 if one or more errors exist. These automatic variables are not written to the output data set.

All variable values, except _N_ and _ERROR_, are initially set to missing. Note that missing numeric values are represented by a period, and missing character values are represented by a blank.

*Figure 3.8   Variable Values Initially Set to Missing*

Input Buffer

```
----+----1----+----2----+----3----+----4----+----5----+----6----+----7
```

Program Data Vector

| IdNumber | Name | Team | StartWeight | EndWeight | Loss |
|----------|------|------|-------------|-----------|------|
| .        |      |      | .           | .         | .    |

The syntax is correct, so the DATA step executes. As the following figure illustrates, the INPUT statement causes SAS to read the first record of raw data into the input buffer. Then, according to the instructions in the INPUT statement, SAS reads the data values in the input buffer and assigns them to variables in the program data vector.

*Figure 3.9   Values Assigned to Variables by the INPUT Statement*

Input Buffer

```
----+----1----+----2----+----3----+----4----+----5----+----6----+----7
1023 David Shaw        red      189 165
```

Program Data Vector

| IdNumber | Name       | Team | StartWeight | EndWeight | Loss |
|----------|------------|------|-------------|-----------|------|
| 1023     | David Shaw | red  | 189         | 165       | .    |

When SAS assigns values to all variables that are listed in the INPUT statement, SAS executes the next statement in the program:

```
Loss = StartWeight - EndWeight;
```

This assignment statement calculates the value for the variable Loss and writes that value to the program data vector, as the following figure shows.

**Figure 3.10** *Value Computed and Assigned to the Variable Loss*

Input Buffer
```
----+----1----+----2----+----3----+----4----+----5----+----6----+----7
1023 David Shaw        red    189 165
```

Program Data Vector

| IdNumber | Name | Team | StartWeight | EndWeight | Loss |
|---|---|---|---|---|---|
| 1023 | David Shaw | red | 189 | 165 | 24 |

SAS has now reached the end of the DATA step, and the program automatically does the following:

- writes the first observation to the data set
- loops back to the top of the DATA step to begin the next iteration
- increments the _N_ automatic variable by 1
- resets the _ERROR_ automatic variable to 0
- except for _N_ and _ERROR_, sets variable values in the program data vector to missing values, as the following figure shows

**Figure 3.11** *Values Set to Missing*

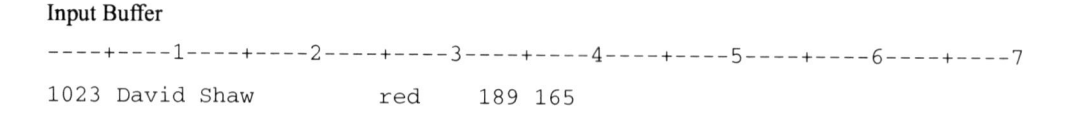

Input Buffer
```
----+----1----+----2----+----3----+----4----+----5----+----6----+----7
1023 David Shaw        red    189 165
```

Program Data Vector

| IdNumber | Name | Team | StartWeight | EndWeight | Loss |
|---|---|---|---|---|---|
| . |  |  |  | . | . | . |

Execution continues. The INPUT statement looks for another record to read. If there are no more records, then SAS closes the data set and the system goes on to the next DATA or PROC step.

In this example, however, more records exist and the INPUT statement reads the second record into the input buffer, as the following figure shows.

*Figure 3.12  Second Record Is Read into the Input Buffer*

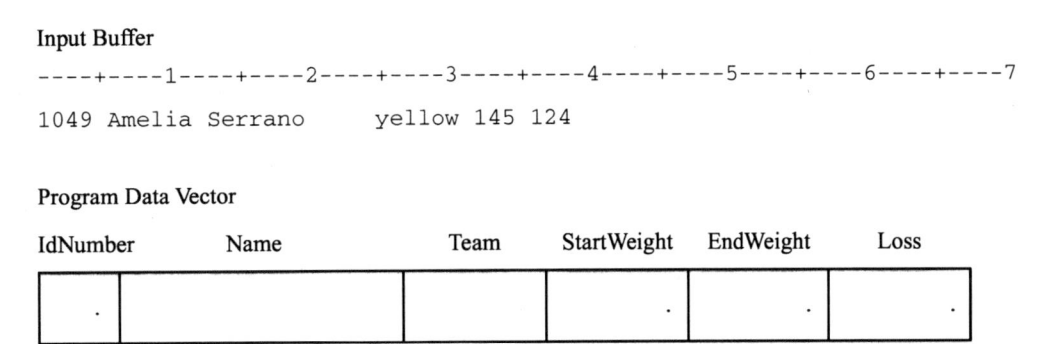

**Input Buffer**

```
----+----1----+----2----+----3----+----4----+----5----+----6----+----7
1049 Amelia Serrano     yellow 145 124
```

**Program Data Vector**

| IdNumber | Name | Team | StartWeight | EndWeight | Loss |
|---|---|---|---|---|---|
| . |  |  | . | . | . |

The following figure shows that SAS assigned values to the variables in the program data vector and calculated the value for the variable Loss. SAS then built the second observation just as it did the first one.

*Figure 3.13  Results of Second Iteration of the DATA Step*

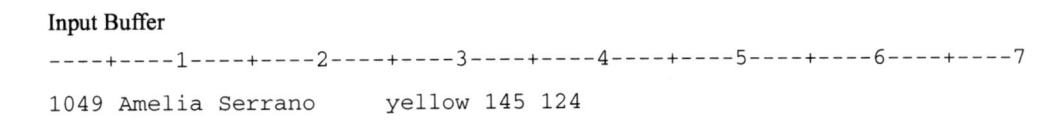

**Input Buffer**

```
----+----1----+----2----+----3----+----4----+----5----+----6----+----7
1049 Amelia Serrano     yellow 145 124
```

**Program Data Vector**

| IdNumber | Name | Team | StartWeight | EndWeight | Loss |
|---|---|---|---|---|---|
| 1049 | Amelia Serrano | yellow | 145 | 124 | 21 |

This entire process continues until SAS detects the end of the file. The DATA step iterates as many times as there are records to read. Then SAS closes the data set WEIGHT_CLUB, and SAS looks for the beginning of the next DATA or PROC step.

Now that SAS has transformed the collected data from raw data into a SAS data set, it can be processed by a SAS procedure.

```
proc print data=weight_club;
   title 'Fitness Center Weight Club';
run;
```

The following output, produced with the PRINT procedure, displays the data set that has just been created.

*Figure 3.14* PROC PRINT Output of the WEIGHT_CLUB Data Set

Fitness Center Weight Club

| Obs | IdNumber | Name | Team | StartWeight | EndWeight | Loss |
|---|---|---|---|---|---|---|
| 1 | 1023 | David Shaw | red | 189 | 165 | 24 |
| 2 | 1049 | Amelia Serrano | yellow | 145 | 124 | 21 |
| 3 | 1219 | Alan Nance | red | 210 | 192 | 18 |
| 4 | 1246 | Ravi Sinha | yellow | 194 | 177 | 17 |
| 5 | 1078 | Ashley McKnight | red | 127 | 118 | 9 |
| 6 | 1221 | Jim Brown | yellow | 220 | | |
| 7 | 1095 | Susan Stewart | blue | 135 | 127 | 8 |
| 8 | 1157 | Rosa Gomez | green | 155 | 141 | 14 |
| 9 | 1331 | Jason Schock | blue | 187 | 172 | 15 |
| 10 | 1067 | Kanoko Nagasaka | green | 135 | 122 | 13 |
| 11 | 1251 | Richard Rose | blue | 181 | 166 | 15 |
| 12 | 1333 | Li-Hwa Lee | green | 141 | 129 | 12 |
| 13 | 1192 | Charlene Armstrong | yellow | 152 | 139 | 13 |
| 14 | 1352 | Bette Long | green | 156 | 137 | 19 |
| 15 | 1262 | Yao Chen | blue | 196 | 180 | 16 |
| 16 | 1087 | Kim Sikorski | red | 148 | 135 | 13 |
| 17 | 1124 | Adrienne Fink | green | 156 | 142 | 14 |
| 18 | 1197 | Lynne Overby | red | 138 | 125 | 13 |
| 19 | 1133 | John VanMeter | blue | 180 | 167 | 13 |
| 20 | 1036 | Becky Redding | green | 135 | 123 | 12 |
| 21 | 1057 | Margie Vanhoy | yellow | 146 | 132 | 14 |
| 22 | 1328 | Hisashi Ito | red | 155 | 142 | 13 |
| 23 | 1243 | Deanna Hicks | blue | 134 | 122 | 12 |
| 24 | 1177 | Holly Choate | red | 141 | 130 | 11 |
| 25 | 1259 | Raoul Sanchez | green | 189 | 172 | 17 |
| 26 | 1017 | Jennifer Brooks | blue | 138 | 127 | 11 |
| 27 | 1099 | Asha Garg | yellow | 148 | 132 | 16 |
| 28 | 1329 | Larry Goss | yellow | 188 | 174 | 14 |

# Supplying Information to Create a SAS Data Set

## Overview of Creating a SAS Data Set

You supply SAS with specific information for reading raw data so that you can create a SAS data set from the raw data. You can use the data set for further processing, data analysis, or report writing. To process raw data in a DATA step, you must:

- use an INPUT statement to tell SAS how to read the data

- define the variables and indicate whether they are character or numeric
- specify the location of the raw data

## *Telling SAS How to Read the Data: Styles of Input*

SAS provides many tools for reading raw data into a SAS data set. These tools include three basic input styles as well as various format modifiers and pointer controls.

List input is used when each field in the raw data is separated by at least one space and does not contain embedded spaces. The INPUT statement simply contains a list of the variable names. List input, however, places numerous restrictions on your data. These restrictions are discussed in detail in Chapter 4, "Starting with Raw Data: The Basics," on page 51. The following example shows list input. Note that there is at least one blank space between each data value.

```
data scores;
   input Name $ Test_1 Test_2 Test_3;
   datalines;
Bill 187 97 103
Carlos 156 76 74
Monique 99 102 129
;
```

Column input enables you to read the same data if it is located in fixed columns:

```
data scores;
   input Name $ 1-7 Test_1 9-11 Test_2 13-15 Test_3 17-19;
   datalines;
Bill     187  97 103
Carlos   156  76  74
Monique   99 102 129
;
```

Formatted input enables you to supply special instructions in the INPUT statement for reading data. For example, to read numeric data that contains special symbols, you need to supply SAS with special instructions so that it can read the data correctly. These instructions, called informats, are discussed in more detail in Chapter 4, "Starting with Raw Data: The Basics," on page 51. In the INPUT statement, you can specify an informat to be used to read a data value, as in the example that follows:

```
data total_sales;
   input Date mmddyy10. +2 Amount comma5.;
   datalines;
09/05/2013  1,382
10/19/2013  1,235
11/30/2013  2,391
;
```

In this example, the MMDDYY10. informat for the variable Date tells SAS to interpret the raw data as a month, day, and year, ignoring the slashes. The COMMA5. informat for the variable Amount tells SAS to interpret the raw data as a number, ignoring the comma. The +2 is a pointer control that tells SAS where to look for the next item. For more information about pointer controls, see Chapter 4, "Starting with Raw Data: The Basics," on page 51.

SAS also enables you to mix these styles of input as required by how values are arranged in the data records. Chapter 4, "Starting with Raw Data: The Basics," on page 51

## Reading Dates with Two-Digit and Four-Digit Year Values

In the previous example, the year values in the dates in the raw data had four digits:

```
09/05/2013
10/19/2013
11/30/2013
```

However, SAS is also capable of reading two-digit year values (for example, 09/05/99). In this example, use the MMDDYY8. informat for the variable Date.

How does SAS know to which century a two-digit year belongs? SAS uses the value of the YEARCUTOFF= SAS system option. In Version 7 and later of SAS, the default value of the YEARCUTOFF= option is 1926. This means that two-digit years from 00 to 19 are assumed to be in the twenty-first century, that is, 2000 to 2019. Two-digit years from 20 to 99 are assumed to be in the twentieth century, that is, 1920 to 1999.

*Note:* The YEARCUTOFF= option and the default setting might be different at your site.

To avoid confusion, you should use four-digit year values in your raw data wherever possible. For more information, see the Dates, Times, and Intervals section of *SAS Language Reference: Concepts*.

## Defining Variables in SAS

So far you have seen that the INPUT statement instructs SAS on how to read raw data lines. At the same time that the INPUT statement provides instructions for reading data, it defines the variables for the data set that come from the raw data. By assuming default values for variable attributes, the INPUT statement does much of the work for you. Later in this documentation that you will learn other statements that enable you to define variables and assign attributes to variables. This section and Chapter 4, "Starting with Raw Data: The Basics," on page 51 concentrate on the use of the INPUT statement.

SAS variables can have these attributes:

- name
- type
- length
- informat
- format
- label
- position in observation
- index type

See the SAS Variables section of *SAS Language Reference: Concepts* for more information about variable attributes.

In an INPUT statement, you must supply each variable name. Unless you also supply an informat, the type is assumed to be numeric, and its length is assumed to be eight bytes. The following INPUT statement creates four numeric variables, each with a length of eight bytes, without requiring you to specify either type or length.

```
input IdNumber Test_1 Test_2 Test_3;
```

The following table summarizes this information.

| Variable Name | Type | Length |
| --- | --- | --- |
| IdNumber | numeric | 8 |
| Test_1 | numeric | 8 |
| Test_2 | numeric | 8 |
| Test_3 | numeric | 8 |

The values of numeric variables can contain only numbers. To store values that contain alphabetic or special characters, you must create a character variable. By following a variable name in an INPUT statement with a dollar sign ($), you create a character variable. The default length of a character variable is also eight bytes. The following statement creates a data set that contains one character variable and four numeric variables, all with a default length of eight bytes.

```
input IdNumber Name $ Test_1 Test_2 Test_3;
```

The following table summarizes this information.

| Variable Name | Type | Length |
| --- | --- | --- |
| IdNumber | numeric | 8 |
| Name | character | 8 |
| Test_1 | numeric | 8 |
| Test_2 | numeric | 8 |
| Test_3 | numeric | 8 |

In addition to specifying the types of variables in the INPUT statement, you can also specify the lengths of character variables. Character variables can be up to 32,767 bytes in length. To specify the length of a character variable in an INPUT statement, you need to supply an informat or use column numbers. For example, following a variable name in the INPUT statement with the informat $20., or with column specifications such as 1-20, creates a character variable that is 20 bytes long.

Note that the length of numeric variables is not affected by informats or column specifications in an INPUT statement. For more information, see *SAS Language Reference: Concepts* for more information about numeric variables and lengths.

Two other variable attributes, format and label, affect how variable values and names are represented when they are printed or displayed. These attributes are assigned with different statements that you will learn about later.

## Indicating the Location of Your Data

### Data Locations

To create a SAS data set, you can read data from one of four locations:

- raw data in the data (job) stream, that is, following a DATALINES statement
- raw data in a file that you specify with an INFILE statement
- data from an existing SAS data set
- data in a database management system (DBMS) file

### Raw Data in the Job Stream

You can place data directly in the job stream with the programming statements that make up the DATA step. The DATALINES statement tells SAS that raw data follows. The single semicolon that follows the last line of data marks the end of the data. The DATALINES statement and data lines must occur last in the DATA step statements:

```
data weight_club;
   input IdNumber 1-4 Name $ 6-24 Team $ StartWeight EndWeight;
   Loss = StartWeight - EndWeight;
   datalines;
1023 David Shaw          red      189 165
1049 Amelia Serrano      yellow   145 124
1219 Alan Nance          red      210 192
1246 Ravi Sinha          yellow   194 177
1078 Ashley McKnight     red      127 118
;
```

### Data in an External File

If your raw data is already stored in a file, then you do not have to bring that file into the data stream. Use an INFILE statement to specify the file containing the raw data. For more information about the INFILE, FILE, and FILENAME statements, see "Using External Files in Your SAS Job" on page 45. The statements in the code that follows demonstrate the same example, this time showing that the raw data is stored in an external file:

```
data weight_club;
   infile 'your-input-file';
   input IdNumber $ 1-4 Name $ 6-23 StartWeight 24-26
         EndWeight 28-30;
   Loss=StartWeight-EndWeight;
run;
```

### Data in a SAS Data Set

You can also use data that is already stored in a SAS data set as input to a new data set. To read data from an existing SAS data set, you must specify the existing data set's name in one of these statements:

- SET statement
- MERGE statement
- MODIFY statement
- UPDATE statement

For example, the statements that follow create a new SAS data set named RED that adds the variable LossPercent:

```
data red;
   set weight_club;
   LossPercent = Loss / StartWeight * 100;
run;
```

The SET statement indicates that the input data is already in the structure of a SAS data set and gives the name of the SAS data set to be read. In this example, the SET statement tells SAS to read the WEIGHT_CLUB data set in the WORK library.

### Data in a DBMS File

If you have data that is stored in another vendor's database management system (DBMS) files, then you can use SAS/ACCESS software to bring this data into a SAS data set. SAS/ACCESS software enables you to assign a libref to a library containing the DBMS file. In this example, a libref is declared, and points to a library containing Oracle data. SAS reads data from an Oracle file into a SAS data set:

```
libname dblib oracle user=scott password=tiger path='hrdept_002';
data employees;
   set dblib.employees;
run;
```

See *SAS/ACCESS for Relational Databases: Reference* for more information about using SAS/ACCESS software to access DBMS files.

## Using External Files in Your SAS Job

Your SAS programs often need to read raw data from a file, or write data or reports to a file that is not a SAS data set. To use a file that is not a SAS data set in a SAS program, you need to tell SAS where to find it. You can do the following:

- Identify the file directly in the INFILE, FILE, or other SAS statement that uses the file.
- Set up a fileref for the file by using the FILENAME statement, and then use the fileref in the INFILE, FILE, or other SAS statement.
- Use operating environment commands to set up a fileref, and then use the fileref in the INFILE, FILE, or other SAS statement.

The first two methods are described here. The third method depends on the operating environment that you use.

For more information, refer to the SAS documentation for your operating environment.

### Identifying an External File Directly

The simplest method for referring to an external file is to use the name of the file in the INFILE, FILE, or other SAS statement that needs to refer to the file. For example, if your raw data is stored in a file in your operating environment, and you want to read the data using a SAS DATA step, then you can tell SAS where to find the raw data by putting the name of the file in the INFILE statement:

```
data temp;
   infile 'your-input-file';
   input IdNumber $ 1-4 Name $ 6-23 StartWeight 24-26
         EndWeight 28-30;
```

```
run;
```

The INFILE statement for this example can appear as follows for various operating environments:

*Table 3.1* Example INFILE Statements for Various Operating Environments

| Operating Environment | INFILE Statement Example |
| --- | --- |
| z/OS | `infile 'fitness.weight.rawdata(club1)';` |
| UNIX | `infile '/usr/local/fitness/club1.dat';` |
| Windows | `infile 'c:\fitness\club1.dat';` |

For more information, refer to the SAS documentation for your operating environment.

### Referencing an External File with a Fileref

An alternate method for referencing an external file is to use the FILENAME statement to set up a fileref for a file. The fileref functions as a shorthand way of referring to an external file. You then use the fileref in later SAS statements that reference the file, such as the FILE or INFILE statement. The advantage of this method is that if the program contains many references to the same external file and the external filename changes, then the program needs to be modified in only one place, rather than in every place where the file is referenced.

Here is the form of the FILENAME statement:

**FILENAME** *fileref* '*your-input-or-output-file*';

The *fileref* must be a valid SAS name, that is, it must

- begin with a letter or an underscore
- contain only letters, numbers, or underscores
- have no more than 8 characters.

Additional restrictions can apply under some operating environments. For more information, refer to the SAS documentation for your operating environment.

For example, you can reference the raw data that is stored in a file in your operating environment by first using the FILENAME statement to specify the name of the file and its fileref, and then using the INFILE statement with the same fileref to reference the file.

```
filename fitclub 'your-input-file';

data temp;
   infile fitclub;
   input IdNumber $ 1-4 Name $ 6-23 StartWeight 24-26 EndWeight 28-30;
run;
```

In this example, the INFILE statement stays the same for all operating environments. The FILENAME statement, however, can appear differently in different operating environments.

The following table shows examples of FILENAME statements for different operating environments.

*Table 3.2  Example FILENAME Statements for Various Operating Environments*

| Operating Environment | FILENAME Statement Example |
| --- | --- |
| z/OS | `filename fitclub 'fitness.weight.rawdata(club1)';` |
| UNIX | `filename fitclub '/usr/local/fitness/club1.dat';` |
| Windows | `filename fitclub 'c:\fitness\club1.dat';` |

If you need to use several files or members from the same directory, partitioned data set (PDS), or MACLIB, then you can use the FILENAME statement to create a fileref that identifies the name of the directory, PDS, or MACLIB. Then you can use the fileref in the INFILE statement and enclose the name of the file, PDS member, or MACLIB member in parentheses immediately after the fileref, as in this example:

```
filename fitclub 'directory-or-PDS-or-MACLIB';

data temp;
   infile fitclub(club1);
   input IdNumber $ 1-4 Name $ 6-23 StartWeight 24-26 EndWeight 28-30;
run;

data temp2;
   infile fitclub(club2);
   input IdNumber $ 1-4 Name $ 6-23 StartWeight 24-26 EndWeight 28-30;
run;
```

In this case, the INFILE statements stay the same for all operating environments. The FILENAME statement, however, can appear differently for different operating environments, as the following table shows:

The following table shows examples of FILENAME statements for different operating environments.

*Table 3.3  Referencing Directories, PDSs, and MACLIBs in Various Operating Environments*

| Operating Environment | FILENAME Statement Example |
| --- | --- |
| z/OS | `filename fitclub 'fitness.weight.rawdata';`* |
| UNIX | `filename fitclub '/usr/local/fitness';` |
| Windows | `filename fitclub 'c:\fitness';` |

\* Under z/OS, the external file must be a PDS or a UFS file.

## Summary

### Statements

DATA <*libref.*>*SAS-data-set;*
　tells SAS to begin creating a SAS data set. If you omit the libref, then SAS creates a temporary SAS data set. (SAS attaches the libref WORK for its internal processing.) If you give a previously defined libref as the first level of the name, then SAS stores the data set permanently in the library referenced by the libref. A SAS program or a portion of a program that begins with a DATA statement and ends with a RUN statement, another DATA statement, or a PROC statement is called a DATA step.

FILENAME *fileref 'your-input-or-output-file';*
　associates a fileref with an external file. Enclose the name of the external file in quotation marks.

INFILE *fileref | 'your-input-file';*
　identifies an external file to be read by an INPUT statement. Specify a fileref that has been assigned with a FILENAME statement or with an appropriate operating environment command, or specify the actual name of the external file.

INPUT *variable* <*$*>;
　reads raw data using list input. At least one blank must occur between any two data values. The $ denotes a character variable.

INPUT *variable* <*$*>*column-range;*
　reads raw data that is aligned in columns. The $ denotes a character variable.

INPUT *variable informat;*
　reads raw data using formatted input. An informat supplies special instructions for reading the data.

LIBNAME *libref 'your-SAS-data-library';*
　associates a libref with a SAS library. Enclose the name of the library in quotation marks. SAS locates a permanent SAS data set by matching the libref in a two-level SAS data set name with the library associated with that libref in a LIBNAME statement. The rules for creating a SAS library depend on your operating environment.

## Learning More

ATTRIB statement
　For information about how the ATTRIB statement enables you to assign attributes to variables, see "ATTRIB Statement" in *SAS Statements: Reference*.

DBMS access
　This documentation explains how to use SAS for reading files of raw data and SAS data sets and writing to SAS data sets. However, SAS documentation for SAS/ACCESS provides complete information about using SAS to read and write information stored in several types of database management system (DBMS) files.

Informats
> For a discussion about informats that you use with dates, see Chapter 15, "Working with Dates in the SAS System," on page 225.

Length of variables
> For more information about how a variable's length affects the values that you can store in the variable, see Chapter 8, "Working with Numeric Variables," on page 117 and Chapter 9, "Working with Character Variables," on page 129.

LINESIZE= *option*
> For information about how to use the LINESIZE= option in an INPUT statement to limit how much of each data line the INPUT statement reads, see "INPUT Statement" in *SAS Statements: Reference.*

MERGE, MODIFY, or UPDATE statements
> In addition to the SET statement, you can read a SAS data set with the MERGE, MODIFY, or UPDATE statements. For more information, see Chapter 19, "Merging SAS Data Sets," on page 289 and Chapter 20, "Updating SAS Data Sets," on page 317.

SET statement
> For information about the SET statement, see Chapter 6, "Starting with SAS Data Sets," on page 91.

USER= *SAS system option*
> You can specify the USER= SAS system option to use one-level names to point to permanent SAS files. (If you specify USER=WORK, then SAS assumes that files referenced with one-level names refer to temporary work files.) For more information, see the *SAS System Options: Reference.*

# Chapter 4
# Starting with Raw Data: The Basics

**Introduction to Raw Data** .................................................. 52
    Purpose ............................................................. 52
    Prerequisites ........................................................ 52
**Examine the Structure of the Raw Data: Factors to Consider** ............... 52
**Reading Unaligned Data** ..................................................... 53
    Understanding List Input ............................................ 53
    Program: Basic List Input ........................................... 53
    Program: When the Data Is Delimited by Characters, Not Blanks ...... 54
    Program: When Consecutive Delimiters Indicate Missing Values ....... 55
    List Input: Points to Remember ...................................... 56
**Reading Data That Is Aligned in Columns** ................................... 57
    Understanding Column Input ......................................... 57
    Program: Reading Data Aligned in Columns ........................... 57
    Understanding Some Advantages of Column Input over Simple List Input ...... 58
    Reading Embedded Blanks and Creating Longer Variables .............. 58
    Program: Skipping Fields When Reading Data Records ................. 59
    Column Input: Points to Remember ................................... 60
**Reading Data That Requires Special Instructions** .......................... 60
    Understanding Formatted Input ...................................... 60
    Program: Reading Data That Requires Special Instructions ........... 61
    Understanding How to Control the Position of the Pointer ........... 62
    Formatted Input: Points to Remember ................................ 63
**Reading Unaligned Data with More Flexibility** .............................. 63
    Understanding How to Make List Input More Flexible ................. 63
    Creating Longer Variables and Reading Numeric Data That
        Contains Special Characters .................................... 64
    Reading Character Data That Contains Embedded Blanks ............... 64
**Mixing Styles of Input** ..................................................... 65
    An Example of Mixed Input .......................................... 65
    Understanding the Effect of Input Style on Pointer Location ........ 66
**Summary** .................................................................... 68
    Statements .......................................................... 68
    Column-Pointer Controls ............................................. 69
**Learning More** .............................................................. 69

## Introduction to Raw Data

### Purpose

To create a SAS data set from raw data, you must examine the data records first to determine how the data values that you want to read are arranged. Then you can look at the styles of reading input that are available in the INPUT statement. SAS provides three basic input styles:

- list
- column
- formatted

You can use these styles individually, in combination with each other, or in conjunction with various line-hold specifiers, line-pointer controls, and column-pointer controls. This section demonstrates various ways of using the INPUT statement to turn your raw data into SAS data sets.

You can enter the data directly in a DATA step or use an existing file of raw data. If your data is machine readable, then you need to learn how to use those tools that enable SAS to read them. If your data is not yet entered, then you can choose the input style that enables you to enter the data most easily.

### Prerequisites

Before continuing, you should understand the concepts presented in the following sections:

- Chapter 1, "What is the SAS System?," on page 3
- Chapter 3, "Introduction to DATA Step Processing," on page 27

## Examine the Structure of the Raw Data: Factors to Consider

Before you can select the appropriate style of input, examine the structure of the raw data that you want to read. Consider some of the following factors:

- how the data is arranged in the input records. For example, are data fields aligned in columns or unaligned? Are they separated by blanks or by other characters?
- whether character values contain embedded blanks
- whether numeric values contain nonnumeric characters such as commas
- whether the data contains time or date values
- whether each input record contains data for more than one observation
- whether data for a single observation is spread over multiple input records

# Reading Unaligned Data

## *Understanding List Input*

The simplest form of the INPUT statement uses list input. List input is used to read data values that are separated by a delimiter character (by default, a blank space). With list input, SAS reads a data value until it encounters a blank space. SAS assumes that the value has ended and assigns the data to the appropriate variable in the program data vector. SAS continues to scan the record until it reaches a non-blank character again. SAS reads a data value until it encounters a blank space or the end of the input record.

## *Program: Basic List Input*

This program uses the health and fitness club data from Chapter 3, "Introduction to DATA Step Processing," on page 27 to illustrate a DATA step that uses list input in an INPUT statement.

```
data club1;
    input IdNumber Name $ Team $ StartWeight EndWeight; 1
    datalines; 2
1023 David red 189 165  3
1049 Amelia yellow 145 124
1219 Alan red 210 192
1246 Ravi yellow 194 177
1078 Ashley red 127 118
1221 Jim yellow 220 .  3
; 2

proc print data=club1;
    title 'Weight of Club Members';
run;
```

The following list corresponds to the numbered items in the preceding program:

1 The variable names in the INPUT statement are specified in exactly the same order as the fields in the raw data records.

2 The DATALINES statement marks the beginning of the data lines. The semicolon that follows the data lines marks the end of the data lines and the end of the DATA step.

3 Each data value in the raw data record is separated from the next by at least one blank space. The last record contains a missing value, represented by a period, for the value of EndWeight.

The following output displays the resulting data set. The PROC PRINT statement that follows the DATA step produces this listing.

*Figure 4.1* Data Set Created with List Input

### Weight of Club Members

| Obs | IdNumber | Name | Team | StartWeight | EndWeight |
|---|---|---|---|---|---|
| 1 | 1023 | David | red | 189 | 165 |
| 2 | 1049 | Amelia | yellow | 145 | 124 |
| 3 | 1219 | Alan | red | 210 | 192 |
| 4 | 1246 | Ravi | yellow | 194 | 177 |
| 5 | 1078 | Ashley | red | 127 | 118 |
| 6 | 1221 | Jim | yellow | 220 | . |

### Program: When the Data Is Delimited by Characters, Not Blanks

This program also uses the health and fitness club data but notice that here the data is delimited by a comma instead of a blank space, the default delimiter.

```
data club1;
   infile datalines dlm=',';  1

   input IdNumber Name $ Team $ StartWeight EndWeight;
   datalines;
1023,David,red,189,165  2
1049,Amelia,yellow,145,124
1219,Alan,red,210,192
1246,Ravi,yellow,194,177
1078,Ashley,red,127,118
1221,Jim,yellow,220,.
;
proc print data=club1;
   title 'Weight of Club Members';
run;
```

The following list corresponds to the numbered items in the preceding program:

1. List input, by default, scans the input records, looking for blank spaces to delimit each data value. The DLM= option enables list input to recognize a character, here a comma, as the delimiter.

2. These data values are separated by commas instead of blanks.

*Note:* This example requires the DLM= option, which is available only in the INFILE statement. Usually, this statement is used only when the input data resides in an external file. The DATALINES specification, however, enables you to take advantage of INFILE statement options, when you are reading data records from the job stream.

**Figure 4.2**  *Reading Data Delimited by Commas*

### Weight of Club Members

| Obs | IdNumber | Name | Team | StartWeight | EndWeight |
|---|---|---|---|---|---|
| 1 | 1023 | David | red | 189 | 165 |
| 2 | 1049 | Amelia | yellow | 145 | 124 |
| 3 | 1219 | Alan | red | 210 | 192 |
| 4 | 1246 | Ravi | yellow | 194 | 177 |
| 5 | 1078 | Ashley | red | 127 | 118 |
| 6 | 1221 | Jim | yellow | 220 | |

## Program: When Consecutive Delimiters Indicate Missing Values

This program also uses the health and fitness club data. Notice that here the data is delimited by a colon (:), and there is a missing value for Alan that is indicated by two consecutive delimiters.

```
data club1;
    infile datalines dsd dlm=':';
    input IdNumber Name $ Team $ StartWeight EndWeight;
    datalines;
1023:David:red:189:165
1049:Amelia:yellow:145:124
1219:Alan::210:192
1246:Ravi:yellow:194:177
1078:Ashley:red:127:118
1221:Jim:yellow:220:.
;
proc print data=club1;
    title 'Weight of Club Members';
run;
```

This example requires the DSD option, which is available only in the INFILE statement. Like the previous example, the DATALINES specification enables you to specify the INFILE statement with the DSD and DLM= options. The DSD option treats two consecutive delimiters as a missing value. (The DSD option also sets the default delimiter to a comma and ignores delimiters that occur within quoted strings.)

If you did not specify the DSD option in this example, then data values would become mismatched and the remaining data would not be read correctly. The following listing shows how the data would have been read if the DSD option were omitted. Notice that the values after Alan's name are incorrect, and the observation for Ravi is omitted. Ravi's IdNumber is read as Alan's EndWeight value and the remainder of that record is ignored.

**Figure 4.3** *DSD Option Incorrectly Omitted*

| Obs | IdNumber | Name | Team | StartWeight | EndWeight |
|---|---|---|---|---|---|
| 1 | 1023 | David | red | 189 | 165 |
| 2 | 1049 | Amelia | yellow | 145 | 124 |
| 3 | 1219 | Alan | 210 | 192 | 1246 |
| 4 | 1078 | Ashley | red | 127 | 118 |
| 5 | 1221 | Jim | yellow | 220 | |

INCORRECT: Weight of Club Members

When you include the DSD option in the INFILE statement, the consecutive delimiters indicate a missing value, and the data is read correctly. The following listing shows the resulting data set. You can see that now the missing Team value for Alan appears as a blank value, and the remaining data set is now read as expected.

**Figure 4.4** *Missing Values Are Read Correctly with the DSD Option*

CORRECT: Weight of Club Members

| Obs | IdNumber | Name | Team | StartWeight | EndWeight |
|---|---|---|---|---|---|
| 1 | 1023 | David | red | 189 | 165 |
| 2 | 1049 | Amelia | yellow | 145 | 124 |
| 3 | 1219 | Alan |  | 210 | 192 |
| 4 | 1246 | Ravi | yellow | 194 | 177 |
| 5 | 1078 | Ashley | red | 127 | 118 |
| 6 | 1221 | Jim | yellow | 220 | |

## List Input: Points to Remember

These are the points to remember when you use list input:

- Use list input when each field is separated by at least one blank space or delimiter. Use the DLM= option in an INFILE statement to indicate delimiters other than spaces. For list input, you must also specify DATALINES in the INFILE statement.
- Specify each field in the order in which they appear in the records of raw data.
- Represent missing values by a placeholder such as a period.

  Under the default behavior, a blank field causes the variable names and values to become mismatched. However, if you use the DSD option with comma delimiters, multiple commas indicate a missing value. To specify a delimiter other than a comma with the DSD option, use the DLM= option.

  *Note:* Specify the DSD and DLM= options in an INFILE statement. For list input, you must also specify DATALINES in the INFILE statement.

- Character values cannot contain embedded blanks, unless you also specify the DSD and DLM= options in an INFILE statement.

- The default length of character variables is eight bytes. SAS truncates a longer value when it writes the value to the program data vector. (To read a character variable that contains more than eight characters with list input, use a LENGTH statement. See "Defining Enough Storage Space for Variables" on page 114.)
- Data must be in standard character or numeric format (that is, it can be read without an informat).

*Note:* List input requires the fewest specifications in the INPUT statement. However, the restrictions that are placed on the data might require that you learn to use other styles of input to read your data. For example, column input, which is discussed in the next section, is less restrictive. This section has introduced only simple list input. See "Understanding How to Make List Input More Flexible" on page 63 to learn about modified list input.

# Reading Data That Is Aligned in Columns

## Understanding Column Input

With column input, data values occupy the same fields within each data record. When you use column input in the INPUT statement, list the variable names and specify column positions that identify the location of the corresponding data fields. You can use column input when your raw data is in fixed columns and does not require the use of informats to be read.

## Program: Reading Data Aligned in Columns

The following program also uses the health and fitness club data, but now two more data values are missing. The data is aligned in columns and SAS reads the data with column input:

```
data club1;
   input IdNumber 1-4 Name $ 6-11 Team $ 13-18 StartWeight 20-22
         EndWeight 24-26;
   datalines;
1023 David  red    189 165
1049 Amelia yellow 145
1219 Alan   red    210 192
1246 Ravi   yellow     177
1078 Ashley red    127 118
1221 Jim    yellow 220
;

proc print data=club1;
   title 'Weight Club Members';
run;
```

The specification that follows each variable name indicates the beginning and ending columns in which the variable value will be found. Note that with column input that you are not required to indicate missing values with a placeholder such as a period.

The following output displays the resulting data set. Missing numeric values occur three times in the data set, and are indicated by periods.

*Figure 4.5  Data Set Created with Column Input*

### Weight Club Members

| Obs | IdNumber | Name | Team | StartWeight | EndWeight |
|---|---|---|---|---|---|
| 1 | 1023 | David | red | 189 | 165 |
| 2 | 1049 | Amelia | yellow | 145 | . |
| 3 | 1219 | Alan | red | 210 | 192 |
| 4 | 1246 | Ravi | yellow | . | 177 |
| 5 | 1078 | Ashley | red | 127 | 118 |
| 6 | 1221 | Jim | yellow | 220 | . |

## Understanding Some Advantages of Column Input over Simple List Input

Here are several advantages of using column input:

- With column input, character variables can contain embedded blanks.

- Column input also enables the creation of variables that are longer than eight bytes. In the preceding example, the variable Name in the data set CLUB1 contains only the members' first names. By using column input, you can read the first and last names as a single value. These differences between input styles are possible for two reasons:

  - Column input uses the columns that you specify to determine the length of character variables.

  - Column input, unlike list input, reads data until it reaches the last specified column, not until it reaches a blank space.

- Column input enables you to skip some data fields when reading records of raw data. It also enables you to read the data fields in any order and reread some fields or parts of fields.

## Reading Embedded Blanks and Creating Longer Variables

This DATA step uses column input to create a new data set named CLUB2. The program still uses the health and fitness club weight data. However, the data has been modified to include members' first and last names. Now the second data field in each record or raw data contains an embedded blank and is 18 bytes long.

```
data club2;
   input IdNumber 1-4 Name $ 6-23 Team $ 25-30 StartWeight 32-34
         EndWeight 36-38;
   datalines;
1023 David Shaw        red    189 165
1049 Amelia Serrano    yellow 145 124
```

```
1219  Alan Nance            red     210 192
1246  Ravi Sinha            yellow  194 177
1078  Ashley McKnight       red     127 118
1221  Jim Brown             yellow  220
;

proc print data=club2;
   title 'Weight Club Members';
run;
```

The following output displays the resulting data set:

*Figure 4.6*  *Data Set Created with Column Input (Embedded Blanks)*

### Weight Club Members

| Obs | IdNumber | Name | Team | StartWeight | EndWeight |
|---|---|---|---|---|---|
| 1 | 1023 | David Shaw | red | 189 | 165 |
| 2 | 1049 | Amelia Serrano | yellow | 145 | 124 |
| 3 | 1219 | Alan Nance | red | 210 | 192 |
| 4 | 1246 | Ravi Sinha | yellow | 194 | 177 |
| 5 | 1078 | Ashley McKnight | red | 127 | 118 |
| 6 | 1221 | Jim Brown | yellow | 220 |  |

## Program: Skipping Fields When Reading Data Records

Column input also enables you to skip over fields or to read the fields in any order. This example uses column input to read the same health and fitness club data, but it reads the value for the variable Team first and omits the variable IdNumber altogether.

You can read or reread part of a value when using column input. For example, because the team names begin with different letters, this program saves storage space by reading only the first character in the field that contains the team name. Note the INPUT statement:

```
data club2;
   input Team $ 25 Name $ 6-23 StartWeight 32-34 EndWeight 36-38;
   datalines;
1023  David Shaw            red     189 165
1049  Amelia Serrano        yellow  145 124
1219  Alan Nance            red     210 192
1246  Ravi Sinha            yellow  194 177
1078  Ashley McKnight       red     127 118
1221  Jim Brown             yellow  220
;

proc print data=club2;
   title 'Weight Club Members';
run;
```

The following output displays the resulting data set. The variable that contains the identification number is no longer in the data set. Instead, Team is the first variable in the new data set, and it contains only one character to represent the team value.

**Figure 4.7** *Data Set Created with Column Input (Skipping Fields)*

### Weight Club Members

| Obs | Team | Name | StartWeight | EndWeight |
|---|---|---|---|---|
| 1 | r | David Shaw | 189 | 165 |
| 2 | y | Amelia Serrano | 145 | 124 |
| 3 | r | Alan Nance | 210 | 192 |
| 4 | y | Ravi Sinha | 194 | 177 |
| 5 | r | Ashley McKnight | 127 | 118 |
| 6 | y | Jim Brown | 220 | . |

### *Column Input: Points to Remember*

Remember the following rules when you use column input:

- Character variables can be up to 32,767 bytes (32KB) in length and are not limited to the default length of eight bytes.
- Character variables can contain embedded blanks.
- You can read fields in any order.
- A placeholder is not required to indicate a missing data value. A blank field is read as missing and does not cause other values to be read incorrectly.
- You can skip over part of the data in the data record.
- You can reread fields or parts of fields.
- You can read standard character and numeric data only. Informats are ignored.

## Reading Data That Requires Special Instructions

### *Understanding Formatted Input*

Sometimes the INPUT statement requires special instructions to read the data correctly. For example, SAS can read numeric data that is in special formats such as binary, packed decimal, or date/time. SAS can also read numeric values that contain special characters such as commas and currency symbols. In these situations, use formatted input. Formatted input combines the features of column input with the ability to read nonstandard numeric or character values. The following data shows formatted input:

- 1,262
- $55.64

- 02JAN2013

## Program: Reading Data That Requires Special Instructions

The data in this program includes numeric values that contain a comma, which is an invalid character for a numeric variable:

```
data january_sales;
   input Item $ 1-16 Amount comma5.;
   datalines;
trucks          1,382
vans            1,235
sedans          2,391
;

proc print data=january_sales;
   title 'January Sales in Thousands';
run;
```

The INPUT statement cannot read the values for the variable Amount as valid numeric values without the additional instructions provided by an informat. The informat COMMA5. enables the INPUT statement to read and store this data as a valid numeric value.

The following figure shows that the informat COMMA5. instructs the program to read five characters of data (the comma counts as part of the length of the data), to remove the comma from the data, and to write the resulting numeric value to the program data vector. Note that the name of an informat always ends in a period (.).

*Figure 4.8  Reading a Value with an Informat*

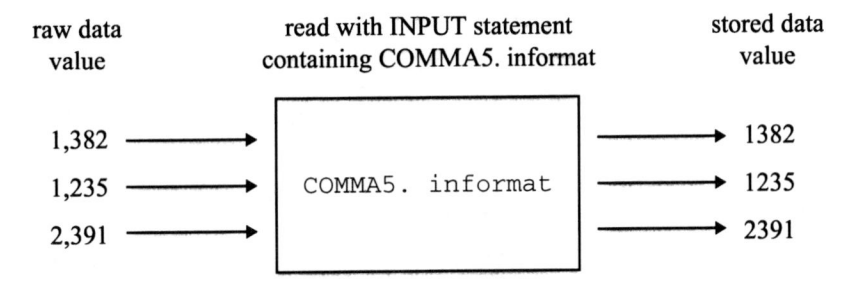

| raw data value | read with INPUT statement containing COMMA5. informat | stored data value |
|---|---|---|
| 1,382 | COMMA5. informat | 1382 |
| 1,235 |  | 1235 |
| 2,391 |  | 2391 |

The following figure shows that the data values are read into the input buffer exactly as they occur in the raw data records. However, they are written to the program data vector

**62** Chapter 4 • *Starting with Raw Data: The Basics*

(and then to the data set as an observation) as valid numeric values without any special characters.

*Figure 4.9* Input Value Compared to Variable Value

**Input Buffer**

```
----+----1----+----2----+----3
trucks          1,382
```

Program Data Vector

| Item | Amount |
|------|--------|
| trucks | 1382 |

The following output displays the resulting data set. The values for Amount contain only numbers. Note that the commas are removed.

*Figure 4.10* Data Set Created with Column and Formatted Input

January Sales in Thousands

| Obs | Item | Amount |
|-----|------|--------|
| 1 | trucks | 1382 |
| 2 | vans | 1235 |
| 3 | sedans | 2391 |

In a report, you might want to include the comma in numeric values to improve readability. Just as the informat gives instructions on how to read a value and to remove the comma, a format gives instructions to add characters to variable values in the output. For an example, see "Writing Output without Creating a Data Set" on page 596.

## Understanding How to Control the Position of the Pointer

As the INPUT statement reads data values, it uses an input pointer to keep track of the position of the data in the input buffer. Column-pointer controls provide additional control over pointer movement and are especially useful with formatted input. Column-pointer controls tell how far to advance the pointer before SAS reads the next value. In this example, SAS reads data lines with a combination of column and formatted input:

```
data january_sales;
   input Item $ 1-16 Amount comma5.;
   datalines;
trucks          1,382
vans            1,235
sedans          2,391
;
```

In the next example, SAS reads data lines by using formatted input with a column-pointer control:

```
data january_sales;
   input Item $10. @17 Amount comma5.;
```

```
        datalines;
trucks          1,382
vans            1,235
sedans          2,391
;
```

After SAS reads the first value for the variable Item, the pointer is left in the next position, column 11. Then the absolute column-pointer control, @17, directs the pointer to move to column 17 in the input buffer. Now, it is in the correct position to read a value for the variable Amount.

In the following program, the relative column-pointer control, +6, instructs the pointer to move six columns to the right before SAS reads the next data value.

```
data january_sales;
    input Item $10. +6 Amount comma5.;
    datalines;
trucks          1,382
vans            1,235
sedans          2,391
;
```

The data in these two programs is aligned in columns. As with column input, you instruct the pointer to move from field to field. You use column specifications with column input. However, with formatted input that you use the length that is specified in the informat together with pointer controls.

## *Formatted Input: Points to Remember*

Remember the following rules when you use formatted input:

- SAS reads formatted input data until it has read the number of columns that the informat indicates. This method of reading the data is different from list input, which reads until a blank space (or other defined delimiter character) is reached.
- You can position the pointer to read the next value by using pointer controls.
- You can read data stored in nonstandard form such as packed decimal, or data that contains commas.
- You have the flexibility of using informats with all the features of column input, as described in "Column Input: Points to Remember" on page 60.

# Reading Unaligned Data with More Flexibility

## *Understanding How to Make List Input More Flexible*

While list input is the simplest to code, remember that it places restrictions on your data. By using format modifiers, you can take advantage of the simplicity of list input without the inconvenience of the usual restrictions. For example, you can use modified list input to do the following:

- Create character variables that are longer than the default length of eight bytes.
- Read numeric data with special characters like commas, hyphens, and currency symbols.

## Creating Longer Variables and Reading Numeric Data That Contains Special Characters

By simply modifying list input with the colon format modifier (:) you can read

- character data that contains more than eight characters
- numeric data that contains special characters

To use the colon format modifier with list input, place the colon between the variable name and the informat. As in simple list input, at least one blank (or other defined delimiter character) must separate each value from the next, and character values cannot contain embedded blanks (or other defined delimiter characters). Consider this DATA step:

```
data january_sales;

   input Item : $12. Amount : comma5.;
   datalines;
Trucks 1,382
Vans 1,235
Sedans 2,391
SportUtility 987
;

proc print data=january_sales;
   title 'January Sales in Thousands';
run;
```

The variable Item has a length of 12, and the variable Amount requires an informat (in this case, COMMA5.) that removes commas from numbers so that they are read as valid numeric values. The data values are not aligned in columns as was required in the last example, which used formatted input to read the data.

The following output displays the resulting data set:

**Figure 4.11** *Data Set Created with Modified List Input (: comma5.)*

January Sales in Thousands

| Obs | Item | Amount |
|---|---|---|
| 1 | trucks | 1382 |
| 2 | vans | 1235 |
| 3 | sedans | 2391 |

## Reading Character Data That Contains Embedded Blanks

Because list input uses a blank space to determine where one value ends and the next one begins, values normally cannot contain blanks. However, with the ampersand format modifier (&) you can use list input to read data that contains single embedded blanks.

The only restriction is that at least two blanks must divide each value from the next data value in the record.

To use the ampersand format modifier with list input, place the ampersand between the variable name and the informat. The following DATA step uses the ampersand format modifier with list input to create the data set CLUB2. Note that the data is not in fixed columns. Therefore, column input is not appropriate.

```
data club2;
    input IdNumber Name & $18. Team $ StartWeight EndWeight;
    datalines;
1023 David Shaw    red 189 165
1049 Amelia Serrano  yellow 145 124
1219 Alan Nance   red 210 192
1246 Ravi Sinha   yellow 194 177
1078 Ashley McKnight  red 127 118
1221 Jim Brown   yellow 220 .
;

proc print data=club2;
    title 'Weight Club Members';
run;
```

The character variable Name, with a length of 18, contains members' first and last names separated by one blank space. The data lines must have two blank spaces between the values for the variable Name and the variable Team for the INPUT statement to correctly read the data.

The following output displays the resulting data set:

*Figure 4.12  Data Set Created with Modified List Input (& $18.)*

### Weight Club Members

| Obs | IdNumber | Name | Team | StartWeight | EndWeight |
|---|---|---|---|---|---|
| 1 | 1023 | David Shaw | red | 189 | 165 |
| 2 | 1049 | Amelia Serrano | yellow | 145 | 124 |
| 3 | 1219 | Alan Nance | red | 210 | 192 |
| 4 | 1246 | Ravi Sinha | yellow | 194 | 177 |
| 5 | 1078 | Ashley McKnight | red | 127 | 118 |
| 6 | 1221 | Jim Brown | yellow | 220 | |

# Mixing Styles of Input

## *An Example of Mixed Input*

When you begin an INPUT statement in a particular style (list, column, or formatted), you are not restricted to using that style alone. You can mix input styles in a single

INPUT statement as long as you mix them in a way that appropriately describes the raw data records. For example, this DATA step uses all three input styles:

```
data club1;
   input IdNumber  ❶
         Name $18.  ❷
         Team $ 25-30  ❸
         StartWeight EndWeight;

   datalines; ❶
1023 David Shaw          red     189 165
1049 Amelia Serrano      yellow  145 124
1219 Alan Nance          red     210 192
1246 Ravi Sinha          yellow  194 177
1078 Ashley McKnight     red     127 118
1221 Jim Brown           yellow  220   .
;

proc print data=club1;
   title 'Weight Club Members';
run;
```

The following list corresponds to the numbered items in the preceding program:

1. The variables IdNumber, StartWeight, and EndWeight are read with list input.
2. The variable Name is read with formatted input.
3. The variable Team is read with column input.

The following output demonstrates that the data is read correctly.

**Figure 4.13** *Data Set Created with Mixed Styles of Input*

Weight Club Members

| Obs | IdNumber | Name | Team | StartWeight | EndWeight |
|---|---|---|---|---|---|
| 1 | 1023 | David Shaw | red | 189 | 165 |
| 2 | 1049 | Amelia Serrano | yellow | 145 | 124 |
| 3 | 1219 | Alan Nance | red | 210 | 192 |
| 4 | 1246 | Ravi Sinha | yellow | 194 | 177 |
| 5 | 1078 | Ashley McKnight | red | 127 | 118 |
| 6 | 1221 | Jim Brown | yellow | 220 | . |

## Understanding the Effect of Input Style on Pointer Location

### Why You Can Get into Trouble By Mixing Input Styles

CAUTION:
**When you mix styles of input in a single INPUT statement, you can get unexpected results if you do not understand where the input pointer is positioned after SAS reads a value in the input buffer.** As the INPUT statement reads data values from the record in the input buffer, it uses a pointer to keep track of its position. Read the following sections so that you understand how the pointer

movement differs between input styles before mixing multiple input styles in a single INPUT statement

### *Pointer Location with Column and Formatted Input*

With column and formatted input, you supply the instructions that determine the exact pointer location. With column input, SAS reads the columns that you specify in the INPUT statement. With formatted input, SAS reads the exact length that you specify with the informat. In both cases, the pointer moves as far as you instruct it and stops. The pointer is left in the column that immediately follows the last column that is read.

Here are two examples of input followed by an explanation of the pointer location. The first DATA step shows column input:

```
data scores;
   input Team $ 1-6 Score 12-13;
   datalines;
red        59
blue       95
yellow     63
green      76
;
```

The second DATA step uses the same data to show formatted input:

```
data scores;
   input Team $6. +5 Score 2.;
   datalines;
red        59
blue       95
yellow     63
green      76
;
```

The following figure shows that the pointer is located in column 7 after the first value is read with either of the two previous INPUT statements.

*Figure 4.14*  Pointer Position: Column and Formatted Input

```
----+----1----+----2
red        59
 ↑
```

Unlike list input, column and formatted input rely totally on your instructions to move the pointer and read the value for the second variable, Score. Column input uses column specifications to move the pointer to each data field. Formatted input uses informats and pointer controls to control the position of the pointer.

This INPUT statement uses column input with the column specifications 12-13 to move the pointer to column 12 and read the value for the variable Score:

```
input Team $ 1-6 Score 12-13;
```

This INPUT statement uses formatted input with the +5 column-pointer control to move the pointer to column 12. Then the value for the variable Score is read with the 2. numeric informat.

```
input Team $6. +5 Score 2.;
```

**68** Chapter 4 • *Starting with Raw Data: The Basics*

Without the use of a pointer control, which moves the pointer to the column where the value begins, this INPUT statement would attempt to read the value for Score in columns 7 and 8, which are blank.

### Pointer Location with List Input

List input, on the other hand, uses a scanning method to determine the pointer location. With list input, the pointer reads until a blank is reached and then stops in the next column. To read the next variable value, the pointer moves automatically to the first nonblank column, discarding any leading blanks that it encounters. Here is the same data that is read with list input:

```
data scores;
   input Team $ Score;
   datalines;
red        59
blue       95
yellow     63
green      76
;
```

The following figure shows that the pointer is located in column 5 after the value `red` is read. Because Score, the next variable, is read with list input, the pointer scans for the next nonblank space before it begins to read a value for Score. Unlike column and formatted input, you do not have to explicitly move the pointer to the beginning of the next field in list input.

**Figure 4.15** Pointer Position: List Input

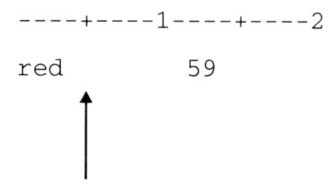

```
----+----1----+----2
red         59
 ↑
```

## Summary

### Statements

DATALINES;
    indicates that data lines immediately follow the DATALINES statement. A semicolon in the line that immediately follows the last data line indicates the end of the data and causes the DATA step to compile and execute.

INFILE DATALINES DLM='*character*';
    identifies the source of the input records as data lines in the job stream rather than as an external file. When your program contains the input data, the data lines directly follow the DATALINES statement. Because you can specify DATALINES in the INFILE statement, you can take advantage of many data-reading options that are available only through the INFILE statement.

    The DLM= option specifies the character that is used to separate data values in the input records. By default, a blank space denotes the end of a data value. This option is useful when you want to use list input to read data records in which a character other than a blank separates data values.

INPUT *variable* <&><$>;
> reads the input data record using list input. The & (ampersand format modifier) enables character values to contain embedded blanks. When you use the ampersand format modifier, two blanks are required to signal the end of a data value. The $ indicates a character variable.

INPUT *variable start-column* <– *end-column*>;
> reads the input data record using column input. You can omit end-column if the data is only 1 byte long. This style of input enables you to skip columns of data that you want to omit.

INPUT *variable* : *informat*;
INPUT *variable* & *informat*;
> reads the input data record using modified list input. The : (colon format modifier) instructs SAS to use the informat that follows to read the data value. The & (ampersand format modifier) instructs SAS to use the informat that follows to read the data value. When you use the ampersand format modifier, two blanks are required to signal the end of a data value.

INPUT <*pointer-control*> *variable informat*;
> reads raw data using formatted input. The informat supplies special instructions to read the data. You can also use a pointer-control to direct SAS to start reading at a particular column.

The syntax given above for the three styles of input shows only one variable. Subsequent variables in the INPUT statement might be described in the same input style as the first one. You can use any of the three styles of input (list, column, and formatted) in a single INPUT statement.

## Column-Pointer Controls

@n
> moves the pointer to the nth column in the input buffer.

+n
> moves the pointer forward n columns in the input buffer.

/
> moves the pointer to the next line in the input buffer.

#n
> moves the pointer to the nth line in the input buffer.

# Learning More

Advanced features
> For some more advanced data-reading features, see Chapter 5, "Starting with Raw Data: Beyond the Basics," on page 71.

Character-delimited data
> For more information about reading data that is delimited by a character other than a blank space, see the DELIMITER= option in "INFILE Statement" in *SAS Statements: Reference*.

Pointer controls
: For a complete discussion and listing of column-pointer controls, line-pointer controls, and line-hold specifiers, see *SAS Statements: Reference*.

Types of input
: For more information about the INPUT statement, see "INPUT Statement" in *SAS Statements: Reference*.

# Chapter 5
# Starting with Raw Data: Beyond the Basics

**Introduction to Beyond the Basics with Raw Data** ......................... 71
   Purpose ................................................................. 71
   Prerequisites ............................................................ 72

**Testing a Condition Before Creating an Observation** ....................... 72

**Creating Multiple Observations from a Single Record** ...................... 73
   Using the Double Trailing @ Line-Hold Specifier ........................ 73
   Understanding How the Double Trailing @ Affects DATA Step Execution ...... 74

**Reading Multiple Records to Create a Single Observation** ................... 77
   How the Data Records Are Structured ..................................... 77
   Method 1: Using Multiple Input Statements ............................... 77
   Method 2: Using the / Line-Pointer Control .............................. 79
   Reading Variables from Multiple Records in Any Order .................... 80
   Understanding How the #n Line-Pointer Control Affects DATA Step Execution .. 81

**Problem Solving: When an Input Record Unexpectedly Does Not Have Enough Values** ............................................................. 84
   Understanding the Default Behavior ...................................... 84
   Methods of Control: Your Options ........................................ 85

**Summary** .................................................................. 87
   Column-Pointer Controls ................................................. 87
   Line-Hold Specifiers .................................................... 87
   Statements .............................................................. 88

**Learning More** ............................................................ 89

## Introduction to Beyond the Basics with Raw Data

### Purpose

To create a SAS data set from raw data, you often need more than the most basic features. In this section, you will learn advanced features for reading raw data that include the following:

- how to understand and then control what happens when a value is unexpectedly missing in an input record

- how to read a record more than once so that you can test a condition before taking action on the current record
- how to create multiple observations from a single input record
- how to read multiple observations to create a single record

## Prerequisites

Before continuing, you should understand the concepts presented in the following sections:

- Chapter 1, "What is the SAS System?," on page 3
- Chapter 3, "Introduction to DATA Step Processing," on page 27

# Testing a Condition Before Creating an Observation

Sometimes you need to read a record, and hold that record in the input buffer while you test for a specified condition before a decision can be made about further processing. For example, the ability to hold a record so that you can read from it again, if necessary, is useful when you need to test for a condition before SAS creates an observation from a data record. To do this, you can use the trailing at-sign (@).

For example, to create a SAS data set that is a subset of a larger group of records, you might need to test for a condition to decide whether a particular record will be used to create an observation. The trailing at-sign placed before the semicolon at the end of an INPUT statement instructs SAS to hold the current data line in the input buffer. This makes the data line available for a subsequent INPUT statement. Otherwise, the next INPUT statement causes SAS to read a new record into the input buffer.

You can set up the process to read each record twice by following these steps:

1. Use an INPUT statement to read a portion of the record.
2. Use a trailing @ at the end of the INPUT statement to hold the record in the input buffer for the execution of the next INPUT statement.
3. Use an IF statement on the portion that is read in to test for a condition.
4. If the condition is met, use another INPUT statement to read the remainder of the record to create an observation.
5. If the condition is not met, the record is released and control passes back to the top of the DATA step.

To read from a record twice, you must prevent SAS from automatically placing a new record into the input buffer when the next INPUT statement executes. Use of a trailing @ in the first INPUT statement serves this purpose. The trailing @ is one of two line-hold specifiers that enable you to hold a record in the input buffer for further processing.

For example, the health and fitness club data contains information about all members. This DATA step creates a SAS data set that contains only members of the red team:

```
data red_team;
    input Team $ 13-18 @;   1
    if Team='red';          2
```

```
         input IdNumber 1-4 StartWeight 20-22 EndWeight 24-26; 3
      datalines;
1023 David   red      189 165
1049 Amelia  yellow   145 124
1219 Alan    red      210 192
1246 Ravi    yellow   194 177
1078 Ashley  red      127 118
1221 Jim     yellow   220  .
;  4

proc print data=red_team;
   title 'Red Team';
run;
```

In this DATA step, these actions occur:

1. The INPUT statement reads a record into the input buffer, reads a data value from columns 13 through 18, and assigns that value to the variable Team in the program data vector. The single trailing @ holds the record in the input buffer.

2. The IF statement enables the current iteration of the DATA step to continue only when the value for Team is **red**. When the value is not red, the current iteration stops and SAS returns to the top of the DATA step, resets values in the program data vector to missing, and releases the held record from the input buffer.

3. The INPUT statement executes only when the value of Team is **red**. It reads the remaining data values from the record held in the input buffer and assigns values to the variables IdNumber, StartWeight, and EndWeight.

4. The record is released from the input buffer when the program returns to the top of the DATA step.

The following output displays the resulting data set:

**Figure 5.1** *Subset Data Set Created with Trailing @*

### Red Team

| Obs | Team | IdNumber | StartWeight | EndWeight |
|---|---|---|---|---|
| 1 | red | 1023 | 189 | 165 |
| 2 | red | 1219 | 210 | 192 |
| 3 | red | 1078 | 127 | 118 |

# Creating Multiple Observations from a Single Record

## Using the Double Trailing @ Line-Hold Specifier

Sometimes you might need to create multiple observations from a single record of raw data. One way to tell SAS how to read such a record is to use the other line-hold specifier, the double trailing at-sign (@@ or "double trailing @"). The double trailing @

not only prevents SAS from reading a new record into the input buffer when a new INPUT statement is encountered. It also prevents the record from being released when the program returns to the top of the DATA step. (Remember that the trailing @ does not hold a record in the input buffer across iterations of the DATA step.)

For example, this DATA step uses the double trailing @ in the INPUT statement:

```
data body_fat;
   input Gender $ PercentFat @@;
   datalines;
m 13.3 f 22
m 22   f 23.2
m 16   m 12
;

proc print data=body_fat;
   title 'Results of Body Fat Testing';
run;
```

The following output displays the resulting data set:

**Figure 5.2** *Data Set Created with Double Trailing @*

Results of Body Fat Testing

| Obs | Gender | PercentFat |
|---|---|---|
| 1 | m | 13.3 |
| 2 | f | 22.0 |
| 3 | m | 22.0 |
| 4 | f | 23.2 |
| 5 | m | 16.0 |
| 6 | m | 12.0 |

## *Understanding How the Double Trailing @ Affects DATA Step Execution*

To understand how the data records in the previous example were read, look at the data lines that were used in the previous DATA step:

```
m 13.3 f 22
m 22   f 23.2
m 16   m 12
```

Each record contains the raw data for two observations instead of one. Consider this example in terms of the flow of the DATA step, as explained in Chapter 3, "Introduction to DATA Step Processing," on page 27.

When SAS reaches the end of the DATA step, it returns to the top of the program and begins the next iteration, executing until there are no more records to read. Each time it returns to the top of the DATA step and executes the INPUT statement, it automatically reads a new record into the input buffer. The second set of data values in each record, therefore, would never be read:

```
m 13.3 f 22
m 22   f 23.2
```

```
m 16    m 12
```

To allow the second set of data values in each record to be read, the double trailing @ tells SAS to hold the record in the input buffer. Each record is held in the input buffer until the end of the record is reached. The program does not automatically place the next record into the input buffer each time the INPUT statement is executed. The current record is not automatically released when it returns to the top of the DATA step. As a result, the pointer location is maintained on the current record, which enables the program to read each value in that record. Each time the DATA step completes an iteration, an observation is written to the data set.

The next five figures demonstrate what happens in the input buffer when a double trailing @ appears in the INPUT statement, as in this example:

```
input Gender $ PercentFat @@;
```

The first figure shows that all values in the program data vector are set to missing. The INPUT statement reads the first record into the input buffer. The program begins to read values from the current pointer location, which is the beginning of the input buffer.

***Figure 5.3***  *First Iteration: First Record Is Read*

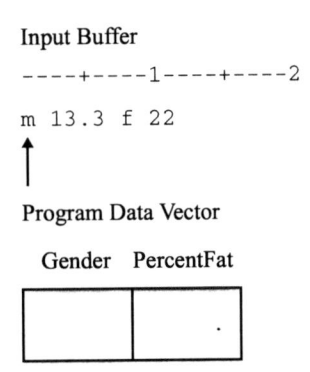

**Input Buffer**

```
----+----1----+----2
m 13.3 f 22
↑
```

**Program Data Vector**

| Gender | PercentFat |
|--------|------------|
|        | .          |

The following figure shows that the value **m** is written to the program data vector. When the pointer reaches the blank space that follows 13.3, the complete value for the variable PercentFat has been read. The pointer stops in the next column, and the value 13.3 is written to the program data vector.

***Figure 5.4***  *First Observation Is Created*

**Input Buffer**

```
----+----1----+----2
m 13.3 f 22
      ↑
```

**Program Data Vector**

| Gender | PercentFat |
|--------|------------|
| m      | 13.3       |

There are no other variables in the INPUT statement and no more statements in the DATA step, so three actions take place:

1. The first observation is written to the data set.

2. The DATA step begins its next iteration.

**76** Chapter 5 • *Starting with Raw Data: Beyond the Basics*

3. The values in the program data vector are set to missing.

The following figure shows the current position of the pointer. SAS is ready to read the next piece of data in the same record.

**Figure 5.5** *Second Iteration: First Record Remains in the Input Buffer*

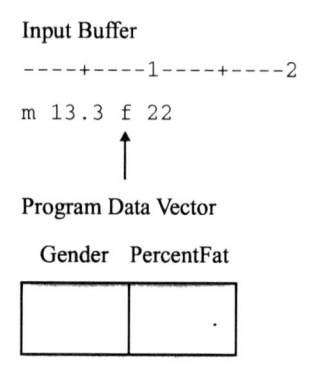

Input Buffer

```
----+----1----+----2
m 13.3 f 22
        ↑
```

Program Data Vector

| Gender | PercentFat |
|--------|------------|
|        | .          |

The following figure shows that the INPUT statement reads the next two values from the input buffer and writes them to the program data vector.

**Figure 5.6** *Second Observation Is Created*

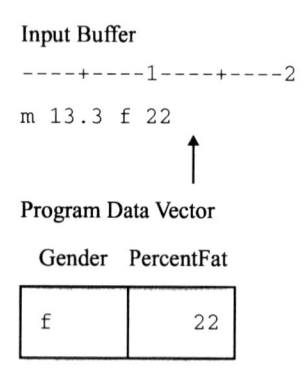

Input Buffer

```
----+----1----+----2
m 13.3 f 22
           ↑
```

Program Data Vector

| Gender | PercentFat |
|--------|------------|
| f      | 22         |

When the DATA step completes the second iteration, the values in the program data vector are written to the data set as the second observation. Then the DATA step begins its third iteration. Values in the program data vector are set to missing, and the INPUT statement executes. The pointer, which is now at column 13 (two columns to the right of the last data value that was read), continues reading. Because this is list input, the pointer scans for the next nonblank character to begin reading the next value. When the pointer reaches the end of the input buffer and fails to find a nonblank character, SAS reads a new record into the input buffer.

The final figure shows that values for the third observation are read from the beginning of the second record.

**Figure 5.7**  Third Iteration: Second Record Is Read into the Input Buffer

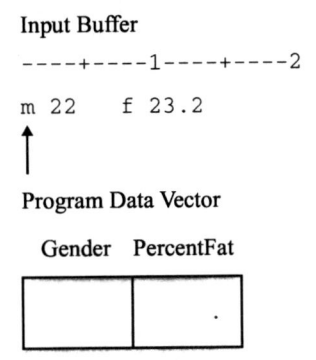

```
Input Buffer
----+----1----+----2
m 22    f 23.2
↑
```

Program Data Vector

| Gender | PercentFat |
|--------|------------|
|        | .          |

The process continues until SAS reads all the records. The resulting SAS data set contains six observations instead of three.

*Note:* Although this program successfully reads all of the data in the input records, SAS writes a message to the log noting that the program had to go to a new line.

# Reading Multiple Records to Create a Single Observation

## How the Data Records Are Structured

An earlier example (see "Reading Character Data That Contains Embedded Blanks" on page 64) shows data for several observations that are contained in a single record of raw data:

```
1023 David Shaw     red 189 165
```

This INPUT statement reads all the data values arranged across a single record:

```
input IdNumber 1-4 Name $ 6-23 Team $ StartWeight EndWeight;
```

Now, consider the opposite situation: when information for a single observation is not contained in a single record of raw data but is scattered across several records. For example, the health and fitness club data could be constructed in such a way that the information about a single member is spread across several records instead of in a single record:

```
1023 David Shaw
red
189 165
```

## Method 1: Using Multiple Input Statements

Multiple INPUT statements, one for each record, can read each record into a single observation, as in this example:

```
input IdNumber 1-4 Name $ 6-23;
```

```
input Team $ 1-6;
input StartWeight 1-3 EndWeight 5-7;
```

To understand how to use multiple INPUT statements, consider what happens as a DATA step executes. Remember that one record is read into the INPUT buffer automatically as each INPUT statement is encountered during each iteration. SAS reads the data values from the input buffer and writes them to the program data vector as variable values. At the end of the DATA step, all the variable values in the program data vector are written automatically as a single observation.

This example uses multiple INPUT statements in a DATA step to read only selected data fields and create a data set containing only the variables IdNumber, StartWeight, and EndWeight.

```
data club2;
    input IdNumber 1-4;   ❶
    input;                ❷
    input StartWeight 1-3 EndWeight 5-7;   ❸
    datalines;
1023 David Shaw
red
189 165
1049 Amelia Serrano
yellow
145 124
1219 Alan Nance
red
210 192
1246 Ravi Sinha
yellow
194 177
1078 Ashley McKnight
red
127 118
1221 Jim Brown
yellow
220  .
;

proc print data=club2;
    title 'Weight Club Members';
run;
```

The following list corresponds to the numbered items in the preceding program:

1  The first INPUT statement reads only one data field in the first record and assigns a value to the variable IdNumber.

2  The second INPUT statement, without arguments, is a null INPUT statement that reads the second record into the input buffer. However, it does not assign a value to a variable.

3  The third INPUT statement reads the third record into the input buffer and assigns values to the variables StartWeight and EndWeight.

The following output displays the resulting data set:

*Figure 5.8  Data Set Created with Multiple INPUT Statements*

## Weight Club Members

| Obs | IdNumber | StartWeight | EndWeight |
|---|---|---|---|
| 1 | 1023 | 189 | 165 |
| 2 | 1049 | 145 | 124 |
| 3 | 1219 | 210 | 192 |
| 4 | 1246 | 194 | 177 |
| 5 | 1078 | 127 | 118 |
| 6 | 1221 | 220 | . |

### *Method 2: Using the / Line-Pointer Control*

Writing a separate INPUT statement for each record is not the only way to create a single observation. You can write a single INPUT statement and use the slash (/) line-pointer control. The slash line-pointer control forces a new record into the input buffer and positions the pointer at the beginning of that record.

This example uses only one INPUT statement to read multiple records:

```
data club2;
   input IdNumber 1-4 / / StartWeight 1-3 EndWeight 5-7;
   datalines;
1023 David Shaw
red
189 165
1049 Amelia Serrano
yellow
145 124
1219 Alan Nance
red
210 192
1246 Ravi Sinha
yellow
194 177
1078 Ashley McKnight
red
127 118
1221 Jim Brown
yellow
220  .
;

proc print data=club2;
   title 'Weight Club Members';
run;
```

**80** Chapter 5 • *Starting with Raw Data: Beyond the Basics*

The / line-pointer control appears exactly where a new INPUT statement begins in the previous example. See "Method 1: Using Multiple Input Statements" on page 77. The sequence of events in the input buffer and the program data vector as this DATA step executes is identical to the previous example in method 1. The / is the signal to read a new record into the input buffer, which happens automatically when the DATA step encounters a new INPUT statement. The preceding example shows two slashes (/ /), indicating that SAS skips a record. SAS reads the first record, skips the second record, and reads the third record.

The following output displays the resulting data set:

*Figure 5.9  Data Set Created with the / Line-Pointer Control*

### Weight Club Members

| Obs | IdNumber | StartWeight | EndWeight |
|---|---|---|---|
| 1 | 1023 | 189 | 165 |
| 2 | 1049 | 145 | 124 |
| 3 | 1219 | 210 | 192 |
| 4 | 1246 | 194 | 177 |
| 5 | 1078 | 127 | 118 |
| 6 | 1221 | 220 | . |

## Reading Variables from Multiple Records in Any Order

You can also read multiple records to create a single observation by pointing to a specific record in a set of input records with the *#n* line-pointer control. As you saw in the last section, the advantage of using the / line-pointer control over multiple INPUT statements is that it requires fewer statements. However, using the *#n* line-pointer control enables you to read the variables in any order, no matter which record contains the data values. It is also useful if you want to skip data lines.

This example uses one INPUT statement to read multiple data lines in a different order:

```
data club2;
   input #2 Team $ 1-6 #1 Name $ 6-23 IdNumber 1-4
       #3 StartWeight 1-3 EndWeight 5-7;
   datalines;
1023 David Shaw
red
189 165
1049 Amelia Serrano
yellow
145 124
1219 Alan Nance
red
210 192
1246 Ravi Sinha
yellow
194 177
```

```
1078 Ashley McKnight
red
127 118
1221 Jim Brown
yellow
220   .
;

proc print data=club2;
   title 'Weight Club Members';
run;
```

The following output displays the resulting data set:

**Figure 5.10**  *Data Set Created with the #n Line-Pointer Control*

**Weight Club Members**

| Obs | Team | Name | IdNumber | StartWeight | EndWeight |
|---|---|---|---|---|---|
| 1 | red | David Shaw | 1023 | 189 | 165 |
| 2 | yellow | Amelia Serrano | 1049 | 145 | 124 |
| 3 | red | Alan Nance | 1219 | 210 | 192 |
| 4 | yellow | Ravi Sinha | 1246 | 194 | 177 |
| 5 | red | Ashley McKnight | 1078 | 127 | 118 |
| 6 | yellow | Jim Brown | 1221 | 220 | . |

The order of the observations is the same as in the raw records (shown in the section "Reading Variables from Multiple Records in Any Order" on page 80). However, the order of the variables in the data set differs from the order of the variables in the raw input data records. This occurs because the order of the variables in the INPUT statements corresponds with their order in the resulting data sets.

## *Understanding How the #n Line-Pointer Control Affects DATA Step Execution*

To understand the importance of the *#n* line-pointer control, remember the sequence of events in the DATA steps that demonstrate the / line-pointer control and multiple INPUT statements. Each record is read into the input buffer sequentially. The data is read, and then a / or a new INPUT statement causes the program to read the next record into the input buffer. It is impossible for the program to read a value from the first record after a value from the second record is read because the data in the first record is no longer available in the input buffer.

To solve this problem, use the *#n* line-pointer control. The *#n* line-pointer control signals the program to create a multiple-line input buffer so that all the data for a single observation is available while the observation is being built in the program data vector. The *#n* line-pointer control also identifies the record in which data for each variable appears. To use the *#n* line-pointer control, the raw data must have the same number of records for each observation; for example, it cannot have three records for one observation and two for the next.

When the program compiles and builds the input buffer, it looks at the INPUT statement and creates an input buffer with as many lines as are necessary to contain the number of records that it needs to read for a single observation. In this example, the highest number of records specified is three, so the input buffer is built to contain three records at one time. The following figures demonstrate the flow of the DATA step in this example.

The following figure shows that the values are set to missing in the program data vector and that the INPUT statement reads the first three records into the input buffer.

*Figure 5.11* Three Records Are Read into the Input Buffer as a Single Observation

Input Buffer

```
----+----1----+----2----+----3----+----4----+----5----+----6
1023 David Shaw

----+----1----+----2----+----3----+----4----+----5----+----6
red

----+----1----+----2----+----3----+----4----+----5----+----6
189 165
```

Program Data Vector

| Team | Name | IdNumber | StartWeight | EndWeight |
|------|------|----------|-------------|-----------|
|      |      | .        | .           | .         |

The INPUT statement for this example is as follows:

```
input #2 Team $ 1-6
      #1 Name $ 6-23 IdNumber 1-4
      #3 StartWeight 1-3 EndWeight 5-7;
```

The first variable is preceded by #2 to indicate that the value in the second record is assigned to the variable Team.

The following figure shows that the pointer advances to the second line in the input buffer, reads the value, and writes it to the program data vector.

*Figure 5.12* Reading from the Second Record First

Input Buffer

```
----+----1----+----2----+----3----+----4----+----5----+----6
1023 David Shaw

----+----1----+----2----+----3----+----4----+----5----+----6
red
↑
----+----1----+----2----+----3----+----4----+----5----+----6
189 165
```

Program Data Vector

| Team | Name | IdNumber | StartWeight | EndWeight |
|------|------|----------|-------------|-----------|
| red  |      | .        | .           | .         |

The following figure shows that the pointer then moves to the sixth column in the first record, reads a value, and assigns it to the variable Name in the program data vector. It then moves to the first column to read the ID number, and assigns it to the variable IdNumber.

*Figure 5.13* Reading from the First Record

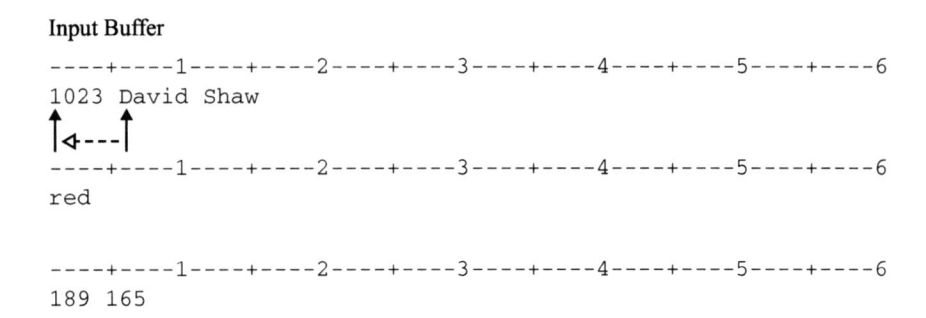

**Input Buffer**

```
----+----1----+----2----+----3----+----4----+----5----+----6
1023 David Shaw
↑ ◁---↑

----+----1----+----2----+----3----+----4----+----5----+----6
red


----+----1----+----2----+----3----+----4----+----5----+----6
189 165
```

**Program Data Vector**

| Team | Name | IdNumber | StartWeight | EndWeight |
|------|------|----------|-------------|-----------|
| red  | David Shaw | 1023 | . | . |

The following figure shows that the process continues with the pointer moving to the third record in the first observation. Values are read and assigned to StartWeight and EndWeight, the last variable that is listed.

*Figure 5.14* Reading from the Third Record

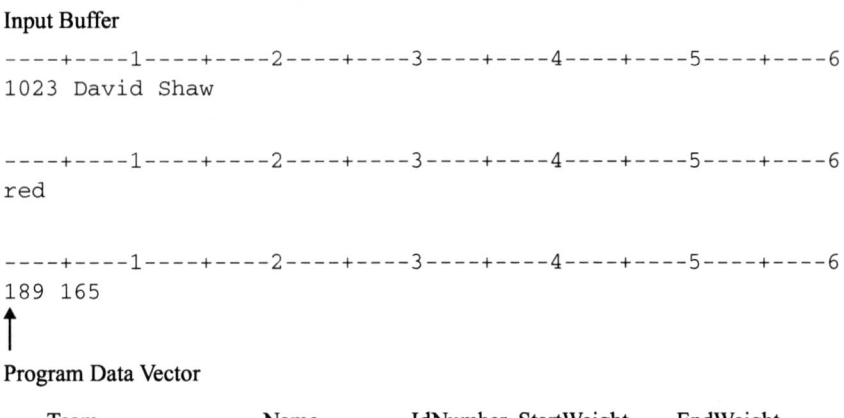

**Input Buffer**

```
----+----1----+----2----+----3----+----4----+----5----+----6
1023 David Shaw

----+----1----+----2----+----3----+----4----+----5----+----6
red

----+----1----+----2----+----3----+----4----+----5----+----6
189 165
↑
```

**Program Data Vector**

| Team | Name | IdNumber | StartWeight | EndWeight |
|------|------|----------|-------------|-----------|
| red  | David Shaw | 1023 | 189 | 165 |

When the bottom of the DATA step is reached, variable values in the program data vector are written as an observation to the data set. The DATA step returns to the top, and values in the program data vector are set to missing. The INPUT statement executes again.

The final figure shows that the next three records are read into the input buffer, ready to create the second observation.

**Figure 5.15** *Reading the Next Three Records into the Input Buffer*

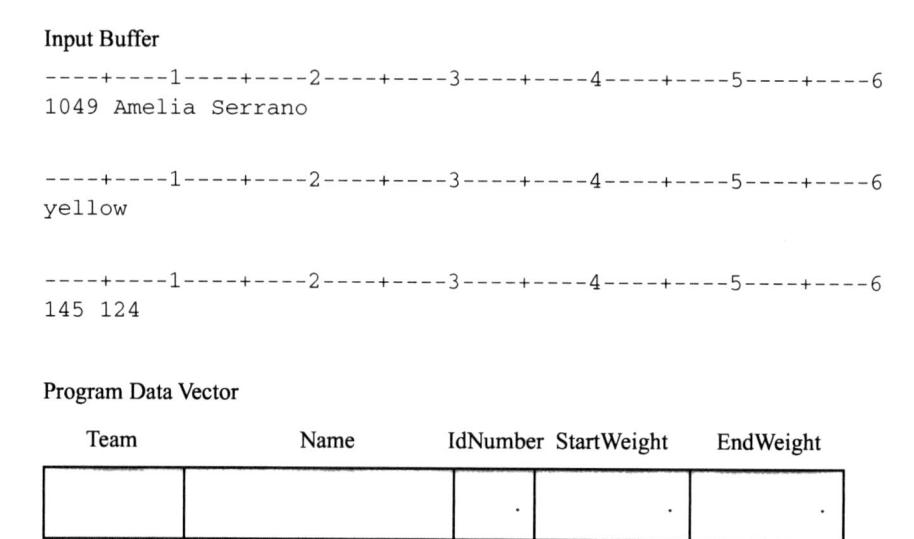

Input Buffer

```
----+----1----+----2----+----3----+----4----+----5----+----6
1049 Amelia Serrano


----+----1----+----2----+----3----+----4----+----5----+----6
yellow


----+----1----+----2----+----3----+----4----+----5----+----6
145 124
```

Program Data Vector

| Team | Name | IdNumber | StartWeight | EndWeight |
|------|------|----------|-------------|-----------|
|      |      | .        | .           | .         |

# Problem Solving: When an Input Record Unexpectedly Does Not Have Enough Values

## Understanding the Default Behavior

When a DATA step reads raw data from an external file, problems can occur when SAS encounters the end of an input line before reading in data for all variables specified in the input statement. This problem can occur when reading variable-length records, records containing missing values, or both.

The following is an example of an external file that contains variable-length records:

```
----+----1----+----2

22
333
4444
55555
```

This DATA step uses the numeric informat 5. to read a single field in each record of raw data and to assign values to the variable TestNumber:

```
data numbers;
   infile 'your-external-file';
   input TestNumber 5.;
run;

proc print data=numbers;
   title 'Test DATA Step';
run;
```

The DATA step reads the first value (22). Because the value is shorter than the 5 characters expected by the informat, the DATA step attempts to finish filling the value with the next record (333). This value is entered into the PDV and becomes the value of the TestNumber variable for the first observation. The DATA step then goes to the next record, but encounters the same problem because the value (4444) is shorter than the value that is expected by the informat. Again, the DATA step goes to the next record, reads the value (55555), and assigns that value to the TestNumber variable for the second observation.

The following output displays the results. After this program runs, the SAS log contains a note to indicate the places where SAS went to the next record to search for data values.

*Figure 5.16* Reading Raw Data Past the End of a Line: Default Behavior

Test DATA Step

| Obs | TestNumber |
|---|---|
| 1 | 333 |
| 2 | 55555 |

## Methods of Control: Your Options

### Four Options: FLOWOVER, STOPOVER, MISSOVER, and TRUNCOVER

To control how SAS behaves after it attempts to read past the end of a data line, you can use the following options in the INFILE statement:

```
infile 'your-external-file' flowover;
```
is the default behavior. The DATA step simply reads the next record into the input buffer, attempting to find values to assign to the rest of the variable names in the INPUT statement.

```
infile 'your-external-file' stopover;
```
causes the DATA step to stop processing if an INPUT statement reaches the end of the current record without finding values for all variables in the statement. Use this option if you expect all of the data in the external file to conform to a given standard, and if you want the DATA step to stop when it encounters a data record that does not conform to the standard.

```
infile 'your-external-file' missover;
```
prevents the DATA step from going to the next line if it does not find values in the current record for all of the variables in the INPUT statement. Instead, the DATA step assigns a missing value for all variables that do not have values.

```
infile 'your-external-file' truncover;
```
causes the DATA step to assign the raw data value to the variable even if the value is shorter than expected by the INPUT statement. If, when the DATA step encounters the end of an input record, there are variables without values, the variables are assigned missing values for that observation.

You can also use these options even when your data lines are in the program itself, that is, when they follow the DATALINES statement. Simply use **datalines** instead of a reference to an external file to indicate that the data records are in the DATA step itself:

- `infile datalines flowover;`

## 86  Chapter 5 • Starting with Raw Data: Beyond the Basics

- `infile datalines stopover;`
- `infile datalines missover;`
- `infile datalines truncover;`

*Note:* The examples in this section show the use of the MISSOVER and TRUNCOVER options with formatted input. You can also use these options with list input and column input.

### Understanding the MISSOVER Option

The MISSOVER option prevents the DATA step from going to the next line if it does not find values in the current record for all of the variables in the INPUT statement. Instead, the DATA step assigns a missing value for all variables that do not have complete values according to any specified informats. The input file contains the following raw data:

```
----+----1----+----2

22
333
4444
55555
```

The following example uses the MISSOVER option:

```
data numbers;
   infile 'your-external-file' missover;
   input TestNumber 5.;
run;

proc print data=numbers;
   title 'Test DATA Step';
run;
```

**Figure 5.17** Output from the MISSOVER Option

```
Test DATA Step

Obs  TestNumber
 1        .
 2        .
 3        .
 4      55555
```

Because the fourth record is the only one whose value matches the informat, it is the only record whose value is assigned to the TestNumber variable. The other observations receive missing values. This result is probably not the desired outcome for this example, but the MISSOVER option can sometimes be valuable. For an example, see "Updating a Data Set" on page 319.

*Note:* If there is a blank line at the end of the last record, the DATA step attempts to load another record into the input buffer. Because there are no more records, the MISSOVER option instructs the DATA step to assign missing values to all variables, and an extra observation is added to the data set. To prevent this situation from occurring, make sure that your input data does not have a blank line at the end of the last record.

## Understanding the TRUNCOVER Option

The TRUNCOVER option causes the DATA step to assign the raw data value to the variable even if the value is shorter that the length that is expected by the INPUT statement. If, when the DATA step encounters the end of an input record, there are variables without values, the variables are assigned missing values for that observation. The following example demonstrated the use of the TRUNCOVER option:

```
data numbers;
    infile 'your-external-file' truncover;
    input TestNumber 5.;
run;

proc print data=numbers;
    title 'Test DATA Step';
run;
```

*Figure 5.18* Output from the TRUNCOVER Option

Test DATA Step

| Obs | TestNumber |
|---|---|
| 1 | 22 |
| 2 | 333 |
| 3 | 4444 |
| 4 | 55555 |

This result shows that all of the values were assigned to the TestNumber variable, despite the fact that three of them did not match the informat. For another example using the TRUNCOVER option, see "Input SAS Data Set for Examples" on page 148.

# Summary

## Column-Pointer Controls

@ *n*
: moves the pointer to the *n* column in the input buffer.

+*n*
: moves the pointer forward *n* columns in the input buffer.

/
: moves the pointer to the next line in the input buffer.

#*n*
: moves the pointer to the *n*th line in the input buffer.

## Line-Hold Specifiers

@
: (trailing @) prevents SAS from automatically reading a new data record into the input buffer when a new INPUT statement is executed within the same iteration of

the DATA step. When used, the trailing @ must be the last item in the INPUT statement.

@@
(double trailing @) prevents SAS from automatically reading a new data record into the input buffer when the next INPUT statement is executed, even if the DATA step returns to the top for another iteration. When used, the double trailing @ must be the last item in the INPUT statement.

## Statements

DATALINES;
indicates that data lines immediately follow. A semicolon in the line that immediately follows the last data line indicates the end of the data and causes the DATA step to compile and execute.

INFILE fileref< FLOWOVER | STOPOVER | MISSOVER | TRUNCOVER>;
INFILE 'external-file' <FLOWOVER | STOPOVER | MISSOVER | TRUNCOVER>;
identifies an external file to be read by an INPUT statement. Specify a fileref that has been assigned with a FILENAME statement or with an appropriate operating environment command. Or you can specify the actual name of the external file.

These options give you control over how SAS behaves if the end of a data record is encountered before all of the variables are assigned values. You can use these options with list, modified list, formatted, and column input.

FLOWOVER
is the default behavior. It causes the DATA step to look in the next record if the end of the current record is encountered before all of the variables are assigned values

MISSOVER
causes the DATA step to assign missing values to any variables that do not have values when the end of a data record is encountered. The DATA step continues processing.

STOPOVER
causes the DATA step to stop execution immediately and write a note to the SAS log.

TRUNCOVER
causes the DATA step to assign values to variables, even if the values are shorter than expected by the INPUT statement, and to assign missing values to any variables that do not have values when the end of a record is encountered.

INPUT *variable* <&> <$>;
reads the input data record using list input. The & (ampersand format modifier) allows character values to contain embedded blanks. When you use the ampersand format modifier, two blanks are required to signal the end of a data value. The $ indicates a character variable.

INPUT variable *start-column* <*end-column*>;
reads the input data record using column input. You can omit end-column if the data is only 1 byte long. This style of input enables you to skip columns of data that you want to omit.

INPUT *variable* : *informat*;
INPUT *variable* & *informat*;
reads the input data record using modified list input. The : (colon format modifier) instructs SAS to use the informat that follows to read the data value. The &

(ampersand format modifier) instructs SAS to use the informat that follows to read the data value. When you use the ampersand format modifier, two blanks are required to signal the end of a data value.

INPUT <*pointer-control*>*variable informat*;
reads raw data using formatted input. The *informat* supplies special instructions to read the data. You can also use a *pointer-control* to direct SAS to start reading at a particular column.

The syntax given above for the three styles of input shows only one *variable*. Subsequent variables in the INPUT statement might be described in the same input style as the first one. You can use any of the three styles of input (list, column, and formatted) in a single INPUT statement.

## Learning More

Handling missing data values
For complete details about the FLOWOVER, STOPOVER, MISSOVER, and TRUNCOVER options in the INFILE statement, see "INFILE Statement" in *SAS Statements: Reference*.

Testing a condition
- For more information about performing conditional processing with the IF statement, see Chapter 10, "Acting on Selected Observations," on page 147 and Chapter 11, "Creating Subsets of Observations," on page 169.

- For a complete discussion and listing of line-pointer controls and line-hold specifiers, see "PUT Statement" in *SAS Statements: Reference*.

# Chapter 6
# Starting with SAS Data Sets

Introduction to Starting with SAS Data Sets ............................. 91
  Purpose ........................................................... 91
  Prerequisites ..................................................... 92
Understanding the Basics ............................................. 92
Input SAS Data Set for Examples ..................................... 92
Reading Selected Observations ....................................... 95
Reading Selected Variables .......................................... 96
  Overview of Reading Selected Variables ........................... 96
  Keeping Selected Variables ....................................... 96
  Dropping Selected Variables ...................................... 97
  Choosing between Data Set Options and Statements ................. 98
  Choosing between the DROP= and KEEP= Data Set Option ............. 98
Creating More Than One Data Set in a Single DATA Step .............. 99
Using the DROP= and KEEP= Data Set Options for Efficiency ......... 101
Summary ............................................................ 102
  Data Set Options ................................................ 102
  Procedures ...................................................... 103
  Statements ...................................................... 103
Learning More ..................................................... 103

## Introduction to Starting with SAS Data Sets

### Purpose

In this section, you will learn how to do the following:

- display information about a SAS data set
- create a new SAS data set from an existing SAS data set rather than creating it from raw data records

Reading a SAS data set in a DATA step is simpler than reading raw data because the work of describing the data to SAS has already been done.

*Prerequisites*

Before continuing, you should understand the concepts presented in the following sections:

- Chapter 1, "What is the SAS System?," on page 3
- Chapter 3, "Introduction to DATA Step Processing," on page 27

## Understanding the Basics

When you use a SAS data set as input into a DATA step, the description of the data set is available to SAS. In your DATA step, use a SET, MERGE, MODIFY, or UPDATE statement to read the SAS data set. Use SAS programming statements to process the data and create an output SAS data set.

In a DATA step, you can create a new data set that is a subset of the original data set. For example, if you have a large data set of personnel data, you might want to look at a subset of observations that meet certain conditions, such as observations for employees hired after a certain date. Alternatively, you might want to see all observations but only a few variables, such as the number of years of education or years of service to the company.

When you use existing SAS data sets, as well as with subsets created from SAS data sets, you can make more efficient use of computer resources than if you use raw data or if you are working with large data sets. Reading fewer variables means that SAS creates a smaller program data vector, and reading fewer observations means that fewer iterations of the DATA step occur. Reading data directly from a SAS data set is more efficient than reading the raw data again, because the work of describing and converting the data has already been done.

One way of looking at a SAS data set is to produce a listing of the data in a SAS data set by using the PRINT procedure. Another way to look at a SAS data set is to display information that describes its structure rather than its data values. To display information about the structure of a data set, use the DATASETS procedure with the CONTENTS statement. If you need to work with a SAS data set that is unfamiliar to you, the CONTENTS statement in the DATASETS procedure displays information such as the name, type, and length of all the variables in the data set. An example that shows the CONTENTS statement in the DATASETS procedure is shown in "Input SAS Data Set for Examples" on page 92.

## Input SAS Data Set for Examples

The examples in this section use a SAS data set named CITY, which contains information about expenditures for a small city. It reports total city expenditures for the years 1980 through 2000 and divides the expenses into two major categories: services and administration.

The following example uses the DATASETS procedure with the NOLIST option to display the CITY data set. The NOLIST option prevents the DATASETS procedure from listing other data sets that are also located in the WORK library:

```
data city;
   input Year 4. @7 ServicesPolice comma6.
        @15 ServicesFire comma6. @22 ServicesWater_Sewer comma6.
        @30 AdminLabor comma6. @39 AdminSupplies comma6.
        @45 AdminUtilities comma6.;
   ServicesTotal=ServicesPolice+ServicesFire+ServicesWater_Sewer;
   AdminTotal=AdminLabor+AdminSupplies+AdminUtilities;
   Total=ServicesTotal+AdminTotal;
   label           Total='Total Outlays'
           ServicesTotal='Services: Total'
          ServicesPolice='Services: Police'
            ServicesFire='Services: Fire'
      ServicesWater_Sewer='Services: Water & Sewer'
              AdminTotal='Administration: Total'
              AdminLabor='Administration: Labor'
           AdminSupplies='Administration: Supplies'
          AdminUtilities='Administration: Utilities';
   datalines;
1993  2,819  1,120    422    391     63     98
1994  2,477  1,160    500    172     47     70
1995  2,028  1,061    510    269     29     79
1996  2,754    893    540    227     21     67
1997  2,195    963    541    214     21     59
1998  1,877    926    535    198     16     80
1999  1,727  1,111    535    213     27     70
2000  1,532  1,220    519    195     11     69
2001  1,448  1,156    577    225     12     58
2002  1,500  1,076    606    235     19     62
2003  1,934    969    646    266     11     63
2004  2,195  1,002    643    256     24     55
2005  2,204    964    692    256     28     70
2006  2,175  1,144    735    241     19     83
2007  2,556  1,341    813    238     25     97
2008  2,026  1,380    868    226     24     97
2009  2,526  1,454    946    317     13     89
2010  2,027  1,486  1,043    226      .     82
2011  2,037  1,667  1,152    244     20     88
2012  2,852  1,834  1,318    270     23     74
2013  2,787  1,701  1,317    307     26     66
;

proc datasets library=work nolist;
   contents data=city;
run;
```

The following outputs display the contents of the CITY data set, as well as information about the data set.

*Figure 6.1* Part 1: The Structure of CITY as Shown by PROC DATASETS

### The SAS System

The DATASETS Procedure

| | | | |
|---|---|---|---|
| Data Set Name | WORK.CITY | Observations | 21 |
| Member Type | DATA | Variables | 10 |
| Engine | V9 | Indexes | 0 |
| Created | 04/30/2013 15:29:12 | Observation Length | 80 |
| Last Modified | 04/30/2013 15:29:12 | Deleted Observations | 0 |
| Protection | | Compressed | NO |
| Data Set Type | | Sorted | NO |
| Label | | | |
| Data Representation | WINDOWS_32 | | |
| Encoding | wlatin1 Western (Windows) | | |

*Figure 6.2* Part 2: The Structure of CITY as Shown by PROC DATASETS

Engine/Host Dependent Information

| | |
|---|---|
| Data Set Page Size | 65536 |
| Number of Data Set Pages | 1 |
| First Data Page | 1 |
| Max Obs per Page | 317 |
| Obs in First Data Page | 21 |
| Number of Data Set Repairs | 0 |
| ExtendObsCounter | YES |
| Filename | C:\Users\userid\AppData\Local\Temp\SAS Temporary Files\_TD1234_D56789_\city.sas7bdat |
| Release Created | 9.0401B0 |
| Host Created | W32_7PRO |

**Figure 6.3** *Part 3: The Structure of CITY as Shown by PROC DATASETS*

| Alphabetic List of Variables and Attributes ||||
|---|---|---|---|
| # | Variable | Type | Len | Label |
| 5 | AdminLabor | Num | 8 | Administration: Labor |
| 6 | AdminSupplies | Num | 8 | Administration: Supplies |
| 9 | AdminTotal | Num | 8 | Administration: Total |
| 7 | AdminUtilities | Num | 8 | Administration: Utilities |
| 3 | ServicesFire | Num | 8 | Services: Fire |
| 2 | ServicesPolice | Num | 8 | Services: Police |
| 8 | ServicesTotal | Num | 8 | Services: Total |
| 4 | ServicesWater_Sewer | Num | 8 | Services: Water &amp; Sewer |
| 10 | Total | Num | 8 | Total Outlays |
| 1 | Year | Num | 8 | |

The following list corresponds to items in the three SAS outputs shown above:

1. The Observations and the Variables fields in Figure 6.1 on page 94 identify the number of observations and the number of variables.

2. The Engine/Host Dependent Information section in Figure 6.2 on page 94 lists detailed information about the data set. This information is generated by the engine, which is the mechanism for reading from and writing to files.

3. The Alphabetic List of Variables and Attributes in Figure 6.3 on page 95 lists the name, type, length, and position of each variable.

4. The Label in Figure 6.3 on page 95 lists the format, informat, and label for each variable, if they exist.

*Operating Environment Information*
   The output in the Engine/Host Dependent Information section might differ, depending on your operating environment. For more information, see the SAS documentation for your operating environment.

# Reading Selected Observations

If you are interested in only part of a large data set, you can use data set options to create a subset of your data. Data set options specify which observations you want the new data set to include. In Chapter 11, "Creating Subsets of Observations," on page 169 you learn how to use the subsetting IF statement to create a subset of a large SAS data set. In this section, you learn how to use the FIRSTOBS= and OBS= data set options to create subsets of a larger data set.

For example, you might not want to read the observations at the beginning of the data set. You can use the FIRSTOBS= data set option to define which observation should be the first one that is processed. For the data set CITY, this example creates a data set that excludes observations that contain data prior to 1991 by specifying FIRSTOBS=12. As a result, SAS does not read the first 11 observations, which contain data prior to 1991. (To see the program that creates the CITY data set, see "DATA Step to Create the CITY Data Set" on page 814.)

The following program creates the data set CITY2, which contains the same number of variables but fewer observations than CITY.

```
data city2;
   set city(firstobs=12);
run;

proc print;
   title 'City Expenditures';
   title2 '1991 - 2000';
run;
```

The following output displays the results:

*Figure 6.4* Subsetting a Data Set by Observations

City Expenditures
2004 - 2013

| Obs | Year | ServicesPolice | ServicesFire | ServicesWater_Sewer | AdminLabor | AdminSupplies | AdminUtilities | ServicesTotal | AdminTotal | Total |
|---|---|---|---|---|---|---|---|---|---|---|
| 1 | 2004 | 2195 | 1002 | 643 | 256 | 24 | 55 | 3840 | 335 | 4175 |
| 2 | 2005 | 2204 | 964 | 692 | 256 | 28 | 70 | 3860 | 354 | 4214 |
| 3 | 2006 | 2175 | 1144 | 735 | 241 | 19 | 83 | 4054 | 343 | 4397 |
| 4 | 2007 | 2556 | 1341 | 813 | 238 | 25 | 97 | 4710 | 360 | 5070 |
| 5 | 2008 | 2026 | 1380 | 868 | 226 | 24 | 97 | 4274 | 347 | 4621 |
| 6 | 2009 | 2526 | 1454 | 946 | 317 | 13 | 89 | 4926 | 419 | 5345 |
| 7 | 2010 | 2027 | 1486 | 1043 | 226 |  | 82 | 4556 |  |  |
| 8 | 2011 | 2037 | 1667 | 1152 | 244 | 20 | 88 | 4856 | 352 | 5208 |
| 9 | 2012 | 2852 | 1834 | 1318 | 270 | 23 | 74 | 6004 | 367 | 6371 |
| 10 | 2013 | 2787 | 1701 | 1317 | 307 | 26 | 66 | 5805 | 399 | 6204 |

You can also specify the last observation that you want to include in a new data set with the OBS= data set option. For example, the next program creates a SAS data set containing only the observations for 1989 (the 10th observation) through 1994 (the 15th observation).

```
data city3;
   set city (firstobs=10 obs=15);
run;
```

# Reading Selected Variables

## Overview of Reading Selected Variables

You can create a subset of a larger data set not only by excluding observations but also by specifying which variables you want the new data set to contain. In a DATA step, you can use the SET statement and the KEEP= or DROP= data set options (or the DROP and KEEP statements) to create a subset from a larger data set by specifying which variables you want the new data set to include.

## Keeping Selected Variables

This example uses the KEEP= data set option in the SET statement to read only the variables that represent the services-related expenditures of the data set CITY.

```
data services;
   set city (keep=Year ServicesTotal ServicesPolice ServicesFire
             ServicesWater_Sewer);
run;

proc print data=services;
   title 'City Services-Related Expenditures';
run;
```

The following output displays the resulting data set. Note that the data set SERVICES contains only those variables that are specified in the KEEP= option.

*Figure 6.5* Selecting Variables with the KEEP= Option

City Services-Related Expenditures

| Obs | Year | ServicesPolice | ServicesFire | ServicesWater_Sewer | ServicesTotal |
|---|---|---|---|---|---|
| 1 | 2004 | 2195 | 1002 | 643 | 3840 |
| 2 | 2005 | 2204 | 964 | 692 | 3860 |
| 3 | 2006 | 2175 | 1144 | 735 | 4054 |
| 4 | 2007 | 2556 | 1341 | 813 | 4710 |
| 5 | 2008 | 2026 | 1380 | 868 | 4274 |
| 6 | 2009 | 2526 | 1454 | 946 | 4926 |
| 7 | 2010 | 2027 | 1486 | 1043 | 4556 |
| 8 | 2011 | 2037 | 1667 | 1152 | 4856 |
| 9 | 2012 | 2852 | 1834 | 1318 | 6004 |
| 10 | 2013 | 2787 | 1701 | 1317 | 5805 |

The following example uses the KEEP statement instead of the KEEP= data set option to read all of the variables from the CITY data set. The KEEP statement creates a new data set (SERVICES) that contains only the variables listed in the KEEP statement. The following program gives results that are identical to those in the previous example:

```
data services;
   set city;
   keep Year ServicesTotal ServicesPolice ServicesFire
        ServicesWater_Sewer;
run;
```

The following example has the same effect as using the KEEP= data set option in the DATA statement. All of the variables are read into the program data vector, but only the specified variables are written to the SERVICES data set:

```
data services (keep=Year ServicesTotal ServicesPolice ServicesFire
                    ServicesWater_Sewer);
   set city;
run;
```

## Dropping Selected Variables

Use the DROP= option to create a subset of a larger data set when you want to specify which variables are being excluded rather than which ones are being included. The following DATA step reads all of the variables from the data set CITY except for those that are specified with the DROP= option. It then creates a data set named SERVICES2:

```
data services2;
```

```
            set city (drop=Total AdminTotal AdminLabor AdminSupplies
                AdminUtilities);
run;

proc print data=services2;
    title 'City Services-Related Expenditures';
run;
```

The following output displays the resulting data set:

**Figure 6.6** *Excluding Variables with the DROP= Option*

City Services-Related Expenditures

| Obs | Year | ServicesPolice | ServicesFire | ServicesWater_Sewer | ServicesTotal |
|---|---|---|---|---|---|
| 1 | 2004 | 2195 | 1002 | 643 | 3840 |
| 2 | 2005 | 2204 | 964 | 692 | 3860 |
| 3 | 2006 | 2175 | 1144 | 735 | 4054 |
| 4 | 2007 | 2556 | 1341 | 813 | 4710 |
| 5 | 2008 | 2026 | 1380 | 868 | 4274 |
| 6 | 2009 | 2526 | 1454 | 946 | 4926 |
| 7 | 2010 | 2027 | 1486 | 1043 | 4556 |
| 8 | 2011 | 2037 | 1667 | 1152 | 4856 |
| 9 | 2012 | 2852 | 1834 | 1318 | 6004 |
| 10 | 2013 | 2787 | 1701 | 1317 | 5805 |

The following example uses the DROP statement instead of the DROP= data set option to read all of the variables from the CITY data set. It also excludes the variables that are listed in the DROP statement from being written to the new data set. The results are identical to those in the previous example:

```
data services2;
    set city;
    drop Total AdminTotal AdminLabor AdminSupplies AdminUtilities;
run;
proc print data=services2;
run;
```

### Choosing between Data Set Options and Statements

When you create only one data set in the DATA step, the data set options to drop and keep variables have the same effect on the output data set as the statements to drop and keep variables. When you want to control which variables are read into the program data vector, use the data set options in the statement (such as a SET statement) that reads the SAS data set. The options are generally more efficient than using the statements. Later topics in this section show you how to use the data set options in some cases where the statements will not work.

### Choosing between the DROP= and KEEP= Data Set Option

In a simple case, you might decide to use the DROP= or KEEP= option, depending on which method enables you to specify fewer variables. If you work with large jobs that read data sets, and you expect that variables might be added between the times your batch jobs run, then you might want to use the KEEP= option to specify which variables are included in the subset data set.

The following figure shows two data sets named SMALL. They have different contents because the new variable F was added to data set BIG before the DATA step ran on Tuesday. The DATA step uses the DROP= option to keep variables D and E from being written to the output data set. The result is that the data sets contain different contents: the second SMALL data set has an extra variable, F. If the DATA step used the KEEP= option to specify A, B, and C, then both of the SMALL data sets would have the same variables (A, B, and C). The addition of variable F to the original data set BIG would have no effect on the creation of the SMALL data set.

*Figure 6.7* Using the DROP= Option

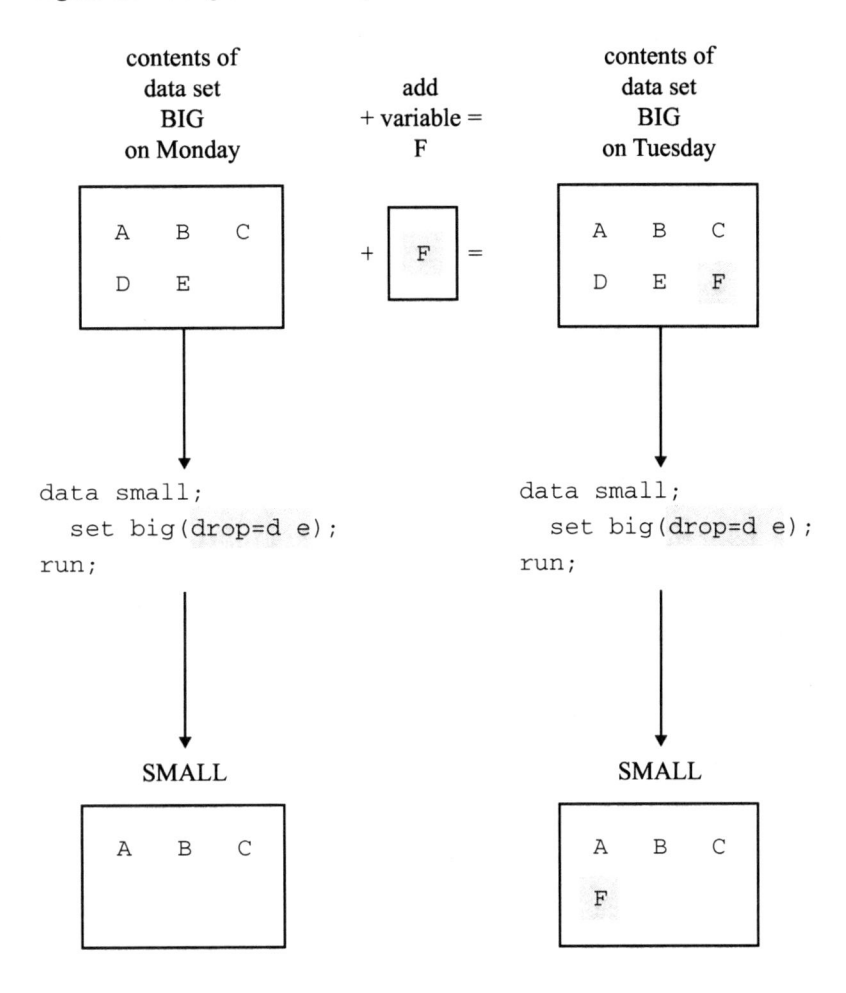

contents of data set BIG on Monday

```
A  B  C
D  E
```

add + variable = F

contents of data set BIG on Tuesday

```
A  B  C
D  E  F
```

```
data small;
  set big(drop=d e);
run;
```

```
data small;
  set big(drop=d e);
run;
```

SMALL

```
A  B  C
```

SMALL

```
A  B  C
F
```

# Creating More Than One Data Set in a Single DATA Step

You can use a single DATA step to create more than one data set at a time. You can create data sets with different contents by using the KEEP= or DROP= data set options. For example, the following DATA step creates two SAS data sets: SERVICES contains variables that show services-related expenditures, and ADMIN contains variables that represent the administration-related expenditures. Use the KEEP= option after each data

**100** Chapter 6 • *Starting with SAS Data Sets*

set name in the DATA statement to determine which variables are written to each SAS data set being created.

```
data services(keep=ServicesTotal ServicesPolice ServicesFire
              ServicesWater_Sewer)
     admin(keep=AdminTotal AdminLabor AdminSupplies
           AdminUtilities);
   set city;
run;

proc print data=services;
   title 'City Expenditures: Services';
run;

proc print data=admin;
   title 'City Expenditures: Administration';
run;
```

The following two output display both data sets. Note that each data set contains only the variables that are specified with the KEEP= option after its name in the DATA statement.

*Figure 6.8* Creating Two Data Sets in One DATA Step, Services

City Expenditures: Services

| Obs | ServicesPolice | ServicesFire | ServicesWater_Sewer | ServicesTotal |
|---|---|---|---|---|
| 1 | 2819 | 1120 | 422 | 4361 |
| 2 | 2477 | 1160 | 500 | 4137 |
| 3 | 2028 | 1061 | 510 | 3599 |
| 4 | 2754 | 893 | 540 | 4187 |
| 5 | 2195 | 963 | 541 | 3699 |
| 6 | 1877 | 926 | 535 | 3338 |
| 7 | 1727 | 1111 | 535 | 3373 |
| 8 | 1532 | 1220 | 519 | 3271 |
| 9 | 1448 | 1156 | 577 | 3181 |
| 10 | 1500 | 1076 | 606 | 3182 |
| 11 | 1934 | 969 | 646 | 3549 |
| 12 | 2195 | 1002 | 643 | 3840 |
| 13 | 2204 | 964 | 692 | 3860 |
| 14 | 2175 | 1144 | 735 | 4054 |
| 15 | 2556 | 1341 | 813 | 4710 |
| 16 | 2026 | 1380 | 868 | 4274 |
| 17 | 2526 | 1454 | 946 | 4926 |
| 18 | 2027 | 1486 | 1043 | 4556 |
| 19 | 2037 | 1667 | 1152 | 4856 |
| 20 | 2852 | 1834 | 1318 | 6004 |
| 21 | 2787 | 1701 | 1317 | 5805 |

**Figure 6.9** *Creating Two Data Sets in One DATA Step, Administration*

City Expenditures: Administration

| Obs | AdminLabor | AdminSupplies | AdminUtilities | AdminTotal |
|---|---|---|---|---|
| 1 | 391 | 63 | 98 | 552 |
| 2 | 172 | 47 | 70 | 289 |
| 3 | 269 | 29 | 79 | 377 |
| 4 | 227 | 21 | 67 | 315 |
| 5 | 214 | 21 | 59 | 294 |
| 6 | 198 | 16 | 80 | 294 |
| 7 | 213 | 27 | 70 | 310 |
| 8 | 195 | 11 | 69 | 275 |
| 9 | 225 | 12 | 58 | 295 |
| 10 | 235 | 19 | 62 | 316 |
| 11 | 266 | 11 | 63 | 340 |
| 12 | 256 | 24 | 55 | 335 |
| 13 | 256 | 28 | 70 | 354 |
| 14 | 241 | 19 | 83 | 343 |
| 15 | 238 | 25 | 97 | 360 |
| 16 | 226 | 24 | 97 | 347 |
| 17 | 317 | 13 | 89 | 419 |
| 18 | 226 | . | 82 | . |
| 19 | 244 | 20 | 88 | 352 |
| 20 | 270 | 23 | 74 | 367 |
| 21 | 307 | 26 | 66 | 399 |

*Note:* In this case, using the KEEP= data set option is necessary, because when you use the KEEP statement, all data sets that are created in the DATA step contain the same variables.

# Using the DROP= and KEEP= Data Set Options for Efficiency

The DROP= and KEEP= data set options are valid in both the DATA statement and the SET statement. However, you can write a more efficient DATA step if you understand the consequences of using these options in the DATA statement rather than the SET statement.

In the DATA statement, these options affect which variables SAS writes from the program data vector to the resulting SAS data set. In the SET statement, these options determine which variables SAS reads from the input SAS data set. Therefore, they determine how the program data vector is built.

When you specify the DROP= or KEEP= option in the SET statement, SAS does not read the excluded variables into the program data vector. If you work with a large data set (perhaps one containing thousands or millions of observations), then you can construct a more efficient DATA step by not reading unneeded variables from the input data set.

Note also that if you use a variable from the input data set to perform a calculation, the variable must be read into the program data vector. If you do not want that variable to appear in the new data set, however, use the DROP= option in the DATA statement to exclude it.

The following DATA step creates the same two data sets as the DATA step in the previous example. It does not read the variable Total into the program data vector. Compare the SET statement here to the one in "Creating More Than One Data Set in a Single DATA Step" on page 99.

```
data services (keep=ServicesTotal ServicesPolice ServicesFire
              ServicesWater_Sewer)
     admin (keep=AdminTotal AdminLabor AdminSupplies
              AdminUtilities);
   set city(drop=Total);
run;

proc print data=services;
   title 'City Expenditures: Services';
run;

proc print data=admin;
   title 'City Expenditures: Administration';
run;
```

In contrast with previous examples, the data set options in this example appear in both the DATA and SET statements. In the SET statement, the DROP= option determines which variables are omitted from the program data vector. In the DATA statement, the KEEP= option controls which variables are written from the program data vector to each data set being created.

*Note:* Using a DROP or KEEP statement is comparable to using a DROP= or KEEP= option in the DATA statement. All variables are included in the program data vector; they are excluded when the observation is written from the program data vector to the new data set. When you create more than one data set in a single DATA step, use the data set options to drop or keep different variables in each of the new data sets. A DROP or KEEP statement, on the other hand, affects all of the data sets that are created.

# Summary

## *Data Set Options*

DROP=*variable(s)*
   specifies the variables to be excluded.

   Used in the SET statement, DROP= specifies the variables that are not to be read from the existing SAS data set into the program data vector. Used in the DATA statement, DROP= specifies the variables to be excluded from the data set that is being created.

FIRSTOBS=*n*
specifies the first observation to be read from the SAS data set that you specify in the SET statement.

KEEP=*variable(s)*
specifies the variables to be included.

Used in the SET statement, KEEP= specifies the variables to be read from the existing SAS data set into the program data vector. Used in the DATA statement, KEEP= specifies which variables in the program data vector are to be written to the data set being created.

OBS=*n*
specifies the last observation to be read from the SAS data set that you specify in the SET statement.

## Procedures

PROC DATASETS <LIBRARY=*SAS-data-library*>;CONTENTS <DATA=*SAS-data set*>;
describes the structure of a SAS data set, including the name, type, and length of all variables in the data set.

## Statements

DATA *SAS-data-set*<(*data-set-options*)>;
begins a DATA step and names the SAS data set or data sets that are being created. You can specify the DROP= or KEEP= data set options in parentheses after each data set name to control which variables are written to the output data set from the program data vector.

DROP *variable(s)*;
specifies the variables to be excluded from the data set that is being created. For more information, see "DROP Statement" in *SAS Statements: Reference*.

KEEP *variable(s)*
specifies the variables to be written to the data set that is being created. For more information, see "KEEP Statement" in *SAS Statements: Reference*.

SET SAS-*data-set*(*data-set-options*);
reads observations from a SAS data set rather than records of raw data. You can specify the DROP= or KEEP= data set options in parentheses after a data set name to control which variables are read into the program data vector from the input data set.

# Learning More

Creating SAS data sets
For a general discussion about creating SAS data sets from other SAS data sets by merging, concatenating, interleaving, and updating, see Chapter 16, "Methods of Combining SAS Data Sets," on page 249.

Data set options
: See the "Data Set Options" section of *SAS Data Set Options: Reference*, and the SAS documentation for your operating environment.

DROP and KEEP statements
: For more information, see "DROP Statement" in *SAS Statements: Reference* and "KEEP Statement" in *SAS Statements: Reference*.

Engines
: *SAS Language Reference: Concepts*.

Subsetting IF statement
: You can use the subsetting IF statement and conditional (IF-THEN) logic when creating a new SAS data set from an existing one. For more information, see Chapter 10, "Acting on Selected Observations," on page 147 and Chapter 11, "Creating Subsets of Observations," on page 169.

# Part 3

# Basic Programming

*Chapter 7*
**Understanding DATA Step Processing** .......................... *107*

*Chapter 8*
**Working with Numeric Variables** .............................. *117*

*Chapter 9*
**Working with Character Variables** ............................ *129*

*Chapter 10*
**Acting on Selected Observations** .............................. *147*

*Chapter 11*
**Creating Subsets of Observations** ............................. *169*

*Chapter 12*
**Working with Grouped or Sorted Observations** ................. *183*

*Chapter 13*
**Using More Than One Observation in a Calculation** ............ *199*

*Chapter 14*
**Finding Shortcuts in Programming** ............................ *215*

*Chapter 15*
**Working with Dates in the SAS System** ........................ *225*

# Chapter 7
# Understanding DATA Step Processing

**Overview of DATA Step Processing** . . . . . . . . . . . . . . . . . . . . . . . . . . . . . . . . . . 107
   Purpose . . . . . . . . . . . . . . . . . . . . . . . . . . . . . . . . . . . . . . . . . . . . . . . . . . . . . . 107
   Prerequisites . . . . . . . . . . . . . . . . . . . . . . . . . . . . . . . . . . . . . . . . . . . . . . . . . . 107

**Input SAS Data Set for Examples** . . . . . . . . . . . . . . . . . . . . . . . . . . . . . . . . . . . 108

**Adding Information to a SAS Data Set** . . . . . . . . . . . . . . . . . . . . . . . . . . . . . . . 109
   Understanding the Assignment Statement . . . . . . . . . . . . . . . . . . . . . . . . . . . . . 109
   Making Uniform Changes to Data By Creating a Variable . . . . . . . . . . . . . . . . . 109
   Adding Information to Some Observations but Not Others . . . . . . . . . . . . . . . . 110
   Making Uniform Changes to Data without Creating Variables . . . . . . . . . . . . . . 111
   Using Variables Efficiently . . . . . . . . . . . . . . . . . . . . . . . . . . . . . . . . . . . . . . . . 112

**Defining Enough Storage Space for Variables** . . . . . . . . . . . . . . . . . . . . . . . . . 114

**Conditionally Deleting an Observation** . . . . . . . . . . . . . . . . . . . . . . . . . . . . . . . 115

**Summary** . . . . . . . . . . . . . . . . . . . . . . . . . . . . . . . . . . . . . . . . . . . . . . . . . . . . . . . 115
   Statements . . . . . . . . . . . . . . . . . . . . . . . . . . . . . . . . . . . . . . . . . . . . . . . . . . . . 115

**Learning More** . . . . . . . . . . . . . . . . . . . . . . . . . . . . . . . . . . . . . . . . . . . . . . . . . . 116

## Overview of DATA Step Processing

### Purpose

To add, modify, and delete information in a SAS data set, you use a DATA step. In this section, you will learn how the DATA step works, the general form of the statements, and some programming techniques.

### Prerequisites

Before continuing, you should understand the concepts presented in the following sections:

- Chapter 3, "Introduction to DATA Step Processing," on page 27
- Chapter 4, "Starting with Raw Data: The Basics," on page 51

## Input SAS Data Set for Examples

Tradewinds Travel Inc. has an external file that they use to manipulate and store data about their tours. The external file contains the following information:

```
1         2    3   4   5
---------------------------
France    8    793 575 Major
Spain     10   805 510 Hispania
India     10    .  489 Royal
Peru      7    722 590 Mundial
```

The columns in the above example contain the following information:

1 the name of the country toured

2 the number of nights on the tour

3 the airfare in US dollars

4 the cost of the land package in US dollars

5 the name of the company that offers the tour

Notice that the cost of the airfare for the tour to India has a missing value, which is indicated by a period.

The following DATA step creates a permanent SAS data set named MYLIB.INTERNATIONALTOURS:

```
libname mylib 'permanent-data-library';

data mylib.internationaltours;
    infile 'input-file';
    input Country $ Nights AirCost LandCost Vendor $;

proc print data = mylib.internationaltours;
   title 'Data Set MYLIB.INTERNATIONALTOURS';
run;
```

The PROC PRINT statement that follows the DATA step produces this display of the MYLIB.INTERNATIONALTOURS data set:

*Figure 7.1  Creating a Permanent SAS Data Set*

### Data Set MYLIB.INTERNATIONALTOURS

| Obs | Country | Nights | AirCost | LandCost | Vendor |
|-----|---------|--------|---------|----------|--------|
| 1 | France | 8 | 793 | 575 | Major |
| 2 | Spain | 10 | 805 | 510 | Hispania |
| 3 | India | 10 | | 489 | Royal |
| 4 | Peru | 7 | 722 | 590 | Mundial |

# Adding Information to a SAS Data Set

## *Understanding the Assignment Statement*

One of the most common reasons for using program statements in the DATA step is to produce new information from the original information or to change the information read by the INPUT or SET/MERGE/MODIFY/UPDATE statement. How do you add information to observations with a DATA step?

The basic method of adding information to a SAS data set is to create a new variable in a DATA step with an assignment statement. An assignment statement has the form:

*variable=expression;*

The variable receives the new information; the expression creates the new information. You specify the calculation necessary to produce the information and write the calculation as the expression. When the expression contains character data, you must enclose the data in quotation marks. SAS evaluates the expression and stores the new information in the variable that you name. It is important to remember that if you need to add the information to only one or two observations out of many, SAS creates that variable for all observations. The SAS data set that is being created must have information in every observation and every variable.

## *Making Uniform Changes to Data By Creating a Variable*

Sometimes you want to make a particular change to every observation. For example, at Tradewinds Travel the airfare must be increased for every tour by $10 because of a new tax. One way to do this is to write an assignment statement that creates a new variable that calculates the new airfare:

```
NewAirCost = AirCost+10;
```

This statement directs SAS to read the value of AirCost, add 10 to it, and assign the result to the new variable, NewAirCost.

When this assignment statement is included in a DATA step, the DATA step looks like this:

```
data newair;
   set mylib.internationaltours;
   NewAirCost = AirCost + 10;

proc print data=newair;
   var Country AirCost NewAirCost;
   title 'Increasing the Air Fare by $10 for All Tours';
run;
```

*Note:* In this example, the VAR statement in the PROC PRINT step determines which variables are displayed in the output.

The following output shows the resulting SAS data set, NEWAIR:

**Figure 7.2** *Adding Information to All Observations By Using a New Variable*

Increasing the Air Fare by $10 for All Tours

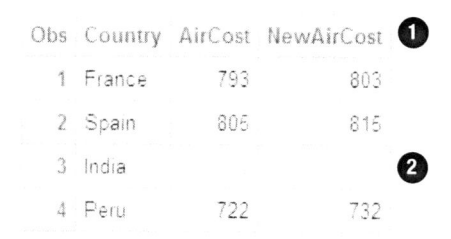

| Obs | Country | AirCost | NewAirCost | ❶ |
|---|---|---|---|---|
| 1 | France | 793 | 803 | |
| 2 | Spain | 805 | 815 | |
| 3 | India | | | ❷ |
| 4 | Peru | 722 | 732 | |

Notice the following in this data set:

1. Because SAS carries out each statement in the DATA step for every observation, NewAirCost is calculated during each iteration of the DATA step.

2. The observation for India contains a missing value for AirCost; SAS therefore assigns a missing value to NewAirCost for that observation.

The SAS data set has information in every observation and every variable.

## Adding Information to Some Observations but Not Others

Often you need to add information to some observations but not to others. For example, some tour operators award bonus points to travel agencies for scheduling particular tours. Two companies, Hispania and Mundial, are offering bonus points this year.

IF-THEN/ELSE statements can cause assignment statements to be carried out only when a condition is met. In the following DATA step, the IF statements check the value of the variable Vendor. If the value is either Hispania or Mundial, information about the bonus points is added to those observations.

```
data bonus;
   set mylib.internationaltours;
   if Vendor = 'Hispania' then BonusPoints = 'For 10+ people';
   else if Vendor = 'Mundial' then BonusPoints = 'Yes';
run;

proc print data=bonus;
   var Country Vendor BonusPoints;
   title1 'Adding Information to Observations for';
   title2 'Vendors Who Award Bonus Points';
run;
```

The following output displays the results:

*Figure 7.3* *Specifying Values for Specific Observations By Using a New Variable*

Adding Information to Observations for
Vendors Who Award Bonus Points

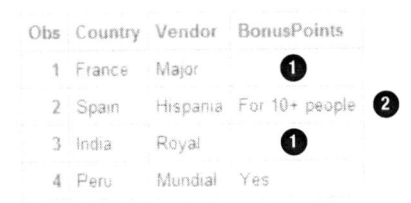

| Obs | Country | Vendor | BonusPoints |
|---|---|---|---|
| 1 | France | Major | ❶ |
| 2 | Spain | Hispania | For 10+ people ❷ |
| 3 | India | Royal | ❶ |
| 4 | Peru | Mundial | Yes |

The new variable BonusPoints has the following information:

1. In the two observations that are not assigned a value for BonusPoints, SAS assigns a missing value, represented by a blank in this case, to indicate the absence of a character value.

2. The first value that SAS encounters for BonusPoints contains 14 characters; therefore, SAS sets aside 14 bytes of storage in each observation for BonusPoints, regardless of the length of the value for that observation.

## Making Uniform Changes to Data without Creating Variables

Sometimes you want to change the value of existing variables without adding new variables. For example, in one DATA step a new variable, NewAirCost, was created to contain the value of the airfare plus the new $10 tax:

```
NewAirCost = AirCost + 10;
```

You can also decide to change the value of an existing variable rather than create a new variable. Following the example, AirCost is changed as follows:

```
AirCost = AirCost + 10;
```

SAS processes this statement just as it does other assignment statements. It evaluates the expression on the right side of the equal sign and assigns the result to the variable on the left side of the equal sign. The fact that the same variable appears on the right and left sides of the equal sign does not matter. SAS evaluates the expression on the right side of the equal sign before looking at the variable on the left side.

The following program contains the new assignment statement:

```
data newair2;
   set mylib.internationaltours;
   AirCost = AirCost + 10;

proc print data=newair2;
   var Country AirCost;
   title 'Adding Tax to the Air Cost Without Adding a New Variable';
run;
```

The following output displays the results:

**Figure 7.4** *Changing the Information in a Variable*

### Adding Tax to the Air Cost Without Adding a New Variable

| Obs | Country | AirCost |
|---|---|---|
| 1 | France | 803 |
| 2 | Spain | 815 |
| 3 | India | |
| 4 | Peru | 732 |

When you change the type of information that a variable contains, you change the meaning of that variable. In this case, you are changing the meaning of AirCost from *airfare without tax* to *airfare with tax*. If you remember the current meaning and if you know that you do not need the original information, then changing a variable's values is useful. However, for many programmers, having separate variables is easier than recalling one variable whose definition changes.

## Using Variables Efficiently

Variables that contain information that applies to only one or two observations use more storage space than necessary. When possible, create fewer variables that apply to more observations in the data set, and allow the different values in different observations to supply the information.

For example, the Major company offers discounts, not bonus points, for groups of 30 or more people. An inefficient program would create separate variables for bonus points and discounts, as follows:

```
   /* inefficient use of variables */
options pagesize=60 linesize=80 pageno=1 nodate;
data tourinfo;
   set mylib.internationaltours;
   if Vendor = 'Hispania' then BonusPoints = 'For 10+ people';
   else if Vendor = 'Mundial' then BonusPoints = 'Yes';
       else if Vendor = 'Major' then Discount = 'For 30+ people';
run;

proc print data=tourinfo;
   var Country Vendor BonusPoints Discount;
   title 'Information About Vendors';
run;
```

The following output displays the results:

*Figure 7.5* *Inefficient: Using Variables That Scatter Information across Multiple Variables*

### Information About Vendors

| Obs | Country | Vendor | BonusPoints | Discount |
|---|---|---|---|---|
| 1 | France | Major | | For 30+ people |
| 2 | Spain | Hispania | For 10+ people | |
| 3 | India | Royal | | |
| 4 | Peru | Mundial | Yes | |

As you can see, storage space is used inefficiently. Both BonusPoints and Discount have a significant number of missing values.

With a little planning, you can make the SAS data set much more efficient. In the following DATA step, the variable Remarks contains information about bonus points, discounts, and any other special features of any tour.

```
   /* efficient use of variables */
data newinfo;
   set mylib.internationaltours;
   if Vendor = 'Hispania' then Remarks = 'Bonus for 10+ people';
   else if Vendor = 'Mundial' then Remarks = 'Bonus points';
      else if Vendor = 'Major' then Remarks = 'Discount: 30+ people';
run;

proc print data=newinfo;
   var Country Vendor Remarks;
   title 'Information About Vendors';
run;
```

The following output displays a more efficient use of variables:

*Figure 7.6* *Efficient: Using Variables to Contain Maximum Information*

### Information About Vendors

| Obs | Country | Vendor | Remarks |
|---|---|---|---|
| 1 | France | Major | Discount: 30+ people |
| 2 | Spain | Hispania | Bonus for 10+ people |
| 3 | India | Royal | |
| 4 | Peru | Mundial | Bonus points |

Remarks has fewer missing values and contains all the information that is used by BonusPoints and Discount in the inefficient example. Using variables efficiently can save storage space and optimize your SAS data set.

## Defining Enough Storage Space for Variables

The first time that a value is assigned to a variable, SAS enables as many bytes of storage space for the variable as there are characters in the first value assigned to it. At times, you might need to specify the amount of storage space that a variable requires.

For example, as shown in the preceding example, the variable Remarks contains miscellaneous information about tours:

```
if Vendor = 'Hispania' then Remarks = 'Bonus for 10+ people';
```

In this assignment statement, SAS enables 20 bytes of storage space for Remarks as there are 20 characters in the first value assigned to it. The longest value might not be the first one assigned, so you specify a more appropriate length for the variable before the first value is assigned to it:

```
length Remarks $ 30;
```

This statement, called a LENGTH statement, applies to the entire data set. It defines the number of bytes of storage that is used for the variable Remarks in every observation. SAS uses the LENGTH statement during compilation, not when it is processing statements on individual observations. The following DATA step shows the use of the LENGTH statement:

```
data newlength;
   set mylib.internationaltours;
   length Remarks $ 30;
   if Vendor = 'Hispania' then Remarks = 'Bonus for 10+ people';
   else if Vendor = 'Mundial' then Remarks = 'Bonus points';
      else if Vendor = 'Major' then Remarks = 'Discount for 30+ people';
run;

proc print data=newlength;
   var Country Vendor Remarks;
   title 'Information About Vendors';
run;
```

The following output displays the NEWLENGTH data set:

*Figure 7.7* Using a LENGTH Statement

### Information About Vendors

| Obs | Country | Vendor | Remarks |
| --- | --- | --- | --- |
| 1 | France | Major | Discount for 30+ people |
| 2 | Spain | Hispania | Bonus for 10+ people |
| 3 | India | Royal | |
| 4 | Peru | Mundial | Bonus points |

Because the LENGTH statement affects variable storage, not the spacing of columns in printed output, the Remarks variable appears the same in Figure 7.7 on page 114 and

Figure 7.6 on page 113. To show the effect of the LENGTH statement on variable storage using the DATASETS procedures, see Chapter 37, "Getting Information about Your SAS Data Sets," on page 701.

## Conditionally Deleting an Observation

If you do not want the program data vector to write to a data set based on a condition, use the DELETE statement in the DATA step. For example, if the tour to Peru has been discontinued, it is no longer necessary to include the observation for Peru in the data set that is being created. The following example uses the DELETE statement to prevent SAS from writing that observation to the output data set:

```
data subset;
   set mylib.internationaltours;
   if Country = 'Peru' then delete;
run;

proc print data=subset;
   title 'Omitting a Discontinued Tour';
run;
```

The following output displays the results:

*Figure 7.8* Deleting an Observation

Omitting a Discontinued Tour

| Obs | Country | Nights | AirCost | LandCost | Vendor |
|---|---|---|---|---|---|
| 1 | France | 8 | 793 | 575 | Major |
| 2 | Spain | 10 | 805 | 510 | Hispania |
| 3 | India | 10 |  | 489 | Royal |

The observation for Peru has been deleted from the data set.

## Summary

### Statements

DELETE;
: prevents SAS from writing a particular observation to the output data set. It usually appears as part of an IF-THEN/ELSE statement.

IF *condition* THEN *action* ELSE *action*;
: tests whether the condition is true. When the condition is true, the THEN statement specifies the action to take. When the condition is false, the ELSE statement provides an alternative action. The action can be one or more statements, including assignment statements.

**LENGTH** *variable* <$> *length*;
> assigns the number of bytes of storage (length) for a variable. Include a dollar sign ($) if the variable is character. The LENGTH statement must appear before the first use of the variable.

*variable = expression*
> is an assignment statement. It causes SAS to evaluate the *expression* on the right side of the equal sign and assign the result to the *variable* on the left. You must select the name of the variable and create the proper expression for calculating its value. The same variable name can appear on the left and right sides of the equal sign because SAS evaluates the right side before assigning the result to the variable on the left side.

# Learning More

Character variables
> For information about expressions involving alphabetic and special characters as well as numbers, see Chapter 9, "Working with Character Variables," on page 129.

DATA step
> For general DATA step information, see Chapter 3, "Introduction to DATA Step Processing," on page 27. Complete information about the DATA step can be found in the "DATA Step Concepts" section of *SAS Language Reference: Concepts*.

IF-THEN/ELSE statements
> The IF-THEN/ELSE statements are discussed in Chapter 10, "Acting on Selected Observations," on page 147.

LENGTH statement
> Additional information about the LENGTH statement can be found in Chapter 8, "Working with Numeric Variables," on page 117 and Chapter 9, "Working with Character Variables," on page 129. To show the effect of the LENGTH statement on variable storage using the DATASETS procedures, see Chapter 37, "Getting Information about Your SAS Data Sets," on page 701.

Missing values
> For more information about missing values, see Chapter 8, "Working with Numeric Variables," on page 117 and Chapter 9, "Working with Character Variables," on page 129.

Numeric variables
> Information about working with numeric variables and expressions can be found in Chapter 8, "Working with Numeric Variables," on page 117.

SAS statements
> For complete reference information about the IF-THEN/ELSE, LENGTH, DELETE, assignment, and comment statements, see *SAS Statements: Reference*.

# Chapter 8
# Working with Numeric Variables

| | |
|---|---|
| **Introduction to Working with Numeric Variables** | **117** |
| Purpose | 117 |
| Prerequisites | 117 |
| **About Numeric Variables in SAS** | **118** |
| **Input SAS Data Set for Examples** | **118** |
| **Calculating with Numeric Variables** | **119** |
| Using Arithmetic Operators in Assignment Statements | 119 |
| Understanding Numeric Expressions and Assignment Statements | 121 |
| Understanding How SAS Handles Missing Values | 121 |
| Calculating Numbers Using SAS Functions | 122 |
| **Comparing Numeric Variables** | **124** |
| **Storing Numeric Variables Efficiently** | **126** |
| **Summary** | **127** |
| Functions | 127 |
| Statements | 127 |
| **Learning More** | **128** |

## Introduction to Working with Numeric Variables

### Purpose

The following concepts are discussed in this section:

- how to perform arithmetic calculations in SAS using arithmetic operators and the SAS functions ROUND and SUM
- how to compare numeric variables using logical operators
- how to store numeric variables efficiently when disk space is limited

### Prerequisites

Before proceeding with this section, you should understand the concepts presented in the following sections:

- Part 1, "Introduction to the SAS System"
- Part 2, "Getting Your Data into Shape"
- Chapter 7, "Understanding DATA Step Processing," on page 107

## About Numeric Variables in SAS

A numeric variable is a variable whose values are numbers.

*Note:* SAS uses double-precision floating-point representation for calculations and, by default, for storing numeric variables in SAS data sets.

SAS accepts numbers in many forms, such as scientific notation, and hexadecimal. For more information, see the discussion on the types of numbers that SAS can read from data lines in *SAS Language Reference: Concepts*. For simplicity, this documentation concentrates on numbers in standard representation, as shown here:

```
1254
 336.05
-243
```

You can use SAS to perform all types of mathematical operations. To perform a calculation in a DATA step, you can write an assignment statement in which the expression contains arithmetic operators, SAS functions, or a combination of the two. To compare numeric variables, you can write an IF-THEN/ELSE statement using logical operators. For more information about numeric functions, see the discussion in the "Functions and CALL Routines" section in *SAS Functions and CALL Routines Reference*.

## Input SAS Data Set for Examples

Tradewinds Travel Inc. has an external file that contains information about their most popular tours:

```
      1              2   3    4    5
---------------------------------------
Japan          8  982 1020 Express
Greece        12    .  748 Express
New Zealand   16 1368 1539 Southsea
Ireland        7  787  628 Express
Venezuela      9  426  505 Mundial
Italy          8  852  598 Express
Russia        14 1106 1024 A-B-C
Switzerland    9  816  834 Tour2000
Australia     12 1299 1169 Southsea
Brazil         8  682  610 Almeida
```

The numbered fields represent

1. the name of the country toured

2. the number of nights on the tour

3. the airfare in US dollars

4   the cost of the land package in US dollars

5   the name of the company that offers the tour

The following program creates a permanent SAS data set named MYLIB.POPULARTOURS:

```
libname mylib 'permanent-data-library';

data mylib.populartours;
   infile 'input-file';
   input Country $ 1-11 Nights AirCost LandCost Vendor $;
run;

proc print data=mylib.populartours;
   title 'Data Set MYLIB.POPULARTOURS';
run;
```

The following output displays the data set:

**Figure 8.1**  *Data Set MYLIB.POPULARTOURS*

Data Set MYLIB.POPULARTOURS

| Obs | Country | Nights | AirCost | LandCost | Vendor |
|-----|---------|--------|---------|----------|--------|
| 1 | Japan | 8 | 982 | 1020 | Express |
| 2 | Greece | 12 |  | 748 | Express |
| 3 | New Zealand | 16 | 1368 | 1539 | Southsea |
| 4 | Ireland | 7 | 787 | 628 | Express |
| 5 | Venezuela | 9 | 426 | 505 | Mundial |
| 6 | Italy | 8 | 852 | 598 | Express |
| 7 | Russia | 14 | 1106 | 1024 | A-B-C |
| 8 | Switzerland | 9 | 816 | 834 | Tour2000 |
| 9 | Australia | 12 | 1299 | 1169 | Southsea |
| 10 | Brazil | 8 | 682 | 610 | Almeida |

In MYLIB.POPULARTOURS, the variables Nights, AirCost, and LandCost contain numbers and are stored as numeric variables. For comparison, variables Country and Vendor contain alphabetic and special characters as well as numbers; they are stored as character variables.

# Calculating with Numeric Variables

## *Using Arithmetic Operators in Assignment Statements*

One way to perform calculations on numeric variables is to write an assignment statement using arithmetic operators. Arithmetic operators indicate addition, subtraction, multiplication, division, and exponentiation (raising to a power). For more information about arithmetic expressions, see the discussion in *SAS Language Reference: Concepts*.

The following table shows operators that you can use in arithmetic expressions.

*Table 8.1* Operators in Arithmetic Expressions

| Operation | Symbol | Example |
|---|---|---|
| addition | + | x = y + z; |
| subtraction | − | x = y - z; |
| multiplication | * | x = y * z |
| division | / | x = y / z |
| exponentiation | ** | x = y ** z |

The following examples show some typical calculations using the Tradewinds Travel sample data.

*Table 8.2* Examples of Using Arithmetic Operators

| Action | SAS Statement |
|---|---|
| Add the airfare and land cost to produce the total cost. | TotalCost = AirCost + Landcost; |
| Calculate the peak season airfares by increasing the basic fare by 10% and adding an $8 departure tax. | PeakAir = (AirCost * 1.10) + 8; |
| Show the cost per night of each land package. | NightCost = LandCost / Nights; |

In each case, the variable on the left side of the equal sign receives the calculated value from the numeric expression on the right side of the equal sign. Including these statements in the following DATA step produces data set NEWTOUR:

```
data newtour;
   set mylib.populartours;
   TotalCost = AirCost + LandCost;
   PeakAir = (AirCost * 1.10) + 8;
   NightCost = LandCost / Nights;
run;

proc print data=newtour;
   var Country Nights AirCost LandCost TotalCost PeakAir NightCost;
   title 'Costs for Tours';
run;
```

The VAR statement in the PROC PRINT step causes only the variables listed in the statement to be displayed in the output.

*Figure 8.2  Creating New Variables By Using Arithmetic Expressions*

## Costs for Tours

| Obs | Country | Nights | AirCost | LandCost | TotalCost | PeakAir | NightCost |
|---|---|---|---|---|---|---|---|
| 1 | Japan | 8 | 982 | 1020 | 2002 | 1088.2 | 127.500 |
| 2 | Greece | 12 |  | 748 |  |  | 62.333 |
| 3 | New Zealand | 16 | 1368 | 1539 | 2907 | 1512.8 | 96.188 |
| 4 | Ireland | 7 | 787 | 628 | 1415 | 873.7 | 89.714 |
| 5 | Venezuela | 9 | 426 | 505 | 931 | 476.6 | 56.111 |
| 6 | Italy | 8 | 852 | 598 | 1450 | 945.2 | 74.750 |
| 7 | Russia | 14 | 1106 | 1024 | 2130 | 1224.6 | 73.143 |
| 8 | Switzerland | 9 | 816 | 834 | 1650 | 905.6 | 92.667 |
| 9 | Australia | 12 | 1299 | 1169 | 2468 | 1436.9 | 97.417 |
| 10 | Brazil | 8 | 682 | 610 | 1292 | 758.2 | 76.250 |

## *Understanding Numeric Expressions and Assignment Statements*

Numeric expressions in SAS share some features with mathematical expressions:

- When an expression contains more than one operator, the operations have the same order of precedence as in a mathematical expression: exponentiation is done first, then multiplication and division, and finally addition and subtraction.

- When operators of equal precedence appear, the operations are performed from left to right (except exponentiation, which is performed right to left).

- Parentheses are used to group parts of an expression; as in mathematical expressions, operations in parentheses are performed first.

*Note:* The equal sign in an assignment statement does not perform the same function as the equal sign in a mathematical equation. The sequence *variable=* in an assignment statement defines the statement, and the variable must appear on the left side of the equal sign. You cannot switch the positions of the result variable and the expression as you can in a mathematical equation.

## *Understanding How SAS Handles Missing Values*

### *Why SAS Assigns Missing Values*

What if an observation lacks a value for a particular numeric variable? For example, in the data set MYLIB.POPULARTOURS, as shown in Figure 8.2 on page 121, the observation for Greece has no value for the variable AirCost. To maintain the rectangular structure of a SAS data set, SAS assigns a missing value to the variable in that observation. A missing value indicates that no information is present for the variable in that observation.

### Rules for Missing Values

The following rules describe missing values in several situations:

- In data lines, a missing numeric value is represented by a period, for example,

    ```
    Greece     8 12   .   748 Express
    ```

    By default, SAS interprets a single period in a numeric field as a missing value. If the INPUT statement reads the value from particular columns, as in column input, a field that contains only blanks also produces a missing value.

- In an expression, a missing numeric value is represented by a period, for example,

    ```
    if AirCost= . then Status = 'Need air cost';
    ```

- In a comparison and in sorting, a missing numeric value is a lower value than any other numeric value.

- In procedure output, SAS by default represents a missing numeric value with a period.

- Some procedures eliminate missing values from their analyses; others do not. Documentation for individual procedures describes how each procedure handles missing values.

### Propagating Missing Values

When you use a missing value in an arithmetic expression, SAS sets the result of the expression to missing. If you use that result in another expression, the next result is also missing. In SAS, this method of treating missing values is called propagation of missing values. For example, Figure 8.2 on page 121 shows that in the data set NEWTOUR, the values for TOTALCOST and PEAKAIR are also missing in the observation for Greece.

*Note:* SAS enables you to distinguish between various types of numeric missing values. See "Missing Values" section of *SAS Language Reference: Concepts*. The SAS language contains 27 special missing values based on the letters A–Z and the underscore (_).

## Calculating Numbers Using SAS Functions

### Rounding Values

In the example data that lists costs of the different tours (Figure 8.1 on page 119), some of the tours have odd prices: $748 instead of $750, $1299 instead of $1300, and so on. Rounded numbers, created by rounding the tour prices to the nearest $10, would be easier to work with.

Programming a rounding calculation with only the arithmetic operators is a lengthy process. However, SAS contains built-in numeric expressions called functions. You can use these functions in expressions just as you do the arithmetic operators. For example, the following assignment statement rounds the value of AirCost to the nearest $50:

```
RoundAir = round(AirCost,50);
```

The following statement calculates the total cost of each tour, rounded to the nearest $100:

```
TotalCostR = round(AirCost + LandCost,100);
```

## Calculating a Cost When There Are Missing Values

As another example, the travel agent can calculate a total cost for the tours based on all nonmissing costs. Therefore, when the airfare is missing (as it is for Greece) the total cost represents the land cost, not a missing value. (Of course, you must decide whether skipping missing values in a particular calculation is a good idea.) The SUM function calculates the sum of its arguments, ignoring missing values. This example illustrates the SUM function:

```
SumCost = sum(AirCost,LandCost);
```

## Combining Functions

It is possible for you to combine functions. The ROUND function rounds the quantity given in the first argument to the nearest unit given in the second argument. The SUM function adds any number of arguments, ignoring missing values. The calculation in the following assignment statement rounds the sum of all nonmissing airfares and land costs to the nearest $100 and assigns the value to RoundSum:

```
RoundSum = round(sum(AirCost,LandCost),100);
```

Using the ROUND and SUM functions in the following DATA step creates the data set MORETOUR:

```
data moretour;
   set mylib.populartours;
   RoundAir = round(AirCost,50);
   TotalCostR = round(AirCost + LandCost,100);
   CostSum = sum(AirCost,LandCost);
   RoundSum = round(sum(AirCost,LandCost),100);
run;

proc print data=moretour;
   var Country AirCost LandCost RoundAir TotalCostR CostSum RoundSum;
   title 'Rounding and Summing Values';
run;
```

The following output displays the results:

*Figure 8.3* Creating New Variables with ROUND and SUM Functions

Rounding and Summing Values

| Obs | Country | AirCost | LandCost | RoundAir | TotalCostR | CostSum | RoundSum |
|---|---|---|---|---|---|---|---|
| 1 | Japan | 982 | 1020 | 1000 | 2000 | 2002 | 2000 |
| 2 | Greece |  | 748 |  |  | 748 | 700 |
| 3 | New Zealand | 1368 | 1539 | 1350 | 2900 | 2907 | 2900 |
| 4 | Ireland | 787 | 628 | 800 | 1400 | 1415 | 1400 |
| 5 | Venezuela | 426 | 505 | 450 | 900 | 931 | 900 |
| 6 | Italy | 852 | 598 | 850 | 1500 | 1450 | 1500 |
| 7 | Russia | 1106 | 1024 | 1100 | 2100 | 2130 | 2100 |
| 8 | Switzerland | 816 | 834 | 800 | 1700 | 1650 | 1700 |
| 9 | Australia | 1299 | 1169 | 1300 | 2500 | 2468 | 2500 |
| 10 | Brazil | 682 | 610 | 700 | 1300 | 1292 | 1300 |

## Comparing Numeric Variables

Often in a program, you need to know whether variables are equal to each other, or whether they are greater than or less than each other. To compare two numeric variables, you can write an IF-THEN/ELSE statement using logical operators.

The following table lists some of the logical operators that you can use for variable comparisons.

*Table 8.3  Logical Operators*

| Symbol | Mnemonic Equivalent | Logical Operation |
| --- | --- | --- |
| = | eq | equal |
| ¬=, ^=, ~= | ne | not equal to ( the ¬=, ^=, or ~= symbol, depending on your keyboard) |
| > | gt | greater than |
| >= | ge | greater than or equal to |
| < | lt | less than |
| <= | le | less than or equal to |

In this example, the total cost of each tour in the POPULARTOURS data set is compared to 2000 using the greater-than logical operator (gt). If the total cost of the tour is greater than 2000, the tour is excluded from the data set. The resulting data set TOURSUNDER2K contains tours that are $2000 or less.

```
data toursunder2K;
   set mylib.populartours;
   TotalCost = AirCost + LandCost;
 if TotalCost gt 2000 then delete;
run;
proc print data=toursunder2K;
   var Country Nights AirCost Landcost TotalCost Vendor;
   title 'Tours $2000 or Less';
run;
```

The following output displays the tours that are less than $2000 in total cost:

*Figure 8.4*  *Comparing Numeric Variables*

### Tours $2000 or Less

| Obs | Country | Nights | AirCost | LandCost | TotalCost | Vendor |
|---|---|---|---|---|---|---|
| 1 | Greece | 12 |  | 748 |  | Express |
| 2 | Ireland | 7 | 787 | 628 | 1415 | Express |
| 3 | Venezuela | 9 | 426 | 505 | 931 | Mundial |
| 4 | Italy | 8 | 852 | 598 | 1450 | Express |
| 5 | Switzerland | 9 | 816 | 834 | 1650 | Tour2000 |
| 6 | Brazil | 8 | 682 | 610 | 1292 | Almeida |

The TotalCost value for Greece is a missing value because any calculation that includes a missing value results in a missing value. In a comparison, missing numeric values are lower than any other numeric value.

If you need to compare a variable to more than one value, you can include multiple comparisons in a condition. To eliminate tours with missing values, a second comparison is added:

```
data toursunder2K2;
   set mylib.populartours;
   TotalCost = AirCost + LandCost;
   if TotalCost gt 2000 or Totalcost = . then delete;
run;

proc print data=toursunder2K2;
   var Country Nights TotalCost Vendor;
   title 'Tours $2000 or Less';
run;
```

The following output displays the results:

*Figure 8.5*  *Multiple Comparisons in a Condition*

### Tours $2000 or Less

| Obs | Country | Nights | TotalCost | Vendor |
|---|---|---|---|---|
| 1 | Ireland | 7 | 1415 | Express |
| 2 | Venezuela | 9 | 931 | Mundial |
| 3 | Italy | 8 | 1450 | Express |
| 4 | Switzerland | 9 | 1650 | Tour2000 |
| 5 | Brazil | 8 | 1292 | Almeida |

Notice that Greece is no longer included in the tours for under $2000.

## Storing Numeric Variables Efficiently

The data sets shown in this section are very small, but data sets are often very large. If you have a large data set, you might need to think about the storage space that your data set occupies. There are ways to save space when you store numeric variables in SAS data sets.

*Note:* The SAS documentation for your operating environment provides information about storing numeric variables whose values are limited to 1 or 0 in the minimum number of bytes used by SAS (either 2 or 3 bytes, depending on your operating environment).

By default, SAS uses 8 bytes of storage in a data set for each numeric variable. Therefore, storing the variables for each observation in the earlier data set MORETOUR requires 75 bytes:

```
56 bytes for numeric variables
   (8 bytes per variable * 7 numeric variables)
11 bytes for Country
 8 bytes for Vendor
_____
75 bytes for all variables
```

When numeric variables contain only integers (whole numbers), you can often shorten them in the data set that is being created. For example, a length of 4 bytes accurately stores all integers up to at least 2,000,000.

*Note:* Under some operating environments, the maximum number of bytes is much greater. For more information, see the documentation for your operating environment.

To change the number of bytes used for each variable, use a LENGTH statement.

A LENGTH statement contains the names of the variables followed by the number of bytes to be used for their storage. For numeric variables, the LENGTH statement affects only the data set that is being created; it does not affect the program data vector. The following program changes the storage space for all numeric variables that are in the data set SHORTER:

```
data shorter;
   set mylib.populartours;
   length Nights AirCost LandCost RoundAir TotalCostR
          Costsum RoundSum 4;
   RoundAir = round(AirCost,50);
   TotalCostR = round(AirCost + LandCost,100);
   CostSum = sum(AirCost,LandCost);
   RoundSum = round(sum(AirCost,LandCost),100);
run;
```

By calculating the storage space that is needed for the variables in each observation of SHORTER, you can see how the LENGTH statement changes the amount of storage space that is used:

```
28 bytes for numeric variables
   (4 bytes per variable in the LENGTH statement X 7 numeric variables)
11 bytes for Country
 8 bytes for Vendor
```

```
47 bytes for all variables
```

Because the 7 variables in SHORTER are shortened by the LENGTH statement, the storage space for the variables in each observation is reduced by almost half.

***CAUTION:***
   **Be careful in shortening the length of numeric variables if your variable values are not integers.** Fractional numbers lose precision permanently if they are truncated. In general, use the LENGTH statement to truncate values only when disk space is limited. Use the default length of 8 bytes to store variables containing fractions.

# Summary

## Functions

ROUND (*expression, round-off-unit*)
> rounds the quantity in *expression* to the figure given in *round-off-unit*. The `expression` can be a numeric variable name, a numeric constant, or an arithmetic expression. Separate *round-off-unit* from *expression* with a comma.

SUM (*expression-1<, expression-2>, . . .*)
> produces the sum of all expressions that you specify in the parentheses. The SUM function ignores missing values as it calculates the sum of the expressions. Each expression can be a numeric variable, a numeric constant, another arithmetic expression, or another numeric function.

## Statements

LENGTH *variable-list number-of-bytes*;
> indicates that the variables in the *variable-list* are to be stored in the data set according to the *number-of-bytes* that you specify. Numeric variables are not affected while they are in the program data vector. The default length for a numeric variable is 8 bytes. In general, the minimum that you should use is 4 bytes for variables that contain integers and 8 bytes for variables that contain fractions. You can assign lengths to both numeric and character variables (discussed in the next section) in a single LENGTH statement.

*variable=expression*;
> is an assignment statement. It causes SAS to calculate the value of the expression on the right side of the equal sign and assign the result to the *variable* on the left. When *variable* is numeric, the expression can be an arithmetic calculation, a numeric constant, or a numeric function.

## Learning More

Abbreviating lists of variables
: Ways to abbreviate lists of variables in function arguments are documented in *SAS Language Reference: Concepts*. Many functions, including the SUM function, accept abbreviated lists of variables as arguments.

DEFAULT= option
: Information about using the DEFAULT= option in the LENGTH statement to assign a default storage length to all newly created numeric variables can be found in *SAS Statements: Reference*.

Logical expressions
: Additional information about the use of logical expressions can be found in *SAS Language Reference: Concepts*.

Numeric precision
: For a discussion about numeric precision, see *SAS Language Reference: Concepts*. Because the computer's hardware determines how a computer stores numbers, the precision with which SAS can store numbers depends on the hardware of the computer system on which it is installed. Specific limits for hardware are discussed in the SAS documentation for each operating environment.

Saving space
: For information about how you can save space by treating some numeric values as character values see Chapter 9, "Working with Character Variables," on page 129.

# Chapter 9
# Working with Character Variables

**Introduction to Working with Character Variables** . . . . . . . . . . . . . . . . . . . . . . . 129
    Purpose . . . . . . . . . . . . . . . . . . . . . . . . . . . . . . . . . . . . . . . . . . . . . . . . . . . . . . . . . . . 129
    Prerequisites . . . . . . . . . . . . . . . . . . . . . . . . . . . . . . . . . . . . . . . . . . . . . . . . . . . . . . . 130
    Character Variables in SAS . . . . . . . . . . . . . . . . . . . . . . . . . . . . . . . . . . . . . . . . . . . 130

**Input SAS Data Set for Examples** . . . . . . . . . . . . . . . . . . . . . . . . . . . . . . . . . . . . . . 130

**Identifying Character Variables and Expressing Character Values** . . . . . . . . . . 131

**Setting the Length of Character Variables** . . . . . . . . . . . . . . . . . . . . . . . . . . . . . . 132

**Handling Missing Values** . . . . . . . . . . . . . . . . . . . . . . . . . . . . . . . . . . . . . . . . . . . . . 134
    Reading Missing Values . . . . . . . . . . . . . . . . . . . . . . . . . . . . . . . . . . . . . . . . . . . . . 134
    Checking for Missing Character Values . . . . . . . . . . . . . . . . . . . . . . . . . . . . . . . . 135
    Setting a Character Variable Value to Missing . . . . . . . . . . . . . . . . . . . . . . . . . . . 136

**Creating New Character Values** . . . . . . . . . . . . . . . . . . . . . . . . . . . . . . . . . . . . . . . 137
    Extracting a Portion of a Character Value . . . . . . . . . . . . . . . . . . . . . . . . . . . . . . 137
    Combining Character Values: Using Concatenation . . . . . . . . . . . . . . . . . . . . . . 139

**Saving Storage Space by Treating Numbers as Characters** . . . . . . . . . . . . . . . . 142

**Summary** . . . . . . . . . . . . . . . . . . . . . . . . . . . . . . . . . . . . . . . . . . . . . . . . . . . . . . . . . . . 143
    Functions . . . . . . . . . . . . . . . . . . . . . . . . . . . . . . . . . . . . . . . . . . . . . . . . . . . . . . . . . 143
    Statements . . . . . . . . . . . . . . . . . . . . . . . . . . . . . . . . . . . . . . . . . . . . . . . . . . . . . . . . 144

**Learning More** . . . . . . . . . . . . . . . . . . . . . . . . . . . . . . . . . . . . . . . . . . . . . . . . . . . . . . 144

## Introduction to Working with Character Variables

### Purpose

In this section, you will learn how to do the following:

- identify character variables
- set the length of character variables
- align character values within character variables
- handle missing values of character variables
- work with character variables, character constants, and character expressions in SAS program statements

- instruct SAS to read fields that contain numbers as character variables in order to save space

## Prerequisites

Before proceeding with this section, you should understand the concepts presented in the following topics:

- Part 1, "Introduction to SAS"
- Part 2, "Getting Your Data into Shape"
- Chapter 7, "Understanding DATA Step Processing," on page 107

## Character Variables in SAS

A character variable is a variable whose value contains letters, numbers, and special characters, and whose length can be from 1 to 32,767 characters long. Character variables can be used in declarative statements, comparison statements, or assignment statements where they can be manipulated to create new character variables.

# Input SAS Data Set for Examples

Tradewinds Travel has an external file with data on flight schedules for tours.

The following DATA step reads the information and stores it in a data set named AIR.DEPARTURES:

```
libname mylib 'permanent-data-library';

data mylib.departures;
   input Country $ 1-9 CitiesInTour 11-12 USGate $ 14-26
         ArrivalDepartureGates $ 28-48;
   datalines;

1         2 3                   4
-------------------------------------------------
Japan     5 San Francisco       Tokyo, Osaka
Italy     8 New York            Rome, Naples
Australia 12 Honolulu           Sydney, Brisbane
Venezuela 4 Miami               Caracas, Maracaibo
Brazil    4                     Rio de Janeiro, Belem
;
proc print data=mylib.departures;
   title 'Data Set AIR.DEPARTURES';
run;
```

The numbered fields represent

1 the name of the country toured

2 the number of cities in the tour

3 the city from which the tour leaves the United States (the gateway city)

4 the cities of arrival and departure in the destination country

The PROC PRINT statement that follows the DATA step produces this display of the AIR.DEPARTURES data set:

*Figure 9.1*   Data Set AIR.DEPARTURES

Data Set AIR.DEPARTURES

| Obs | Country | CitiesInTour | USGate | ArrivalDepartureGates |
|---|---|---|---|---|
| 1 | Japan | 5 | San Francisco | Tokyo, Osaka |
| 2 | Italy | 8 | New York | Rome, Naples |
| 3 | Australia | 12 | Honolulu | Sydney, Brisbane |
| 4 | Venezuela | 4 | Miami | Caracas, Maracaibo |
| 5 | Brazil | 4 | | Rio de Janeiro, Belem |

In AIR.DEPARTURES, the variables Country, USGate, and ArrivalDepartureGates contain information other than numbers, so they must be stored as character variables. The variable CitiesInTour contains only numbers. Therefore, it can be created and stored as either a character or numeric variable.

# Identifying Character Variables and Expressing Character Values

To store character values in a SAS data set, you need to create a character value. One way to create a character variable is to define it in an input statement. Simply place a dollar sign after the variable name in the INPUT statement, as shown in the DATA step that created AIR.DEPARTURES:

```
input Country $ 1-9 CitiesInTour 11-12 USGate $ 14-26
      ArrivalDepartureGates $ 28-48;
```

You can also create a character variable and assign a value to it in an assignment statement. Simply enclose the value in quotation marks:

```
Schedule = '3-4 tours per season';
```

Either single quotation marks (apostrophes) or double quotation marks are acceptable. If the value itself contains a single quotation mark, then enclose the value in double quotation marks, as in

```
Remarks = "See last year's schedule";
```

*Note:* Matching quotation marks properly is important. Missing or extraneous quotation marks cause SAS to misread both the erroneous statement and the statements following it.

When you specify a character value in an expression, you must also enclose the value in quotation marks. For example, the following statement compares the value of USGate to San Francisco and, when a match occurs, assigns the airport code SFO to the variable Airport:

```
if USGate = 'San Francisco' then Airport =
'SFO';
```

In character values, SAS distinguishes uppercase letters from lowercase letters. For example, in the data set AIR.DEPARTURES, the value of USGate in the observation for

**132** Chapter 9 • *Working with Character Variables*

Australia is `Honolulu`. The following IF condition is true; therefore, SAS assigns to Airport the value HNL:

```
else if USGate = 'Honolulu' then Airport = 'HNL';
```

However, the following condition is false:

```
if USGate = 'HONOLULU' then Airport = 'HNL';
```

SAS does not select that observation because the characters in Honolulu and HONOLULU are not equivalent.

The following program places these shaded statements in a DATA step:

```
data charvars;
   set mylib.departures;
   Schedule = '3-4 tours per season';
   Remarks = "See last year's schedule";
   if USGate = 'San Francisco' then Airport = 'SFO';
     else if USGate = 'Honolulu' then Airport = 'HNL';
run;

proc print data=charvars noobs; 1
   var Country Schedule Remarks USGate Airport;
   title 'Tours By City of Departure';
run;
```

1 The NOOBS option in the PROC PRINT statement suppresses the display of observation numbers in the output.

The following output displays the character variables in the data set CHARVARS:

*Figure 9.2* Examples of Character Variables

**Tours By City of Departure**

| Country | Schedule | Remarks | USGate | Airport |
|---|---|---|---|---|
| Japan | 3-4 tours per season | See last year's schedule | San Francisco | SFO |
| Italy | 3-4 tours per season | See last year's schedule | New York | |
| Australia | 3-4 tours per season | See last year's schedule | Honolulu | HNL |
| Venezuela | 3-4 tours per season | See last year's schedule | Miami | |
| Brazil | 3-4 tours per season | See last year's schedule | | |

## Setting the Length of Character Variables

This example illustrates why you might want to specify a length for a character variable, rather than let the first assigned value determine the length. Because New York City has two airports, both the abbreviations for John F. Kennedy International Airport and La Guardia Airport can be assigned to the Airport variable as in the DATA step.

*Note:* When you create character variables, SAS determines the length of the variable from its first occurrence in the DATA step. Therefore, you must allow for the longest possible value in the first statement that mentions the variable. If you do not assign the longest value the first time the variable is assigned, then data can be truncated.

```
   /* first attempt */
data aircode;
   set mylib.departures;
   if USGate = 'San Francisco' then Airport = 'SFO';
   else if USGate = 'Honolulu' then Airport = 'HNL';
      else if USGate = 'New York' then Airport = 'JFK or LGA';
run;

proc print data=aircode;
   var Country USGate Airport;
   title 'Country by US Point of Departure';
run;
```

The following output displays the results:

*Figure 9.3* Truncation of Character Values

### Country by US Point of Departure

| Obs | Country | USGate | Airport |
|---|---|---|---|
| 1 | Japan | San Francisco | SFO |
| 2 | Italy | New York | JFK |
| 3 | Australia | Honolulu | HNL |
| 4 | Venezuela | Miami | |
| 5 | Brazil | | |

Only the characters JFK appear in the observation for New York. SAS first encounters Airport in the statement that assigns the value SFO. Therefore, SAS creates Airport with a length of three bytes and uses only the first three characters in the New York observation.

To allow space to write JFK or LGA, use a LENGTH statement as the first reference to Airport. The LENGTH statement is a declarative statement and has the form

**LENGTH** *variable-list* $ *number-of-bytes*;

*variable-list*
: specifies the variable or variables to which you are assigning the length *number-of-bytes*. The dollar sign ($) indicates that the variable is a character variable. The LENGTH statement determines the length of a character variable in both the program data vector and the data set that are being created. (In contrast, a LENGTH statement determines the length of a numeric variable only in the data set that is being created.) The maximum length of any character value in SAS is 32,767 bytes.

This LENGTH statement assigns a length of 10 to the character variable Airport:

```
length Airport $ 10;
```

*Note:* If you use a LENGTH statement to assign a length to a character variable, then it must be the first reference to the character variables in the DATA step. Therefore, the best position in the DATA step for a LENGTH statement is immediately after the DATA statement.

The following DATA step includes the LENGTH statement for Airport. Remember that you can use the DATASETS procedure to display the length of variables in a SAS data set.

```
   /* correct method */
data aircode2;
   length Airport $ 10;
   set mylib.departures;
   if USGate = 'San Francisco' then Airport = 'SFO';
   else if USGate = 'Honolulu' then Airport = 'HNL';
      else if USGate = 'New York' then Airport = 'JFK or LGA';
         else if USGate = 'Miami' then Airport = 'MIA';
run;

proc print data=aircode2;
   var Country USGate Airport;
   title 'Country by US Point of Departure';
run;
```

The following output displays the results:

**Figure 9.4** *Using a LENGTH Statement to Capture Complete Variable Information*

Country by US Point of Departure

| Obs | Country | USGate | Airport |
|---|---|---|---|
| 1 | Japan | San Francisco | SFO |
| 2 | Italy | New York | JFK or LGA |
| 3 | Australia | Honolulu | HNL |
| 4 | Venezuela | Miami | MIA |
| 5 | Brazil | | |

# Handling Missing Values

## Reading Missing Values

SAS uses a blank to represent a missing value of a character variable. For example, the data line for Brazil lacks the departure city from the United States:

```
Japan      5 San Francisco      Tokyo, Osaka
Italy      8 New York           Rome, Naples
Australia 12 Honolulu           Sydney, Brisbane
Venezuela  4 Miami              Caracas, Maracaibo
Brazil     4                    Rio de Janeiro, Belem
```

As Figure 9.1 on page 131 shows, when the INPUT statement reads the data line for Brazil and determines that the value for USGate in columns 14-26 is missing, SAS assigns a missing value to USGate for that observation. The missing value is represented by a blank when printing.

One special case occurs when you read character data values with list input. In that case, you must use a period to represent a missing value in data lines. (Blanks in list input separate values. Therefore, SAS interprets blanks as a signal to keep searching for the value, not as a missing value.) In the following DATA step, the TourGuide information for Venezuela is missing and is represented with a period:

```
data missingval;
   length Country $ 10 TourGuide $ 10;
   input Country TourGuide;
   datalines;
Japan Yamada
Italy Militello
Australia Edney
Venezuela .
Brazil Cardoso
;

proc print data=missingval;
   title 'Missing Values for Character List Input Data';
run;
```

The following output displays the results:

*Figure 9.5*  Using a Period in List Input for Missing Character Data

Missing Values for Character List Input Data

| Obs | Country | TourGuide |
|-----|---------|-----------|
| 1 | Japan | Yamada |
| 2 | Italy | Militello |
| 3 | Australia | Edney |
| 4 | Venezuela | |
| 5 | Brazil | Cardoso |

SAS recognized the period as a missing value in the fourth data line. Therefore, it recorded a missing value for the character variable TourGuide in the resulting data set.

## Checking for Missing Character Values

When you want to check for missing character values, compare the character variable to a blank enclosed in quotation marks:

```
if USGate = ' ' then GateInformation = 'Missing';
```

The following DATA step includes this statement to check USGate for missing information. The results are recorded in GateInformation:

```
data checkgate;
   length GateInformation $ 15;
   set mylib.departures;
   if USGate = ' ' then GateInformation = 'Missing';
   else GateInformation = 'Available';
run;
proc print data=checkgate;
   var Country CitiesIntour USGate ArrivalDepartureGates GateInformation;
   title 'Checking For Missing Gate Information';
run;
```

The following output displays the results:

*Figure 9.6* Checking for Missing Character Values

**Checking For Missing Gate Information**

| Obs | Country | CitiesInTour | USGate | ArrivalDepartureGates | GateInformation |
|---|---|---|---|---|---|
| 1 | Japan | 5 | San Francisco | Tokyo, Osaka | Available |
| 2 | Italy | 8 | New York | Rome, Naples | Available |
| 3 | Australia | 12 | Honolulu | Sydney, Brisbane | Available |
| 4 | Venezuela | 4 | Miami | Caracas, Maracaibo | Available |
| 5 | Brazil | 4 |  | Rio de Janeiro, Belem | Missing |

## Setting a Character Variable Value to Missing

You can assign missing character values in assignment statements by setting the character variable to a blank enclosed in quotation marks. For example, the following statement sets the day of departure based on the number of days in the tour. If the number of cities in the tour is a week or less, then the day of departure is a Sunday. Otherwise, the day of departure is not known and is set to a missing value.

```
if Cities <=7 then DayOfDeparture = 'Sunday';
else DayOfDeparture = ' ';
```

The following DATA step includes these statements:

```
data departuredays;
   set mylib.departures;
   length DayOfDeparture $ 8;
   if CitiesInTour <=7 then DayOfDeparture = 'Sunday';
   else DayOfDeparture = ' ';
run;

proc print data=departuredays;
   var Country CitiesInTour DayOfDeparture;
   title 'Departure Day is Sunday or Missing';
run;
```

The following output displays the results:

*Figure 9.7* Assigning Missing Character Values

**Departure Day is Sunday or Missing**

| Obs | Country | CitiesInTour | DayOfDeparture |
|---|---|---|---|
| 1 | Japan | 5 | Sunday |
| 2 | Italy | 8 |  |
| 3 | Australia | 12 |  |
| 4 | Venezuela | 4 | Sunday |
| 5 | Brazil | 4 | Sunday |

# Creating New Character Values

## *Extracting a Portion of a Character Value*

### *Understanding the SCAN Function*

Some character values might contain multiple pieces of information that need to be isolated and assigned to separate character variables. For example, the value of ArrivalDepartureGates contains two cities: the city of arrival and the city of departure. How can the individual values be isolated so that separate variables can be created for the two cities?

The SCAN function returns a character string when it is given the source string, the position of the desired character string, and a character delimiter:

**SCAN** (*source, n<,list-of-delimiters>*)

*source* is the value that you want to examine. It can be any type of character expression, including character variables, character constants, and so on. *n* is the position of the term to be selected from *source*. *list-of-delimiters* can list one, multiple, or no delimiters. If you specify more than one delimiter, then SAS uses any of them. If you omit the delimiter, then SAS divides words according to a default list of delimiters (including the blank and some special characters).

For example, to select the first term in the value of ArrivalDepartureGates and assign it to a new variable named ArrivalGate, write

```
ArrivalGate = scan(ArrivalDepartureGates,1,',');
```

The SCAN function examines the value of ArrivalDepartureGates and selects the first string as identified by a comma.

Although default values can be used for the delimiter, it is a good idea to specify the delimiter to be used. If the default delimiter is used in the SCAN function when the observation for Brazil is processed, then SAS recognizes a blank space as the delimiter and selects `Rio` rather than `Rio de Janeiro` as the first term. Specifying the delimiter enables you to control where the division of the term occurs.

To select the second term from ArrivalDepartureGates and assign it to a new variable term named DEPARTUREGATE, specify the following:

```
DepartureGate = scan(ArrivalDepartureGates,2,',');
```

*Note:* Within a DATA step, the default length of a target variable where the expression contains the SCAN function is the same as the length of the first SCAN argument. In the SQL procedure or in any WHERE clause, the maximum length of a string that is returned by the SCAN function is 200 characters. For more information about the SCAN function, see *SAS Functions and CALL Routines: Reference*.

### *Aligning New Values*

Remember that SAS maintains the existing alignment of a character value used in an expression; it does not perform any automatic realignment. This example creates the values for a new variable DepartureGate from the values of ArrivalDepartureGates. The value of ArrivalDepartureGates contains a comma and a blank between the two city names as shown in the following output:

*Figure 9.8* Dividing Values into Separate Words Using the SCAN Function

Data Set Air.Departure

| Obs | Country | CitiesInTour | USGate | ArrivalDepartureGates |
|---|---|---|---|---|
| 1 | Japan | 6 | San Francisco | Tokyo, Osaka |
| 2 | Italy | 8 | New York | Rome, Naples |
| 3 | Australia | 12 | Honolulu | Sydney, Brisbane |
| 4 | Venezuela | 4 | Miami | Caracas, Maracaibo |
| 5 | Brazil | 4 | | Rio de Janeiro, Belem |

When the SCAN function divides the names at the comma, the second term begins with a blank. Therefore, all the values that are assigned to DepartureGate begin with a blank.

To left-align the values, use the LEFT function:

**LEFT** (*source*)

The LEFT function produces a value that has all leading blanks in *source* moved to the right side of the value. Therefore, the result is left aligned. *source* can be any type of character expression, including a character variable, a character constant enclosed in quotation marks, or another character function.

This example uses the LEFT function in the second assignment statement:

```
DepartureGate = scan(ArrivalDepartureGates,2,',');
DepartureGate = left(DepartureGate);
```

You can also nest the two functions:

```
DepartureGate = left(scan(ArrivalDepartureGates,2,','));
```

When you nest functions, SAS performs the action in the innermost function first. It uses the result of that function as the argument of the next function, and so on.

The following DATA step creates separate variables for the arrival gates and the departure gates:

```
data gates;
   set mylib.departures;
   ArrivalGate = scan(ArrivalDepartureGates,1,',');
   DepartureGate = left(scan(ArrivalDepartureGates,2,','));
run;

proc print data=gates;
   var Country ArrivalDepartureGates ArrivalGate DepartureGate;
   title 'Arrival and Departure Gates';
run;
```

The following output displays the results:

*Figure 9.9* Dividing Values into Separate Words with the SCAN Function

### Arrival and Departure Gates

| Obs | Country | ArrivalDepartureGates | ArrivalGate | DepartureGate |
|---|---|---|---|---|
| 1 | Japan | Tokyo, Osaka | Tokyo | Osaka |
| 2 | Italy | Rome, Naples | Rome | Naples |
| 3 | Australia | Sydney, Brisbane | Sydney | Brisbane |
| 4 | Venezuela | Caracas, Maracaibo | Caracas | Maracaibo |
| 5 | Brazil | Rio de Janeiro, Belem | Rio de Janeiro | Belem |

### *Saving Storage Space When Using the SCAN Function*

In a DATA step, the SCAN function assigns the length of the first argument to the target variable in an assignment statement, if the length of the target variable is not already assigned. In the SQL procedure or in a WHERE clause in any procedure, the maximum length of a word that is returned by the SCAN function is 200 characters. If you use the SCAN function in an expression that contains operators or other functions, the maximum length of a word returned by the SCAN function is 32,767 characters.

Setting the lengths of ArrivalGate and DepartureGate to the needed values rather than to the default lengths might save storage space. Because SAS sets the length of a character variable the first time SAS encounters it, the LENGTH statement must appear before the assignment statements that create values for the variables:

```
data gatelength;
   length ArrivalGate $ 14 DepartureGate $ 9;
   set mylib.departures;
   ArrivalGate = scan(ArrivalDepartureGate,1,',');
   DepartureGate = left(scan(ArrivalDepartureGate,2,','));
run;
```

## *Combining Character Values: Using Concatenation*

### *Understanding Concatenation of Variable Values*

SAS enables you to combine character values into longer ones using an operation known as concatenation. Concatenation combines character values by placing them one after the other and assigning them to a variable. In SAS programming, the concatenation operator is a pair of vertical bars (||). If your keyboard does not have a solid vertical bar, use two broken vertical bars (¦¦) or two exclamation points (!!). The length of the new variable is the sum of the lengths of each variable or number of characters that is specified in a LENGTH statement for the new variable.

The following figure illustrates concatenation:

**Figure 9.10** *Concatenation of Two Values*

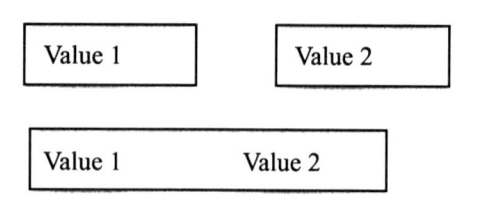

| Value 1 |    | Value 2 |

| Value 1 | Value 2 |

### Performing a Simple Concatenation

The following statement uses the CAT function to combine all the cities named as gateways into a single variable named AllGates:

```
AllGates = cat(USGate,ArrivalDepartureGates);
```

*Note:* The CAT function does not use the || concatenation operator that is described in "Understanding Concatenation of Variable Values" on page 139.

SAS attaches the beginning of each value of ArrivalDepartureGates to the end of each value of USGate and assigns the results to AllGates. The following DATA step includes this statement:

```
data all;
   set mylib.departures;
   AllGates = cat(USGate,ArrivalDepartureGates);
run;

proc print data=all;
   var Country USGate ArrivalDepartureGates AllGates;
   title 'All Tour Gates';
run;
```

The following output displays the results:

**Figure 9.11** *Simple Concatenation*

All Tour Gates

| Obs | Country | USGate | ArrivalDepartureGates | AllGates |
|---|---|---|---|---|
| 1 | Japan | San Francisco | Tokyo, Osaka | San FranciscoTokyo, Osaka |
| 2 | Italy | New York | Rome, Naples | New York Rome, Naples |
| 3 | Australia | Honolulu | Sydney, Brisbane | Honolulu Sydney, Brisbane |
| 4 | Venezuela | Miami | Caracas, Maracaibo | Miami Caracas, Maracaibo |
| 5 | Brazil |  | Rio de Janeiro, Belem | Rio de Janeiro, Belem |

The length of USGate in the above output is 13 bytes, but only San Francisco uses all of them. Therefore, the other values contain blanks at the end, and the value for Brazil is entirely blank.

When a character value that is to be concatenated is shorter than the length of the variable to which it belongs, SAS pads the value with trailing blanks during concatenation. However, HTML is the default output style for SAS, and it ignores the trailing blanks in the variable values when it displays the concatenated output. Instead, HTML places one blank between the concatenated values. Therefore, the values for AllGates in the above output contain one blank between the concatenated values of USGate and ArrivalDepartureGates. For example, the value in AllGates for the fourth observation contains one blank between Miami and Caracas. HTML output ignores the eight trailing blanks that follow the value of Miami in the USGATE variable when it displays the concatenated variables.

*Note:* You can use the NOBREAKSPACE style attribute to keep the trailing blanks between the concatenated variables. For information about the NOBREAKSPACE style attribute, see "Style Attributes" in *SAS Output Delivery System: User's Guide*.

### *Adding Additional Characters*

The TRIM function enables you to remove blanks between concatenated variables. It also enables you to add delimiters such as commas, colons, or blanks between concatenated variables.

*Note:* The TRIM function uses the || concatenation operator that is described in "Understanding Concatenation of Variable Values" on page 139.

If you use the TRIM function to delete the blanks between concatenated variables, then the results might be difficult to read. To make the result easier to read, you can concatenate a comma and blank between the trimmed value of USGate and the value of ArrivalDepartureGates. Also, to align the AllGates2 value for Brazil with all other values of AllGates2, use an IF-THEN statement to equate the value of AllGates2 with the value of ArrivalDepartureGates in that observation.

```
AllGates2 = trim(USGate)||', '||ArrivalDepartureGates;
if Country = 'Brazil' then AllGates2 = ArrivalDepartureGates;
```

This DATA step includes these statements:

```
data all2;
   set mylib.departures;
   AllGates2 = trim(USGate)||', '||ArrivalDepartureGates;
   if Country = 'Brazil' then AllGates2 = ArrivalDepartureGates;
run;

proc print data=all2;
   var Country USGate ArrivalDepartureGates AllGates2;
   title 'All Tour Gates';
run;
```

*Note:* In the ALLGATES2 statement, ', ' is a literal value. SAS adds the size of the two variables and the length of the literal to calculate the size of the concatenated variable. HTML ignores any additional trailing blanks and inserts only the specified comma and one blank.

The following output displays the results:

*Figure 9.12  Concatenating Additional Characters for Readability*

**All Tour Gates**

| Obs | Country | USGate | ArrivalDepartureGates | AllGates2 |
|---|---|---|---|---|
| 1 | Japan | San Francisco | Tokyo, Osaka | San Francisco, Tokyo, Osaka |
| 2 | Italy | New York | Rome, Naples | New York, Rome, Naples |
| 3 | Australia | Honolulu | Sydney, Brisbane | Honolulu, Sydney, Brisbane |
| 4 | Venezuela | Miami | Caracas, Maracaibo | Miami, Caracas, Maracaibo |
| 5 | Brazil | | Rio de Janeiro, Belem | Rio de Janeiro, Belem |

For more information about removing trailing blanks or adding delimiters between concatenated variables, see "TRIM Function" in *SAS Functions and CALL Routines: Reference*.

## Saving Storage Space by Treating Numbers as Characters

Remember that SAS uses eight bytes of storage for every numeric value in the DATA step. By default, SAS also uses eight bytes of storage for each numeric value in an output data set. However, a character value can contain a minimum of one character. In that case, SAS uses one byte for the character variable, both in the program data vector and in the output data set. In addition, SAS treats the digits 0 through 9 in a character value like any other character. When you are not going to perform calculations on a variable, you can save storage space by treating a value that contains digits as a character value.

For example, some tours offer various prices, depending on the quality of the hotel room. The brochures rank the rooms as two stars, three stars, and so on. In this case the values 2, 3, and 4 are really the names of categories, and arithmetic operations are not expected to be performed on them. Therefore, the values can be read into a character variable. The following DATA step reads HotelRank as a character variable and assigns it a length of one byte:

```
data hotels;
   input Country $ 1-9 HotelRank $ 11 LandCost;
   datalines;
Italy     2  498
Italy     4  698
Australia 2  915
Australia 3 1169
Australia 4 1399
;

proc print data=hotels;
   title 'Hotel Rankings';
run;
```

In the previous example, the INPUT statement assigns HotelRank a length of one byte because the INPUT statement reads one column to find the value (shown by the use of column input). If you are using list input, then place a LENGTH statement before the INPUT statement to set the length to one byte.

If you read a number as a character value and then discover that you need to use it in a numeric expression, then you can do so without making changes in your program. SAS automatically produces a numeric value from the character value for use in the expression; it also issues a note in the log that the conversion occurred. (Of course, the conversion causes the DATA step to use slightly more computer resources.) The original variable remains unchanged.

The following output displays the results:

*Figure 9.13* *Saving Storage Space By Creating a Character Variable*

Hotel Rankings

| Obs | Country | HotelRank | LandCost |
|---|---|---|---|
| 1 | Italy | 2 | 498 |
| 2 | Italy | 4 | 698 |
| 3 | Australia | 2 | 915 |
| 4 | Australia | 3 | 1169 |
| 5 | Australia | 4 | 1399 |

*Note:* Note that the width of the column is not the default width of eight.

# Summary

## *Functions*

LEFT (*source*)
: left-aligns *source* by moving any leading blanks to the end of the value. *source* can be any type of character expression, including a character variable, a character constant enclosed in quotation marks, or another character function. Because any blanks removed from the left are added to the right, the length of the result matches the length of *source*. For more information, see "LEFT Function" in *SAS Functions and CALL Routines: Reference*.

SCAN (*source*, n <,*list-of-delimiters*>)
: selects the nth term from *source*. *source* can be any type of character expression, including a character variable, a character constant enclosed in quotation marks, or another character function. To choose the character that divides the terms, use a delimiter. If you omit the delimiter, then SAS divides the terms using a default list of delimiters (the blank and some special characters). For more information, see "SCAN Function" in *SAS Functions and CALL Routines: Reference*.

TRIM (*source*)
: trims trailing blanks from *source*. *source* can be any type of character expression, including a character variable, a character constant enclosed in quotation marks, or another character function. The TRIM function does not affect how a variable is stored. If you use the TRIM function to remove trailing blanks and assign the trimmed value to a variable that is longer than that value, then SAS pads the value with new trailing blanks to make the value match the length of the new variable. For more information, see "TRIM Function" in *SAS Functions and CALL Routines: Reference*.

## Statements

LENGTH *variable-list$number-of-bytes*;
   assigns a length that you specify in *number-of-bytes* to the character variable or variables in *variable-list*. You can assign any number of lengths in a single LENGTH statement, and you can assign lengths to both character and numeric variables in the same statement. Place a dollar sign ($) before the length of any character variable. For more information, see "LENGTH Statement" in *SAS Statements: Reference*.

# Learning More

Character values
   This section illustrates the flexibility that SAS provides for manipulating character values. In addition to the functions that are described in this section, the following character functions are also frequently used:

COMPBL
   removes multiple blanks from a character string.

COMPRESS
   removes specified character(s) from the source.

INDEX
   searches the source data for a pattern of characters.

LOWCASE
   converts all letters in an argument to lowercase.

RIGHT
   right-aligns the source.

SUBSTR
   extracts a group of characters.

TRANSLATE
   replaces specific characters in a character expression.

UPCASE
   returns the source data in uppercase.

The INDEX and UPCASE functions are discussed in Chapter 10, "Acting on Selected Observations," on page 147. Complete descriptions of all character functions appear in *SAS Functions and CALL Routines: Reference*.

Character variables
   Detailed information about character variables is found in *SAS Language Reference: Concepts*.

   Additional information about aligning character variables is explained in the TEMPLATE procedure in *SAS Output Delivery System: User's Guide*, and in the REPORT procedure in *Base SAS Procedures Guide*.

Comparing uppercase and lowercase characters
   How to compare uppercase and lowercase characters is shown in Chapter 10, "Acting on Selected Observations," on page 147.

Concatenation operator
: Information about the concatenation operator can be found in *SAS Language Reference: Concepts*.

DATASETS procedure
: Using the DATASETS procedure to display the length of variables in a SAS data set is explained in Chapter 37, "Getting Information about Your SAS Data Sets," on page 701.

IF-THEN statements
: A detailed explanation of the IF-THEN statements can be found in Chapter 10, "Acting on Selected Observations," on page 147.

Informats and formats
: Complete information about the numerous SAS informats and formats for reading and writing character variables is found in *SAS Formats and Informats: Reference*.

Missing values
: Detailed information about missing values is found in *SAS Language Reference: Concepts*.

VLENGTH function
: The VLENGTH function is explained in detail in *SAS Functions and CALL Routines: Reference*.

# Chapter 10
# Acting on Selected Observations

Introduction to Acting on Selected Observations . . . . . . . . . . . . . . . . . . . . . . . . 147
    Purpose . . . . . . . . . . . . . . . . . . . . . . . . . . . . . . . . . . . . . . . . . . . . . . . . . . . . . 147
    Prerequisites . . . . . . . . . . . . . . . . . . . . . . . . . . . . . . . . . . . . . . . . . . . . . . . . . 148

Input SAS Data Set for Examples . . . . . . . . . . . . . . . . . . . . . . . . . . . . . . . . . . . . 148

Selecting Observations . . . . . . . . . . . . . . . . . . . . . . . . . . . . . . . . . . . . . . . . . . . . 149
    Understanding the Selection Process . . . . . . . . . . . . . . . . . . . . . . . . . . . . . . 149
    Selecting Observations Based on a Simple Condition . . . . . . . . . . . . . . . . . . 150
    Providing an Alternative Action . . . . . . . . . . . . . . . . . . . . . . . . . . . . . . . . . . 151
    Creating a Series of Mutually Exclusive Conditions . . . . . . . . . . . . . . . . . . . 152

Constructing Conditions . . . . . . . . . . . . . . . . . . . . . . . . . . . . . . . . . . . . . . . . . . . 154
    Understanding Construct Conditions . . . . . . . . . . . . . . . . . . . . . . . . . . . . . . 154
    Selecting an Observation Based on Simple Conditions . . . . . . . . . . . . . . . . . 155
    Using More Than One Comparison in a Condition . . . . . . . . . . . . . . . . . . . . 155

Comparing Characters . . . . . . . . . . . . . . . . . . . . . . . . . . . . . . . . . . . . . . . . . . . . 161
    Types of Character Comparisons . . . . . . . . . . . . . . . . . . . . . . . . . . . . . . . . . 161
    Comparing Uppercase and Lowercase Characters . . . . . . . . . . . . . . . . . . . . . 161
    Selecting All Values That Begin with the Same Group of Characters . . . . . . . . 162
    Selecting a Range of Character Values . . . . . . . . . . . . . . . . . . . . . . . . . . . . . 163
    Finding a Value Anywhere within Another Character Value . . . . . . . . . . . . . . 164

Summary . . . . . . . . . . . . . . . . . . . . . . . . . . . . . . . . . . . . . . . . . . . . . . . . . . . . . . 165
    Statements . . . . . . . . . . . . . . . . . . . . . . . . . . . . . . . . . . . . . . . . . . . . . . . . . . 165
    Functions . . . . . . . . . . . . . . . . . . . . . . . . . . . . . . . . . . . . . . . . . . . . . . . . . . 166

Learning More . . . . . . . . . . . . . . . . . . . . . . . . . . . . . . . . . . . . . . . . . . . . . . . . . . 166

## Introduction to Acting on Selected Observations

### Purpose

One of the most useful features of SAS is its ability to perform an action on only the observations that you select. The following concepts are discussed in this section:

- how the selection process works
- how to write statements that select observations based on a condition
- some special points about selecting numeric and character variables

*Prerequisites*

You should understand the concepts presented in all previous sections before proceeding with this section.

## Input SAS Data Set for Examples

Tradewinds Travel offers tours to art museums and galleries in various cities. The company decided that in order to make its process more efficient, additional information is needed. For example, if the tour covers too many museums and galleries within a time period, then the number of museums visited must be decreased or the number of days for the tour needs to change. If the guide who is assigned to the tour is not available, then another guide must be assigned. Most of the process involves selecting observations that meet or that do not meet various criteria and then taking the required action.

The Tradewinds Travel tour data is stored in an external file that contains the following information:

```
1             2 3    4 5              6         7
-------------------------------------------------------
Rome          3 1550 7 4 M, 3 G       D'Amico   Torres
Brasilia      8 3360 6 5 M, 1 other   Lucas     Lucas
London        6 2460 5 3 M, 2 G       Wilson    Lucas
Warsaw        6    . 8 5 M, 1 G, 2 other Lucas  D'Amico
Madrid        3  740 5 3 M, 2 other   Torres    D'Amico
Amsterdam     4 1160 6 3 M, 3 G                 Vandever
```

The following list explains the numbered items in the preceding file:

1   identifies the name of the destination city

2   specifies the number of nights in the city

3   specifies the cost of the land package in U.S. dollars

4   specifies the number of events the trip offers (such as visits to museums and galleries)

5   provides a brief description of events. **M** indicates a museum, **G** indicates a gallery, and **other** indicates another type of event

6   provides the name of the tour guide

7   provides the name of the backup tour guide

The following DATA step creates MYLIB.ARTTOURS:

```
libname mylib 'SAS-library';

data mylib.arttours;
   infile 'input-file' truncover;
   input City $ 1-9 Nights 11 LandCost 13-16 NumberOfEvents 18
         EventDescription $ 20-36 TourGuide $ 38-45
         BackUpGuide $ 47-54;
run;
```

```
proc print data=mylib.arttours;
   title 'Data Set MYLIB.ARTTOURS';
run;
```

*Note:* When the TRUNCOVER option is specified in the INFILE statement, and when the record is shorter than what the INPUT statement expects, SAS reads a variable length record.

The PROC PRINT statement that follows the DATA step produces this display of the MYLIB.ARTTOURS data set:

*Figure 10.1* Data Set MYLIB.ARTTOURS

Data Set MYLIB.ARTTOURS

| Obs | City | Nights | LandCost | NumberOfEvents | EventDescription | TourGuide | BackUpGuide |
|---|---|---|---|---|---|---|---|
| 1 | Rome | 3 | 1550 | 7 | 4 M, 3 G | D'Amico | Torres |
| 2 | Brasilia | 8 | 3360 | 6 | 5 M, 1 other | Lucas | Lucas |
| 3 | London | 6 | 2460 | 5 | 3 M, 2 G | Wilson | Lucas |
| 4 | Warsaw | 6 | . | 8 | 5 M, 1 G, 2 other | Lucas | D'Amico |
| 5 | Madrid | 3 | 740 | 5 | 3 M, 2 other | Torres | D'Amico |
| 6 | Amsterdam | 4 | 1160 | 6 | 3 M, 3 G |  | Vandever |

The following list describes several of the fields in the output:

- NumberOfEvents contains the number of attractions that were visited during the tour.
- EventDescription lists the number of museums (M), art galleries (G), and other attractions (other) that were visited.
- TourGuide lists the name of the tour guide that is assigned to the tour.
- BackUpGuide lists the alternate tour guide in case the original tour guide is unavailable.

# Selecting Observations

## Understanding the Selection Process

The most common way that SAS selects observations for action in a DATA step is through the IF-THEN statement:

**IF** *condition* **THEN** *action*;

The condition is one or more comparisons, for example:

- `City = 'Rome'`
- `NumberOfEvents > Nights`
- `TourGuide = 'Lucas' and Nights > 7`

The symbol > means greater than. How to use symbols as comparison operators is explained in "Understanding Construct Conditions" on page 154.

For a given observation, a comparison is either true or false. In the first example, the value of City is either **Rome** or it is not. In the second example, the value of NumberOfEvents in the current observation is either greater than the value of Nights in the same observation or it is not. If the condition contains more than one comparison, as in the third example, then SAS evaluates all of the conditions according to its rules (discussed later) and declares the entire condition to be true or false.

When the condition is true, SAS takes the action in the THEN clause. The action must be expressed as a SAS statement that can be executed in an individual iteration of the DATA step. Such statements are called executable statements. The most common executable statements are assignment statements, as shown in the following examples:

- `LandCost=LandCost + 30;`
- `Calendar='Check schedule';`
- `TourGuide='Torres';`

This section concentrates on assignment statements in the THEN clause, but examples in other sections show other types of statements that are used with the THEN clause.

Statements that provide information about a data set are not executable. Such statements are called declarative statements. For example, the LENGTH statement affects a variable as a whole, and not how the variable is treated in a particular observation. Therefore, you cannot use a LENGTH statement in a THEN clause.

When the condition is false, SAS ignores the THEN clause and proceeds to the next statement in the DATA step.

## Selecting Observations Based on a Simple Condition

The following DATA step uses the previous example conditions and actions in IF-THEN statements:

```
data revise;
   set mylib.arttours;
   if City='Rome' then LandCost=LandCost + 30;
   if NumberOfEvents > Nights then Calendar='Check schedule';
   if TourGuide='Lucas' and Nights > 7 then TourGuide='Torres';
run;

proc print data=revise;
   var City Nights LandCost NumberOfEvents TourGuide Calendar;
   title 'Tour Information';
run;
```

The following output displays the results.

*Figure 10.2* Selecting Observations with IF-THEN Statements

Tour Information

| Obs | City | Nights | LandCost | NumberOfEvents | TourGuide | Calendar |
|---|---|---|---|---|---|---|
| 1 | Rome | 3 | 1580 | 7 | D'Amico | Check schedule |
| 2 | Brasilia | 8 | 3360 | 6 | Torres | |
| 3 | London | 6 | 2460 | 5 | Wilson | |
| 4 | Warsaw | 6 | . | 8 | Lucas | Check schedule |
| 5 | Madrid | 3 | 740 | 5 | Torres | Check schedule |
| 6 | Amsterdam | 4 | 1160 | 6 | | Check schedule |

In the output, you can see several changes:

- The land cost was increased by $30 in the observation for Rome.
- Four observations have a greater number of events than the number of days in the tour.
- The tour guide for Brasilia is replaced by Torres because the original tour guide is Lucas, and the number of nights in the tour is greater than 7.

## Providing an Alternative Action

Remember that SAS creates a variable in all observations, even if you do not assign the variable a value in all observations. In the previous output, the value of Calendar is blank in two observations. A second IF-THEN statement can assign a different value, as in these examples:

```
if NumberOfEvents > Nights then Calendar='Check schedule';
if NumberOfEvents <= Nights then Calendar='No problems';
```

The symbol <= means less than or equal to. In this case, SAS compares the values of Events and Nights twice, once in each IF condition. A more efficient way to provide an alternative action is to use an ELSE statement:

**ELSE** *action*;

An ELSE statement names an alternative action to be taken when the IF condition is false. It must immediately follow the corresponding IF-THEN statement, as shown here:

```
if NumberOfEvents > Nights then Calendar='Check schedule';
else Calendar='No problems';
```

The REVISE2 DATA step adds the preceding ELSE statement to the previous DATA step:

```
data revise2;
   set mylib.arttours;
   if City='Rome' then LandCost=LandCost + 30;
   if NumberOfEvents > Nights then Calendar='Check schedule';
   else Calendar='No problems';
```

```
        if TourGuide='Lucas' and Nights > 7 then TourGuide='Torres';
run;

proc print data=revise2;
    var City Nights LandCost NumberOfEvents TourGuide Calendar;
    title 'Tour Information';
run;
```

The following output displays the results.

*Figure 10.3  Providing an Alternative Action with the ELSE Statement*

Tour Information

| Obs | City | Nights | LandCost | NumberOfEvents | TourGuide | Calendar |
|---|---|---|---|---|---|---|
| 1 | Rome | 3 | 1580 | 7 | D'Amico | Check schedule |
| 2 | Brasilia | 8 | 3360 | 6 | Torres | No problems |
| 3 | London | 6 | 2460 | 5 | Wilson | No problems |
| 4 | Warsaw | 6 | . | 8 | Lucas | Check schedule |
| 5 | Madrid | 3 | 740 | 5 | Torres | Check schedule |
| 6 | Amsterdam | 4 | 1160 | 6 |  | Check schedule |

## Creating a Series of Mutually Exclusive Conditions

Using an ELSE statement after an IF-THEN statement provides one alternative action when the IF condition is false. However, many cases involve a series of mutually exclusive conditions, each of which requires a separate action. In this example, tour prices can be classified as high, medium, or low. A series of IF-THEN and ELSE statements classifies the tour prices appropriately:

```
if LandCost >= 2500 then Price='High';
else if LandCost >= 1500 then Price='Medium';
    else Price='Low';
```

The symbol >= means greater than or equal to. To see how SAS executes this series of statements, consider two observations: Amsterdam, whose value of LandCost is 1160, and Brasilia, whose value is 3360.

When the value of LandCost is 1160, SAS processes the program in the following way:

1. SAS tests whether 1160 is equal to or greater than 2500, determines that the comparison is false, ignores the THEN clause, and proceeds to the ELSE statement.

2. The action in the ELSE statement is to evaluate another condition. SAS tests whether 1160 is equal to or greater than 1500, determines that the comparison is false, ignores the THEN clause, and proceeds to the accompanying ELSE statement.

3. SAS executes the action in the ELSE statement and assigns Price the value Low.

When the value of LandCost is 3360, SAS processes the program in the following way:

1. SAS tests whether 3360 is greater than or equal to 2500, determines that the comparison is true, and executes the action in the THEN clause. The value of Price becomes High.
2. SAS ignores the ELSE statement. Because the entire remaining series is part of the first ELSE statement, SAS skips all remaining actions in the series.

A simple way to think of these actions is to remember that when an observation satisfies one condition in a series of mutually exclusive IF-THEN/ELSE statements, SAS processes the THEN action and skips the rest of the statements. Therefore, you can increase the efficiency of a program by ordering the IF-THEN/ELSE statements so that the most common conditions appear first.

The following DATA step includes the preceding series of statements:

```
data prices;
   set mylib.arttours;
   if LandCost >= 2500 then Price='High  ';
   else if LandCost >= 1500 then Price='Medium';
        else Price='Low';
run;

proc print data=prices;
   var City LandCost Price;
   title 'Tour Prices';
run;
```

The following output displays the results.

*Figure 10.4* Assigning Mutually Exclusive Values with IF-THEN/ELSE Statements

Tour Prices

| Obs | City | LandCost | Price |
|---|---|---|---|
| 1 | Rome | 1550 | Medium |
| 2 | Brasilia | 3360 | High |
| 3 | London | 2460 | Medium |
| 4 | Warsaw | . | Low |
| 5 | Madrid | 740 | Low |
| 6 | Amsterdam | 1160 | Low |

Note the value of Price in the fourth observation. The Price value is Low because the LandCost value for the Warsaw trip is a missing value. Remember that a missing value is the lowest possible numeric value.

## Constructing Conditions

### Understanding Construct Conditions

When you use an IF-THEN statement, you ask SAS to make a comparison. SAS must determine whether a value is equal to another value, greater than another value, and so on.

SAS has the following six main comparison operators:

*Table 10.1  Comparison Operators*

| Symbol | Mnemonic Operator | Meaning |
| --- | --- | --- |
| = | EQ | equal to |
| ¬=, ^=, ~= | NE | not equal to (the ¬, ^, or ~ symbol, depending on your keyboard) |
| > | GT | greater than |
| < | LT | less than |
| >= | GE | greater than or equal to |
| <= | LE | less than or equal to |

The symbols in the table are based on mathematical symbols. The letter abbreviations, known as mnemonic operators, have the same effect as the symbols. Use the form that you prefer, but remember that you can use the mnemonic operators only in comparisons. For example, the equal sign in an assignment statement must be represented by the symbol =, not the mnemonic operator. Both of the following statements compare the number of nights in the tour to six:

- `if Nights >= 6 then Stay='Week+';`
- `if Nights ge 6 then Stay='Week+';`

The terms on each side of the comparison operator can be variables, expressions, or constants. The side a particular term appears on does not matter, as long as you use the correct operator. All of the following comparisons are constructed correctly for use in SAS statements:

- `Guide=' '`
- `LandCost ne .`
- `LandCost lt 1200`
- `600 ge LandCost`
- `NumberOfEvents / Nights > 2`
- `2 <= NumberOfEvents / Nights`

### Selecting an Observation Based on Simple Conditions

The following DATA step illustrates some of these conditions:

```
data changes;
   set mylib.arttours;
   if Nights >= 6 then Stay='Week+';
   else Stay = 'Days';
   if LandCost ne . then Remarks='OK   ';
   else Remarks = 'Redo';
   if LandCost lt 1200 then Budget='Low    ';
   else Budget = 'Medium';
   if NumberOfEvents / Nights > 2 then Pace='Too fast';
   else Pace='OK';
run;

proc print data=changes;
   var City Nights LandCost NumberOfEvents Stay Remarks Budget Pace;
   title 'Tour Information';
run;
```

The following output displays the results:

*Figure 10.5* *Assigning Values to Variables According to Specific Conditions*

Tour Information

| Obs | City | Nights | LandCost | NumberOfEvents | Stay | Remarks | Budget | Pace |
|---|---|---|---|---|---|---|---|---|
| 1 | Rome | 3 | 1550 | 7 | Days | OK | Medium | Too fast |
| 2 | Brasilia | 8 | 3360 | 6 | Week+ | OK | Medium | OK |
| 3 | London | 6 | 2460 | 5 | Week+ | OK | Medium | OK |
| 4 | Warsaw | 6 | . | 8 | Week+ | Redo | Low | OK |
| 5 | Madrid | 3 | 740 | 5 | Days | OK | Low | OK |
| 6 | Amsterdam | 4 | 1160 | 6 | Days | OK | Low | OK |

### Using More Than One Comparison in a Condition

#### Specifying Multiple Comparisons

You can specify more than one comparison in a condition with these operators:

- & or AND
- | or OR

A condition can contain any number of AND operators, OR operators, or both.

### Making Comparisons When All of the Conditions Must Be True

When comparisons are connected by AND, all of the comparisons must be true for the condition to be true. Consider this example:

```
if City='Brasilia' and TourGuide='Lucas' then Remarks='Bilingual';
```

The comparison is true for observations in which the value of City is **Brasilia** and the value of TourGuide is **Lucas**.

A common comparison is to determine whether a value is between two quantities, greater than one quantity and less than another quantity. For example, to select observations in which the value of LandCost is greater than or equal to 2000, and less than or equal to 2500, you can write a comparison with AND:

```
if LandCost >= 2000 and LandCost <= 2500 then Price = '2000-2500';
```

This is a simpler way to write this comparison:

```
if 2000 <= LandCost <= 2500 then Price = '2000-2500';
```

This comparison has the same meaning as the previous one. You can use any of the operators <, <=, >, >=, or their mnemonic equivalents in this way.

The following DATA step includes these multiple comparison statements:

```
data showand;
   set mylib.arttours;
   if City='Brasilia' and TourGuide='Lucas' then Remarks='Bilingual';
   if 2000 <= LandCost <= 2500 then Price='2000-2500';
run;

proc print data=showand;
   var City LandCost TourGuide Remarks Price;
   title 'Tour Information';
run;
```

The following output displays the results.

**Figure 10.6** Using AND When Making Multiple Comparisons

Tour Information

| Obs | City | LandCost | TourGuide | Remarks | Price |
|---|---|---|---|---|---|
| 1 | Rome | 1550 | D'Amico | | |
| 2 | Brasilia | 3360 | Lucas | Bilingual | |
| 3 | London | 2460 | Wilson | | 2000-2500 |
| 4 | Warsaw | . | Lucas | | |
| 5 | Madrid | 740 | Torres | | |
| 6 | Amsterdam | 1160 | | | |

### When Only One Condition Must Be True

When comparisons are connected by OR, only one of the comparisons needs to be true for the condition to be true. Consider the following example:

```
if LandCost gt 1500 or LandCost / Nights gt 400 then Level='Deluxe';
```

Any observation in which the land cost is over $1500, the cost per night is over $200, or both, satisfies the condition. The following DATA step shows this condition:

```
data showor;
   set mylib.arttours;
   if LandCost gt 1500 or LandCost / Nights gt 400 then Level='Deluxe';
run;
proc print data=showor;
   var City LandCost Nights Level;
   title 'Tour Information';
run;
```

The following output displays the results.

*Figure 10.7* Using OR When Making Multiple Comparisons

Tour Information

| Obs | City | LandCost | Nights | Level |
|---|---|---|---|---|
| 1 | Rome | 1550 | 3 | Deluxe |
| 2 | Brasilia | 3360 | 8 | Deluxe |
| 3 | London | 2460 | 6 | Deluxe |
| 4 | Warsaw | . | 6 | |
| 5 | Madrid | 740 | 3 | |
| 6 | Amsterdam | 1160 | 4 | |

## *Using Negative Operators with AND or OR*

Be careful when you combine negative operators with OR. Often, the operator that you really need is AND. For example, the variable TourGuide contains some problems with the data. In the observation for Brasilia, the tour guide and the backup tour guide are both `Lucas`. In the observation for Amsterdam, the name of the tour guide is missing. You want to label the observations that have no problems with TourGuide as OK. Should you write the IF condition with OR or with AND?

The following DATA step shows both conditions:

```
data showorand;
   set mylib.arttours;
   if TourGuide ne BackUpGuide or TourGuide ne ' ' then GuideCheckUsingOR='OK';
   else GuideCheckUsingOR='No';
   if TourGuide ne BackUpGuide and TourGuide ne ' ' then GuideCheckUsingAND='OK';
   else GuideCheckUsingAND='No';
run;

proc print data = showorand;
   var City TourGuide BackUpGuide GuideCheckUsingOR GuideCheckUsingAND;
   title 'Negative Operators with OR and AND';
run;
```

The following output displays the results.

*Figure 10.8* Using Negative Operators When Making Comparisons

Negative Operators with OR and AND

| Obs | City | TourGuide | BackUpGuide | GuideCheckUsingOR | GuideCheckUsingAND |
|---|---|---|---|---|---|
| 1 | Rome | D'Amico | Torres | OK | OK |
| 2 | Brasilia | Lucas | Lucas | OK | No |
| 3 | London | Wilson | Lucas | OK | OK |
| 4 | Warsaw | Lucas | D'Amico | OK | OK |
| 5 | Madrid | Torres | D'Amico | OK | OK |
| 6 | Amsterdam |  | Vandever | OK | No |

In the IF-THEN/ELSE statements that create GuideCheckUsingOR, only one comparison needs to be true to make the condition true. Note that the following conditions are true for the Brasilia and Amsterdam observations in the data set MYLIB.ARTTOURS:

- In the observation for Brasilia, TourGuide does not have a missing value and the comparison `TourGuide NE ' '` is true.

- For Amsterdam, the comparison `TourGuide NE BackUpGuide` is true.

Because one OR comparison is true in each observation, GuideCheckUsingOR is labeled OK for all observations. The IF-THEN/ELSE statements that create GuideCheckUsingAND achieve better results. That is, the AND operator selects the observations in which the value of TourGuide is not the same as BackUpGuide and is not missing.

### *Using Complex Comparisons That Require AND and OR*

A condition can contain both AND operators and OR operators. When it does, SAS evaluates the AND operators before the OR operators. The following example specifies a list of cities and a list of guides:

```
/* first attempt */
if City='Brasilia' or City='Rome' and TourGuide='Lucas' or
   TourGuide="D'Amico" then Topic= 'Art history';
```

SAS first joins the items that are connected by AND:

`City='Rome' and TourGuide='Lucas'`

Then SAS makes the following OR comparisons:

```
City='Brasilia'
    or
City='Rome' and TourGuide='Lucas'
    or
TourGuide="D'Amico"
```

To group the City comparisons and the TourGuide comparisons, use parentheses:

```
/* correct method */
if (City='Brasilia' or City='Rome') and
```

```
    (TourGuide='Lucas' or TourGuide="D'Amico") then
Topic='Art history';
```

SAS evaluates the comparisons within parentheses first and uses the results as the terms of the larger comparison. You can use parentheses in any condition to control the grouping of comparisons or to make the condition easier to read.

The following DATA step illustrates these conditions:

```
data combine;
   set mylib.arttours;
   if (City='Brasilia' or City='Rome') and
      (TourGuide='Lucas' or TourGuide="D'Amico") then
      Topic='Art history';
run;

proc print data=combine;
   var City TourGuide Topic;
   title 'Tour Information';
run;
```

The following output displays the results.

*Figure 10.9  Using Parentheses to Combine Comparisons with AND and OR*

Tour Information

| Obs | City | TourGuide | Topic |
|---|---|---|---|
| 1 | Rome | D'Amico | Art history |
| 2 | Brasilia | Lucas | Art history |
| 3 | London | Wilson | |
| 4 | Warsaw | Lucas | |
| 5 | Madrid | Torres | |
| 6 | Amsterdam | | |

### *Abbreviating Numeric Comparisons*

Two considerations about numeric comparisons are especially helpful to know:

- An abbreviated form of comparison is possible.

- Abbreviated comparisons with OR require you to use caution.

In computing terms, a value of TRUE is 1 and a value of FALSE is 0. In SAS, the following rules apply:

- Any numeric value other than 0 is true.

- A value of 0 or missing is false.

Therefore, a numeric variable or expression can stand alone in a condition. If its value is a number other than 0, then the condition is true. If its value is 0 or missing, then the condition is false.

The following example assigns a value to the variable Remarks only if the value of LandCost is present for a given observation:

```
if LandCost then Remarks='Ready to budget';
```

This statement is equivalent to

```
if LandCost ne . and LandCost ne 0 then Remarks='Ready to budget';
```

Be careful when you abbreviate comparisons with OR. It is easy to produce unexpected results. For example, this IF-THEN statement selects tours that last six or eight nights:

```
/* first try */
if Nights=6 or 8 then Stay='Medium';
```

SAS treats the condition as the following comparisons:

```
Nights=6
   or
   8
```

The second comparison does not use the values of Nights. The comparison uses the number 8 standing alone. Because the number 8 is neither 0 nor a missing value, it always has the value TRUE. Because only one comparison in a series of OR comparisons needs to be true to make the condition true, this condition is true for all observations.

The following comparisons correctly select observations that have six or eight nights:

```
/* correct way */
if Nights=6 or Nights=8 then Stay='Medium';
```

The following DATA step includes these IF-THEN statements:

```
data morecomp;
   set mylib.arttours;
   if LandCost then Remarks='Ready to budget';
   else Remarks='Need land cost';
   if Nights=6 or Nights=8 then Stay='Medium';
   else Stay='Short';
run;

proc print data=morecomp;
   var City Nights LandCost Remarks Stay;
   title 'Tour Information';
run;
```

The following output displays the results.

*Figure 10.10  Abbreviating Numeric Comparisons*

Tour Information

| Obs | City | Nights | LandCost | Remarks | Stay |
|---|---|---|---|---|---|
| 1 | Rome | 3 | 1550 | Ready to budget | Short |
| 2 | Brasilia | 8 | 3360 | Ready to budget | Medium |
| 3 | London | 6 | 2460 | Ready to budget | Medium |
| 4 | Warsaw | 6 | . | Need land cost | Medium |
| 5 | Madrid | 3 | 740 | Ready to budget | Short |
| 6 | Amsterdam | 4 | 1160 | Ready to budget | Short |

# Comparing Characters

## Types of Character Comparisons

Some special situations occur when you make character comparisons. You might need to perform the following tasks:

- compare uppercase and lowercase characters
- select all values that begin with a particular group of characters
- select all values that begin with a particular range of characters
- find a particular value anywhere within another character value

## Comparing Uppercase and Lowercase Characters

SAS distinguishes between uppercase and lowercase letters in comparisons. For example, the values `Madrid` and `MADRID` are not equivalent. To compare values that might occur in different cases, use the UPCASE function to produce an uppercase value. Then make the comparison between two uppercase values, as shown here:

```
data newguide;
   set mylib.arttours;
   if upcase(City)='MADRID' then TourGuide='Balarezo';
run;

proc print data=newguide;
   var City TourGuide;
   title 'Tour Guides';
run;
```

Within the comparison, SAS produces an uppercase version of the value of City and compares it to the uppercase constant MADRID. The value of City in the observation remains in its original case.

The following output displays the results.

**Figure 10.11** *Data Set Produced by an Uppercase Comparison*

Tour Guides

| Obs | City | TourGuide |
|---|---|---|
| 1 | Rome | D'Amico |
| 2 | Brasilia | Lucas |
| 3 | London | Wilson |
| 4 | Warsaw | Lucas |
| 5 | Madrid | Balarezo |
| 6 | Amsterdam | |

Now `Balarezo` is assigned as the tour guide for `Madrid` because the UPCASE function compares the uppercase value of `Madrid` with the value `MADRID`. The UPCASE function enables SAS to read the two values as equal.

### Selecting All Values That Begin with the Same Group of Characters

Sometimes you need to select a group of character values, such as all tour guides whose names begin with the letter D.

By default, SAS compares values of different lengths by adding blanks to the end of the shorter value and testing the result against the longer value. Here is an example:

```
   /* first attempt */
if Tourguide='D' then Chosen='Yes';
else Chosen='No';
```

SAS interprets the comparison as

```
TourGuide='D       '
```

D is followed by seven blanks because TourGuide, a character variable created by column input, has a length of eight bytes. Because the value of TourGuide never consists of the single letter D, the comparison is never true.

To compare a long value to a shorter standard, put a colon (:) after the operator, as in this example:

```
   /* correct method */
if TourGuide=:'D' then Chosen='Yes';
else Chosen='No';
```

The colon causes SAS to compare the same number of characters in the shorter value and the longer value. In this case, the shorter string contains one character. Therefore, SAS tests only the first character from the longer value. All names that begin with a D make the comparison true. (If you are not sure that all the values of TourGuide begin with a capital letter, then use the UPCASE function.) The following DATA step selects names that begin with D:

```
data dguide;
   set mylib.arttours;
   if TourGuide=:'D' then Chosen='Yes';
   else Chosen='No';
run;

proc print data=dguide;
   var City TourGuide Chosen;
   title 'Guides Whose Names Begin with D';
run;
```

The following output displays the results.

*Figure 10.12  Selecting All Values Beginning with a Particular String*

Guides Whose Names Begin with D

| Obs | City | TourGuide | Chosen |
|---|---|---|---|
| 1 | Rome | D'Amico | Yes |
| 2 | Brasilia | Lucas | No |
| 3 | London | Wilson | No |
| 4 | Warsaw | Lucas | No |
| 5 | Madrid | Torres | No |
| 6 | Amsterdam |  | No |

## *Selecting a Range of Character Values*

You might want to select values that begin with a range of characters, such as all names that begin with A through L or M through Z. To select a range of character values, you need to understand the following points:

- In computer processing, letters have magnitude. A is the smallest letter in the alphabet and Z is the largest. Therefore, the comparison A<B is true; so is the comparison D>C.[1]
- A blank is smaller than any letter.

The following statements divide the names of the guides into two groups, A-L and M-Z, by combining the comparison operator with the colon:

```
if TourGuide <=: 'L' then TourGuideGroup='A-L';
   else TourGuideGroup='M-Z';
```

The following DATA step creates the groups:

---

[1] The magnitude of letters in the alphabet is true for all operating environments under which SAS runs. Other points, such as whether uppercase or lowercase letters are larger and how to treat numbers in character values, depend on your operating environment. For more information about how character values are sorted under various operating environments, see Chapter 12, "Working with Grouped or Sorted Observations," on page 183.

```
data guidegrp;
   set mylib.arttours;
   if TourGuide <=: 'L' then TourGuideGroup'A-L';
   else TourGuideGroup='M-Z';
run;

proc print data=guidegrp;
   var City TourGuide TourGuideGroup;
   title 'Tour Guide Groups';
run;
```

The following output displays the results.

*Figure 10.13* Selecting All Values Beginning with a Range of Characters

Tour Guide Groups

| Obs | City | TourGuide | TourGuideGroup |
|---|---|---|---|
| 1 | Rome | D'Amico | A-L |
| 2 | Brasilia | Lucas | A-L |
| 3 | London | Wilson | M-Z |
| 4 | Warsaw | Lucas | A-L |
| 5 | Madrid | Torres | M-Z |
| 6 | Amsterdam |  | A-L |

All names that begin with A through L, as well as the missing value, go into group A-L. The missing value goes into that group because a blank is smaller than any letter.

### *Finding a Value Anywhere within Another Character Value*

A data set is needed that lists tours that visit other attractions in addition to museums and galleries. In the data set MYLIB.ARTTOURS, the variable EventDescription refers to those events as `other`. However, the position of the word `other` varies in different observations. How can it be determined that `other` exists anywhere in the value of EventDescription for a given observation?

The INDEX function determines whether a specified character string (the excerpt) is present within a particular character value (the source):

**INDEX** (*source, excerpt*)

Both *source* and *excerpt* can be any type of character expression, including character strings that are enclosed in quotation marks, character variables, and other character functions. If *excerpt* does occur within *source*, then the function returns the position of the first character of *excerpt*, which is a positive number. If it does not, then the function returns a 0. By testing for a value greater than 0, you can determine whether a particular character string is present in another character value.

The following statements select observations containing the string `other`:

```
if index(EventDescription,'other') > 0 then OtherEvents='Yes';
else OtherEvents='No';
```

You can also write the condition in the following way:

```
if index(EventDescription,'other') then OtherEvents='Yes';
else OtherEvents='No';
```

The second example uses the fact that any value other than 0 or missing makes the condition true. This statement is included in the following DATA step:

```
data otherevent;
   set mylib.arttours;
   if index(EventDescription,'other') then OtherEvents='Yes';
   else OtherEvents='No';
run;

proc print data=otherevent;
   var City EventDescription OtherEvents;
   title 'Tour Events';
run;
```

The following output displays the results.

*Figure 10.14* Finding a Character String within Another Value

Tour Events

| Obs | City | EventDescription | OtherEvents |
|---|---|---|---|
| 1 | Rome | 4 M, 3 G | No |
| 2 | Brasilia | 5 M, 1 other | Yes |
| 3 | London | 3 M, 2 G | No |
| 4 | Warsaw | 5 M, 1 G, 2 other | Yes |
| 5 | Madrid | 3 M, 2 other | Yes |
| 6 | Amsterdam | 3 M, 3 G | No |

In the observations for Brasilia and Madrid, the INDEX function returns the value 8 because the string `other` is found beginning in position eight of the variable (`5 M, 1 other` for Brasilia and `3 M, 2 other` for Madrid). For New York, it returns the value 13 because the string `other` is found beginning in position thirteen of the variable (`5 M, 1 G, 2 other`). In the remaining observations, the function does not find the string `other` and returns a 0.

# Summary

### Statements

IF *condition* THEN *action*;
<ELSE *action*>;

tests whether *condition* is true. If the condition is true, the action in the THEN clause is carried out. If *condition* is false and an ELSE statement is present, then the ELSE action is carried out. If *condition* is false and no ELSE statement is present, then the next statement in the DATA step is processed. *Condition* specifies one or more numeric or character comparisons. *action* must be an executable statement. That is, one that can be processed in an individual iteration of the DATA step. (Statements that affect the entire DATA step, such as LENGTH, are not executable.)

In SAS processing, any numeric value other than 0 or missing is true. The values 0 and missing are false. Therefore, a numeric value can stand alone in a comparison. If its value is 0 or missing, then the comparison is false. Otherwise, the comparison is true.

## Functions

INDEX(*source, excerpt*)
: searches *source* for the string given in *excerpt*. Both *source* and *excerpt* can be any type of character expressions, such as character variables, character strings that are enclosed in quotation marks, other character functions, and so on. When *excerpt* is present in *source*, the function returns the position of the first character of *excerpt* (a positive number). When *excerpt* is not present, the function returns a 0.

UPCASE(*argument*)
: produces an uppercase value of *argument*, which can be any type of character expression, such as a character variable, character string enclosed in quotation marks, other character functions, and so on.

# Learning More

Base SAS functions
: Base SAS functions are documented in *SAS Functions and CALL Routines: Reference*.

Comparison and logical operators
: For more information, see "SAS Operators in Expressions" in *SAS Language Reference: Concepts*.

Executable statements
: You can issue only executable statements in IF-THEN/ELSE statements. For a complete list of executable and nonexecutable statements, see "DATA Step Statements" in *SAS Statements: Reference*.

IF-THEN/ELSE statement and clauses
: For more information, see "IF-THEN/ELSE Statement" in *SAS Statements: Reference*.

IN operator
: You can use the IN operator to shorten a comparison when you are comparing a value to a series of numeric or character constants (not variables or expressions). For more information, see "The IN Operator in Character Comparisons" in *SAS Language Reference: Concepts* and "The IN Operator in Numeric Comparisons" in *SAS Language Reference: Concepts*.

SELECT statement
: The SELECT statement, which selects observations based on a condition, is equivalent to a series of IF-THEN/ELSE statements. If you have a long series of conditions and actions, then the DATA step can be easier to read if you write the

conditions in a SELECT group. For more information, see "SELECT Statement" in *SAS Statements: Reference*.

TRUNCOVER option
The TRUNCOVER option in the INFILE statement is described in the note in "Input SAS Data Set for Examples" on page 148. For more information, see "INFILE Statement" in *SAS Statements: Reference*.

# Chapter 11
# Creating Subsets of Observations

Introduction to Creating Subsets of Observations ..................... 169
    Purpose ......................................................... 169
    Prerequisites .................................................... 169

Input SAS Data Set for Examples ..................................... 170

Selecting Observations for a New SAS Data Set ........................ 171
    Deleting Observations Based on a Condition ....................... 171
    Accepting Observations Based on a Condition ...................... 173
    Comparing the DELETE and Subsetting IF Statements ................ 173

Conditionally Writing Observations to One or More SAS Data Sets ...... 175
    Understanding the OUTPUT Statement ............................... 175
    Example for Conditionally Writing Observations to Multiple Data Sets ....... 175
    Avoiding a Common Mistake When Writing to Multiple Data Sets ..... 176
    Understanding Why the Placement of the OUTPUT Statement Is Important .... 177
    Writing an Observation Multiple Times to One or More Data Sets ..... 179

Summary ............................................................ 180
    Statements ...................................................... 180

Learning More ...................................................... 181

## Introduction to Creating Subsets of Observations

### Purpose

In this section, you will learn to select specific observations from existing SAS data sets to create new data sets. Specifically, you will learn about the following concepts:

- how to create a new SAS data set that includes only some of the observations from the input data source

- how to create several new SAS data sets by writing observations from an input data source, using a single DATA step

### Prerequisites

Before proceeding with this section, you should understand the concepts presented in the following sections:

- Chapter 3, "Introduction to DATA Step Processing," on page 27
- Chapter 4, "Starting with Raw Data: The Basics," on page 51
- Chapter 5, "Starting with Raw Data: Beyond the Basics," on page 71
- Chapter 6, "Starting with SAS Data Sets," on page 91
- Chapter 10, "Acting on Selected Observations," on page 147
- Chapter 7, "Understanding DATA Step Processing," on page 107

## Input SAS Data Set for Examples

Tradewinds Travel has a schedule for tours to various art museums and galleries. It would be convenient to keep different SAS data sets that contain different information about the tours. The tour data is stored in an external file that contains the following information:

```
1          2 3    4      5
--------------------------------
Rome       3 1550 Medium D'Amico
Brasilia   8 3360 High   Lucas
London     6 2460 Medium Wilson
Warsaw     6    .        Lucas
Madrid     3  740 Low    Torres
Amsterdam  4 1160 Low
```

The following list describes the fields in the input file:

1  provides the name of the destination city

2  specifies the number of nights on the tour

3  specifies the cost of the land package in U.S. dollars

4  specifies a rating for the budget

5  provides the name of the tour guide

The following program creates a permanent SAS data set named MYLIB.ARTS:

```
libname mylib 'SAS-library';

data mylib.arts;
   infile 'input-file' truncover;
   input City $ 1-9 Nights 11 LandCost 13-16 Budget $ 18-23
         TourGuide $ 25-32;
run;

proc print data=mylib.arts;
   title 'Data Set MYLIB.ARTS';
run;
```

The following output displays the results.

*Figure 11.1* Data Set MYLIB.ARTS

Data Set MYLIB.ARTS

| Obs | City | Nights | LandCost | Budget | TourGuide |
|---|---|---|---|---|---|
| 1 | Rome | 3 | 1550 | Medium | D'Amico |
| 2 | Brasilia | 8 | 3360 | High | Lucas |
| 3 | London | 6 | 2460 | Medium | Wilson |
| 4 | Warsaw | 6 | | | Lucas |
| 5 | Madrid | 3 | 740 | Low | Torres |
| 6 | Amsterdam | 4 | 1160 | Low | |

# Selecting Observations for a New SAS Data Set

## Deleting Observations Based on a Condition

There are two ways to select specific observations in a SAS data set when you create a new SAS data set:

1. Delete the observations that do not meet a condition, keeping only the ones that you want.
2. Accept only the observations that meet a condition.

To delete an observation, first identify it with an IF condition, and then use a DELETE statement in the THEN clause:

**IF** *condition* **THEN DELETE**;

Processing the DELETE statement for an observation causes SAS to return immediately to the beginning of the DATA step for a new observation without writing the current observation to the output DATA set. The DELETE statement does not include the observation in the output data set, but it does not delete the observation from the input data set. For example, the following statement deletes observations that contain a missing value for LandCost:

```
if LandCost=. then delete;
```

The following DATA step includes this statement:

```
data remove;
   set mylib.arts;
   if LandCost=. then delete;
run;

proc print data=remove;
   title 'Tours With Complete Land Costs';
```

```
run;
```

The following output displays the results.

*Figure 11.2* Deleting Observations That Have a Particular Value

Tours With Complete Land Costs

| Obs | City | Nights | LandCost | Budget | TourGuide |
|---|---|---|---|---|---|
| 1 | Rome | 3 | 1550 | Medium | D'Amico |
| 2 | Brasilia | 8 | 3360 | High | Lucas |
| 3 | London | 6 | 2460 | Medium | Wilson |
| 4 | Madrid | 3 | 740 | Low | Torres |
| 5 | Amsterdam | 4 | 1160 | Low | |

Warsaw, the observation that is missing a value for LandCost, is not included in the resulting data set, REMOVE.

You can also delete observations as you enter data from an external file. The following DATA step produces the same SAS data set as the REMOVE data set.

```
data remove2;
   infile 'input-file' truncover;
   input City $ 1-9 Nights 11 LandCost 13-16 Budget $ 18-23
         TourGuide $ 25-32;
   if LandCost=. then delete;
run;

proc print data=remove2;
   title 'Tours With Complete Land Costs';
run;
```

The following output displays the results.

*Figure 11.3* Deleting Observations While Reading from an External File

Tours With Complete Land Costs

| Obs | City | Nights | LandCost | Budget | TourGuide |
|---|---|---|---|---|---|
| 1 | Rome | 3 | 1550 | Medium | D'Amico |
| 2 | Brasilia | 8 | 3360 | High | Lucas |
| 3 | London | 6 | 2460 | Medium | Wilson |
| 4 | Madrid | 3 | 740 | Low | Torres |
| 5 | Amsterdam | 4 | 1160 | Low | |

### Accepting Observations Based on a Condition

One data set that is needed by the travel agency contains observations for tours that last only six nights. One way to make the selection is to delete observations in which the value of Nights is not equal to 6:

```
if Nights ne 6 then delete;
```

A more straightforward way is to select only observations meeting the criterion. The subsetting IF statement selects the observations that you specify. It contains only a condition:

**IF** *condition*;

The implicit action in a subsetting IF statement is always the same: if the condition is true, then continue processing the observation. If it is false, then stop processing the observation and return to the top of the DATA step for a new observation. The statement is called subsetting because the result is a subset of the original observations. For example, if you want to select only observations in which the value of Nights is equal to 6, then you specify the following statement:

```
if Nights = 6;
```

The following DATA step includes the subsetting IF statement:

```
data subset6;
   set mylib.arts;
   if nights=6;
run;

proc print data=subset6;
   title 'Six-Night Tours';
run;
```

The following output displays the results.

*Figure 11.4* Selecting Observations with a Subsetting IF Statement

Six-Night Tours

| Obs | City | Nights | LandCost | Budget | TourGuide |
|---|---|---|---|---|---|
| 1 | London | 6 | 2460 | Medium | Wilson |
| 2 | Warsaw | 6 | . | . | Lucas |

Two observations met the criteria for a six-night tour.

### Comparing the DELETE and Subsetting IF Statements

These are the main reasons to consider when choosing between a DELETE statement and a subsetting IF statement:

- It is usually easier to choose the statement that requires the fewest comparisons to identify the condition.
- It is usually easier to think in positive terms than negative ones (this favors the subsetting IF).

One additional situation favors the subsetting IF: it is the safer method to use if your data has missing or misspelled values. Consider the following situation.

Tradewinds Travel needs a SAS data set of low-priced to medium-priced tours. Knowing that the values of Budget are `Low`, `Medium`, and `High`, a first thought would be to delete observations with a value of `High`. The following program creates a SAS data set by deleting observations that have a Budget value of `HIGH`:

```
   /* first attempt */
data lowmed;
   set mylib.arts;
   if upcase(Budget)='HIGH' then delete;
run;

proc print data=lowmed;
   title 'Medium and Low Priced Tours';
run;
```

The following output displays the results.

**Figure 11.5** *Producing a Subset by Deletion*

Medium and Low Priced Tours

| Obs | City      | Nights | LandCost | Budget | TourGuide |
|-----|-----------|--------|----------|--------|-----------|
| 1   | Rome      | 3      | 1550     | Medium | D'Amico   |
| 2   | London    | 6      | 2460     | Medium | Wilson    |
| 3   | Warsaw    | 6      |          |        | Lucas     |
| 4   | Madrid    | 3      | 740      | Low    | Torres    |
| 5   | Amsterdam | 4      | 1160     | Low    |           |

The data set LOWMED contains both the tours that you want and the tour to Warsaw. The inclusion of the tour to Warsaw is erroneous because the value of Budget for the Warsaw observation is missing. Using a subsetting IF statement ensures that the data set contains exactly the observations that you want. This DATA step creates the subset with a subsetting IF statement:

```
   /* a safer method */
data lowmed2;
   set mylib.arts;
   if upcase(Budget)='MEDIUM' or upcase(Budget)='LOW';
run;

proc print data=lowmed2;
   title 'Medium and Low Priced Tours';
run;
```

The following output displays the results.

*Figure 11.6* Producing an Exact Subset with the Subsetting IF Statement

Medium and Low Priced Tours

| Obs | City | Nights | LandCost | Budget | TourGuide |
|---|---|---|---|---|---|
| 1 | Rome | 3 | 1550 | Medium | D'Amico |
| 2 | London | 6 | 2460 | Medium | Wilson |
| 3 | Madrid | 3 | 740 | Low | Torres |
| 4 | Amsterdam | 4 | 1160 | Low | |

The result is a SAS data set with no missing values for Budget.

# Conditionally Writing Observations to One or More SAS Data Sets

### Understanding the OUTPUT Statement

SAS enables you to create multiple SAS data sets in a single DATA step using an OUTPUT statement:

**OUTPUT** <*SAS-data-set(s)*>;

When you use an OUTPUT statement without specifying a data set name, SAS writes the current observation to all data sets that are named in the DATA statement. If you want to write observations to a selected data set, then you specify that data set name directly in the OUTPUT statement. Any data set name appearing in the OUTPUT statement must also appear in the DATA statement.

### Example for Conditionally Writing Observations to Multiple Data Sets

One of the SAS data sets contains tours that are guided by the tour guide Lucas and the other contains tours led by other guides. Writing to multiple data sets is accomplished by performing the following tasks:

- naming both data sets in the DATA statement
- selecting the observations using an IF condition
- using an OUTPUT statement in the THEN and ELSE clauses to output the observations to the appropriate data sets

The following DATA step shows how to write to multiple data sets:

```
data lucastour othertours;
   set mylib.arts;
   if TourGuide='Lucas' then output lucastour;
   else output othertours;
run;
```

```
proc print data=lucastour;
   title "Data Set with TourGuide='Lucas'";
run;

proc print data=othertours;
   title "Data Set with Other Guides";
run;
```

The following output displays the results.

*Figure 11.7  Creating Two Data Sets with One DATA Step*

Data Set with TourGuide = 'Lucas'

| Obs | City | Nights | LandCost | Budget | TourGuide |
|---|---|---|---|---|---|
| 1 | Brasilia | 8 | 3360 | High | Lucas |
| 2 | Warsaw | 6 | . | | Lucas |

Data Set with Other Guides

| Obs | City | Nights | LandCost | Budget | TourGuide |
|---|---|---|---|---|---|
| 1 | Rome | 3 | 1550 | Medium | D'Amico |
| 2 | London | 6 | 2460 | Medium | Wilson |
| 3 | Madrid | 3 | 740 | Low | Torres |
| 4 | Amsterdam | 4 | 1160 | Low | |

## Avoiding a Common Mistake When Writing to Multiple Data Sets

If you use an OUTPUT statement, then you suppress the automatic output of observations at the end of the DATA step. Therefore, if you plan to use any OUTPUT statements in a DATA step, then you must program all output for that step with OUTPUT statements. For example, in the previous DATA step you sent output to both LUCASTOUR and OTHERTOURS. For comparison, the following program shows what would happen if you omit the ELSE statement in the DATA step:

```
data lucastour2 othertour2;
   set mylib.arts;
   if TourGuide='Lucas' then output lucastour2;
run;

proc print data=lucastour2;
   title "Data Set with Guide='Lucas'";
run;

proc print data=othertour2;
   title "Data Set with Other Guides";
run;
```

The following output displays the results.

*Figure 11.8  Failing to Direct Output to a Second Data Set*

Data Set with Guide = 'Lucas'

| Obs | City | Nights | LandCost | Budget | TourGuide |
|---|---|---|---|---|---|
| 1 | Brasilia | 8 | 3360 | High | Lucas |
| 2 | Warsaw | 6 | . |  | Lucas |

No observations are written to OTHERTOUR2 because output was not directed to it.

## Understanding Why the Placement of the OUTPUT Statement Is Important

By default SAS writes an observation to the output data set at the end of each iteration. When you use an OUTPUT statement, you override the automatic output feature. Where you place the OUTPUT statement, therefore, is very important. For example, if a variable value is calculated after the OUTPUT statement executes, then that value is not available when the observation is written to the output data set.

For example, in the following DATA step, an assignment statement is placed after the IF-THEN/ELSE group:

```
/* first attempt to combine assignment and OUTPUT statements */
data lucasdays otherdays;
   set mylib.arts;
   if TourGuide='Lucas' then output lucasdays;
   else output otherdays;
   Days=Nights+1;
run;

proc print data=lucasdays;
   title "Number of Days in Lucas's Tours";
run;

proc print data=otherdays;
   title "Number of Days in Other Guides' Tours";
run;
```

The following output displays the results.

*Figure 11.9  Unintended Results: Writing Observations Before Assigning Values*

Number of Days in Lucas's Tours

| Obs | City | Nights | LandCost | Budget | TourGuide | Days |
|---|---|---|---|---|---|---|
| 1 | Brasilia | 8 | 3360 | High | Lucas | . |
| 2 | Warsaw | 6 | . |  | Lucas | . |

## Number of Days in Other Guides' Tours

| Obs | City | Nights | LandCost | Budget | TourGuide | Days |
|---|---|---|---|---|---|---|
| 1 | Rome | 3 | 1550 | Medium | D'Amico | . |
| 2 | London | 6 | 2460 | Medium | Wilson | . |
| 3 | Madrid | 3 | 740 | Low | Torres | . |
| 4 | Amsterdam | 4 | 1160 | Low | | . |

The value of Days is missing in all observations because the OUTPUT statement writes the observation to the SAS data sets before the assignment statement is processed. If you want the value of Day to appear in the data sets, then use the assignment statement before you use the OUTPUT statement. The following program shows the correct position:

```
/* correct position of assignment statement */
data lucasdays2 otherdays2;
   set mylib.arts;
   Days=Nights + 1;
   if TourGuide='Lucas' then output lucasdays2;
   else output otherdays2;
run;

proc print data=lucasdays2;
   title "Number of Days in Lucas's Tours";
run;
proc print data=otherdays2;
   title "Number of Days in Other Guides' Tours";
run;
```

The following output displays the results.

*Figure 11.10* Intended Results: Assigning Values After Writing Observations

## Number of Days in Lucas's Tours

| Obs | City | Nights | LandCost | Budget | TourGuide | Days |
|---|---|---|---|---|---|---|
| 1 | Brasilia | 8 | 3360 | High | Lucas | 9 |
| 2 | Warsaw | 6 | . | | Lucas | 7 |

Number of Days in Other Guides' Tours

| Obs | City | Nights | LandCost | Budget | TourGuide | Days |
|---|---|---|---|---|---|---|
| 1 | Rome | 3 | 1550 | Medium | D'Amico | 4 |
| 2 | London | 6 | 2460 | Medium | Wilson | 7 |
| 3 | Madrid | 3 | 740 | Low | Torres | 4 |
| 4 | Amsterdam | 4 | 1160 | Low |  | 5 |

## *Writing an Observation Multiple Times to One or More Data Sets*

After SAS processes an OUTPUT statement, the observation remains in the program data vector and you can continue programming with it. You can even write it again to the same SAS data set or to a different one. The following example creates two pairs of data sets, one pair based on the name of the tour guide and one pair based on the number of nights.

```
data lucastour othertour weektour daytour;
   set mylib.arts;
   if TourGuide='Lucas' then output lucastour;
   else output othertour;
   if nights >= 6 then output weektour;
   else output daytour;
run;

proc print data=lucastour;
   title "Lucas's Tours";
run;

proc print data=othertour;
   title "Other Guides' Tours";
run;
proc print data=weektour;
   title 'Tours Lasting a Week or More';
run;

proc print data=daytour;
   title 'Tours Lasting Less Than a Week';
run;
```

The following output displays the results.

*Figure 11.11  Assigning Observations to More Than One Data Set*

Lucas's Tours

| Obs | City | Nights | LandCost | Budget | TourGuide |
|---|---|---|---|---|---|
| 1 | Brasilia | 8 | 3360 | High | Lucas |
| 2 | Warsaw | 6 |  |  | Lucas |

Other Guides' Tours

| Obs | City | Nights | LandCost | Budget | TourGuide |
|---|---|---|---|---|---|
| 1 | Rome | 3 | 1550 | Medium | D'Amico |
| 2 | London | 6 | 2460 | Medium | Wilson |
| 3 | Madrid | 3 | 740 | Low | Torres |
| 4 | Amsterdam | 4 | 1160 | Low | |

Tours Lasting a Week or More

| Obs | City | Nights | LandCost | Budget | TourGuide |
|---|---|---|---|---|---|
| 1 | Brasilia | 8 | 3360 | High | Lucas |
| 2 | London | 6 | 2460 | Medium | Wilson |
| 3 | Warsaw | 6 | . | | Lucas |

Tours Lasting Less Than a Week

| Obs | City | Nights | LandCost | Budget | TourGuide |
|---|---|---|---|---|---|
| 1 | Rome | 3 | 1550 | Medium | D'Amico |
| 2 | Madrid | 3 | 740 | Low | Torres |
| 3 | Amsterdam | 4 | 1160 | Low | |

The first IF-THEN/ELSE group writes all observations to either data set LUCASTOUR or OTHERTOUR. The second IF-THEN/ELSE group writes the same observations to a different pair of data sets, WEEKTOUR and DAYTOUR. This repetition is possible because each observation remains in the program data vector after the first OUTPUT statement is processed and can be written again.

# Summary

## Statements

DATA *<libref-1> SAS-data-set-1< . . .<libref-n> SAS-data-set-n>*;
names the SAS data sets to be created in the DATA step.

DELETE;
deletes the current observation. The DELETE statement is usually used as part of an IF-THEN/ELSE group.

IF *condition*;
: tests whether *condition* is true. If it is true, then SAS continues to process the current observation. If it is not true, then SAS stops processing the observation, does not add it to the SAS data set, and returns to the top of the DATA step. The conditions that are used are the same as in the IF-THEN/ELSE statements. This type of IF statement is called a subsetting IF statement because it produces a subset of the original observations.

OUTPUT <*SAS-data-set*>;
: immediately writes the current observation to *SAS-data-set*. The observation remains in the program data vector. You can continue programming with observation, including writing it again to a SAS data set. When an OUTPUT statement appears in a DATA step, SAS does not automatically output observations to the SAS data set. You must specify the destination for all output in the DATA step with OUTPUT statements. Any SAS data set that you specify in an OUTPUT statement must also appear in the DATA statement.

# Learning More

Comparison and logical operators
: For more information, see "SAS Operators in Expressions" in *SAS Language Reference: Concepts* and Chapter 10, "Acting on Selected Observations," on page 147.

DROP= and KEEP= data set options
: Using the DROP= and KEEP= data set options to output a subset of variables to a SAS data set are discussed in Chapter 6, "Starting with SAS Data Sets," on page 91. For more information, see "DROP= Data Set Option" in *SAS Data Set Options: Reference* and "KEEP= Data Set Option" in *SAS Data Set Options: Reference*.

FIRSTOBS= and OBS= data set options
: Using these data set options to select observations from the beginning, middle, or end of a SAS data set are discussed in Chapter 6, "Starting with SAS Data Sets," on page 91. For more information, see "FIRSTOBS= Data Set Option" in *SAS Data Set Options: Reference* and "OBS= Data Set Option" in *SAS Data Set Options: Reference*.

IF-THEN/ELSE statement
: For more information, see "IF-THEN/ELSE Statement" in *SAS Statements: Reference*.

DELETE and OUTPUT statements
: For more information, see "DELETE Statement" in *SAS Statements: Reference* and "OUTPUT Statement" in *SAS Statements: Reference*.

WHERE statement
: The WHERE statement selects observations based on a condition. Its action is similar to that of a subsetting IF statement. The WHERE statement is extremely useful in PROC steps, and it can also be useful in some DATA steps. The WHERE statement selects observations before they enter the program data vector. In contrast, the subsetting IF statement selects observations already in the program data vector.

    For more information, see "Selecting Observations" on page 444. See also "WHERE Statement" in *SAS Statements: Reference*.

    *Note:* In some cases, the same condition in a WHERE statement in the DATA step and in a subsetting IF statement produces different subsets. The difference is described in the discussion of the WHERE statement in "WHERE Statement" in *SAS Statements: Reference*. Be sure you understand the difference before you use the WHERE statement in the DATA step. With that caution in mind, a WHERE statement can increase the efficiency of the DATA step considerably.

# Chapter 12
# Working with Grouped or Sorted Observations

**Introduction to Working with Grouped or Sorted Observations** .............. **183**
    Purpose ................................................. 183
    Prerequisites ............................................ 184

**Input SAS Data Set for Examples** ................................. **184**

**Working with Grouped Data** ..................................... **185**
    Understanding the Basics of Grouping Data ...................... 185
    Grouping Observations with the SORT Procedure .................. 186
    Grouping by More Than One Variable ........................... 187
    Arranging Groups in Descending Order .......................... 188
    Finding the First or Last Observation in a Group ................... 189

**Working with Sorted Data** ...................................... **192**
    Understanding Sorted Data .................................. 192
    Sorting Data ............................................. 192
    Deleting Duplicate Observations .............................. 193
    Understanding Collating Sequences ............................ 195
    ASCII Collating Sequence ................................... 196
    EBCDIC Collating Sequence ................................. 196

**Summary** .................................................. **197**
    Procedures .............................................. 197
    Statements .............................................. 197

**Learning More** .............................................. **197**

## Introduction to Working with Grouped or Sorted Observations

### Purpose

Sometimes you need to create reports where observations are grouped according to the values of a particular variable, or where observations are sorted alphabetically. The following concepts are discussed in this section:

- how to group observations by variables and how to work with grouped observations
- how to sort the observations and how to work with sorted observations

*Prerequisites*

Before proceeding with this section, you should understand the concepts presented in the following sections:

- Chapter 1, "What is the SAS System?," on page 3
- "Overview of DATA Step Processing" on page 107
- Chapter 4, "Starting with Raw Data: The Basics," on page 51
- Chapter 5, "Starting with Raw Data: Beyond the Basics," on page 71
- Chapter 6, "Starting with SAS Data Sets," on page 91
- Chapter 7, "Understanding DATA Step Processing," on page 107

## Input SAS Data Set for Examples

Tradewinds Travel has an external file that contains data about tours that emphasize either architecture or scenery. After the data is created in a SAS data set and the observations for those tours are grouped together, SAS can produce reports for each group separately. In addition, if the observations need to be alphabetized by country, SAS can sort them. The external file looks like this:

```
   1              2               3   4     5
   ----------------------------------------------
   Spain          architecture    10  1020  World
   Japan          architecture     8  1440  Express
   Switzerland    scenery          9  1468  World
   Brazil         architecture     8  1150  World
   Ireland        scenery          7  1116  Express
   New Zealand    scenery         16  2978  Southsea
   Italy          architecture     8   936  Express
   Greece         scenery         12  1396  Express
```

The following list describes the fields in the input file:

1 provides the name of the destination country

2 identifies the tour's area of emphasis

3 specifies the number of nights on the tour

4 specifies the cost of the land package in U.S. dollars

5 lists the name of the tour vendor

The following DATA step creates the permanent SAS data set MYLIB.ARCH_OR_SCEN:

```
libname mylib 'SAS-library';

data mylib.arch_or_scen;
   infile 'input-file' truncover;
   input Country $ 1-11 TourType $ 13-24 Nights LandCost Vendor $;
run;
```

```
proc print data=mylib.arch_or_scen;
   title 'Data Set MYLIB.ARCH_OR_SCEN';
run;
```

The PROC PRINT statement that follows the DATA step produces this display of the MYLIB.ARCH_OR_SCEN data set:

*Figure 12.1* Data Set MYLIB.ARCH_OR_SCEN

Data Set MYLIB.ARCH_OR_SCEN

| Obs | Country | TourType | Nights | LandCost | Vendor |
|-----|---------|----------|--------|----------|--------|
| 1 | Spain | architecture | 10 | 1020 | World |
| 2 | Japan | architecture | 8 | 1440 | Express |
| 3 | Switzerland | scenery | 9 | 1468 | World |
| 4 | Brazil | architecture | 8 | 1150 | World |
| 5 | Ireland | scenery | 7 | 1116 | Express |
| 6 | New Zealand | scenery | 16 | 2978 | Southsea |
| 7 | Italy | architecture | 8 | 936 | Express |
| 8 | Greece | scenery | 12 | 1396 | Express |

# Working with Grouped Data

## Understanding the Basics of Grouping Data

The basic method for grouping data is to use a BY statement:

**BY** *list-of-variables*;

The BY statement can be used in a DATA step with a SET, MERGE, MODIFY, or UPDATE statement, or it can be used in SAS procedures.

To work with grouped data using the SET, MERGE, MODIFY, or UPDATE statements, the data must meet these conditions:

- The observations must be in a SAS data set, not an external file.
- The variables that define the groups must appear in the BY statement.
- All observations in the input data set must be in ascending or descending numeric or character order, or grouped in some way, such as by calendar month or by a formatted value, according to the variables that are specified in the BY statement.

   *Note:* If you use the MODIFY statement, the input data does not need to be in any order. However, ordering the data can improve performance.

If the third condition is not met, the data is stored in a SAS data set but is not arranged in the groups that you want. You can order the data using the SORT procedure (discussed in the next section).

After the SAS data set is arranged in some order, you can use the BY statement to group values of one or more common variables.

## Grouping Observations with the SORT Procedure

All observations in the input data set must be in a particular order. To meet this condition, the observations in MYLIB.ARCH_OR_SCEN can be ordered by the values of TourType, which are **architecture** or **scenery**. Use the SORT procedure to sort the observations by TourType:

```
proc sort data=mylib.arch_or_scen out=tourorder;
   by TourType;
run;
```

The SORT procedure sorts the data set MYLIB.ARCH_OR_SCEN alphabetically according to the values of TourType. The sorted observations go into a new data set specified by the OUT= option. In this example, TOURORDER is the sorted data set. If the OUT= option is omitted, the sorted version of the data set replaces the data set MYLIB.ARCH_OR_SCEN.

The SORT procedure does not produce output other than the sorted data set. A message in the SAS log says that the SORT procedure was executed:

*Log 12.1  Message That the SORT Procedure Has Executed Successfully*

```
880  proc sort data=mylib.arch_or_scen out=tourorder;
881     by TourType;
882  run;

NOTE: There were 8 observations read from the data set MYLIB.ARCH_OR_SCEN.
NOTE: The data set WORK.TOURORDER has 8 observations and 5 variables.
NOTE: PROCEDURE SORT used (Total process time):
      real time           0.20 seconds
      cpu time            0.04 seconds
```

To see the sorted data set, add a PROC PRINT step to the program:

```
proc sort data=mylib.arch_or_scen out=tourorder;
   by TourType;
run;

proc print data=tourorder;
   var TourType Country Nights LandCost Vendor;
   title 'Tours Sorted by Architecture or Scenery';
run;
```

The following output displays the results.

*Figure 12.2  Displaying the Sorted Output*

**Tours Sorted by Architecture or Scenery**

| Obs | TourType | Country | Nights | LandCost | Vendor |
|---|---|---|---|---|---|
| 1 | architecture | Spain | 10 | 1020 | World |
| 2 | architecture | Japan | 8 | 1440 | Express |
| 3 | architecture | Brazil | 8 | 1150 | World |
| 4 | architecture | Italy | 8 | 936 | Express |
| 5 | scenery | Switzerland | 9 | 1468 | World |
| 6 | scenery | Ireland | 7 | 1116 | Express |
| 7 | scenery | New Zealand | 16 | 2978 | Southsea |
| 8 | scenery | Greece | 12 | 1396 | Express |

By default, SAS arranges groups in ascending order of the BY values, smallest to largest. Sorting a data set does not change the order of the variables within it. However, most examples in this section use a VAR statement in the PRINT procedure to display the BY variable in the first column. (The PRINT procedure and other procedures used in this documentation can also produce a separate report for each BY group.)

## Grouping by More Than One Variable

You can group observations by as many variables as you want. This example groups observations by TourType, Vendor, and LandCost:

```
proc sort data=mylib.arch_or_scen out=tourorder2;
   by TourType Vendor LandCost;
run;

proc print data=tourorder2;
   var TourType Vendor LandCost Country Nights;
   title 'Tours Grouped by Type of Tour, Vendor, and Price';
run;
```

The following output displays the results.

*Figure 12.3* Grouping by Several Variables

Tours Grouped by Type of Tour, Vendor, and Price

| Obs | TourType | Vendor | LandCost | Country | Nights |
|---|---|---|---|---|---|
| 1 | architecture | Express | 936 | Italy | 8 |
| 2 | architecture | Express | 1440 | Japan | 8 |
| 3 | architecture | World | 1020 | Spain | 10 |
| 4 | architecture | World | 1150 | Brazil | 8 |
| 5 | scenery | Express | 1116 | Ireland | 7 |
| 6 | scenery | Express | 1396 | Greece | 12 |
| 7 | scenery | Southsea | 2978 | New Zealand | 16 |
| 8 | scenery | World | 1468 | Switzerland | 9 |

As this example shows, SAS groups the observations by the first variable that is named within those groups, by the second variable named, and so on. The groups defined by all variables contain only one observation each. In this example, no two variables have the same values for all observations. In other words, this example does not have any duplicate entries.

## Arranging Groups in Descending Order

In the data sets that are grouped by TourType, the group for `architecture` comes before the group for `scenery` because `architecture` begins with an "a", and "a" is smaller than "s" in computer processing. (The order of characters, known as their collating sequence, is discussed later in this section.) To produce a descending order for a particular variable, place the DESCENDING option before the name of the variable in the BY statement of the SORT procedure. In the next example, the observations are grouped in descending order by TourType, but in ascending order by Vendor and LandCost:

```
proc sort data=mylib.arch_or_scen out=tourorder3;
   by descending TourType Vendor LandCost;
run;

proc print data=tourorder3;
   var TourType Vendor LandCost Country Nights;
   title 'Descending Order of TourType';
run;
```

The following output displays the results.

*Figure 12.4  Combining Descending and Ascending Sorted Observations*

Descending Order of TourType

| Obs | TourType | Vendor | LandCost | Country | Nights |
|---|---|---|---|---|---|
| 1 | scenery | Express | 1116 | Ireland | 7 |
| 2 | scenery | Express | 1396 | Greece | 12 |
| 3 | scenery | Southsea | 2978 | New Zealand | 16 |
| 4 | scenery | World | 1468 | Switzerland | 9 |
| 5 | architecture | Express | 936 | Italy | 8 |
| 6 | architecture | Express | 1440 | Japan | 8 |
| 7 | architecture | World | 1020 | Spain | 10 |
| 8 | architecture | World | 1150 | Brazil | 8 |

## Finding the First or Last Observation in a Group

If you do not want to display the entire data set, you can create a data set that contains only the least expensive tour that features architecture, and the least expensive tour that features scenery:

First, sort the data set by TourType and LandCost:

```
proc sort data=mylib.arch_or_scen out=tourorder4;
   by TourType LandCost;
run;

proc print data=tourorder4;
   var TourType LandCost Country Nights Vendor;
   title 'Tours Arranged by TourType and LandCost';
run;
```

The following output displays the results.

*Figure 12.5* Sorting to Find the Least Expensive Tours

Tours Arranged by TourType and LandCost

| Obs | TourType | LandCost | Country | Nights | Vendor |
|---|---|---|---|---|---|
| 1 | architecture | 936 | Italy | 8 | Express |
| 2 | architecture | 1020 | Spain | 10 | World |
| 3 | architecture | 1150 | Brazil | 8 | World |
| 4 | architecture | 1440 | Japan | 8 | Express |
| 5 | scenery | 1116 | Ireland | 7 | Express |
| 6 | scenery | 1396 | Greece | 12 | Express |
| 7 | scenery | 1468 | Switzerland | 9 | World |
| 8 | scenery | 2978 | New Zealand | 16 | Southsea |

You sorted LandCost in ascending order, so the first observation in each value of TourType has the lowest value of LandCost. If you can locate the first observation in each BY group in a DATA step, you can use a subsetting IF statement to select that observation. SAS provides a way to locate the first observation with each value of TourType.

When you use a BY statement in a DATA step, SAS automatically creates two additional variables for each variable in the BY statement. One variable is named FIRST.*variable*, where *variable* is the name of the BY variable. The other variable is named LAST.*variable*. Their values are either 1 or 0. They exist in the program data vector and are available for DATA step programming, but SAS does not add them to the SAS data set that is being created. For example, the DATA step begins with these statements:

```
data lowcost;
   set tourorder4;
   by TourType;
   ...more SAS statements...
run;
```

The BY statement causes SAS to create one variable called FIRST.TOURTYPE and another variable called LAST.TOURTYPE. When SAS processes the first observation with the value `architecture`, the value of FIRST.TOURTYPE is 1. In other observations with the value `architecture`, it is 0. Similarly, when SAS processes the last observation with the value `architecture`, the value of LAST.TOURTYPE is 1. In other observations with the value `architecture`, it is 0. The same result occurs in the scenery group with the observations.

SAS does not write FIRST. and LAST. variables to the output data set, so you cannot display their values with the PRINT procedure. Therefore, the simplest method of displaying the values of FIRST. and LAST. variables is to assign their values to other variables. This example assigns the value of FIRST.TOURTYPE to a variable named FirstTour and the value of LAST.TOURTYPE to a variable named LastTour:

```
data temp;
   set tourorder4;
```

```
    by TourType;
    FirstTour=first.TourType;
    LastTour=last.TourType;
run;
proc print data=temp;
    var Country Tourtype FirstTour LastTour;
    title 'Specifying FIRST.TOURTYPE and LAST.TOURTYPE';
run;
```

The following output displays the results.

*Figure 12.6  Demonstrating FIRST. and LAST. Values*

Specifying FIRST.TOURTYPE and LAST.TOURTYPE

| Obs | Country | TourType | FirstTour | LastTour |
|---|---|---|---|---|
| 1 | Italy | architecture | 1 | 0 |
| 2 | Spain | architecture | 0 | 0 |
| 3 | Brazil | architecture | 0 | 0 |
| 4 | Japan | architecture | 0 | 1 |
| 5 | Ireland | scenery | 1 | 0 |
| 6 | Greece | scenery | 0 | 0 |
| 7 | Switzerland | scenery | 0 | 0 |
| 8 | New Zealand | scenery | 0 | 1 |

In this data set, Italy is the first observation with the value `architecture`. For that observation, the value of FIRST.TOURTYPE is 1. Italy is not the last observation with the value `architecture`, so the value of LAST.TOURTYPE is 0. The observations for Spain and Brazil are neither the first nor the last with the value `architecture`. Both FIRST.TOURTYPE and LAST.TOURTYPE are 0 for them. Japan is the last observation with the value `architecture`. The value of LAST.TOURTYPE is 1. The same rules apply to observations in the `scenery` group.

Now you are ready to use FIRST.TOURTYPE in a subsetting IF statement. When the data is sorted by TourType and LandCost, selecting the first observation in each type of tour gives you the lowest price of any tour in that category:

```
proc sort data=mylib.arch_or_scen out=tourorder4;
    by TourType LandCost;
run;

data lowcost;
    set tourorder4;
    by TourType;
    if first.TourType;
run;

proc print data=lowcost;
    title 'Least Expensive Tour for Each Type of Tour';
run;
```

The following output displays the results.

*Figure 12.7* Selecting One Observation from Each BY Group

Least Expensive Tour for Each Type of Tour

| Obs | Country | TourType | Nights | LandCost | Vendor |
|---|---|---|---|---|---|
| 1 | Italy | architecture | 8 | 936 | Express |
| 2 | Ireland | scenery | 7 | 1116 | Express |

# Working with Sorted Data

## Understanding Sorted Data

By default, groups appear in ascending order of the BY values. In some cases you want to emphasize the order in which the observations are sorted, not the fact that they can be grouped. For example, you might want to alphabetize the tours by country.

To sort your data in a particular order, use the SORT procedure just as you do for grouped data. When the sorted order is more important than the grouping, you usually want only one observation with a given BY value in the resulting data set. Therefore, you might need to remove duplicate observations.

*Operating Environment Information*
The SORT procedure accesses either a sorting utility that is supplied as part of SAS, or a sorting utility that is supplied by the host operating environment. All examples in this documentation use the SAS sorting utility. Some operating environment utilities do not accept particular options, including the NODUPRECS option described later in this section. The default sorting utility is set by your site. For more information about the utilities available to you, see the documentation for your operating environment.

## Sorting Data

The following example sorts data set MYLIB.ARCH_OR_SCEN by Country:

```
proc sort data=mylib.arch_or_scen out=bycountry;
   by Country;
run;

proc print data=bycountry;
   title 'Tours in Alphabetical Order by Country';
run;
```

The following output displays the results.

*Figure 12.8* Sorting Data

Tours in Alphabetical Order by Country

| Obs | Country | TourType | Nights | LandCost | Vendor |
|---|---|---|---|---|---|
| 1 | Brazil | architecture | 8 | 1150 | World |
| 2 | Greece | scenery | 12 | 1396 | Express |
| 3 | Ireland | scenery | 7 | 1116 | Express |
| 4 | Italy | architecture | 8 | 936 | Express |
| 5 | Japan | architecture | 8 | 1440 | Express |
| 6 | New Zealand | scenery | 16 | 2978 | Southsea |
| 7 | Spain | architecture | 10 | 1020 | World |
| 8 | Switzerland | scenery | 9 | 1468 | World |

## Deleting Duplicate Observations

You can eliminate duplicate observations in a SAS data set by using the NODUPRECS option with the SORT procedure. The following programs show you how to create a SAS data set and then remove duplicate observations.

The external file shown below contains a duplicate observation for Switzerland:

```
Spain        architecture  10 1020 World
Japan        architecture   8 1440 Express
Switzerland  scenery        9 1468 World
Brazil       architecture   8 1150 World
Switzerland  scenery        9 1468 World
Ireland      scenery        7 1116 Express
New Zealand  scenery       16 2978 Southsea
Italy        architecture   8  936 Express
Greece       scenery       12 1396 Express
```

The following DATA step creates a permanent SAS data set named MYLIB.ARCH_OR_SCEN2.

```
libname mylib 'SAS-library';

data mylib.arch_or_scen2;
   infile 'input-file';
   input Country $ 1-11 TourType $ 13-24 Nights LandCost Vendor $;
run;

proc print data=mylib.arch_or_scen2;
   title 'Data Set MYLIB.ARCH_OR_SCEN2';
run;
```

The following output shows that this data set contains a duplicate observation for Switzerland.

*Figure 12.9* Data Set MYLIB.ARCH_OR_SCEN2

Data Set MYLIB.ARCH_OR_SCEN2

| Obs | Country | TourType | Nights | LandCost | Vendor |
|---|---|---|---|---|---|
| 1 | Spain | architecture | 10 | 1020 | World |
| 2 | Japan | architecture | 8 | 1440 | Express |
| 3 | Switzerland | scenery | 9 | 1468 | World |
| 4 | Brazil | architecture | 8 | 1150 | World |
| 5 | Switzerland | scenery | 9 | 1468 | World |
| 6 | Ireland | scenery | 7 | 1116 | Express |
| 7 | New Zealand | scenery | 16 | 2978 | Southsea |
| 8 | Italy | architecture | 8 | 936 | Express |
| 9 | Greece | scenery | 12 | 1396 | Express |

The following program uses the NODUPRECS option in the SORT procedure to delete duplicate observations. The program creates a new data set called FIXED:

```
proc sort data=mylib.arch_or_scen2 out=fixed noduprecs;
    by Country;
run;

proc print data=fixed;
    title 'Data Set FIXED: MYLIB.ARCH_OR_SCEN2 With Duplicates Removed';
run;
```

The following output displays messages that appear in the SAS log.

*Log 12.2* *SAS Log Indicating Deleted Duplicate Observations*

```
697  proc sort data=mylib.arch_or_scen2 out=fixed noduprecs;
698     by Country;
699  run;

NOTE: There were 9 observations read from the data set MYLIB.ARCH_OR_SCEN2.
NOTE: 1 duplicate observations were deleted.
NOTE: The data set WORK.FIXED has 8 observations and 5 variables.
NOTE: PROCEDURE SORT used (Total process time):
      real time           0.01 seconds
      cpu time            0.01 seconds


700
701  proc print data=fixed;
702     title 'Data Set FIXED: MYLIB.ARCH_OR_SCEN2 With Duplicates Removed';
703  run;

NOTE: There were 8 observations read from the data set WORK.FIXED.
NOTE: PROCEDURE PRINT used (Total process time):
      real time           0.03 seconds
      cpu time            0.03 seconds
```

The following output shows the results of the NODUPRECS option.

*Figure 12.10* *Data Set FIXED with No Duplicate Observations*

Data Set FIXED: MYLIB.ARCH_OR_SCEN2 With Duplicates Removed

| Obs | Country | TourType | Nights | LandCost | Vendor |
|---|---|---|---|---|---|
| 1 | Brazil | architecture | 8 | 1150 | World |
| 2 | Greece | scenery | 12 | 1396 | Express |
| 3 | Ireland | scenery | 7 | 1116 | Express |
| 4 | Italy | architecture | 8 | 936 | Express |
| 5 | Japan | architecture | 8 | 1440 | Express |
| 6 | New Zealand | scenery | 16 | 2978 | Southsea |
| 7 | Spain | architecture | 10 | 1020 | World |
| 8 | Switzerland | scenery | 9 | 1468 | World |

## Understanding Collating Sequences

Both numeric and character variables can be sorted into ascending or descending order. For numeric variables, ascending or descending order is easy to understand. For character variables, ascending or descending order is more complex. Character values include uppercase and lowercase letters, special characters, and the digits 0 through 9 when they are treated as characters rather than as numbers.

The order in which character values are sorted is called a collating sequence. By default, SAS sorts characters in one of two sequences: EBCDIC or ASCII, depending on the operating environment under which SAS is running. For reference, both sequences are displayed here.

As long as you work under a single operating environment, you seldom need to think about the details of collating sequences. However, when you transfer files from an operating environment that uses EBCDIC to an operating environment that uses ASCII or vice versa, character values that are sorted in one operating environment are not necessarily in the correct order for the other operating environment. The simplest solution to the problem is to sort the character data (not numeric data) again in the destination operating environment. For detailed information about collating sequences, see the documentation for your operating environment.

## *ASCII Collating Sequence*

The following operating systems use the ASCII collating sequence:

- UNIX and its derivatives
- Windows

Here is the English-language ASCII sequence from the smallest to the largest character that you can display:

- blank!"#$%&'()*+,− ./0123456789:;<=>?@
- ABCDEFGHIJKLMNOPQRSTUVWXYZ [ \ ] ° _
- abcdefghijklmnopqrstuvwxyz{}~

The main features of the ASCII sequence are that digits are smaller than uppercase letters and uppercase letters are smaller than lowercase letters. The blank is the smallest character that you can display, followed by the other types of characters:

blank < digits < uppercase letters < lowercase letters

## *EBCDIC Collating Sequence*

The z/OS operating system uses the EBCDIC collating sequence.

Here is the English-language EBCDIC sequence from the smallest to largest character that you can display:

- blank.<(+|&!$*);¬ − /,%_>?:#@'="
- abcdefghijklmnopqr~stuvwxyz
- {ABCDEFGHI}JKLMNOPQR\ STUVWXYZ
- 0123456789

The main features of the EBCDIC sequence are that lowercase letters are smaller than uppercase letters and uppercase letters are smaller than digits. The blank is the smallest character that you can display, followed by the other types of characters:

blank < lowercase letters < uppercase letters < digits

# Summary

## Procedures

PROC SORT <DATA=*SAS-data-set*> <OUT=*SAS-data-set*> <NODUPRECS>;
   sorts a SAS data set by the values of variables that are listed in the BY statement. If you specify the OUT= option, the sorted data is stored in a different SAS data set than the input data. The NODUPRECS option tells PROC SORT to eliminate identical observations.

## Statements

BY <DESCENDING> *variable-1* < . . . <DESCENDING> *variable-n*>;
   in a DATA step, causes SAS to create FIRST. and LAST. variables for each variable named in the BY statement. The value of FIRST.*variable-1* is 1 for the first observation with a given BY value, and 0 for other observations. Similarly, the value of LAST.*variable-1* is 1 for the last observation for a given BY value, and 0 for other observations. The BY statement can follow a SET, MERGE, MODIFY, or UPDATE statement in the DATA step. It cannot be used with an INPUT statement. By default, SAS assumes that the data that is being read with a BY statement is in ascending order of the BY values. The DESCENDING option indicates that values of the variable that follow are in the opposite order, that is, largest to smallest.

# Learning More

Alternative to sorting observations
   You can use an alternative method to sort observations by creating an index that identifies the observations with particular values of a variable. For more information, see "SAS Data Files" in *SAS Language Reference: Concepts*.

BY statement and BY-group processing
   For more information, see "BY Statement" in *SAS Statements: Reference* and "BY-Group Processing in the DATA Step" in *SAS Language Reference: Concepts*.

Interleaving, merging, and updating SAS data sets
   For more information, see Chapter 18, "Interleaving SAS Data Sets," on page 281, Chapter 19, "Merging SAS Data Sets," on page 289, and Chapter 20, "Updating SAS Data Sets," on page 317.

   These operations depend on the BY statement in the DATA step. Interleaving combines data sets in sorted order. Match-merging joins observations that are identified by the value of a BY variable. Updating uses a data set that contains transactions to change values in a master file.

   For more information about SAS statements, see "MERGE Statement" in *SAS Statements: Reference* and "UPDATE Statement" in *SAS Statements: Reference*.

NOTSORTED option
   The NOTSORTED option can be used in both DATA and PROC steps, except for the SORT procedure. The NOTSORTED option is useful when data is grouped

according to the values of a variable, but the groups are not in ascending or descending order. Using the NOTSORTED option in the BY statement enables SAS to process them. For more information, see Chapter 32, "Writing Lines to the SAS Log or to an Output File," on page 595.

SORT procedure

For more information about the SORT procedure and the role of the BY statement, see "SORT" in *Base SAS Procedures Guide*. For information about the NOTSORTED option in the BY statement, see "BY Statement" in *SAS Statements: Reference*.

The following items refer to sorting SAS data sets:

- When you work with large data sets, plan your work so that you sort the data set as few times as possible. For example, if you need to sort a data set by State at the beginning of a program, and by City within State later, then sort the data set by State and City at the beginning of the program.

- To eliminate observations whose BY values duplicate BY values in other observations (but not necessarily values of other variables), use the NODUPKEY option in the SORT procedure.

- SAS can sort data in sequences other than English-language EBCDIC or ASCII. Examples include the Danish-Norwegian and Finnish or Swedish sequences.

The SAS documentation for your operating environment presents operating-environment-specific information about the SORT procedure. In general, many considerations about sorting data depend on the operating environment and other local conditions at your site, such as whether various operating environment utilities are available.

# Chapter 13
# Using More Than One Observation in a Calculation

Introduction to Using More Than One Observation in a Calculation . . . . . . . . . . 199
    Purpose . . . . . . . . . . . . . . . . . . . . . . . . . . . . . . . . . . . . . . . . . . . . . . . . . . . . . . . . . . 199
    Prerequisites . . . . . . . . . . . . . . . . . . . . . . . . . . . . . . . . . . . . . . . . . . . . . . . . . . . . . . 200

Input File and SAS Data Set for Examples . . . . . . . . . . . . . . . . . . . . . . . . . . . . . . 200

Accumulating a Total for an Entire Data Set . . . . . . . . . . . . . . . . . . . . . . . . . . . . . 201
    Creating a Running Total . . . . . . . . . . . . . . . . . . . . . . . . . . . . . . . . . . . . . . . . . . . . 201
    Writing Only the Total . . . . . . . . . . . . . . . . . . . . . . . . . . . . . . . . . . . . . . . . . . . . . . 203

Obtaining a Total for Each BY Group . . . . . . . . . . . . . . . . . . . . . . . . . . . . . . . . . . 204

Writing to Separate Data Sets . . . . . . . . . . . . . . . . . . . . . . . . . . . . . . . . . . . . . . . . . 206
    Writing Observations to Separate Data Sets . . . . . . . . . . . . . . . . . . . . . . . . . . . . . 206
    Writing Totals to Separate Data Sets . . . . . . . . . . . . . . . . . . . . . . . . . . . . . . . . . . . 207
    The Program . . . . . . . . . . . . . . . . . . . . . . . . . . . . . . . . . . . . . . . . . . . . . . . . . . . . . 207

Using a Value in a Later Observation . . . . . . . . . . . . . . . . . . . . . . . . . . . . . . . . . . 209

Summary . . . . . . . . . . . . . . . . . . . . . . . . . . . . . . . . . . . . . . . . . . . . . . . . . . . . . . . . . . 212
    Statements . . . . . . . . . . . . . . . . . . . . . . . . . . . . . . . . . . . . . . . . . . . . . . . . . . . . . . . 212

Learning More . . . . . . . . . . . . . . . . . . . . . . . . . . . . . . . . . . . . . . . . . . . . . . . . . . . . . 213

## Introduction to Using More Than One Observation in a Calculation

### Purpose

In this section, you will learn about calculations that require more than one observation. The following types of examples are included in this section:

- accumulating a total across a data set or a BY group
- saving a value from one observation in order to compare it to a value in a later observation

*Prerequisites*

Before proceeding with this section, you should understand the concepts presented in the following sections:

- Chapter 7, "Understanding DATA Step Processing," on page 107
- Chapter 12, "Working with Grouped or Sorted Observations," on page 183

## Input File and SAS Data Set for Examples

Tradewinds Travel needs to know how much business the company did with various tour vendors during the peak season. The data that the company wants to look at is the total number of people that are scheduled on tours with various vendors. It also wants to look at the total value of the tours that are scheduled.

The following external file contains data about Tradewinds Travel tours:

```
1              2    3        4
---------------------------
Germany      1150 Express   10
Spain        1020 World     12
Brazil       1080 World      6
India         978 Express    .
Japan        1440 Express   10
Greece       1396 Express   20
New Zealand  2978 Southsea   6
Venezuela     850 World      8
Italy         936 Express    9
Russia       1848 World      6
Switzerland  1468 World     20
Australia    2158 Southsea  10
Ireland      1116 Express    9
```

The following list describes the fields in the input file:

1   provides the name of the destination country for the tour

2   specifies the cost of the land package in U.S. dollars

3   specifies the name of the trip vendor

4   specifies the number of people that were booked on that tour

The first step is to create a permanent SAS data set. The following program creates the data set MYLIB.TOURREVENUE:

```
libname mylib 'SAS-library';

data mylib.tourrevenue;
   infile 'input-file' truncover;
   input Country $ 1-11 LandCost Vendor $ NumberOfBookings;
run;

proc print data=mylib.tourrevenue;
```

```
        title 'SAS Data Set MYLIB.TOURREVENUE';
run;
```

The PROC PRINT statement that follows the DATA step produces this display of the MYLIB.TOURREVENUE data set.

*Figure 13.1   Data Set MYLIB.TOURREVENUE*

SAS Data Set MYLIB.TOURREVENUE

| Obs | Country | LandCost | Vendor | NumberOfBookings |
|---|---|---|---|---|
| 1 | Germany | 1150 | Express | 10 |
| 2 | Spain | 1020 | World | 12 |
| 3 | Brazil | 1080 | World | 6 |
| 4 | India | 978 | Express | . |
| 5 | Japan | 1440 | Express | 10 |
| 6 | Greece | 1396 | Express | 20 |
| 7 | New Zealand | 2978 | Southsea | 6 |
| 8 | Venezuela | 850 | World | 8 |
| 9 | Italy | 936 | Express | 9 |
| 10 | Russia | 1848 | World | 6 |
| 11 | Switzerland | 1468 | World | 20 |
| 12 | Australia | 2158 | Southsea | 10 |
| 13 | Ireland | 1116 | Express | 9 |

Each observation in the data set MYLIB.TOURREVENUE contains the cost of a tour and the number of people who booked that tour. The tasks of Tradewinds Travel are as follows:

- to determine how much money was spent with each vendor and with all vendors together
- to store the totals in a SAS data set that is separate from the individual vendors' records
- to find the tour that produced the most revenue, which is determined by the land cost times the number of people who booked the tour

# Accumulating a Total for an Entire Data Set

## Creating a Running Total

The first task in performing calculations on the data set MYLIB.TOURREVENUE is to find the total number of people who booked tours with Tradewinds Travel. Therefore, a variable is needed whose value starts at 0 and increases by the number of bookings in each observation. The Sum statement gives you that capability:

*variable* + *expression*

In a Sum statement, the value of *variable* on the left side of the plus sign is 0 before the statement is processed for the first time. Processing the statement adds the value of *expression* on the right side of the plus sign to the initial value. The Sum variable retains the new value until the next processing of the statement. The Sum statement ignores a missing value for the expression. The previous total remains unchanged.

The following statement creates the total number of bookings:

```
TotalBookings + NumberOfBookings;
```

The following DATA step includes the Sum statement shown above:

```
data total;
   set mylib.tourrevenue;
   TotalBookings + NumberOfBookings;
run;

proc print data=total;
   var Country NumberOfBookings TotalBookings;
   title 'Total Tours Booked';
run;
```

The following output displays the results.

**Figure 13.2** *Accumulating a Total for a Data Set*

Total Tours Booked

| Obs | Country | NumberOfBookings | TotalBookings |
|---|---|---|---|
| 1 | Germany | 10 | 10 |
| 2 | Spain | 12 | 22 |
| 3 | Brazil | 6 | 28 |
| 4 | India | . | 28 |
| 5 | Japan | 10 | 38 |
| 6 | Greece | 20 | 58 |
| 7 | New Zealand | 6 | 64 |
| 8 | Venezuela | 8 | 72 |
| 9 | Italy | 9 | 81 |
| 10 | Russia | 6 | 87 |
| 11 | Switzerland | 20 | 107 |
| 12 | Australia | 10 | 117 |
| 13 | Ireland | 9 | 126 |

The TotalBookings variable in the last observation of the TOTAL data set contains the total number of bookings for the year.

## Writing Only the Total

If the total is the only information that is needed from the data set, a data set that contains only one observation and one variable (the TotalBookings variable) can be created by writing a DATA step that performs all of the following functions:

- specifies the END= option in the SET statement to determine whether the current observation is the last observation
- uses a subsetting IF statement to write only the last observation to the SAS data set
- specifies the KEEP= option in the DATA step to keep only the variable that totals the bookings

When the END= option in the SET statement is specified, the variable that is named in the END= option is set to 1 when the DATA step is processing the last observation. The variable that is named in the END= option is set to 0 for other observations:

**SET** *SAS-data-set* <END=*variable*>;

SAS does not add the END= variable to the data set that is being created. By testing the value of the END= variable, you can determine which observation is the last observation.

The following program selects the last observation with a subsetting IF statement and uses a KEEP= data set option to keep only the variable TotalBookings in the data set:

```
data total2(keep=TotalBookings);
   set mylib.tourrevenue end=Lastobs;
   TotalBookings + NumberOfBookings;
   if Lastobs;
run;

proc print data=total2;
   title 'Total Number of Tours Booked';
run;
```

The following output displays the results.

*Figure 13.3* Selecting the Last Observation in a Data Set

Total Number of Tours Booked

| Obs | TotalBookings |
|---|---|
| 1 | 126 |

The condition in the subsetting IF statement is true when Lastobs has a value of 1. When SAS is processing the last observation from MYLIB.TOURREVENUE, it assigns to Lastobs the value 1. Therefore, the subsetting IF statement accepts only the last observation from MYLIB.TOURREVENUE, and SAS writes the last observation to the data set TOTAL2.

## Obtaining a Total for Each BY Group

An additional requirement of Tradewinds Travel is to determine the number of tours that are booked with each vendor. In order to accomplish this task, a program must group the data by a variable. That is, the program must organize the data set into groups of observations, with one group for each vendor. In this case, the program must group the data by the Vendor variable. Each group is known generically as a BY- group. The variable that is used to determine the groupings is called a BY variable.

In order to group the data by the Vendor variable:

- include a PROC SORT step to group the observations by the Vendor variable
- use a BY statement in the DATA step
- use a Sum statement to total the bookings
- reset the Sum variable to 0 at the beginning of each group of observations

The following program sorts the data set by Vendor and sums the total bookings for each vendor.

```
proc sort data=mylib.tourrevenue out=mylib.sorttour;
    by Vendor;
run;

data totalby;
   set mylib.sorttour;
   by Vendor;
   if First.Vendor then VendorBookings=0;
   VendorBookings + NumberOfBookings;
run;

proc print data=totalby;
   title 'Summary of Bookings by Vendor';
run;
```

In the preceding program, the FIRST.Vendor variable is used in an IF-THEN statement to set the Sum variable (VendorBookings) to 0 in the first observation of each BY group. (For more information about the FIRST.*variable* and LAST.*variable* temporary variables, see "Finding the First or Last Observation in a Group" on page 189.)

The following output displays the results.

*Figure 13.4* Creating Totals for BY Groups

Summary of Bookings by Vendor

| Obs | Country | LandCost | Vendor | NumberOfBookings | VendorBookings |
|---|---|---|---|---|---|
| 1 | Germany | 1150 | Express | 10 | 10 |
| 2 | India | 978 | Express | . | 10 |
| 3 | Japan | 1440 | Express | 10 | 20 |
| 4 | Greece | 1396 | Express | 20 | 40 |
| 5 | Italy | 936 | Express | 9 | 49 |
| 6 | Ireland | 1116 | Express | 9 | 58 |
| 7 | New Zealand | 2978 | Southsea | 6 | 6 |
| 8 | Australia | 2158 | Southsea | 10 | 16 |
| 9 | Spain | 1020 | World | 12 | 12 |
| 10 | Brazil | 1080 | World | 6 | 18 |
| 11 | Venezuela | 850 | World | 8 | 26 |
| 12 | Russia | 1848 | World | 6 | 32 |
| 13 | Switzerland | 1468 | World | 20 | 52 |

Notice that while this output does in fact include the total number of bookings for each vendor, it also includes a great deal of extraneous information. Reporting the total bookings for each vendor requires only the variables Vendor and VendorBookings from the last observation for each vendor. Therefore, the program can use the following elements:

- the DROP= or KEEP= data set options to eliminate the variables Country, LandCost, and NumberOfBookings from the output data set
- the LAST.Vendor variable in a subsetting IF statement to write only the last observation in each group to the data set TOTALBY

The following program creates data set TOTALBY:

```
proc sort data=mylib.tourrevenue out=mylib.sorttour;
   by Vendor;
run;

data totalby(drop=country landcost NumberOfBookings);
   set mylib.sorttour;
   by Vendor;
   if First.Vendor then VendorBookings=0;
   VendorBookings + NumberOfBookings;
   if Last.Vendor;
run;

proc print data=totalby;
```

```
         title 'Total Bookings by Vendor';
run;
```

The following output displays the results.

*Figure 13.5*  Putting Totals for Each BY Group in a New Data Set

```
       Total Bookings by Vendor

   Obs   Vendor     VendorBookings
    1    Express          58
    2    Southsea         16
    3    World            52
```

# Writing to Separate Data Sets

## Writing Observations to Separate Data Sets

Tradewinds Travel wants overall information about the tours that were conducted this year. One SAS data set is needed to contain detailed information about each tour, including the total amount that was spent on that tour. Another SAS data set is needed to contain the total number of bookings with each vendor and the total amount spent with that vendor. Both of these data sets can be created using the techniques that you have learned so far.

Begin the program by creating two SAS data sets from the SAS data set MYLIB.SORTTOUR using the following DATA and SET statements:

```
data tourdetails vendordetails;
   set mylib.sorttour;
```

The data set TOURDETAILS will contain the individual records, and VENDORDETAILS will contain the information about vendors. The observations do not need to be grouped for TOURDETAILS, but they need to be grouped by Vendor for VENDORDETAILS.

If the data is not already grouped by Vendor, first use the SORT procedure. Add a BY statement to the DATA step for use with VENDORDETAILS.

```
proc sort data=mylib.tourrevenue out=mylib.sorttour;
   by Vendor;
run;

data tourdetails vendordetails;
   set mylib.sorttour;
   by Vendor;
run;
```

The only calculation that is needed for the individual tours is the amount of money that was spent on each tour. Therefore, calculate the amount in an assignment statement and write the observation to TOURDETAILS.

```
Money=LandCost * NumberOfBookings;
output tourdetails;
```

The portion of the DATA step that builds TOURDETAILS is now complete.

## Writing Totals to Separate Data Sets

Because observations remain in the program data vector after an OUTPUT statement executes, you can continue using the observations in programming statements. The rest of the DATA step creates information for the VENDORDETAILS data set.

Use the FIRST.Vendor variable to determine when SAS is processing the first observation in each group.

Then set the Sum variables VendorBookings and VendorMoney to 0 in that observation. VendorBookings accumulates the total number the bookings for each vendor, and VendorMoney accumulates the total costs. Add the following statements to the DATA step:

```
if First.Vendor then
   do;
      VendorBookings=0;
      VendorMoney=0;
   end;
VendorBookings + NumberOfBookings;
VendorMoney + Money;
```

*Note:* The program uses a DO group. Using DO groups enables the program to evaluate a condition once and take more than one action as a result. For more information about DO groups, see "Performing More Than One Action in an IF-THEN Statement" on page 217.

The last observation in each BY–group contains the totals for that vendor. Therefore, use the following statement to output the last observation to the data set VENDORDETAILS:

```
if Last.Vendor then output vendordetails;
```

As a final step, use KEEP= and DROP= data set options to remove extraneous variables from the two data sets so that each data set has just the variables that are wanted.

```
data tourdetails(drop=VendorBookings VendorMoney)
     vendordetails(keep=Vendor VendorBookings VendorMoney);
```

## The Program

The following is the complete program that creates the VENDORDETAILS and TOURDETAILS data sets:

```
proc sort data=mylib.tourrevenue out=mylib.sorttour;
   by Vendor;
run;

data tourdetails(drop=VendorBookings VendorMoney)
     vendordetails(keep=Vendor VendorBookings VendorMoney);
   set mylib.sorttour;
   by Vendor;
   Money=LandCost * NumberOfBookings;
   output tourdetails;
```

```
            if First.Vendor then
               do;
                  VendorBookings=0;
                  VendorMoney=0;
               end;
            VendorBookings + NumberOfBookings;
            VendorMoney + Money;
            if Last.Vendor then output vendordetails;
         run;

         proc print data=tourdetails;
            title 'Detail Records: Dollars Spent on Individual Tours';
         run;

         proc print data=vendordetails;
            title 'Vendor Totals: Dollars Spent and Bookings by Vendor';
         run;
```

The following output displays detail tour records in one SAS data set and vendor totals in another.

*Figure 13.6* Detail Tour Records in the TOURDETAILS Data Set

Detail Records: Dollars Spent on Individual Tours

| Obs | Country | LandCost | Vendor | NumberOfBookings | Money |
|---|---|---|---|---|---|
| 1 | Germany | 1150 | Express | 10 | 11500 |
| 2 | India | 978 | Express | . | . |
| 3 | Japan | 1440 | Express | 10 | 14400 |
| 4 | Greece | 1396 | Express | 20 | 27920 |
| 5 | Italy | 936 | Express | 9 | 8424 |
| 6 | Ireland | 1116 | Express | 9 | 10044 |
| 7 | New Zealand | 2978 | Southsea | 6 | 17868 |
| 8 | Australia | 2158 | Southsea | 10 | 21580 |
| 9 | Spain | 1020 | World | 12 | 12240 |
| 10 | Brazil | 1080 | World | 6 | 6480 |
| 11 | Venezuela | 850 | World | 8 | 6800 |
| 12 | Russia | 1848 | World | 6 | 11088 |
| 13 | Switzerland | 1468 | World | 20 | 29360 |

**Figure 13.7** *Vendor Totals in the VENDORDETAILS Data Set*

Vendor Totals: Dollars Spent and Bookings by Vendor

| Obs | Vendor | VendorBookings | VendorMoney |
|---|---|---|---|
| 1 | Express | 58 | 72288 |
| 2 | Southsea | 16 | 39448 |
| 3 | World | 52 | 65968 |

# Using a Value in a Later Observation

A further requirement of Tradewinds Travel is a separate SAS data set that contains the tour that generated the most revenue. (The revenue total equals the price of the tour multiplied by the number of bookings.) One method of creating the new data set might be to follow these three steps:

1. Calculate the revenue in a DATA step.

2. Sort the data set in descending order by the revenue.

3. Use another DATA step with the OBS= data set option to write that observation.

A more efficient method compares the revenue from all observations in a single DATA step. SAS can retain a value from the current observation to use in future observations. When the processing of the DATA step reaches the next observation, the held value represents information from the previous observation.

The RETAIN statement causes a variable that is created in the DATA step to retain its value from the current observation into the next observation. The variable is not set to missing at the beginning of each iteration of the DATA step. RETAIN is a declarative statement, not an executable statement. This statement has the following form:

**RETAIN** *variable-1* <... *variable-n*>;

To compare the Revenue value in one observation to the Revenue value in the next observation, create a retained variable named HoldRevenue and assign the value of the current Revenue variable to it. In the next observation, the HoldRevenue variable contains the Revenue value from the previous observation, and its value can be compared to that of Revenue in the current observation.

To see how the RETAIN statement works, look at the next example. The following DATA step writes observations to data set TEMP before SAS assigns the current revenue to HoldRevenue:

```
data temp;
   set mylib.tourrevenue;
   retain HoldRevenue;
   Revenue=LandCost * NumberOfBookings;
   output;
   HoldRevenue=Revenue;
run;
```

**210** *Chapter 13* • *Using More Than One Observation in a Calculation*

```
proc print data=temp;
   var Country LandCost NumberOfBookings Revenue HoldRevenue;
   title 'Tour Revenue';
run;
```

The following output displays the results.

*Figure 13.8* Retaining a Value By Using the Retain Statement

Tour Revenue

| Obs | Country | LandCost | NumberOfBookings | Revenue | HoldRevenue |
|---|---|---|---|---|---|
| 1 | Germany | 1150 | 10 | 11500 | . |
| 2 | Spain | 1020 | 12 | 12240 | 11500 |
| 3 | Brazil | 1080 | 6 | 6480 | 12240 |
| 4 | India | 978 | . | . | 6480 |
| 5 | Japan | 1440 | 10 | 14400 | . |
| 6 | Greece | 1396 | 20 | 27920 | 14400 |
| 7 | New Zealand | 2978 | 6 | 17868 | 27920 |
| 8 | Venezuela | 850 | 8 | 6800 | 17868 |
| 9 | Italy | 936 | 9 | 8424 | 6800 |
| 10 | Russia | 1848 | 6 | 11088 | 8424 |
| 11 | Switzerland | 1468 | 20 | 29360 | 11088 |
| 12 | Australia | 2158 | 10 | 21580 | 29360 |
| 13 | Ireland | 1116 | 9 | 10044 | 21580 |

The value of HoldRevenue is missing at the beginning of the first observation. It is still missing when the OUTPUT statement writes the first observation to TEMP. After the OUTPUT statement, an assignment statement assigns the value of Revenue to HoldRevenue. Because HoldRevenue is retained, that value is present at the beginning of the next iteration of the DATA step. When the OUTPUT statement executes again, the value of HoldRevenue still contains that value.

To find the largest value of Revenue, assign the value of Revenue to HoldRevenue only when Revenue is larger than HoldRevenue, as shown in the following program:

```
data mostrevenue;
   set mylib.tourrevenue;
   retain HoldRevenue;
   Revenue=LandCost * NumberOfBookings;
   if Revenue > HoldRevenue then HoldRevenue=Revenue;
run;

proc print data=mostrevenue;
   var Country LandCost NumberOfBookings Revenue HoldRevenue;
   title 'Tour Revenue';
run;
```

The following output displays the results.

*Figure 13.9* Holding the Largest Value in a Retained Variable

Tour Revenue

| Obs | Country | LandCost | NumberOfBookings | Revenue | HoldRevenue |
|---|---|---|---|---|---|
| 1 | Germany | 1150 | 10 | 11500 | 11500 |
| 2 | Spain | 1020 | 12 | 12240 | 12240 |
| 3 | Brazil | 1080 | 6 | 6480 | 12240 |
| 4 | India | 978 | . | . | 12240 |
| 5 | Japan | 1440 | 10 | 14400 | 14400 |
| 6 | Greece | 1396 | 20 | 27920 | 27920 |
| 7 | New Zealand | 2978 | 6 | 17868 | 27920 |
| 8 | Venezuela | 850 | 8 | 6800 | 27920 |
| 9 | Italy | 936 | 9 | 8424 | 27920 |
| 10 | Russia | 1848 | 6 | 11088 | 27920 |
| 11 | Switzerland | 1468 | 20 | 29360 | 29360 |
| 12 | Australia | 2158 | 10 | 21580 | 29360 |
| 13 | Ireland | 1116 | 9 | 10044 | 29360 |

The value of HoldRevenue in the last observation represents the largest revenue that is generated by any tour. To determine which observation the value came from, create a variable named HoldCountry to hold the name of the country from the observation with the largest revenue. Include HoldCountry in the RETAIN statement to retain its value until explicitly changed. Then use the END= data set option to select the last observation, and use the KEEP= data set option to keep only HoldRevenue and HoldCountry in MOSTREVENUE:

```
data mostrevenue (keep=HoldCountry HoldRevenue);
   set mylib.tourrevenue  end=LastOne;
   retain HoldRevenue HoldCountry;
   Revenue=LandCost * NumberOfBookings;
   if Revenue > HoldRevenue then
      do;
         HoldRevenue=Revenue;
         HoldCountry=Country;
      end;
   if LastOne;
run;
proc print data=mostrevenue;
   title 'Country with the Largest Value of Revenue';
run;
```

*Note:* The program uses a DO group. Using DO groups enables the program to evaluate a condition once and take more than one action as a result. For more information about DO groups, see "Performing More Than One Action in an IF-THEN Statement" on page 217.

The following output displays the results.

*Figure 13.10* *Using the RETAIN and Subsetting IF Statements to Find the Most Revenue*

Country with the Largest Value of Revenue

| Obs | HoldRevenue | HoldCountry |
|---|---|---|
| 1 | 29360 | Switzerland |

# Summary

## *Statements*

RETAIN *variable-1* < . . . *variable-n*>;
: retains the value of *variable* for use in a subsequent observation. The RETAIN statement prevents the value of the variable from being reinitialized to missing when control returns to the top of the DATA step.

  The RETAIN statement affects variables that are created in the current DATA step (for example, variables that are created with an INPUT or assignment statement). Variables that are read with a SET, MERGE, or UPDATE statement are retained automatically. Naming them in a RETAIN statement has no effect.

  The RETAIN statement can assign an initial value to a variable. If a variable needs to have the same value in all observations of a DATA step, then it is more efficient to put the value in a RETAIN statement rather than in an assignment statement. SAS assigns the value in the RETAIN statement when it is compiling the DATA step, but it executes the assignment statement during each execution of the DATA step.

SET *SAS-data-set* <END=*variable*>;
: reads from the specified *SAS-data-set*. The *variable* that is specified in the END= option has the value 0 until SAS is processing the last observation in the data set. Then the variable has the value 1. SAS does not include the END= variable in the data set that is being created.

*variable* + *expression*;
: is called a Sum statement. It adds the result of *expression* on the right side of the plus sign to *variable* on the left side of the plus sign, and holds the new value of *variable* for use in subsequent observations. The expression can be a numeric variable or expression. The value of *variable* is retained. If the expression is a missing value, *variable* maintains its previous value. Before the Sum statement is executed for the first time, the default value of *variable* is 0.

  The plus sign is required in the Sum statement. To subtract successive values from a starting value, add negative values to the Sum variable.

# Learning More

Automatic variable _N_
: provides a way to count the number of times SAS executes a DATA step. For more information, see Chapter 32, "Writing Lines to the SAS Log or to an Output File," on page 595.

    SAS creates _N_ in each DATA step. The first time SAS begins to execute the DATA step, the value of _N_ is 1. The second time SAS begins to execute, the value is 2, and so on. SAS does not add _N_ to the output data set. Using _N_ is more efficient than using a Sum statement.

DO groups
: specifies a group of statements that are executed as a unit. For information about DO group processing, see Chapter 14, "Finding Shortcuts in Programming," on page 215. For more information about the DO statement, see "DO Statement" in *SAS Statements: Reference*.

END= option
: can be used in a SET statement. For an example, see Chapter 22, "Conditionally Processing Observations from Multiple SAS Data Sets," on page 349.

KEEP= and DROP= data set options
: specify which observations should be kept or dropped from a data set. These options can be used on an input or output data set. For more information, see Chapter 6, "Starting with SAS Data Sets," on page 91. See also "DROP= Data Set Option" in *SAS Data Set Options: Reference*, and "KEEP= Data Set Option" in *SAS Data Set Options: Reference*.

LAG family of functions
: provide another way to retain a value from one observation for use in a subsequent observation. LAG functions can retain a value for up to 100 observations. For more information, see "LAG Function" in *SAS Functions and CALL Routines: Reference*.

RETAIN, Sum, and SET statements
: For more information, see "RETAIN Statement" in *SAS Statements: Reference*, "Sum Statement" in *SAS Statements: Reference*, and "SET Statement" in *SAS Statements: Reference*.

Sum and SUMBY statements
: can be used in the PRINT procedure if the only purpose in getting a total is to display it in a report. The Sum and SUMBY statements in the PRINT procedure are discussed in Chapter 27, "Producing Detail Reports with the PRINT Procedure," on page 433.

SUMMARY and MEANS procedures
: can also be used to compute totals. For more information, see "SUMMARY" in *Base SAS Procedures Guide* and "MEANS" in *Base SAS Procedures Guide*.

# Chapter 14
# Finding Shortcuts in Programming

Introduction to Shortcuts .................................................. 215
    Purpose ............................................................. 215
    Prerequisites ....................................................... 215

Input File and SAS Data Set ............................................... 216

Performing More Than One Action in an IF-THEN Statement .................... 217

Performing the Same Action for a Series of Variables ...................... 219
    Using a Series of IF-THEN Statements ............................... 219
    Grouping Variables into Arrays ..................................... 219
    Repeating the Action ............................................... 220
    Selecting the Current Variable ..................................... 220

Summary ................................................................... 222
    Statements ......................................................... 222

Learning More ............................................................. 223

## Introduction to Shortcuts

### Purpose

In this section you will learn two DATA step programming techniques that make the code easier to write and read:

- using a DO group to perform more than one action after evaluating an IF condition
- using arrays to perform the same action on more than one variable with a single group of statements

### Prerequisites

Before proceeding with this section, you should understand the topics presented in the following sections:

- Chapter 7, "Understanding DATA Step Processing," on page 107
- Chapter 10, "Acting on Selected Observations," on page 147

## Input File and SAS Data Set

In the following example, Tradewinds Travel is making adjustments to their data about tours to art museums and galleries. The data for the tours is as follows:

```
1             2 3 4 5       6
-------------------------
Rome          4 3 . D'Amico 2
Paris         5 . 1 Lucas   5
London        3 2 . Wilson  3
New York      5 1 2 Lucas   5
Madrid        . . 5 Torres  4
Amsterdam 3 3 .             .
```

The following list explains the numbered items in the preceding file:

1 This column provides the name of the destination country.

2 This column provides the number of museums to be visited.

3 This column provides the number of art galleries in the tour.

4 This column provides the number of other attractions to be toured.

5 This column lists the last name of the tour guide.

6 This column lists the number of years of experience for the tour guide.

The following program creates the permanent SAS data set MYLIB.ATTRACTIONS:

```
libname mylib 'permanent-data-library';

data mylib.attractions;
   infile 'input-file';
   input City $ 1-9 Museums 11 Galleries 13
         Other 15 TourGuide $ 17-24 YearsExperience 26;
run;

proc print data=mylib.attractions;
   title 'Data Set MYLIB.ATTRACTIONS';
run;
```

The following output shows the results.

*Figure 14.1* Data Set MYLIB.ATTRACTIONS

### Data Set MYLIB.ATTRACTIONS

| Obs | City | Museums | Galleries | Other | TourGuide | YearsExperience |
|---|---|---|---|---|---|---|
| 1 | Rome | 4 | 3 | . | D'Amico | 2 |
| 2 | Paris | 5 | . | 1 | Lucas | 5 |
| 3 | London | 3 | 2 | . | Wilson | 3 |
| 4 | New York | 5 | 1 | 2 | Lucas | 5 |
| 5 | Madrid | . | . | 5 | Torres | 4 |
| 6 | Amsterdam | 3 | 3 | . | | . |

# Performing More Than One Action in an IF-THEN Statement

Several changes are needed in the observations for Madrid and Amsterdam. One way to select those observations is to evaluate an IF condition in a series of IF-THEN statements, as follows:

```
/* multiple actions based on the same condition */
data updatedattractions;
   set mylib.attractions;
   if City = 'Madrid' then Museums = 3;
   if City = 'Madrid' then Other = 2;
   if City = 'Amsterdam' then TourGuide = 'Vandever';
   if City = 'Amsterdam' then YearsExperience = 4;
run;
```

To avoid writing the IF condition twice for each city, use a DO group in the THEN clause, for example:

IF *condition* THEN

**DO**;

...*more SAS statements*...

**END**;

The DO statement causes all statements following it to be treated as a unit until a matching END statement appears. A group of SAS statements that begin with DO and end with END is called a DO group.

The following DATA step replaces the multiple IF-THEN statements with DO groups:

```
/* a more efficient method */
data updatedattractions2;
   set mylib.attractions;
```

```
            if City = 'Madrid' then
               do;
                  Museums = 3;
                  Other = 2;
               end;
            else if City = 'Amsterdam' then
               do;
                  TourGuide = 'Vandever';
                  YearsExperience = 4;
               end;
      run;

      proc print data=updatedattractions2;
         title 'Data Set MYLIB.UPDATEDATTRACTIONS';
      run;
```

The following output displays the results.

*Figure 14.2* Using DO Groups to Produce a Data Set

### Data Set MYLIB.UPDATEDATTRACTIONS

| Obs | City | Museums | Galleries | Other | TourGuide | YearsExperience |
|---|---|---|---|---|---|---|
| 1 | Rome | 4 | 3 | | D'Amico | 2 |
| 2 | Paris | 5 | | 1 | Lucas | 5 |
| 3 | London | 3 | 2 | | Wilson | 3 |
| 4 | New York | 5 | 1 | 2 | Lucas | 5 |
| 5 | Madrid | 3 | | 2 | Torres | 4 |
| 6 | Amsterdam | 3 | 3 | | Vandever | 4 |

Using DO groups makes the program faster to write and easier to read. It also makes the program more efficient for SAS in two ways:

1. The IF condition is evaluated fewer times. (Although there are more statements in this DATA step than in the preceding one, the DO, and END statements require very few computer resources.)

2. The conditions `City = 'Madrid'` and `City = 'Amsterdam'` are mutually exclusive, as condensing the multiple IF-THEN statements into two statements shows. You can make the second IF-THEN statement part of an ELSE statement. Therefore, the second IF condition is not evaluated when the first IF condition is true.

# Performing the Same Action for a Series of Variables

### Using a Series of IF-THEN Statements

In the data set MYLIB.ATTRACTIONS, the variables Museums, Galleries, and Other contain missing values when the tour does not feature that type of attraction. To change the missing values to 0, you can write a series of IF-THEN statements with assignment statements, as the following program illustrates:

```
   /* same action for different variables */
data changes;
   set mylib.attractions;
   if Museums = . then Museums = 0;
   if Galleries = . then Galleries = 0;
   if Other = . then Other = 0;
run;
```

The pattern of action is the same in the three IF-THEN statements. Only the variable name is different. To make the program easier to read, you can write SAS statements that perform the same action several times, changing only the variable that is affected. This technique is called array processing, and consists of the following three steps:

1. grouping variables into arrays
2. repeating the action
3. selecting the current variable to be acted upon

### Grouping Variables into Arrays

In DATA step programming, you can put variables into a temporary group called an array. To define an array, use an ARRAY statement. A simple ARRAY statement has the following form:

**ARRAY** *array-name*{*number-of-variables*} *variable-1* <. . . *variable-n*>;

*Array-name* is a SAS name that you choose to identify the group of variables. *Number-of-variables*, enclosed in braces, tells SAS how many variables you are grouping, and *variable-1*<. . . *variable-n*> lists their names.

*Note:* If you have worked with arrays in other programming languages, note that arrays in SAS are different from those in many other languages. In SAS, an array is simply a convenient way of temporarily identifying a group of variables by assigning an alias to them. It is not a permanent data structure. It exists only for the duration of the DATA step. The *array-name* identifies the array and distinguishes it from any other arrays in the same DATA step. It is not a variable.

The following ARRAY statement lists the three variables Museums, Galleries, and Other:

```
array changelist{3} Museums Galleries Other;
```

This statement tells SAS to do the following:

- make a group named CHANGLIST for the duration of this DATA step

- put three variable names in CHANGELIST: Museums, Galleries, and Other

In addition, by listing a variable in an ARRAY statement, you assign the variable an extra name with the form *array-name {position}*, where *position* is the position of the variable in the list (1, 2, or 3 in this case). The position can be a number, or the name of a variable whose value is the number. This additional name is called an array reference, and the position is called the subscript. The previous ARRAY statement assigns CHANGELIST{1} to Museums, CHANGELIST{2} to Galleries, and CHANGELIST{3} to Other. From that point in the DATA step, you can refer to the variable by either its original name or by its array reference. For example, the names Museums and CHANGELIST{1} are equivalent.

## *Repeating the Action*

To tell SAS to perform the same action several times, use an iterative DO loop of the following form:

**DO** *index-variable*=1 TO *number-of-variables-in-array*;

...*SAS statements*...

**END**;

An iterative DO loop begins with an iterative DO statement, contains other SAS statements, and ends with an END statement. The loop is processed repeatedly (iterated) according to the directions in the iterative DO statement. The iterative DO statement contains an *index-variable* whose name you choose and whose value changes in each iteration of the loop. In array processing, you usually want the loop to execute as many times as there are variables in the array. Therefore, you specify that the values of *index-variable* are 1 to *number-of-variables-in-array*. By default, SAS increases the value of *index-variable* by 1 before each new iteration of the loop. When the value becomes greater than *number-of-variables-in-array*, SAS stops processing the loop. By default, SAS adds the index variable to the data set that is being created.

An iterative DO loop that processes three times and has an index variable named Count looks like this:

```
do Count = 1 to 3;
   ...SAS statements...
end;
```

The first time the loop is processed, the value of Count is 1; the second time, the value is 2; and the third time, the value is 3. At the beginning of the fourth execution, the value of Count is 4, exceeding the specified range of 1 to 3. SAS stops processing the loop.

## *Selecting the Current Variable*

Now that you have grouped the variables and you know how many times the loop is processed, you must tell SAS which variable in the array to use in each iteration of the loop. Recall that variables in an array can be identified by their array references, and that the subscript of the reference can be a variable name as well as a number. Therefore, you can write programming statements in which the index variable of the DO loop is the subscript of the array reference:

*array-name {index-variable}*

When the value of the index variable changes, the subscript of the array reference (and, therefore, the variable that is referenced) also changes.

The following statement uses the index variable Count as the subscript of array references:

```
if changelist{Count} = . then changelist{Count} = 0;
```

You can place this statement inside an iterative DO loop. When the value of Count is 1, SAS reads the array reference as CHANGELIST{1} and processes the IF-THEN statement on CHANGELIST{1}, that is, Museums. When Count has the value 2 or 3, SAS processes the statement on CHANGELIST{2}, Galleries, or CHANGELIST{3}, Other. The complete iterative DO loop with array references looks like this:

```
do Count = 1 to 3;
   if changelist{Count} = . then changelist{Count} = 0;
end;
```

These statements tell SAS to do the following processing:

- perform the actions in the loop three times
- replace the array subscript Count with the current value of Count for each iteration of the IF-THEN statement
- locate the variable with that array reference and process the IF-THEN statement on that variable

The following DATA step uses the ARRAY statement and iterative DO loop:

```
libname mylib 'permanent-data-library';

data changes;
   set mylib.attractions;
   array changelist{3} Museums Galleries Other;
   do Count = 1 to 3;
      if changelist{Count} = . then changelist{Count} = 0;
   end;
run;

proc print data=changes;
   title 'Tour Attractions';
run;
```

The following output displays the results.

*Figure 14.3* Using an Array and an Iterative DO Loop to Produce a Data Set

## Tour Attractions

| Obs | City | Museums | Galleries | Other | TourGuide | YearsExperience | Count |
|---|---|---|---|---|---|---|---|
| 1 | Rome | 4 | 3 | 0 | D'Amico | 2 | 4 |
| 2 | Paris | 5 | 0 | 1 | Lucas | 5 | 4 |
| 3 | London | 3 | 2 | 0 | Wilson | 3 | 4 |
| 4 | New York | 5 | 1 | 2 | Lucas | 5 | 4 |
| 5 | Madrid | 0 | 0 | 5 | Torres | 4 | 4 |
| 6 | Amsterdam | 3 | 3 | 0 | | | 4 |

The data set CHANGES shows that the missing values for the variables Museums, Galleries, and Other are now zero. In addition, the data set contains the variable Count

**222** Chapter 14 • *Finding Shortcuts in Programming*

with the value 4 (the value that caused processing of the loop to cease in each observation).

To exclude Count from the data set, use a DROP= data set option:

```
data changes2 (drop=Count);
   set mylib.attractions;
   array changelist{3} Museums Galleries Other;
   do Count = 1 to 3;
      if changelist{Count} = . then changelist{count} = 0;
   end;
run;

proc print data=changes2;
   title 'Tour Attractions';
run;
```

The following output displays the results.

***Figure 14.4*** *Dropping the Index Variable from a Data Set*

### Tour Attractions

| Obs | City | Museums | Galleries | Other | TourGuide | YearsExperience |
|---|---|---|---|---|---|---|
| 1 | Rome | 4 | 3 | 0 | D'Amico | 2 |
| 2 | Paris | 5 | 0 | 1 | Lucas | 5 |
| 3 | London | 3 | 2 | 0 | Wilson | 3 |
| 4 | New York | 5 | 1 | 2 | Lucas | 5 |
| 5 | Madrid | 0 | 0 | 5 | Torres | 4 |
| 6 | Amsterdam | 3 | 3 | 0 | | |

## Summary

### Statements

ARRAY *array-name{number-of-variables} variable-1 <. . . variable-n>*;
   creates a named, ordered, list of variables that exists for processing of the current DATA step. *Array-name* must be a valid SAS name. Each variable is the name of a variable to be included in the array. *Number-of-variables* is the number of variables listed.

   When you place a variable in an array, the variable can also be accessed by *array-name {position}*, where *position* is the position of the variable in the list (from 1 to *number-of-variables*). In this way of accessing the variable is called an array reference, and the position is known as the subscript of the array reference. After you list a variable in an ARRAY statement, programming statements in the same DATA step can use either the original name of the variable or the array reference.

This documentation uses braces around the subscript. Parentheses ( ) are also acceptable, and square brackets [ ] are acceptable in operating environments that support those characters. Refer to the documentation for your operating environment to determine the supported characters.

DO;
... *SAS statements* ...
END;

treats the enclosed *SAS statements* as a unit. A group of statements beginning with DO and ending with END is called a DO group. DO groups usually appear in THEN clauses or ELSE statements.

DO *index-variable*=1 TO *number-of-variables-in-array*;
... *SAS statements* ...
END;

is known as an iterative DO loop. In each execution of the DATA step, an iterative DO loop is processed repeatedly (is iterated) based on the value of *index-variable*. To create an index variable, use a SAS variable name in an iterative DO statement.

When you use iterative DO loops for array processing, the value of *index-variable* usually starts at 1 and increases by 1 before each iteration of the loop. When the value becomes greater than the *number-of-variables-in-array* (usually the number of variables in the array being processed), SAS stops processing the loop and proceeds to the next statement in the DATA step.

In array processing, the SAS statements in an iterative DO loop usually contain array references whose subscript is the name of the index variable (as in *array-name* {*index-variable*}). In each iteration of the loop, SAS replaces the subscript in the reference with the index variable's current value. Therefore, successive iterations of the loop cause SAS to process the statements on the first variable in the array, then on the second variable, and so on.

# Learning More

Arrays

Arrays can be single or multidimensional. For more information, see "Array Processing" in *SAS Language Reference: Concepts*, "ARRAY Statement" in *SAS Statements: Reference*, and "Array Reference Statement" in *SAS Statements: Reference*.

DO groups

Iterative DO statements are flexible and powerful. They are useful in many situations other than array processing. The range of the index variable can start and stop with any number, and the increment can be any positive or negative number. The range of the index variable can be given as starting and stopping values, the values of the DIM, LBOUND, and HBOUND functions, a list of values separated by commas, or a combination of these. A range can also contain a WHILE or UNTIL clause. The index variable can also be a character variable (in that case, the range must be given as a list of character values). For more information, see:

- "DO Statement" in *SAS Statements: Reference*
- "DO Statement, Iterative" in *SAS Statements: Reference*
- "DIM Function" in *SAS Functions and CALL Routines: Reference*
- "HBOUND Function" in *SAS Functions and CALL Routines: Reference*

- "LBOUND Function" in *SAS Functions and CALL Routines: Reference*

DO UNTIL and DO WHILE statements

A DO WHILE statement processes a loop as long as a condition is true. A DO UNTIL statement processes a loop until a condition is true. (A DO UNTIL loop always processes at least once. A DO WHILE loop is not processed at all if the condition is initially false.) For more information, see "DO UNTIL Statement" in *SAS Statements: Reference* and "DO WHILE Statement" in *SAS Statements: Reference*.

# Chapter 15
# Working with Dates in the SAS System

| | |
|---|---:|
| **Introduction to Working with Dates** | 226 |
| Purpose | 226 |
| Prerequisites | 226 |
| **Understanding How SAS Handles Dates** | 226 |
| How SAS Stores Date Values | 226 |
| Determining the Century for Dates with Two-Digit Years | 227 |
| **Input File and SAS Data Set for Examples** | 228 |
| **Entering Dates** | 228 |
| Understanding Informats for Date Values | 228 |
| Reading a Date Value | 229 |
| Using Good Programming Practices to Read Dates | 230 |
| Using Dates as Constants | 232 |
| **Displaying Dates** | 233 |
| Understanding How SAS Displays Values | 233 |
| Formatting a Date Value | 234 |
| Assigning Permanent Date Formats to Variables | 235 |
| Changing Formats Temporarily | 237 |
| **Using Dates in Calculations** | 237 |
| Sorting Dates | 237 |
| Creating New Date Variables | 238 |
| **Using SAS Date Functions** | 239 |
| Finding the Day of the Week | 239 |
| **Comparing Durations and SAS Date Values** | 241 |
| **Summary** | 243 |
| Statements | 243 |
| Formats and Informats for Dates | 243 |
| Functions | 243 |
| System Options | 244 |
| **Learning More** | 244 |

# Introduction to Working with Dates

## Purpose

SAS stores dates as single, unique numbers so that they can be used in programs like any other numeric variable. In this section, you will learn how to do the following:

- make SAS read dates in raw data files and store them as SAS date values
- indicate which calendar form SAS should use to display SAS date values
- calculate with dates, that is, determine the number of days between dates, find the day of the week on which a date falls, and use today's date in calculations

## Prerequisites

You should understand the following topics before proceeding with this section:

- Chapter 7, "Understanding DATA Step Processing," on page 107
- Chapter 11, "Creating Subsets of Observations," on page 169
- Chapter 12, "Working with Grouped or Sorted Observations," on page 183

# Understanding How SAS Handles Dates

## How SAS Stores Date Values

Dates are written in many different ways. Some dates contain only numbers. Others contain various combinations of numbers, letters, and characters. For example, all the following forms represent the date July 26, 2013:

*Table 15.1* How Dates Are Formatted

| 072613 | 26JUL13 | 132607 |
| 7/26/13 | 26JUL2013 | July 26, 2013 |

With so many different forms of dates, there must be a way to store dates and use them in calculations, regardless of how dates are entered or displayed.

The common ground that SAS uses to represent dates is called a SAS data value. No matter which form you use to write a date, SAS can convert and store that date as the number of days between January 1, 1960, and the date that you enter.

The following figure shows some dates written in calendar form and as SAS date values:

*Figure 15.1* Comparing Calendar Dates to SAS Date Values

Calendar Date

Jul 4 1776   Jan 1 1959   Jan 1 1960   Jan 1 1961   Jul 26 2000

-67019       -365         0            366          15547

SAS Date Value

In SAS, every date is a unique number on a number line. Dates before January 1, 1960, are negative numbers; those after January 1, 1960, are positive. Because SAS date values are numeric variables, you can sort them easily, determine time intervals, and use dates as constants, as arguments in SAS functions, or in calculations.

*Note:* SAS date values are valid for dates based on the Gregorian calendar from A.D. 1582 through A.D. 19,900. Use caution when working with historical dates. Although the Gregorian calendar was used throughout most of Europe from 1582, Great Britain and the American colonies did not adopt the calendar until 1752.

## Determining the Century for Dates with Two-Digit Years

If dates in your external data sources or SAS program statements contain two-digit years, then you can determine which century prefix should be assigned to them by using the YEARCUTOFF= system option. The YEARCUTOFF= system option specifies the first year of the 100-year span that is used to determine the century of a two-digit year. For example, YEARCUTOFF=1950 means that two-digit years 50 through 99 correspond to 1950 through 1999. Two-digit years 00 through 49 correspond to 2000 through 2049. The default value of the YEARCUTOFF= system option is 1926, but you can change the YEARCUTOFF= value in an OPTIONS statement to accommodate the range of date values that you are working with.

Before you use the YEARCUTOFF= system option, examine the dates in your data:

- If the dates in your data fall within a 100-year span, then you can use the YEARCUTOFF= system option.

- If the dates in your data do not fall within a 100-year span, then you must either convert the two-digit years to four-digit years or use a DATA step with conditional logic to assign the proper century prefix.

After you determine that the YEARCUTOFF= system option is appropriate for your range of data, you can determine the setting to use. The best setting for YEARCUTOFF= is the year before the lowest year in your data. For example, if you have data in a range from 2013 to 2112, then set YEARCUTOFF= to 2013. This is the result of setting YEARCUTOFF= to 2013:

- SAS interprets all two-digit dates in the range of 13 through 99 as 2013 through 2099.

- SAS interprets all two-digit dates in the range of 00 through 12 as 2100 through 2112.

With YEARCUTOFF= set to 2013, a two-digit year of 13 would be interpreted as 2013 and a two-digit year of 05 would be interpreted as 2105.

## Input File and SAS Data Set for Examples

In the travel industry, some of the most important data about a tour includes dates such as:

- when the tour leaves and returns
- when payments are due
- when refunds are allowed, and so on

Tradewinds Travel has data that contains dates of past and upcoming popular tours as well as the number of nights spent on the tour. The raw data is stored in an external file that looks like this:

```
Japan        13may2000  8
Greece       17oct99    12
New Zealand  03feb2001  16
Brazil       28feb2001  8
Venezuela    10nov00    9
Italy        25apr2001  8
Russia       03jun1997  14
Switzerland  14jan2001  9
Australia    24oct98    12
Ireland      27aug2000  7
```

The first column lists the name of the country that was toured. The second column lists the departure date. The third column lists the number of nights on the tour.

## Entering Dates

### Understanding Informats for Date Values

In order for SAS to read a value as a SAS date value, you must give it a set of directions called an informat. By default, SAS reads numeric variables with a standard numeric informat that does not include letters or special characters. When a field that contains data does not match the standard patterns, you specify the appropriate informat in the INPUT statement.

SAS provides many informats. Four informats that are commonly used to read date values are:

MMDDYY8.
    reads dates written as mm/dd/yy.

MMDDYY10.
    reads dates written as mm/dd/yyyy.

DATE7.
    reads dates in the form ddMMMyy.

DATE9.
    reads dates in the form ddMMMyyyy.

Note that each informat name ends with a period and contains a width specification that tells SAS how many columns to read.

## Reading a Date Value

To create a SAS data set for the Tradewinds Travel data, the DATE9. informat is used in the INPUT statement to read the variable DepartureDate.

```
input Country $ 1-11 @13 DepartureDate date9. Nights;
```

Using an informat in the INPUT statement is called formatted input. The formatted input in this example contains the following items:

- a pointer to indicate the column in which the value begins (@13)
- the name of the variable to be read (DepartureDate)
- the name of the informat to use (DATE9.)

The following DATA step creates MYLIB.TOURDATES using the DATE9. informat to create SAS date values:

```
options yearcutoff=1920;
libname mylib 'permanent-data-library';

data mylib.tourdates;
   infile 'input-file';
   input Country $ 1-11 @13 DepartureDate date9. Nights;
run;

proc print data=mylib.tourdates;
   title 'Tour Departure Dates as SAS Date Values';
run;
```

230  Chapter 15  •  Working with Dates in the SAS System

The following output displays the results:

*Figure 15.2   Creating SAS Date Values from Calendar Dates*

```
            Tour Departure Dates as SAS Date Values

            Obs   Country         DepartureDate   Nights

             1    Japan               14743          8
             2    Greece              14534         12
             3    New Zealand         15009         16
             4    Brazil              15034          8
             5    Venezuela           14924          9
             6    Italy               15090          8
             7    Russia              13668         14
             8    Switzerland         14989          9
             9    Australia           14176         12
            10    Ireland             14849          7
```

Compare the SAS values of the variable DepartureDate with the values of the raw data shown in the previous section. The data set MYLIB.TOURDATES shows that SAS read the departure dates and created SAS date values. Now you need a way to display the dates in a recognizable form.

## Using Good Programming Practices to Read Dates

When reading dates, it is good programming practice to always use the DATE9. or MMDDYY10. informats to make sure that the data is read correctly. If you use the DATE7. or MMDDYY8. informat, then SAS reads only the first two digits of the year. If the data contains four-digit years, then SAS reads the century and not the year.

In the example, the PRINT procedure uses the FORMAT option to format the dates.

Consider the Tradewinds Travel external file with both two-digit years and four-digit years:

```
Japan        13may2000  8
Greece       17oct99    12
New Zealand  03feb2001  16
Brazil       28feb2001  8
Venezuela    10nov00    9
Italy        25apr2001  8
France       03jun1997  14
Switzerland  14jan2001  9
Australia    24oct98    12
Ireland      27aug2000  7
```

The following DATA step creates a SAS data set MYLIB.TOURDATES7 by using the DATE7. informat:

```
options yearcutoff=1920;

data mylib.tourdates7;
   infile 'input-file';
   input Country $ 1-11 @13 DepartureDate date7. Nights;
run;

proc print data=mylib.tourdates7;
   title 'Tour Departure Dates Using the DATE7. Informat';
   title2 'Displayed as Two-Digit Calendar Dates';
   format DepartureDate date7.;
run;

proc print data=mylib.tourdates7;
   title 'Tour Departure Dates Using the DATE7. Informat';
   title2 'Displayed as Four-Digit Calendar Dates';
   format DepartureDate date9.;
run;
```

The PRINT procedures format DepartureDate using two-digit year (DATE7.) and four-digit year (DATE9.) calendar dates. The following output shows how using the wrong date informat can produce invalid SAS data sets. This is the output from the program:

*Figure 15.3  Using the Wrong Date Informat*

Tour Departure Dates Using the DATE7. Informat
Displayed as Two-Digit Calendar Dates

| Obs | Country | DepartureDate | Nights |
|---|---|---|---|
| 1 | Japan | 13MAY20 | 0 |
| 2 | Greece | 17OCT99 | 12 |
| 3 | New Zealand | 03FEB20 | 1 |
| 4 | Brazil | 28FEB20 | 1 |
| 5 | Venezuela | 10NOV00 | 9 |
| 6 | Italy | 25APR20 | 1 |
| 7 | Russia | 03JUN19 | 97 |
| 8 | Switzerland | 14JAN20 | 1 |
| 9 | Australia | 24OCT98 | 12 |
| 10 | Ireland | 27AUG20 | 0 |

**Figure 15.4** *Using the Wrong Informat Can Produce Invalid SAS Data Sets*

Tour Departure Dates Using the DATE7. Informat
Displayed as Four-Digit Calendar Dates

| Obs | Country | DepartureDate | Nights |
|---|---|---|---|
| 1 | Japan | 13MAY1920 | 0 |
| 2 | Greece | 17OCT1999 | 12 |
| 3 | New Zealand | 03FEB1920 | 1 |
| 4 | Brazil | 28FEB1920 | 1 |
| 5 | Venezuela | 10NOV2000 | 9 |
| 6 | Italy | 25APR1920 | 1 |
| 7 | Russia | 03JUN2019 | 97 |
| 8 | Switzerland | 14JAN1920 | 1 |
| 9 | Australia | 24OCT1998 | 12 |
| 10 | Ireland | 27AUG1920 | 0 |

Notice that the four-digit years in the input file do not match the years in MYLIB.TOURDATES7 for observations 1, 3, 4, 6, 7, 8, and 10:

- SAS stopped reading the date after seven characters. It read the first two digits, the century, and not the complete four-digit year.

- To read the data for the next variable, SAS moved the pointer one column and read the next two numeric characters (the years 00, 01, and 97) as the value for the variable Nights. The data for Nights in the input file was ignored.

- When the dates were formatted for four-digit calendar dates, SAS used the YEARCUTOFF= 1920 system option to determine the century for the two-digit year. What was originally 1997 in observation 7 became 2019, and what was originally 2000 and 2001 in observations 1, 3, 4, 6, 8, and 10 became 1920.

## Using Dates as Constants

If the tour of Switzerland leaves on January 21, 2001 instead of January 14, then you can use the following assignment statement to make the update:

```
if Country = 'Switzerland' then DepartureDate = '21jan2001'd;
```

The value '21jan2001'D is a SAS date constant. To write a SAS date constant, enclose a date in quotation marks in the standard SAS form ddMMMyyyy and immediately follow the final quotation mark with the letter D. The D suffix tells SAS to convert the calendar date to a SAS date value. The following DATA step includes the use of the SAS date constant. The FORMAT option in the PRINT procedure writes the date with the DATE9. format:

```
data correctdates;
   set mylib.tourdates;
```

```
        if Country = 'Switzerland' then DepartureDate = '21jan2001'd;
run;

proc print data=correctdates;
    title 'Corrected Departure Date for Switzerland';
    format DepartureDate date9.;
run;
```

The following output displays the results:

*Figure 15.5* Changing a Date By Using a SAS Date Constant

**Corrected Departure Date for Switzerland**

| Obs | Country | DepartureDate | Nights |
|-----|---------|---------------|--------|
| 1 | Japan | 13MAY2000 | 8 |
| 2 | Greece | 17OCT1999 | 12 |
| 3 | New Zealand | 03FEB2001 | 16 |
| 4 | Brazil | 28FEB2001 | 8 |
| 5 | Venezuela | 10NOV2000 | 9 |
| 6 | Italy | 25APR2001 | 8 |
| 7 | Russia | 03JUN1997 | 14 |
| 8 | Switzerland | 21JAN2001 | 9 |
| 9 | Australia | 24OCT1998 | 12 |
| 10 | Ireland | 27AUG2000 | 7 |

# Displaying Dates

## Understanding How SAS Displays Values

To understand how to display the departure dates, you need to understand how SAS displays values in general. SAS displays all data values with a set of directions called a format. By default, SAS uses a standard numeric format with no commas, letters, or other special notation to display the values of numeric variables. Figure 15.2 on page 230 shows that writing SAS date values with the standard numeric format produces numbers that are difficult to recognize. To display these numbers as calendar dates, you need to specify a SAS date format for the variable.

SAS date formats are available for the most common ways of writing calendar dates. The DATE9. format represents dates in the form ddMMMyyyy. If you want the month, day, and year to be spelled out, then use the WORDDATE18. format. The WEEKDATE29. format includes the day of the week. There are also formats available for number representations such as the format MMDDYY8., which displays the calendar date in the form mm/dd/yy, or the format MMDDYY10., which displays the calendar date in the form mm/dd/yyyy. Like informat names, each format name ends with a period and contains a width specification that tells SAS how many columns to use when displaying the date value.

## Formatting a Date Value

You tell SAS which format to use by specifying the variable and the format name in a FORMAT statement. The following FORMAT statement assigns the MMDDYY10. format to the variable DepartureDate:

```
format DepartureDate mmddyy10.;
```

In this example, the FORMAT statement contains the following items:

- the name of the variable (DepartureDate)
- the name of the format to be used (MMDDYY10.)

The following PRINT procedures format the variable DepartureDate in both the two-digit year calendar format and the four-digit year calendar format:

```
proc print data=mylib.tourdates;
   title 'Departure Dates in Two-Digit Calendar Format';
   format DepartureDate mmddyy8.;
run;

proc print data=mylib.tourdates;
   title 'Departure Dates in Four-Digit Calendar Format';
   format DepartureDate mmddyy10.;
run;
```

The following output displays the results:

*Figure 15.6* Displaying a Formatted Date Value: Two-Digit Calendar Format

Departure Dates in Two-Digit Calendar Format

| Obs | Country | DepartureDate | Nights |
|---|---|---|---|
| 1 | Japan | 05/13/00 | 8 |
| 2 | Greece | 10/17/99 | 12 |
| 3 | New Zealand | 02/03/01 | 16 |
| 4 | Brazil | 02/28/01 | 8 |
| 5 | Venezuela | 11/10/00 | 9 |
| 6 | Italy | 04/25/01 | 8 |
| 7 | Russia | 06/03/97 | 14 |
| 8 | Switzerland | 01/14/01 | 9 |
| 9 | Australia | 10/24/98 | 12 |
| 10 | Ireland | 08/27/00 | 7 |

*Figure 15.7* *Displaying a Formatted Date Value: Four-Digit Calendar Format*

Departure Dates in Four-Digit Calendar Format

| Obs | Country | DepartureDate | Nights |
|---|---|---|---|
| 1 | Japan | 05/13/2000 | 8 |
| 2 | Greece | 10/17/1999 | 12 |
| 3 | New Zealand | 02/03/2001 | 16 |
| 4 | Brazil | 02/28/2001 | 8 |
| 5 | Venezuela | 11/10/2000 | 9 |
| 6 | Italy | 04/25/2001 | 8 |
| 7 | Russia | 06/03/1997 | 14 |
| 8 | Switzerland | 01/14/2001 | 9 |
| 9 | Australia | 10/24/1998 | 12 |
| 10 | Ireland | 08/27/2000 | 7 |

Placing a FORMAT statement in a PROC step associates the format with the variable only for that step. To associate a format with a variable permanently, use the FORMAT statement in a DATA step.

## Assigning Permanent Date Formats to Variables

The next example creates a new permanent SAS data set and assigns the DATE9. format in the DATA step. Now all subsequent procedures and DATA steps that use the variable DepartureDate will use the DATE9. format by default. The PROC CONTENTS step displays the characteristics of the data set MYLIB.TOURDATE.

```
options yearcutoff=1920;

data mylib.fmttourdate;
   set mylib.tourdates;
   format DepartureDate date9.;
run;

proc contents data=mylib.fmttourdate nodetails;
run;
```

**236** Chapter 15 • *Working with Dates in the SAS System*

The following output displays that the DATE9. format is permanently associated with DepartureDate:

*Figure 15.8* *Assigning a Format in a DATA Step*

The CONTENTS Procedure

| Data Set Name | MYLIB.FMTTOURDATE | Observations | 10 |
|---|---|---|---|
| Member Type | DATA | Variables | 3 |
| Engine | V9 | Indexes | 0 |
| Created | 04/30/2013 10:14:06 | Observation Length | 32 |
| Last Modified | 04/30/2013 10:14:06 | Deleted Observations | 0 |
| Protection | | Compressed | NO |
| Data Set Type | | Sorted | NO |
| Label | | | |
| Data Representation | WINDOWS_32 | | |
| Encoding | wlatin1 Western (Windows) | | |

| Engine/Host Dependent Information ||
|---|---|
| Data Set Page Size | 65536 |
| Number of Data Set Pages | 1 |
| First Data Page | 1 |
| Max Obs per Page | 2039 |
| Obs in First Data Page | 10 |
| Number of Data Set Repairs | 0 |
| ExtendObsCounter | YES |
| Filename | c:\Users\lirezn\mylib\fmttourdate.sas7bdat |
| Release Created | 9.0401B0 |
| Host Created | W32_7PRO |

| Alphabetic List of Variables and Attributes ||||| 
|---|---|---|---|---|
| # | Variable | Type | Len | Format |
| 1 | Country | Char | 11 | |
| 2 | DepartureDate | Num | 8 | DATE9. |
| 3 | Nights | Num | 8 | |

### Changing Formats Temporarily

If you are preparing a report that requires the date in a different format, then you can override the permanent format by using a FORMAT statement in a PROC step. For example, to display the value for DepartureDate in the data set MYLIB.TOURDATES in the form of month-name dd, yyyy, you can issue a FORMAT statement in a PROC PRINT step. The following program specifies the WORDDATE18. format for the variable DepartureDate:

```
proc print data=mylib.tourdates;
   title 'Tour Departure Dates';
   format DepartureDate worddate18.;
run;
```

The following output displays the results:

**Figure 15.9** *Overriding a Previously Specified Format*

Tour Departure Dates

| Obs | Country | DepartureDate | Nights |
|---|---|---|---|
| 1 | Japan | May 13, 2000 | 8 |
| 2 | Greece | October 17, 1999 | 12 |
| 3 | New Zealand | February 3, 2001 | 16 |
| 4 | Brazil | February 28, 2001 | 8 |
| 5 | Venezuela | November 10, 2000 | 9 |
| 6 | Italy | April 25, 2001 | 8 |
| 7 | Russia | June 3, 1997 | 14 |
| 8 | Switzerland | January 14, 2001 | 9 |
| 9 | Australia | October 24, 1998 | 12 |
| 10 | Ireland | August 27, 2000 | 7 |

The format DATE9. is still permanently assigned to DepartureDate. Calendar dates in the remaining examples are in the form ddMMMyyyy unless a FORMAT statement is included in the PROC PRINT step.

## Using Dates in Calculations

### Sorting Dates

Because SAS date values are numeric variables, you can sort them and use them in calculations. The following example uses the data set MYLIB.TOURDATES to extract other information about the Tradewinds Travel data.

To help determine how frequently tours are scheduled, you can print a report with the tours listed in chronological order. The first step is to specify the following BY statement in a PROC SORT step to tell SAS to arrange the observations in ascending order of the date variable DepartureDate:

```
by DepartureDate;
```

By using a VAR statement in the following PROC PRINT step, you can list the departure date as the first column in the report:

```
proc sort data=mylib.fmttourdate out=sortdate;
   by DepartureDate;
run;

proc print data=sortdate;
   var DepartureDate Country Nights;
   title 'Departure Dates Listed in Chronological Order';
run;
```

The following output displays the results:

*Figure 15.10* Sorting by SAS Date Values

Departure Dates Listed in Chronological Order

| Obs | DepartureDate | Country | Nights |
|---|---|---|---|
| 1 | 03JUN1997 | Russia | 14 |
| 2 | 24OCT1998 | Australia | 12 |
| 3 | 17OCT1999 | Greece | 12 |
| 4 | 13MAY2000 | Japan | 8 |
| 5 | 27AUG2000 | Ireland | 7 |
| 6 | 10NOV2000 | Venezuela | 9 |
| 7 | 14JAN2001 | Switzerland | 9 |
| 8 | 03FEB2001 | New Zealand | 16 |
| 9 | 28FEB2001 | Brazil | 8 |
| 10 | 25APR2001 | Italy | 8 |

The observations in the data set SORTDATE are now arranged in chronological order. Note that there are no FORMAT statements in this example, so the dates are displayed in the DATE9. format that you assigned to DepartureDate when you created the data set MYLIB.FMTTOURDATE.

## Creating New Date Variables

Because you know the departure date and the number of nights spent on each tour, you can calculate the return date for each tour. To start, create a new variable by adding the number of nights to the departure date, as follows:

```
Return = DepartureDate + Nights;
```

The result is a SAS date value for the return date that you can display by assigning it the DATE9. format, as follows:

```
options yearcutoff=1920;
data home;
   set mylib.tourdates;
   Return = DepartureDate + Nights;
   format Return date9.;
run;

proc print data=home;
   title 'Dates of Departure and Return';
run;
```

The following output displays the results:

*Figure 15.11  Adding Days to a Date Value*

Dates of Departure and Return

| Obs | Country | DepartureDate | Nights | Return |
|---|---|---|---|---|
| 1 | Japan | 14743 | 8 | 21MAY2000 |
| 2 | Greece | 14534 | 12 | 29OCT1999 |
| 3 | New Zealand | 15009 | 16 | 19FEB2001 |
| 4 | Brazil | 15034 | 8 | 08MAR2001 |
| 5 | Venezuela | 14924 | 9 | 19NOV2000 |
| 6 | Italy | 15090 | 8 | 03MAY2001 |
| 7 | Russia | 13668 | 14 | 17JUN1997 |
| 8 | Switzerland | 14989 | 9 | 23JAN2001 |
| 9 | Australia | 14176 | 12 | 05NOV1998 |
| 10 | Ireland | 14849 | 7 | 03SEP2000 |

Note that because the variable DepartureDate in the data set MYLIB.TOURDATES has no permanent format, you see a numeric value instead of a readable calendar date for that variable.

# Using SAS Date Functions

## Finding the Day of the Week

SAS has various functions that produce calendar dates from SAS date values. SAS date functions enable you to do such things as derive partial date information or use the current date in calculations.

If the final payment for a tour is due 30 days before the tour leaves, then the final payment date can be calculated using subtraction. However, Tradewinds Travel is closed on Sundays. If the payment is due on a Sunday, then an additional day must be subtracted to make the payment due on Saturday. The WEEKDAY function, which returns the day of the week as a number from 1 through 7 (Sunday through Saturday), can be used to determine whether the return day is a Sunday.

The following statements determine the final payment date by

- subtracting 30 from the departure date
- checking the value returned by the WEEKDAY function
- subtracting an additional day if necessary

```
DueDate=DepartureDate - 30;
if Weekday(DueDate)=1 then DueDate=DueDate - 1;
```

Constructing a data set with these statements produces a list of payment due dates. The following program includes these statements and assigns the format WEEKDATE29. to the new variable DueDate:

```
options yearcutoff=1920;
data pay;
   set mylib.tourdates;
   DueDate = DepartureDate - 30;
   if Weekday(DueDate) = 1 then DueDate = DueDate - 1;
   format DueDate weekdate29.;
run;

proc print data=pay;
   var Country DueDate;
   title 'Date and Day of Week Payment Is Due';
run;
```

The following output displays the results:

*Figure 15.12  Using the WEEKDAY Function*

**Date and Day of Week Payment Is Due**

| Obs | Country | DueDate |
|---|---|---|
| 1 | Japan | Thursday, April 13, 2000 |
| 2 | Greece | Friday, September 17, 1999 |
| 3 | New Zealand | Thursday, January 4, 2001 |
| 4 | Brazil | Monday, January 29, 2001 |
| 5 | Venezuela | Wednesday, October 11, 2000 |
| 6 | Italy | Monday, March 26, 2001 |
| 7 | Russia | Saturday, May 3, 1997 |
| 8 | Switzerland | Friday, December 15, 2000 |
| 9 | Australia | Thursday, September 24, 1998 |
| 10 | Ireland | Friday, July 28, 2000 |

## Comparing Durations and SAS Date Values

You can use SAS date values to find the units of time between dates. Tradewinds Travel was founded on February 8, 1982. On November 23, 1999, you decide to find out how old Tradewinds Travel is, and you write the following program:

```
options yearcutoff=1920;
   /* Calculating a duration in days */
data ttage;
   Start = '08feb82'd;
   RightNow = today();
   Age = RightNow - Start;
   format Start RightNow date9.;
run;

proc print data=ttage;
   title 'Age of Tradewinds Travel';
run;
```

The following output displays the results:

**Figure 15.13** Calculating a Duration in Days

```
            Age of Tradewinds Travel

        Obs     Start     RightNow    Age

         1    08FEB1982   30APR2013  11404
```

The value of Age is 11404, a number that looks like an unformatted SAS date value. However, Age is actually the difference between February 8, 1982, and April 30, 2013, and represents a duration in days, not a SAS date value. To make the value of Age more understandable, divide the number of days by 365 (more precisely, 365.25) to produce a duration in years. The following DATA step calculates the age of Tradewinds Travel in years:

```
options yearcutoff=1920;
   /* Calculating a duration in years */
data ttage2;
   Start = '08feb82'd;
   RightNow = today();
   AgeInDays = RightNow - Start;
   AgeInYears = AgeInDays / 365.25;
   format AgeInYears 4.1 Start RightNow date9.;
run;

proc print data=ttage2;
   title 'Age in Years of Tradewinds Travel';
run;
```

The following output displays the results:

**Figure 15.14** Calculating a Duration in Years

```
            Age in Years of Tradewinds Travel

     Obs     Start     RightNow   AgeInDays  AgeInYears

      1    08FEB1982   30APR2013    11404       31.2
```

To show a portion of a year, the value for AgeInYears is assigned a numeric format of 4.1 in the FORMAT statement of the DATA step. The 4 tells SAS that the number contains up to four characters. The 1 tells SAS that the number includes one digit after the decimal point.

# Summary

## Statements

*date-variable*='*ddMMMyy*'D;
: is an assignment statement that tells SAS to convert the date in quotation marks to a SAS date value and assign it to *date-variable*. The SAS date constant '*ddMMMyy*'D specifies a particular date (for example, '23NOV00'D), and can be used in many SAS statements and expressions, not only assignment statements.

FORMAT *date-variable date-format*;
: tells SAS to format the values of *date-variable* using *date-format*. A FORMAT statement within a DATA step permanently associates a format with *date-variable*.

INPUT *date-variable date-informat*;
: tells SAS how to read the values for the date-variable from an external file. *date-informat* is an instruction that tells SAS the form of the date in the external file.

## Formats and Informats for Dates

DATE9.
: the form of the *date-variable* is ddMMMyyyy (for example, 30APR2013).

DATE7.
: the form of the *date-variable* is ddMMMyy (for example, 23NOV00).

MMDDYY10.
: the form of the *date-variable* is mm/dd/yyyy (for example, 11/23/2012).

MMDDYY8.
: the form of the *date-variable* is mm/dd/yy (for example, 11/23/00).

WORDDATE18.
: the form of the *date-variable* is month-name dd, yyyy (for example, November 23, 2013).

WEEKDATE29.
: the form of the *date-variable* is day-of-the-week, month-name dd, yyyy (for example, Thursday, November 23, 2013).

## Functions

WEEKDAY (*SAS-date-value*)
: is a function that returns the day of the week for a SAS date value. Values are numbers between 1 and 7, with Sunday assigned the value 1.

TODAY()
: is a function that returns a SAS date value corresponding to the date on which the SAS program is initiated.

## System Options

YEARCUTOFF=
specifies the first year of a 100-year span that is used by informats and functions to read two-digit years, and used by formats to display two-digit years. The value that is specified in YEARCUTOFF= can result in a range of years that span two centuries. If YEARCUTOFF=1950, then any two-digit value between 50 and 99 inclusive refers to the first half of the 100-year span, which is in the 1900s. Any two-digit value between 00 and 49 inclusive refers to the second half of the 100-year span, which is in the 2000s. YEARCUTOFF= has no effect on existing SAS dates or dates that are read from input data that include a four-digit year.

# Learning More

ATTRIB statement
Information about using the ATTRIB statement to assign or change a permanent format can be found in *SAS Statements: Reference*.

DATASETS procedure
To assign or change a variable to a permanent format see the DATASETS procedure in Chapter 36, "Managing SAS Libraries," on page 695.

PUT and INPUT functions
The PUT and INPUT functions can be used for correcting two common errors in working with SAS dates: treating date values that contain letters or symbols as character variables, or storing dates written as numbers as ordinary numeric variables. Neither method enables you to use dates in calculations. For more information, see "PUT Statement" in *SAS Statements: Reference* and "INPUT Statement" in *SAS Statements: Reference*.

SAS date values
Documentation on informats, formats, and functions for working with SAS date values, SAS time, and SAS datetime values can be found in *SAS Language Reference: Concepts*. This documentation includes the following date and time information:

- SAS stores a time as the number of seconds since midnight of the current day. For example, 9:30 am is 34200. A number of this type is known as a SAS time value. A SAS time value is independent of the date; the count begins at 0 each midnight.

- When a date and a time are both present, SAS stores the value as the number of seconds since midnight, January 1, 1960. For example, 9:30 am, November 23, 2000, is 1290591000. This type of number is known as a SAS datetime value.

- SAS date and time informats read fields of different widths. SAS date and time formats can display date variables in different ways according to the widths that you specify in the format name. The number at the end of the format or informat name indicates the number of columns that SAS can use. For example, the DATE9. informat reads up to nine columns (as in 23NOV2013). The WEEKDATE8. format displays eight columns, as in Thursday, and WEEKDATE27. displays 27 columns, as in Thursday, November 23, 2013.

- SAS provides date, time, and datetime intervals for counting different periods of elapsed time, such as MONTH, which represents an interval from the beginning of one month to the next, not a period of 30 or 31 days.
- International date, time, and datetime formats.

SYSDATE9

To include the current date in a title, you can use the macro variable SYSDATE9, which is explained in Chapter 27, "Producing Detail Reports with the PRINT Procedure," on page 433.

## Part 4

# Combining SAS Data Sets

*Chapter 16*
  **Methods of Combining SAS Data Sets** ............................ *249*

*Chapter 17*
  **Concatenating SAS Data Sets** .................................. *257*

*Chapter 18*
  **Interleaving SAS Data Sets** .................................... *281*

*Chapter 19*
  **Merging SAS Data Sets** ........................................ *289*

*Chapter 20*
  **Updating SAS Data Sets** ....................................... *317*

*Chapter 21*
  **Modifying SAS Data Sets** ...................................... *335*

*Chapter 22*
  **Conditionally Processing Observations from
  Multiple SAS Data Sets** ........................................ *349*

# Chapter 16
# Methods of Combining SAS Data Sets

Introduction to Combining SAS Data Sets . . . . . . . . . . . . . . . . . . . . . . . . . . . . . . 249
   Purpose . . . . . . . . . . . . . . . . . . . . . . . . . . . . . . . . . . . . . . . . . . . . . . . . . . . . . . . 249
   Prerequisites . . . . . . . . . . . . . . . . . . . . . . . . . . . . . . . . . . . . . . . . . . . . . . . . . . . 249
Definition of Concatenating . . . . . . . . . . . . . . . . . . . . . . . . . . . . . . . . . . . . . . . . . 250
Definition of Interleaving . . . . . . . . . . . . . . . . . . . . . . . . . . . . . . . . . . . . . . . . . . . 250
Definition of Merging . . . . . . . . . . . . . . . . . . . . . . . . . . . . . . . . . . . . . . . . . . . . . . 251
Definition of Updating . . . . . . . . . . . . . . . . . . . . . . . . . . . . . . . . . . . . . . . . . . . . . 252
Definition of Modifying . . . . . . . . . . . . . . . . . . . . . . . . . . . . . . . . . . . . . . . . . . . . 253
Comparing Modifying, Merging, and Updating Data Sets . . . . . . . . . . . . . . . . . . 254
Learning More . . . . . . . . . . . . . . . . . . . . . . . . . . . . . . . . . . . . . . . . . . . . . . . . . . . . 255

## Introduction to Combining SAS Data Sets

### Purpose

SAS provides several methods for combining SAS data sets. In this section, you will be introduced to five methods of combining data sets:

- concatenating
- interleaving
- merging
- updating
- modifying

Subsequent sections teach you how to use these methods.

### Prerequisites

Before continuing with this section, you should understand the concepts that are presented in the following sections:

- Chapter 7, "Understanding DATA Step Processing," on page 107

- Chapter 6, "Starting with SAS Data Sets," on page 91
- Chapter 7, "Understanding DATA Step Processing," on page 107

## Definition of Concatenating

Concatenating combines two or more SAS data sets, one after the other, into a single SAS data set. You concatenate data sets by using either the SET statement in a DATA step or the APPEND procedure.

The following figure shows the results of concatenating two SAS data sets, and the DATA step that produces the results.

*Figure 16.1* Concatenating Two SAS Data Sets

| DATA1 | DATA2 | COMBINED |
|---|---|---|
| Year | Year | Year |
| 2008 | 2008 | 2008 |
| 2009 | 2009 | 2009 |
| 2010 + | 2010 = | 2010 |
| 2011 | 2011 | 2011 |
| 2012 | 2012 | 2012 |
|  |  | 2008 |
|  |  | 2009 |
|  |  | 2010 |
|  |  | 2011 |
|  |  | 2012 |

```
data combined;
    set data1 data2;
run;
```

## Definition of Interleaving

Interleaving combines individual, sorted SAS data sets into one sorted SAS data set. For each observation, the following figure shows the value of the variable by which the data sets are sorted. You interleave data sets using a SET statement along with a BY statement.

In the following example, the data sets are sorted by the variable Year.

*Figure 16.2* Interleaving SAS Data Sets

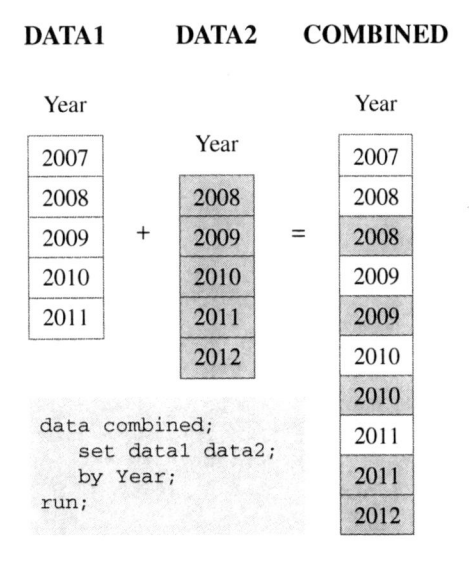

**DATA1**    **DATA2**    **COMBINED**

| Year |   | Year |   | Year |
|------|---|------|---|------|
| 2007 |   |      |   | 2007 |
| 2008 |   | 2008 |   | 2008 |
| 2009 | + | 2009 | = | 2008 |
| 2010 |   | 2010 |   | 2009 |
| 2011 |   | 2011 |   | 2009 |
|      |   | 2012 |   | 2010 |
|      |   |      |   | 2010 |
|      |   |      |   | 2011 |
|      |   |      |   | 2011 |
|      |   |      |   | 2012 |

```
data combined;
   set data1 data2;
   by Year;
run;
```

# Definition of Merging

Merging combines observations from two or more SAS data sets into a single observation in a new data set.

A one-to-one merge, shown in the following figure, combines observations based on their position in the data sets. You use the MERGE statement for one-to-one merging.

*Figure 16.3* One-to-One Merging

**DATA1**    **DATA2**    **COMBINED**
VarX         VarY         VarX  VarY

| X1 |   | Y1 |   | X1 | Y1 |
|----|---|----|---|----|----|
| X2 |   | Y2 |   | X2 | Y2 |
| X3 | + | Y3 | = | X3 | Y3 |
| X4 |   | Y4 |   | X4 | Y4 |
| X5 |   | Y5 |   | X5 | Y5 |

```
data combined;
   merge data1 data2;
run;
```

A match-merge, shown in the following figure, combines observations based on the values of one or more common variables. If you are performing a match-merge, use the MERGE statement along with a BY statement.

In the following example, two data sets are match-merged by the value of the variable Year.

**Figure 16.4** *Match-Merging Two SAS Data Sets*

| DATA1 |  |   | DATA2 |  |   | COMBINED |  |  |
|---|---|---|---|---|---|---|---|---|
| Year | VarX |   | Year | VarY |   | Year | VarX | VarY |
| 2008 | X1 |   | 2008 | Y1 |   | 2008 | X1 | Y1 |
| 2009 | X2 |   | 2008 | Y2 |   | 2008 | X1 | Y2 |
| 2010 | X3 | + | 2010 | Y3 | = | 2009 | X2 | . |
| 2011 | X4 |   | 2011 | Y4 |   | 2010 | X3 | Y3 |
| 2012 | X5 |   | 2012 | Y5 |   | 2011 | X4 | Y4 |
|   |   |   |   |   |   | 2012 | X5 | Y5 |

```
data combined;
   merge data1 data2;
   by Year;
run;
```

Notice that there is a missing value for the variable VarY in the COMBINED data set. The value is missing because the DATA2 data set has no observations for the year 2009.

# Definition of Updating

Updating a SAS data set replaces the values of variables in one data set (the master data set) with values from another data set (the transaction data set). If the UPDATEMODE= option in the UPDATE statement is set to MISSINGCHECK, then missing values in a transaction data set do not replace existing values in a master data set. If the UPDATEMODE= option is set to NOMISSINGCHECK, then missing values in a transaction data set replace existing values in a master data set. The default setting is MISSINGCHECK.

You update a data set by using the UPDATE statement along with a BY statement. Both of the input data sets must be sorted by the variable that you use in the BY statement.

The following figure shows the results of updating a SAS data set.

*Figure 16.5*  Updating a Master Data Set

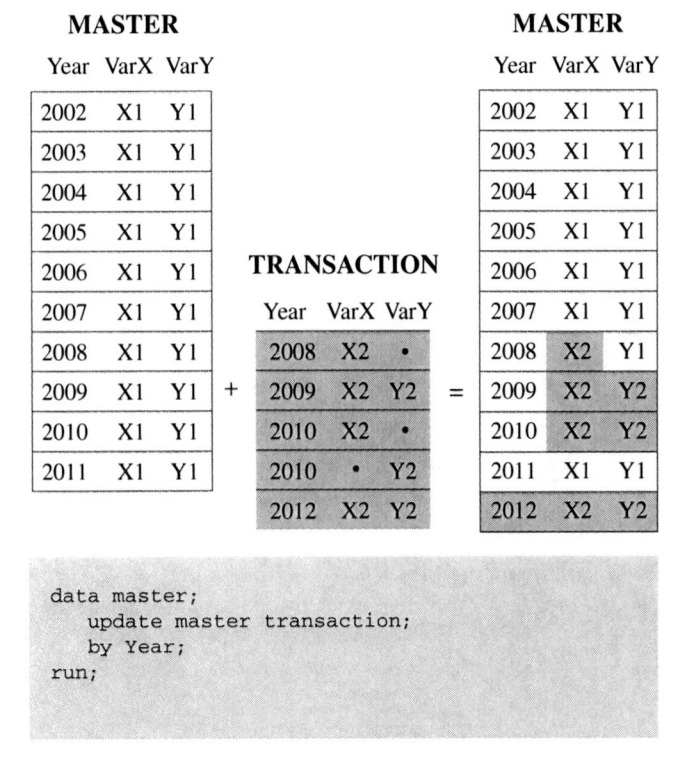

| MASTER | | |
|---|---|---|
| Year | VarX | VarY |
| 2002 | X1 | Y1 |
| 2003 | X1 | Y1 |
| 2004 | X1 | Y1 |
| 2005 | X1 | Y1 |
| 2006 | X1 | Y1 |
| 2007 | X1 | Y1 |
| 2008 | X1 | Y1 |
| 2009 | X1 | Y1 |
| 2010 | X1 | Y1 |
| 2011 | X1 | Y1 |

+

| TRANSACTION | | |
|---|---|---|
| Year | VarX | VarY |
| 2008 | X2 | • |
| 2009 | X2 | Y2 |
| 2010 | X2 | • |
| 2010 | • | Y2 |
| 2012 | X2 | Y2 |

=

| MASTER | | |
|---|---|---|
| Year | VarX | VarY |
| 2002 | X1 | Y1 |
| 2003 | X1 | Y1 |
| 2004 | X1 | Y1 |
| 2005 | X1 | Y1 |
| 2006 | X1 | Y1 |
| 2007 | X1 | Y1 |
| 2008 | X2 | Y1 |
| 2009 | X2 | Y2 |
| 2010 | X2 | Y2 |
| 2011 | X1 | Y1 |
| 2012 | X2 | Y2 |

```
data master;
   update master transaction;
   by Year;
run;
```

Notice that the TRANSACTION data set contains missing values. When the update occurs, the new MASTER data set retains the values from the original MASTER data set, and no missing values appear.

# Definition of Modifying

Modifying a SAS data set replaces, deletes, or appends observations in an existing data set. Modifying a SAS data set is similar to updating a SAS data set, but the following differences exist:

- Modifying cannot create a new data set, but updating can.
- Unlike updating, modifying does not require that the master data set or the transaction data set be sorted.

You use a MODIFY statement along with a BY statement to change an existing data set.

In the following example, the MASTER data set is updated by YEAR.

**Figure 16.6**  *Modifying a Data Set*

**MASTER**

| Year | VarX | VarY |
|------|------|------|
| 2003 | X1 | Y1 |
| 2004 | X1 | Y1 |
| 2005 | X1 | Y1 |
| 2006 | X1 | Y1 |
| 2007 | X1 | Y1 |
| 2008 | X1 | Y1 |
| 2009 | X1 | Y1 |
| 2010 | X1 | Y1 |
| 2011 | X1 | Y1 |
| 2012 | X1 | Y1 |

+

**TRANSACTION**

| Year | VarX | VarY |
|------|------|------|
| 2011 | X2 | . |
| 2011 | . | Y2 |
| 2009 | X2 | . |
| 2012 | X2 | Y2 |
| 2010 | X2 | Y2 |

=

**MASTER**

| Year | VarX | VarY |
|------|------|------|
| 2003 | X1 | Y1 |
| 2004 | X1 | Y1 |
| 2005 | X1 | Y1 |
| 2006 | X1 | Y1 |
| 2007 | X1 | Y1 |
| 2008 | X1 | Y1 |
| 2009 | X2 | Y1 |
| 2010 | X2 | Y2 |
| 2011 | X2 | Y2 |
| 2012 | X2 | Y2 |

```
data master;
   modify master transaction;
   by Year;
run;
```

Notice that the TRANSACTION data set contains missing values. When the MASTER data set is modified, the new MASTER data set retains the values from the original MASTER data set, and no missing values appear.

# Comparing Modifying, Merging, and Updating Data Sets

The table that follows summarizes several differences among the MERGE, UPDATE, and MODIFY statements.

**Table 16.1**  *Differences among the MERGE, UPDATE, and MODIFY Statements*

| Criterion | MERGE | UPDATE | MODIFY |
|---|---|---|---|
| Data sets must be sorted or indexed | Match-merge: Yes<br>One-to-one merge: No | Yes | No |
| BY values must be unique | No | Master data set: Yes<br>Transaction data set: No | No |
| Can create or delete variables | Yes | Yes | No |

| Criterion | MERGE | UPDATE | MODIFY |
|---|---|---|---|
| Number of data sets combined | Any number | 2 | 2 |
| Processing missing values | Overwrites nonmissing values from first data set with missing values from second data set | Default behavior: missing values in the transaction data set do not replace values in the master data set | Depends on the value of the UPDATEMODE= option (see "Handling Missing Values" on page 345) Default: MISSINGCHECK |

# Learning More

Concatenating data sets
  For more information, see Chapter 17, "Concatenating SAS Data Sets," on page 257.

Interleaving data sets
  For more information, see Chapter 18, "Interleaving SAS Data Sets," on page 281.

Manipulating data sets
  You can manipulate data sets as you combine them. For example, you can select certain observations from each data set and determine which data set an observation came from. For more information, see Chapter 22, "Conditionally Processing Observations from Multiple SAS Data Sets," on page 349.

MERGE, MODIFY, and UPDATE statements
  For more information, see "MERGE Statement" in *SAS Statements: Reference*, "MODIFY Statement" in *SAS Statements: Reference*, and "UPDATE Statement" in *SAS Statements: Reference*.

Merging data sets
  For more information, see Chapter 19, "Merging SAS Data Sets," on page 289.

Modifying data sets
  For more information, see Chapter 21, "Modifying SAS Data Sets," on page 335, and Chapter 22, "Conditionally Processing Observations from Multiple SAS Data Sets," on page 349.

Updating data sets
  For more information, see Chapter 20, "Updating SAS Data Sets," on page 317.

# Chapter 17
# Concatenating SAS Data Sets

**Introduction to Concatenating SAS Data Sets** ............................. 257
    Purpose .................................................................. 257
    Prerequisites ............................................................. 258

**Concatenating Data Sets with the SET Statement** ......................... 258
    Understanding the SET Statement ........................................ 258
    Using the SET Statement: The Simplest Case .............................. 258
    Using the SET Statement When Data Sets Contain Different Variables ........ 261
    Using the SET Statement When Variables Have Different Attributes .......... 262

**Concatenating Data Sets By Using the APPEND Procedure** .................. 274
    Understanding the APPEND Procedure .................................... 274
    Using the APPEND Procedure: The Simplest Case ......................... 274
    Using the APPEND Procedure When Data Sets Contain Different Variables .... 275
    Using the APPEND Procedure When Variables Have Different Attributes ...... 277

**Choosing between the SET Statement and the APPEND Procedure** .......... 278

**Summary** ................................................................ 279
    Statements ............................................................... 279
    Procedures ............................................................... 279

**Learning More** .......................................................... 280

## Introduction to Concatenating SAS Data Sets

### Purpose

Concatenating combines two or more SAS data sets, one after the other, into a single data set. The number of observations in the new data set is the sum of the number of observations in the original data sets.

You can concatenate SAS data sets by using one of the following methods:

- the SET statement in a DATA step
- the APPEND procedure

If the data sets that you concatenate contain the same variables, and each variable has the same attributes in all data sets, then the results of the SET statement and PROC APPEND are the same. In other cases, the results differ. In this section you will learn both of these methods and their differences so that you can decide which one to use.

*Prerequisites*

Before continuing with this section, you should understand the concepts presented in the following sections:

- Chapter 6, "Starting with SAS Data Sets," on page 91
- Chapter 7, "Understanding DATA Step Processing," on page 107
- Chapter 8, "Working with Numeric Variables," on page 117
- Chapter 9, "Working with Character Variables," on page 129

# Concatenating Data Sets with the SET Statement

## Understanding the SET Statement

The SET statement reads observations from one or more SAS data sets and uses them to build a new data set.

The SET statement for concatenating data sets has the following form:

**SET** *SAS-data-sets*;

*SAS-data-sets* specifies the two or more SAS data sets to concatenate. The observations from the first data set that you name in the SET statement appear first in the new data set. The observations from the second data set follow those from the first data set, and so on. The list can contain any number of data sets.

## Using the SET Statement: The Simplest Case

In the simplest situation, the data sets that you concatenate contain the same variables (variables with the same name). In addition, the type, length, informat, format, and label of each variable match across all data sets. In this case, SAS copies all observations from the first data set into the new data set. It then copies all observations from the second data set into the new data set, and so on. Each observation is an exact copy of the original.

In the following example, a company that uses SAS to maintain personnel records for six separate departments decided to combine all personnel records. Two departments, Sales and Customer Support, store their data in the same form. Each observation in both data sets contains values for these variables:

EmployeeID
: specifies a character variable that contains the employee's identification number.

Name
: specifies a character variable that contains the employee's name in the form last name, comma, first name.

HireDate
: specifies a numeric variable that contains the date on which the employee was hired. This variable has a format of DATE9.

Salary
: specifies a numeric variable that contains the employee's annual salary in US dollars.

HomePhone
: specifies a character variable that contains the employee's home telephone number.

The following program creates the SALES and CUSTOMER_SUPPORT data sets:

```
data sales;
    input EmployeeID $ 1-9 Name $ 11-29 @30 HireDate date9.
        Salary HomePhone $;
    format HireDate date9.;
    datalines;
429685482 Martin, Virginia    09aug2002 45000 493-0824
244967839 Singleton, MaryAnn  24apr2004 34000 929-2623
996740216 Leighton, Maurice   16dec2001 57000 933-6908
675443925 Freuler, Carl       15feb2010 54500 493-3993
845729308 Cage, Merce         19oct2009 64000 286-0519
;
run;

proc print data=sales;
    title 'Sales Department Employees';
run;

data customer_support;
    input EmployeeID $ 1-9 Name $ 11-29 @30 HireDate date9.
        Salary HomePhone $;
    format HireDate date9.;
    datalines;
324987451 Sayre, Jay          15nov2005 66000 933-2998
596771321 Tolson, Andrew      18mar2000 54000 929-4800
477562122 Jensen, Helga       01feb2004 70300 286-2816
894724859 Kulenic, Marie      24jun2004 54800 493-1472
988427431 Zweerink, Anna      07jul2011 59000 929-3885
;
run;

proc print data=customer_support;
    title 'Customer Support Department Employees';
run;
```

The following output displays the results of both DATA steps:

*Figure 17.1* *The SALES Data Set*

Sales Department Employees

| Obs | EmployeeID | Name | HireDate | Salary | HomePhone |
|---|---|---|---|---|---|
| 1 | 429685482 | Martin, Virginia | 09AUG2002 | 45000 | 493-0824 |
| 2 | 244967839 | Singleton, MaryAnn | 24APR2004 | 34000 | 929-2623 |
| 3 | 996740216 | Leighton, Maurice | 16DEC2001 | 57000 | 933-6908 |
| 4 | 675443925 | Freuler, Carl | 15FEB2010 | 54500 | 493-3993 |
| 5 | 845729308 | Cage, Merce | 19OCT2009 | 64000 | 286-0519 |

*Figure 17.2* The CUSTOMER_SUPPORT Data Set

Customer Support Department Employees

| Obs | EmployeeID | Name | HireDate | Salary | HomePhone |
|---|---|---|---|---|---|
| 1 | 324987451 | Sayre, Jay | 15NOV2005 | 66000 | 933-2998 |
| 2 | 596771321 | Tolson, Andrew | 18MAR2000 | 54000 | 929-4800 |
| 3 | 477562122 | Jensen, Helga | 01FEB2004 | 70300 | 286-2816 |
| 4 | 894724859 | Kulenic, Marie | 24JUN2004 | 54800 | 493-1472 |
| 5 | 988427431 | Zweerink, Anna | 07JUL2011 | 59000 | 929-3885 |

To concatenate the two data sets, list them in the SET statement. Use the PRINT procedure to display the resulting DEPT1_2 data set.

```
data dept1_2;
   set sales customer_support;
run;

proc print data=dept1_2;
   title 'Employees in Sales and Customer Support Departments';
run;
```

The following output displays the new DEPT1_2 data set. The data set contains all observations from SALES followed by all observations from CUSTOMER_SUPPORT:

*Figure 17.3* The Concatenated DEPT1_2 Data Set

Employees in Sales and Customer Support Departments

| Obs | EmployeeID | Name | HireDate | Salary | HomePhone |
|---|---|---|---|---|---|
| 1 | 429685482 | Martin, Virginia | 09AUG2002 | 45000 | 493-0824 |
| 2 | 244967839 | Singleton, MaryAnn | 24APR2004 | 34000 | 929-2623 |
| 3 | 996740216 | Leighton, Maurice | 16DEC2001 | 57000 | 933-6908 |
| 4 | 675443925 | Freuler, Carl | 15FEB2010 | 54500 | 493-3993 |
| 5 | 845729308 | Cage, Merce | 19OCT2009 | 64000 | 286-0519 |
| 6 | 324987451 | Sayre, Jay | 15NOV2005 | 66000 | 933-2998 |
| 7 | 596771321 | Tolson, Andrew | 18MAR2000 | 54000 | 929-4800 |
| 8 | 477562122 | Jensen, Helga | 01FEB2004 | 70300 | 286-2816 |
| 9 | 894724859 | Kulenic, Marie | 24JUN2004 | 54800 | 493-1472 |
| 10 | 988427431 | Zweerink, Anna | 07JUL2011 | 59000 | 929-3885 |

## Using the SET Statement When Data Sets Contain Different Variables

The two data sets in the previous example contain the same variables, and each variable is defined the same way in both data sets. However, you might want to concatenate data sets when not all variables are common to the data sets that are named in the SET statement. In this case, each observation in the new data set includes all variables from the SAS data sets that are named in the SET statement.

The examples in this section show the SECURITY data set, and the concatenation of this data set to the SALES and the CUSTOMER_SUPPORT data sets. Not all variables are common to the three data sets. The personnel records for the Security department do not include the variable HomePhone. The variable Gender does not appear in the SALES or the CUSTOMER_SUPPORT data sets.

The following program creates the SECURITY data set.

```
data security;
   input EmployeeID $ 1-9 Name $ 11-29 Gender $ 30
         @32 HireDate date9. Salary;
   format HireDate date9.;
   datalines;
744289612 Saparilas, Theresa  F 09may2005 45000
824904032 Brosnihan, Dylan    M 04jan2009 49000
242779184 Chao, Daeyong       M 28sep2004 48500
544382887 Slifkin, Leah       F 24jul2011 54000
933476520 Perry, Marguerite   F 19apr2010 49500
;
run;

proc print data=security;
   title 'Security Department Employees';
run;
```

The following output displays the SECURITY data set:

*Figure 17.4* The SECURITY Data Set

**Security Department Employees**

| Obs | EmployeeID | Name | Gender | HireDate | Salary |
|---|---|---|---|---|---|
| 1 | 744289612 | Saparilas, Theresa | F | 09MAY2005 | 45000 |
| 2 | 824904032 | Brosnihan, Dylan | M | 04JAN2009 | 49000 |
| 3 | 242779184 | Chao, Daeyong | M | 28SEP2004 | 48500 |
| 4 | 544382887 | Slifkin, Leah | F | 24JUL2011 | 54000 |
| 5 | 933476520 | Perry, Marguerite | F | 19APR2010 | 49500 |

The following program concatenates the SALES, CUSTOMER_SUPPORT, and SECURITY data sets, and creates the new data set, DEPT1_3:

```
data dept1_3;
   set sales customer_support security;
```

```
run;

proc print data=dept1_3;
    title 'Employees in Sales, Customer Support,';
    title2 'and Security Departments';
run;
```

The following output displays the concatenated DEPT1_3 data set:

*Figure 17.5* The Concatenated DEPT1_3 Data Set

Employees in Sales, Customer Support,
and Security Departments

| Obs | EmployeeID | Name | HireDate | Salary | HomePhone | Gender |
|---|---|---|---|---|---|---|
| 1 | 429685482 | Martin, Virginia | 09AUG2002 | 45000 | 493-0824 | |
| 2 | 244967839 | Singleton, MaryAnn | 24APR2004 | 34000 | 929-2623 | |
| 3 | 996740216 | Leighton, Maurice | 16DEC2001 | 57000 | 933-6908 | |
| 4 | 675443925 | Freuler, Carl | 15FEB2010 | 54500 | 493-3993 | |
| 5 | 845729308 | Cage, Merce | 19OCT2009 | 64000 | 286-0519 | |
| 6 | 324987451 | Sayre, Jay | 15NOV2005 | 66000 | 933-2998 | |
| 7 | 596771321 | Tolson, Andrew | 18MAR2000 | 54000 | 929-4800 | |
| 8 | 477562122 | Jensen, Helga | 01FEB2004 | 70300 | 286-2816 | |
| 9 | 894724859 | Kulenic, Marie | 24JUN2004 | 54800 | 493-1472 | |
| 10 | 988427431 | Zweerink, Anna | 07JUL2011 | 59000 | 929-3885 | |
| 11 | 744289612 | Saparilas, Theresa | 09MAY2005 | 45000 | | F |
| 12 | 824904032 | Brosnihan, Dylan | 04JAN2009 | 49000 | | M |
| 13 | 242779184 | Chao, Daeyong | 28SEP2004 | 48500 | | M |
| 14 | 544382887 | Slifkin, Leah | 24JUL2011 | 54000 | | F |
| 15 | 933476520 | Perry, Marguerite | 19APR2010 | 49500 | | F |

All observations in the data set DEPT1_3 have values for both the variable Gender and the variable HomePhone. Observations from data sets SALES and CUSTOMER_SUPPORT, the data sets that do not contain the variable Gender, have missing values for Gender (indicated by blanks under the variable name). Observations from SECURITY, the data set that does not contain the variable HomePhone, have missing values for HomePhone (indicated by blanks under the variable name).

## Using the SET Statement When Variables Have Different Attributes

### Understanding Attributes

Each variable in a SAS data set can have as many as six attributes that are associated with it. The following is a list of the attributes:

name
> identifies a variable. When SAS looks at two or more data sets, it considers variables with the same name to be the same variable.

type
> identifies a variable as character or numeric.

length
> refers to the number of bytes that SAS uses to store each of the variable's values in a SAS data set. Length is an especially important consideration when you use character variables, because the default length of character variables is eight bytes. If your data values are greater than eight bytes, then you can use a LENGTH statement to specify the number of bytes of storage that you need so that your data is not truncated.

informat
> refers to the instructions that SAS uses when reading data values. These instructions specify the form of an input value.

format
> refers to the instructions that SAS uses when writing data values. These instructions specify the form of an output value.

label
> refers to descriptive text that is associated with a specific variable.

If the data sets that you name in the SET statement contain variables with the same names and types, then you can concatenate the data sets without modification. However, if variable types differ, then you must modify one or more data sets before concatenating them. When lengths, formats, informats, or labels differ, you might want to modify one or more data sets before proceeding.

## *Using the SET Statement When Variables Have Different Types*

If a variable is defined as a character variable in one data set that is named in the SET statement, and as a numeric variable in another, SAS issues an error message and does not concatenate the data sets.

In the following example, the Accounting department in the company treats the employee identification number (EmployeeID) as a numeric variable, whereas all other departments treat it as a character variable.

The following program creates the ACCOUNTING data set, which is concatenated along with other data sets:

```
data accounting;
    input EmployeeID 1-9 Name $ 11-29 Gender $ 30
        @32 HireDate date9. Salary;
    format HireDate date9.;
    datalines;
634875680 Gardinski, Barbara  F 29may2001 59000
824576630 Robertson, Hannah   F 14mar2010 65500
744826703 Gresham, Jean       F 28apr1999 67000
824447605 Kruize, Ronald      M 23may2001 58000
988674342 Linzer, Fritz       M 23jul2007 63500
;
run;

proc print data=accounting;
    title 'Accounting Department Employees';
run;
```

The following output displays the ACCOUNTING data set:

**Figure 17.6** *The ACCOUNTING Data Set*

Accounting Department Employees

| Obs | EmployeeID | Name | Gender | HireDate | Salary |
|---|---|---|---|---|---|
| 1 | 634875680 | Gardinski, Barbara | F | 29MAY2001 | 59000 |
| 2 | 824576630 | Robertson, Hannah | F | 14MAR2010 | 65500 |
| 3 | 744826703 | Gresham, Jean | F | 28APR1999 | 67000 |
| 4 | 824447605 | Kruize, Ronald | M | 23MAY2001 | 58000 |
| 5 | 988674342 | Linzer, Fritz | M | 23JUL2007 | 63500 |

The following program attempts to concatenate the data sets for all four departments:

```
data dept1_4;
   set sales customer_support security accounting;
run;
```

The program fails because of the difference in variable type among the four departments. SAS writes the following error message to the log:

```
ERROR: Variable EmployeeID has been defined as both character
       and numeric.
```

## *Changing the Type of a Variable*

One way to correct the error in the previous example is to change the type of the variable EmployeeID in ACCOUNTING from numeric to character. Because performing calculations on employee identification numbers is unlikely, EmployeeID can be a character variable.

You can change the type of the variable EmployeeID in the following ways:

- re-create the data set, changing the INPUT statement so that it identifies EmployeeID as a character variable

- use the PUT function to create a new variable, and data set options to rename and drop variables

The following program uses the PUT function and data set options to change the variable type of EmployeeID from numeric to character.

```
data new_accounting (rename=(TempVar=EmployeeID)drop=EmployeeID);   １
   set accounting;   ２
   TempVar=put(EmployeeID, 9.);   ３
run;

proc datasets library=work;   ４
   contents data=new_accounting;
run;
```

The following list corresponds to the numbered items in the preceding program:

1. The RENAME= data set option renames the variable TempVar to EmployeeID when SAS writes an observation to the output data set. The DROP= data set option is

applied before the RENAME= option. The result is a change in the variable type for EmployeeID from numeric to character.

*Note:* Although this example creates a new data set called NEW_ACCOUNTING, you can create a data set that has the same name as the data set that is listed in the SET statement. If you do this, then the type attribute for EmployeeID is permanently altered in the ACCOUNTING data set.

2  The SET statement reads observations from the ACCOUNTING data set.

3  The PUT function converts a numeric value to a character value, and applies a format to the variable EmployeeID. The assignment statement assigns the result of the PUT function to the variable TempVar.

4  The DATASETS procedure enables you to verify the new attribute type for EmployeeID.

The following output displays a partial listing from PROC DATASETS. Note that EmployeeID is now a character variable:

*Figure 17.7*  PROC DATASETS Output for the NEW_ACCOUNTING Data Set

| # | Variable | Type | Len | Format |
|---|----------|------|-----|--------|
| 5 | EmployeeID | Char | 9 | |
| 2 | Gender | Char | 1 | |
| 3 | HireDate | Num | 8 | DATE9. |
| 1 | Name | Char | 19 | |
| 4 | Salary | Num | 8 | |

Alphabetic List of Variables and Attributes

Now that the types of all variables match, you can easily concatenate all four data sets using the following program:

```
data dept1_4;
   set sales customer_support security new_accounting;
run;

proc print data=dept1_4;
   title 'Employees in Sales, Customer Support, Security,';
   title2 'and Accounting Departments';
run;
```

The following output displays the concatenated DEPT1_4 data set:

*Figure 17.8* *The Concatenated Dept1_4 Data Set*

Employees in Sales, Customer Support, Security, and Accounting Departments

| Obs | EmployeeID | Name | HireDate | Salary | HomePhone | Gender |
|---|---|---|---|---|---|---|
| 1 | 429685482 | Martin, Virginia | 09AUG2002 | 45000 | 493-0824 | |
| 2 | 244967839 | Singleton, MaryAnn | 24APR2004 | 34000 | 929-2623 | |
| 3 | 996740216 | Leighton, Maurice | 16DEC2001 | 57000 | 933-6908 | |
| 4 | 675443925 | Freuler, Carl | 15FEB2010 | 54500 | 493-3993 | |
| 5 | 845729308 | Cage, Merce | 19OCT2009 | 64000 | 286-0519 | |
| 6 | 324987451 | Sayre, Jay | 15NOV2005 | 66000 | 933-2998 | |
| 7 | 596771321 | Tolson, Andrew | 18MAR2000 | 54000 | 929-4800 | |
| 8 | 477562122 | Jensen, Helga | 01FEB2004 | 70300 | 286-2816 | |
| 9 | 894724859 | Kulenic, Marie | 24JUN2004 | 54800 | 493-1472 | |
| 10 | 988427431 | Zweerink, Anna | 07JUL2011 | 59000 | 929-3885 | |
| 11 | 744289612 | Saparilas, Theresa | 09MAY2005 | 45000 | | F |
| 12 | 824904032 | Brosnihan, Dylan | 04JAN2009 | 49000 | | M |
| 13 | 242779184 | Chao, Daeyong | 28SEP2004 | 48500 | | M |
| 14 | 544382887 | Slifkin, Leah | 24JUL2011 | 54000 | | F |
| 15 | 933476520 | Perry, Marguerite | 19APR2010 | 49500 | | F |
| 16 | 634875680 | Gardinski, Barbara | 29MAY2001 | 59000 | | F |
| 17 | 824576630 | Robertson, Hannah | 14MAR2010 | 65500 | | F |
| 18 | 744826703 | Gresham, Jean | 28APR1999 | 67000 | | F |
| 19 | 824447605 | Kruize, Ronald | 23MAY2001 | 58000 | | M |
| 20 | 988674342 | Linzer, Fritz | 23JUL2007 | 63500 | | M |

## Using the SET Statement When Variables Have Different Formats, Informats, or Labels

When you concatenate data sets with the SET statement, the following rules determine which formats, informats, and labels are associated with variables in the new data set.

- An explicitly defined format, informat, or label overrides a default, regardless of the position of the data sets in the SET statement.

- If two or more data sets explicitly define different formats, informats, or labels for the same variable, then the variable in the new data set assumes the attribute from the first data set in the SET statement that explicitly defines that attribute.

Returning to the examples, you might have noticed that the DATA steps that created the SALES, CUSTOMER_SUPPORT, SECURITY, and ACCOUNTING data sets use a FORMAT statement to explicitly assign a format of DATE9. to the variable HireDate. Therefore, although HireDate is a numeric variable, it appears in all displays as DDMMMYYYY (for example, 13DEC2000). The SHIPPING data set that is created in the following example, however, uses a format of DATE11. for HireDate. The DATE11. format is displayed as DD-MMM-YYYY (for example, 13–DEC–2012).

In addition, the SALES, CUSTOMER_SUPPORT, SECURITY, and ACCOUNTING data sets contain a default format for Salary, whereas the SHIPPING data set contains an explicitly defined format, COMMA6., for the same variable. The COMMA6. format inserts a comma in the appropriate place when SAS displays the numeric variable Salary.

The following program creates the data set for the Shipping department:

```
data shipping;
   input employeeID $ 1-9 Name $ 11-29 Gender $ 30
         @32 HireDate date11.
         @42 Salary;
   format HireDate date11.
          Salary comma6.;
   datalines;
688774609 Carlton, Susan      F 28jan2012 41000
922448328 Hoffmann, Gerald    M 12oct2012 40500
544909752 DePuis, David       M 23aug2011 43500
745609821 Hahn, Kenneth       M 23aug2011 45500
634774295 Landau, Jennifer    F 30apr2012 43500
;
run;

proc print data=shipping;
   title 'Shipping Department Employees';
run;
```

The following output displays the SHIPPING data set:

*Figure 17.9* The SHIPPING Data Set

**Shipping Department Employees**

| Obs | employeeID | Name | Gender | HireDate | Salary |
|---|---|---|---|---|---|
| 1 | 688774609 | Carlton, Susan | F | 28-JAN-2012 | 41,000 |
| 2 | 922448328 | Hoffmann, Gerald | M | 12-OCT-2012 | 40,500 |
| 3 | 544909752 | DePuis, David | M | 23-AUG-2011 | 43,500 |
| 4 | 745609821 | Hahn, Kenneth | M | 23-AUG-2011 | 45,500 |
| 5 | 634774295 | Landau, Jennifer | F | 30-APR-2012 | 43,500 |

Now consider what happens when you concatenate SHIPPING with the previous four data sets.

```
data dept1_5;
   set sales customer_support security new_accounting shipping;
run;
```

```
proc print data=dept1_5;
   title 'Employees in Sales, Customer Support, Security,';
   title2 'Accounting, and Shipping Departments';
run;
```

The following output displays the concatenation of five data sets:

*Figure 17.10*    *The DEPT1_5 Data Set: Concatenation of Five Data Sets*

**Employees in Sales, Customer Support, Security, Accounting, and Shipping Departments**

| Obs | EmployeeID | Name | HireDate | Salary | HomePhone | Gender |
|---|---|---|---|---|---|---|
| 1 | 429685482 | Martin, Virginia | 09AUG2002 | 45,000 | 493-0824 | |
| 2 | 244967839 | Singleton, MaryAnn | 24APR2004 | 34,000 | 929-2623 | |
| 3 | 996740216 | Leighton, Maurice | 16DEC2001 | 57,000 | 933-6908 | |
| 4 | 675443925 | Freuler, Carl | 15FEB2010 | 54,500 | 493-3993 | |
| 5 | 845729308 | Cage, Merce | 19OCT2009 | 64,000 | 286-0519 | |
| 6 | 324987451 | Sayre, Jay | 15NOV2005 | 66,000 | 933-2998 | |
| 7 | 596771321 | Tolson, Andrew | 18MAR2000 | 54,000 | 929-4800 | |
| 8 | 477562122 | Jensen, Helga | 01FEB2004 | 70,300 | 286-2816 | |
| 9 | 894724859 | Kulenic, Marie | 24JUN2004 | 54,800 | 493-1472 | |
| 10 | 988427431 | Zweerink, Anna | 07JUL2011 | 59,000 | 929-3885 | |
| 11 | 744289612 | Saparilas, Theresa | 09MAY2005 | 45,000 | | F |
| 12 | 824904032 | Brosnihan, Dylan | 04JAN2009 | 49,000 | | M |
| 13 | 242779184 | Chao, Daeyong | 28SEP2004 | 48,500 | | M |
| 14 | 544382887 | Slifkin, Leah | 24JUL2011 | 54,000 | | F |
| 15 | 933476520 | Perry, Marguerite | 19APR2010 | 49,500 | | F |
| 16 | 634875680 | Gardinski, Barbara | 29MAY2001 | 59,000 | | F |
| 17 | 824576630 | Robertson, Hannah | 14MAR2010 | 65,500 | | F |
| 18 | 744826703 | Gresham, Jean | 28APR1999 | 67,000 | | F |
| 19 | 824447605 | Kruize, Ronald | 23MAY2001 | 58,000 | | M |
| 20 | 988674342 | Linzer, Fritz | 23JUL2007 | 63,500 | | M |
| 21 | 688774609 | Carlton, Susan | 28JAN2012 | 41,000 | | F |
| 22 | 922448328 | Hoffmann, Gerald | 12OCT2012 | 40,500 | | M |
| 23 | 544909752 | DePuis, David | 23AUG2011 | 43,500 | | M |
| 24 | 745609821 | Hahn, Kenneth | 23AUG2011 | 45,500 | | M |
| 25 | 634774295 | Landau, Jennifer | 30APR2012 | 43,500 | | F |

In this concatenation, the input data sets contain the variable HireDate, which was explicitly defined using two different formats. The data sets also contain the variable Salary, which has both a default and an explicit format. You can see from the output that SAS creates the new data set according to the rules mentioned earlier:

- In the case of HireDate, SAS uses the format that is defined in the first data set that is named in the SET statement (DATE9. in SALES).

- In the case of Salary, SAS uses the explicit format (COMMA6.) that is defined in the SHIPPING data set. In this case, SAS does not use the default format.

Notice the difference if you perform a similar concatenation but reverse the order of the data sets in the SET statement.

```
data dept5_1;
    set shipping new_accounting security customer_support sales;
run;

proc print data=dept5_1;
    title 'Employees in Shipping, Accounting, Security,';
    title2 'Customer Support, and Sales Departments';
run;
```

**270** Chapter 17 • Concatenating SAS Data Sets

The following output displays the DEPT5_1 data set, but with a different order of concatenation:

*Figure 17.11* The DEPT5_1 Data Set: Changing the Order of Concatenation

**Employees in Shipping, Accounting, Security, Customer Support, and Sales Departments**

| Obs | employeeID | Name | Gender | HireDate | Salary | HomePhone |
|---|---|---|---|---|---|---|
| 1 | 688774609 | Carlton, Susan | F | 28-JAN-2012 | 41,000 | |
| 2 | 922448328 | Hoffmann, Gerald | M | 12-OCT-2012 | 40,500 | |
| 3 | 544909752 | DePuis, David | M | 23-AUG-2011 | 43,500 | |
| 4 | 745609821 | Hahn, Kenneth | M | 23-AUG-2011 | 45,500 | |
| 5 | 634774295 | Landau, Jennifer | F | 30-APR-2012 | 43,500 | |
| 6 | 634875680 | Gardinski, Barbara | F | 29-MAY-2001 | 59,000 | |
| 7 | 824576630 | Robertson, Hannah | F | 14-MAR-2010 | 65,500 | |
| 8 | 744826703 | Gresham, Jean | F | 28-APR-1999 | 67,000 | |
| 9 | 824447605 | Kruize, Ronald | M | 23-MAY-2001 | 58,000 | |
| 10 | 988674342 | Linzer, Fritz | M | 23-JUL-2007 | 63,500 | |
| 11 | 744289612 | Saparilas, Theresa | F | 09-MAY-2005 | 45,000 | |
| 12 | 824904032 | Brosnihan, Dylan | M | 04-JAN-2009 | 49,000 | |
| 13 | 242779184 | Chao, Daeyong | M | 28-SEP-2004 | 48,500 | |
| 14 | 544382887 | Slifkin, Leah | F | 24-JUL-2011 | 54,000 | |
| 15 | 933476520 | Perry, Marguerite | F | 19-APR-2010 | 49,500 | |
| 16 | 324987451 | Sayre, Jay | | 15-NOV-2005 | 66,000 | 933-2998 |
| 17 | 596771321 | Tolson, Andrew | | 18-MAR-2000 | 54,000 | 929-4800 |
| 18 | 477562122 | Jensen, Helga | | 01-FEB-2004 | 70,300 | 286-2816 |
| 19 | 894724859 | Kulenic, Marie | | 24-JUN-2004 | 54,800 | 493-1472 |
| 20 | 988427431 | Zweerink, Anna | | 07-JUL-2011 | 59,000 | 929-3885 |
| 21 | 429685482 | Martin, Virginia | | 09-AUG-2002 | 45,000 | 493-0824 |
| 22 | 244967839 | Singleton, MaryAnn | | 24-APR-2004 | 34,000 | 929-2623 |
| 23 | 996740216 | Leighton, Maurice | | 16-DEC-2001 | 57,000 | 933-6908 |
| 24 | 675443925 | Freuler, Carl | | 15-FEB-2010 | 54,500 | 493-3993 |
| 25 | 845729308 | Cage, Merce | | 19-OCT-2009 | 64,000 | 286-0519 |

Compared with the output in Figure 17.10 on page 268, this example shows that not only does the order of the observations change, but in the case of HireDate, the DATE11. format specified in SHIPPING now prevails because that data set now appears first in the SET statement. The COMMA6. format prevails for the variable Salary because SHIPPING is the only data set that explicitly specifies a format for the variable.

## *Using the SET Statement When Variables Have Different Lengths*

If you use the SET statement to concatenate data sets in which the same variable has different lengths, the outcome of the concatenation depends on whether the variable is character or numeric. The SET statement determines the length of variables as follows:

- For a character or numeric variable, an explicitly defined length overrides a default, regardless of the position of the data sets in the SET statement.

- If two or more data sets explicitly define different lengths for the same numeric variable, the variable in the new data set has the same length as the variable in the data set that appears first in the SET statement.

- If the length of a character variable differs among data sets, whether the differences are explicit, the variable in the new data set has the same length as the variable in the data set that appears first in the SET statement.

The following program creates the RESEARCH data set for the sixth department, Research. Notice that the INPUT statement for this data set creates the variable Name with a length of 27. In all other data sets, Name has a length of 19. In this example, if Name had a length of 19, the values of Name would be truncated.

```
data research;
   input EmployeeID $ 1-9 Name $ 11-37 Gender $ 38
         @40 HireDate date9. Salary;
   format HireDate date9.;
   datalines;
922854076 Schoenberg, Marguerite    F 19nov2004 60500
770434994 Addison-Hardy, Jonathon   M 23feb2011 63500
242784883 McNaughton, Elizabeth     F 24jul2001 65000
377882806 Tharrington, Catherine    F 28sep2004 60000
292450691 Frangipani, Christopher   M 12aug2008 63000
;
run;

proc print data=research;
   title 'Research Department Employees';
run;
```

The following output displays the RESEARCH data set:

*Figure 17.12  The RESEARCH Data Set*

### Research Department Employees

| Obs | EmployeeID | Name | Gender | HireDate | Salary |
|---|---|---|---|---|---|
| 1 | 922854076 | Schoenberg, Marguerite | F | 19NOV2004 | 60500 |
| 2 | 770434994 | Addison-Hardy, Jonathon | M | 23FEB2011 | 63500 |
| 3 | 242784883 | McNaughton, Elizabeth | F | 24JUL2001 | 65000 |
| 4 | 377882806 | Tharrington, Catherine | F | 28SEP2004 | 60000 |
| 5 | 292450691 | Frangipani, Christopher | M | 12AUG2008 | 63000 |

If you concatenate all six data sets, naming RESEARCH in any position except the first in the SET statement, SAS defines Name with a length of 19.

If you want your program to use the Name variable that has a length of 27, you have two options:

- change the order of data sets in the SET statement
- change the length of Name in the new data set

For the first option, list the data set (RESEARCH) that uses the longer length first:

```
data dept6_1;
   set research shipping new_accounting
       security customer_support sales;
run;
```

For the second option, include a LENGTH statement in the DATA step that creates the new data set. If you change the length of a numeric variable, then the LENGTH statement can appear anywhere in the DATA step. However, if you change the length of a character variable, then the LENGTH statement must precede the SET statement.

The following program creates the data set DEPT1_6A. The LENGTH statement gives the character variable Name a length of 27, even though the first data set in the SET statement (SALES) assigns it a length of 19.

```
data dept1_6a;
   length Name $ 27;
   set sales customer_support security
       new_accounting shipping research;
run;

proc print data=dept1_6a;
   title 'Employees in All Departments';
run;
```

The following output shows that all values of Name are complete. Note that the order of the variables in the new data set changes because Name is the first variable encountered in the DATA step.

*Figure 17.13*  *The DEPT1_6A Data Set: Using a LENGTH Statement for the Name Variable*

**Employees in All Departments**

| Obs | Name | EmployeeID | HireDate | Salary | HomePhone | Gender |
|---|---|---|---|---|---|---|
| 1 | Martin, Virginia | 429685482 | 09AUG2002 | 45,000 | 493-0824 | |
| 2 | Singleton, MaryAnn | 244967839 | 24APR2004 | 34,000 | 929-2623 | |
| 3 | Leighton, Maurice | 996740216 | 16DEC2001 | 57,000 | 933-6908 | |
| 4 | Freuler, Carl | 675443925 | 15FEB2010 | 54,500 | 493-3993 | |
| 5 | Cage, Merce | 845729308 | 19OCT2009 | 64,000 | 286-0519 | |
| 6 | Sayre, Jay | 324987451 | 15NOV2005 | 66,000 | 933-2998 | |
| 7 | Tolson, Andrew | 596771321 | 18MAR2000 | 54,000 | 929-4800 | |
| 8 | Jensen, Helga | 477562122 | 01FEB2004 | 70,300 | 286-2816 | |
| 9 | Kulenic, Marie | 894724859 | 24JUN2004 | 54,800 | 493-1472 | |
| 10 | Zweerink, Anna | 988427431 | 07JUL2011 | 59,000 | 929-3885 | |
| 11 | Saparilas, Theresa | 744289612 | 09MAY2005 | 45,000 | | F |
| 12 | Brosnihan, Dylan | 824904032 | 04JAN2009 | 49,000 | | M |
| 13 | Chao, Daeyong | 242779184 | 28SEP2004 | 48,500 | | M |
| 14 | Slifkin, Leah | 544382887 | 24JUL2011 | 54,000 | | F |
| 15 | Perry, Marguerite | 933476520 | 19APR2010 | 49,500 | | F |
| 16 | Gardinski, Barbara | 634875680 | 29MAY2001 | 59,000 | | F |
| 17 | Robertson, Hannah | 824576630 | 14MAR2010 | 65,500 | | F |
| 18 | Gresham, Jean | 744826703 | 28APR1999 | 67,000 | | F |
| 19 | Kruize, Ronald | 824447605 | 23MAY2001 | 58,000 | | M |
| 20 | Linzer, Fritz | 988674342 | 23JUL2007 | 63,500 | | M |
| 21 | Carlton, Susan | 688774609 | 28JAN2012 | 41,000 | | F |
| 22 | Hoffmann, Gerald | 922448328 | 12OCT2012 | 40,500 | | M |
| 23 | DePuis, David | 544909752 | 23AUG2011 | 43,500 | | M |
| 24 | Hahn, Kenneth | 745609821 | 23AUG2011 | 45,500 | | M |
| 25 | Landau, Jennifer | 634774295 | 30APR2012 | 43,500 | | F |
| 26 | Schoenberg, Marguerite | 922854076 | 19NOV2004 | 60,500 | | F |
| 27 | Addison-Hardy, Jonathon | 770434994 | 23FEB2011 | 63,500 | | M |
| 28 | McNaughton, Elizabeth | 242784883 | 24JUL2001 | 65,000 | | F |
| 29 | Tharrington, Catherine | 377882806 | 28SEP2004 | 60,000 | | F |
| 30 | Frangipani, Christopher | 292450691 | 12AUG2008 | 63,000 | | M |

## Concatenating Data Sets By Using the APPEND Procedure

### Understanding the APPEND Procedure

The APPEND procedure adds the observations from one SAS data set to the end of another SAS data set. PROC APPEND does not process the observations in the first data set. It adds the observations in the second data set directly to the end of the original data set.

The APPEND procedure has the following form:

**PROC APPEND** BASE=*base-SAS-data-set* <DATA=*SAS-data-set-to-append*> <FORCE>;

*base-SAS-data-set*
> names the SAS data set to which you want to append the observations. If this data set does not exist, then SAS creates it. At the completion of PROC APPEND, the value of *base-SAS-data-set* becomes the current (most recently created) SAS data set.

*SAS-data-set-to-append*
> names the SAS data set that contains the observations to add to the end of the base data set. If you omit this option, then PROC APPEND adds the observations in the most recently created SAS data set to the end of the base data set.

FORCE
> forces PROC APPEND to concatenate the files in some situations in which the procedure would normally fail.

### Using the APPEND Procedure: The Simplest Case

The following program appends the data set CUSTOMER_SUPPORT to the data set SALES. Both data sets contain the same variables and each variable has the same attributes in both data sets.

```
proc append base=sales data=customer_support;
run;

proc print data=sales;
   title 'Employees in Sales and Customer Support Departments';
run;
```

The following output shows the results:

*Figure 17.14  Output from PROC APPEND*

**Employees in Sales and Customer Support Departments**

| Obs | EmployeeID | Name | HireDate | Salary | HomePhone |
|---|---|---|---|---|---|
| 1 | 429685482 | Martin, Virginia | 09AUG2002 | 45000 | 493-0824 |
| 2 | 244967839 | Singleton, MaryAnn | 24APR2004 | 34000 | 929-2623 |
| 3 | 996740216 | Leighton, Maurice | 16DEC2001 | 57000 | 933-6908 |
| 4 | 675443925 | Freuler, Carl | 15FEB2010 | 54500 | 493-3993 |
| 5 | 845729308 | Cage, Merce | 19OCT2009 | 64000 | 286-0519 |
| 6 | 324987451 | Sayre, Jay | 15NOV2005 | 66000 | 933-2998 |
| 7 | 596771321 | Tolson, Andrew | 18MAR2000 | 54000 | 929-4800 |
| 8 | 477562122 | Jensen, Helga | 01FEB2004 | 70300 | 286-2816 |
| 9 | 894724859 | Kulenic, Marie | 24JUN2004 | 54800 | 493-1472 |
| 10 | 988427431 | Zweerink, Anna | 07JUL2011 | 59000 | 929-3885 |

The resulting data set is identical to the data set that was created by naming SALES and CUSTOMER_SUPPORT in the SET statement. (See Figure 17.3 on page 260.) It is important to realize that PROC APPEND permanently alters the SALES data set, which is the data set for the BASE= option. SALES now contains observations from both the Sales and the Customer Support departments.

## Using the APPEND Procedure When Data Sets Contain Different Variables

Recall that the SECURITY data set contains the variable Gender, which is not in the SALES data set, and lacks the variable HomePhone, which is present in the SALES data set. What happens if you try to use PROC APPEND to concatenate data sets that contain different variables?

If you try to append SECURITY to SALES using the following program, then the concatenation fails:

```
proc append base=sales data=security;
run;
```

**276** Chapter 17 • Concatenating SAS Data Sets

SAS writes the following messages to the log:

*Log 17.1  SAS Log: PROC APPEND Error*

```
338  proc append base=sales data=security;
339  run;

NOTE: Appending WORK.SECURITY to WORK.SALES.
WARNING: Variable Gender was not found on BASE file. The variable will not be
added to the BASE
         file.
WARNING: Variable HomePhone was not found on DATA file.
ERROR: No appending done because of anomalies listed above.
       Use FORCE option to append these files.
NOTE: 0 observations added.
NOTE: The data set WORK.SALES has 5 observations and 5 variables.
NOTE: Statements not processed because of errors noted above.
NOTE: PROCEDURE APPEND used (Total process time):
      real time           0.01 seconds
      cpu time            0.03 seconds

NOTE: The SAS System stopped processing this step because of errors.
```

You must use the FORCE option with PROC APPEND when the DATA= data set contains a variable that is not in the BASE= data set. If you modify the program to include the FORCE option, then it successfully concatenates the files.

```
data sales;
    input EmployeeID $ 1-9 Name $ 11-29 @30 HireDate date9.
          Salary HomePhone $;
    format HireDate date9.;
    datalines;
429685482 Martin, Virginia    09aug2002 45000 493-0824
244967839 Singleton, MaryAnn  24apr2004 34000 929-2623
996740216 Leighton, Maurice   16dec2001 57000 933-6908
675443925 Freuler, Carl       15feb2010 54500 493-3993
845729308 Cage, Merce         19oct2009 64000 286-0519
;
run;

data security;
    input EmployeeID $ 1-9 Name $ 11-29 Gender $ 30
          @32 HireDate date9. Salary;
    format HireDate date9.;
    datalines;
744289612 Saparilas, Theresa  F 09may2005 45000
824904032 Brosnihan, Dylan    M 04jan2009 49000
242779184 Chao, Daeyong       M 28sep2004 48500
544382887 Slifkin, Leah       F 24jul2011 54000
933476520 Perry, Marguerite   F 19apr2010 49500
;
run;

proc append base=sales data=security force;
run;

proc print data=sales;
    title 'Employees in the Sales and the Security Departments';
```

```
run;
```

The following output displays the results:

**Figure 17.15** *Results of Using FORCE with PROC APPEND*

Employees in the Sales and the Security Departments

| Obs | EmployeeID | Name | HireDate | Salary | HomePhone |
|---|---|---|---|---|---|
| 1 | 429685482 | Martin, Virginia | 09AUG2002 | 45000 | 493-0824 |
| 2 | 244967839 | Singleton, MaryAnn | 24APR2004 | 34000 | 929-2623 |
| 3 | 996740216 | Leighton, Maurice | 16DEC2001 | 57000 | 933-6908 |
| 4 | 675443925 | Freuler, Carl | 15FEB2010 | 54500 | 493-3993 |
| 5 | 845729308 | Cage, Merce | 19OCT2009 | 64000 | 286-0519 |
| 6 | 744289612 | Saparilas, Theresa | 09MAY2005 | 45000 | |
| 7 | 824904032 | Brosnihan, Dylan | 04JAN2009 | 49000 | |
| 8 | 242779184 | Chao, Daeyong | 28SEP2004 | 48500 | |
| 9 | 544382887 | Slifkin, Leah | 24JUL2011 | 54000 | |
| 10 | 933476520 | Perry, Marguerite | 19APR2010 | 49500 | |

This output illustrates two important points about using PROC APPEND to concatenate data sets with different variables:

- If the BASE= data set contains a variable that is not in the DATA= data set (for example, HomePhone), then PROC APPEND concatenates the data sets and assigns a missing value to that variable in the observations that are taken from the DATA= data set.

- If the DATA= data set contains a variable that is not in the BASE= data set (for example, Gender), then the FORCE option in PROC APPEND forces the procedure to concatenate the two data sets. But because that variable is not in the descriptor portion of the BASE= data set, the procedure cannot include it in the concatenated data set.

*Note:* In the current example, each data set contains a variable that is not in the other. It is only the case of a variable in the DATA= data set that is not in the BASE= data set that requires the use of the FORCE option. However, both cases display a warning in the log.

## Using the APPEND Procedure When Variables Have Different Attributes

When you use PROC APPEND with variables that have different attributes, the following rules apply:

- If a variable has different attributes in the BASE= data set than it does in the DATA= data set, then the attributes in the BASE= data set prevail. In the cases of different formats, informats, and labels, the concatenation succeeds.

- If the length of a variable is longer in the BASE= data set than in the DATA= data set, then the concatenation succeeds.

- If the length of a variable is longer in the DATA= data set than in the BASE= data set, or if the same variable is a character variable in one data set and a numeric variable in the other, then PROC APPEND fails to concatenate the files unless you specify the FORCE option.

Using the FORCE option has these consequences:

- The length that is specified in the BASE= data set prevails. Therefore, SAS truncates values from the DATA= data set to fit them into the length that is specified in the BASE= data set.

- The type that is specified in the BASE= data set prevails. The procedure replaces values of the wrong type (all values for the variable in the DATA= data set) with missing values.

# Choosing between the SET Statement and the APPEND Procedure

If two data sets contain the same variables and the variables have the same attributes, then the file that results from concatenating them with the SET statement is the same as the file that results from concatenating them with the APPEND procedure. The APPEND procedure concatenates much faster than the SET statement, particularly when the BASE= data set is large, because the APPEND procedure does not process the observations from the BASE= data set. However, the two methods of concatenating are sufficiently different when the variables or their attributes differ between data sets. In this case, you must consider the differences in behavior before you decide which method to use.

The following table summarizes the major differences between using the SET statement and using the APPEND procedure to concatenate files.

*Table 17.1* Differences between the SET Statement and the APPEND Procedure

| Criterion | SET Statement | APPEND Procedure |
|---|---|---|
| Number of data sets that you can concatenate | Uses any number of data sets. | Uses two data sets. |
| Handling of data sets that contain different variables | Uses all variables and assigns missing values where appropriate. | Uses all variables in the BASE= data set and assigns missing values to observations from the DATA= data set where appropriate. Requires the FORCE option to concatenate data sets if the DATA= data set contains variables that are not in the BASE= data set. Cannot include variables found only in the DATA= data set when concatenating the data sets. |

| Criterion | SET Statement | APPEND Procedure |
|---|---|---|
| Handling of different formats, informats, or labels | Uses explicitly defined formats, informats, and labels rather than defaults. If two or more data sets explicitly define the format, informat, or label, then SAS uses the definition from the data set you name first in the SET statement. | Uses formats, informats, and labels from the BASE= data set. |
| Handling of different variable lengths | If the same variable has a different length in two or more data sets, then SAS uses the length from the data set you name first in the SET statement. | Requires the FORCE option if the length of a variable is longer in the DATA= data set. Truncates the values of the variable to match the length in the BASE= data set. |
| Handling of different variable types | Does not concatenate the data sets. | Requires the FORCE option to concatenate data sets. Uses the type attribute from the BASE= data set and assigns missing values to the variable in observations from the DATA= data set. |

# Summary

## Statements

LENGTH *variables* <$> *length*;
   specifies the number of bytes that are used for storing variables.

SET *SAS-data-sets*;
   reads one or more SAS data sets and creates a single SAS data set that you specify in the DATA statement.

## Procedures

PROC APPEND BASE=*base-SAS-data-set* <DATA=*SAS-data-set-to-append*> <FORCE>;
   appends the DATA= data set to the BASE= data set. *Base-SAS-data-set* names the SAS data set to which you want to append the observations. If this data set does not exist, then SAS creates it. At the completion of PROC APPEND, the base data set becomes the current (most recently created) SAS data set. *SAS-data-set-to-append* names the SAS data set that contains the observations to add to the end of the base data set. If you omit this option, then PROC APPEND adds the observations in the

current SAS data set to the end of the base data set. The FORCE option forces PROC APPEND to concatenate the files in situations in which the procedure would otherwise fail.

## Learning More

CONTENTS statement
: displays information about a data set, including the names and attributes of all variables. This information reveals any problems that you might have when you try to concatenate data sets, and helps you decide whether to use the SET statement or PROC APPEND. For more information about using the CONTENTS statement in the DATASETS procedure, see "DATASETS" in *Base SAS Procedures Guide*.

END= statement option
: enables you to determine when SAS is processing the last observation in the DATA step. For more information, see Chapter 22, "Conditionally Processing Observations from Multiple SAS Data Sets," on page 349.

IN= data set option
: enables you to process observations from each data set differently. For more information about using the IN= option in the SET statement, see Chapter 22, "Conditionally Processing Observations from Multiple SAS Data Sets," on page 349.

Variable attributes
: For more information, see "SAS Variable Attributes" in *SAS Language Reference: Concepts*.

# Chapter 18
# Interleaving SAS Data Sets

Introduction to Interleaving SAS Data Sets ............................... 281
    Purpose ................................................................. 281
    Prerequisites ............................................................ 281

**Understanding BY-Group Processing Concepts** ........................... 282

**Interleaving Data Sets** ................................................. 282
    Preparing to Interleave Data Sets ........................................ 282
    Understanding the Interleaving Process .................................. 284
    Using the Interleaving Process ........................................... 285

**Summary** .............................................................. 286
    Statements ............................................................. 286

**Learning More** ......................................................... 287

## Introduction to Interleaving SAS Data Sets

### Purpose

Interleaving combines individual sorted SAS data sets into one sorted data set. You interleave data sets by using a SET statement and a BY statement in a DATA step. The number of observations in the new data set is the sum of the number of observations in the original data sets.

In this section, you will learn how to use the BY statement, how to sort data sets to prepare for interleaving, and how to use the SET and BY statements together to interleave observations.

### Prerequisites

Before continuing with this section, you should understand the concepts that are presented in the following sections:

- Chapter 4, "Starting with Raw Data: The Basics," on page 51
- Chapter 6, "Starting with SAS Data Sets," on page 91

## Understanding BY-Group Processing Concepts

The BY statement specifies the variable or variables by which you want to interleave the data sets. In order to understand interleaving, you must understand BY variables, BY values, and BY groups.

BY variable
: specifies a variable that is named in a BY statement and by which the data is sorted or needs to be sorted.

BY value
: specifies the value of a BY variable.

BY group
: specifies the set of all observations with the same value for a BY variable (when only one BY variable is specified). If you use more than one variable in a BY statement, then a BY group is a group of observations with a unique combination of values for those variables. In discussions of interleaving, BY groups commonly span more than one data set.

## Interleaving Data Sets

### Preparing to Interleave Data Sets

Before you can interleave data sets, the data must be sorted by the same variable or variables that will be used with the BY statement that accompanies your SET statement.

For example, the Research and Development division and the Publications division of a company both maintain data sets containing information about each project currently under way. Each data set includes these variables:

Project
: specifies a unique code that identifies the project.

Department
: specifies the name of a department that is involved in the project.

Manager
: specifies the last name of the manager from Department.

StaffCount
: specifies the number of people working for Manager on this project.

Senior management for the company wants to combine the data sets by Project so that the new data set shows the resources that both divisions are devoting to each project. Both data sets must be sorted by Project before they can be interleaved.

The following program creates and displays the data set RESEARCH_DEVELOPMENT. Note that the input data is already sorted by Project.

```
data research_development;
   length Department Manager $ 10;
   input Project $ Department $ Manager $ StaffCount;
   datalines;
```

```
MP971 Designing Daugherty 10
MP971 Coding Newton 8
MP971 Testing Miller 7
SL827 Designing Ramirez 8
SL827 Coding Cho 10
SL827 Testing Baker 7
WP057 Designing Hascal 11
WP057 Coding Constant 13
WP057 Testing Slivko 10
;
run;

proc print data=research_development;
    title 'Research and Development Project Staffing';
run;
```

The following output displays the RESEARCH_DEVELOPMENT data set:

*Figure 18.1* The RESEARCH_DEVELOPMENT Data Set

Research and Development Project Staffing

| Obs | Department | Manager | Project | StaffCount |
|---|---|---|---|---|
| 1 | Designing | Daugherty | MP971 | 10 |
| 2 | Coding | Newton | MP971 | 8 |
| 3 | Testing | Miller | MP971 | 7 |
| 4 | Designing | Ramirez | SL827 | 8 |
| 5 | Coding | Cho | SL827 | 10 |
| 6 | Testing | Baker | SL827 | 7 |
| 7 | Designing | Hascal | WP057 | 11 |
| 8 | Coding | Constant | WP057 | 13 |
| 9 | Testing | Slivko | WP057 | 10 |

The following program creates, sorts, and displays the second data set, PUBLICATIONS. Note that the output data set is sorted by Project.

```
data publications;
    length Department Manager $ 10;
    input Manager $ Department $ Project $ StaffCount;
    datalines;
Cook Writing WP057 5
Deakins Writing SL827 7
Franscombe Editing MP971 4
Henry Editing WP057 3
King Production SL827 5
Krysonski Production WP057 3
Lassiter Graphics SL827 3
Miedema Editing SL827 5
Morard Writing MP971 6
```

```
Posey Production MP971 4
Spackle Graphics WP057 2
;
run;

proc sort data=publications;
   by Project;
run;

proc print data=publications;
   title 'Publications Project Staffing';
run;
```

The following output displays the PUBLICATIONS data set:

*Figure 18.2   The PUBLICATIONS Data Set*

Publications Project Staffing

| Obs | Department | Manager | Project | StaffCount |
|---|---|---|---|---|
| 1 | Editing | Franscombe | MP971 | 4 |
| 2 | Writing | Morard | MP971 | 6 |
| 3 | Production | Posey | MP971 | 4 |
| 4 | Writing | Deakins | SL827 | 7 |
| 5 | Production | King | SL827 | 5 |
| 6 | Graphics | Lassiter | SL827 | 3 |
| 7 | Editing | Miedema | SL827 | 5 |
| 8 | Writing | Cook | WP057 | 5 |
| 9 | Editing | Henry | WP057 | 3 |
| 10 | Production | Krysonski | WP057 | 3 |
| 11 | Graphics | Spackle | WP057 | 2 |

## Understanding the Interleaving Process

When you interleave data sets, SAS creates a new data set as follows:

1. Before executing the SET statement, SAS reads the descriptor portion of each data set that you name in the SET statement. Then SAS creates a program data vector that, by default, contains all the variables from all data sets as well as any variables created by the DATA step. SAS sets the value of each variable to missing.

2. SAS looks at the first BY group in each data set in the SET statement in order to determine which BY group should appear first in the new data set.

3. SAS copies to the new data set all observations in that BY group from each data set that contains observations in the BY group. SAS copies from the data sets in the same order as they appear in the SET statement.

4. SAS looks at the next BY group in each data set to determine which BY group should appear next in the new data set.

5. SAS sets the value of each variable in the program data vector to missing.

6. SAS repeats steps 3 through 5 until it has copied all observations to the new data set.

### *Using the Interleaving Process*

The following program uses the SET and BY statements to interleave the data sets RESEARCH_DEVELOPMENT and PUBLICATIONS.

```
data rnd_pubs;
   set research_development publications;
   by Project;
run;

proc print data=rnd_pubs;
   title 'Project Participation by Research and Development';
   title2 'and Publications Departments';
   title3 'Sorted by Project';
run;
```

The new data set, RND_PUBS, includes all observations from both data sets. Each BY group in the new data set contains observations from RESEARCH_DEVELOPMENT followed by observations from PUBLICATIONS.

*Figure 18.3* Interleaving Two Data Sets

Project Participation by Research and Development
and Publications Departments
Sorted by Project

| Obs | Department | Manager | Project | StaffCount |
|---|---|---|---|---|
| 1 | Designing | Daugherty | MP971 | 10 |
| 2 | Coding | Newton | MP971 | 8 |
| 3 | Testing | Miller | MP971 | 7 |
| 4 | Editing | Franscombe | MP971 | 4 |
| 5 | Writing | Morard | MP971 | 6 |
| 6 | Production | Posey | MP971 | 4 |
| 7 | Designing | Ramirez | SL827 | 8 |
| 8 | Coding | Cho | SL827 | 10 |
| 9 | Testing | Baker | SL827 | 7 |
| 10 | Writing | Deakins | SL827 | 7 |
| 11 | Production | King | SL827 | 5 |
| 12 | Graphics | Lassiter | SL827 | 3 |
| 13 | Editing | Miedema | SL827 | 5 |
| 14 | Designing | Hascal | WP057 | 11 |
| 15 | Coding | Constant | WP057 | 13 |
| 16 | Testing | Slivko | WP057 | 10 |
| 17 | Writing | Cook | WP057 | 5 |
| 18 | Editing | Henry | WP057 | 3 |
| 19 | Production | Krysonski | WP057 | 3 |
| 20 | Graphics | Spackle | WP057 | 2 |

# Summary

## Statements

**SET** *SAS-data-sets*;

**BY** *variable-list*;

The SET statement reads multiple sorted SAS data sets and creates one sorted SAS data set. *SAS-data-sets* is a list of the SAS data sets to interleave.

The BY statement is used with a SET statement to perform BY-group processing. *Variable-list* contains the names of one or more variables (BY variables) by which to interleave the data sets. All of the data sets must be sorted by the same variables before you can interleave them.

# Learning More

Indexes
: You do not need to sort unordered data sets before interleaving them if the data sets have an index on the variable or variables by which you want to interleave. For more information, see "Understanding SAS Indexes" in *SAS Language Reference: Concepts*.

Combining SAS data sets
: For information about combining SAS data sets using the SET statement when the data sets contain different variables, attributes, types, or lengths, see "Concatenating Data Sets with the SET Statement" on page 258.

  For information about combining SAS data sets with the APPEND procedure, see "Concatenating Data Sets By Using the APPEND Procedure" on page 274.

  The same rules apply to interleaving data sets as to concatenating them.

SORT procedure and the BY statement
: For more information, see Chapter 12, "Working with Grouped or Sorted Observations," on page 183.

# Chapter 19
# Merging SAS Data Sets

**Introduction to Merging SAS Data Sets** .................................... **289**
    Purpose .................................................................. 289
    Prerequisites ............................................................. 290

**Understanding the MERGE Statement** ..................................... **290**

**One-to-One Merging** ........................................................ **290**
    Definition of One-to-One Merging ........................................ 290
    Performing a Simple One-to-One Merge .................................... 290
    Performing a One-to-One Merge on Data Sets with the Same Variables ........ 293

**Match-Merging** ............................................................. **296**
    Merging with a BY Statement ............................................. 296
    Input SAS Data Set for Examples ......................................... 297
    The Program ............................................................. 299
    Explanation .............................................................. 300
    Match-Merging Data Sets with Multiple Observations in a BY Group ......... 300
    Match-Merging Data Sets with Dropped Variables .......................... 305
    Match-Merging Data Sets with the IN= Data Set Option .................... 307
    Match-Merging Data Sets with the Same Variables ......................... 308
    Match-Merging Data Sets That Lack a Common Variable .................... 308

**Choosing between One-to-One Merging and Match-Merging** ............... **310**
    Comparing Match-Merge Methods ......................................... 310
    Input SAS Data Set for Examples ......................................... 311
    When to Use a One-to-One Merge ......................................... 312
    When to Use a Match-Merge .............................................. 313

**Summary** .................................................................. **315**
    Statements ............................................................... 315

**Learning More** ............................................................. **315**

## Introduction to Merging SAS Data Sets

### Purpose

Merging combines observations from two or more SAS data sets into a single observation in a new SAS data set. The new data set contains all variables from all the original data sets unless you specify otherwise.

In this section, you will learn about two types of merging: one-to-one merging and match merging. In one-to-one merging, you do not use a BY statement. Observations are combined based on their positions in the input data sets. In match merging, you use a BY statement to combine observations from the input data sets based on common values of the variable by which you merge the data sets.

## Prerequisites

Before continuing with this section, you should understand the concepts that are presented in the following sections:

- Chapter 4, "Starting with Raw Data: The Basics," on page 51
- Chapter 6, "Starting with SAS Data Sets," on page 91

# Understanding the MERGE Statement

You merge data sets using the MERGE statement in a DATA step. The form of the MERGE statement that is used in this section is the following:

**MERGE** *SAS-data-set-list*;

**BY** *variable-list*;

*SAS-data-set-list*
specifies the names of two or more SAS data sets to merge. The list can contain any number of data sets.

*variable-list*
specifies one or more variables by which to merge the data sets. If you use a BY statement, then the data sets must be sorted by the same BY variables before you can merge them.

# One-to-One Merging

## Definition of One-to-One Merging

When you use the MERGE statement without a BY statement, SAS combines the first observation in all data sets you name in the MERGE statement into the first observation in the new data set, the second observation in all data sets into the second observation in the new data set, and so on. In a one-to-one merge, the number of observations in the new data set is equal to the number of observations in the largest data set that you name in the MERGE statement.

## Performing a Simple One-to-One Merge

### Input SAS Data Set for Examples
The instructor of a college acting class wants to schedule a conference with each student. One data set, CLASS, contains these variables:

Name
> specifies the student's name.

Year
> specifies the student's year: first, second, third, or fourth.

Major
> specifies the student's area of specialization. This value is always missing for first-year and second-year students, who have not yet selected a major subject to study.

The following program creates and displays the CLASS data set:

```
data class;
    input Name $ 1-25 Year $ 26-34 Major $ 36-50;
    datalines;
Abbott, Jennifer         first
Carter, Tom              third      Theater
Mendoza, Elissa          fourth     Mathematics
Tucker, Rachel           first
Uhl, Roland              second
Wacenske, Maurice        third      Theater
;
run;

proc print data=class;
    title 'Acting Class Roster';
run;
```

The following output displays the CLASS data set:

*Figure 19.1* *Acting Class Data Set*

Acting Class Roster

| Obs | Name | Year | Major |
|---|---|---|---|
| 1 | Abbott, Jennifer | first | |
| 2 | Carter, Tom | third | Theater |
| 3 | Mendoza, Elissa | fourth | Mathematics |
| 4 | Tucker, Rachel | first | |
| 5 | Uhl, Roland | second | |
| 6 | Wacenske, Maurice | third | Theater |

A second data set contains a list of the dates and times the instructor scheduled conferences, and the rooms in which the conferences are to take place. The following program creates and displays the TIME_SLOT data set. Note the use of the date format and informat.

```
data time_slot;
    input Date date9.  @12 Time $ @19 Room $;
    format date date9.;
    datalines;
14sep2012  10:00  103
14sep2012  10:30  103
```

```
   14sep2012  11:00  207
   15sep2012  10:00  105
   15sep2012  10:30  105
   17sep2012  11:00  207
   ;
run;

proc print data=time_slot;
   title 'Dates, Times, and Locations of Conferences';
run;
```

The following output displays the TIME_SLOT data set:

**Figure 19.2** *The TIME_SLOT Data Set*

Dates, Times, and Locations of Conferences

| Obs | Date | Time | Room |
|---|---|---|---|
| 1 | 14SEP2012 | 10:00 | 103 |
| 2 | 14SEP2012 | 10:30 | 103 |
| 3 | 14SEP2012 | 11:00 | 207 |
| 4 | 15SEP2012 | 10:00 | 105 |
| 5 | 15SEP2012 | 10:30 | 105 |
| 6 | 17SEP2012 | 11:00 | 207 |

## The Program

The following program performs a one-to-one merge of these data sets, assigning a time slot for a conference to each student in the class.

```
data schedule;
   merge class time_slot;
run;

proc print data=schedule;
   title 'Student Conference Assignments';
run;
```

The following output displays the SCHEDULE data set:

*Figure 19.3  One-to-One Merge: The Conference Schedule Data Set*

### Student Conference Assignments

| Obs | Name | Year | Major | Date | Time | Room |
|---|---|---|---|---|---|---|
| 1 | Abbott, Jennifer | first |  | 14SEP2012 | 10:00 | 103 |
| 2 | Carter, Tom | third | Theater | 14SEP2012 | 10:30 | 103 |
| 3 | Mendoza, Elissa | fourth | Mathematics | 14SEP2012 | 11:00 | 207 |
| 4 | Tucker, Rachel | first |  | 15SEP2012 | 10:00 | 105 |
| 5 | Uhl, Roland | second |  | 15SEP2012 | 10:30 | 105 |
| 6 | Wacenske, Maurice | third | Theater | 17SEP2012 | 11:00 | 207 |

### *Explanation*

The preceding output, One-to-One Merge. the SCHEDULE Data Set shows that the new data set combines the first observation from CLASS with the first observation from TIME_SLOT. It then combines the second observation from CLASS with the second observation from TIME_SLOT, and so on.

## Performing a One-to-One Merge on Data Sets with the Same Variables

### *Input SAS Data Set for Examples*

The previous example illustrates the simplest case of a one-to-one merge: the data sets contain the same number of observations, all variables have unique names, and you want to keep all variables from both data sets in the new data set. This example merges data sets that contain variables that have the same names. Also, the second data set in this example contains one more observation than the first data set. Each data set contains data about a separate acting class.

In addition to the CLASS data set, the instructor also uses the CLASS2 data set, which contains the same variables as CLASS but one more observation. The following program creates and displays the CLASS2 data set:

```
data class2;
    input Name $ 1-25 Year $ 26-34 Major $ 36-50;
    datalines;
Hitchcock-Tyler, Erin     second
Keil, Deborah             third     Theater
Nacewicz, Chester         third     Theater
Norgaard, Rolf            second
Prism, Lindsay            fourth    Anthropology
Singh, Rajiv              second
Wittich, Stefan           third     Physics
;
run;

proc print data=class2;
```

```
         title 'Acting Class Roster';
         title2 '(second section)';
      run;
```

The following output displays the CLASS2 data set:

*Figure 19.4*  *The CLASS2 Data Set*

```
                      Acting Class Roster
                         (second section)

     Obs  Name                   Year     Major

       1  Hitchcock-Tyler, Erin  second
       2  Keil, Deborah          third    Theater
       3  Nacewicz, Chester      third    Theater
       4  Norgaard, Rolf         second
       5  Prism, Lindsay         fourth   Anthropology
       6  Singh, Rajiv           second
       7  Wittich, Stefan        third    Physics
```

## *The Program*

Instead of scheduling conferences for one class, the instructor wants to schedule acting exercises for pairs of students, one student from each class. The instructor wants to create a data set in which each observation contains the name of one student from each class and the date, time, and location of the exercise. The variables Year and Major should not be in the new data set.

This new data set can be created by merging the data sets CLASS, CLASS2, and TIME_SLOT. Because Year and Major are not wanted in the new data set, the DROP= data set option can be used to drop them. Notice that the data sets CLASS and CLASS2 both contain the variable Name, but the values for Name are different in each data set. To preserve both sets of values, the RENAME= data set option must be used to rename the variable in one of the data sets.

The following program uses the DROP and RENAME data set options to merge the three data sets:

```
data exercise;
   merge class  (drop=Year Major)
         class2 (drop=Year Major rename=(Name=Name2))
         time_slot;
run;

proc print data=exercise;
   title 'Acting Class Exercise Schedule';
run;
```

The following output displays the merged data set:

*Figure 19.5* Merging Three Data Sets

Acting Class Exercise Schedule

| Obs | Name | Name2 | Date | Time | Room |
|---|---|---|---|---|---|
| 1 | Abbott, Jennifer | Hitchcock-Tyler, Erin | 14SEP2012 | 10:00 | 103 |
| 2 | Carter, Tom | Keil, Deborah | 14SEP2012 | 10:30 | 103 |
| 3 | Mendoza, Elissa | Nacewicz, Chester | 14SEP2012 | 11:00 | 207 |
| 4 | Tucker, Rachel | Norgaard, Rolf | 15SEP2012 | 10:00 | 105 |
| 5 | Uhl, Roland | Prism, Lindsay | 15SEP2012 | 10:30 | 105 |
| 6 | Wacenske, Maurice | Singh, Rajiv | 17SEP2012 | 11:00 | 207 |
| 7 |  | Wittich, Stefan |  |  |  |

## *Explanation*

The following steps describe how SAS merges the data sets:

1. Before executing the DATA step, SAS reads the descriptor portion of each data set that you name in the MERGE statement. Then SAS creates a program data vector for the new data set that, by default, contains all of the variables from all of the data sets. It also contains variables that are created by the DATA step. However, in this case the DROP= data set option excludes the variables Year and Major from the program data vector. The RENAME= data set option adds the variable Name2 to the program data vector. Therefore, the program data vector contains the variables Name, Name2, Date, Time, and Room.

2. SAS sets the value of each variable in the program data vector to missing, as the following figure illustrates.

*Figure 19.6* Program Data Vector Before Reading from Data Sets

| Name | Name2 | Date | Time | Room |
|---|---|---|---|---|
|  |  | . |  |  |

3. Next, SAS reads and copies the first observation from each data set into the program data vector (reading the data sets in the same order they appear in the MERGE statement), as the following figure illustrates.

**Figure 19.7** Program Data Vector After Reading from Each Data Set

| Name | Name2 | Date | Time | Room |
|---|---|---|---|---|
| Abbott, Jennifer | | | . | |

| Name | Name2 | Date | Time | Room |
|---|---|---|---|---|
| Abbott, Jennifer | Hitchcock-Tyler, Erin | | . | |

| Name | Name2 | Date | Time | Room |
|---|---|---|---|---|
| Abbott, Jennifer | Hitchcock-Tyler, Erin | 14SEP2000 | 10:00 | 103 |

4. After processing the first observation from the last data set and executing any other statements in the DATA step, SAS writes the contents of the program data vector to the new data set. If the DATA step attempts to read past the end of a data set, then the values of all variables from that data set in the program data vector are set to missing.

   This behavior has two important consequences:

   - If a variable exists in more than one data set, then the value from the last data set SAS reads is the value that goes into the new data set, even if that value is missing. If you want to keep all the values for like-named variables from different data sets, then you must rename one or more of the variables with the RENAME= data set option so that each variable has a unique name.

   - After SAS processes all observations in a data set, the program data vector and all subsequent observations in the new data set have missing values for the variables unique to that data set. So, as the next figure shows, the program data vector for the last observation in the new data set contains missing values for all variables except Name2.

   **Figure 19.8** Program Data Vector for the Last Observation

| Name | Name2 | Date | Time | Room |
|---|---|---|---|---|
| | Wittich, Stefan | | . | |

5. SAS continues to merge observations until it has copied all observations from all data sets.

# Match-Merging

## Merging with a BY Statement

Merging with a BY statement enables you to match observations according to the values of the BY variables that you specify. Before you can perform a match-merge, all data sets must be sorted by the variables that you want to use for the merge.

In order to understand match-merging, you must understand three key concepts:

## Match-Merging

BY *variable*
: specifies a variable that is named in a BY statement.

BY *value*
: specifies the value of a BY variable.

BY *group*
: specifies the set of all observations with the same value for the BY variable (if there is only one BY variable). If you use more than one variable in a BY statement, then a BY group is the set of observations with a unique combination of values for those variables. In discussions of match-merging, BY groups commonly span more than one data set.

### Input SAS Data Set for Examples

The director of a small repertory theater company, the Little Theater, maintains company records in two SAS data sets, COMPANY and FINANCE.

*Table 19.1  Variables in the COMPANY and FINANCE Data Sets*

| Data Set | Variable | Description |
| --- | --- | --- |
| COMPANY | Name | player's name |
|  | Age | player's age |
|  | Gender | player's gender |
| FINANCE | Name | player's name |
|  | IdNumber | player's employee ID number |
|  | Salary | player's annual salary |

The following program creates, sorts, and displays the COMPANY and FINANCE data sets:

```
data company;
    input Name $ 1-25 Age 27-28 Gender $ 30;
    datalines;
Vincent, Martina          34 F
Phillipon, Marie-Odile    28 F
Gunter, Thomas            27 M
Harbinger, Nicholas       36 M
Benito, Gisela            32 F
Rudelich, Herbert         39 M
Sirignano, Emily          12 F
Morrison, Michael         32 M
;
run;

proc sort data=company;
    by Name;
run;
```

```
data finance;
   input IdNumber $ 1-11 Name $ 13-37 Salary;
   datalines;
074-53-9892 Vincent, Martina          35000
776-84-5391 Phillipon, Marie-Odile    29750
929-75-0218 Gunter, Thomas            27500
446-93-2122 Harbinger, Nicholas       33900
228-88-9649 Benito, Gisela            28000
029-46-9261 Rudelich, Herbert         35000
442-21-8075 Sirignano, Emily          5000
;
run;

proc sort data=finance;
   by Name;
run;

proc print data=company;
   title 'Little Theater Company Roster';
run;

proc print data=finance;
   title 'Little Theater Employee Information';
run;
```

The following output displays the COMPANY and FINANCE data sets. Notice that the FINANCE data set does not contain an observation for Michael Morrison:

*Figure 19.9  The COMPANY Data Set*

Little Theater Company Roster

| Obs | Name | Age | Gender |
|-----|------|-----|--------|
| 1 | Benito, Gisela | 32 | F |
| 2 | Gunter, Thomas | 27 | M |
| 3 | Harbinger, Nicholas | 36 | M |
| 4 | Morrison, Michael | 32 | M |
| 5 | Phillipon, Marie-Odile | 28 | F |
| 6 | Rudelich, Herbert | 39 | M |
| 7 | Sirignano, Emily | 12 | F |
| 8 | Vincent, Martina | 34 | F |

*Figure 19.10* The FINANCE Data Set

Little Theater Employee Information

| Obs | IdNumber | Name | Salary |
|---|---|---|---|
| 1 | 228-88-9649 | Benito, Gisela | 28000 |
| 2 | 929-75-0218 | Gunter, Thomas | 27500 |
| 3 | 446-93-2122 | Harbinger, Nicholas | 33900 |
| 4 | 776-84-5391 | Phillipon, Marie-Odile | 29750 |
| 5 | 029-46-9261 | Rudelich, Herbert | 35000 |
| 6 | 442-21-8075 | Sirignano, Emily | 5000 |
| 7 | 074-53-9892 | Vincent, Martina | 35000 |

## The Program

To avoid having to maintain two separate data sets, the director wants to merge the records for each player from both data sets into a new data set that contains all of the variables. The variable that is common to both data sets is Name. Therefore, Name is the appropriate BY variable.

The data sets are already sorted by Name, so no further sorting is required. The following program merges them by Name:

```
data employee_info;
   merge company finance;
   by name;
run;

proc print data=employee_info;
   title 'Little Theater Employee Information';
   title2 '(including personal and financial information)';
run;
```

The following output displays the merged EMPLOYEE_INFO data set:

*Figure 19.11    Match-Merging: The EMPLOYEE_INFO Data Set*

Little Theater Employee Information
(including personal and financial information)

| Obs | Name | Age | Gender | IdNumber | Salary |
|---|---|---|---|---|---|
| 1 | Benito, Gisela | 32 | F | 228-88-9649 | 28000 |
| 2 | Gunter, Thomas | 27 | M | 929-75-0218 | 27500 |
| 3 | Harbinger, Nicholas | 36 | M | 446-93-2122 | 33900 |
| 4 | Morrison, Michael | 32 | M | | |
| 5 | Phillipon, Marie-Odile | 28 | F | 776-84-5391 | 29750 |
| 6 | Rudelich, Herbert | 39 | M | 029-46-9261 | 35000 |
| 7 | Sirignano, Emily | 12 | F | 442-21-8075 | 5000 |
| 8 | Vincent, Martina | 34 | F | 074-53-9892 | 35000 |

## *Explanation*

The new data set contains one observation for each player in the company. Each observation contains all the variables from both data sets. Notice in particular the fourth observation. The data set FINANCE does not have an observation for Michael Morrison. In this case, the values of the variables that are unique to FINANCE (IdNumber and Salary) are missing.

## *Match-Merging Data Sets with Multiple Observations in a BY Group*

### *Input SAS Data Set for Examples*

The Little Theater has a third data set, REPERTORY, that tracks the casting assignments in each of the season's plays. REPERTORY contains these variables:

Play
  specifies the name of one of the plays in the repertory.

Role
  specifies the name of a character in Play.

IdNumber
  specifies the employee ID number of the player playing Role.

The following program creates and displays the REPERTORY data set:

```
data repertory;
   input Play $ 1-23 Role $ 25-48 IdNumber $ 50-60;
   datalines;
No Exit                Estelle                  074-53-9892
No Exit                Inez                     776-84-5391
No Exit                Valet                    929-75-0218
```

```
    No Exit                 Garcin              446-93-2122
    Happy Days              Winnie              074-53-9892
    Happy Days              Willie              446-93-2122
    The Glass Menagerie     Amanda Wingfield    228-88-9649
    The Glass Menagerie     Laura Wingfield     776-84-5391
    The Glass Menagerie     Tom Wingfield       929-75-0218
    The Glass Menagerie     Jim O'Connor        029-46-9261
    The Dear Departed       Mrs. Slater         228-88-9649
    The Dear Departed       Mrs. Jordan         074-53-9892
    The Dear Departed       Henry Slater        029-46-9261
    The Dear Departed       Ben Jordan          446-93-2122
    The Dear Departed       Victoria Slater     442-21-8075
    The Dear Departed       Abel Merryweather   929-75-0218
    ;
run;

proc print data=repertory;
    title 'Little Theater Season Casting Assignments';
run;
```

The following output displays the REPERTORY data set:

*Figure 19.12   The REPERTORY Data Set*

### Little Theater Season Casting Assignments

| Obs | Play | Role | IdNumber |
|---|---|---|---|
| 1 | No Exit | Estelle | 074-53-9892 |
| 2 | No Exit | Inez | 776-84-5391 |
| 3 | No Exit | Valet | 929-75-0218 |
| 4 | No Exit | Garcin | 446-93-2122 |
| 5 | Happy Days | Winnie | 074-53-9892 |
| 6 | Happy Days | Willie | 446-93-2122 |
| 7 | The Glass Menagerie | Amanda Wingfield | 228-88-9649 |
| 8 | The Glass Menagerie | Laura Wingfield | 776-84-5391 |
| 9 | The Glass Menagerie | Tom Wingfield | 929-75-0218 |
| 10 | The Glass Menagerie | Jim O'Connor | 029-46-9261 |
| 11 | The Dear Departed | Mrs. Slater | 228-88-9649 |
| 12 | The Dear Departed | Mrs. Jordan | 074-53-9892 |
| 13 | The Dear Departed | Henry Slater | 029-46-9261 |
| 14 | The Dear Departed | Ben Jordan | 446-93-2122 |
| 15 | The Dear Departed | Victoria Slater | 442-21-8075 |
| 16 | The Dear Departed | Abel Merryweather | 929-75-0218 |

**302** Chapter 19 • *Merging SAS Data Sets*

To maintain confidentiality during preliminary casting, this data set identifies players by employee ID number. However, casting decisions are now final, and the manager wants to replace each employee ID number with the player's name. Of course, it is possible to re-create the data set, entering each player's name instead of the employee ID number in the raw data. However, it is more efficient to make use of the FINANCE data set, which already contains the name and employee ID number of all players.

When the data sets are merged, SAS adds the players' names to the data set. Of course, before you can merge the data sets, you must sort them by IdNumber.

```
proc sort data=finance;
   by IdNumber;
run;

proc sort data=repertory;
   by IdNumber;
run;

proc print data=finance;
   title 'Little Theater Employee Information';
   title2 '(sorted by employee ID number)';
run;

proc print data=repertory;
   title 'Little Theater Season Casting Assignments';
   title2 '(sorted by employee ID number)';
run;
```

The following output displays the FINANCE and REPERTORY data sets, sorted by IdNumber:

*Figure 19.13* The FINANCE Data Set Sorted by IdNumber

```
         Little Theater Employee Information
              (sorted by employee ID number)

      Obs   IdNumber       Name                   Salary

       1    029-46-9261    Rudelich, Herbert       35000
       2    074-53-9892    Vincent, Martina        35000
       3    228-88-9649    Benito, Gisela          28000
       4    442-21-8075    Sirignano, Emily         5000
       5    446-93-2122    Harbinger, Nicholas     33900
       6    776-84-5391    Phillipon, Marie-Odile  29750
       7    929-75-0218    Gunter, Thomas          27500
```

*Figure 19.14 The REPERTORY Data Set Sorted by IdNumber*

Little Theater Season Casting Assignments
(sorted by employee ID number)

| Obs | Play | Role | IdNumber |
|-----|------|------|----------|
| 1 | The Glass Menagerie | Jim O'Connor | 029-46-9261 |
| 2 | The Dear Departed | Henry Slater | 029-46-9261 |
| 3 | No Exit | Estelle | 074-53-9892 |
| 4 | Happy Days | Winnie | 074-53-9892 |
| 5 | The Dear Departed | Mrs. Jordan | 074-53-9892 |
| 6 | The Glass Menagerie | Amanda Wingfield | 228-88-9649 |
| 7 | The Dear Departed | Mrs. Slater | 228-88-9649 |
| 8 | The Dear Departed | Victoria Slater | 442-21-8075 |
| 9 | No Exit | Garcin | 446-93-2122 |
| 10 | Happy Days | Willie | 446-93-2122 |
| 11 | The Dear Departed | Ben Jordan | 446-93-2122 |
| 12 | No Exit | Inez | 776-84-5391 |
| 13 | The Glass Menagerie | Laura Wingfield | 776-84-5391 |
| 14 | No Exit | Valet | 929-75-0218 |
| 15 | The Glass Menagerie | Tom Wingfield | 929-75-0218 |
| 16 | The Dear Departed | Abel Merryweather | 929-75-0218 |

These two data sets contain seven BY groups. That is, among the 23 observations are seven different values for the BY variable, IdNumber. The first BY group has a value of 029-46-9261 for IdNumber. FINANCE has one observation in this BY group; REPERTORY has two. The last BY group has a value of 929-75-0218 for IdNumber. FINANCE has one observation in this BY group; REPERTORY has three.

## The Program

The following program merges the data sets FINANCE and REPERTORY. It also illustrates what happens when a BY group in one data set has more observations in it than the same BY group in the other data set.

The resulting data set contains all variables from both data sets.

```
data repertory_name;
   merge finance repertory;
   by IdNumber;
run;

proc print data=repertory_name;
   title 'Little Theater Season Casting Assignments';
   title2 'with employee financial information';
run;
```

The following output displays the merged data set:

*Figure 19.15  Match-Merge with Multiple Observations in a BY Group*

### Little Theater Season Casting Assignments
### with employee financial information

| Obs | IdNumber | Name | Salary | Play | Role |
|---|---|---|---|---|---|
| 1 | 029-46-9261 | Rudelich, Herbert | 35000 | The Glass Menagerie | Jim O'Connor |
| 2 | 029-46-9261 | Rudelich, Herbert | 35000 | The Dear Departed | Henry Slater |
| 3 | 074-53-9892 | Vincent, Martina | 35000 | No Exit | Estelle |
| 4 | 074-53-9892 | Vincent, Martina | 35000 | Happy Days | Winnie |
| 5 | 074-53-9892 | Vincent, Martina | 35000 | The Dear Departed | Mrs. Jordan |
| 6 | 228-88-9649 | Benito, Gisela | 28000 | The Glass Menagerie | Amanda Wingfield |
| 7 | 228-88-9649 | Benito, Gisela | 28000 | The Dear Departed | Mrs. Slater |
| 8 | 442-21-8075 | Sirignano, Emily | 5000 | The Dear Departed | Victoria Slater |
| 9 | 446-93-2122 | Harbinger, Nicholas | 33900 | No Exit | Garcin |
| 10 | 446-93-2122 | Harbinger, Nicholas | 33900 | Happy Days | Willie |
| 11 | 446-93-2122 | Harbinger, Nicholas | 33900 | The Dear Departed | Ben Jordan |
| 12 | 776-84-5391 | Phillipon, Marie-Odile | 29750 | No Exit | Inez |
| 13 | 776-84-5391 | Phillipon, Marie-Odile | 29750 | The Glass Menagerie | Laura Wingfield |
| 14 | 929-75-0218 | Gunter, Thomas | 27500 | No Exit | Valet |
| 15 | 929-75-0218 | Gunter, Thomas | 27500 | The Glass Menagerie | Tom Wingfield |
| 16 | 929-75-0218 | Gunter, Thomas | 27500 | The Dear Departed | Abel Merryweather |

### Explanation

Carefully examine the first few observations in the new data set and consider how SAS creates them.

1. Before executing the DATA step, SAS reads the descriptor portion of the two data sets and creates a program data vector that contains all variables from both data sets:

   - IdNumber, Name, and Salary from FINANCE
   - Play and Role from REPERTORY.

   IdNumber is already in the program data vector because it is in the FINANCE data set. SAS sets the values of all variables to missing, as the following figure illustrates.

*Figure 19.16  Program Data Vector Before Reading from Data Sets*

| IdNumber | Name | Salary | Play | Role |
|---|---|---|---|---|
|  |  | . |  |  |

2. SAS looks at the first BY group in each data set to determine which BY group should appear first. In this case, the first BY group, observations with the value 029-46-9261 for IdNumber, is the same in both data sets.

3. SAS reads and copies the first observation from FINANCE into the program data vector, as the next figure illustrates.

*Figure 19.17* Program Data Vector After Reading FINANCE Data Set

| IdNumber | Name | Salary | Play | Role |
|---|---|---|---|---|
| 029-46-9261 | Rudelich, Herbert | 35000 | | |

4. SAS reads and copies the first observation from REPERTORY into the program data vector, as the next figure illustrates. If a data set does not have any observations in a BY group, then the program data vector contains missing values for the variables that are unique to that data set.

*Figure 19.18* Program Data Vector After Reading REPERTORY Data Set

| IdNumber | Name | Salary | Play | Role |
|---|---|---|---|---|
| 029-46-9261 | Rudelich, Herbert | 35000 | The Glass Menagerie | Jim O'Connor |

5. SAS writes the observation to the new data set and retains the values in the program data vector. (If the program data vector contained variables created by the DATA step, then SAS would set them to missing after writing to the new data set.)

6. SAS looks for a second observation in the BY group in each data set. REPERTORY has one; FINANCE does not. The MERGE statement reads the second observation in the BY group from REPERTORY. Because FINANCE has only one observation in the BY group, the statement uses the values of Name (**Rudelich, Herbert**) and Salary (**35000**) that were retained in the program data vector for the second observation in the new data set. The next figure illustrates this behavior.

*Figure 19.19* Program Data Vector with Second Observation in the BY Group

| IdNumber | Name | Salary | Play | Role |
|---|---|---|---|---|
| 029-46-9261 | Rudelich, Herbert | 35000 | The Dear Departed | Henry Slater |

7. SAS writes the observation to the new data set. Neither data set contains any more observations in this BY group. Therefore, as the final figure illustrates, SAS sets all values in the program data vector to missing and begins processing the next BY group. It continues processing observations until it exhausts all observations in both data sets.

*Figure 19.20* Program Data Vector before New BY Groups

| IdNumber | Name | Salary | Play | Role |
|---|---|---|---|---|
| | | . | | |

## Match-Merging Data Sets with Dropped Variables

Now that casting decisions are final, the director wants to post the casting list, but does not want to include salary or employee ID information. As the next program illustrates,

Salary and IdNumber can be eliminated by using the DROP= data set option when creating the new data set.

```
data newrep (drop=IdNumber);
   merge finance (drop=Salary) repertory;
   by IdNumber;
run;

proc print data=newrep;
   title 'Final Little Theater Season Casting Assignments';
run;
```

*Note:* The difference in placement of the two DROP= data set options is crucial. Dropping IdNumber in the DATA statement means that the variable is available to the MERGE and BY statements (to which it is essential), but that it does not go into the new data set. Dropping Salary in the MERGE statement means that the MERGE statement does not even read this variable, so Salary is unavailable to the program statements. Because the variable Salary is not needed for processing, it is more efficient to prevent it from being read into the PDV in the first place.

The following output displays the merged data set without the IdNumber and Salary variables:

*Figure 19.21* Match-Merging Data Sets with Dropped Variables

Final Little Theater Season Casting Assignments

| Obs | Name | Play | Role |
|---|---|---|---|
| 1 | Rudelich, Herbert | The Glass Menagerie | Jim O'Connor |
| 2 | Rudelich, Herbert | The Dear Departed | Henry Slater |
| 3 | Vincent, Martina | No Exit | Estelle |
| 4 | Vincent, Martina | Happy Days | Winnie |
| 5 | Vincent, Martina | The Dear Departed | Mrs. Jordan |
| 6 | Benito, Gisela | The Glass Menagerie | Amanda Wingfield |
| 7 | Benito, Gisela | The Dear Departed | Mrs. Slater |
| 8 | Sirignano, Emily | The Dear Departed | Victoria Slater |
| 9 | Harbinger, Nicholas | No Exit | Garcin |
| 10 | Harbinger, Nicholas | Happy Days | Willie |
| 11 | Harbinger, Nicholas | The Dear Departed | Ben Jordan |
| 12 | Phillipon, Marie-Odile | No Exit | Inez |
| 13 | Phillipon, Marie-Odile | The Glass Menagerie | Laura Wingfield |
| 14 | Gunter, Thomas | No Exit | Valet |
| 15 | Gunter, Thomas | The Glass Menagerie | Tom Wingfield |
| 16 | Gunter, Thomas | The Dear Departed | Abel Merryweather |

## Match-Merging Data Sets with the IN= Data Set Option

You can merge data sets after checking to see if an observation occurs in one or more data sets by using the IN= data set option. Suppose that you create a data set called HIGH_SALARY that includes only members of the company that make more than $28,000. You want to see a list of all the roles for these higher paid actors. You can create this list if you merge the REPERTORY data set with the HIGH_SALARY data set. Use the IN= data set option to include only roles that were portrayed by a higher paid actor. That is, you want to include a role in the merged data set only if the actor was also in the HIGH_SALARY data set.

First, create the HIGH_SALARY data set.

```
data high_salary;
   set finance;
   if (salary > 28000);
run;
```

Sort the data sets by IdNumber and then merge them. Use the IN= data set option to include only the roles that were portrayed by actors in the HIGH_SALARY data set. Print the list of HIGH_PAID_ROLES to see the resulting data set.

```
proc sort data=high_salary;
   by IdNumber;
run;

proc sort data=repertory;
   by IdNumber;
run;

data high_paid_roles(drop=IdNumber);
   merge high_salary(in=in_high_sal) repertory;
   by IdNumber;
   if in_high_sal then output;
run;

proc print data=high_paid_roles;
title 'Roles of Highest Paid Performers';
run;
```

The following output shows the list of roles for the higher paid actors. The list includes 10 observations. This is shorter than the entire list of 16 observations in the REPERTORY data set (see Figure 19.14 on page 303).

**Figure 19.22** Roles for Actors in HIGH_SALARY

| Obs | Name | Salary | Play | Role |
|---|---|---|---|---|
| 1 | Rudelich, Herbert | 35000 | The Glass Menagerie | Jim O'Connor |
| 2 | Rudelich, Herbert | 35000 | The Dear Departed | Henry Slater |
| 3 | Vincent, Martina | 35000 | No Exit | Estelle |
| 4 | Vincent, Martina | 35000 | Happy Days | Winnie |
| 5 | Vincent, Martina | 35000 | The Dear Departed | Mrs. Jordan |
| 6 | Harbinger, Nicholas | 33900 | No Exit | Garcin |
| 7 | Harbinger, Nicholas | 33900 | Happy Days | Willie |
| 8 | Harbinger, Nicholas | 33900 | The Dear Departed | Ben Jordan |
| 9 | Phillipon, Marie-Odile | 29750 | No Exit | Inez |
| 10 | Phillipon, Marie-Odile | 29750 | The Glass Menagerie | Laura Wingfield |

*Roles of Highest Paid Performers*

## Match-Merging Data Sets with the Same Variables

You can match-merge data sets that contain the same variables (variables with the same name) by using the RENAME= data set option, just as you would when performing a one-to-one merge (see "Performing a One-to-One Merge on Data Sets with the Same Variables" on page 293).

If you do not use the RENAME= option and a variable exists in more than one data set, then the value of that variable in the last data set that is read is the value that goes into the new data set.

## Match-Merging Data Sets That Lack a Common Variable

You can name any number of data sets in the MERGE statement. However, if you are match-merging the data sets, then you must be sure they all have a common variable and are sorted by that variable. If the data sets do not have a common variable, then you might be able to use another data set that has variables common to the original data sets to merge them.

For example, consider the data sets that are used in the match-merge examples. The following table displays the names of the data sets and the names of the variables in each data set.

**Table 19.2** Data Sets and Variables That Are Used in Match-Merge Examples

| Data Set | Variables |
|---|---|
| COMPANY | Name, Age, Gender |
| FINANCE | Name, IdNumber, Salary |
| REPERTORY | Play, Role, IdNumber |

These data sets do not share a common variable. However, COMPANY and FINANCE share the variable Name. Similarly, FINANCE and REPERTORY share the variable IdNumber. Therefore, as the next program shows, you can merge the data sets into one with two separate DATA steps. As usual, you must sort the data sets by the appropriate BY variable. (REPERTORY is already sorted by IdNumber.)

```
   /* Sort FINANCE and COMPANY by Name */
proc sort data=finance;
   by Name;
run;

proc sort data=company;
   by Name;
run;

   /* Merge COMPANY and FINANCE into a */
   /* temporary data set.              */
data temp;
   merge company finance;
   by Name;
run;

proc sort data=temp;
   by IdNumber;
run;

   /* Merge the temporary data set with REPERTORY */
data all;
   merge temp repertory;
   by IdNumber;
run;

proc print data=all;
   title 'Little Theater Complete Casting Information';
run;
```

In order to merge the three data sets, this program performs the following tasks:

- sorts FINANCE and COMPANY by Name
- merges COMPANY and FINANCE into a temporary data set, TEMP
- sorts TEMP by IdNumber
- merges TEMP and REPERTORY by IdNumber

The following output displays the resulting data set, ALL:

*Figure 19.23*  *Match-Merging Data Sets That Lack a Common Variable*

### Little Theater Complete Casting Information

| Obs | Name | Age | Gender | IdNumber | Salary | Play | Role |
|---|---|---|---|---|---|---|---|
| 1 | Morrison, Michael | 32 | M | | | | |
| 2 | Rudelich, Herbert | 39 | M | 029-46-9261 | 35000 | The Glass Menagerie | Jim O'Connor |
| 3 | Rudelich, Herbert | 39 | M | 029-46-9261 | 35000 | The Dear Departed | Henry Slater |
| 4 | Vincent, Martina | 34 | F | 074-53-9892 | 35000 | No Exit | Estelle |
| 5 | Vincent, Martina | 34 | F | 074-53-9892 | 35000 | Happy Days | Winnie |
| 6 | Vincent, Martina | 34 | F | 074-53-9892 | 35000 | The Dear Departed | Mrs. Jordan |
| 7 | Benito, Gisela | 32 | F | 228-88-9649 | 28000 | The Glass Menagerie | Amanda Wingfield |
| 8 | Benito, Gisela | 32 | F | 228-88-9649 | 28000 | The Dear Departed | Mrs. Slater |
| 9 | Sirignano, Emily | 12 | F | 442-21-8075 | 5000 | The Dear Departed | Victoria Slater |
| 10 | Harbinger, Nicholas | 36 | M | 446-93-2122 | 33900 | No Exit | Garcin |
| 11 | Harbinger, Nicholas | 36 | M | 446-93-2122 | 33900 | Happy Days | Willie |
| 12 | Harbinger, Nicholas | 36 | M | 446-93-2122 | 33900 | The Dear Departed | Ben Jordan |
| 13 | Phillipon, Marie-Odile | 28 | F | 776-84-5391 | 29750 | No Exit | Inez |
| 14 | Phillipon, Marie-Odile | 28 | F | 776-84-5391 | 29750 | The Glass Menagerie | Laura Wingfield |
| 15 | Gunter, Thomas | 27 | M | 929-75-0218 | 27500 | No Exit | Valet |
| 16 | Gunter, Thomas | 27 | M | 929-75-0218 | 27500 | The Glass Menagerie | Tom Wingfield |
| 17 | Gunter, Thomas | 27 | M | 929-75-0218 | 27500 | The Dear Departed | Abel Merryweather |

# Choosing between One-to-One Merging and Match-Merging

## Comparing Match-Merge Methods

Use one-to-one merging when you want to combine one observation from each data set, but it is not important to match observations. For example, when merging an observation that contains a student's name, year, and major with an observation that contains a date, time, and location for a conference, it does not matter which student gets which time slot. Therefore, a one-to-one merge is appropriate.

In cases where you must merge certain observations, use a match-merge. For example, when merging employee information from two different data sets, it is crucial that you merge observations that relate to the same employee. Therefore, you must use a match-merge.

Sometimes you might want to merge by a particular variable, and your data is arranged in such a way that you can see that a one-to-one merge will work. The next example illustrates a case when you could use a one-to-one merge for matching observations because you are certain that your data is ordered correctly. However, as a subsequent example shows, it is risky to use a one-to-one merge in such situations.

## Input SAS Data Set for Examples

Consider the COMPANY2 data set. Each observation in this data set corresponds to an observation with the same value of Name in FINANCE. The program that follows creates and displays the COMPANY2 data set. It also displays the FINANCE data set for comparison.

```
data company2;
    input name $ 1-25 age 27-28 gender $ 30;
    datalines;
Benito, Gisela            32 F
Gunter, Thomas            27 M
Harbinger, Nicholas       36 M
Phillipon, Marie-Odile    28 F
Rudelich, Herbert         39 M
Sirignano, Emily          12 F
Vincent, Martina          34 F
;
run;

proc print data=company2;
    title 'Little Theater Company Roster';
run;

proc print data=finance;
    title 'Little Theater Employee Information';
run;
```

The following output displays the two data sets:

**Figure 19.24** The COMPANY2 Data Set

Little Theater Company Roster

| Obs | name | age | gender |
| --- | --- | --- | --- |
| 1 | Benito, Gisela | 32 | F |
| 2 | Gunter, Thomas | 27 | M |
| 3 | Harbinger, Nicholas | 36 | M |
| 4 | Phillipon, Marie-Odile | 28 | F |
| 5 | Rudelich, Herbert | 39 | M |
| 6 | Sirignano, Emily | 12 | F |
| 7 | Vincent, Martina | 34 | F |

*Figure 19.25* The FINANCE Data Set

Little Theater Employee Information

| Obs | IdNumber | Name | Salary |
|---|---|---|---|
| 1 | 228-88-9649 | Benito, Gisela | 28000 |
| 2 | 929-75-0218 | Gunter, Thomas | 27500 |
| 3 | 446-93-2122 | Harbinger, Nicholas | 33900 |
| 4 | 776-84-5391 | Phillipon, Marie-Odile | 29750 |
| 5 | 029-46-9261 | Rudelich, Herbert | 35000 |
| 6 | 442-21-8075 | Sirignano, Emily | 5000 |
| 7 | 074-53-9892 | Vincent, Martina | 35000 |

## When to Use a One-to-One Merge

The following program shows that because both data sets are sorted by Name and because each observation in one data set has a corresponding observation in the other data set, a one-to-one merge has the same result as merging by Name.

```
   /* One-to-one merge */
data one_to_one;
   merge company2 finance;
run;

proc print data=one_to_one;
   title 'Using a One-to-One Merge to Combine';
   title2 'COMPANY2 and FINANCE';
run;

   /* Match-merge */
data match;
   merge company2 finance;
   by name;
run;

proc print data=match;
   title 'Using a Match-Merge to Combine';
   title2 'COMPANY2 and FINANCE';
run;
```

The following output displays the results of the two merges. You can see that they are identical:

**Figure 19.26** *Using a One-to-One Merge to Combine Observations When Observations Correspond*

Using a One-to-One Merge to Combine
COMPANY2 and FINANCE

| Obs | name | age | gender | IdNumber | Salary |
|---|---|---|---|---|---|
| 1 | Benito, Gisela | 32 | F | 228-88-9649 | 28000 |
| 2 | Gunter, Thomas | 27 | M | 929-75-0218 | 27500 |
| 3 | Harbinger, Nicholas | 36 | M | 446-93-2122 | 33900 |
| 4 | Phillipon, Marie-Odile | 28 | F | 776-84-5391 | 29750 |
| 5 | Rudelich, Herbert | 39 | M | 029-46-9261 | 35000 |
| 6 | Sirignano, Emily | 12 | F | 442-21-8075 | 5000 |
| 7 | Vincent, Martina | 34 | F | 074-53-9892 | 35000 |

**Figure 19.27** *Using a Match-Merge to Combine Observations When Observations Correspond*

Using a Match-Merge to Combine
COMPANY2 and FINANCE

| Obs | name | age | gender | IdNumber | Salary |
|---|---|---|---|---|---|
| 1 | Benito, Gisela | 32 | F | 228-88-9649 | 28000 |
| 2 | Gunter, Thomas | 27 | M | 929-75-0218 | 27500 |
| 3 | Harbinger, Nicholas | 36 | M | 446-93-2122 | 33900 |
| 4 | Phillipon, Marie-Odile | 28 | F | 776-84-5391 | 29750 |
| 5 | Rudelich, Herbert | 39 | M | 029-46-9261 | 35000 |
| 6 | Sirignano, Emily | 12 | F | 442-21-8075 | 5000 |
| 7 | Vincent, Martina | 34 | F | 074-53-9892 | 35000 |

Even though the resulting data sets are identical, it is not wise to use a one-to-one merge when it is essential to merge a particular observation from one data set with a particular observation from another data set.

## When to Use a Match-Merge

In the previous example, you can easily determine that the data sets contain the same values for Name and that the values appear in the same order. However, if the data sets contained hundreds of observations, then it would be difficult to determine that all the values match. If the observations do not match, then serious problems can occur. The next example illustrates why you should not use a one-to-one merge for matching observations.

Consider the original data set, COMPANY, which contains an observation for Michael Morrison (see Figure 19.9 on page 298). The FINANCE data set has no corresponding observation. If you do not realize this difference and try to use the following program to perform a one-to-one merge with FINANCE, then several problems could appear.

```
data badmerge;
   merge company finance;
run;

proc print data=badmerge;
   title 'Using a One-to-One Merge Instead of a Match-Merge';
run;
```

The following output shows the potential problems:

**Figure 19.28** *One-to-One Merge with Unequal Numbers of Observations in Each Data Set*

Using a One-to-One Merge Instead of a Match-Merge

| Obs | Name | Age | Gender | IdNumber | Salary |
|-----|------|-----|--------|----------|--------|
| 1 | Benito, Gisela | 32 | F | 228-88-9649 | 28000 |
| 2 | Gunter, Thomas | 27 | M | 929-75-0218 | 27500 |
| 3 | Harbinger, Nicholas | 36 | M | 446-93-2122 | 33900 |
| 4 | Phillipon, Marie-Odile | 32 | M | 776-84-5391 | 29750 |
| 5 | Rudelich, Herbert | 28 | F | 029-46-9261 | 35000 |
| 6 | Sirignano, Emily | 39 | M | 442-21-8075 | 5000 |
| 7 | Vincent, Martina | 12 | F | 074-53-9892 | 35000 |
| 8 | Vincent, Martina | 34 | F | | |

The first three observations merge correctly. However, FINANCE does not have an observation for Michael Morrison. A one-to-one merge makes no attempt to match parts of the observations from the different data sets. It simply combines observations based on their positions in the data sets that you name in the MERGE statement. Therefore, the fourth observation in BADMERGE combines the fourth observation in COMPANY (Michael's name, age, and gender) with the fourth observation in FINANCE (Marie-Odile's name, employee ID number, and salary). As SAS combines the observations, Marie-Odile's name overwrites Michael's. After writing this observation to the new data set, SAS processes the next observation in each data set. These observations are similarly mismatched.

This type of mismatch continues until the seventh observation when the MERGE statement exhausts the observations in the smaller data set, FINANCE. After writing the seventh observation to the new data set, SAS begins the next iteration of the DATA step. Because SAS has read all observations in FINANCE, it sets the values for variables from that data set to missing in the program data vector. Then it reads the values for Name, Age, and Gender from COMPANY and writes the contents of the program data vector to the new data set. Therefore, the last observation has the same value for NAME as the previous observation and contains missing values for IdNumber and Salary.

These missing values and the duplication of the value for Name might make you suspect that the observations did not merge as you intended them to. However, if instead of being an additional observation, the observation for Michael Morrison replaced another observation in COMPANY2, then no observations would have missing values, and the problem would not be as easy to detect. Therefore, you are safer using a match-merge in situations that call for it even if you think the data is arranged so that a one-to-one merge has the same results.

# Summary

## Statements

MERGE *SAS-data-set-list*;
BY *variable-list*;

The MERGE statement reads observations in multiple SAS data sets and combines them into one observation in one new SAS data set. *SAS-data-set-list* is a list of the SAS data sets to merge. The list can contain any number of data sets.

*Variable-list* is the name of one or more variables by which to merge the data sets. If you use a BY statement, then the data sets must be sorted by the same BY variables before you can merge them. If you do not use a BY statement, then SAS merges observations based on their positions in the original data sets.

# Learning More

Indexes

If a data set has an index on the variable or variables named in the BY statement that accompanies the MERGE statement, then you do not need to sort that data set. For more information, see "Understanding SAS Indexes" in *SAS Language Reference: Concepts*.

SAS date and time formats and informats

The examples in this section read Time as a character variable, and they read Date with a SAS date informat. You could read Time using special SAS time informats. For more information about SAS date and time informats, see "Working with SAS Dates and Times" in *SAS Language Reference: Concepts*.

# Chapter 20
# Updating SAS Data Sets

Introduction to Updating SAS Data Sets ................................ 317
    Purpose ............................................................. 317
    Prerequisites ........................................................ 317

Understanding the UPDATE Statement ................................. 318

Understanding How to Select BY Variables ............................ 318

Updating a Data Set .................................................... 319

Updating with Incremental Values ...................................... 324

Understanding the Differences between Updating and Merging ......... 326
    General Comparisons between Updating and Merging ................. 326
    How the UPDATE and MERGE Statements Process Missing Values Differently . 328
    How the UPDATE and MERGE Statements Process Multiple
      Observations in a BY Group Differently ............................ 329

Handling Missing Values ............................................... 329

Summary .............................................................. 333
    Statements .......................................................... 333

Learning More ......................................................... 333

## Introduction to Updating SAS Data Sets

### Purpose

Updating replaces the values of variables in one data set with nonmissing values from another data set. In this section, you will learn about the following concepts:

- difference between master data sets and transaction data sets
- using the UPDATE statement
- how to choose between updating and merging

### Prerequisites

Before using this section, you should understand the concepts that are presented in the following sections:

- Chapter 4, "Starting with Raw Data: The Basics," on page 51
- Chapter 6, "Starting with SAS Data Sets," on page 91
- Chapter 19, "Merging SAS Data Sets," on page 289

## Understanding the UPDATE Statement

When you update, you work with two SAS data sets. The data set that contains the original information is the master data set. The data set that contains the new information is the transaction data set. Many applications, such as maintaining mailing lists and inventories, call for periodic updates of information.

In a DATA step, the UPDATE statement reads observations from the transaction data set and updates corresponding observations (observations with the same value of all BY variables) from the master data set. All nonmissing values for variables in the transaction data set replace the corresponding values that are read from the master data set. SAS writes the modified observations to the data set that you name in the DATA statement without modifying either the master or the transaction data set.

The general form of the UPDATE statement follows:

**UPDATE** *master-SAS-data-set transaction-SAS-data-set*;

**BY** *identifier-list*;

*master-SAS-data-set*
   specifies the SAS data set that contains information that you want to update.

*transaction-SAS-data-set*
   specifies the SAS data set that contains information for updating the master data set.

*identifier-list*
   specifies the list of BY variables by which you identify corresponding observations.

If the master data set contains an observation that does not correspond to an observation in the transaction data set, then the DATA step writes that observation to the new data set without modification. An observation from the transaction data set that does not correspond to any observation in the master data set becomes the basis for a new observation. The new observation can be modified by other observations from the transaction data set before it is written to the new data set.

## Understanding How to Select BY Variables

The master data set and the transaction data set must be sorted by the same variable or variables that you specify in the BY statement. Select a variable that meets these criteria:

- The value of the variable is unique for each observation in the master data set. If you use more than one BY variable, no two observations in the master data set should have the same values for all BY variables.
- The variable or variables never need to be updated.

Some examples of variables that you can use in the BY statement include employee or student identification numbers, stock numbers, and the names of objects in an inventory.

If you are updating a data set, you probably do not want duplicate values of BY variables in the master data set. For example, if you update by Name, each observation in the master data set should have a unique value of Name. If you update by Name and Age, two or more observations can have the same value for either Name or Age but should not have the same values for both. SAS warns you if it finds duplicates but proceeds with the update. It applies all transactions only to the first observation in the BY group in the master data set.

# Updating a Data Set

In this example, the circulation department of a magazine maintains a mailing list that contains tens of thousands of names. Each issue of the magazine contains a form for readers to fill out when they change their names or addresses. To simplify the maintenance job, the form requests that readers send only new information. New subscribers can start a subscription by completing the entire form. When a form is received, a data entry operator enters the information about the form into a raw data file. The mailing list is updated once per month from the raw data file.

The mailing list includes these variables for each subscriber:

SubscriberId
> specifies a unique number that is assigned to the subscriber at the time the subscription begins. A subscriber's SubscriberId never changes.

Name
> specifies the subscriber's name. The last name appears first, followed by a comma and the first name.

StreetAddress
> specifies the subscriber's street address.

City
> specifies the subscriber's city.

StateProv
> specifies the subscriber's state or province. This variable is missing for addresses outside the United States and Canada.

PostalCode
> specifies the subscriber's postal code (ZIP code for addresses in the United States).

Country
> specifies the subscriber's country.

The following program creates and displays the first part of this data set. The raw data are already sorted by SubscriberId.

```
data mail_list;
   input SubscriberId 1-8 Name $ 9-27 StreetAddress $ 28-47 City $ 48-62
         StateProv $ 63-64 PostalCode $ 67-73 Country $ ;
   datalines;
1001    Ericson, Jane      111 Clancey Court   Chapel Hill    NC   27514   USA
1002    Dix, Martin        4 Shepherd St.      Vancouver      BC   V6C 3E8 Canada
1003    Gabrielli, Theresa Via Pisanelli, 25   Roma                00196   Italy
1004    Clayton, Aria      14 Bridge St.       San Francisco  CA   94124   USA
1005    Archuleta, Ruby    Box 108             Milagro        NM   87429   USA
1006    Misiewicz, Jeremy  43-C Lakeview Apts. Madison        WI   53704   USA
```

```
                    1007    Ahmadi, Hafez      52 Rue Marston      Paris                    75019   France
                    1008    Jacobson, Becky    1 Lincoln St.       Tallahassee     FL       32312   USA
                    1009    An, Ing            95 Willow Dr.       Toronto         ON       M5J 2T3 Canada
                    1010    Slater, Emily      1009 Cherry St.     York            PA       17407   USA
                    ;
                 run;

                 proc print data=mail_list;
                    title 'Magazine Master Mailing List';
                 run;
```

The following output displays the master MAIL_LIST data set:

*Figure 20.1* Magazine Master Mailing List

Magazine Master Mailing List

| Obs | SubscriberId | Name | StreetAddress | City | StateProv | PostalCode | Country |
|---|---|---|---|---|---|---|---|
| 1 | 1001 | Ericson, Jane | 111 Clancey Court | Chapel Hill | NC | 27514 | USA |
| 2 | 1002 | Dix, Martin | 4 Shepherd St. | Vancouver | BC | V6C 3E8 | Canada |
| 3 | 1003 | Gabrielli, Theresa | Via Pisanelli, 25 | Roma | | 00196 | Italy |
| 4 | 1004 | Clayton, Aria | 14 Bridge St. | San Francisco | CA | 94124 | USA |
| 5 | 1005 | Archuleta, Ruby | Box 108 | Milagro | NM | 87429 | USA |
| 6 | 1006 | Misiewicz, Jeremy | 43-C Lakeview Apts. | Madison | WI | 53704 | USA |
| 7 | 1007 | Ahmadi, Hafez | 52 Rue Marston | Paris | | 75019 | France |
| 8 | 1008 | Jacobson, Becky | 1 Lincoln St. | Tallahassee | FL | 32312 | USA |
| 9 | 1009 | An, Ing | 95 Willow Dr. | Toronto | ON | M5J 2T3 | Canada |
| 10 | 1010 | Slater, Emily | 1009 Cherry St. | York | PA | 17407 | USA |

This month the information that follows is received for updating the mailing list:

- Martin Dix changed his name to Martin Dix-Rosen.

- Jane Ericson's postal code changed.

- Jeremy Misiewicz moved to a new street address. His city, state, and postal code remain the same.

- Ing An moved from Toronto, Ontario, to Calgary, Alberta.

- Martin Dix-Rosen, shortly after changing his name, moved from Vancouver, British Columbia, to Seattle, Washington.

- Two new subscribers joined the list. They are given SubscriberID numbers 1011 and 1012.

Each change is entered into the raw data file as soon as it is received. In each case, only the customer's SubscriberId and the new information are entered. The raw data file looks like this:

```
                 1002    Dix-Rosen, Martin
                 1001                                                            27516
                 1006                       932 Webster St.
                 1009                       2540 Pleasant St.   Calgary    AB    T2P 4H2
                 1011    Mitchell, Wayne    28 Morningside Dr.  New York   NY    10017   USA
```

*Updating a Data Set* **321**

```
1002                        P.O. Box 1850      Seattle       WA  98101  USA
1012    Stavros, Gloria     212 Northampton Rd. South Hadley  MA  01075  USA
```

The data is in fixed columns, matching the INPUT statement that created MAIL_LIST.

First, you must transform the raw data into a SAS data set and sort that data set by SubscriberId so that you can use it to update the master list.

```
data mail_trans;
   infile datalines missover;
   input SubscriberId 1-8 Name $ 9-27 StreetAddress $ 28-47 City $ 48-62
         StateProv $ 63-64 PostalCode $ 67-73 Country $ 75-80;
   datalines;
1002    Dix-Rosen, Martin
1001                                                           27516
1006                        932 Webster St.
1009                        2540 Pleasant St.  Calgary       AB  T2P 4H2
1011    Mitchell, Wayne     28 Morningside Dr. New York      NY  10017  USA
1002                        P.O. Box 1850      Seattle       WA  98101  USA
1012    Stavros, Gloria     212 Northampton Rd. South Hadley  MA  01075  USA
;
run;

proc sort data=mail_trans;
   by SubscriberId;
run;

proc print data=mail_trans;
   title 'Magazine Mailing List Changes';
   title2 '(for current month)';
run;
```

Note the MISSOVER option in the INFILE statement. The MISSOVER option prevents the INPUT statement from going to a new line to search for values for variables that have not received values. Instead, any variables that have not received values are set to missing. For example, when the first record is read, the end of the record is encountered before any value has been assigned to the Country variable. Instead of going to the next record to search for a value for Country, the Country variable is assigned a missing value. For more information about the MISSOVER option, see "Methods of Control: Your Options" on page 85.

The following output displays the sorted MAIL_TRANS transaction data set, which lists the magazine changes for the current month:

**Figure 20.2** *Magazine Mailing List Changes: Sorted*

Magazine Mailing List Changes
(for current month)

| Obs | SubscriberId | Name | StreetAddress | City | StateProv | PostalCode | Country |
|---|---|---|---|---|---|---|---|
| 1 | 1001 | | | | | 27516 | |
| 2 | 1002 | Dix-Rosen, Martin | | | | | |
| 3 | 1002 | | P.O. Box 1850 | Seattle | WA | 98101 | USA |
| 4 | 1006 | | 932 Webster St. | | | | |
| 5 | 1009 | | 2540 Pleasant St. | Calgary | AB | T2P 4H2 | |
| 6 | 1011 | Mitchell, Wayne | 28 Morningside Dr. | New York | NY | 10017 | USA |
| 7 | 1012 | Stavros, Gloria | 212 Northampton Rd. | South Hadley | MA | 01075 | USA |

Now that the new data are in a sorted SAS data set, the following program updates the mailing list.

```
data mail_newlist;
   update mail_list mail_trans;
   by SubscriberId;
run;

proc print data=mail_newlist;
   title 'Magazine Mailing List';
   title2 '(updated for current month)';
run;
```

The following output displays the MAIL_NEWLIST data set:

**Figure 20.3** *The Updated Mailing List*

Magazine Mailing List
(updated for current month)

| Obs | SubscriberId | Name | StreetAddress | City | StateProv | PostalCode | Country |
|---|---|---|---|---|---|---|---|
| 1 | 1001 | Ericson, Jane | 111 Clancey Court | Chapel Hill | NC | 27516 | USA |
| 2 | 1002 | Dix-Rosen, Martin | P.O. Box 1850 | Seattle | WA | 98101 | USA |
| 3 | 1003 | Gabrielli, Theresa | via Pisanelli, 25 | Roma | | 00196 | Italy |
| 4 | 1004 | Clayton, Aria | 14 Bridge St. | San Francisco | CA | 94124 | USA |
| 5 | 1005 | Archuleta, Ruby | Box 108 | Milagro | NM | 87429 | USA |
| 6 | 1006 | Misiewicz, Jeremy | 932 Webster St. | Madison | WI | 53704 | USA |
| 7 | 1007 | Ahmadi, Hafez | 52 Rue Marston | Paris | | 75019 | France |
| 8 | 1008 | Jacobson, Becky | 1 Lincoln St. | Tallahassee | FL | 32312 | USA |
| 9 | 1009 | An, Ing | 2540 Pleasant St. | Calgary | AB | T2P 4H2 | Canada |
| 10 | 1010 | Slater, Emily | 1009 Cherry St. | York | PA | 17407 | USA |
| 11 | 1011 | Mitchell, Wayne | 28 Morningside Dr. | New York | NY | 10017 | USA |
| 12 | 1012 | Stavros, Gloria | 212 Northampton Rd. | South Hadley | MA | 01075 | USA |

The data for subscriber 1002, who has two update transactions (see the MAIL_TRANS data set), is used below to show what happens when you update an observation in the master data set with corresponding observations from the transaction data set.

1. Before executing the DATA step, SAS reads the descriptor portion of each data set that is named in the UPDATE statement and. By default, SAS then creates a program data vector that contains all of the variables from all data sets. As the following figure shows, SAS sets the value of each variable to missing. (Use the DROP= or KEEP= data set option to exclude one or more variables.)

*Figure 20.4* Program Data Vector before Execution of the DATA Step

| SubscriberId | Name | Street Address | City | StateProv | PostalCode | Country |
|---|---|---|---|---|---|---|
| . | | | | | | |

2. Next, SAS reads the first observation from the master data set and copies it into the program data vector, as the following figure shows.

*Figure 20.5* Program Data Vector After Reading the First Observation from the Master Data Set

| SubscriberId | Name | Street Address | City | StateProv | PostalCode | Country |
|---|---|---|---|---|---|---|
| 1002 | Dix, Martin | 4 Shepherd St. | Vancouver | BC | V6C 3E8 | Canada |

3. SAS applies the first transaction by copying all nonmissing values (the value of Name) from the first observation in this BY group (ID=1002) into the program data vector, as the following figure shows.

*Figure 20.6* Program Data Vector After Applying the First Transaction

| SubscriberId | Name | Street Address | City | StateProv | PostalCode | Country |
|---|---|---|---|---|---|---|
| 1002 | Dix-Rosen, Martin | 4 Shepherd St. | Vancouver | BC | V6C 3E8 | Canada |

4. After completing this transaction, SAS looks for another observation in the same BY group in the transaction data set. If it finds a second observation with the same value for ID, then it also applies the second transaction (new values for StreetAddress, City, StateProv, PostalCode, and Country). Now the observation contains the new values from both transactions, as the following figure shows.

*Figure 20.7* Program Data Vector After Applying the Second Transaction

| SubscriberId | Name | Street Address | City | StateProv | Postal Code | Country |
|---|---|---|---|---|---|---|
| 1002 | Dix-Rosen, Martin | P.O. Box 1850 | Seattle | WA | 98101 | USA |

5. After completing the second transaction, SAS looks for a third observation in the same BY group. Because no such observation exists, it writes the observation in its current form to the new data set and sets the values in the program data vector to missing.

As the DATA step iterates, the UPDATE statement continues processing observations in this way until it reaches the end of the master and transaction data sets. The two observations in the transaction data set that describe new subscribers (and therefore have no corresponding observation in the master data set) become observations in the new data set.

Remember that if there are duplicate observations in the master data set, all matching observations in the transaction data set are applied only to the first of the duplicate observations in the master data set.

## Updating with Incremental Values

Some applications do not update a data set by overwriting values in the master data set with new values from a transaction data set. Instead, they update a variable by mathematically manipulating its value based on the value of a variable in the transaction data set.

In this example, a bookstore uses SAS to keep track of weekly sales and year-to-date sales. The program that follows creates, sorts by Title, and displays the data set, YEAR_SALES, which contains the year-to-date information.

```
data year_sales;
   input Title $ 1-25 Author $ 27-50 Sales;
   datalines;
The Milagro Beanfield War Nichols, John            303
The Stranger              Camus, Albert            150
Always Coming Home        LeGuin, Ursula            79
Falling through Space     Gilchrist, Ellen         128
Don Quixote               Cervantes, Miguel de      87
The Handmaid's Tale       Atwood, Margaret          64
;

proc sort data=year_sales;
   by title;
run;

proc print data=year_sales;
   title 'Bookstore Sales, Year-to-Date';
   title2 'By Title';
run;
```

The following output displays the YEAR_SALES data set:

*Figure 20.8* *The YEAR_SALES Data Set, Sorted by Title*

Bookstore Sales, Year-to-Date
By Title

| Obs | Title | Author | Sales |
|---|---|---|---|
| 1 | Always Coming Home | LeGuin, Ursula | 79 |
| 2 | Don Quixote | Cervantes, Miguel de | 87 |
| 3 | Falling through Space | Gilchrist, Ellen | 128 |
| 4 | The Handmaid's Tale | Atwood, Margaret | 64 |
| 5 | The Milagro Beanfield War | Nichols, John | 303 |
| 6 | The Stranger | Camus, Albert | 150 |

Every Saturday a SAS data set is created containing information about all the books that were sold during the past week. The following program creates, sorts by Title, and displays the data set WEEK_SALES, which contains the current week's information.

```
data week_sales;
    input Title $ 1-25 Author $ 27-50 Sales;
    datalines;
The Milagro Beanfield War Nichols, John          32
The Stranger              Camus, Albert          17
Always Coming Home        LeGuin, Ursula         10
Falling through Space     Gilchrist, Ellen       12
The Accidental Tourist    Tyler, Anne            15
The Handmaid's Tale       Atwood, Margaret        8
;
proc sort data=week_sales;
    by title;
run;

proc print data=week_sales;
    title 'Bookstore Sales for Current Week';
    title2 'By Title';
run;
```

The following output displays the WEEK_SALES data set, which contains the same variables as the YEAR_SALES data set, but the variable Sales represents sales for only one week:

*Figure 20.9* The WEEK_SALES Data Set, Sorted by Title

**Bookstore Sales for Current Week**
**By Title**

| Obs | Title | Author | Sales |
|---|---|---|---|
| 1 | Always Coming Home | LeGuin, Ursula | 10 |
| 2 | Falling through Space | Gilchrist, Ellen | 12 |
| 3 | The Accidental Tourist | Tyler, Anne | 15 |
| 4 | The Handmaid's Tale | Atwood, Margaret | 8 |
| 5 | The Milagro Beanfield War | Nichols, John | 32 |
| 6 | The Stranger | Camus, Albert | 17 |

*Note:* If the transaction data set is updating only titles that are already in YEAR_SALES, it does not need to contain the variable Author. However, because this variable is there, the transaction data set can be used to add complete observations to the master data set.

The program that follows uses the weekly information to update the YEAR_SALES data set and displays the new data set.

```
data total_sales;
    drop NewSales;    ❶
    update year_sales week_sales (rename=(Sales=NewSales));    ❷
    by Title;
```

```
        sales=sum(Sales,NewSales);   ❸
   run;

   proc print data=total_sales;
      title 'Updated Year-to-Date Sales';
   run;
```

The following list corresponds to the numbered items in the preceding program:

1. The program drops the variable NewSales because it is not needed in the new data set.

2. The RENAME= data set option in the UPDATE statement changes the name of the variable Sales in the transaction data set (WEEK_SALES) to NewSales. As a result, these values do not replace the value of Sales that are read from the master data set (YEAR_SALES).

3. The Sales value that is in the updated data set (TOTAL_SALES) is the sum of the year-to-date sales and the weekly sales.

The following output displays the TOTAL_SALES data set. In addition to updating sales information for the titles already in the master data set, the UPDATE statement has added a new title, The Accidental Tourist:

**Figure 20.10**  *Updated Year-to-Date Sales with Weekly Sales*

Updated Year-to-Date Sales

| Obs | Title | Author | Sales |
|---|---|---|---|
| 1 | Always Coming Home | LeGuin, Ursula | 89 |
| 2 | Don Quixote | Cervantes, Miguel de | 87 |
| 3 | Falling through Space | Gilchrist, Ellen | 140 |
| 4 | The Accidental Tourist | Tyler, Anne | 15 |
| 5 | The Handmaid's Tale | Atwood, Margaret | 72 |
| 6 | The Milagro Beanfield War | Nichols, John | 335 |
| 7 | The Stranger | Camus, Albert | 167 |

# Understanding the Differences between Updating and Merging

## General Comparisons between Updating and Merging

The MERGE statement and the UPDATE statement both match observations from two SAS data sets. However, the two statements differ significantly. It is important to distinguish between the two processes and to choose the one that is appropriate for your application.

The most straightforward differences are as follows:

- The UPDATE statement uses only two data sets. The number of data sets that the MERGE statement can use is limited only by machine-dependent factors such as memory and disk space.

- A BY statement must accompany an UPDATE statement. The MERGE statement performs a one-to-one merge if no BY statement follows it.

- The two statements also process observations differently when a data set contains missing values or multiple observations in a BY group.

To illustrate the differences, compare the results of updating and merging the SAS data set MAIL_LIST and the data set MAIL_TRANS. You have already seen the results of updating in the example that created the MAIL_NEWLIST data set. That output is displayed again for easy comparison.

The following output displays the MAIL_NEWLIST data set:

*Figure 20.11* The Updated Magazine Mailing List

Magazine Mailing List
(updated for current month)

| Obs | SubscriberId | Name | StreetAddress | City | StateProv | PostalCode | Country |
|---|---|---|---|---|---|---|---|
| 1 | 1001 | Ericson, Jane | 111 Clancey Court | Chapel Hill | NC | 27516 | USA |
| 2 | 1002 | Dix-Rosen, Martin | P.O. Box 1850 | Seattle | WA | 98101 | USA |
| 3 | 1003 | Gabrielli, Theresa | Via Pisanelli, 25 | Roma | | 00196 | Italy |
| 4 | 1004 | Clayton, Aria | 14 Bridge St. | San Francisco | CA | 94124 | USA |
| 5 | 1005 | Archuleta, Ruby | Box 108 | Milagro | NM | 87429 | USA |
| 6 | 1006 | Misiewicz, Jeremy | 932 Webster St. | Madison | WI | 53704 | USA |
| 7 | 1007 | Ahmadi, Hafez | 52 Rue Marston | Paris | | 75019 | France |
| 8 | 1008 | Jacobson, Becky | 1 Lincoln St. | Tallahassee | FL | 32312 | USA |
| 9 | 1009 | An, Ing | 2540 Pleasant St. | Calgary | AB | T2P 4H2 | Canada |
| 10 | 1010 | Slater, Emily | 1009 Cherry St. | York | PA | 17407 | USA |
| 11 | 1011 | Mitchell, Wayne | 28 Morningside Dr. | New York | NY | 10017 | USA |
| 12 | 1012 | Stavros, Gloria | 212 Northampton Rd. | South Hadley | MA | 01075 | USA |

In contrast, the following program merges the two data sets.

```
data mail_merged;
   merge mail_list mail_trans;
   by SubscriberId;
run;

proc print data=mail_merged;
   title 'Magazine Mailing List';
run;
```

The following output displays the MAIL_MERGED data set:

**Figure 20.12** *Magazine Mailing List: Merging the Master and Transaction Data Sets*

Magazine Mailing List

| Obs | SubscriberId | Name | StreetAddress | City | StateProv | PostalCode | Country |
|---|---|---|---|---|---|---|---|
| 1 | 1001 | | | | | 27516 | |
| 2 | 1002 | Dix-Rosen, Martin | | | | | |
| 3 | 1002 | | P.O. Box 1850 | Seattle | WA | 98101 | USA |
| 4 | 1003 | Gabrielli, Theresa | Via Pisanelli, 25 | Roma | | 00196 | Italy |
| 5 | 1004 | Clayton, Aria | 14 Bridge St. | San Francisco | CA | 94124 | USA |
| 6 | 1005 | Archuleta, Ruby | Box 108 | Milagro | NM | 87429 | USA |
| 7 | 1006 | | 932 Webster St. | | | | |
| 8 | 1007 | Ahmadi, Hafez | 52 Rue Marston | Paris | | 75019 | France |
| 9 | 1008 | Jacobson, Becky | 1 Lincoln St. | Tallahassee | FL | 32312 | USA |
| 10 | 1009 | | 2540 Pleasant St. | Calgary | AB | T2P 4H2 | |
| 11 | 1010 | Slater, Emily | 1009 Cherry St. | York | PA | 17407 | USA |
| 12 | 1011 | Mitchell, Wayne | 28 Morningside Dr. | New York | NY | 10017 | USA |
| 13 | 1012 | Stavros, Gloria | 212 Northampton Rd. | South Hadley | MA | 01075 | USA |

The MERGE statement produces a data set that contains 13 observations, whereas UPDATE produces a data set containing 12 observations. In addition, merging the data sets results in several missing values, whereas updating does not. Obviously, using the wrong statement can result in incorrect data. The differences between the merged and updated data sets result from the ways the two statements handle missing values and multiple observations in a BY group.

## How the UPDATE and MERGE Statements Process Missing Values Differently

During an update, if a value for a variable is missing in the transaction data set, SAS uses the value from the master data set when it writes the observation to the new data set. When merging the same observations, SAS overwrites the value in the program data vector with the missing value. For example, the following observation exists in data set MAILING_MASTER.

```
1001     ERICSON, JANE    111 CLANCEY COURT    CHAPEL HILL    NC    27514
```

The following corresponding observation exists in MAILING_TRANS.

```
1001                                                                27516
```

Updating combines the two observations and creates the following observation:

```
1001     ERICSON, JANE    111 CLANCEY COURT    CHAPEL HILL    NC    27516
```

Merging combines the two observations and creates this observation:

```
1001                                                                27516
```

### How the UPDATE and MERGE Statements Process Multiple Observations in a BY Group Differently

SAS does not write an updated observation to the new data set until it has applied all the transactions in a BY group. When merging data sets, SAS writes one new observation for each observation in the data set with the largest number of observations in the BY group. For example, consider this observation from MAILING_MASTER:

```
1002    DIX, MARTIN         4 SHEPHERD ST.    NORWICH       VT    05055
```

Consider the corresponding observations from MAILING_TRANS:

```
1002    DIX-ROSEN, MARTIN
1002                        R.R. 2, BOX 1850  HANOVER       NH    03755
```

The UPDATE statement applies both transactions and combines these observations into a single one:

```
1002    DIX-ROSEN, MARTIN   R.R. 2, BOX 1850  HANOVER       NH    03755
```

The MERGE statement, on the other hand, first merges the observation from MAILING_MASTER with the first observation in the corresponding BY group in MAILING_TRANS. All values of variables from the observation in MAILING_TRANS are used, even if they are missing. Then SAS writes the observation to the new data set:

```
1002    DIX-ROSEN, MARTIN
```

Next, SAS looks for other observations in the same BY group in each data set. Because more observations are in the BY group in MAILING_TRANS, all the values in the program data vector are retained. SAS merges them with the second observation in the BY group from MAILING_TRANS and writes the result to the new data set:

```
1002                        R.R. 2, BOX 1850  HANOVER       NH    03755
```

Therefore, merging creates two observations for the new data set, whereas updating creates only one.

## Handling Missing Values

If you update a master data set with a transaction data set, and the transaction data set contains missing values, then you can use the UPDATEMODE= option in the UPDATE statement to tell SAS how you want to handle the missing values. The UPDATEMODE= option specifies whether missing values in a transaction data set replace existing values in a master data set.

The syntax for using the UPDATEMODE= option with the UPDATE statement follows:

**UPDATE** *master-SAS-data-set transaction-SAS-data-set*
<UPDATEMODE=MISSINGCHECK | NOMISSINGCHECK>;
    **BY** *by-variable*;

The MISSINGCHECK value in the UPDATEMODE= option prevents missing values in a transaction data set from replacing values in a master data set. This is the default. The NOMISSINGCHECK value in the UPDATEMODE= option enables missing values in a transaction data set to replace values in a master data set by preventing the check for missing data from being performed.

**330** Chapter 20 • *Updating SAS Data Sets*

The following examples show how SAS handles missing values when you use the UPDATEMODE= option in the UPDATE statement.

The following example creates and sorts a master data set:

```
data inventory;
   input PartNumber $ Description $ Stock @17
   ReceivedDate date9. @27 Price;
   format  ReceivedDate date9.;
   datalines;
K89R seal    34   27jul2004 245.00
M4J7 sander 98   20jun2004 45.88
LK43 filter 121  19may2005 10.99
MN21 brace 43    10aug2005 27.87
BC85 clamp 80    16aug2005 9.55
NCF3 valve 198   20mar2005 24.50
;
run;

proc sort data=inventory;
   by PartNumber;
run;

proc print data=inventory;
   title 'Master Data Set';
   title2 'Tool Warehouse Inventory';
run;
```

The following output displays the master INVENTORY data set:

*Figure 20.13* The Master Inventory Data Set

```
                    Master Data Set
                 Tool Warehouse Inventory

       Obs  PartNumber  Description  Stock  ReceivedDate   Price

        1    BC85         clamp        80    16AUG2005      9.55

        2    K89R         seal         34    27JUL2004    245.00

        3    LK43         filter      121    19MAY2005     10.99

        4    M4J7         sander       98    20JUN2004     45.88

        5    MN21         brace        43    10AUG2005     27.87

        6    NCF3         valve       198    20MAR2005     24.50
```

The following example creates and sorts a transaction data set:

```
data add_inventory;
   input PartNumber $ 1-4 Description $ 6-11 Stock 13-15 @17 Price;
   datalines;
K89R seal       245.00
M4J7 sander 121 45.88
LK43 filter 34  10.99
MN21 brace      28.87
```

```
BC85 clamp   57  11.64
NCF3 valve  121    .
;
run;

proc sort data=add_inventory;
   by PartNumber;
run;

proc print data=add_inventory;
   title 'Transaction Data Set';
   title2 'Tool Warehouse Inventory';
run;
```

The following output displays the transaction ADD_INVENTORY data set:

**Figure 20.14** *The Transaction Data Set*

Transaction Data Set
Tool Warehouse Inventory

| Obs | PartNumber | Description | Stock | Price |
|---|---|---|---|---|
| 1 | BC85 | clamp | 57 | 11.64 |
| 2 | K89R | seal | . | 245.00 |
| 3 | LK43 | filter | 34 | 10.99 |
| 4 | M4J7 | sander | 121 | 45.88 |
| 5 | MN21 | brace | . | 28.87 |
| 6 | NCF3 | valve | 121 | . |

In the following example, SAS uses the NOMISSINGCHECK value of the UPDATEMODE= option in the UPDATE statement:

```
data new_inventory;
   update inventory add_inventory updatemode=nomissingcheck;
   by PartNumber;
   ReceivedDate=today();
run;

proc print data=new_inventory;
   title 'Updated Master Data Set';
   title2 'Tool Warehouse Inventory';
run;
```

The following output displays the results of using the NOMISSINGCHECK value. Observations 2 and 5 contain missing values for STOCK because the transaction data set contains missing values for STOCK for these items. Because checking for missing values in the transaction data set is not done, the original value in STOCK is replaced by missing values. In the sixth observation, the original value of PRICE is replaced by a missing value.

*Figure 20.15* Updated Master Data Set: UPDATEMODE=NOMISSINGCHECK

**Updated Master Data Set**
**Tool Warehouse Inventory**

| Obs | PartNumber | Description | Stock | ReceivedDate | Price |
|-----|------------|-------------|-------|--------------|-------|
| 1 | BC85 | clamp | 57 | 09APR2013 | 11.64 |
| 2 | K89R | seal | . | 09APR2013 | 245.00 |
| 3 | LK43 | filter | 34 | 09APR2013 | 10.99 |
| 4 | M4J7 | sander | 121 | 09APR2013 | 45.88 |
| 5 | MN21 | brace | . | 09APR2013 | 28.87 |
| 6 | NCF3 | valve | 121 | 09APR2013 | . |

The following output displays the results of using the MISSINGCHECK value. Note that no missing values are written to the updated master data set. The missing data in observations 2, 5, and 6 of the transaction data set is ignored, and the original data from the master data set remains.

*Figure 20.16* Updated Master Data Set: UPDATEMODE=MISSINGCHECK

**Updated Master Data Set**
**Tool Warehouse Inventory**

| Obs | PartNumber | Description | Stock | ReceivedDate | Price |
|-----|------------|-------------|-------|--------------|-------|
| 1 | BC85 | clamp | 57 | 09APR2013 | 11.64 |
| 2 | K89R | seal | 34 | 09APR2013 | 245.00 |
| 3 | LK43 | filter | 34 | 09APR2013 | 10.99 |
| 4 | M4J7 | sander | 121 | 09APR2013 | 45.88 |
| 5 | MN21 | brace | 43 | 09APR2013 | 28.87 |
| 6 | NCF3 | valve | 121 | 09APR2013 | 24.50 |

For more information, see "UPDATE Statement" in *SAS Statements: Reference*.

# Summary

## Statements

UPDATE *master-SAS-data-set transaction-SAS-data-set*;
BY *identifier-list*;
> The UPDATE statement replaces the values of variables in one SAS data set with nonmissing values from another SAS data set. *Master-SAS-data-set* is the SAS data set that contains information that you want to update. *Transaction-SAS-data-set* is the SAS data set that contains information for updating the master data set.
>
> *Identifier-list* in the BY statement is the list of BY variables by which you identify corresponding observations.

# Learning More

DATASETS procedure
> When you update a data set, you create a new data set containing the updated information. Typically, you want to use PROC DATASETS to delete the old master data set and rename the new one so that you can use the same program the next time you update the information. For more information, see "Understanding the DATASETS Procedure" on page 696.

Indexes
> If a data set has an index on the variable or variables named in the BY statement that accompanies the UPDATE statement, you do not need to sort that data set. For more information, see "Understanding SAS Indexes" in *SAS Language Reference: Concepts*.

MERGE statement
> For more information, see Chapter 19, "Merging SAS Data Sets," on page 289.

# Chapter 21
# Modifying SAS Data Sets

Introduction to Modifying SAS Data Sets ............................... 335
    Purpose ............................................................ 335
    Prerequisites ....................................................... 336

Input SAS Data Set for Examples ....................................... 336

Modifying a SAS Data Set: The Simplest Case .......................... 337

Modifying a Master Data Set with Observations from a Transaction Data Set .. 338
    Understanding the MODIFY Statement ................................ 338
    Adding New Observations to the Master Data Set ..................... 339
    Checking for Program Errors ........................................ 339
    The Program ...................................................... 339

Understanding How Duplicate BY Variables Affect File Update ............. 343
    How the DATA Step Processes Duplicate BY Variables ................ 343
    The Program ...................................................... 343

Handling Missing Values ............................................... 345

Summary ............................................................. 347
    Statements ........................................................ 347

Learning More ........................................................ 348

## Introduction to Modifying SAS Data Sets

### Purpose

Modifying changes observations directly in the original master file. It does not create a copy of the file. In this section, you will learn how to use the MODIFY statement in a DATA step to do the following:

- replace values in a data set

- replace values in a master data set with values from a transaction data set

- append observations to an existing SAS data set

- delete observations from an existing SAS data set

## Prerequisites

Before continuing with this section, you should understand the concepts that are presented in the following sections:

- Chapter 4, "Starting with Raw Data: The Basics," on page 51
- Chapter 6, "Starting with SAS Data Sets," on page 91
- Chapter 19, "Merging SAS Data Sets," on page 289
- Chapter 20, "Updating SAS Data Sets," on page 317

# Input SAS Data Set for Examples

This section looks at examples from an inventory tracking system that is used by a tool vendor. The examples use the SAS data set INVENTORY as input. The data set contains these variables:

PartNumber
    is a character variable that contains a unique value that identifies each item.

Description
    is a character variable that contains the text description of each item.

InStock
    is a numeric variable that contains a value that describes how many units of each tool the warehouse has in stock.

ReceivedDate
    is a numeric variable that contains the SAS date value that is the day for which InStock values are current.

Price
    is a numeric variable that contains the price of each item.

The following program creates and displays the INVENTORY_TOOL data set:

```
data inventory_tool;
   input PartNumber $ Description $ InStock @17
        ReceivedDate date9. @27 Price;
   format ReceivedDate date9.;
   datalines;
K89R seal    34  27jul2010 245.00
M4J7 sander  98  20jun2011 45.88
LK43 filter  121 19may2011 10.99
MN21 brace   43  10aug2012 27.87
BC85 clamp   80  16aug2012 9.55
NCF3 valve   198 20mar2012 24.50
KJ66 cutter  6   18jun2010 19.77
UYN7 rod     211 09sep2010 11.55
JD03 switch  383 09jan2013 13.99
BV1E timer   26  03aug2013 34.50
;
run;

proc print data=inventory_tool;
```

```
        title 'Tool Warehouse Inventory';
run;
```

The following output displays the INVENTORY_TOOL data set:

*Figure 21.1   The INVENTORY_TOOL Data Set*

**Tool Warehouse Inventory**

| Obs | PartNumber | Description | InStock | ReceivedDate | Price |
|---|---|---|---|---|---|
| 1 | K89R | seal | 34 | 27JUL2010 | 245.00 |
| 2 | M4J7 | sander | 98 | 20JUN2011 | 45.88 |
| 3 | LK43 | filter | 121 | 19MAY2011 | 10.99 |
| 4 | MN21 | brace | 43 | 10AUG2012 | 27.87 |
| 5 | BC85 | clamp | 80 | 16AUG2012 | 9.55 |
| 6 | NCF3 | valve | 198 | 20MAR2012 | 24.50 |
| 7 | KJ66 | cutter | 6 | 18JUN2010 | 19.77 |
| 8 | UYN7 | rod | 211 | 09SEP2010 | 11.55 |
| 9 | JD03 | switch | 383 | 09JAN2013 | 13.99 |
| 10 | BV1E | timer | 26 | 03AUG2013 | 34.50 |

## Modifying a SAS Data Set: The Simplest Case

You can use the MODIFY statement to replace all values for a specific variable or variables in a data set. The syntax for using the MODIFY statement for this purpose is

**MODIFY** *SAS-data-set*;

In the following program, the price of each part in the inventory is increased by 15%. The new values for PRICE replace the old values on all records in the original INVENTORY_TOOL data set. The FORMAT statement in the print procedure writes the price of each item with two-digit decimal precision.

```
data inventory_tool;
   modify inventory_tool;
   price=price+(price*.15);
run;

proc print data=inventory_tool;
   title 'Tool Warehouse Inventory';
   title2 '(Price reflects 15% increase)';
   format price 8.2;
run;
```

The following output displays the modified INVENTORY_TOOL data set:

*Figure 21.2  The INVENTORY_TOOL Data Set with Updated Prices*

Tool Warehouse Inventory
(Price reflects 15% increase)

| Obs | PartNumber | Description | InStock | ReceivedDate | Price |
|---|---|---|---|---|---|
| 1 | K89R | seal | 34 | 27JUL2010 | 281.75 |
| 2 | M4J7 | sander | 98 | 20JUN2011 | 52.76 |
| 3 | LK43 | filter | 121 | 19MAY2011 | 12.64 |
| 4 | MN21 | brace | 43 | 10AUG2012 | 32.05 |
| 5 | BC85 | clamp | 80 | 16AUG2012 | 10.98 |
| 6 | NCF3 | valve | 198 | 20MAR2012 | 28.18 |
| 7 | KJ66 | cutter | 6 | 18JUN2010 | 22.74 |
| 8 | UYN7 | rod | 211 | 09SEP2010 | 13.28 |
| 9 | JD03 | switch | 383 | 09JAN2013 | 16.09 |
| 10 | BV1E | timer | 26 | 03AUG2013 | 39.68 |

# Modifying a Master Data Set with Observations from a Transaction Data Set

## Understanding the MODIFY Statement

The MODIFY statement replaces data in a master data set with data from a transaction data set, and makes the changes in the original master data set. You can use a BY statement to match observations from the transaction data set with observations in the master data set. The syntax for using the MODIFY statement and the BY statement follows:

**MODIFY** *master-SAS-data-set transaction-SAS-data-set*;

**BY** *by-variable*;

*Master-SAS-data-set* specifies the SAS data set that you want to modify. *Transaction-SAS-data-set* specifies the SAS data set that provides the values for updating the master data set. *By-variable* specifies one or more variables by which you identify corresponding observations.

When you use a BY statement with the MODIFY statement, the DATA step uses dynamic WHERE processing to find observations in the master data set. Neither the master data set nor the transaction data set needs to be sorted. However, for large data sets, sorting the data before you modify it can enhance performance significantly.

### Adding New Observations to the Master Data Set

You can use the MODIFY statement to add observations to an existing master data set. If the transaction data set contains an observation that does not match an observation in the master data set, then SAS enables you to write a new observation to the master data set if you use an explicit OUTPUT statement in your program. When you specify an explicit OUTPUT statement, you must also specify a REPLACE statement if you want to replace observations in place. All new observations append to the end of the master data set.

### Checking for Program Errors

You can use the _IORC_ automatic variable for error checking in your DATA step program. The _IORC_ automatic variable contains the return code for each I/O operation that the MODIFY statement attempts to perform.

The best way to test the values of _IORC_ is with the mnemonic codes that are provided by the SYSRC autocall macro. Each mnemonic code describes one condition. The mnemonics provide an easy method for testing problems in a DATA step program. The following is a partial list of codes:

_DSENMR
: specifies that the transaction data set observation does not exist in the master data set (used only with MODIFY and BY statements). If consecutive observations with different BY values do not find a match in the master data set, then both of them return _DSENMR.

_DSEMTR
: specifies that multiple transaction data set observations with a given BY value do not exist in the master data set (used only with MODIFY and BY statements). If consecutive observations with the same BY values do not find a match in the master data set, then the first observation returns _DSENMR and the subsequent observations return _DSEMTR.

_DSENOM
: specifies that no matching observation was found.

_SOK
: specifies that the observation was located in the master data set and indicates that the MODIFY statement executed successfully.

### The Program

The program in this section updates values in a master data set with values from a transaction data set. If a transaction does not exist in the master data set, then the program adds the transaction to the master data set.

In this example, a warehouse received a shipment of new items, and the INVENTORY_TOOL master data set must be modified to reflect the changes. The master data set contains a complete list of the inventory items. The transaction data set contains items that are on the master inventory as well as new inventory items.

The following program creates the ADD_INVENTORY transaction data set, which contains items for updating the master data set. The PartNumber variable contains the part number for the item and corresponds to PartNumber in the INVENTORY_TOOL data set. The Description variable names the item. The NewStock variable contains the number of each item in the current shipment. The NewPrice variable contains the new price of the item.

The program attempts to update the master data set INVENTORY_TOOL (see Figure 21.1 on page 337) according to the values in the ADD_INVENTORY transaction data set. The program uses the _IORC_ automatic variable to detect errors.

```
data inventory_tool;
   input PartNumber $ Description $ InStock @17
         ReceivedDate date9. @27 Price;
   format ReceivedDate date9.;
   datalines;
K89R seal    34  27jul2010 245.00
M4J7 sander  98  20jun2011 45.88
LK43 filter  121 19may2011 10.99
MN21 brace   43  10aug2012 27.87
BC85 clamp   80  16aug2012 9.55
NCF3 valve   198 20mar2012 24.50
KJ66 cutter  6   18jun2010 19.77
UYN7 rod     211 09sep2010 11.55
JD03 switch  383 09jan2013 13.99
BV1E timer   26  03aug2013 34.50
;
run;

data add_inventory;   ❶
   input PartNumber $ Description $ NewStock @16 NewPrice;
   datalines;
K89R seal    6  247.50
AA11 hammer  55 32.26
BB22 wrench  21 17.35
KJ66 cutter  10 24.50
CC33 socket  7  22.19
BV1E timer   30 36.50
;
run;

data inventory_tool;
   modify inventory_tool add_inventory;   ❷
   by PartNumber;
   select (_iorc_);   ❸
      /* The observation exists in the master data set. */
      when (%sysrc(_sok)) do;   ❹
         InStock=InStock+NewStock;
         ReceivedDate=today();
         Price=NewPrice;
         replace;   ❺
      end;
      /* The observation does not exist in the master data set. */
      when (%sysrc(_dsenmr)) do;   ❻
         InStock=NewStock;
         ReceivedDate=today();
         Price=NewPrice;
         output;   ❼
         _error_=0;
      end;
      otherwise do;   ❽
```

## Modifying a Master Data Set with Observations from a Transaction Data Set 341

```
                put 'An unexpected I/O error has occurred.'/
                    'Check you data and your program.';   8
                _error_=0;
                stop;
            end;
        end;
    run;

    proc print data=inventory_tool;
        title 'Tool Warehouse Inventory';
    run;
```

The following list corresponds to the numbered items in the preceding program:

1 The DATA statement creates the ADD_INVENTORY transaction data set.

2 The MODIFY statement loads the data from the INVENTORY_TOOL and ADD_INVENTORY data sets.

3 The _IORC_ automatic variable is used for error checking. The value of _IORC_ is a numeric return code that indicates the status of the most recent I/O operation.

4 The SYSRC autocall macro checks to see whether the value of _IORC_ is _SOK. If the value is _SOK, then an observation in the transaction data set matches an observation in the master data set.

5 The REPLACE statement updates the INVENTORY master data set by replacing the observation in the master data set with the observation from the transaction data set.

6 The SYSRC autocall macro checks to see whether the value of _IORC_ is _DSENMR. If the value is _DSENMR, then an observation in the transaction data set does not exist in the master data set.

7 The OUTPUT statement writes the current observation to the end of the master data set.

8 If neither condition is met, the PUT statement writes a message to the log.

**342** Chapter 21 • Modifying SAS Data Sets

The following output displays the updated INVENTORY_TOOL data set:

*Figure 21.3* The Updated INVENTORY_TOOL Data Set

Tool Warehouse Inventory

| Obs | PartNumber | Description | InStock | ReceivedDate | Price |
|---|---|---|---|---|---|
| 1 | K89R | seal | 40 | 09APR2013 | 247.50 |
| 2 | M4J7 | sander | 98 | 20JUN2011 | 45.88 |
| 3 | LK43 | filter | 121 | 19MAY2011 | 10.99 |
| 4 | MN21 | brace | 43 | 10AUG2012 | 27.87 |
| 5 | BC85 | clamp | 80 | 16AUG2012 | 9.55 |
| 6 | NCF3 | valve | 198 | 20MAR2012 | 24.50 |
| 7 | KJ66 | cutter | 16 | 09APR2013 | 24.50 |
| 8 | UYN7 | rod | 211 | 09SEP2010 | 11.55 |
| 9 | JD03 | switch | 383 | 09JAN2013 | 13.99 |
| 10 | BV1E | timer | 56 | 09APR2013 | 36.50 |
| 11 | AA11 | hammer | 55 | 09APR2013 | 32.26 |
| 12 | BB22 | wrench | 21 | 09APR2013 | 17.35 |
| 13 | CC33 | socket | 7 | 09APR2013 | 22.19 |

SAS writes the following message to the log:

```
NOTE: The data set WORK.INVENTORY_TOOL has been updated.  There were 3
observations rewritten, 3 observations added and 0 observations deleted.
```

*CAUTION:*
   **If you execute your program without the OUTPUT and REPLACE statements, then your master file might not update correctly.** Using OUTPUT or REPLACE in a DATA step overrides the default replacement of observations. If you use these statements in a DATA step, then you must explicitly program each action that you want to take.

For more information, see "OUTPUT Statement" in *SAS Statements: Reference* and "REPLACE Statement" in *SAS Statements: Reference*.

# Understanding How Duplicate BY Variables Affect File Update

## *How the DATA Step Processes Duplicate BY Variables*

When you use a BY statement with MODIFY, both the master and the transaction data sets can have observations with duplicate values of BY variables. Neither the master nor the transaction data set needs to be sorted, because BY-group processing uses dynamic WHERE processing to find an observation in the master data set.

The DATA step processes duplicate observations in the following ways:

- If duplicate BY values exist in the master data set, then MODIFY applies the current transaction to the first occurrence in the master data set.

- If duplicate BY values exist in the transaction data set, then the observations are applied one on top of another so that the values overwrite each other. The value in the last transaction is the final value in the master data set.

- If both the master and the transaction data sets contain duplicate BY values, then MODIFY applies each transaction to the first occurrence in the group in the master data set.

## *The Program*

The program in this section updates the INVENTORY_2 master data set with observations from the ADD_INVENTORY_2 transaction data set. Both data sets contain consecutive and nonconsecutive duplicate values of the BY variable PartNumber.

The following program creates the master data set INVENTORY_2. Note that the data set contains three observations for PartNumber M4J7.

```
data inventory_2;
   input PartNumber $ Description $ InStock @17
         ReceivedDate date9. @27 Price;
   format ReceivedDate date9.;
   datalines;
K89R seal   34  27jul1998 245.00
M4J7 sander 98  20jun2012 45.88
M4J7 sander 98  20jun2012 45.88
LK43 filter 121 19may2013 10.99
MN21 brace  43  10aug2013 27.87
M4J7 sander 98  20jun2012 45.88
BC85 clamp  80  16aug2013 9.55
NCF3 valve  198 20mar2013 24.50
KJ66 cutter 6   18jun2013 19.77
;
run;

proc print data=inventory_2;
   title 'INVENTORY_2 Data Set';
run;
```

The following output displays the INVENTORY_2 data set:

*Figure 21.4* *The INVENTORY_2 Data Set*

INVENTORY_2 Data Set

| Obs | PartNumber | Description | InStock | ReceivedDate | Price |
|---|---|---|---|---|---|
| 1 | K89R | seal | 34 | 27JUL1998 | 245.00 |
| 2 | M4J7 | sander | 98 | 20JUN2012 | 45.88 |
| 3 | M4J7 | sander | 98 | 20JUN2012 | 45.88 |
| 4 | LK43 | filter | 121 | 19MAY2013 | 10.99 |
| 5 | MN21 | brace | 43 | 10AUG2013 | 27.87 |
| 6 | M4J7 | sander | 98 | 20JUN2012 | 45.88 |
| 7 | BC85 | clamp | 80 | 16AUG2013 | 9.55 |
| 8 | NCF3 | valve | 198 | 20MAR2013 | 24.50 |
| 9 | KJ66 | cutter | 6 | 18JUN2013 | 19.77 |

The following program creates the transaction data set ADD_INVENTORY_2, and then modifies the master data set INVENTORY_2. Note that the data set ADD_INVENTORY_2 contains three observations for PartNumber M4J7.

```
data add_inventory_2;
   input PartNumber $ Description $ NewStock;
   datalines;
K89R abc 17
M4J7 def 72
M4J7 ghi 66
LK43 jkl 311
M4J7 mno 43
BC85 pqr 75
;
run;

data inventory_2;
   modify inventory_2 add_inventory_2;
   by PartNumber;
   ReceivedDate=today();
   InStock=InStock+NewStock;
run;

proc print data=inventory_2;
   title 'Tool Warehouse Inventory';
run;
```

The following output displays the updated INVENTORY_2 data set:

*Figure 21.5* The Updated INVENTORY_2 Data Set: Duplicate BY Variables

Tool Warehouse Inventory

| Obs | PartNumber | Description | InStock | ReceivedDate | Price |
|---|---|---|---|---|---|
| 1 | K89R | abc | 51 | 09APR2013 | 245.00 |
| 2 | M4J7 | mno | 279 | 09APR2013 | 45.88 |
| 3 | M4J7 | sander | 98 | 20JUN2012 | 45.88 |
| 4 | LK43 | jkl | 432 | 09APR2013 | 10.99 |
| 5 | MN21 | brace | 43 | 10AUG2013 | 27.87 |
| 6 | M4J7 | sander | 98 | 20JUN2012 | 45.88 |
| 7 | BC85 | pqr | 155 | 09APR2013 | 9.55 |
| 8 | NCF3 | valve | 198 | 20MAR2013 | 24.50 |
| 9 | KJ66 | cutter | 6 | 18JUN2013 | 19.77 |

# Handling Missing Values

By default, if the transaction data set contains missing values for a variable that is common to both the master and the transaction data sets, then the MODIFY statement does not replace values in the master data set with missing values.

If you want to replace values in the master data set with missing values, then you use the UPDATEMODE= option in the MODIFY statement. UPDATEMODE= specifies whether missing values in a transaction data set replace existing values in a master data set.

The syntax for using the UPDATEMODE= option with the MODIFY statement follows:

**MODIFY** *master-SAS-data-set transaction-SAS-data-set*
<UPDATEMODE=MISSINGCHECK | NOMISSINGCHECK>;

**BY** *by-variable*;

MISSINGCHECK prevents missing values in a transaction data set from replacing values in a master data set. This is the default. NOMISSINGCHECK enables missing values in a transaction data set to replace values in a master data set by preventing the check for missing data from being performed.

The following example creates the master data set Event_List, which contains the schedule and codes for athletic events. The example then updates Event_List with the transaction data set Event_Change, which contains new information about the schedule. Because the MODIFY statement uses the NOMISSINGCHECK value of the UPDATEMODE= option, values in the master data set are replaced by missing values from the transaction data set.

The following program creates the EVENT_LIST master data set:

```
data Event_List;
   input Event $ 1-10 Weekday $ 12-20 TimeofDay $ 22-30 Fee Code;
   datalines;
Basketball Monday     evening    10 58
Soccer     Tuesday    morning     5 33
Yoga       Wednesday  afternoon  15 92
Swimming   Wednesday  morning    10 63
;
run;

proc print data=Event_List;
   title 'EVENT_LIST Data Set';
run;
```

The following output displays the EVENT_LIST data set:

*Figure 21.6* The EVENT_LIST Data Set

EVENT_LIST Data Set

| Obs | Event | Weekday | TimeofDay | Fee | Code |
|---|---|---|---|---|---|
| 1 | Basketball | Monday | evening | 10 | 58 |
| 2 | Soccer | Tuesday | morning | 5 | 33 |
| 3 | Yoga | Wednesday | afternoon | 15 | 92 |
| 4 | Swimming | Wednesday | morning | 10 | 63 |

The following program creates the EVENT_CHANGE transaction data set:

```
data Event_Change;
   input Event $ 1-10 Weekday $ 12-20 Fee Code;
   datalines;
Basketball Wednesday 10 .
Yoga       Monday     . 63
Swimming              . .
;
run;

proc print data=Event_Change;
   title 'EVENT_CHANGE Data Set';
run;
```

The following output displays the EVENT_CHANGE transaction data set:

*Figure 21.7*  *The EVENT_CHANGE Transaction Data Set*

EVENT_CHANGE Data Set

| Obs | Event | Weekday | Fee | Code |
|---|---|---|---|---|
| 1 | Basketball | Wednesday | 10 | . |
| 2 | Yoga | Monday | . | 63 |
| 3 | Swimming | | . | . |

The following program modifies and writes the master data set:

```
data Event_List;
   modify Event_List Event_Change updatemode=nomissingcheck;
   by Event;
run;

proc print data=Event_List;
   title 'Schedule of Athletic Events';
run;
```

The following output displays the modified EVENT_LIST data set:

*Figure 21.8*  *The EVENT_LIST Master Data Set: Missing Values*

Schedule of Athletic Events

| Obs | Event | Weekday | TimeofDay | Fee | Code |
|---|---|---|---|---|---|
| 1 | Basketball | Wednesday | evening | 10 | . |
| 2 | Soccer | Tuesday | morning | 5 | 33 |
| 3 | Yoga | Monday | afternoon | . | 63 |
| 4 | Swimming | | morning | . | . |

# Summary

## Statements

BY *by-variable*;
: identifies corresponding observations in a master data set and a transaction data set. *By-variable* specifies one or more variables to use with the BY statement.

MODIFY *master-SAS-data-set transaction-SAS-data-set*
<UPDATEMODE=MISSINGCHECK|NOMISSINGCHECK>;
> replaces the values of variables in one SAS data set with values from another SAS data set. *Master-SAS-data-set* contains data that you want to update. *Transaction-SAS-data-set* contains observations for updating the master data set.
>
> The UPDATEMODE= argument determines whether missing values in the transaction data set overwrite values in the master data set. The MISSINGCHECK option prevents missing values in a transaction data set from replacing values in a master data set. This is the default. The NOMISSINGCHECK option enables missing values in a transaction data set to replace values in a master data set by preventing the check for missing data from being performed.

MODIFY *SAS-data-set*;
> replaces the values of variables in a data set with values that you specify in your program.

OUTPUT;
> if a MODIFY statement is present, writes the current observation to the end of the master data set.

REPLACE;
> if a MODIFY statement is present, writes the current observation to the same physical location from which it was read in a data set that is named in the DATA statement.

## Learning More

MERGE statement
> For more information, see "MERGE Statement" in *SAS Statements: Reference*.

MODIFY statement
> For more information about the various applications of the MODIFY statement, see "MODIFY Statement" in *SAS Statements: Reference*.

UPDATE statement
> For more information, see "UPDATE Statement" in *SAS Statements: Reference*.

# Chapter 22
# Conditionally Processing Observations from Multiple SAS Data Sets

**Introduction to Conditional Processing from Multiple SAS Data Sets** ......... 349
    Purpose .................................................................... 349
    Prerequisites .............................................................. 350

**Input SAS Data Sets for Examples** ........................................ 350

**Determining Which Data Set Contributed the Observation** ............... 353
    Understanding the IN= Data Set Option ............................... 353
    The Program ............................................................. 354

**Combining Selected Observations from Multiple Data Sets** .............. 358

**Performing a Calculation Based on the Last Observation** ................ 359
    Understanding When the Last Observation Is Processed ............. 359
    The Program ............................................................. 360

**Summary** ................................................................... 361
    Statements ............................................................... 361

**Learning More** .............................................................. 362

## Introduction to Conditional Processing from Multiple SAS Data Sets

### Purpose

When combining SAS data sets, you can process observations conditionally, based on which data set contributed that observation. You can do the following:

- determine which data set contributed each observation in the combined data set

- create a new data set that includes only selected observations from the data sets that you combine

- determine when SAS is processing the last observation in the DATA step so that you can execute conditional operations, such as creating totals

You have seen some of these concepts in earlier topics, but in this section you will apply them to the processing of multiple data sets. The examples use the SET statement, but you can also use all of the features that are discussed here with the MERGE, MODIFY, and UPDATE statements.

*Prerequisites*

Before using this section, you should understand the concepts that are presented in the following sections:

- Chapter 4, "Starting with Raw Data: The Basics," on page 51
- Chapter 6, "Starting with SAS Data Sets," on page 91
- Chapter 18, "Interleaving SAS Data Sets," on page 281

# Input SAS Data Sets for Examples

The following program creates two SAS data sets, SOUTHAMERICAN and EUROPEAN. Each data set contains the following variables:

Year
: is the year in which South American and European countries competed in the World Cup Finals from 1954 to 1998.

Country
: is the name of the competing country.

Score
: is the final score of the game.

Result
: is the result of the game. The value for winners is **won**; the value for losers is `lost`.

The PROC SORT statements in the example below sort the data sets in ascending order according to the BY variable. To create the interleaved data set in the next example, the data must be in ascending order.

```
data southamerican;
   title 'South American World Cup Finalists from 1954 to 1998';
   input  Year $  Country $ 9-23 Score $ 25-28 Result $ 32-36;
   datalines;
1998      Brazil           0-3      lost
1994      Brazil           3-2      won
1990      Argentina        0-1      lost
1986      Argentina        3-2      won
1978      Argentina        3-1      won
1970      Brazil           4-1      won
1962      Brazil           3-1      won
1958      Brazil           5-2      won
;

data european;
   title 'European World Cup Finalists From 1954 to 1998';
   input  Year $  Country $ 9-23 Score $ 25-28 Result $ 32-36;
   datalines;
1998      France           3-0      won
1994      Italy            2-3      lost
1990      West Germany     1-0      won
1986      West Germany     2-3      lost
1982      Italy            3-1      won
```

```
    1982    West Germany      1-3     lost
    1978    Netherlands       1-3     lost
    1974    West Germany      2-1     won
    1974    Netherlands       1-2     lost
    1970    Italy             1-4     lost
    1966    England           4-2     won
    1966    West Germany      2-4     lost
    1962    Czechoslovakia    1-3     lost
    1958    Sweden            2-5     lost
    1954    West Germany      3-2     won
    1954    Hungary           2-3     lost
;
run;

proc sort data=southamerican;
   by year;

proc print data=southamerican;
   title 'World Cup Finalists:';
   title2 'South American Countries';
   title3 'from 1954 to 1998';
run;

proc sort data=european;
   by year;
run;

proc print data=european;
   title 'World Cup Finalists:';
   title2 'European Countries';
   title3 'from 1954 to 1998';
run;
```

The PROC SORT statement sorts the data set in ascending order according to the BY variable. To create the interleaved data set in the next example, the data must be in ascending order.

The following output displays the SOUTHAMERICAN data set:

*Figure 22.1* World Cup Finalists by Continent: South America

World Cup Finalists:
South American Countries
from 1954 to 1998

| Obs | Year | Country | Score | Result |
|---|---|---|---|---|
| 1 | 1958 | Brazil | 5-2 | won |
| 2 | 1962 | Brazil | 3-1 | won |
| 3 | 1970 | Brazil | 4-1 | won |
| 4 | 1978 | Argentina | 3-1 | won |
| 5 | 1986 | Argentina | 3-2 | won |
| 6 | 1990 | Argentina | 0-1 | lost |
| 7 | 1994 | Brazil | 3-2 | won |
| 8 | 1998 | Brazil | 0-3 | lost |

The following output displays the EUROPEAN data set:

*Figure 22.2* World Cup Finalists by Continent: Europe

World Cup Finalists:
European Countries
from 1954 to 1998

| Obs | Year | Country | Score | Result |
|---|---|---|---|---|
| 1 | 1954 | West Germany | 3-2 | won |
| 2 | 1954 | Hungary | 2-3 | lost |
| 3 | 1958 | Sweden | 2-5 | lost |
| 4 | 1962 | Czechoslovakia | 1-3 | lost |
| 5 | 1966 | England | 4-2 | won |
| 6 | 1966 | West Germany | 2-4 | lost |
| 7 | 1970 | Italy | 1-4 | lost |
| 8 | 1974 | West Germany | 2-1 | won |
| 9 | 1974 | Netherlands | 1-2 | lost |
| 10 | 1978 | Netherlands | 1-3 | lost |
| 11 | 1982 | Italy | 3-1 | won |
| 12 | 1982 | West Germany | 1-3 | lost |
| 13 | 1986 | West Germany | 2-3 | lost |
| 14 | 1990 | West Germany | 1-0 | won |
| 15 | 1994 | Italy | 2-3 | lost |
| 16 | 1998 | France | 3-0 | won |

# Determining Which Data Set Contributed the Observation

## Understanding the IN= Data Set Option

When you create a new data set by combining observations from two or more data sets, knowing which data set an observation came from can be useful. For example, you might want to perform a calculation based on which data set contributed an observation. Otherwise, you might lose important contextual information that you need for later processing. You can determine which data set contributed a particular observation by using the IN= data set option.

The IN= data set option enables you to determine which data sets have contributed to the observation that is currently in the program data vector. The syntax for this option in the SET statement follows:

SET *SAS-data-set-1* (IN=*variable*) *SAS-data-set-2*;

BY *a-common-variable*;

When you use the IN= option with a data set in a SET, MERGE, MODIFY, or UPDATE statement, SAS creates a temporary variable associated with that data set. The value of *variable* is 1 if the data set has contributed to the observation currently in the program data vector. The value is 0 if it has not contributed. You can use the IN= option with any or all the data sets that you name in a SET, MERGE, MODIFY, or UPDATE statement. But, you must use a different variable name in each case.

*Note:* The IN= variable exists during the execution of the DATA step only. It is not written to the output data set that is created.

## The Program

The original data sets, SOUTHAMERICAN and EUROPEAN, do not need a variable that identifies the countries' continent because all observations in SOUTHAMERICAN pertain to the South American continent, and all observations in EUROPEAN pertain to the European continent. However, when you combine the data sets, you lose the context, which in this case is the relevant continent for each observation. The following example uses the SET statement with a BY statement to combine the two data sets into one data set that contains all the observations in chronological order:

```
data finalists;
   set southamerican european;
   by year;
run;

proc print data=finalists;
   title 'World Cup Finalists';
   title2 'from 1954 to 1998';
run;
```

The following output displays the FINALISTS data set:

*Figure 22.3* *World Cup Finalists Grouped by Year*

World Cup Finalists
from 1954 to 1998

| Obs | Year | Country | Score | Result |
|---|---|---|---|---|
| 1 | 1954 | West Germany | 3-2 | won |
| 2 | 1954 | Hungary | 2-3 | lost |
| 3 | 1958 | Brazil | 5-2 | won |
| 4 | 1958 | Sweden | 2-5 | lost |
| 5 | 1962 | Brazil | 3-1 | won |
| 6 | 1962 | Czechoslovakia | 1-3 | lost |
| 7 | 1966 | England | 4-2 | won |
| 8 | 1966 | West Germany | 2-4 | lost |
| 9 | 1970 | Brazil | 4-1 | won |
| 10 | 1970 | Italy | 1-4 | lost |
| 11 | 1974 | West Germany | 2-1 | won |
| 12 | 1974 | Netherlands | 1-2 | lost |
| 13 | 1978 | Argentina | 3-1 | won |
| 14 | 1978 | Netherlands | 1-3 | lost |
| 15 | 1982 | Italy | 3-1 | won |
| 16 | 1982 | West Germany | 1-3 | lost |
| 17 | 1986 | Argentina | 3-2 | won |
| 18 | 1986 | West Germany | 2-3 | lost |
| 19 | 1990 | Argentina | 0-1 | lost |
| 20 | 1990 | West Germany | 1-0 | won |
| 21 | 1994 | Brazil | 3-2 | won |
| 22 | 1994 | Italy | 2-3 | lost |
| 23 | 1998 | Brazil | 0-3 | lost |
| 24 | 1998 | France | 3-0 | won |

Notice that this output would be more useful if it showed from which data set each observation originated. To solve this problem, the following program uses the IN= data set option in conjunction with IF-THEN/ELSE statements. By determining which data set contributed an observation, the conditional statement executes and assigns the appropriate value to the variable Continent in each observation in the new data set FINALISTS.

```
data finalists;
   set southamerican (in=S) european;    ①
   by Year;
   if S then Continent='South America';  ②
   else Continent='Europe';
run;

proc print data=finalists;
   title 'World Cup Finalists';
   title2 'from 1954 to 1998';
run;
```

The following list corresponds to the numbered items in the preceding program:

1   The IN= option in the SET statement tells SAS to create a variable named S.

2   When the current observation comes from the data set SOUTHAMERICAN, the value of S is 1. Otherwise, the value is 0. The IF-THEN/ELSE statements execute one of two assignment statements, depending on the value of S. If the observation comes from the data set SOUTHAMERICAN, then the value that is assigned to Continent is South America. If the observation comes from the data set EUROPEAN, then the value that is assigned to Continent is Europe.

The following output displays the updated FINALISTS data set:

**Figure 22.4** World Cup Finalists with Continent

### World Cup Finalists from 1954 to 1998

| Obs | Year | Country | Score | Result | Continent |
|---|---|---|---|---|---|
| 1 | 1954 | West Germany | 3-2 | won | Europe |
| 2 | 1954 | Hungary | 2-3 | lost | Europe |
| 3 | 1958 | Brazil | 5-2 | won | South America |
| 4 | 1958 | Sweden | 2-5 | lost | Europe |
| 5 | 1962 | Brazil | 3-1 | won | South America |
| 6 | 1962 | Czechoslovakia | 1-3 | lost | Europe |
| 7 | 1966 | England | 4-2 | won | Europe |
| 8 | 1966 | West Germany | 2-4 | lost | Europe |
| 9 | 1970 | Brazil | 4-1 | won | South America |
| 10 | 1970 | Italy | 1-4 | lost | Europe |
| 11 | 1974 | West Germany | 2-1 | won | Europe |
| 12 | 1974 | Netherlands | 1-2 | lost | Europe |
| 13 | 1978 | Argentina | 3-1 | won | South America |
| 14 | 1978 | Netherlands | 1-3 | lost | Europe |
| 15 | 1982 | Italy | 3-1 | won | Europe |
| 16 | 1982 | West Germany | 1-3 | lost | Europe |
| 17 | 1986 | Argentina | 3-2 | won | South America |
| 18 | 1986 | West Germany | 2-3 | lost | Europe |
| 19 | 1990 | Argentina | 0-1 | lost | South America |
| 20 | 1990 | West Germany | 1-0 | won | Europe |
| 21 | 1994 | Brazil | 3-2 | won | South America |
| 22 | 1994 | Italy | 2-3 | lost | Europe |
| 23 | 1998 | Brazil | 0-3 | lost | South America |
| 24 | 1998 | France | 3-0 | won | Europe |

## Combining Selected Observations from Multiple Data Sets

To create a data set that contains only the observations that are selected according to a particular criterion, you can use the subsetting IF statement and a SET statement that specifies multiple data sets. The following DATA step reads two input data sets to create a combined data set that lists only the winning teams:

```
data champions(drop=result); ❶
   set southamerican (in=S) european; ❷
   by Year;
   if result='won';  ❸
   if S then Continent='South America';  ❹
   else Continent='Europe';
run;

proc print data=champions;
   title 'World Cup Champions from 1954 to 1998';
   title2 'including Countries'' Continent';
run;
```

The following list corresponds to the numbered items in the preceding program:

1. The DROP= data set option drops the variable Result from the new data set CHAMPIONS because all values for this variable are the same.

2. The SET statement reads observations from two data sets: SOUTHAMERICAN and EUROPEAN. The S= data option creates the variable S, which is set to 1 each time an observation is contributed by the SOUTHAMERICAN data set.

3. A subsetting IF statement writes the observation to the output data set CHAMPIONS only if the value of the Result variable is **won**.

4. When the current observation comes from the data set SOUTHAMERICAN, the value of S is 1. Otherwise, the value is 0. The IF-THEN/ELSE statements execute one of two assignment statements, depending on the value of S. If the observation comes from the data set SOUTHAMERICAN, then the value assigned to Continent is South America. If the observation comes from the data set EUROPEAN, then the value assigned to Continent is Europe.

The following output displays the CHAMPIONS data set:

*Figure 22.5* Combining Selected Observations

World Cup Champions from 1954 to 1998
including Countries' Continent

| Obs | Year | Country | Score | Continent |
|---|---|---|---|---|
| 1 | 1954 | West Germany | 3-2 | Europe |
| 2 | 1958 | Brazil | 5-2 | South America |
| 3 | 1962 | Brazil | 3-1 | South America |
| 4 | 1966 | England | 4-2 | Europe |
| 5 | 1970 | Brazil | 4-1 | South America |
| 6 | 1974 | West Germany | 2-1 | Europe |
| 7 | 1978 | Argentina | 3-1 | South America |
| 8 | 1982 | Italy | 3-1 | Europe |
| 9 | 1986 | Argentina | 3-2 | South America |
| 10 | 1990 | West Germany | 1-0 | Europe |
| 11 | 1994 | Brazil | 3-2 | South America |
| 12 | 1998 | France | 3-0 | Europe |

# Performing a Calculation Based on the Last Observation

## Understanding When the Last Observation Is Processed

Many applications require that you determine when the DATA step processes the last observation in the input data set. For example, you might want to perform calculations only on the last observation in a data set, or you might want to write an observation only after the last observation has been processed. For this purpose, you can use the END= option for the SET, MERGE, MODIFY, or UPDATE statements. The syntax for this option is:

**SET** *SAS-data-set-list* END=*variable*;

The END= option defines a temporary variable whose value is 1 when the DATA step is processing the last observation. At all other times, the value of variable is 0. Although the DATA step can use the END= variable, SAS does not add it to the resulting data set.

*Note:* Chapter 13, "Using More Than One Observation in a Calculation," on page 199 explains how to use the END= option in the SET statement with a single data set. The END= option works the same way with multiple data sets. However, it is

**360** Chapter 22 • *Conditionally Processing Observations from Multiple SAS Data Sets*

important to note that END= is set to 1 only when the last observation from all input data sets is being processed.

## The Program

This example uses the data in SOUTHAMERICAN and EUROPEAN to calculate how many years a team from each continent won the World Cup from 1954 to 1998.

To perform this calculation, this program must perform the following tasks:

1. Identify on which continent a country is located.

2. Keep a running total of how many times a team from each continent won the World Cup.

3. After processing all observations, multiply the final total for each continent by 4 (the length of time between World Cups) to determine the length of time each continent has been a World Cup champion.

4. Write only the final observation to the output data set. The variables that contain the totals do not contain the final total until the last observation is processed.

The following DATA step calculates the running totals and produces the output data set that contains only those totals.

```
data timespan (keep=YearsSouthAmerican keep=YearsEuropean);   1
    set southamerican (in=S) european end=LastYear;   2
    by Year;
    if result='won' then
        do;
            if S then SouthAmericanWins+1;   3
            else EuropeanWins+1;   3
        end;
    if lastyear then   4
        do;
            YearsSouthAmerican=SouthAmericanWins*4;
            YearsEuropean=EuropeanWins*4;
            output;   5
        end;
run;

proc print data=timespan;
    title 'Total Years as Reigning World Cup Champions';
    title2 'from 1954 to 1998';
run;
```

The following list corresponds to the numbered items in the preceding program:

1 The KEEP= option writes only the YearsSouthAmerican and YearsEuropean variables to the TIMESPAN data set.

2 The END= option creates the temporary variable LastYear. The value of LastYear is 0 until the DATA step begins processing the last observation. At that point, the value of LastYear is set to 1.

3 Two new variables, SouthAmericanWins and EuropeanWins, keep a running total of the number of victories each continent achieved. For each observation in which the value of the variable Result is **won,** a different sum statement executes, based on the data set that the observation came from:

```
SouthAmericanWins+1;
```

or

```
EuropeanWins+1
```

4   When the DATA step begins processing the last observation, the value of LASTYEAR changes from 0 to 1. When this change occurs, the conditional statement **IF LastYear** becomes true, and the statements that follow it are executed. The assignment statement multiplies the total number of victories for each continent by 4 and assigns the result to the appropriate variable, YearsSouthAmerican or YearsEuropean.

5   The OUTPUT statement writes the observation to the newly created data set. Remember that the DATA step automatically writes an observation at the end of each iteration. However, the OUTPUT statement turns off this automatic feature. The DATA step writes only the last observation to TIMESPAN. When the DATA step writes the observation from the program data vector to the output data set, it writes only two variables, YearsSouthAmerican and YearsEuropean, as directed by the KEEP= data set option in the DATA statement.

The following output displays the TIMESPAN data set:

**Figure 22.6**  *Using the END= Option to Perform a Calculation Based on the Last Observation in the Data Sets*

Total Years as Reigning World Cup Champions from 1954 to 1998

| Obs | YearsSouthAmerican | YearsEuropean |
|---|---|---|
| 1 | 24 | 24 |

# Summary

## Statements

IF *condition*;
   tests whether the condition is true. If it is true, then SAS continues processing the current observation. If it is false, then SAS stops processing the observation and returns to the beginning of the DATA step. This type of IF statement is called a subsetting IF statement because it produces a subset of the original observations.

IF *condition* THEN *action*;
<ELSE *action*;>
   tests whether condition is true. If it is true, then the action in the THEN clause is executed. If the condition is false and an ELSE statement is present, the ELSE action is executed. If the condition is false and no ELSE statement is present, then execution proceeds to the next statement in the DATA step.

SET *SAS-data-set* (IN=*variable*) *SAS-data-set-list*;
   creates a variable that is associated with a SAS data set. The value of *variable* is 1 if the data set has contributed to the observation currently in the program data vector.

The value is 0 if it has not. The IN= variable exists only while the DATA step executes. It is not written to the output data set.

You can use the IN= option with any data set that you name in the SET, MERGE, MODIFY, or UPDATE statement, but use a different variable name for each one.

SET *SAS-data-set-list* END=*variable*;
creates a variable whose value is 0 until the DATA step starts to process its last observation. When processing of the last observation begins, the value of *variable* changes to 1. The END= variable exists only while the DATA step executes. It is not written to the output data set.

You can also use the END= option with the MERGE, MODIFY, and UPDATE statements.

# Learning More

DATA set options
For an introduction to data set options, see "Reading Selected Variables" on page 96.

DO statement
For information about DO-loop processing, see Chapter 14, "Finding Shortcuts in Programming," on page 215.

IF statements
For more information about both the subsetting and conditional IF statements, see Chapter 10, "Acting on Selected Observations," on page 147.

OUTPUT and subsetting IF statement
For information about using the OUTPUT and subsetting IF statements, see Chapter 10, "Acting on Selected Observations," on page 147.

SUM statement and END= option
For information about accumulating totals and using the END= option, see "Accumulating a Total for an Entire Data Set" on page 201 and "Obtaining a Total for Each BY Group" on page 204.

## Part 5

# Debugging SAS Programs

*Chapter 23*
**Analyzing Your SAS Session with the SAS Log** .................. *365*

*Chapter 24*
**Directing SAS Output and the SAS Log** ....................... *391*

*Chapter 25*
**Diagnosing and Avoiding Errors** .............................. *399*

*Chapter 26*
**Finding Logic Errors in Your Program** ......................... *415*

# Chapter 23
# Analyzing Your SAS Session with the SAS Log

**Introduction to Analyzing Your SAS Session with the SAS Log** .............. 366
   Purpose ........................................................................ 366
   Prerequisites .................................................................. 366

**Understanding the SAS Log** ........................................................ 366
   Understanding the Role of the SAS Log ................................ 366
   Resolving Errors with the Log ............................................ 368

**Locating the SAS Log** ............................................................... 369

**Understanding the Log Structure** ................................................ 369
   The Components of a SAS Log ........................................... 369
   Messages in the SAS Log .................................................. 370
   Notes in the SAS Log ....................................................... 370
   Warning Messages in the SAS Log ...................................... 371
   Error Messages in the SAS Log .......................................... 372
   Detecting a Syntax Error ................................................... 373

**Writing to the SAS Log** ............................................................. 375
   Default Output to the SAS Log ........................................... 375
   Writing Messages to the SAS Log: The PUT Statement ............ 375
   Writing the Contents of an Input Record: The LIST Statement .... 377
   Writing Messages to the SAS Log: The %PUT Macro Statement .. 380

**Suppressing Information in the SAS Log** ...................................... 382
   Using SAS System Options to Suppress Log Output ................ 382
   Suppressing SAS Statements ............................................. 382
   Suppressing System Notes ................................................ 383
   Limiting the Number of Error Messages ............................... 383
   Controlling the Level of Detail in the SAS Log ...................... 383
   Suppressing SAS Statements, Notes, and Error Messages ......... 384

**Changing the Appearance of the Log** ........................................... 387

**Summary** ............................................................................... 387
   Statements .................................................................... 387
   System Options .............................................................. 388

**Learning More** ....................................................................... 388

# Introduction to Analyzing Your SAS Session with the SAS Log

## Purpose

The SAS log is a useful tool for analyzing your SAS session and programs. In this section, you will learn about the following concepts:

- the log in relation to output
- types of messages in the SAS log
- the log structure
- writing to the SAS log
- suppressing information in the SAS log

## Prerequisites

You should understand the basic SAS programming concepts that are presented in the following sections:

- Chapter 1, "What is the SAS System?," on page 3
- Chapter 7, "Understanding DATA Step Processing," on page 107
- Chapter 4, "Starting with Raw Data: The Basics," on page 51

# Understanding the SAS Log

## Understanding the Role of the SAS Log

The SAS log results from executing a SAS program, and in that sense, it is output. The SAS log provides a record of everything that you do in your SAS session or with your SAS program, from the names of the data sets that you create to the number of observations and variables in those data sets. This record can tell you what statements were executed, how much time the DATA and PROC steps required, and whether your program contains errors.

As with SAS output, the destination of the SAS log varies depending on your method of running SAS and on your operating environment. The content of the SAS log varies according to the DATA and PROC steps that are executed and the options that are used.

The sample log in the following example was created by a SAS program that contains a DATA step and two PROC steps. The DATA step uses the file that is located in "The UNIVERSITY_TEST_SCORES Data Set" on page 815. Use the LIBNAME statement to create a libref and identify a location for your data set. The data set is stored in a SAS library that is referenced by the libref OUT throughout the rest of this section.

Use the following form of the LIBNAME statement:

```
libname libref 'your-data-library';
```

The sample log shown below is created by executing the following SAS program:

```
libname out 'your-data-library';

data out.university_test_scores;
   infile out 'your-input-file';
   input Test $ Gender $ Year TestScore;
run;

proc sort data=out.university_test_scores;
   by test;
run;

proc print data=out.university_test_scores;
   by test;
   label TestScore='Test Score';
   title1 'University Test Scores by Year, 2005-2011';
   title3 'Separate Statistics by Test Type';
run;
```

**Log 23.1** Example of a SAS Log

```
1    libname out 'your-data-library';
NOTE: Libref OUT was successfully assigned as follows:
     Engine:        V9
     Physical Name: your-data-library
2
3    data out.university_test_scores;
4       infile 'your-input-file';
5       input Test $ Gender $ Year TestScore;
6    run;

NOTE: The infile 'your-input-file' is:
     Filename=your-input-file
     RECFM=V,LRECL=32767,File Size (bytes)=544,
     Last Modified=03May2013:07:18:05,
     Create Time=01May2013:10:41:25

NOTE: 28 records were read from the infile 'your-input-file'.
     The minimum record length was 17.
     The maximum record length was 18.
NOTE: The data set OUT.UNIVERSITY_TEST_SCORES has 28 observations and 4
variables.
NOTE: DATA statement used (Total process time):
     real time           0.03 seconds
     cpu time            0.03 seconds


7
8    proc sort data=out.university_test_scores;
9       by test;
10   run;

NOTE: There were 28 observations read from the data set
OUT.UNIVERSITY_TEST_SCORES.
NOTE: The data set OUT.UNIVERSITY_TEST_SCORES has 28 observations and 4
variables.
NOTE: PROCEDURE SORT used (Total process time):
     real time           0.15 seconds
     cpu time            0.04 seconds


11
12   proc print data=out.university_test_scores;
NOTE: Writing HTML Body file: sashtml.htm
13      by test;
14      label TestScore='Test Score';
15      title1 'University Test Scores by Year, 2005-2011';
16      title3 'Separate Statistics by Test Type';
17   run;

NOTE: There were 28 observations read from the data set
OUT.UNIVERSITY_TEST_SCORES.
NOTE: PROCEDURE PRINT used (Total process time):
     real time           0.51 seconds
     cpu time            0.39 seconds
```

The SAS program that created this log ran without errors. The notes are informational messages that indicate how SAS processed your program.

## Resolving Errors with the Log

If the previous program contained errors, then those errors would be reflected, as part of the session, in the SAS log. SAS generates messages for data errors, syntax errors, and programming errors. You can browse those messages, make necessary changes to your program, and then rerun the program successfully.

SAS does not always identify all of the program errors the first time you execute your program. Fixing one error might uncover other errors in your program.

## Locating the SAS Log

The destination of your log depends on the method that you are using to start, run, and exit SAS. It also depends on your operating environment and on the setting of SAS system options. The following table shows the default destination for each method of operation:

*Table 23.1 Default Destinations for the SAS Log*

| Method of Operation | Destination of SAS Log |
| --- | --- |
| SAS windowing environment (interactive full-screen) | Log window |
| Interactive line mode | On the terminal display, as statements are entered |
| Noninteractive SAS programs | Depends on the operating environment |
| Batch jobs | Printer or disk file, depending on your operating environment |

## Understanding the Log Structure

### The Components of a SAS Log

The SAS log provides valuable information that helps you identify problems when you execute your programs. The SAS log is especially helpful if you have questions and need to contact on-site SAS personnel or SAS Technical Support. The contents of the log aids in diagnosing your problem.

The log contains the following elements:

- SAS statements for the DATA and PROC steps
- error messages
- notes and warning messages
- notes that contain the number of observations and variables for each data set that is created

## Messages in the SAS Log

SAS performs error processing during the compilation and execution phases of your program, and writes note, warning, and error messages to the log. If your program has an error, SAS attempts to interpret the intended meaning of the error. If SAS cannot correct the error, SAS writes a message to the log. Some errors are explained fully by the message in the log. Other messages are not as easy to interpret because SAS is not always able to detect exactly where the error occurred. You can debug SAS programs by understanding processing messages in the SAS log and then modifying your code.

The SAS log displays the following types of messages that can help you debug your programs:

- notes
- warnings
- errors

Each of these messages gives you different information about the program that is being processed. Reviewing all messages in the log is a good practice to follow.

## Notes in the SAS Log

A SAS note is an informational message and does not stop your program from executing. A note can indicate that part of your code is programmatically incorrect.

A SAS note can also provide information about such items as processing time, the successful or unsuccessful completion of a DATA or PROC step, the number of records that were read from a file, and the name of the input file that is used in your program.

The following are examples of SAS notes:

- NOTE: The data set WORK.TEST has 50 observations and five variables.
- NOTE: 35 records were read from the INFILE *file-name*.
- NOTE: Variable ABCD is uninitialized.
- NOTE: The SAS System stopped processing this step because of errors.
- NOTE: No variables in data set WORK.TEST.
- NOTE: Invalid data for *variable-name* at line *n*.

The following program creates the STOCK data set that is used as input to the second DATA step. The STOCK data set has an invalid value for Inventory. The INPUT statement identifies Inventory as a numeric value, but the fourth observation has alphabetic characters in this field. The program uses PUT statements to write user-supplied messages to the log. (For more information about the PUT statement, see "Writing Messages to the SAS Log: The PUT Statement" on page 375.) Notice that the first record has a missing value in the Product field. This missing value does not cause the program to stop executing.

```
data stock;
    input inventory QuantitySold Idnum 8-11 Product $ 13-18 cost 20-24;
    datalines;
100 52 1001         67.45
345 49 1020 saw     99.99
237 55 2003 wrench  34.97
abc 65 3015 shovel  25.99
```

```
          932 38 4215 rake    22.50
          ;
       run;

       data stock2;
          set stock;
          if inventory < 300 and QuantitySold > 50 then
             put 'Time to order product: ' @27 product= @42 idnum=;
             else;
                if inventory > 300 then
                   put 'No need to order product: ' @27 product= @42 idnum=;
       run;
```

**Log 23.2**  *Log Output from the Stock Inventory Program*

```
94    data stock;
95       input inventory QuantitySold Idnum 8-11 Product $ 13-18;
96       datalines;

NOTE: Invalid data for inventory in line 100 1-3.
      RULE:        ----+----1----+----2----+----3----+----4----+----5----+----6----
+----7----+----8----+-
100           abc 65 3015 shovel
inventory=. QuantitySold=65 Idnum=3015 Product=shovel _ERROR_=1 _N_=4
NOTE: The data set WORK.STOCK has 5 observations and 4 variables.
NOTE: DATA statement used (Total process time):
      real time           0.00 seconds
      cpu time            0.00 seconds


102   ;
103   run;
104
105   data stock2;
106      set stock;
107      if inventory < 300 and QuantitySold > 50 then
108         put 'Time to order product: ' @27 product= @42 idnum=;
109      if inventory > 300 then
110         put 'No need to order product: ' @27 product= @42 idnum=;
111   run;

Time to order product:    Product=         Idnum=1001
No need to order product: Product=saw      Idnum=1020
Time to order product:    Product=wrench   Idnum=2003
Time to order product:    Product=shovel   Idnum=3015
No need to order product: Product=rake     Idnum=4215
NOTE: There were 5 observations read from the data set WORK.STOCK.
NOTE: The data set WORK.STOCK2 has 5 observations and 4 variables.
NOTE: DATA statement used (Total process time):
      real time           0.00 seconds
      cpu time            0.00 seconds
```

The NOTE identifies invalid data for an inventory item. SAS was expecting a numeric value for Inventory, and instead the value of Inventory for the fourth observation is a character. SAS continues to execute the program.

## Warning Messages in the SAS Log

A SAS warning message alerts you to potential problems with your code but does not stop program execution. For example, warning messages are issued when you enter a word incorrectly and SAS is able to interpret the word, and when a program produces no output.

It is important to view the warning messages in the SAS log to verify whether your program executed as expected. The following are examples of warning messages in the SAS log:

- WARNING: The data set WORK.TEST might be incomplete. When this step was stopped, there were 0 observations and 0 variables.
- WARNING: Data set WORK.TEST was not replaced because this step was stopped.
- WARNING: Assuming that the symbol DATA was misspelled as date.

Executing the following DATA step results in an error message, a note, and a warning in the SAS log. In the example, the SET statement identifies the input data set. However, SAS could not find the data set:

```
data test;
   set mydataset;
run;
```

**Log 23.3** *Log Output with Warning Messages*

```
114  data test;
115     set mydataset;
ERROR: File WORK.MYDATASET.DATA does not exist.
NOTE: The SAS System stopped processing this step because of errors.
WARNING: The data set WORK.TEST may be incomplete.  When this step was stopped
there were 0 observations and 0 variables.
WARNING: Data set WORK.TEST was not replaced because this step was stopped.
NOTE: DATA statement used (Total process time):
      real time           0.03 seconds
      cpu time            0.00 seconds
```

The WARNING messages in the log give you information about the temporary output data set, WORK.TEST, that the program was attempting to create.

To correct the problem, make sure that the input data set in the SET statement already exists. The error might be as simple as a misspelled data set name. Execute the program again and review your log messages.

## Error Messages in the SAS Log

A SAS error message alerts you to a significant problem with your code. SAS either stops program processing or flags errors and continues to process your program. An error message is written to the log.

One of the most common errors is a missing semicolon at the end of a statement. SAS can write multiple messages to the log when it encounters a missing semicolon, depending on where in the program the error occurred. It might be difficult to determine the location of the error because there is no specific message that identifies the error as a missing semicolon.

The following are examples of error messages in the SAS log:

- ERROR: Variable ITEM2 not found.
- ERROR: Illegal reference to array ALL.
- ERROR: LIBNAME MYLIB is not assigned.
- ERROR: Syntax error, statement will be ignored.

For an example of an error message in a SAS log, see "Warning Messages in the SAS Log" on page 371. Often, multiple types of messages occur in the same log.

### Detecting a Syntax Error

The following SAS program contains one DATA step and two PROC steps. However, the DATA step has a syntax error. It does not end with a semicolon. The SAS log that is located below the program explains program processing in detail.

```
libname out 'your-data-library';

   /* omitted semicolon */
data out.university_test_scores2
   infile 'your-input-file';
   input test $ gender $ year TestScore;
run;

proc sort data=out.university_test_scores2;
   by test;
run;

proc print data=out.university_test_scores2;
   by test;
run;
```

**374** Chapter 23 • *Analyzing Your SAS Session with the SAS Log*

The following log shows how SAS processes the program step by step:

**Log 23.4** *Log Output That Identifies a Syntax Error*

```
16   libname out 'your-data-library';
NOTE: Libref OUT was successfully assigned as follows:   1
      Engine:        V9
      Physical Name: your-data-library
17
18
19      /* omitted semicolon */
20   data out.university_test_scores2   2
21      infile 'your-input-file';
22      input test $ gender $ year TestScore;
23   run;

ERROR: No DATALINES or INFILE statement.   3
ERROR: Extension for physical file name "your-input-file" does not
       correspond to a valid member type.   3
NOTE: The SAS System stopped processing this step because of errors.   3
WARNING: The data set OUT.UNIVERSITY_TEST_SCORES2 may be incomplete.  When this
step was stopped
         there were 0 observations and 4 variables.   4
WARNING: The data set WORK.INFILE may be incomplete.  When this step was stopped
there were 0
         observations and 4 variables.   4
NOTE: DATA statement used (Total process time):
      real time           0.01 seconds
      cpu time            0.01 seconds


24
25   proc sort data=out.university_test_scores2;   5
26      by test;
27   run;

NOTE: Input data set is empty.   5
NOTE: The data set OUT.UNIVERSITY_TEST_SCORES2 has 0 observations and 4
variables.   5
NOTE: PROCEDURE SORT used (Total process time):
      real time           0.01 seconds
      cpu time            0.01 seconds


28
29   proc print data=out.university_test_scores2;   6
30      by test;
31   run;

NOTE: No observations in data set OUT.UNIVERSITY_TEST_SCORES2.   6
NOTE: PROCEDURE PRINT used (Total process time):
      real time           0.00 seconds
      cpu time            0.00 seconds
```

The following list corresponds to the numbered items in the preceding log:

1 The LIBNAME statement successfully associates the libref OUT with your library.

2 The DATA statement is missing a semicolon, causing an error in the program.

3 The error messages identify the error, and the note shows that the program stopped executing.

4 Warning messages provide information about the OUT.UNIVERSITY_TEST_SCORES2 and WORK.INFILE data sets. SAS reads

past the DATA statement until it encounters a semicolon and interprets WORK.INFILE as a data set that is to be created.

5   SAS attempts to sort the output data set.

6   SAS attempts to write the OUT.UNIVERSITY_TEST_SCORES2 data set.

To correct the error, add a semicolon after the DATA statement and execute the program again.

# Writing to the SAS Log

## Default Output to the SAS Log

The previous sample logs show the information that appears in the log by default. You can also write to the log by using the PUT statement or the LIST statement within a DATA step. You can also use the %PUT macro statement anywhere in your program. These statements can be used to debug your SAS programs.

## Writing Messages to the SAS Log: The PUT Statement

### Introduction to the PUT Statement

The PUT statement enables you to write information that you specify, including text strings and variable values, to the log. Each time the DATA step iterates, the PUT statement is executed, and a message is written to the log. Values can be written in column, list, formatted, or named output style. (For more information, see the different forms of the INPUT statement in *SAS Statements: Reference*.) The PUT statement can be a useful debugging tool. You can write messages from different parts of your program to the log, and track the execution of your program.

This is a simplified form of the PUT statement:

**PUT** <'*message*'> | <*variable-name*>;

*message* specifies the message that you would like to see displayed in the SAS log. Character literals must be enclosed in quotation marks. Your message text can include a word or phrase, such as "Notice This", that helps you identify the message more easily.

*variable-name* specifies a variable whose value is written to the log. If you use an equal sign after the variable name, both the variable name and the value will be written to the log.

By using the following PUT statement, you can write the values of all variables, including the automatic variables _ERROR_ and _N_, that are defined in the current DATA step:

   PUT _ALL_;

For more information about the PUT statement and how it can be used, see "PUT Statement" in *SAS Statements: Reference*.

### Example: Writing to the Log with the PUT Statement

The first DATA step in the following example creates a temporary data set called INVENTORY. This data set is used as input to the SET statement in the second DATA step. The PUT statements write messages to the log indicating whether items need to be

**376** Chapter 23 • Analyzing Your SAS Session with the SAS Log

ordered. By entering the variable name Item followed by an equal sign in the PUT statement, the PUT statement writes both the variable name and its value to the log. This example shows the log output for the PUT statement.

```
data inventory;
    input InStock QuantitySold Idnum 8-11 Item $ 13-18;
    datalines;
100 52 1001 hammer
345 49 1020 saw
237 55 2003 wrench
864 65 3015 shovel
932 38 4215 rake
;
run;

data inventory2;
   set inventory;
   if InStock < 300 and QuantitySold > 50 then
      put 'Time to order product: ' Item=;
   if InStock > 300 then
      put 'No need to order product: ' Item=;
run;
```

**Log 23.5** *Log Output Using the PUT Statement*

```
1     data inventory;
2        input InStock QuantitySold Idnum 8-11 Item $ 13-18;
3        datalines;

NOTE: The data set WORK.INVENTORY has 5 observations and 4 variables.
NOTE: DATA statement used (Total process time):
      real time           0.30 seconds
      cpu time            0.09 seconds


9     ;
10    run;
11
12    data inventory2;
13       set inventory;
14       if InStock < 300 and QuantitySold > 50 then
15          put 'Time to order product: ' Item=;
16       if InStock > 300 then
17          put 'No need to order product: ' Item=;
18    run;

Time to order product: Item=hammer
No need to order product: Item=saw
Time to order product: Item=wrench
No need to order product: Item=shovel
No need to order product: Item=rake
NOTE: There were 5 observations read from the data set WORK.INVENTORY.
NOTE: The data set WORK.INVENTORY2 has 5 observations and 4 variables.
NOTE: DATA statement used (Total process time):
      real time           0.01 seconds
      cpu time            0.01 seconds
```

The PUT statement writes five lines to the log.

Your program code can include instructions for writing output in specific columns. If you do this, it might make your output more readable. In the example above, you can change the IF statements in the following way:

```
    if Inventory < 300 and QuantitySold > 50 then
       put 'Time to order product: ' @27 Item= @39 Idnum=;
    if Inventory > 300 then
       putlog 'No need to order product: ' @27 Item= @39 Idnum=;
```

The @27 indicates that the variable Item begins in column 27. The @39 indicates that the variable Idnum begins in column 39. The following partial SAS log shows the output:

```
Time to order product:      Item=hammer Idnum=1001
No need to order product:   Item=saw    Idnum=1020
Time to order product:      Item=wrench Idnum=2003
No need to order product:   Item=shovel Idnum=3015
No need to order product:   Item=rake   Idnum=4215
```

### Writing the Contents of an Input Record: The LIST Statement

#### Introduction to the LIST Statement

Use the LIST statement in the DATA step to write the current input record to the log. You can use the LIST statement only on data that is read from an INPUT statement. It has no effect on data that is read with the SET, MERGE, MODIFY, or UPDATE statements. The LIST statement writes observations at the end of each iteration of the DATA step.

#### Example: Listing the Current Input Record

The following program shows that the LIST statement, like the PUT statement, can be very effective when combined with conditional processing to write selected information to the log.

```
libname out 'your-data-library';

data out.university_test_scores3;
   infile 'your-data-file';
   input test $ gender $ year TestScore;
   if TestScore < 525 then delete;
   else list;
run;
```

**Log 23.6** *Listing the Contents of a Record*

```
46   libname out 'your-data-library';
NOTE: Libref OUT was successfully assigned as follows:
      Engine:        V9
      Physical Name: your-data-library
47
48
49
50   data out.university_test_scores3;
51      infile 'your-input-file';
52      input test $ gender $ year TestScore;
53      if TestScore < 525 then delete;
54      else list;
55   run;

NOTE: The infile 'your-input-file' is:
      Filename=your-input-file,
      RECFM=V,LRECL=32767,File Size (bytes)=544,
      Last Modified=03May2013:07:18:05,
      Create Time=01May2013:10:41:25

RULE:       ----+----1----+----2----+----3----+----4----+----5----+----6----
+----7----+----8----+--
21          Math    m 2008 525   18
23          Math    m 2009 527   18
25          Math    m 2010 530   18
27          Math    m 2011 531   18
NOTE: 28 records were read from the infile 'your-input-file'.
      The minimum record length was 17.
      The maximum record length was 18.
NOTE: The data set OUT.UNIVERSITY_TEST_SCORES3 has 4 observations and 4
variables.
NOTE: DATA statement used (Total process time):
      real time           0.01 seconds
      cpu time            0.01 seconds
```

When the LIST statement is executed, SAS causes the current input buffer to be written to the log. Note the presence of the columns ruler before the first line of data. The ruler indicates that input data was written to the log. The ruler can be used to reference column positions in the input buffer. Notice that the LIST statement causes the record length to be written at the end of each line (in this case, each record has a length of 18). This feature of the LIST statement works only in operating environments that support variable-length input records.

### *Example: Listing Records That Have Missing Data*

The following is another example of using the LIST statement to write records to the log. The example creates the EMPLOYEE data set. The program uses an INPUT statement to describe the arrangement of values in the input record. The input data for each employee is three lines long and begins with an employee ID number. The third line contains the value for Salary. After reading the ID number, the #3 notation instructs SAS to go to the third line of each record to read the value for Salary. The IF statement instructs SAS to list the record for any employees that have a missing value for Salary.

```
data employee;
   input IdNum 1-9 #3 Salary 1-8;
   if salary=. then list;
   datalines;
234567890
James Smith
70356.79
```

```
345678912
Jeffery Feldenstern
.
382623454
Sandy Lineman
75724.96
346521145
Jose Garcia
.
;
run;

proc print data=employee;
   format salary dollar10.2;
   title 'Employee Salary';
run;
```

You can see that two observations have missing values for Salary.

*Log 23.7  Listing the Contents of a Record That Has Missing Values*

```
70    data employee;
71       input IdNum 1-9 #3 Salary 1-8;
72       if salary=. then list;
73       datalines;
RULE:       ----+----1----+----2----+----3----+----4----+----5----+----6----
+----7----+----8----+-
77          345678912
78          Jeffery Feldenstern
79          .
83          346521145
84          Jose Garcia
85          .
NOTE: The data set WORK.EMPLOYEE has 4 observations and 2 variables.
NOTE: DATA statement used (Total process time):
      real time           0.00 seconds
      cpu time            0.00 seconds


86    ;
87    run;
88
89    proc print data=employee;
90       format salary dollar10.2;
91       title 'Employee Salary';
92    run;

NOTE: There were 4 observations read from the data set WORK.EMPLOYEE.
NOTE: PROCEDURE PRINT used (Total process time):
      real time           0.00 seconds
      cpu time            0.01 seconds
```

The following output shows the missing values.

*Figure 23.1  Employee Salary with Missing Values*

Employee Salary

| Obs | IdNum | Salary |
|---|---|---|
| 1 | 234567890 | $70,356.79 |
| 2 | 345678912 | . |
| 3 | 382623454 | $75,724.96 |
| 4 | 346521145 | . |

## Writing Messages to the SAS Log: The %PUT Macro Statement

### Introduction to the %PUT Macro Statement

The %PUT statement is a macro statement that is independent of the DATA step. It can be used anywhere in your program to write a message to the log.

### Example: Writing Messages to the Log

The following example shows how the %PUT macro statement can be used in a SAS program:

```
data professions;
    input Name $ 1-17 Gender $ 19 Occupation $ 21-33;
    datalines;
Shirley Grayson    F attorney
Kristen Hagshould  F doctor
Matthew Rodriguez  M
Michael Wu         M mathematician
Sophie Majkut      F physicist
;

%put Notice: This is the end of the first DATA step.;

data professions2;
    set professions;
    %put Notice: Testing for missing values.;
    if gender = 'M' and Occupation = ' ' then Occupation='MISSING';
run;

%put Notice: This is the end of the second DATA step.;

proc print data=professions2;
    title 'Staff Occupations';
run;
%put Notice: This is the end of the program.;
```

**Log 23.8** *Output from Using the %PUT Macro Statement*

```
11   data professions;
12      input Name $ 1-17 Gender $ 19 Occupation $ 21-33;
13      datalines;

NOTE: The data set WORK.PROFESSIONS has 5 observations and 3 variables.
NOTE: DATA statement used (Total process time):
      real time           0.07 seconds
      cpu time            0.00 seconds


19   ;
20
21   %put Notice: This is the end of the first DATA step.;
Notice: This is the end of the first DATA step.
22
23   data professions2;
24      set professions;
25      %put Notice: Testing for missing values.;
Notice: Testing for missing values.
26      if gender = 'M' and Occupation = ' ' then Occupation='MISSING';
27   run;

NOTE: There were 5 observations read from the data set WORK.PROFESSIONS.
NOTE: The data set WORK.PROFESSIONS2 has 5 observations and 3 variables.
NOTE: DATA statement used (Total process time):
      real time           0.01 seconds
      cpu time            0.01 seconds


28
29   %put Notice: This is the end of the second DATA step.;
Notice: This is the end of the second DATA step.
30
31   proc print data=professions2;
32      title 'Staff Occupations';
33   run;

NOTE: There were 5 observations read from the data set WORK.PROFESSIONS2.
NOTE: PROCEDURE PRINT used (Total process time):
      real time           0.04 seconds
      cpu time            0.01 seconds


34   %put Notice: This is the end of the program.;
Notice: This is the end of the program.
```

SAS produces the following results:

**Figure 23.2** *Staff Occupations*

Staff Occupations

| Obs | Name | Gender | Occupation |
|---|---|---|---|
| 1 | Shirley Grayson | F | attorney |
| 2 | Kristen Hagshould | F | doctor |
| 3 | Matthew Rodriguez | M | MISSING |
| 4 | Michael Wu | M | mathematician |
| 5 | Sophie Majkut | F | physicist |

# Suppressing Information in the SAS Log

## Using SAS System Options to Suppress Log Output

There might be times when you want to prevent some information from being written to the SAS log. You can suppress SAS statements, system messages, and error messages with the NOSOURCE, NONOTES, and ERRORS= SAS system options. You can specify these options when you invoke SAS, in the OPTIONS window, or in an OPTIONS statement. In this section, the options are specified in OPTIONS statements.

*Note:* All SAS system options remain in effect for the duration of your session or until you change the options.

## Suppressing SAS Statements

If you regularly execute large SAS programs without making changes, then you can use the NOSOURCE system option as follows to suppress the listing of the SAS statements to the log:

```
options nosource;
```

The NOSOURCE system option causes only source lines that contain errors to be written to the log. You can return to the default by specifying the SOURCE system option as follows:

```
options source;
```

The SOURCE system option causes all subsequent source lines to be written to the log.

You can also control whether secondary source statements (from files that are included with a %INCLUDE statement) are written to the log. Use the NOSOURCE2 system option to suppress secondary statements:

```
options nosource2;
```

Use the SOURCE2 system option to include secondary source statements in the log:

```
options source2;
```

## Suppressing System Notes

Much of the information that is supplied by the log appears as notes, including the following items:

- copyright information
- licensing and site information
- number of observations and variables in the data set

SAS also issues a note when it has stopped processing a step because of errors.

If you do not want the notes to appear in the log, then use the NONOTES system option to suppress them:

```
options nonotes;
```

All messages starting with NOTE are suppressed. You can return to the default by specifying the NOTES system option:

```
options notes;
```

## Limiting the Number of Error Messages

SAS writes messages for data input errors that appear in your SAS program. The default number for the error messages is usually 20, but might vary from site to site. Use the ERRORS= system option to specify the maximum number of observations for which error messages are written to the log.

Note that this option limits only the error messages that are produced for incorrect data. This type of error is caused primarily by trying to read character values for a variable that the INPUT statement defines as numeric.

If data errors are detected in more observations than the number that you specify, then processing continues, but error messages are not written for the additional errors. For example, the following OPTIONS statement specifies writing for a maximum of five observations:

```
options errors=5;
```

However, as discussed in "Suppressing SAS Statements, Notes, and Error Messages" on page 384, it might not be to your advantage to suppress error messages.

*Note:* No option is available to eliminate warning messages.

## Controlling the Level of Detail in the SAS Log

The MSGLEVEL= system option controls the level of detail in messages that are written to the SAS log. The values for MSGLEVEL= are either N or I. N indicates that SAS writes notes, warning, and error messages only. This is the default. I indicates that SAS writes additional messages that relate to index usage, merge processing, and sort utilities.

The PRINTMSGLIST system option controls the writing of extended lists of messages to the SAS log. PRINTMSGLIST is the default, and writes the entire list of messages to the SAS log. NOPRINTMSGLIST writes the top-level messages only.

## Suppressing SAS Statements, Notes, and Error Messages

The following SAS program reads the test score data as in other examples, but in this example the character symbol for the variable Gender is omitted. Also, the data is not sorted before using a BY statement with PROC PRINT. For efficiency, statements, notes, and error messages are suppressed.

```
libname out 'your-data-library';
options nosource nonotes errors=0;

data out.university_test_scores4;
   infile 'your-input-file';
   input test $ gender year TestScore 25-27;
run;

proc print;
   by test;
run;
```

The SAS log that appears is shown in the following output. Because the SAS system option ERRORS=0 is specified, the error limit is reached immediately, and the errors that result from trying to read Gender as a numeric value are not written to the log. Also, specifying the NOSOURCE and NONOTES system options causes the log to contain no SAS statements that can be verified and no notes to explain SAS processing. The log does contain an error message that explains that OUT.UNIVERSITY_TEST_SCORES4 is not sorted in ascending sequence. This error is not caused by invalid input data, so the ERRORS=0 option has no effect on this error.

*Log 23.9  Suppressing Information in the SAS Log*

```
1    libname out 'your-data-library';
NOTE: Libref OUT was successfully assigned as follows:
      Engine:        V9
      Physical Name: your-data-library
2    options nosource nonotes errors=0;
ERROR: Data set OUT.UNIVERSITY_TEST_SCORES4 is not sorted in ascending sequence.
The current BY
       group has test = Verbal and the next BY group has test = Math.
```

*Note:* The NOSOURCE, NONOTES, and ERRORS= system options are used to save space. They are most useful with a program that has already been tested, such as one that is run regularly. However, as demonstrated in this section, using these system options is not always appropriate. During development of a new program, the error messages in the log might be essential for debugging, and should not be limited. Similarly, notes should not be suppressed because they can help you identify problems with a program. They are especially important if you seek help in debugging your program from someone who is unfamiliar with it. You should not suppress any information in the log until you have already executed the program without errors.

The following partial output shows the results if the previous example is executed again with the SOURCE, NOTES, and ERRORS= options:

```
options source notes errors=4;

data out.university_test_scores5;
   infile 'your-input-file';
```

```
      input test $ gender year TestScore;
run;

proc print data=out.university_test_scores5;
   by test;
run;
```

**Log 23.10** SAS Log with Output: ERRORS=4

```
13   libname out 'your-data-library';
NOTE: Libref OUT was successfully assigned as follows:
      Engine:        V9
      Physical Name: your-data-library
14
15   options source notes errors=4;
16
17   data out.university_test_scores5;
18      infile 'your-input-file';
19      input test $ gender year TestScore;
20   run;

NOTE: The infile 'your-input-file' is:
      Filename=your-input-file,
      RECFM=V,LRECL=32767,File Size (bytes)=544,
      Last Modified=03May2013:07:18:05,
      Create Time=01May2013:10:41:25

NOTE: Invalid data for gender in line 1 8-8.
RULE:      ----+----1----+----2----+----3----+----4----+----5----+----6----
+----7----+----8----+--
1          Verbal m 2005 504   18
test=Verbal gender=. year=2005 TestScore=504 _ERROR_=1 _N_=1
NOTE: Invalid data for gender in line 2 8-8.
2          Verbal f 2005 496   17
test=Verbal gender=. year=2005 TestScore=496 _ERROR_=1 _N_=2
NOTE: Invalid data for gender in line 3 8-8.
3          Verbal m 2006 504   18
test=Verbal gender=. year=2006 TestScore=504 _ERROR_=1 _N_=3
NOTE: Invalid data for gender in line 4 8-8.
WARNING: Limit set by ERRORS= option reached.  Further errors of this type will
not be printed.
4          Verbal f 2006 497   17
test=Verbal gender=. year=2006 TestScore=497 _ERROR_=1 _N_=4
NOTE: 28 records were read from the infile 'your-input-file'.
      The minimum record length was 17.
      The maximum record length was 18.
NOTE: The data set OUT.UNIVERSITY_TEST_SCORES5 has 28 observations and 4
variables.
NOTE: DATA statement used (Total process time):
      real time           0.04 seconds
      cpu time            0.04 seconds


21
22   proc print data=out.university_test_scores5;
23      by test;
24   run;

ERROR: Data set OUT.UNIVERSITY_TEST_SCORES5 is not sorted in ascending sequence.
The current BY
      group has test = Verbal and the next BY group has test = Math.
NOTE: The SAS System stopped processing this step because of errors.
NOTE: There were 15 observations read from the data set
OUT.UNIVERSITY_TEST_SCORES5.
NOTE: PROCEDURE PRINT used (Total process time):
      real time           0.06 seconds
      cpu time            0.04 seconds
```

With this execution of the program, the log is a more effective problem-solving tool. The log includes all the SAS statements from the program, as well as many informative notes. Specifically, it includes enough messages about the invalid data for the variable

Gender so that the problem can be identified. With this information, the program can be modified and rerun successfully.

## Changing the Appearance of the Log

Except in an interactive session, you can customize the log by using the PAGE and SKIP statements. Use the PAGE statement to move to a new page in the log. Use the SKIP statement to skip lines in the log. With the SKIP statement, you can specify the number of lines that you want to skip. If you do not specify a number, then SAS skips one line. If the number that you specify exceeds the number of lines remaining on the page, then SAS treats the SKIP statement like a PAGE statement and skips to the top of the next page. The PAGE and SKIP statements do not appear in the log.

## Summary

### Statements

LIST;
: lists in the SAS log the contents of the input buffer for the observation that is being processed.

SAS Log
: *SAS Language Reference: Concepts* provides additional information about the SAS log.

PAGE;
: skips to a new page in the log.

PUT '*message*' <*variable-name*>;
PUT <*variable-list*> | <_ALL_>;
: writes lines to the SAS log, the output file, or any file that is specified in a FILE statement. If no FILE statement was executed in the current iteration of the DATA step, then the PUT statement writes to the SAS log.

*message* specifies the text that you enter to be written to the log.

*variable-name* specifies the variable whose value is written to the log.

*variable-list* names a list of variables whose values are written to the log.

_ALL_ specifies that the values of all variables, including _ERROR_ and _N_, are to be written to the log.

The PUT statement is valid in a DATA step.

For more information, see "PUT Statement" in *SAS Statements: Reference*.

%PUT '*message*';
: writes a message to the SAS log.

*message* specifies the text that you enter to be written to the log.

You can use the %PUT macro statement anywhere in your program.

For more information, see "%PUT Statement" in *SAS Macro Language: Reference*.

SKIP <*n*>;
> in the SAS log, skips the number of lines that you specify with the value *n*. If you do not specify a value, then SAS writes a blank line to the log. If you specify a number that is greater than the number of lines remaining on the page, then SAS treats the SKIP statement like a PAGE statement and skips to the top of the next page.

## *System Options*

ERRORS=*n*
> specifies the maximum number of observations for which error messages about data input errors are written to the log.

NOTES | NONOTES
> controls whether notes are written to the log.

SOURCE | NOSOURCE
> controls whether SAS statements are written to the log.

SOURCE2 | NOSOURCE2
> controls whether secondary SAS statements from files included by %INCLUDE statements are written to the log.

# Learning More

Automatic variables
> For information about the automatic variables _N_ and _ERROR_, see Chapter 25, "Diagnosing and Avoiding Errors," on page 399.

Debugging
> For more information about debugging, see Chapter 25, "Diagnosing and Avoiding Errors," on page 399.

FILE and PUT statements
> For more information, see Chapter 33, "Understanding and Customizing SAS Output: The Basics," on page 613.

Log window
> For more information about the Log window, see "Working with SAS Windows" on page 761 and "The SAS Log" in *SAS Language Reference: Concepts*.

Operating environment-specific information
> The SAS documentation for your operating environment contains information about the appearance and destination of the SAS log, as well as for routing output.

The SAS environment
> For more information about starting a SAS session and the different ways that you can execute your programs, see "Introduction to the SAS Environment" on page 741 and "Introduction to the SAS Windowing Environment" in *SAS Language Reference: Concepts*.

SAS statements
> *SAS Statements: Reference* provides complete reference information about SAS statements that work across all operating environments. Refer to the SAS documentation for your operating environment for information about operating-environment specific options.

SAS system options

*SAS System Options: Reference* provides complete reference information about SAS system options that work across all operating environments. Refer to the SAS documentation for your operating environment for information about operating-environment specific options.

# Chapter 24
# Directing SAS Output and the SAS Log

Introduction to Directing SAS Output and the SAS Log .................... 391
    Purpose ........................................................................ 391
    Prerequisites ................................................................... 392

Input File and SAS Data Set for Examples .............................. 392

Routing the Output and the SAS Log with PROC PRINTTO ................ 393
    Routing Output to an Alternate Location ................................. 393
    Routing the SAS Log to an Alternate Location .......................... 394
    Restoring the Default Destination ....................................... 395

Storing the Output and the SAS Log in the SAS Windowing Environment ..... 395
    Understanding the Default Destination .................................. 395
    Storing the Contents of the Output and Log Windows .................... 396

Redefining the Default Destination in a Batch or Noninteractive Environment . 396
    Determining the Default Destination .................................... 396
    Changing the Default Destination ....................................... 396
    Understanding the Configuration File ................................... 397

Summary ........................................................................ 397
    PROC PRINTTO Statement Options ..................................... 397
    SAS Windowing Environment Commands ............................... 397
    SAS System Options ....................................................... 398

Learning More ................................................................. 398

## Introduction to Directing SAS Output and the SAS Log

### Purpose

SAS provides several methods to direct SAS output and the SAS log to different destinations. In this section, you learn how to use the following SAS language elements:

- PRINTTO procedure from within a program or session to route DATA step output, the SAS log, or procedure output from their default destinations to another destination

- FILE command, in the SAS windowing environment, to store the contents of the Log and Output windows in files

- PRINT= and LOG= system options when you invoke SAS to redefine the destination of the log and output for an entire SAS session

*Prerequisites*

Before proceeding with this section, you should understand the following features and concepts:

- creating DATA step or PROC step output
- locating the log and procedure output
- referencing external files

# Input File and SAS Data Set for Examples

The examples in this section are based on data from university test scores. The data contains test scores for entering university classes from 2005 to 2011. For a complete listing of the input data, see "The UNIVERSITY_TEST_SCORES Data Set" on page 815. The input file has the following structure:

```
Verbal f 2009 503
Verbal m 2010 507
Verbal f 2010 503
Verbal m 2011 509
Verbal f 2011 502
Math   m 2005 521
Math   f 2005 484
Math   m 2006 524
Math   f 2006 484
```

The input file contains the following values from left to right:

- type of test
- gender of student
- year of the exam
- average score

The following program creates the data set that this section uses:

```
data test_scores;
   input Test $ Gender $ Year TestScore;
   datalines;
Verbal m 2005 504
Verbal f 2005 496
Verbal m 2006 504
Verbal f 2006 497
Verbal m 2007 501
Verbal f 2007 497
   ... more data lines ...
Math   f 2009 492
Math   m 2010 530
Math   f 2010 494
Math   m 2011 531
```

```
Math    f  2011  496
;
```

# Routing the Output and the SAS Log with PROC PRINTTO

## *Routing Output to an Alternate Location*

You can use the PRINTTO procedure to redirect SAS procedure output from the HTML destination to an alternate location. These locations are:

- a permanent file
- a SAS catalog entry
- a dummy file, which serves to suppress the output

After PROC PRINTTO executes, all procedure output is sent to the alternate location until you execute another PROC PRINTTO statement or until your program or session ends.

The default destination for the procedure output depends on how you configure SAS to handle output. For more information, see the discussion of SAS output in Chapter 34, "Understanding and Customizing SAS Output: The Output Delivery System (ODS)," on page 643 and Chapter 33, "Understanding and Customizing SAS Output: The Basics," on page 613.

*Note:* If you used the Output Delivery System (ODS) to close the HTML destination, then PROC PRINTTO does not receive any output to redirect. However, the procedure results still go to the destination that you specified with ODS.

You use the PRINT= option in the PROC PRINTTO statement to specify the name of the file or SAS catalog that will contain the procedure output. If you specify a file, then either use the complete name of the file in quotation marks or use a fileref for the file. (See "Using External Files in Your SAS Job" on page 45 for more information about filerefs and filenames.) You can also specify the NEW option in the PROC PRINTTO statement so that SAS replaces the previous contents of the output file. Otherwise, SAS appends the output to any output that is currently in the file.

To route output to an alternate file, insert a PROC PRINTTO step in the program before the PROC step that generates the procedure output. The following program routes the output from PROC PRINT to an external file:

```
proc printto print='alternate-output-file' new;
run;

proc print data=test_scores;
    title 'Test Scores for Entering University Classes';
run;

proc printto;
run;
```

After the PROC PRINT step executes, *alternate-output-file* contains the procedure output. The second PROC PRINTTO step redirects output back to its default location.

**394** *Chapter 24* • *Directing SAS Output and the SAS Log*

The PRINTTO procedure does not produce the output. Instead, it tells SAS to route the results of all subsequent procedures until another PROC PRINTTO statement executes. Therefore, the PROC PRINTTO statement must precede the procedure whose output you want to route.

The figure below shows how SAS uses PROC PRINTTO to route procedure output. You can also use PROC PRINTTO multiple times in a program so that output from different steps of a SAS job is stored in different files.

*Figure 24.1* Using PROC PRINTTO Route Output

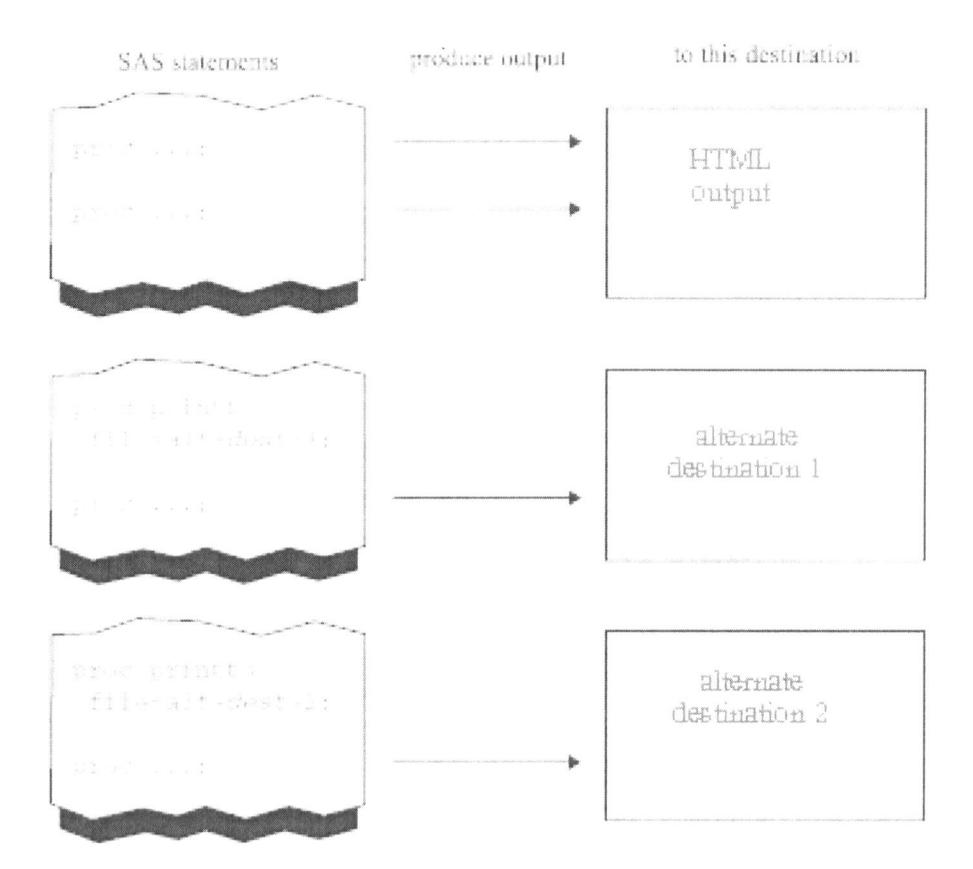

## Routing the SAS Log to an Alternate Location

You can use the PRINTTO procedure to redirect the SAS log to an alternate location. The location can be one of the following:

- a permanent file
- a SAS catalog entry
- a dummy file to suppress the log

After PROC PRINTTO executes, the log is sent either to a permanent external file or to a SAS catalog entry until you execute another PROC PRINTTO statement, or until your program or session ends.

You use the LOG= option in the PROC PRINTTO statement to specify the name of the file or SAS catalog that will contain the log. If you specify a file, then either use the complete name of the file in quotation marks or use a fileref for the file. You can also specify the NEW option in the PROC PRINTTO statement so that SAS replaces the previous contents of the file. Otherwise, SAS appends the log to any log that is currently in the file.

The following program routes the SAS log to an alternate file:

```
proc printto log='alternate-log-file';
run;
```

After the PROC PRINT step executes, *alternate-log-file* contains the SAS log.

### Restoring the Default Destination

Specify the PROC PRINTTO statement with no argument when you want to route the log and the output back to their default destinations:

```
proc printto;
run;
```

You might want to return only the log or only the procedure output to its default destination. The following PROC PRINTTO statement routes only the log back to the default destination:

```
proc printto log=log;
run;
```

The following PROC PRINTTO statement routes only the procedure output to the default destination:

```
proc printto print=print;
run;
```

## Storing the Output and the SAS Log in the SAS Windowing Environment

### Understanding the Default Destination

Within the SAS windowing environment, the default destination for most procedure output is HTML output that appears in the Results Viewer window. However, you can use the Output Delivery System (ODS) to change the format of your output.

Each time you execute a procedure within a single session, SAS appends the output to the existing output. To view the results, you can use one of the following methods:

- scroll the Results Viewer window, which contains the output in the order in which you generated it

- use the Results window to select a pointer that is a link to the procedure output

The SAS windowing environment interacts with certain aspects of ODS to format, control, and manage your output.

In the SAS windowing environment, the default destination for the SAS log messages is the Log window. When you execute a procedure, SAS appends the log messages to the existing log messages in the Log window. You can scroll the Log window to see the results. To print your log messages, execute the PRINT command. To clear the contents of the Log window, execute the CLEAR command. When your session ends, SAS automatically clears the window.

Within the SAS windowing environment, you can use the PRINTTO procedure to route log messages or procedure output to a location other than the default location, just as you can in other methods of operation. For details, see "Routing the Output and the SAS Log with PROC PRINTTO" on page 393. You can also use ODS to change the format of the procedure output.

For additional information about using ODS, viewing procedure output, and changing the format of the procedure output, see Chapter 33, "Understanding and Customizing SAS Output: The Basics," on page 613.

### Storing the Contents of the Output and Log Windows

If you want to store a copy of the contents of the Results Viewer or Log window in a file, then use the FILE command. On the command line, specify the FILE command followed by the name of the file:

file '*file-to-store-contents-of-window*'

SAS has a built-in safeguard that prevents you from accidentally overwriting a file. If you inadvertently specify an existing file, then a dialog box appears. The dialog box asks you to choose a course of action, provides you with information, and might prevent you from overwriting the file by mistake. You are asked whether to:

- replace the contents of the file
- append the contents of the file
- cancel the FILE command

# Redefining the Default Destination in a Batch or Noninteractive Environment

### Determining the Default Destination

Usually, in a batch or noninteractive environment, SAS routes procedure output to the listing file and routes the SAS log to a log file. These files are usually defined by your installation and are created automatically when you invoke SAS. Contact your on-site SAS support personnel if you have questions pertaining to your site.

### Changing the Default Destination

If you want to redefine the default destination for procedure output, then use the PRINT= system option. If you want to redefine the default destination for the SAS log, then use the LOG= system option. You specify these options only at initialization.

*Operating Environment Information*
The way that you specify output destinations when you use SAS system options depends on your operating environment. For details, see the SAS documentation for your operating environment.

Options that you must specify at initialization are called configuration options. The configuration options affect the following:

- the initialization of the SAS System
- the hardware interface

- the operating system interface

In contrast to other SAS system options, which affect the appearance of output, file handling, use of system variables, or processing of observations, you cannot change configuration options in the middle of a program. You specify configuration options when SAS is invoked, either in the configuration file or in the SAS command.

### *Understanding the Configuration File*

The configuration file is a special file that contains configuration options as well as other SAS system options and their settings. Each time you invoke SAS, the settings of the configuration file are examined. You can specify the options in the configuration file in the same format as they are used in the SAS command for your operating environment. For example, under UNIX this file's contents might include the following:

```
WORK=WORK
SASUSER=SASUSER
EXPLORER
```

SAS automatically sets the options as they appear in the configuration file. If you specify options both in the configuration file and in the SAS command, then the options are concatenated. If you specify the same option in the SAS command and in the configuration file, then the setting in the SAS command overrides the setting in the file. For example, specifying the NOEXPLORER option in the SAS command overrides the EXPLORER option in the configuration file and tells SAS to start your session without displaying the Explorer window.

## Summary

### *PROC PRINTTO Statement Options*

**PROC PRINTTO** <PRINT='*alternate-output-file*'> <LOG='*alternate-log-file*'> <NEW>;

PRINT='*alternate-output-file*'
 identifies the location and routes the procedure output to this alternate location.

LOG='*alternate-log-file*'
 identifies the location and routes the SAS log to this alternate location.

NEW
 specifies that the current log or procedure output writes over the previous contents of the file.

### *SAS Windowing Environment Commands*

CLEAR
 clears the contents of a window, as specified.

FILE <*file-to-store-contents-of-window*>
 routes a copy of the contents of a window to the file that you specify. The original contents remain in place.

PRINT
 writes the contents of the window.

## SAS System Options

LOG=*system-filename*
: redefines the default destination for the SAS log to the file named *system-filename*.

PRINT=*system-filename*
: redefines the default destination for procedure output to the file named *system-filename*.

# Learning More

Output Delivery System
: For complete reference documentation about the Output Delivery System, see *SAS Output Delivery System: User's Guide*.

PROC PRINTTO
: For complete reference documentation, see "PRINTTO" in *Base SAS Procedures Guide*.

SAS environment
: For details about the methods of operating SAS and interactive processing in the windowing environment, see Chapter 40, "Introducing the SAS Environment," on page 741, Chapter 41, "Using the SAS Windowing Environment," on page 753, and Chapter 42, "Customizing the SAS Environment," on page 793.

SAS log
: For complete reference information about the SAS log and procedure output, see *SAS Language Reference: Concepts*.

SAS system options
: For details about SAS system options, including configuration options, see *SAS System Options: Reference*.

  For operating-environment specific information about routing output, the PRINT= option, LOG= option, and other SAS system options, see the SAS documentation for your operating environment.

# Chapter 25
# Diagnosing and Avoiding Errors

**Introduction to Diagnosing and Avoiding Errors** .......................... **399**
    Purpose ........................................................ 399
    Prerequisites ................................................... 400

**Understanding How the SAS Supervisor Checks a Job** ..................... **400**

**Understanding How SAS Processes Errors** ............................... **400**

**Distinguishing Types of Errors** ......................................... **401**
    SAS Programming Errors ........................................ 401
    Syntax Errors .................................................. 401
    Execution-time Errors ........................................... 401
    Data Errors .................................................... 401
    Semantic Errors ................................................ 402

**Diagnosing Errors** .................................................... **402**
    Examples in This Section ........................................ 402
    Diagnosing Syntax Errors ........................................ 402
    Diagnosing Execution-time Errors ................................. 406
    Diagnosing Data Errors .......................................... 408
    Diagnosing Semantic Errors ...................................... 411

**Using a Quality Control Checklist** ...................................... **412**

**Learning More** ....................................................... **412**

## Introduction to Diagnosing and Avoiding Errors

### Purpose

In this section, you will learn how to diagnose errors in your programs by learning about the following concepts:

- how the SAS Supervisor checks a program for errors
- how to distinguish among the types of errors
- how to interpret the notes, warning messages, and error messages in the log
- what to check for as you develop a program

*Prerequisites*

You should understand the concepts that are presented in the following sections:

- Chapter 3, "Introduction to DATA Step Processing," on page 27
- Chapter 4, "Starting with Raw Data: The Basics," on page 51
- Chapter 7, "Understanding DATA Step Processing," on page 107
- Chapter 23, "Analyzing Your SAS Session with the SAS Log," on page 365

# Understanding How the SAS Supervisor Checks a Job

To better understand the errors that you make so that you can avoid them in the future, it is important to understand how the SAS Supervisor checks a job. The SAS Supervisor is the part of SAS that is responsible for executing SAS programs. To check the syntax of a SAS program, the SAS Supervisor does the following:

- reads the SAS statements and data
- translates the program statements into executable machine code or intermediate code
- creates data sets
- calls SAS procedures, as requested
- writes error messages
- ends the job

The SAS Supervisor knows the following information about DATA and PROC steps:

- the forms and types of statements that can be present in a DATA step
- the types of statements and the options that can be present in a PROC step

To process a program, the SAS Supervisor scans all the SAS statements and breaks each statement into words. Each word is processed separately. When all the words in a step are processed, the step is executed. If the SAS Supervisor detects an error, then it flags the error at its location and writes an explanation in the log. The SAS Supervisor assumes that anything that it does not recognize is an error.

# Understanding How SAS Processes Errors

When SAS detects an error, it usually underlines the error in the log or underlines the point at which it detects the error, identifying the error with a number. Each number is uniquely associated with an error message. Then SAS enters syntax check mode, and reads the remaining program statements, checks their syntax, and underlines additional errors if necessary.

In a batch or noninteractive program, an error in a DATA step statement causes SAS to remain in syntax check mode for the rest of the program. It does not execute any more DATA or PROC steps that create external files or SAS data sets. Procedures that are read from SAS data sets execute with 0 observations, and procedures that do not read SAS data sets execute normally. A syntax error in a PROC step usually affects only that step. At the end of the step, SAS writes a message in the log for each error that is detected.

# Distinguishing Types of Errors

## SAS Programming Errors

The following types of errors can occur when SAS compiles and executes your program:

- syntax
- execution-time
- data
- semantic

## Syntax Errors

Syntax errors are errors that are made in the SAS statements of a program. They occur when program statements do not conform to the rules of the SAS language. SAS detects syntax errors as it compiles each DATA and PROC step. These are some of the types of syntax errors:

- misspelled SAS keywords
- unmatched quotation marks
- missing or invalid punctuation
- invalid statement or data set options

## Execution-time Errors

Execution-time errors cause a program to fail when it is submitted for execution. Most execution-time errors that are not serious produce notes in the SAS log, but the program is allowed to run to completion. However, for more serious errors, SAS issues error messages and stops all processing. These are some of the types of execution-time errors:

- invalid arguments to functions
- invalid mathematical operations, such as division by 0
- observations in the wrong order for BY-group processing
- references to a nonexistent member of an array
- INPUT statements that do not match the data lines
- an incorrect reference in an INFILE statement (for example, misspelling or otherwise incorrectly stating the external file)

## Data Errors

Data errors are a type of execution-time error. Data errors occur when the raw data that you are analyzing with a SAS program contains invalid values. For example, a data error occurs if you specify a numeric variable in an INPUT statement when the data is character data. Data errors do not cause a program to stop. Instead, they generate notes in the SAS log. These are some of the types of data errors:

- defining a variable as numeric when the data value is actually character
- generating missing values as a result of performing an operation on missing values
- reading a variable with an INPUT statement when the variable is not in the correct position in a file
- using the Sum statement with character variables

### Semantic Errors

Semantic errors are another type of an execution-time error. They occur when the form of a SAS statement is correct, but some elements are not valid in a particular usage. These are some of the types of semantic errors:

- specifying the wrong number of arguments for a function
- using a numeric variable name where only character variables are valid
- using invalid references to an array
- using a libref that has not yet been assigned

# Diagnosing Errors

### Examples in This Section

Many of the programs in this section use university test scores to illustrate errors in the SAS log. Other programs in this section use other data.

### Diagnosing Syntax Errors

#### When SAS Detects a Syntax Error
The SAS Supervisor detects syntax errors as it compiles each step, and then SAS does the following:

- writes the word ERROR to the log
- identifies the error's location
- writes an explanation of the error

#### Example: Missing Semicolon and Misspelled Keyword
In the following program, the CHART procedure is used to analyze data. Note that a semicolon in the DATA statement is omitted, and the keyword INFILE is misspelled.

```
libname out 'your-data-library';

data out.error1
   infill 'your-input-file';
   input test $ gender $ year TestScore;
```

```
run;

proc chart data=out.error1;
   hbar test / sumvar=TestScore type=mean group=gender discrete;
run;
```

The following output shows the result of the two syntax errors:

*Log 25.1  Diagnosing Syntax Errors: Missing Semicolon and Misspelled Keyword*

```
11    libname out 'your-data-library';
NOTE: Libref OUT was successfully assigned as follows:
      Engine:        V9
      Physical Name: your-data-library
12
13    data out.error1
14       infill 'your-input-file';
15       input test $ gender $ year TestScore;
16    run;

ERROR: No DATALINES or INFILE statement.
ERROR: Extension for physical file name "your-input-file" does not
       correspond to a valid member type.
NOTE: The SAS System stopped processing this step because of errors.
WARNING: The data set OUT.ERROR1 may be incomplete.  When this step was stopped
there were 0
         observations and 4 variables.
WARNING: Data set OUT.ERROR1 was not replaced because this step was stopped.
WARNING: The data set WORK.INFILL may be incomplete.  When this step was stopped
there were 0
         observations and 4 variables.
WARNING: Data set WORK.INFILL was not replaced because this step was stopped.
NOTE: DATA statement used (Total process time):
      real time           0.00 seconds
      cpu time            0.00 seconds


17
18    proc chart data=out.error1;
19       hbar test / sumvar=TestScore type=mean group=gender discrete;
20    run;

NOTE: No observations in data set OUT.ERROR1.
NOTE: PROCEDURE CHART used (Total process time):
      real time           0.00 seconds
      cpu time            0.00 seconds
```

As the log indicates, SAS recognizes the keyword DATA and attempts to process the DATA step. Because the DATA statement must end with a semicolon, SAS assumes that INFILL is a data set name, and that two data sets are being created: OUT.ERROR1 and WORK.INFILL. Because SAS considers INFILL the name of a data set, it does not recognize it as part of another statement, and therefore does not detect the spelling error. Because the string in quotation marks is invalid in a DATA statement, SAS stops processing here and created no observations for either data set.

SAS attempts to execute the program logically based on the statements that it contains, according to the steps outlined earlier in this section. The second syntax error, the misspelled keyword, is never recognized because SAS considers the DATA statement to be in effect until a semicolon ends the statement. The point to remember is that when multiple errors are made in the same program, not all of them might be detected the first time the program is executed. Errors might also be flagged differently in a group than if they were made alone. You might find that one correction uncovers another error or at least changes its explanation in the log.

To illustrate this point, the previous program is executed again with the semicolon added to the DATA statement. An attempt to correct the misspelled keyword introduces a different spelling error, as shown below:

```
libname out 'your-data-library';

data out.error2;
   unfile 'your-input-file';
   input test $ gender $ year TestScore;
run;

proc chart data=out.error2;
   hbar test / sumvar=TestScore type=mean group=gender discrete;
run;
```

**Log 25.2** *Diagnosing Syntax Errors: Misspelled Keyword*

```
19   libname out 'your-data-library';
NOTE: Libref OUT was successfully assigned as follows:
     Engine:         V9
     Physical Name:  your-data-library
20

21   data out.error2;
22      unfile 'your-input-file';
        ------
        180
ERROR 180-322: Statement is not valid or it is used out of proper order.

23      input test $ gender $ year TestScore;
24   run;

ERROR: No DATALINES or INFILE statement.
NOTE: The SAS System stopped processing this step because of errors.
WARNING: The data set OUT.ERROR2 may be incomplete.  When this step was stopped
there were 0
         observations and 4 variables.
NOTE: DATA statement used (Total process time):
     real time           0.01 seconds
     cpu time            0.01 seconds


25
26   proc chart data=out.error2;
27      hbar test / sumvar=TestScore type=mean group=gender discrete;
28   run;

NOTE: No observations in data set OUT.ERROR2.
NOTE: PROCEDURE CHART used (Total process time):
     real time           0.01 seconds
     cpu time            0.01 seconds
```

With the semicolon added, SAS now attempts to create only one data set. From then on, SAS reads the SAS statements as it did before, and issues many of the same messages. However, this time SAS considers the UNFILE statement invalid or out of proper order, and it creates no observations for the data set.

If this example is rerun with the correct spelling for INFILE but with a misspelling of the filename in the INFILE statement, then the error is detected at execution time and the data is not read.

### *Example: Missing Semicolon in SET Statement*

The following example shows the SAS log when you use a SET statement without a semicolon. The first DATA step creates the data set INSURANCE. The SET statement in the second DATA step uses INSURANCE as input. The SAS log shows that an error was encountered.

```
data insurance;
   input PolicyNum $ Name $ Amount 23-27 District $;
   datalines;
4356 Susan Bellingham 45000 North
2678 James Hastings   35000 West
4967 Jan Spiro        49000 North
1367 Robert Hernandez 63000 South
7366 Walter Peters    66000 East
;
run;

data policy;
   set insurance
run;

proc print data=policy;
run;
```

**Log 25.3** *Log Output for a Missing Semicolon in a SET Statement*

```
44    data insurance;
45        input PolicyNum $ Name $ Amount 23-27 District $;
46        datalines;

NOTE: The data set WORK.INSURANCE has 5 observations and 4 variables. [1]
NOTE: DATA statement used (Total process time):
      real time           0.00 seconds
      cpu time            0.00 seconds


52    ;
53    run;
54
55    data policy;
56        set insurance
57    run;
ERROR: File WORK.RUN.DATA does not exist. [2]
58

NOTE: The SAS System stopped processing this step because of errors. [3]
WARNING: The data set WORK.POLICY may be incomplete.  When this step was stopped
there were 0
         observations and 4 variables. [4]
WARNING: Data set WORK.POLICY was not replaced because this step was stopped.
NOTE: DATA statement used (Total process time):
      real time           0.01 seconds
      cpu time            0.01 seconds


59    proc print data=policy;
60    run;

NOTE: No observations in data set WORK.POLICY. [5]
NOTE: PROCEDURE PRINT used (Total process time):
      real time           0.01 seconds
      cpu time            0.01 seconds
```

The following list corresponds to the numbered items in the preceding log:

1 SAS successfully creates the INSURANCE data set.

2 SAS encounters an error when the SET statement in the second DATA step is read (the semicolon is missing as the end of the statement). SAS continues to read the next line until it encounters a semicolon.

3 The note indicates that the program stopped processing.

4 The warning message indicates that there are no observations in the POLICY data set.

5 The PRINT procedure produces no output because the POLICY data set is empty.

To correct the program error, add a semicolon after the SET statement. Execute your program again.

## Diagnosing Execution-time Errors

### When SAS Detects an Execution-time Error

Several types of errors are detected at execution time. Errors occur when the language element is correct, but the element might not be valid for a particular usage.

When the SAS Supervisor encounters an execution-time error, it does the following:

- writes a note, warning, or error message to the log, depending on the seriousness of the error
- in some cases, lists the values that are stored in the program data vector
- continues or stops processing, depending on the seriousness of the error

### *Example: Misspelled Input File*

If the following program executes with the correct spelling for INFILE but with a misspelling of the filename in the INFILE statement, then the error is detected at execution time and the data is not read.

```
   /* misspelled filename in the INFILE statement */
libname out 'your-data-library';

data out.error3;
   infile 'an-incorrect-filename';
   input test $ gender $ year TestScore;
run;

proc chart data=out.error3;
   hbar test / sumvar=TestScore type=mean group=gender discrete;
run;
```

**Log 25.4**  *Log Output for a Misspelled Input File*

```
29      /* misspelled filename in the INFILE statement */
30    libname out 'your-data-library';
NOTE: Libref OUT was successfully assigned as follows:
      Engine:        V9
      Physical Name: your-data-library
31
32    data out.error3;
33       infile 'your-input-file';
34       input test $ gender $ year TestScore @@;
35    run;

ERROR: Physical file does not exist, your-input-file.
NOTE: The SAS System stopped processing this step because of errors.
WARNING: The data set OUT.ERROR3 may be incomplete.  When this step was stopped
there were 0
         observations and 4 variables.
NOTE: DATA statement used (Total process time):
      real time           0.00 seconds
      cpu time            0.00 seconds


36
37    proc chart data=out.error3;
38       hbar test / sumvar=TestScore type=mean group=gender discrete;
39    run;

NOTE: No observations in data set OUT.ERROR3.
NOTE: PROCEDURE CHART used (Total process time):
      real time           0.00 seconds
      cpu time            0.00 seconds
```

As the log indicates, SAS cannot find the file. SAS stops processing because of errors and creates no observations in the data set.

## Diagnosing Data Errors

### When SAS Detects a Data Error

Data errors occur when data values are invalid. When SAS detects data errors during execution, it continues processing, and writes the following information to the SAS log:

- a note that describes the error
- a list of values that are stored in the input buffer
- a list of values that are stored in the program data vector

Note that the values that are listed in the program data vector include two variables that are created automatically by SAS:

_N_
   counts the number of times the DATA step iterates.

_ERROR_
   indicates the occurrence of an error during an execution of the DATA step. The value that is assigned to the variable _ERROR_ is 0 when no error is encountered and 1 when an error is encountered.

These automatic variables are assigned temporarily to each observation, and are not stored with the data set.

### Example: Invalid Raw Data

The following program uses raw input data that is not aligned correctly in the columns that are described in the INPUT statement. The sixth data line is shifted two spaces to the right, and the rest of the data lines, except for the first, are shifted one space to the right. The program uses formats to determine how variable values are written:

```
    /* data in wrong columns */
proc format;
    value xscore . ='accurate scores unavailable';
run;

data out.error4;
    input test $ 1-8 gender $ 18 year 20-23 TestScore 25-27;
    format TestScore xscore.;
    datalines;
verbal         m 2008 463
 verbal        f 2008 468
 verbal        m 2011 459
 verbal        f 2011 461
 math          m 2008 514
  math          f 2008 467
 math          m 2011 509
 math          f 2011 509
 ;

proc print data=out.error4;
    title 'Viewing Incorrect Output';
run;
```

The following output shows the results of the SAS program.

*Figure 25.1* Detecting Data Errors in Raw Data

### Viewing Incorrect Output

| Obs | test | gender | year | TestScore |
|---|---|---|---|---|
| 1 | verbal | m | 2008 | 463 |
| 2 | verbal |  | 200 | 46 |
| 3 | verbal |  | 201 | 45 |
| 4 | verbal |  | 201 | 46 |
| 5 | math |  | 200 | 51 |
| 6 | math |  | . | accurate scores unavailable |
| 7 | math |  | 201 | 50 |
| 8 | math |  | 201 | 50 |

This program generates output, but it is not the expected output. The first observation appears to be correct, but subsequent observations have the following problems:

- The values for the variable Gender are missing.
- Only the first three digits of the value for the variable Year are shown, except in the sixth observation, where a missing value is indicated.
- The third digit of the value for the variable TestScore is missing, except in the sixth observation, which does show the assigned value for the missing value.

**410** Chapter 25 • Diagnosing and Avoiding Errors

The following SAS log explains program processing:

**Log 25.5** *Diagnosing Data Errors*

```
61      /* data in wrong columns */
62    proc format;
63       value xscore . ='accurate scores unavailable';
NOTE: Format XSCORE has been output.
64    run;

NOTE: PROCEDURE FORMAT used (Total process time):
      real time           0.01 seconds
      cpu time            0.00 seconds


65
66    data out.error4;
67       input test $ 1-8 gender $ 18 year 20-23 TestScore 25-27;
68       format TestScore xscore.;
69       datalines;

NOTE: Invalid data for year in line 75 20-23.
NOTE: Invalid data for TestScore in line 75 25-27.
RULE:       ----+----1----+----2----+----3----+----4----+----5----+----6----
+----7----+----8----+-
75         math            f 2008 467
test=math gender=  year=. TestScore=accurate scores unavailable _ERROR_=1 _N_=6
NOTE: The data set OUT.ERROR4 has 8 observations and 4 variables.
NOTE: DATA statement used (Total process time):
      real time           0.01 seconds
      cpu time            0.01 seconds


78        ;
79
80    proc print data=out.error4;
81       title 'Viewing Incorrect Output';
82    run;

NOTE: There were 8 observations read from the data set OUT.ERROR4.
NOTE: PROCEDURE PRINT used (Total process time):
      real time           0.03 seconds
      cpu time            0.03 seconds
```

The errors are flagged, starting with the first message that line 75 contains invalid data for the variable Year. The rule indicates that input data has been written to the log. SAS lists in the log the values that are stored in the program data vector. The following lines from the log indicate that SAS encountered an error:

```
NOTE: Invalid data for year in line 75 20-23.
NOTE: Invalid data for TestScore in line 75 25-27.
RULE:       ----+----1----+----2----+----3----+----4----+----5----+----6
----+----7----+----8----+-
16         math            f 2008 467
test=math gender=  year=. TestScore=accurate scores unavailable _ERROR_=1 _N_=6
```

Missing values are shown for the variables Gender and Year. The notes in the log indicate that the sixth line of input contained the error. The _ERROR_ automatic variable indicates that an error has occurred, and the _N_ automatic variable identifies the sixth iteration of the DATA step as having the error.

To debug the program, either the raw data can be repositioned or the INPUT statement can be rewritten. Remember that all data lines, except the first, were shifted at least one space to the right. The variable Test was unaffected, but the variable Gender was completely removed from its designated field, except in the first observation. Therefore, SAS reads the variable Gender as a missing value for the second through the eighth observation. In the sixth observation, for which the data was shifted to the right an additional space, the character value for Gender occupied part of the field for the numeric variable Year. When SAS encounters the invalid data, it treats the value as a missing value but also notes in the log that the data is invalid. It is important to remember that SAS can use only the information that you provide, not what you intended to provide. For valid output, your input data must be valid.

## Diagnosing Semantic Errors

### When SAS Detects a Semantic Error

Semantic errors occur when the language element is correct, but the element might not be valid for a particular usage.

### Example: Missing Argument in a Function

The following example shows the SAS log when SAS encounters a semantic error. The example uses the COUNT function to find the number of times that a substring, in this case "th", appears within the character string "This is the thistle." The COUNT function has two arguments. The first argument contains the string that is to be searched. The second argument contains the string that is searched for. An error occurs if you use only one argument with the COUNT function, as the following example shows.

```
data _null_;
   x='This is the thistle.';
   y='th';
   occurrences=count(x);
   put occurrences=;
run;
```

*Log 25.6* Missing Argument in a Function

```
249  data _null_;
250     x='This is the thistle.';
251     y='th';
252     occurrences=count(x);
                     -----
                      71
ERROR 71-185: The COUNT function call does not have enough arguments.

253     put occurrences=;
254  run;

NOTE: The SAS System stopped processing this step because of errors.
NOTE: DATA statement used (Total process time):
      real time           0.01 seconds
      cpu time            0.01 seconds
```

An error occurred because one of the arguments for the COUNT function is missing. To correct the error, add the second argument to COUNT. (The syntax for the COUNT function is `count(string, substring);`.) Execute the program again. SAS searches the string for any character string that matches "th". Because the COUNT function identifies two occurrences of "th", the value of Occurrences is 2. Notice that the first word in the string is uppercase and therefore does not match "th".

## Using a Quality Control Checklist

If you follow some basic guidelines as you develop a program, then you can avoid common errors. Use the following checklist to flag and correct common mistakes before you submit your program.

- Check the syntax of your program. In particular, check that the following items are correct:
  - All SAS statements must end with a semicolon. Make sure that you have not omitted any semicolons or accidentally entered the wrong character.
  - Any starting and ending quotation marks must match. You can use either single or double quotation marks.
  - Most SAS statements begin with a SAS keyword. (Exceptions are assignment statements and Sum statements.) Make sure that you have not misspelled or omitted any of the keywords.
  - Every DO and SELECT statement must be followed by an END statement.
- Check the order of your program.

  SAS usually executes the statements in a DATA step one by one, in the order in which they appear. After executing the DATA step, SAS moves to the next step and continues in the same way. Make sure that all the SAS statements appear in order so that SAS can execute them properly. For example, an INFILE statement, if used, must precede an INPUT statement.

  Also, be sure to end steps with the RUN statement. This is especially important at the end of your program because the RUN statement causes the previous step to be executed.
- Check your INPUT statement and your data.

  SAS classifies all variables as either character or numeric. The assignment in the INPUT statement as either character or numeric must correspond to the actual values of variables in your data. Also, SAS allows for list, column, formatted, or named input. The method of input that you specify in the INPUT statement must correspond with the actual arrangement of raw data.

## Learning More

INFILE statement options
  The "INFILE Statement" in *SAS Statements: Reference* contains information about using the MISSOVER and STOPOVER options in the INFILE statement as debugging tools.

  The MISSOVER option prevents a SAS program from going past the end of a line to read values with list input if it does not find values in the current line for all INPUT statement variables. SAS then assigns missing values to variables for which no values appear in the current input line.

  The STOPOVER option stops processing the DATA step when an INPUT statement that uses list input reaches the end of the current record without finding values for all variables in the statement. SAS then sets _ERROR_ to 1, stops building the data set, and writes an incomplete data line.

PUT statement
For information about the PUT statement, see "PUT Statement" in *SAS Statements: Reference*.

Program data vector and input buffer
"Introduction to DATA Step Processing" on page 27, and "Introduction to Raw Data" on page 52 contain information about the program data vector and the input buffer.

The SAS log
Chapter 23, "Analyzing Your SAS Session with the SAS Log," on page 365 contains information about the SAS log and its structure. "The SAS Log" in *SAS Language Reference: Concepts* contains more information about the SAS log.

SAS output
"SAS Output" in *SAS Language Reference: Concepts* contains more information about SAS output.

Your SAS session
Other sections provide more information about your SAS session. "Introduction to Analyzing Your SAS Session with the SAS Log" on page 366 discusses log structure, types of messages in the log, and the log in relation to output.

# Chapter 26
# Finding Logic Errors in Your Program

Finding Logic Errors in Your Program . . . . . . . . . . . . . . . . . . . . . . . . . . . . . . . . . 415
    Purpose . . . . . . . . . . . . . . . . . . . . . . . . . . . . . . . . . . . . . . . . . . . . . . . . . . . . . . 415
    Prerequisites . . . . . . . . . . . . . . . . . . . . . . . . . . . . . . . . . . . . . . . . . . . . . . . . . . 415

Using the DATA Step Debugger . . . . . . . . . . . . . . . . . . . . . . . . . . . . . . . . . . . . . 416

Basic Usage . . . . . . . . . . . . . . . . . . . . . . . . . . . . . . . . . . . . . . . . . . . . . . . . . . . . . 416
    How a Debugger Session Works . . . . . . . . . . . . . . . . . . . . . . . . . . . . . . . . . . . 416
    Using the Windows . . . . . . . . . . . . . . . . . . . . . . . . . . . . . . . . . . . . . . . . . . . . . 417
    Entering Commands . . . . . . . . . . . . . . . . . . . . . . . . . . . . . . . . . . . . . . . . . . . . 417
    Working with Expressions . . . . . . . . . . . . . . . . . . . . . . . . . . . . . . . . . . . . . . . 417
    Assigning Commands to Function Keys . . . . . . . . . . . . . . . . . . . . . . . . . . . . . 417

Using the Macro Facility with the Debugger . . . . . . . . . . . . . . . . . . . . . . . . . . . 418
    Using Macros as Debugging Tools . . . . . . . . . . . . . . . . . . . . . . . . . . . . . . . . . 418
    Creating Customized Debugging Commands with Macros . . . . . . . . . . . . . . . . 418
    Debugging a DATA Step Generated by a Macro . . . . . . . . . . . . . . . . . . . . . . . 418

Examples . . . . . . . . . . . . . . . . . . . . . . . . . . . . . . . . . . . . . . . . . . . . . . . . . . . . . . 419
    Example 1: Debugging a Simple DATA Step When Output Is Missing . . . . . . . . 419
    Example 2: Working with Formats . . . . . . . . . . . . . . . . . . . . . . . . . . . . . . . . . 424
    Example 3: Debugging DO Loops . . . . . . . . . . . . . . . . . . . . . . . . . . . . . . . . . 429
    Example 4: Examining Formatted Values of Variables . . . . . . . . . . . . . . . . . . . 430

## Finding Logic Errors in Your Program

### Purpose

The DATA step debugger, which consists of windows and a group of commands, helps you to interactively identify logic errors, and sometimes data errors, in your SAS programs. Unlike syntax errors, logic errors do not stop a program from running. Instead, logic errors cause the program to produce unexpected results. For example, if you create a DATA step that keeps track of inventory, and your program shows that you are out of stock but your warehouse is full, you have a logic error in your program.

### Prerequisites

Before continuing with this session, you should understand the concepts that are presented in the following sections:

- Chapter 3, "Introduction to DATA Step Processing," on page 27
- Chapter 7, "Understanding DATA Step Processing," on page 107

## Using the DATA Step Debugger

By issuing commands, you can execute DATA step statements one by one and pause to display the resulting variable values in a window. By observing the results that are displayed, you can determine where the logic error lies. Because the debugger is interactive, you can repeat the process of issuing commands and observing the results as many times as needed in a single debugging session. To invoke the debugger, add the DEBUG option to the DATA statement and execute the program.

The DATA step debugger enables you to perform these tasks:

- execute statements one by one or in groups
- bypass the execution of one or more statements
- suspend execution at selected statements, either in each iteration of DATA step statements or on a condition that you specify, and resume execution on command
- monitor the values of selected variables and suspend execution at the point a value changes
- display the values of variables and assign new values to them
- display the attributes of variables
- receive help for individual debugger commands
- assign debugger commands to function keys
- use the macro facility to generate customized debugger commands

The following section provides usage information and examples.

For a list of debugger commands and their descriptions, see Appendix 2, "DATA Step Debugger Commands," on page 823.

## Basic Usage

### How a Debugger Session Works

When you submit a DATA step with the DEBUG option, SAS compiles the step, displays the debugger windows, and pauses until you enter a debugger command to begin execution. For example, if you begin execution with the GO command, SAS executes each statement in the DATA step. To suspend execution at a particular line in the DATA step, use the BREAK command to set breakpoints at statements that you select. Then issue the GO command. The GO command starts or resumes execution until the breakpoint is reached.

To execute the DATA step one statement at a time or a few statements at a time, use the STEP command. By default, the STEP command is mapped to the ENTER key.

In a debugging session, statements in a DATA step can iterate as many times as they would outside the debugging session. When the last iteration has finished, a message appears in the DEBUGGER LOG window.

You cannot restart DATA step execution in a debugging session after the DATA step finishes executing. You must resubmit the DATA step in your SAS session. However, you can examine the final values of variables after execution has ended.

You can debug only one DATA step at a time. You can use the debugger only with a DATA step, and not with a PROC step.

## Using the Windows

The DATA step debugger contains two primary windows, the DEBUGGER LOG and the DEBUGGER SOURCE windows. The windows appear when you execute a DATA step with the DEBUG option.

The DEBUGGER LOG window records the debugger commands that you issue and their results. The last line is the debugger command line, where you issue debugger commands. The debugger command line is marked with a greater than (>) prompt.

The DEBUGGER SOURCE window contains the SAS statements that comprise the DATA step that you are debugging. The window enables you to view your position in the DATA step as you debug your program. In the window, the SAS statements have the same line numbers as they do in the SAS log.

You can enter windowing environment commands on the window command lines. You can also execute commands by using function keys.

## Entering Commands

For a list of commands and their descriptions, see Appendix 2, "DATA Step Debugger Commands," on page 823.

Enter DATA step debugger commands on the debugger command line. Follow these rules when you enter a command:

- A command can occupy only one line (except for a DO group).
- A DO group can extend over more than one line.
- To enter multiple commands, separate the commands with semicolons:

```
examine _all_; set letter='bill'; examine letter
```

## Working with Expressions

All SAS operators that are described in "SAS Operators in Expressions" in *SAS Language Reference: Concepts* are valid in debugger expressions. Debugger expressions cannot contain functions.

A debugger expression must fit on one line. You cannot continue an expression on another line.

## Assigning Commands to Function Keys

To assign debugger commands to function keys, open the Keys window. Position your cursor in the Definitions column of the function key that you want to assign, and begin the command with the term DSD. To assign more than one command to a function key, enclose the commands (separated by semicolons) in quotation marks. Be sure to save your changes. These examples show commands assigned to function keys:

- `dsd step3`
- `dsd 'examine cost saleprice; go 120;'`

# Using the Macro Facility with the Debugger

### Using Macros as Debugging Tools

You can use the SAS macro facility with the debugger to invoke macros from the DEBUGGER LOG command line. You can also define macros and use macro program statements, such as %LET, on the debugger command line.

Macros are useful for storing a series of debugger commands. Executing the macro at the DEBUGGER LOG command line then generates the entire series of debugger commands. You can also use macros with parameters to build different series of debugger commands based on various conditions.

### Creating Customized Debugging Commands with Macros

You can create a customized debugging command by defining a macro on the DEBUGGER LOG command line. Then invoke the macro from the command line. For example, to examine the variable COST, to execute five statements, and then to examine the variable DURATION, define the following macro (in this case the macro is called EC). Note that the example uses the alias for the EXAMINE command.

`%macro ec; ex cost; step 5; ex duration; %mend ec;`

To issue the commands, invoke macro EC from the DEBUGGER LOG command line:

`%ec`

The DEBUGGER LOG displays the value of COST, executes the next five statements, and then displays the value of DURATION.

*Note:* Defining a macro on the DEBUGGER LOG command line enables you to use the macro only during the current debugging session, because the macro is not permanently stored. To create a permanently stored macro, use the Program Editor.

### Debugging a DATA Step Generated by a Macro

You can use a macro to generate a DATA step, but debugging a DATA step that is generated by a macro can be difficult. The SAS log displays a copy of the macro, but not the DATA step that the macro generated. If you use the DEBUG option at this point, the text that the macro generates appears as a continuous stream to the debugger. As a result, there are no line breaks where execution can pause.

To debug a DATA step that is generated by a macro:

1. Use the MPRINT and MFILE system options when you execute your program.

2. Assign the fileref MPRINT to an existing external file. MFILE routes the program output to the external file. Note that if you rerun your program, current output appends to the previous output in your file.

3. Invoke the macro from a SAS session.

4. In the Editor window, issue the INCLUDE command or use the File menu to open your external file.

5. Add the DEBUG option to the DATA statement and begin a debugging session.

6. When you locate the logic error, correct the portion of the macro that generated that statement or statements.

# Examples

## Example 1: Debugging a Simple DATA Step When Output Is Missing

### Discovering a Problem
This program creates information about a travel tour group. The data files contain two types of records. One type contains the tour code, and the other type contains customer information. The program creates a report listing tour number, name, age, and gender for each customer.

```
/* first execution */
data tours (drop=type);
   input @1 type $ @;
   if type='H' then do;
      input @3 Tour $20.;
      return;
      end;
   else if type='P' then do;
      input @3 Name $10. Age 2. +1 Sex $1.;
      output;
      end;
   datalines;
H Tour 101
P Mary E     21 F
P George S   45 M
P Susan K     3 F
H Tour 102
P Adelle S   79 M
P Walter P   55 M
P Fran I     63 F
;

proc print data=tours;
   title 'Tour List';
run;
```

**Tour List**

| Obs | Tour | Name | Age | Sex |
|-----|------|---------|-----|-----|
| 1 | | Mary E | 21 | F |
| 2 | | George S | 45 | M |
| 3 | | Susan K | 3 | F |
| 4 | | Adelle S | 79 | M |
| 5 | | Walter P | 55 | M |
| 6 | | Fran I | 63 | F |

The program executes without error, but the output is unexpected. The output does not contain values for the variable Tour. Viewing the SAS log will not help you debug the program because the data are valid and no errors appear in the log. To help identify the logic error, run the DATA step again using the DATA step debugger.

### Examining Data Values after the First Iteration

To debug a DATA step, create a hypothesis about the logic error and test it by examining the values of variables at various points in the program. For example, issue the EXAMINE command from the debugger command line to display the values of all variables in the program data vector before execution begins:

```
examine _all_
```

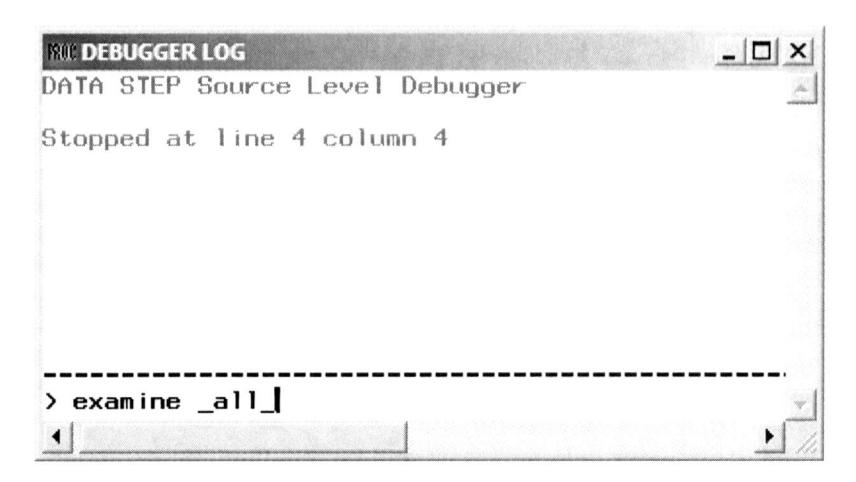

```
DEBUGGER LOG
DATA STEP Source Level Debugger
Stopped at line 4 column 4

---------------------------------------------
> examine _all_
```

*Note:* Most debugger commands have abbreviations, and you can assign commands to function keys. The examples in this section, however, show the full command. For a list of all commands, see "DATA Step Debugger Commands by Category" in *Base SAS Utilities: Reference*.

When you press ENTER, the following display appears:

```
DEBUGGER LOG
DATA STEP Source Level Debugger

Stopped at line 4 column 4
> examine _all_
type =
Tour =
Name =
Age = .
Sex =
_ERROR_ = 0
_N_ = 1
------------------------------------------------
>
```

The values of all variables appear in the DEBUGGER LOG window. SAS has compiled, but not yet executed, the INPUT statement.

Use the STEP command to execute the DATA step statements one at a time. By default, the STEP command is assigned to the ENTER key. Press ENTER repeatedly to step through the first iteration of the DATA step, and stop when the RETURN statement in the program is highlighted in the DEBUGGER SOURCE window.

Because Tour information was missing in the program output, enter the EXAMINE command to view the value of the variable Tour for the first iteration of the DATA step.

```
examine tour
```

The following display shows the results:

```
DEBUGGER LOG
Sex =
_ERROR_ = 0
_N_ = 1
>
Stepped to line 5 column 4
>
Stepped to line 6 column 7
>
Stepped to line 7 column 7
> examine tour
Tour = Tour 101
------------------------------------------------
>
```

```
DEBUGGER SOURCE
 3  data tours (drop=type) /debug;
 4     input @1 type $ @;
 5     if type='H' then do;
 6        input @3 Tour $20.;
 7        return;
 8     end;
 9     else if type='P' then do;
10        input @3 Name $10. Age 2. +1 Sex $1.;
11        output;
12     end;
13  datalines;
```

The variable Tour contains the value Tour 101, showing you that Tour was read. The first iteration of the DATA step worked as intended. Press ENTER to reach the top of the DATA step.

**422** Chapter 26 • Finding Logic Errors in Your Program

### *Examining Data Values after the Second Iteration*

You can use the BREAK command (also known as setting a breakpoint) to suspend DATA step execution at a particular line that you designate. In this example, suspend execution before executing the ELSE statement by setting a breakpoint at line 9.

```
break 9
```

When you press ENTER, an exclamation point appears at line 9 in the DEBUGGER SOURCE window to mark the breakpoint:

```
DEBUGGER SOURCE                                          _ □ ×
  3 data tours (drop=type) /debug;
  4    input @1 type $ @;
  5    if type='H' then do;
  6       input @3 Tour $20.;
  7       return;
  8       end;
! 9    else if type='P' then do;
 10       input @3 Name $10. Age 2. +1 Sex $1.;
 11       output;
 12       end;
 13    datalines;
```

Execute the GO command to continue DATA step execution until it reaches the breakpoint (in this case, line 9):

```
go
```

The following display shows the result:

```
DEBUGGER LOG                                             _ □ ×
Stepped to line 6 column 7
>
Stepped to line 7 column 7
> examine tour
Tour = Tour 101
>
Stepped to line 4 column 4
> break 9
Breakpoint 1 set at line 9
> go
Break at line 9 column 9
------------------------------------------------------------
>
```

```
DEBUGGER SOURCE
  3 data tours (drop=type) /debug;
  4    input @1 type $ @;
  5    if type='H' then do;
  6       input @3 Tour $20.;
  7       return;
  8       end;
! 9    else if type='P' then do;
 10       input @3 Name $10. Age 2. +1 Sex $1.;
 11       output;
 12       end;
 13    datalines;
```

SAS suspended execution just before the ELSE statement in line 7. Examine the values of all the variables to see their status at this point.

```
examine _all_
```

The following display shows the values:

```
 DEBUGGER LOG                                    _ □ ×
Breakpoint 1 set at line 9
> go
Break at line 9 column 9
>      examine _all_
type = P
Tour =
Name =
Age = .
Sex =
_ERROR_ = 0
_N_ = 2
```

You expect to see a value for Tour, but it does not appear. The program data vector gets reset to missing values at the beginning of each iteration and therefore does not retain the value of Tour. To solve the logic problem, you need to include a RETAIN statement in the SAS program.

### Ending the Debugger

To end the debugging session, issue the QUIT command on the debugger command line:

```
quit
```

The debugging windows disappear, and the original SAS session resumes.

### Correcting the DATA Step

Correct the original program by adding the RETAIN statement. Delete the DEBUG option from the DATA step, and resubmit the program:

```
      /* corrected version */
   data tours (drop=type);
      retain Tour;
      input @1 type $ @;
      if type='H' then do;
         input @3 Tour $20.;
         return;
         end;
      else if type='P' then do;
         input @3 Name $10. Age 2. +1 Sex $1.;
         output;
         end;
   datalines;
   H Tour 101
   P Mary E      21 F
   P George S    45 M
   P Susan K      3 F
   H Tour 102
   P Adelle S    79 M
   P Walter P    55 M
   P Fran I      63 F
   ;

   run;

   proc print;
      title 'Tour List';
   run;
```

The values for Tour now appear in the output:

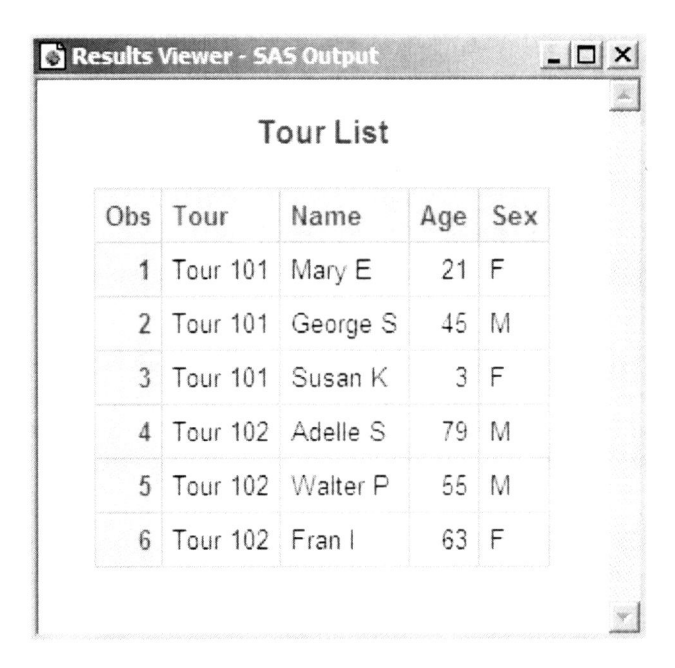

| Obs | Tour | Name | Age | Sex |
|---|---|---|---|---|
| 1 | Tour 101 | Mary E | 21 | F |
| 2 | Tour 101 | George S | 45 | M |
| 3 | Tour 101 | Susan K | 3 | F |
| 4 | Tour 102 | Adelle S | 79 | M |
| 5 | Tour 102 | Walter P | 55 | M |
| 6 | Tour 102 | Fran I | 63 | F |

Tour List

## *Example 2: Working with Formats*

This example shows how to debug a program when you use format statements to format dates. The following program creates a report that lists travel tour dates for specific countries.

```
data tours;
   length Country $ 10;
   input Country $10. Start : mmddyy. End : mmddyy.;
   Duration=end-start;
datalines;
Italy        033012 041312
Brazil       021912 022812
Japan        052212 061512
Venezuela    110312 11801
Australia    122112 011513
;

proc print data=tours;
   format start end date9.;
   title 'Tour Duration';
run;
```

## Tour Duration

| Obs | Country | Start | End | Duration |
|---|---|---|---|---|
| 1 | Italy | 30MAR2012 | 13APR2012 | 14 |
| 2 | Brazil | 19FEB2012 | 28FEB2012 | 9 |
| 3 | Japan | 22MAY2012 | 15JUN2012 | 24 |
| 4 | Venezuela | 03NOV2012 | 18JAN2012 | -290 |
| 5 | Australia | 21DEC2012 | 15JAN2013 | 25 |

The value of Duration for the tour to Venezuela shows a negative number, -290 days. To help identify the error, run the DATA step again using the DATA step debugger. SAS displays the following debugger windows:

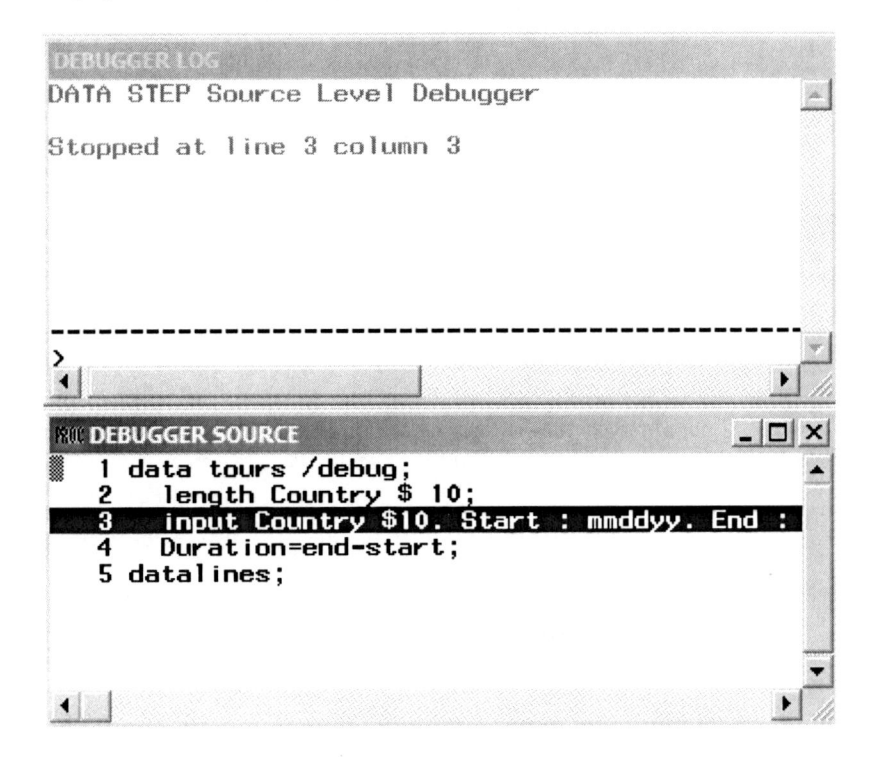

```
DEBUGGER LOG
DATA STEP Source Level Debugger

Stopped at line 3 column 3

------------------------------------------------
>
```

```
DEBUGGER SOURCE
  1  data tours /debug;
  2     length Country $ 10;
  3     input Country $10. Start : mmddyy. End :
  4     Duration=end-start;
  5  datalines;
```

At the DEBUGGER LOG command line, issue the EXAMINE command to display the values of all variables in the program data vector before execution begins:

```
examine _all_
```

Initial values of all variables appear in the DEBUGGER LOG window. SAS has not yet executed the INPUT statement.

Press ENTER to issue the STEP command. SAS executes the INPUT statement, and the assignment statement is now highlighted.

Issue the EXAMINE command to display the current value of all variables:

```
examine _all_
```

The following display shows the results:

```
DEBUGGER LOG                                    _ □ ×
>
Stepped to line 4 column 3
> examine _all_
Country = Italy
Start = 19082
End = 19096
Duration = .
_ERROR_ = 0
_N_ = 1
------------------------------------------------
>
```

```
DEBUGGER SOURCE
    1  data tours /debug;
    2     length Country $ 10;
    3     input Country $10. Start : mmddyy. End :
    4     Duration=end-start;
    5  datalines;
```

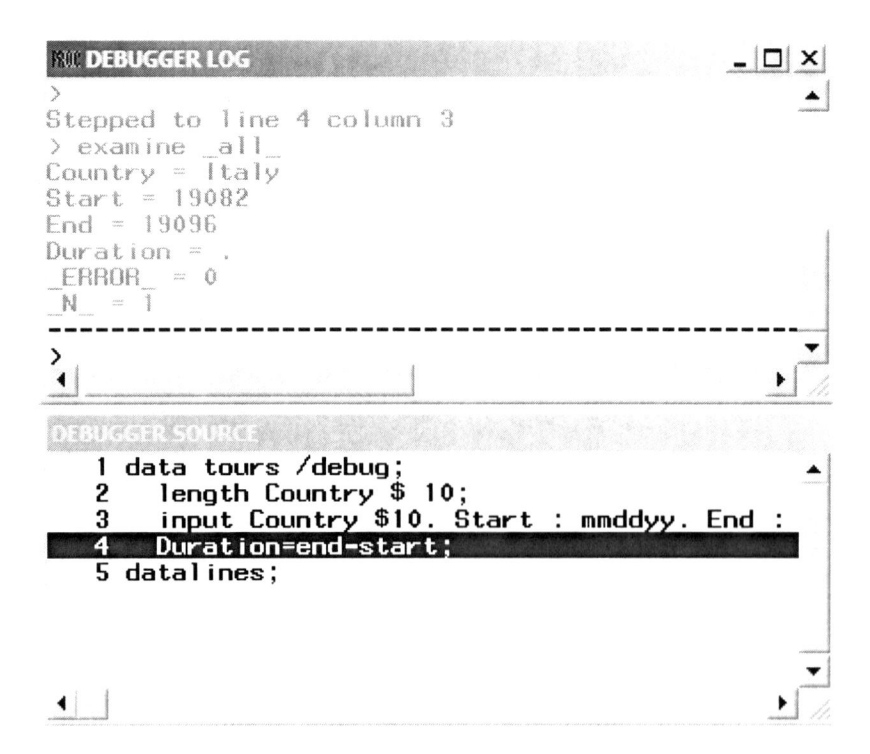

Because a problem exists with the Venezuela tour, suspend execution before the assignment statement when the value of Country equals Venezuela. Set a breakpoint to do this:

```
break 4 when country='Venezuela'
```

Execute the GO command to resume program execution:

```
go
```

SAS stops execution when the country name is Venezuela. You can examine Start and End tour dates for the Venezuela trip. Because the assignment statement is highlighted (indicating that SAS has not yet executed that statement), there will be no value for Duration.

Execute the EXAMINE command to view the value of the variables after execution:

```
examine _all_
```

The following display shows the results:

```
    DEBUGGER LOG                                    _ □ ×
> go
Break at line 4 column 3
> examine _all_
Country = Venezuela
Start = 19300
End = 19010
Duration = .
_ERROR_ = 0
_N_ = 4
------------------------------------------------
>

    DEBUGGER SOURCE
    1  data tours /debug;
    2    length Country $ 10;
    3    input Country $10. Start : mmddyy. End :
  ! 4    Duration=end-start;
    5  datalines;
```

To view formatted SAS dates, issue the EXAMINE command using the DATEw. format:

`examine start date7. end date7.`

The following display shows the results:

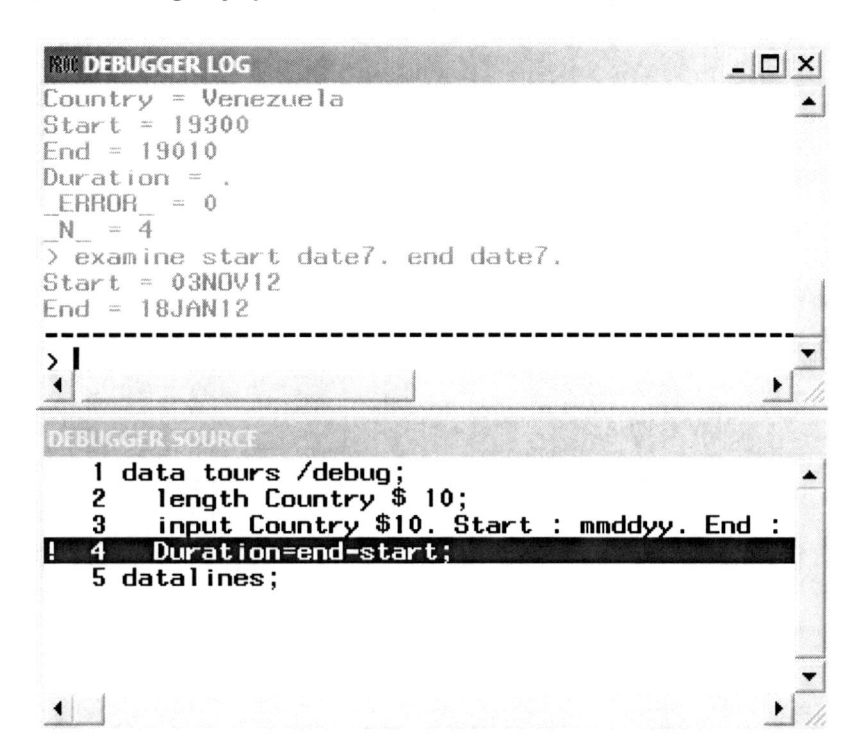

```
    DEBUGGER LOG                                    _ □ ×
Country = Venezuela
Start = 19300
End = 19010
Duration = .
_ERROR_ = 0
_N_ = 4
> examine start date7. end date7.
Start = 03NOV12
End = 18JAN12
------------------------------------------------
>

    DEBUGGER SOURCE
    1  data tours /debug;
    2    length Country $ 10;
    3    input Country $10. Start : mmddyy. End :
  ! 4    Duration=end-start;
    5  datalines;
```

Because the tour ends on November 18, 2012, and not on January 18, 2012, there is an error in the variable End. Examine the source data in the program and notice that the value for End has a typographical error. By using the SET command, you can temporarily set the value of End to November 18 to see whether you get the anticipated result. Issue the SET command using the DDMMMYY*w*. format:

```
set end='18nov12'd
```

Press ENTER to issue the STEP command and execute the assignment statement.

Issue the EXAMINE command to view the tour date and Duration fields:

```
examine start date7. end date7. duration
```

The following display shows the results:

```
DEBUGGER LOG
> examine start date7. end date7.
Start = 03NOV12
End = 18JAN12
> set end='18nov12'd
>
Stepped to line 5 column 1
> examine start date7. end date7. duration
Start = 03NOV12
End = 18NOV12
Duration = 15
--------------------------------------------
>
```

```
DEBUGGER SOURCE
  1  data tours /debug;
  2    length Country $ 10;
  3    input Country $10. Start : mmddyy. End
! 4    Duration=end-start;
  5  datalines;
```

The Start, End, and Duration fields contain correct data.

End the debugging session by issuing the QUIT command on the DEBUGGER LOG command line. Correct the original data in the SAS program, delete the DEBUG option, and resubmit the program.

```
   /* corrected version */

data tours;
  length Country $ 10;
  input Country $10. Start : mmddyy. End : mmddyy.;
  duration=end-start;
datalines;
Italy       033012 041312
Brazil      021912 022812
Japan       052212 061512
Venezuela   110312 111812
Australia   122112 011513
;

proc print data=tours;
   format start end date9.;
   title 'Tour Duration';
run;
```

### Tour Duration

| Obs | Country | Start | End | Duration |
|---|---|---|---|---|
| 1 | Italy | 30MAR2012 | 13APR2012 | 14 |
| 2 | Brazil | 19FEB2012 | 28FEB2012 | 9 |
| 3 | Japan | 22MAY2012 | 15JUN2012 | 24 |
| 4 | Venezuela | 03NOV2012 | 18NOV2012 | 15 |
| 5 | Australia | 21DEC2012 | 15JAN2013 | 25 |

## Example 3: Debugging DO Loops

An iterative DO, DO WHILE, or DO UNTIL statement can iterate many times during a single iteration of the DATA step. When you debug DO loops, you can examine several iterations of the loop by using the AFTER option in the BREAK command. The AFTER option requires a number that indicates how many times the loop will iterate before it reaches the breakpoint. The BREAK command then suspends program execution. For example, consider this data set:

```
data new / debug;
   set old;
   do i=1 to 20;
      newtest=oldtest+i;
      output;
   end;
run;
```

To set a breakpoint at the assignment statement (line 4 in this example) after every five iterations of the DO loop, issue this command:

```
break 4 after 5
```

When you issue the GO commands, the debugger suspends execution when $i$ has the values of 5, 10, 15, and 20 in the DO loop iteration.

In an iterative DO loop, select a value for the AFTER option that can be divided evenly into the number of iterations of the loop. For example, in this DATA step, 5 can be evenly divided into 20. When the DO loop iterates the second time, $i$ again has the values of 5, 10, 15, and 20.

If you do not select a value that can be evenly divided (such as 3 in this example), the AFTER option causes the debugger to suspend execution when $i$ has the values of 3, 6, 9, 12, 15, and 18. When the DO loop iterates the second time, $i$ has the values of 1, 4, 7, 10, 13, and 16.

## Example 4: Examining Formatted Values of Variables

You can use a SAS format or a user-created format when you display a value with the EXAMINE command. For example, assume that the variable BEGIN contains a SAS date value. To display the day of the week and date, use the WEEKDATE*w.* format with EXAMINE:

```
examine begin weekdate17.
```

When the value of BEGIN is 033012, the debugger displays the following:

```
Sun, Mar 30, 2012
```

As another example, you can create a format named SIZE:

```
proc format;
   value size 1-5='small'
              6-10='medium'
              11-high='large';
run;
```

To debug a DATA step that applies the format SIZE. to the variable STOCKNUM, use the format with EXAMINE:

```
examine stocknum size.
```

For example, when the value of STOCKNUM is 7, the debugger displays the following:

```
STOCKNUM = medium
```

## Part 6

# Producing Reports

*Chapter 27*
**Producing Detail Reports with the PRINT Procedure** ............. *433*

*Chapter 28*
**Creating Summary Tables with the TABULATE Procedure** ....... *473*

*Chapter 29*
**Creating Detail and Summary Reports with the REPORT Procedure** ............................................... *501*

# Chapter 27
# Producing Detail Reports with the PRINT Procedure

**Introduction to Producing Reports with the PRINT Procedure** .............. 434
    Purpose .......................................................... 434
    Prerequisites ..................................................... 434

**Input File and SAS Data Sets for Examples** ............................. 434

**Creating Simple Reports** ............................................. 436
    Showing All the Variables .......................................... 436
    Labeling the Observation Column .................................... 437
    Suppressing the Observation Column ................................. 438
    Emphasizing a Key Variable ........................................ 439
    Reporting the Values of Selected Variables ........................... 442
    Selecting Observations ............................................ 444

**Creating Enhanced Reports** .......................................... 446
    Ways to Enhance a Report ......................................... 446
    Specifying Formats for the Variables ................................ 447
    Summing Numeric Variables ........................................ 448
    Grouping Observations by Variable Values ............................ 449
    Grouping Observations in Multiple Sections ........................... 456

**Creating Customized Reports** ......................................... 458
    Ways to Customize a Report ....................................... 458
    Understanding Titles and Footnotes ................................. 458
    Adding Titles and Footnotes ....................................... 459
    Defining Labels .................................................. 460
    Splitting Labels across Two or More Lines ........................... 461
    Adding Blanks Lines .............................................. 462
    Changing the Report Style ......................................... 463

**Making Your Reports Easy to Change** .................................. 465
    Understanding the SAS Macro Facility ............................... 465
    Using Automatic Macro Variables ................................... 465
    Using Your Own Macro Variables ................................... 466
    Defining Macro Variables .......................................... 467
    Referring to Macro Variables ....................................... 467

**Summary** .......................................................... 469
    PROC PRINT Statements .......................................... 469
    PROC SORT Statements ........................................... 471
    SAS Macro Language ............................................. 471

**Learning More** ..................................................... 472

**434** Chapter 27 • *Producing Detail Reports with the PRINT Procedure*

# Introduction to Producing Reports with the PRINT Procedure

## Purpose

Reports that you create with the PRINT procedure contain one row for every observation that is selected for inclusion in the report. A report provides information about every record that is processed. For example, a report for a sales company can include all of the information about every sale made during a particular quarter of the year. The PRINT procedure is one of several report writing tools that you can use to create a variety of reports.

In this section, you will learn how to do the following:

- produce simple reports by using a few basic PROC PRINT options and statements
- produce enhanced reports by adding additional statements that format values, sum columns, group observations, and compute totals
- customize the appearance of reports by adding titles, footnotes, column labels, and the style of the report output
- substitute text by using macro variables

## Prerequisites

Before proceeding with this section, you should be familiar with the following features and concepts:

- the assignment statement
- the SORT procedure
- the BY statement
- the location of the procedure output

# Input File and SAS Data Sets for Examples

The examples in this section use one input file and five SAS data sets. For a complete listing of the input data, see "The YEAR_SALES Data Set" on page 816.

The input file contains sales records for a company, TruBlend Coffee Makers, that distributes the coffee machines. The file has the following structure:

```
01     1     Hollingsworth  Deluxe      260     49.50
01     1     Garcia         Standard     41     30.97
01     1     Hollingsworth  Deluxe      330     49.50
01     1     Jensen         Standard   1110     30.97
01     1     Garcia         Standard    715     30.97
01     1     Jensen         Deluxe      675     49.50
02     1     Jensen         Standard     45     30.97
02     1     Garcia         Deluxe       10     49.50
```

*...more data lines...*

```
12      4       Hollingsworth   Deluxe      125     49.50
12      4       Jensen          Standard    1254    30.97
12      4       Hollingsworth   Deluxe      175     49.50
```

The input file contains the following values from left to right:

- the month in which a sale was made
- the quarter of the year in which a sale was made
- the name of the sales representative
- the type of coffee maker sold (standard or deluxe)
- the number of units sold
- the price of each unit in US dollars

The first of the five SAS data sets is named YEAR_SALES. This data set contains all the sales data from the input file, and a new variable named AmountSold, which is created by multiplying Units by Price. The other four data sets are created from the YEAR_SALES data set. Each of the four data sets contains a subset of the data for each quarter. The data sets are QTR01, QTR02, QTR03, and QTR04.

The following program creates the five SAS data sets:

```
data year_sales;
   infile 'your-input-file';
   input Month $ Quarter $ SalesRep $14. Type $ Units Price;
   AmountSold = Units * Price;
run;

data qtr01;
   set year_sales(where=(quarter='1'));
run;

data qtr02;
   set year_sales(where=(quarter='2'));
run;

data qtr03;
   set year_sales(where=(quarter='3'));
run;

data qtr04;
   set year_sales(where=(quarter='4'));
run;
```

## Creating Simple Reports

### Showing All the Variables

By default, the PRINT procedure generates a simple report that shows the values of all the variables and the observations in the data set. For example, the following PROC PRINT step creates a report for the first sales quarter:

```
proc print data=qtr01;
   title 'TruBlend Coffee Makers First Quarter Sales Report';
run;
```

The following output shows the values of all the variables for all the observations in QTR01:

*Figure 27.1* Showing All Variables and All Observations

### TruBlend Coffee Makers First Quarter Sales Report

| Obs | Month | Quarter | SalesRep | Type | Units | Price | AmountSold |
|---|---|---|---|---|---|---|---|
| 1 | 01 | 1 | Hollingsworth | Deluxe | 260 | 49.50 | 12870.00 |
| 2 | 01 | 1 | Garcia | Standard | 41 | 30.97 | 1269.77 |
| 3 | 01 | 1 | Hollingsworth | Standard | 330 | 30.97 | 10220.10 |
| 4 | 01 | 1 | Jensen | Standard | 110 | 30.97 | 3406.70 |
| 5 | 01 | 1 | Garcia | Deluxe | 715 | 49.50 | 35392.50 |
| 6 | 01 | 1 | Jensen | Standard | 675 | 30.97 | 20904.75 |
| 7 | 02 | 1 | Garcia | Standard | 2045 | 30.97 | 63333.65 |
| 8 | 02 | 1 | Garcia | Deluxe | 10 | 49.50 | 495.00 |
| 9 | 02 | 1 | Garcia | Standard | 40 | 30.97 | 1238.80 |
| 10 | 02 | 1 | Hollingsworth | Standard | 1030 | 30.97 | 31899.10 |
| 11 | 02 | 1 | Jensen | Standard | 153 | 30.97 | 4738.41 |
| 12 | 02 | 1 | Garcia | Standard | 98 | 30.97 | 3035.06 |
| 13 | 03 | 1 | Hollingsworth | Standard | 125 | 30.97 | 3871.25 |
| 14 | 03 | 1 | Jensen | Standard | 154 | 30.97 | 4769.38 |
| 15 | 03 | 1 | Garcia | Standard | 118 | 30.97 | 3654.46 |
| 16 | 03 | 1 | Hollingsworth | Standard | 25 | 30.97 | 774.25 |
| 17 | 03 | 1 | Jensen | Standard | 525 | 30.97 | 16259.25 |
| 18 | 03 | 1 | Garcia | Standard | 310 | 30.97 | 9600.70 |

The Obs column identifies each observation by number. By default, SAS displays the observation number at the beginning of each row.

The top of the report has a title. The TITLE statement in the PROC PRINT step produces the title. For more information about the TITLE statement, see "Creating Customized Reports" on page 458. For now, remember that all the examples include at least one TITLE statement that produces a descriptive title similar to the one in this example.

The content of the report is very similar to the contents of the original data set QTR01. However, the report is easy to produce and to enhance.

## *Labeling the Observation Column*

A quick way to modify the report is to label the observation number (Obs column). The following SAS program includes the OBS= option in the PROC PRINT statement to change the column label for the Obs column:

```
proc print data=qtr01 obs='Observation Number';
   title 'TruBlend Coffee Makers First Quarter Sales Report';
run;
```

The following output shows the report:

*Figure 27.2* Labeling the Observation Column

### TruBlend Coffee Makers First Quarter Sales Report

| Observation Number | Month | Quarter | SalesRep | Type | Units | Price | AmountSold |
|---|---|---|---|---|---|---|---|
| 1 | 01 | 1 | Hollingsworth | Deluxe | 260 | 49.50 | 12870.00 |
| 2 | 01 | 1 | Garcia | Standard | 41 | 30.97 | 1269.77 |
| 3 | 01 | 1 | Hollingsworth | Standard | 330 | 30.97 | 10220.10 |
| 4 | 01 | 1 | Jensen | Standard | 110 | 30.97 | 3406.70 |
| 5 | 01 | 1 | Garcia | Deluxe | 715 | 49.50 | 35392.50 |
| 6 | 01 | 1 | Jensen | Standard | 675 | 30.97 | 20904.75 |
| 7 | 02 | 1 | Garcia | Standard | 2045 | 30.97 | 63333.65 |
| 8 | 02 | 1 | Garcia | Deluxe | 10 | 49.50 | 495.00 |
| 9 | 02 | 1 | Garcia | Standard | 40 | 30.97 | 1238.80 |
| 10 | 02 | 1 | Hollingsworth | Standard | 1030 | 30.97 | 31899.10 |
| 11 | 02 | 1 | Jensen | Standard | 153 | 30.97 | 4738.41 |
| 12 | 02 | 1 | Garcia | Standard | 98 | 30.97 | 3035.06 |
| 13 | 03 | 1 | Hollingsworth | Standard | 125 | 30.97 | 3871.25 |
| 14 | 03 | 1 | Jensen | Standard | 154 | 30.97 | 4769.38 |
| 15 | 03 | 1 | Garcia | Standard | 118 | 30.97 | 3654.46 |
| 16 | 03 | 1 | Hollingsworth | Standard | 25 | 30.97 | 774.25 |
| 17 | 03 | 1 | Jensen | Standard | 525 | 30.97 | 16259.25 |
| 18 | 03 | 1 | Garcia | Standard | 310 | 30.97 | 9600.70 |

## Suppressing the Observation Column

A quick way to simplify the report is to suppress the observation number (Obs column). Usually, it is unnecessary to identify each observation by number. (In some cases, you might want to show the observation numbers.) The following SAS program includes the NOOBS option in the PROC PRINT statement to suppress the Obs column:

```
proc print data=qtr01 noobs;
   title 'TruBlend Coffee Makers First Quarter Sales Report';
run;
```

The following output shows the report:

*Figure 27.3  Suppressing the Observation Column*

**TruBlend Coffee Makers First Quarter Sales Report**

| Month | Quarter | SalesRep | Type | Units | Price | AmountSold |
|---|---|---|---|---|---|---|
| 01 | 1 | Hollingsworth | Deluxe | 260 | 49.50 | 12870.00 |
| 01 | 1 | Garcia | Standard | 41 | 30.97 | 1269.77 |
| 01 | 1 | Hollingsworth | Standard | 330 | 30.97 | 10220.10 |
| 01 | 1 | Jensen | Standard | 110 | 30.97 | 3406.70 |
| 01 | 1 | Garcia | Deluxe | 715 | 49.50 | 35392.50 |
| 01 | 1 | Jensen | Standard | 675 | 30.97 | 20904.75 |
| 02 | 1 | Garcia | Standard | 2045 | 30.97 | 63333.65 |
| 02 | 1 | Garcia | Deluxe | 10 | 49.50 | 495.00 |
| 02 | 1 | Garcia | Standard | 40 | 30.97 | 1238.80 |
| 02 | 1 | Hollingsworth | Standard | 1030 | 30.97 | 31899.10 |
| 02 | 1 | Jensen | Standard | 153 | 30.97 | 4738.41 |
| 02 | 1 | Garcia | Standard | 98 | 30.97 | 3035.06 |
| 03 | 1 | Hollingsworth | Standard | 125 | 30.97 | 3871.25 |
| 03 | 1 | Jensen | Standard | 154 | 30.97 | 4769.38 |
| 03 | 1 | Garcia | Standard | 118 | 30.97 | 3654.46 |
| 03 | 1 | Hollingsworth | Standard | 25 | 30.97 | 774.25 |
| 03 | 1 | Jensen | Standard | 525 | 30.97 | 16259.25 |
| 03 | 1 | Garcia | Standard | 310 | 30.97 | 9600.70 |

## Emphasizing a Key Variable

### Understanding the ID Statement

To emphasize a key variable in a data set, you can use the ID statement in the PROC PRINT step. When you identify a variable in the ID statement, PROC PRINT displays the values of this variable in the first column of each row of the report. Highlighting a key variable in this way can help answer questions about your data. For example, the report can answer this question: "For each sales representative, what are the sales figures for the first quarter of the year?" The following two examples demonstrate how to answer this question quickly using data that is unsorted and sorted.

## Using an Unsorted Key Variable

To produce a report that emphasizes the sales representative, the PROC PRINT step includes an ID statement that specifies the variable SalesRep. Here is the revised program:

```
proc print data=qtr01;
   id SalesRep;
   title 'TruBlend Coffee Makers First Quarter Sales Report';
run;
```

Because the ID statement automatically suppresses the observation numbers, the NOOBS option is not needed in the PROC PRINT statement.

The following output shows the report:

**Figure 27.4** *Using the ID Statement with an Unsorted Variable*

### TruBlend Coffee Makers First Quarter Sales Report

| SalesRep | Month | Quarter | Type | Units | Price | AmountSold |
|---|---|---|---|---|---|---|
| Hollingsworth | 01 | 1 | Deluxe | 260 | 49.50 | 12870.00 |
| Garcia | 01 | 1 | Standard | 41 | 30.97 | 1269.77 |
| Hollingsworth | 01 | 1 | Standard | 330 | 30.97 | 10220.10 |
| Jensen | 01 | 1 | Standard | 110 | 30.97 | 3406.70 |
| Garcia | 01 | 1 | Deluxe | 715 | 49.50 | 35392.50 |
| Jensen | 01 | 1 | Standard | 675 | 30.97 | 20904.75 |
| Garcia | 02 | 1 | Standard | 2045 | 30.97 | 63333.65 |
| Garcia | 02 | 1 | Deluxe | 10 | 49.50 | 495.00 |
| Garcia | 02 | 1 | Standard | 40 | 30.97 | 1238.80 |
| Hollingsworth | 02 | 1 | Standard | 1030 | 30.97 | 31899.10 |
| Jensen | 02 | 1 | Standard | 153 | 30.97 | 4738.41 |
| Garcia | 02 | 1 | Standard | 98 | 30.97 | 3035.06 |
| Hollingsworth | 03 | 1 | Standard | 125 | 30.97 | 3871.25 |
| Jensen | 03 | 1 | Standard | 154 | 30.97 | 4769.38 |
| Garcia | 03 | 1 | Standard | 118 | 30.97 | 3654.46 |
| Hollingsworth | 03 | 1 | Standard | 25 | 30.97 | 774.25 |
| Jensen | 03 | 1 | Standard | 525 | 30.97 | 16259.25 |
| Garcia | 03 | 1 | Standard | 310 | 30.97 | 9600.70 |

Notice that the names of the sales representatives are not in any particular order. The report will be easier to read when the observations are grouped together in alphabetical order by sales representative.

## Using a Sorted Key Variable

If your data is not already ordered by the key variable, then use PROC SORT to sort the observations by this variable. If you do not specify an output data set, then PROC SORT permanently changes the order of the observations in the input data set.

The following program shows how to alphabetically order the observations by sales representative:

```
proc sort data=qtr01;  1
   by SalesRep;  2
run;

proc print data=qtr01;
   id SalesRep;  3
   title 'TruBlend Coffee Makers First Quarter Sales Report';
run;
```

The following list corresponds to the numbered items in the preceding program:

1  A PROC SORT step precedes the PROC PRINT step. PROC SORT orders the observations in the data set alphabetically by the values of the BY variable and overwrites the input data set.

2  A BY statement sorts the observations alphabetically by SalesRep.

3  An ID statement identifies the observations with the value of SalesRep rather than with the observation number. PROC PRINT uses the sorted order of SalesRep to create the report.

The following output shows the report:

*Figure 27.5* *Using the ID Statement with a Sorted Key Variable*

### TruBlend Coffee Makers First Quarter Sales Report

| SalesRep | Month | Quarter | Type | Units | Price | AmountSold |
|---|---|---|---|---|---|---|
| Garcia | 01 | 1 | Standard | 41 | 30.97 | 1269.77 |
| Garcia | 01 | 1 | Deluxe | 715 | 49.50 | 35392.50 |
| Garcia | 02 | 1 | Standard | 2045 | 30.97 | 63333.65 |
| Garcia | 02 | 1 | Deluxe | 10 | 49.50 | 495.00 |
| Garcia | 02 | 1 | Standard | 40 | 30.97 | 1238.80 |
| Garcia | 02 | 1 | Standard | 98 | 30.97 | 3035.06 |
| Garcia | 03 | 1 | Standard | 118 | 30.97 | 3654.46 |
| Garcia | 03 | 1 | Standard | 310 | 30.97 | 9600.70 |
| Hollingsworth | 01 | 1 | Deluxe | 260 | 49.50 | 12870.00 |
| Hollingsworth | 01 | 1 | Standard | 330 | 30.97 | 10220.10 |
| Hollingsworth | 02 | 1 | Standard | 1030 | 30.97 | 31899.10 |
| Hollingsworth | 03 | 1 | Standard | 125 | 30.97 | 3871.25 |
| Hollingsworth | 03 | 1 | Standard | 25 | 30.97 | 774.25 |
| Jensen | 01 | 1 | Standard | 110 | 30.97 | 3406.70 |
| Jensen | 01 | 1 | Standard | 675 | 30.97 | 20904.75 |
| Jensen | 02 | 1 | Standard | 153 | 30.97 | 4738.41 |
| Jensen | 03 | 1 | Standard | 154 | 30.97 | 4769.38 |
| Jensen | 03 | 1 | Standard | 525 | 30.97 | 16259.25 |

Now, the report clearly shows what each sales representative sold during the first three months of the year.

## Reporting the Values of Selected Variables

By default, the PRINT procedure reports the values of all the variables in the data set. However, to control which variables are shown and in what order, add a VAR statement to the PROC PRINT step.

For example, the information for the variables Quarter, Type, and Price is unnecessary. Therefore, the report needs to show only the values of the variables that are specified in the following order:

```
SalesRep  Month  Units  AmountSold
```

The following program adds the VAR statement to create a report that lists the values of the four variables in a specific order:

```
proc print data=qtr01 noobs;
   var SalesRep Month Units AmountSold;
   title 'TruBlend Coffee Makers First Quarter Sales Report';
run;
```

This program does not include the ID statement. It is unnecessary to identify the observations because the variable SalesRep is the first variable that is specified in the VAR statement. The NOOBS option in the PROC PRINT statement suppresses the observation numbers so that the sales representative appears in the first column of the report.

*Note:* If the ID statement is used and it names one of the variables in the VAR statement, the information is duplicated and two columns for the same variable appear in the report.

The following output shows the report:

*Figure 27.6* Showing Selected Variables

### TruBlend Coffee Makers First Quarter Sales Report

| SalesRep | Month | Units | AmountSold |
|---|---|---|---|
| Garcia | 01 | 41 | 1269.77 |
| Garcia | 01 | 715 | 35392.50 |
| Garcia | 02 | 2045 | 63333.65 |
| Garcia | 02 | 10 | 495.00 |
| Garcia | 02 | 40 | 1238.80 |
| Garcia | 02 | 98 | 3035.06 |
| Garcia | 03 | 118 | 3654.46 |
| Garcia | 03 | 310 | 9600.70 |
| Hollingsworth | 01 | 260 | 12870.00 |
| Hollingsworth | 01 | 330 | 10220.10 |
| Hollingsworth | 02 | 1030 | 31899.10 |
| Hollingsworth | 03 | 125 | 3871.25 |
| Hollingsworth | 03 | 25 | 774.25 |
| Jensen | 01 | 110 | 3406.70 |
| Jensen | 01 | 675 | 20904.75 |
| Jensen | 02 | 153 | 4738.41 |
| Jensen | 03 | 154 | 4769.38 |
| Jensen | 03 | 525 | 16259.25 |

The report is concise because it contains only those variables that are specified in the VAR statement.

The next example revises the report to show only those observations that satisfy a particular condition.

## Selecting Observations

### Understanding the WHERE Statement

To select observations that meet a particular condition from a data set, use a WHERE statement. The WHERE statement subsets the input data by specifying certain conditions that each observation must meet before it is available for processing.

The condition that you define in a WHERE statement is an arithmetic or logical expression that generally consists of a sequence of operands and operators.[1] To compare character values, you must enclose them in single or double quotation marks and the values must match exactly, including capitalization. You can also specify multiple comparisons that are joined by logical operators in the WHERE statement.

Using the WHERE statement might improve the efficiency of your SAS programs because SAS is not required to read all the observations in the input data set.

### Making a Single Comparison

You can select observations based on a single comparison by using the WHERE statement. The following program uses a single comparison in a WHERE statement to produce a report that shows the sales activity for a sales representative named Garcia:

```
proc print data=qtr01 noobs;
   var SalesRep Month Units AmountSold;
   where SalesRep='Garcia';
   title 'TruBlend Coffee Makers Quarterly Sales for Garcia';
run;
```

In the WHERE statement, the value `Garcia` is enclosed in quotation marks because SalesRep is a character variable. In addition, the letter G in the value `Garcia` is uppercase so that it matches exactly the value in the data set QTR01.

---

[1] The construction of the WHERE statement is similar to the construction of IF and IF-THEN statements.

The following output shows the report:

*Figure 27.7*  *Making a Single Comparison*

### TruBlend Coffee Makers First Quarter Sales Report

| SalesRep | Month | Units | AmountSold |
|---|---|---|---|
| Garcia | 01 | 41 | 1269.77 |
| Garcia | 01 | 715 | 35392.50 |
| Garcia | 02 | 2045 | 63333.65 |
| Garcia | 02 | 10 | 495.00 |
| Garcia | 02 | 40 | 1238.80 |
| Garcia | 02 | 98 | 3035.06 |
| Garcia | 03 | 118 | 3654.46 |
| Garcia | 03 | 310 | 9600.70 |

### *Making Multiple Comparisons*

You can also select observations based on two or more comparisons by using the WHERE statement. However, when you use multiple WHERE statements in a PROC step, then only the last statement is used. You can create a compound comparison by using the AND operator. For example, the following WHERE statement selects observations where Garcia sold only the deluxe coffee maker:

```
where SalesRep = 'Garcia' and Type='Deluxe';
```

The following program uses two comparisons in a WHERE statement to produce a report that shows sales activities for a sales representative (Garcia) during the first month of the year:

```
proc print data=year_sales noobs;
   var SalesRep Month Units AmountSold;
   where SalesRep='Garcia' and Month='01';
   title 'TruBlend Coffee Makers First Month Sales Report for Garcia';
run;
```

The WHERE statement uses the logical AND operator. Therefore, both comparisons must be true for PROC PRINT to include an observation in the report.

The following output shows the report:

*Figure 27.8*  *Making Two Comparisons*

### TruBlend Coffee Makers First Month Sales Report for Garcia

| SalesRep | Month | Units | AmountSold |
|---|---|---|---|
| Garcia | 01 | 41 | 1269.77 |
| Garcia | 01 | 715 | 35392.50 |

You might also want to select observations that meet at least one of several conditions. The following program uses two comparisons in the WHERE statement to create a report that shows every sale during the first quarter of the year that was greater than 500 units or more than $20,000:

```
proc print data=qtr01 noobs;
   var SalesRep Month Units AmountSold;
   where Units>500 or AmountSold>20000;
      title 'Sales Rep Q1 Monthly Report for Sales Above 500 Units or $20,000';
run;
```

Notice this WHERE statement uses the logical OR operator. Therefore, only one of the comparisons must be true for PROC PRINT to include an observation in the report.

The following output shows the report:

*Figure 27.9* *Making Comparisons for One Condition or Another*

Sales Rep Q1 Monthly Report for Sales Above 500 Units or $20,000

| SalesRep | Month | Units | AmountSold |
|---|---|---|---|
| Garcia | 01 | 715 | 35392.50 |
| Jensen | 01 | 675 | 20904.75 |
| Garcia | 02 | 2045 | 63333.65 |
| Hollingsworth | 02 | 1030 | 31899.10 |
| Jensen | 03 | 525 | 16259.25 |

# Creating Enhanced Reports

## Ways to Enhance a Report

With just a few PROC PRINT statements and options, you can produce a variety of detail reports. By using additional statements and options that enhance the reports, you can do the following using PROC PRINT:

- format the columns
- sum the numeric variables
- group the observations based on variable values
- sum the groups of variable values
- group the observations in separate sections

The examples in this section use the SAS data set QTR02, which was created in "Input File and SAS Data Sets for Examples" on page 434.

### Specifying Formats for the Variables

Specifying the formats of variables is a simple yet effective way to enhance the readability of your reports. By adding the FORMAT statement to your program, you can specify formats for variables. The format of a variable is a pattern that SAS uses to write the values of the variables. For example, SAS contains formats that add commas to numeric values, that add dollar signs to figures, or that report values as Roman numerals.

Using a format can make the values of the variables Units and AmountSold easier to read than in the previous reports. Specifically, Units can use a COMMA format with a total field width of 7, which includes commas to separate every three digits and omits decimal values. AmountSold can use a DOLLAR format with a total field width of 14, which includes commas to separate every three digits, a decimal point, two decimal places, and a dollar sign.

The following program illustrates how to apply these formats using a FORMAT statement:

```
proc print data=qtr02 noobs;
   var SalesRep Month Units AmountSold;
   where Units>500 or AmountSold>20000;
   format Units comma7. AmountSold dollar14.2;
   title 'Sales Rep Q2 Monthly Report for Sales Above 500 Units or $20,000';
run;
```

PROC PRINT applies the COMMA7. format to the values of the variable Units and the DOLLAR14.2 format to the values of the variable AmountSold.

The following output shows the report:

*Figure 27.10* Formatting Numeric Variables

**Sales Rep Q2 Monthly Report for Sales Above 500 Units or $20,000**

| SalesRep | Month | Units | AmountSold |
|---|---|---|---|
| Hollingsworth | 04 | 530 | $16,414.10 |
| Jensen | 04 | 1,110 | $34,376.70 |
| Garcia | 04 | 1,715 | $53,113.55 |
| Jensen | 04 | 675 | $20,904.75 |
| Hollingsworth | 05 | 1,120 | $34,686.40 |
| Hollingsworth | 05 | 1,030 | $31,899.10 |
| Garcia | 06 | 512 | $15,856.64 |
| Garcia | 06 | 1,000 | $30,970.00 |

AmountSold uses the DOLLAR14.2 format. The maximum column width is 14 spaces. Two spaces are reserved for the decimal part of a value. The remaining 12 spaces include the decimal point, whole numbers, the dollar sign, commas, and a minus sign if a value is negative.

Units uses the COMMA7. format. The maximum column width is seven spaces. The column width includes the numeric value, commas, and a minus sign if a value is negative.

The formats do not affect the internal data values that are stored in the SAS data set. The formats change only how the current PROC step displays the values in the report.

*Note:* Be sure to specify enough columns in the format to contain the largest value. If the format that you specify is not wide enough to contain the largest value, including special characters such as commas and dollar signs, then SAS applies the most appropriate format.

### Summing Numeric Variables

In addition to reporting the values in a data set, you can add the SUM statement to compute subtotals and totals for the numeric variables. The SUM statement enables you to request totals for one or more variables.

The following program produces a report that shows totals for the two numeric variables Units and AmountSold:

```
proc print data=qtr02 noobs;
   var SalesRep Month Units AmountSold;
   where Units>500 or AmountSold>20000;
   format Units comma7. AmountSold dollar14.2;
   sum Units AmountSold;
   title 'Sales Rep Q2 Monthly Report for Sales above 500 Units or $20,000';
run;
```

The following output shows the report:

*Figure 27.11* Summing Numeric Variables

### Sales Rep Q2 Monthly Report for Sales Above 500 Units or $20,000

| SalesRep | Month | Units | AmountSold |
|---|---|---|---|
| Hollingsworth | 04 | 530 | $16,414.10 |
| Jensen | 04 | 1,110 | $34,376.70 |
| Garcia | 04 | 1,715 | $53,113.55 |
| Jensen | 04 | 675 | $20,904.75 |
| Hollingsworth | 05 | 1,120 | $34,686.40 |
| Hollingsworth | 05 | 1,030 | $31,899.10 |
| Garcia | 06 | 512 | $15,856.64 |
| Garcia | 06 | 1,000 | $30,970.00 |
|  |  | 7,692 | $238,221.24 |

The totals for Units and AmountSold are computed by summing the values for each sale made by all the sales representatives. As the next example shows, the PRINT procedure can also separately compute subtotals for each sales representative.

## Grouping Observations by Variable Values

### Overview of Grouping Observations by Variable Values

The BY statement enables you to obtain separate analyses on groups of observations. The previous example used the SUM statement to compute totals for the variables Units and AmountSold. However, the totals were for all three sales representatives as one group. The following examples show how to use the BY, ID, and SUMBY statements as a part of the PROC PRINT step to separate the sales representatives into three groups with three separate subtotals and one grand total.

### Computing Group Subtotals

To obtain separate subtotals for specific numeric variables, add a BY statement to the PROC PRINT step. When you use a BY statement, the PRINT procedure expects that you have already sorted the data set by using the BY variables. Therefore, if your data is not sorted in the proper order, then you must add a PROC SORT step before the PROC PRINT step.

The BY statement produces a separate table for each BY group in the report. Above each table is a BY line that is a heading for the BY group.

*Note:* Do not specify in the VAR statement the variable that you use in the BY statement. Otherwise, the values of the BY variable appear twice in the report, as a header across the page and in columns down the page.

The following program uses the BY statement in the PROC PRINT step to obtain separate subtotals of the variables Units and AmountSold for each sales representative:

```
proc sort data=qtr02;
   by SalesRep;  1
run;

proc print data=qtr02 noobs;
   var Month Units AmountSold;  2
   where Units>500 or AmountSold>20000;
   format Units comma7. AmountSold dollar14.2;
   sum Units AmountSold;
   by SalesRep;  2
   title1 'Sales Rep Q2 Totals for Sales above 500 Units or $20,000';
run;
```

The following list corresponds to the numbered items in the preceding program:

1  The BY statement in the PROC SORT step sorts the data.

2  The variable SalesRep becomes part of the BY statement instead of the VAR statement.

The following output shows the report:

*Figure 27.12* *Grouping Observations with the BY Statement*

### Sales Rep Q2 Monthly Report for Sales Above 500 Units or $20,000

#### SalesRep=Garcia

| Month | Units | AmountSold |
|---|---|---|
| 04 | 1,715 | $53,113.55 |
| 06 | 512 | $15,856.64 |
| 06 | 1,000 | $30,970.00 |
| SalesRep | 3,227 | $99,940.19 |

#### SalesRep=Hollingsworth

| Month | Units | AmountSold |
|---|---|---|
| 04 | 530 | $16,414.10 |
| 05 | 1,120 | $34,686.40 |
| 05 | 1,030 | $31,899.10 |
| SalesRep | 2,680 | $82,999.60 |

#### SalesRep=Jensen

| Month | Units | AmountSold |
|---|---|---|
| 04 | 1,110 | $34,376.70 |
| 04 | 675 | $20,904.75 |
| SalesRep | 1,785 | $55,281.45 |
|  | 7,692 | $238,221.24 |

### *Labeling Subtotals and the Grand Total*

The previous example uses the default labels for the subtotal and the grand total labels. The subtotal default label is the BY variable, and the grand total default label is no label. You can use the SUMLABEL= option to replace the default subtotal label and the GRANDTOTAL_LABEL= option to replace a blank grand total label.

The following example adds the SUMLABEL= option and the GRANDTOTAL_LABEL= option to the PROC PRINT statement. This program assumes that QTR02 data has been previously sorted by the variables SalesRep.

```
proc print data=qtr02 noobs sumlabel="Total" grandtotal_label="Grand Total";;
   var Month Units AmountSold;
   where Units>500 or AmountSold>20000;
   format Units comma7. AmountSold dollar14.2;
```

```
    sum Units AmountSold;
 by SalesRep;
    title1 'Sales Rep Q2 Monthly Report for Sales Above 500 Units or $20,000';
 run;
```

The following output shows the report:

**Sales Rep Q2 Monthy Report for Sales Above 500 Units or $20,000**

SalesRep=Garcia

| Month | Units | AmountSold |
|---|---|---|
| 04 | 1,715 | $53,113.55 |
| 06 | 512 | $15,856.64 |
| 06 | 1,000 | $30,970.00 |
| Total | 3,227 | $99,940.19 |

SalesRep=Hollingsworth

| Month | Units | AmountSold |
|---|---|---|
| 04 | 530 | $16,414.10 |
| 05 | 1,120 | $34,686.40 |
| 05 | 1,030 | $31,899.10 |
| Total | 2,680 | $82,999.60 |

SalesRep=Jensen

| Month | Units | AmountSold |
|---|---|---|
| 04 | 1,110 | $34,376.70 |
| 04 | 675 | $20,904.75 |
| Total | 1,785 | $55,281.45 |
| Grand Total | 7,692 | $238,221.24 |

## *Identifying Group Subtotals*

You can use both the BY and ID statements in the PROC PRINT step to modify the appearance of your report. When you specify the same variables in both the BY and ID statements, the PRINT procedure uses the ID variable to identify the start of the BY group.

The following example uses the data set that was sorted in the last example and adds the ID statement to the PROC PRINT step:

```
proc print data=qtr02 sumlabel="Total" grandtotal_label="Grand Total";
   var Month Units AmountSold;
```

```
where Units>500 or AmountSold>20000;
format Units comma7. AmountSold dollar14.2;
sum Units AmountSold;
by SalesRep;
id SalesRep;
title1 'Sales Rep Q2 Monthly Report for Sales Above 500 Units or $20,000';
run;
```

The following output shows the report:

**Figure 27.13** *Grouping Observations with the BY and ID Statements*

**Sales Rep Q2 Monthly Report for Sales Above 500 Units or $20,000**

| SalesRep | Month | Units | AmountSold |
|---|---|---|---|
| Garcia | 04 | 1,715 | $53,113.55 |
|  | 06 | 512 | $15,856.64 |
|  | 06 | 1,000 | $30,970.00 |
| Total |  | 3,227 | $99,940.19 |

| SalesRep | Month | Units | AmountSold |
|---|---|---|---|
| Hollingsworth | 04 | 530 | $16,414.10 |
|  | 05 | 1,120 | $34,686.40 |
|  | 05 | 1,030 | $31,899.10 |
| Total |  | 2,680 | $82,999.60 |

| SalesRep | Month | Units | AmountSold |
|---|---|---|---|
| Jensen | 04 | 1,110 | $34,376.70 |
|  | 04 | 675 | $20,904.75 |
| Total |  | 1,785 | $55,281.45 |
| Grand Total |  | 7,692 | $238,221.24 |

The report has two distinct features. PROC PRINT separates the report into groups and suppresses the repetitive values of the BY and ID variables. The BY line does not appear above the group because the BY and ID statements are used together in the PROC PRINT step.

Remember these general rules about the SUM, BY, and ID statements:

- You can specify a variable in the SUM statement while omitting it in the VAR statement. PROC PRINT simply adds the variable to the list of variables in the VAR statement.

- You do not specify variables in the SUM statement that you used in the ID or BY statement.

- When you use a BY statement and you specify only one BY variable, PROC PRINT subtotals the SUM variable for each BY group that contains more than one observation.
- When you use a BY statement and you specify multiple BY variables, PROC PRINT shows a subtotal for a BY variable only when the value changes and when there are multiple observations with that value.

### Computing Multiple Group Subtotals

You can also use two or more variables in a BY statement to define groups and subgroups. The following program produces a report that groups observations first by sales representative and then by month:

```
proc sort data=qtr02;
   by SalesRep Month;  1
run;

proc print data=qtr02 noobs n='Sales Transactions:' 2
                              'Total Sales Transactions:' 2 ;
   var Units AmountSold; 3
   where Units>500 or AmountSold>20000;
   format Units comma7. AmountSold dollar14.2;
   sum Units AmountSold;
   by SalesRep Month 3 ;
   title1 'Monthly Sales Rep Totals for Sales Above 500 Units or $20,000';
run;
```

The following list corresponds to the numbered items in the preceding program:

1 The BY statement in the PROC SORT step sorts the data by SalesRep and Month.

2 The N= option in the PROC PRINT statement reports the number of observations in a BY group. Because of the SUM statement, it also reports the overall total number of observations at the end of the report. The first piece of explanatory text that N= provides precedes the number for each BY group. The second piece of explanatory text that N= provides precedes the number for the overall total.

3 The variables SalesRep and Month are omitted in the VAR statement because the variables are specified in the BY statement. This prevents PROC PRINT from reporting the values for these variables twice.

The following output shows the report:

**Figure 27.14** *Grouping Observations with Multiple BY Variables*

### Sales Rep Q2 Monthly Report for Sales Above 500 Units or $20,000

**SalesRep=Garcia Month=04**

| Units | AmountSold |
|---|---|
| 1,715 | $53,113.55 |

Sales Transactions 1

**SalesRep=Garcia Month=06**

| Units | AmountSold |
|---|---|
| 512 | $15,856.64 |
| 1,000 | $30,970.00 |
| 1,512 | $46,826.64 |
| 3,227 | $99,940.19 |

Sales Transactions 2

**SalesRep=Hollingsworth Month=04**

| Units | AmountSold |
|---|---|
| 530 | $16,414.10 |

Sales Transactions 1

**SalesRep=Hollingsworth Month=05**

| Units | AmountSold |
|---|---|
| 1,120 | $34,686.40 |
| 1,030 | $31,899.10 |
| 2,150 | $66,585.50 |
| 2,680 | $82,999.60 |

Sales Transactions 2

**SalesRep=Jensen Month=04**

| Units | AmountSold |
|---|---|
| 1,110 | $34,376.70 |
| 675 | $20,904.75 |
| 1,785 | $55,281.45 |
| 1,785 | $55,281.45 |
| 7,692 | $238,221.24 |

Sales Transactions 2
Total Sales Transactions 8

## Computing Group Totals

When you use multiple BY variables as in the previous example, you can suppress the subtotals every time a change occurs for the value of the BY variables. Use the SUMBY statement to control which BY variable causes subtotals to appear.

You can specify only one SUMBY variable, and this variable must also be specified in the BY statement. PROC PRINT computes sums when a change occurs to the following values:

- the value of the SUMBY variable
- the value of any variable in the BY statement that is specified before the SUMBY variable

For example, consider the following statements:

```
by Quarter SalesRep Month;
sumby SalesRep;
```

SalesRep is the SUMBY variable. In the BY statement, Quarter comes before SalesRep while Month comes after SalesRep. Therefore, these statements cause PROC PRINT to compute totals when either Quarter or SalesRep changes value, but not when Month changes value.

The following program omits the monthly subtotals for each sales representative by designating SalesRep as the variable to sum by:

```
proc print data=qtr02 sumlabel="Total" grandtotal_label="Grand Total";
   var Units AmountSold;
   where Units>500 or AmountSold>20000;
   format Units comma7. AmountSold dollar14.2;
   sum Units AmountSold;
   by SalesRep Month;
   id SalesRep Month;
   sumby SalesRep;
   title1 'Sales Rep Q2 Monthly Report for Sales Above 500 Units or $20,000';
run;
```

This program assumes that QTR02 data has been previously sorted by the variables SalesRep and Month.

The following output shows the report:

**Figure 27.15** Combining Subtotals for Groups of Observations

### Sales Rep Q2 Monthly Report for Sales Above 500 Units or $20,000

| SalesRep | Month | Units | AmountSold |
|---|---|---|---|
| Garcia | 04 | 1,715 | $53,113.55 |

| SalesRep | Month | Units | AmountSold |
|---|---|---|---|
| Garcia | 06 | 512 | $15,856.64 |
|  |  | 1,000 | $30,970.00 |
| Total |  | 3,227 | $99,940.19 |

| SalesRep | Month | Units | AmountSold |
|---|---|---|---|
| Hollingsworth | 04 | 530 | $16,414.10 |

| SalesRep | Month | Units | AmountSold |
|---|---|---|---|
| Hollingsworth | 05 | 1,120 | $34,686.40 |
|  |  | 1,030 | $31,899.10 |
| Total |  | 2,680 | $82,999.60 |

| SalesRep | Month | Units | AmountSold |
|---|---|---|---|
| Jensen | 04 | 1,110 | $34,376.70 |
|  |  | 675 | $20,904.75 |
| Total |  | 1,785 | $55,281.45 |
| Grand Total |  | 7,692 | $238,221.24 |

## Grouping Observations in Multiple Sections

You can also create a report with multiple sections by using the PAGEBY statement with the BY statement. The PAGEBY statement identifies a variable in the BY statement that causes the PRINT procedure to begin the report in a new section when a change occurs to the following values:

- the value of the BY variable
- the value of any BY variable that precedes it in the BY statement

The following program uses a PAGEBY statement with the BY statement to create a report with multiple sections. This program assumes that QTR02 data has been previously sorted by the variables SalesRep and Month.

```
proc print data=qtr02 sumlabel="Total" grandtotal_label="Grand Total";
   var Units AmountSold;
   where Units>500 or AmountSold>20000;
   format Units comma7. AmountSold dollar14.2;
   sum Units AmountSold;
   by SalesRep Month;
   id SalesRep Month;
   sumby SalesRep;
   pageby SalesRep;
   title1 'Sales Rep Quarterly Totals for Sales above 500 Units or $20,000';
run;
```

The following output shows the report:

*Figure 27.16   Grouping Observations on Separate Pages*

**Sales Rep Q2 Monthly Report for Sales Above 500 Units or $20,000**

| SalesRep | Month | Units | AmountSold |
|---|---|---|---|
| Garcia | 04 | 1,715 | $53,113.55 |

| SalesRep | Month | Units | AmountSold |
|---|---|---|---|
| Garcia | 06 | 512 | $15,856.64 |
|  |  | 1,000 | $30,970.00 |
| Total |  | 3,227 | $99,940.19 |

**Sales Rep Q2 Monthly Report for Sales Above 500 Units or $20,000**

| SalesRep | Month | Units | AmountSold |
|---|---|---|---|
| Hollingsworth | 04 | 530 | $16,414.10 |

| SalesRep | Month | Units | AmountSold |
|---|---|---|---|
| Hollingsworth | 05 | 1,120 | $34,686.40 |
|  |  | 1,030 | $31,899.10 |
| Total |  | 2,680 | $82,999.60 |

**Sales Rep Q2 Monthly Report for Sales Above 500 Units or $20,000**

| SalesRep | Month | Units | AmountSold |
|---|---|---|---|
| Jensen | 04 | 1,110 | $34,376.70 |
|  |  | 675 | $20,904.75 |
| Total |  | 1,785 | $55,281.45 |
| Grand Total |  | 7,692 | $238,221.24 |

A new section occurs in the report when the value of the variable SalesRep changes from **Garcia** to `Hollingsworth` and from **Hollingsworth** to **Jensen**.

# Creating Customized Reports

## Ways to Customize a Report

As you have seen from the previous examples, the PRINT procedure produces simple reports quickly and easily. With additional statements and options, you can enhance the readability of your reports. For example, you can do the following:

- Add descriptive titles and footnotes.
- Define and split labels across multiple lines.
- Add double spacing.
- Ensure that the column widths are uniform across the pages of the report.
- Change the style of the output.

## Understanding Titles and Footnotes

Adding descriptive titles and footnotes is one of the easiest and most effective ways to improve the appearance of a report. You can use the TITLE statement to include from 1 to 10 lines of text at the top of the report. You can use the FOOTNOTE statement to include from 1 to 10 lines of text at the bottom of the report.

In the TITLE statement, you can specify *n* immediately following the keyword TITLE, to indicate the level of the TITLE statement. *n* is a number from 1 to 10 that specifies the line number of the TITLE. You must enclose the text of each title in single or double quotation marks.

Skipping over some values of *n* indicates that those lines are blank. For example, if you specify TITLE1 and TITLE3 statements but skip TITLE2, then a blank line occurs between the first and third lines.

When you specify a title, SAS uses that title for all subsequent output until you cancel it or define another title for that line. A TITLE statement for a given line cancels the previous TITLE statement for that line and for all lines below it, that is, for those with larger *n* values.

To cancel all existing titles, specify a TITLE statement without the *n* value:

`title;`

To suppress the *n*th title and all titles below it, use the following statement:

`titlen;`

Footnotes work the same way as titles. In the FOOTNOTE statement, you can specify *n* immediately following the keyword FOOTNOTE, to indicate the level of the FOOTNOTE statement. *n* is a number from 1 to 10 that specifies the line number of the FOOTNOTE. You must enclose the text of each footnote in single or double quotation marks. As with the TITLE statement, skipping over some values of *n* indicates that those lines are blank.

Remember that the footnotes are pushed up from the bottom of the report. In other words, the FOOTNOTE statement with the largest number appears on the bottom line.

When you specify a footnote, SAS uses that footnote for all subsequent output until you cancel it or define another footnote for that line. You cancel and suppress footnotes in the same way that you cancel and suppress titles.

*Note:* The maximum title length and footnote length that is allowed depends on your operating environment and the value of the LINESIZE= system option. Refer to the SAS documentation for your operating environment for more information.

## *Adding Titles and Footnotes*

The following program includes titles and footnotes in a report of second quarter sales during the month of April:

```
proc sort data=qtr02;
   by SalesRep;
run;

proc print data=qtr02 noobs;
   var SalesRep Month Units AmountSold;
   where Month='04';
   format Units comma7. AmountSold dollar14.2;
   sum Units AmountSold;
   title1 'TruBlend Coffee Makers, Inc.';
   title3 'Quarterly Sales Report';
   footnote1 'April Sales Totals';
   footnote2 'COMPANY CONFIDENTIAL INFORMATION';
run;
```

The report includes three title lines and two footnote lines. The program omits the TITLE2 statement so that the second title line is blank.

The following output shows the report:

*Figure 27.17* Adding Titles and Footnotes

**TruBlend Coffee Makers, Inc.**

**Quarterly Sales Report**

| SalesRep | Month | Units | AmountSold |
|---|---|---|---|
| Garcia | 04 | 150 | $4,645.50 |
| Garcia | 04 | 1,715 | $53,113.55 |
| Hollingsworth | 04 | 260 | $8,052.20 |
| Hollingsworth | 04 | 530 | $16,414.10 |
| Jensen | 04 | 1,110 | $34,376.70 |
| Jensen | 04 | 675 | $20,904.75 |
|  |  | 4,440 | $137,506.80 |

April Sales Totals
COMPANY CONFIDENTIAL INFORMATION

## Defining Labels

By default, SAS uses variable names for column headings. However, to improve the appearance of a report, you can specify your own column headings.

To override the default headings, you need to do the following:

- Add the LABEL option to the PROC PRINT statement.
- Define the labels in the LABEL statement.

The LABEL option causes the report to display labels, instead of variable names, for the column headings. You use the LABEL statement to assign the labels for the specific variables. A label can be up to 256 characters long, including blanks, and must be enclosed in single or double quotation marks. If you assign labels when you created the SAS data set, then you can omit the LABEL statement from the PROC PRINT step.

The following program modifies the previous program and defines labels for the variables SalesRep, Units, and AmountSold:

```
proc sort data=qtr02;
   by SalesRep;
run;

proc print data=qtr02 noobs label;
   var SalesRep Month Units AmountSold;
   where Month='04';
   format Units comma7. AmountSold dollar14.2;
   sum Units AmountSold;
```

```
      label SalesRep   = 'Sales Rep.'
            Units      = 'Units Sold'
            AmountSold = 'Amount Sold';
      title 'TruBlend Coffee Maker Sales Report for April';
      footnote;
run;
```

The TITLE statement redefines the first title and cancels any additional titles that might have been previously defined. The FOOTNOTE statement cancels any footnotes that might have been previously defined.

The following output shows the report:

*Figure 27.18  Defining Labels*

**TruBlend Coffee Maker Sales Report for April**

| Sales Rep. | Month | Units Sold | Amount Sold |
|---|---|---|---|
| Garcia | 04 | 150 | $4,645.50 |
| Garcia | 04 | 1,715 | $53,113.55 |
| Hollingsworth | 04 | 260 | $8,052.20 |
| Hollingsworth | 04 | 530 | $16,414.10 |
| Jensen | 04 | 1,110 | $34,376.70 |
| Jensen | 04 | 675 | $20,904.75 |
|  |  | 4,440 | $137,506.80 |

## Splitting Labels across Two or More Lines

Sometimes labels are too long to fit on one line, or you might want to split a label across two or more lines. You can use the SPLIT= option to control where the labels are separated into multiple lines.

The SPLIT= option replaces the LABEL option in the PROC PRINT statement. (You do not need to use both SPLIT= and LABEL because SPLIT= implies that PROC PRINT use labels.) In the SPLIT= option, you specify an alphanumeric character that indicates where to split labels. To use the SPLIT= option, you need to do the following:

- Define the split character as a part of the PROC PRINT statement.
- Define the labels with a split character in the LABEL statement.

The following PROC PRINT step defines the slash (/) as the split character and includes slashes in the LABEL statements to split the labels Sales Representative, Units Sold, and Amount Sold into two lines each:

```
proc sort data=qtr02;
   by SalesRep;
run;

proc print data=qtr02 noobs split='/';
   var SalesRep Month Units AmountSold;
```

```
        where Month='04';
        format Units comma7. AmountSold dollar14.2;
        sum Units AmountSold;
        title 'TruBlend Coffee Maker Sales Report for April';
        label SalesRep   = 'Sales/Representative'
              Units      = 'Units/Sold'
              AmountSold = 'Amount/Sold';
    run;
```

The following output shows the report:

*Figure 27.19  Reporting: Splitting Labels into Two Lines*

TruBlend Coffee Maker Sales Report for April

| Sales Representative | Month | Units Sold | Amount Sold |
|---|---|---|---|
| Garcia | 04 | 150 | $4,645.50 |
| Garcia | 04 | 1,715 | $53,113.55 |
| Hollingsworth | 04 | 260 | $8,052.20 |
| Hollingsworth | 04 | 530 | $16,414.10 |
| Jensen | 04 | 1,110 | $34,376.70 |
| Jensen | 04 | 675 | $20,904.75 |
|  |  | 4,440 | $137,506.80 |

## Adding Blanks Lines

You might want to improve the appearance of a report by adding one or more blank lines between the rows of the report. The following program uses the BLANKLINE option in the PROC PRINT statement to add lines between the observations for sales representatives in the report:

```
proc sort data=qtr02;
   by SalesRep;
run;

proc print data=qtr02 noobs split='/' blankline=2;
   var SalesRep Month Units AmountSold;
   where Month='04';
   format Units comma7. AmountSold dollar14.2;
   sum Units AmountSold;
   title 'TruBlend Coffee Maker Sales Report for April';
   label SalesRep   = 'Sales/Representative'
         Units      = 'Units/Sold'
         AmountSold = 'Amount/Sold';
run;
```

The following output shows the report:

*Figure 27.20* Adding Blank Lines

### TruBlend Coffee Maker Sales Report for April

| Sales Representative | Month | Units Sold | Amount Sold |
|---|---|---|---|
| Garcia | 04 | 150 | $4,645.50 |
| Garcia | 04 | 1,715 | $53,113.55 |
|  |  |  |  |
| Hollingsworth | 04 | 260 | $8,052.20 |
| Hollingsworth | 04 | 530 | $16,414.10 |
|  |  |  |  |
| Jensen | 04 | 1,110 | $34,376.70 |
| Jensen | 04 | 675 | $20,904.75 |
|  |  | 4,440 | $137,506.80 |

## *Changing the Report Style*

By default, the Output Delivery System (ODS) creates reports for the HTML destination using the default style of HTMLBlue. You can change the report style for all ODS destinations except for the LISTING destination and any destination that does not create output. Instead of explicitly using ODS to change a report style, you can use the STYLE= option in the PROC PRINT statement. The STYLE= option interfaces with ODS to change the report style. You can change report attributes such as fonts, colors, text alignment, and table cell size. Here is the syntax of the STYLE= option:

STYLE <*(location(s))*>=<*style-element-name*> <[*style-attribute-name=style-attribute-value*]>

*location(s)* identifies the part of the report that the STYLE= option affects. Report locations include the table structure, column headings, data in the cells, BY labels, the SUM line that contains BY group totals and the report grand totals, N= option values, and the OBS column heading and data.

*style-element-name* is the name of a style element that is registered with ODS. A style element is a collection of style attributes that apply to a particular part of the output for a SAS program. For example, a style element might contain instructions for the presentation of column headings or for the presentation of the data inside table cells. Each location has a default style element.

*style-attribute-name=style-attribute-value* describes the style attribute to change and its value.

For a list of style elements and style attributes, see *SAS Output Delivery System: User's Guide*.

**464** Chapter 27 • *Producing Detail Reports with the PRINT Procedure*

The following program modifies the previous program to change the cell spacing in the tables, the color and font style for the header, and the grand total. The style for header is based on the headerstrong style element.

```
proc sort data=qtr02;
  by SalesRep;
run;

proc print data=qtr02 noobs split='/'
           style(table)={cellpadding=10}
           style(header)=headerstrong{backgroundcolor=very light green
                                      color=green fontstyle=italic}
           style(grandtotal)={backgroundcolor=very light green color=green};
   var SalesRep Month Units AmountSold;
   where Month='04';
   format Units comma7. AmountSold dollar14.2;
   sum Units AmountSold;
   label SalesRep= 'Sales/Representative'
       Units = 'Units/Sold'
           AmountSold= 'Amount/Sold';
   title 'TruBlend Coffee Maker Sales Report for April';
run;
```

The following output shows the report:

*Figure 27.21 Changing the Report Style*

### TruBlend Coffee Maker Sales Report for April

| Sales Representative | Month | Units Sold | Amount Sold |
|---|---|---|---|
| Garcia | 04 | 150 | $4,645.50 |
| Garcia | 04 | 1,715 | $53,113.55 |
| Hollingsworth | 04 | 260 | $8,052.20 |
| Hollingsworth | 04 | 530 | $16,414.10 |
| Jensen | 04 | 1,110 | $34,376.70 |
| Jensen | 04 | 675 | $20,904.75 |
|  |  | 4,440 | $137,506.80 |

# Making Your Reports Easy to Change

### *Understanding the SAS Macro Facility*

Base SAS includes the macro facility as a tool to customize SAS and to reduce the amount of text that you must enter to do common tasks. The macro facility enables you to assign a name to character strings or groups of SAS programming statements.

Thereafter, you can work with the names rather than with the text itself. When you use a macro facility name in a SAS program, the macro facility generates SAS statements and commands as needed. The rest of SAS receives those statements and uses them in the same way it uses the ones that you enter in the standard manner.

The macro facility enables you to create macro variables to substitute text in SAS programs. One of the major advantages of using macro variables is that it enables you to change the value of a variable in one place in your program and then have the change appear in multiple references throughout your program. You can substitute text by using automatic macro variables or by using your own macro variables, which you define and assign values to.

### *Using Automatic Macro Variables*

The SAS macro facility includes many automatic macro variables. Some of the values associated with the automatic macro variables depend on your operating environment. You can use automatic macro variables to provide the time, the day of the week, and the date based on your computer's internal clock as well as other processing information.

To include a second title on a report that displays the text string "Produced on" followed by today's date, add the following TITLE statement to your program:

```
title2 "Produced on &SYSDATE9";
```

Notice the syntax for this statement. First, the ampersand that precedes SYSDATE9 tells the SAS macro facility to replace the reference with its assigned value. In this case, the assigned value is the date on which the SAS session started and is expressed as *ddmmmyyyy*, where

- *dd* is a two-digit date
- *mmm* is the first three letters of the month name
- *yyyy* is a four-digit year

Second, the text of the TITLE statement is enclosed in double quotation marks because the SAS macro facility resolves macro variable references in the TITLE statement and the FOOTNOTE statement only if they are in double quotation marks.

The following program, which includes a PROC SORT step and the TITLE statement, demonstrates how to use the SYSDATE9 automatic macro variable:

```
proc sort data=qtr04;
   by SalesRep;
run;

proc print data=qtr04 noobs split='/';
   var SalesRep Month Units AmountSold;
   format Units comma7. AmountSold dollar14.2;
```

```
         sum Units AmountSold;
         title1 'TruBlend Coffee Maker Quarterly Sales Report';
         title2 "Produced on &SYSDATE9";
         label SalesRep    = 'Sales/Rep.'
               Units       = 'Units/Sold'
               AmountSold  = 'Amount/Sold';
run;
```

The following output shows the report:

*Figure 27.22  Using Automatic Macro Variables*

**TruBlend Coffee Maker Quarterly Sales Report**
**Produced on 09APR2013**

| Sales Rep. | Month | Units Sold | Amount Sold |
|---|---|---|---|
| Garcia | 10 | 250 | $7,742.50 |
| Garcia | 10 | 365 | $11,304.05 |
| Garcia | 11 | 198 | $6,132.06 |
| Garcia | 11 | 120 | $3,716.40 |
| Garcia | 12 | 1,000 | $30,970.00 |
| Hollingsworth | 10 | 530 | $16,414.10 |
| Hollingsworth | 10 | 265 | $8,207.05 |
| Hollingsworth | 11 | 1,230 | $38,093.10 |
| Hollingsworth | 11 | 150 | $7,425.00 |
| Hollingsworth | 12 | 125 | $6,187.50 |
| Hollingsworth | 12 | 175 | $5,419.75 |
| Jensen | 10 | 975 | $30,195.75 |
| Jensen | 10 | 55 | $1,703.35 |
| Jensen | 11 | 453 | $14,029.41 |
| Jensen | 11 | 70 | $2,167.90 |
| Jensen | 12 | 876 | $27,129.72 |
| Jensen | 12 | 1,254 | $38,836.38 |
|  |  | 8,091 | $255,674.02 |

## Using Your Own Macro Variables

In addition to using automatic macro variables, you can use the %LET statement to define your own macro variables and refer to them with the ampersand prefix. Defining

*Making Your Reports Easy to Change* **467**

macro variables at the beginning of your program enables you to change other parts of the program easily. The example in this section shows how to define two macro variables, Quarter and Year, and how to refer to them in a TITLE statement.

## *Defining Macro Variables*

To use two macro variables that produce flexible report titles, first define the macro variables. The following %LET statements define the two macro variables:

```
%let Quarter=Fourth;
%let Year=2011;
```

The name of the first macro variable is Quarter and it is assigned the value Fourth. The name of the second macro variable is Year and it is assigned the value 2011.

Macro variable names such as these conform to the following rules for SAS names:

- macro variable names are one to 32 characters long
- macro variable names begin with a letter or an underscore
- letters, numbers, and underscores follow the first character.

In these simple situations, do not assign values to macro variables that contain unmatched quotation marks or semicolons. If the values contain leading or trailing blanks, then SAS removes the blanks.

## *Referring to Macro Variables*

To refer to the value of a macro variable, place an ampersand prefix in front of the name of the variable. The following TITLE statement contains references to the values of the macro variables Quarter and Year, which were previously defined in %LET statements:

```
title3 "&Quarter Quarter &Year Sales Totals";
```

The complete program, which includes the two %LET statements and the TITLE3 statement, follows:

```
%let Quarter=Fourth;    1
%let Year=2013;         2

proc sort data=qtr04;
   by SalesRep;
run;

proc print data=qtr04 noobs split='/' width=uniform;
   var SalesRep Month Units AmountSold;
   format Units comma7. AmountSold dollar14.2;
   sum Units AmountSold;
   title1 'TruBlend Coffee Maker Quarterly Sales Report';
   title2 "Produced on &SYSDATE9";
   title3 "&Quarter Quarter &Year Sales Totals";    3
   label SalesRep   = 'Sales/Rep.'
         Units      = 'Units/Sold'
         AmountSold = 'Amount/Sold';
run;
```

1  The %LET statement creates a macro variable with the sales quarter. When an ampersand precedes Quarter, the SAS macro facility knows to replace any reference to &Quarter with the assigned value of Fourth.

2 The %LET statement creates a macro variable with the year. When ampersand precedes Year, the SAS macro facility knows to replace any reference to &Year with the assigned value of 2011.

3 The text of the TITLE2 and TITLE3 statements are enclosed in double quotation marks so that the SAS macro facility can resolve them.

The following list corresponds to the numbered items in the preceding program:

The following output shows the report:

*Figure 27.23  Using Your Own Macro Variables*

**TruBlend Coffee Maker Quarterly Sales Report**
**Produced on 09APR2013**
**Fourth Quarter 2012 Sales Totals**

| Sales Rep. | Month | Units Sold | Amount Sold |
|---|---|---|---|
| Garcia | 10 | 250 | $7,742.50 |
| Garcia | 10 | 365 | $11,304.05 |
| Garcia | 11 | 198 | $6,132.06 |
| Garcia | 11 | 120 | $3,716.40 |
| Garcia | 12 | 1,000 | $30,970.00 |
| Hollingsworth | 10 | 530 | $16,414.10 |
| Hollingsworth | 10 | 265 | $8,207.05 |
| Hollingsworth | 11 | 1,230 | $38,093.10 |
| Hollingsworth | 11 | 150 | $7,425.00 |
| Hollingsworth | 12 | 125 | $6,187.50 |
| Hollingsworth | 12 | 175 | $5,419.75 |
| Jensen | 10 | 975 | $30,195.75 |
| Jensen | 10 | 55 | $1,703.35 |
| Jensen | 11 | 453 | $14,029.41 |
| Jensen | 11 | 70 | $2,167.90 |
| Jensen | 12 | 876 | $27,129.72 |
| Jensen | 12 | 1,254 | $38,836.38 |
|  |  | 8,091 | $255,674.02 |

Using macro variables can make your programs easy to modify. For example, if the previous program contained many references to Quarter and Year, then changes in only three places will produce an entirely different report:

- the two values in the %LET statements

- the data set name in the PROC PRINT statement

# Summary

## PROC PRINT Statements

**PROC PRINT** <DATA=*SAS-data-set*> <*option(s)*>;
    **BY** *variable(s)*;
    **FOOTNOTE**<*n*> <*'footnote'*>;
    **FORMAT** *variable(s) format-name*;
    **ID** *variable(s)*;
    **LABEL** *variable='label'*;
    **PAGEBY** *variable*;
    **SUM** *variable(s)*;
    **SUMBY** *variable*;
    **TITLE**<*n*> <*'title'*>;
    **VAR** *variable(s)*;
**WHERE** *where-expression*;

PROC PRINT <DATA=*SAS-data-set*> <*option(s)*>;
: starts the procedure and, when used alone, shows all variables for all observations in the SAS-data-set in the report. Other statements, that are listed below, enable you to control what to report.

You can specify the following options in the PROC PRINT statement:

BLANKLINE=*n* | BLANKLINE=(COUNT=*n* <STYLE=[*style-attribute-specification(s)*]>)
: specifies to insert a blank line after every n observations. The observation count is reset to 0 at the beginning of every BY group for all ODS destinations.

DATA=*SAS-data-set*
: names the SAS data set that PROC PRINT uses. If you omit DATA=, then PROC PRINT uses the most recently created data set.

GRANDTOTAL_LABEL="*label*"
: displays a label on the grand total line.

LABEL
: uses variable labels instead of variable names as column headings for any variables that have labels defined. Variable labels appear only if you use the LABEL option or the SPLIT= option. You can specify labels in LABEL statements in the DATA step that creates the data set or in the PROC PRINT step. If you do not specify the LABEL option or if there is no label for a variable, then PROC PRINT uses the variable name.

N<="*string-1*" <"*string-2*">>
: shows the number of observations in the data set, in BY groups, or both. It can also specify explanatory text to include with the number.

NOOBS
: suppresses the observation numbers in the output. This option is useful when you omit an ID statement and do not want to show the observation numbers.

OBS="*column-header*"
: specifies a column heading for the column that identifies each observation by number.

SPLIT='*split-character*'
: specifies the split character, which controls line breaks in column headings. PROC PRINT breaks a column heading when it reaches the split character and continues the header on the next line. The split character is not part of the column heading.

    PROC PRINT uses variable labels only when you use the LABEL option or the SPLIT= option. It is not necessary to use both the LABEL and SPLIT= options because SPLIT= implies to use labels.

STYLE=<*location(s)*>=<*style-element-name*> <[*style-attribute-specification(s)*]>
: specify one or more style elements for the Output Delivery System to use for different parts of the report.

SUMLABEL="*label*"
: displays a label on the summary line in place of the BY variable name.

WIDTH=UNIFORM
: uses each variable's formatted width as its column width on all pages. If the variable does not have a format that explicitly specifies a field width, then PROC PRINT uses the widest data value as the column width. Without this option, PROC PRINT fits as many variables and observations on a page as possible. Therefore, the report might contain a different number of columns on each page.

BY *variable(s)*;
: produces a separate section of the report for each BY group. The BY group consists of the *variables* that you specify. When you use a BY statement, the procedure expects that the input data set is sorted by the *variables*.

FOOTNOTE<*n*> <'*footnote*'>;
: specifies a footnote. The argument *n* is a number from 1 to 10 that immediately follows the word FOOTNOTE, with no intervening blank, and specifies the line number of the FOOTNOTE. The text of each footnote must be enclosed in single or double quotation marks. The maximum footnote length that is allowed depends on your operating environment and the value of the LINESIZE= system option. Refer to the SAS documentation for your operating environment for more information.

FORMAT *variable(s) format-name*;
: enables you to report the value of a *variable* using a special pattern that you specify as *format-name*.

ID *variable(s)*;
: specifies one or more variables that PROC PRINT uses instead of observation numbers to identify observations in the report.

LABEL *variable*='*label*';
: specifies to use labels for column headings. *Variable* names the variable to label, and *label* specifies a string of up to 256 characters, which includes blanks. The *label* must be enclosed in single or double quotation marks.

PAGEBY *variable*;
: causes PROC PRINT to begin a new page when the *variable* that you specify changes value or when any variable that you list before it in the BY statement changes value. You must use a BY statement with the PAGEBY statement.

SUM *variable(s)*;
: identifies the numeric variables to total in the report. You can specify a variable in the SUM statement and omit it in the VAR statement because PROC PRINT will add

the variable to the VAR list. PROC PRINT ignores requests to total the BY and ID variables. In general, when you also use the BY statement, the SUM statement produces subtotals each time the value of a BY variable changes.

SUMBY *variable*;

limits the number of sums that appear in the report. PROC PRINT reports totals only when *variable* changes value or when any variable that is listed before it in the BY statement changes value. You must use a BY statement with the SUMBY statement.

TITLE<*n*> <'*title*'>;

specifies a title. The argument *n* is a number from 1 to 10 that immediately follows the word TITLE, with no intervening blank, and specifies the level of the TITLE. The text of each *title* must be enclosed in single or double quotation marks. The maximum title length that is allowed depends on your operating environment and the value of the LINESIZE= system option. Refer to the SAS documentation for your operating environment for more information.

VAR *variable(s)*;

identifies one or more variables that appear in the report. The variables appear in the order in which you list them in the VAR statement. If you omit the VAR statement, then all the variables appear in the report.

WHERE *where-expression*;

subsets the input data set by identifying certain conditions that each observation must meet before an observation is available for processing. *Where-expression* defines the condition. The condition is a valid arithmetic or logical expression that generally consists of a sequence of operands and operators.

## *PROC SORT Statements*

**PROC SORT** <DATA=*SAS-data-set*>;

   **BY** *variable(s)*;

PROC SORT DATA=*SAS-data-set*;

sorts a SAS data set by the values of variables that you list in the BY statement.

BY *variable(s)*;

specifies one or more variables by which PROC SORT sorts the observations. By default, PROC SORT arranges the data set by the values in ascending order (smallest value to largest).

## *SAS Macro Language*

%LET *macro-variable=value*;

is a macro statement that defines a macro-variable and assigns it a value. The value that you define in the %LET statement is substituted for the macro-variable in output. To use the macro-variable in a program, include an ampersand (&) prefix before it.

SYSDATE9

is an automatic macro variable that contains the date on which a SAS job or session began to execute. SYSDATE9 contains a SAS date value in the DATE9 format (ddmmmyyyy). The date displays a two-digit date, the first three letters of the month name, and a four-digit year. To use it in a program, you include an ampersand (&) prefix before SYSDATE9.

## Learning More

Data Set Indexes
: For information about indexing data sets, see *SAS Data Set Options: Reference*. You do not need to sort data sets before using a BY statement in the PRINT procedure if the data sets have an index for the variable or variables that are specified in the BY statement.

Style Elements and Style Attributes
: For complete information about style elements and style attributes, see *SAS Output Delivery System: User's Guide*.

PROC PRINT
: For complete documentation, see *Base SAS Procedures Guide*.

PROC SORT
: For more information, see Chapter 12, "Working with Grouped or Sorted Observations," on page 183. For complete reference documentation about the SORT procedure, see *Base SAS Procedures Guide*.

SAS formats
: For complete documentation, see *SAS Formats and Informats: Reference*. Formats that are available with SAS software include fractions, hexadecimal values, roman numerals, Social Security numbers, date and time values, and numbers written as words.

SAS macro facility
: For complete reference documentation, see *SAS Macro Language: Reference*.

WHERE statement
: For complete reference documentation, see *SAS Statements: Reference*. For a complete discussion of WHERE processing, see *SAS Language Reference: Concepts*.

## Chapter 28
# Creating Summary Tables with the TABULATE Procedure

Introduction to Creating Summary Tables with the TABULATE Procedure ... **474**
    Purpose ... 474
    Prerequisites ... 474

**Understanding Summary Table Design** ... **474**

**Understanding the Basics of the TABULATE Procedure** ... **476**
    Required Statements for the TABULATE Procedure ... 476
    Begin with the PROC TABULATE Statement ... 477
    Specify Class Variables with the CLASS Statement ... 477
    Specify Analysis Variables with the VAR Statement ... 477
    Define the Table Structure with the TABLE Statement ... 477
    Identifying Missing Values for Class Variables ... 478

**Input File and SAS Data Set for Examples** ... **479**

**Creating Simple Summary Tables** ... **480**
    Creating a Basic One-Dimensional Summary Table ... 480
    Creating a Basic Two-Dimensional Summary Table ... 481
    Creating a Basic Three-Dimensional Summary Table ... 482
    Producing Multiple Tables in a Single PROC TABULATE Step ... 484

**Creating More Sophisticated Summary Tables** ... **485**
    Creating Hierarchical Tables to Report on Subgroups ... 485
    Formatting Output ... 487
    Calculating Descriptive Statistics ... 487
    Reporting on Multiple Statistics ... 488
    Reducing Code and Applying a Single Label to Multiple Elements ... 489
    Getting Summaries for All Variables ... 490
    Defining Labels ... 491
    Using Styles and the Output Delivery System ... 492
    Ordering Class Variables ... 495

**Summary** ... **496**
    Global Statement ... 496
    TABULATE Procedure Statements ... 496

**Learning More** ... **499**

# Introduction to Creating Summary Tables with the TABULATE Procedure

## Purpose

Summary tables display the relationships that exist among the variables in a data set. The variables in the data set form the columns, rows, and pages of summary tables. The data at each intersection of a column and row (that is, each cell) shows a relationship between the variables. The TABULATE procedure enables you to create a variety of summary tables.

In this section, you learn how to do the following:

- produce simple summary tables by using a few basic PROC TABULATE options and statements
- produce enhanced summary tables by summarizing more complex relationships between and across variables, applying formats to variables, and calculating statistics for variables
- add the finishing touches to tables by using labels, by specifying fonts and colors with the Output Delivery System, and by ordering class variables

## Prerequisites

To understand the examples in this section, you should be familiar with the following features and concepts:

- "Locating Procedure Output" on page 617
- "Adding Titles and Footnotes" on page 459

# Understanding Summary Table Design

If you design your summary table in advance, then you can save time and write simpler SAS code to produce the summary table. The basic steps of summary table design and construction are listed next. For a detailed step-by-step example of the design process, see *PROC TABULATE by Example*.

Before designing a summary table, it is important to understand that the summary table produces summary data wherever values for two or more variables intersect. The point of intersection is a cell. When values for two or more variables intersect, the variables are said to be crossed. The process of crossing variables to form intersections is called crosstabulation. Variables in columns, rows, and pages can be crossed to produce summary data. The following program creates a summary table that shows how two variables are crossed. For the input file and SAS Data set, see "Input File and SAS Data Set for Examples" on page 479.

```
proc format;
   value $slrpclr 'Hollingsworth'='cx00BBBB'
                  other         ='cx00DDDD';
```

```
      value $slrpmsk 'Hollingsworth'='value Y'
                     'Garcia'        ='value X'
                     'Jensen'        ='value Z';
      value $quarclr '3'='cx00BBBB'
                     other ='cx00DDDD';
      value $quarmsk '1'='value A'
                     '2'='value B'
                     '3'='value C'
                     '4'='value D';
      value sumclr   222290-222291 ='cx00BBBB'
                     225326-225327 ='cx00BBBB'
                     108900-110000 ='cx2E7371'
                      81000- 82000 ='cx00BBBB'
                      96000- 97000 ='cx00BBBB'
                      59000- 59650 ='cx00BBBB'
                     other ='w';
run;

ods html body='/dept/pub/doc/701misc/authoring/TW6025/sgml/delete.htm';
proc tabulate data=year_sales format=comma10.;
   title 'Crossing Value C with Value Y';
   class SalesRep Quarter/ style=[background=cx00DDDD];
   classlev salesrep / style=[background=$slrpclr. just=right];
   classlev quarter / style=[background=$quarclr.];
   var AmountSold / style=[background=cx00DDDD];
   table SalesRep='Variable 2',
         (AmountSold='Variable 1'*sum=' '*Quarter=' ')*
         [s=[background=sumclr. foreground=sumclr.]] /
         box={style={background=cx00DDDD}};
   format salesrep $slrpmsk. Quarter $quarmsk.;
run;
ods html close;
```

The following summary table displays how two variables are crossed by highlighting a single value for each variable:

*Figure 28.1* Crossing Variables

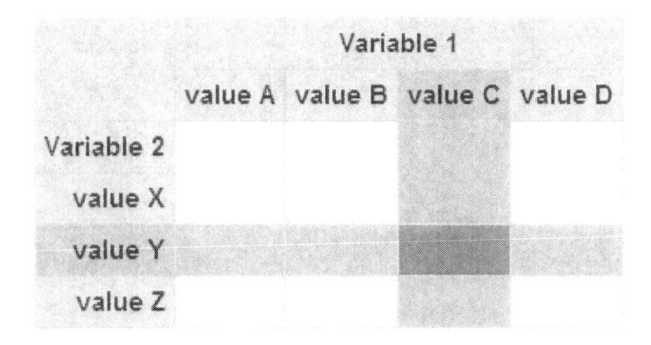

Here are the basic steps for designing and constructing a summary table:

1. Start with a question that you want to answer with a summary table.
2. Identify the variables necessary to answer your question.

- See whether any of the data sets that you are using already use the variables that you identified. If they do not, then you might be able to use the FORMAT procedure to reclassify the variable values in these data sets. This enables you produce the data that you need.

  For example, you can apply a new format to values for a variable MONTH so that they become values for a variable QUARTER. To do this, assign the values representing the first three months to a value for quarter one, values representing the second set of three months to a value for quarter two, and so on.

- If possible, use discrete variables rather than continuous variables for categories or headings. If you must use continuous variables, then it might be helpful to create categories. For example, you can group ages into categories such as ages 15-19, 20-35, 36-55, and 56-higher. This creates four categories rather than a possible 56+ categories. You can use PROC FORMAT to categorize the data.

- Choose formats for the variables and the data that you want to display in your summary table. See whether the data in your data sets is in a format that you can use. You might need to create new formats with PROC FORMAT, or copy the formats of variables from another data set so that the data is formatted in the same way.

3. Review the data for anything that might cause discrepancies in your report.
   - Remove data that does not relate to your needs.
   - Identify missing data.
   - Make sure that the data overall seems to make logical sense.

4. Choose statistics that help answer your question. For a complete list of statistics, see "Statistics Available in PROC TABULATE" in the *Base SAS Procedures Guide*.

5. Decide on the basic structure of the table. Use the variables that you have identified to determine the headings for the columns, rows, and pages. The values of the variables are the subheadings. Statistics are usually represented as subheadings, but are sometimes represented as headings. is an example of a template for a very basic table.

6. Decide on the style of the table. You can use ODS to customize the appearance of your output. For more information about the Output Delivery System, see *SAS Output Delivery System: User's Guide*.

# Understanding the Basics of the TABULATE Procedure

## Required Statements for the TABULATE Procedure

The TABULATE procedure requires three statements, usually in the following order:

1. PROC TABULATE statement
2. CLASS statements or VAR statements or both
3. TABLE statements

Note that there can be multiple CLASS statements, VAR statements, and TABLE statements.

## Begin with the PROC TABULATE Statement

The TABULATE procedure begins with a PROC TABULATE statement. Many options are available with the PROC TABULATE statement. However, most of the examples in this section use only two options, the DATA= option and the FORMAT= option. The PROC TABULATE statement that follows is used for all of the examples in this section:

```
proc tabulate data=year_sales format=comma10.;
```

You can direct PROC TABULATE to use a specific SAS data set with the DATA= option. If you omit the DATA= option in the current job or session, then the TABULATE procedure uses the SAS data set that was created most recently.

You can specify a default format for PROC TABULATE to apply to the value in each cell in the table with the FORMAT= option. You can specify any valid SAS numeric format or user-defined format.

## Specify Class Variables with the CLASS Statement

Use the CLASS statement to specify which variables are class variables. Class variables (that is, classification variables) contain values that are used to form categories. In summary tables, the categories are used as the column, row, and page headings. The categories are crossed to obtain descriptive statistics. See Figure 28.1 on page 475 for an example of crossing categories (variable values).

Class variables can be either character or numeric. The default statistic for class variables is N, which is the frequency or number of observations in the data set for which there are nonmissing variable values.

The following CLASS statement specifies the variables SalesRep and Type as class variables:

```
class SalesRep Type;
```

For important information about how PROC TABULATE behaves when class variables that have missing values are listed in a CLASS statement but are not used in a TABLE statement, see "Identifying Missing Values for Class Variables" on page 478.

## Specify Analysis Variables with the VAR Statement

Use the VAR statement to specify which variables are analysis variables. Analysis variables contain numeric values for which you want to compute statistics. The default statistic for analysis variables is SUM.

The following VAR statement specifies the variable AmountSold as an analysis variable:

```
var AmountSold;
```

## Define the Table Structure with the TABLE Statement

### Syntax of a TABLE Statement

Use the TABLE statement to define the structure of the table that you want PROC TABULATE to produce. A TABLE statement consists of one to three dimension expressions, separated by commas. Dimension expressions define the columns, rows, and pages of a summary table. Options can follow dimension expressions. You must specify at least one TABLE statement, because there is no default table in a PROC TABULATE step. Here are three variations of the syntax for a basic TABLE statement:

```
TABLE column-expression;
TABLE row-expression, column-expression;
TABLE page-expression, row-expression, column-expression;
```

In this syntax

- a column expression is required
- a row expression is optional
- a page expression is optional
- the order of the expressions must be page expression, row expression, and then column expression

Here is an example of a basic TABLE statement with three dimension expressions:

```
table SalesRep, Type, AmountSold;
```

This TABLE statement defines a three-dimensional summary table that places:

- the values of the variable AmountSold in the column dimension
- the values of the variable Type in the row dimension
- the values of the variable SalesRep in the page dimension

### Restrictions on a TABLE Statement

Here are restrictions on the TABLE statement:

- A TABLE statement must have a column dimension.
- Every variable that is used in a dimension expression in a TABLE statement must appear in either a CLASS statement or a VAR statement, but not both.
- All analysis variables must be in the same dimension and cannot be crossed. Therefore, only one dimension of any TABLE statement can contain analysis variables.

### Identifying Missing Values for Class Variables

You can identify missing values for class variables with the MISSING option. By default, if an observation contains a missing value for any class variable, that observation is excluded from all tables. The value is excluded even if the variable does not appear in the TABLE statement for one or more tables. Therefore, it is helpful to run your program at least once with the MISSING option to identify missing values.

The MISSING option creates a separate category in the summary table for missing values. It can be used with the PROC TABULATE statement or the CLASS statement. If you specify the MISSING option in the PROC TABULATE statement, the procedure considers missing values as valid levels for all class variables:

```
proc tabulate data=year_sales format=comma10. missing;
   class SalesRep;
   class Month Quarter;
   var AmountSold;
```

Because the MISSING option is in the PROC TABULATE statement in this example, observations with missing values for SalesRep, Month, or Quarter are displayed in the summary table.

If you specify the MISSING option in a CLASS statement, PROC TABULATE considers missing values as valid levels for the class variable(s) that are specified in that CLASS statement:

```
proc tabulate data=year_sales format=comma10.;
   class SalesRep;
   class Month Quarter / missing;
   var AmountSold;
```

Because the MISSING option is in the second CLASS statement, observations with missing values for Month or Quarter are displayed in the summary table, but observations with a missing value for SalesRep are not displayed.

If you have class variables with missing values in your data set, then you must decide whether the observations with the missing values should be omitted from every table. If the observations should not be omitted, then you can fill in the missing values where appropriate or continue to run the PROC TABULATE step with the MISSING option. For other options for handling missing values, see "Handling Missing Data" in *PROC TABULATE by Example*. For general information about missing values, see "Missing Values" in *SAS Language Reference: Concepts*.

## Input File and SAS Data Set for Examples

The examples in this section use one input file and one SAS data set. For more information, see "The YEAR_SALES Data Set" on page 816.

The input file contains sales records for a company, TruBlend Coffee Makers, that distributes the coffee machines. The file has the following structure:

```
01    1    Hollingsworth  Deluxe      260    49.50
01    1    Garcia         Standard     41    30.97
01    1    Hollingsworth  Deluxe      330    49.50
01    1    Jensen         Standard   1110    30.97
01    1    Garcia         Standard    715    30.97
01    1    Jensen         Deluxe      675    49.50
02    1    Jensen         Standard     45    30.97
02    1    Garcia         Deluxe       10    49.50

...more data lines...

12    4    Hollingsworth  Deluxe      125    49.50
12    4    Jensen         Standard   1254    30.97
12    4    Hollingsworth  Deluxe      175    49.50
```

The input file contains the following data from left to right:

- the month in which a sale was made
- the quarter of the year in which a sale was made
- the name of the sales representative
- the type of coffee maker sold (standard or deluxe)
- the number of units sold
- the price of each unit in US dollars

The SAS data set is named YEAR_SALES. This data set contains all the sales data from the input file and data from a new variable named AmountSold, which is created by multiplying Units by Price.

The following program creates the SAS data set that is used in this section:

```
data year_sales;
   infile 'your-input-file';
   input Month $ Quarter $ SalesRep $14. Type $ Units Price;
   AmountSold = Units * Price;
run;
```

# Creating Simple Summary Tables

## Creating a Basic One-Dimensional Summary Table

The simplest summary table contains multiple columns but only a single row. It is called a one-dimensional summary table because it has only a column dimension. The PROC TABULATE step that follows creates a one-dimensional summary table that answers the question, "How many times did each sales representative make a sale?"

```
options linesize=84 pageno=1 nodate;

proc tabulate data=year_sales format=comma10.;
   title1 'TruBlend Coffee Makers, Inc.';
   title2 'Number of Sales by Each Sales Representative';
   class SalesRep; 1
   table SalesRep; 2
run;
```

The numbered items in the previous program correspond to the following:

1  The variable SalesRep is specified as a class variable in the CLASS statement. A category is created for each value of SalesRep wherever SalesRep is used in a TABLE statement.

2  The variable SalesRep is specified in the column dimension of the TABLE statement. A column is created for each category of SalesRep. Each column shows the number of times (N) that values belonging to the category appear in the data set.

The following summary table displays the results of this program:

**Figure 28.2** Basic One-Dimensional Summary Table

TruBlend Coffee Makers, Inc.
Number of Sales by Each Sales Representative

| SalesRep |||
|---|---|---|
| Garcia | Hollingsworth | Jensen |
| N | N | N |
| 40 | 32 | 38 |

The values 40, 32, and 38 are the frequency with which each sales representative's name (Garcia, Hollingsworth, and Jensen) occurs in the data set. For this data set, each occurrence of the sales representative's name in the data set represents a sale.

### *Creating a Basic Two-Dimensional Summary Table*

The most commonly used form of a summary table has at least one column and multiple rows, and is called a two-dimensional summary table. The PROC TABULATE step that follows creates a two-dimensional summary table that answers the question, "What was the amount that was sold by each sales representative?"

```
options linesize=84 pageno=1 nodate;

proc tabulate data=year_sales format=comma10.;
   title1 'TruBlend Coffee Makers, Inc.';
   title2 'Amount Sold by Each Sales Representative';
   class SalesRep; 1
   var AmountSold; 2
   table SalesRep, 3
         AmountSold; 4 ;
run;
```

The shaded areas in the previous program correspond to the following:

1   The variable SalesRep is specified as a class variable in the CLASS statement. A category is created for each value of SalesRep wherever SalesRep is used in a TABLE statement.

2   The variable AmountSold is specified as an analysis variable in the VAR statement. The values of AmountSold are used to compute statistics wherever AmountSold is used in a TABLE statement.

3   The variable SalesRep is in the row dimension of the TABLE statement. A row is created for each value or category of SalesRep.

4   The variable AmountSold is in the column dimension of the TABLE statement. The default statistic for analysis variables, SUM, is used to summarize the values of AmountSold.

The following summary table displays the results of this program: The variable AmountSold has been crossed with the variable SalesRep to produce each cell of the summary table. The column heading AmountSold includes the subheading SUM. The values that are displayed in the column dimension are sums of the amount sold by each sales representative.

*Figure 28.3* Basic Two-Dimensional Summary Table

```
              TruBlend Coffee Makers, Inc.
          Amount Sold by Each Sales Representative

                              AmountSold

                                 Sum

              SalesRep

              Garcia          512,071

              Hollingsworth   347,246

              Jensen          461,163
```

### Creating a Basic Three-Dimensional Summary Table

Three-dimensional summary tables produce the output on separate pages with rows and columns on each page. The PROC TABULATE step that follows creates a three-dimensional summary table that answers the question, "What was the amount that was sold during each quarter of the year by each sales representative?"

```
options linesize=84 pageno=1 nodate;

proc tabulate data=year_sales format=comma10.;
   title1 'TruBlend Coffee Makers, Inc.';
   title2 'Quarterly Sales by Each Sales Representative';
   class SalesRep Quarter; 1
   var AmountSold; 2
   table SalesRep, 3
         Quarter, 4
         AmountSold; 5
run;
```

The numbered items in the previous program correspond to the following:

1. The variables SalesRep and Quarter are specified as class variables in the CLASS statement. A category is created for each value of SalesRep wherever SalesRep is used in the TABLE statement. Similarly, a category is created for each value of Quarter wherever Quarter is used in a TABLE statement.

2. The variable AmountSold is specified as an analysis variable in the VAR statement. The values of AmountSold are used to compute statistics wherever AmountSold is used in a TABLE statement.

3. The variable SalesRep is used in the page dimension of the TABLE statement. A page is created for each value or category of SalesRep.

4. The variable Quarter is used in the row dimension of the TABLE statement. A row is created for each value or category of Quarter.

5 The variable AmountSold is used in the column dimension of the TABLE statement. The default statistic for analysis variables, SUM, is used to summarize the values of AmountSold.

The following summary table displays the results of this program. This summary table has a separate page for each sales representative. For each sales representative, the amount sold is reported for each quarter. The column heading AmountSold includes the subheading SUM. The values that are displayed in this column indicate the total amount sold in US dollars for each quarter by each sales representative.

*Figure 28.4* Basic Three-Dimensional Summary Table

TruBlend Coffee Makers, Inc.
Quarterly Sales by Each Sales Representative

SalesRep Garcia

| Quarter | AmountSold Sum |
|---|---|
| 1 | 118,020 |
| 2 | 108,860 |
| 3 | 225,326 |
| 4 | 59,865 |

TruBlend Coffee Makers, Inc.
Quarterly Sales by Each Sales Representative

SalesRep Hollingsworth

| Quarter | AmountSold Sum |
|---|---|
| 1 | 59,635 |
| 2 | 96,161 |
| 3 | 109,704 |
| 4 | 81,747 |

TruBlend Coffee Makers, Inc.
Quarterly Sales by Each Sales Representative

SalesRep Jensen

| Quarter | AmountSold Sum |
|---|---|
| 1 | 50,078 |
| 2 | 74,731 |
| 3 | 222,291 |
| 4 | 114,063 |

### Producing Multiple Tables in a Single PROC TABULATE Step

You can produce multiple tables in a single PROC TABULATE step. However, you cannot change how a variable is used or defined in the middle of the step. In other words, the variables in the CLASS or VAR statements are defined only once for all TABLE statements in the PROC TABULATE step. If you need to change how a variable is used or defined for different TABLE statements, then you must place the TABLE statements, and define the variables, in multiple PROC TABULATE steps. The program that follows produces three summary tables during one execution of the TABULATE procedure:

```
options linesize=84 pageno=1 nodate;

proc tabulate data=year_sales format=comma10.;
   title1 'TruBlend Coffee Makers, Inc.';
   title2 'Sales of Deluxe Model Versus Standard Model';
   class SalesRep Type;
   var AmountSold Units;
   table Type; 1
   table Type, Units; 2
   table SalesRep, Type, AmountSold; 3
run;
```

The numbered items in the previous program correspond to the following:

1. The first TABLE statement produces a one-dimensional summary table with the values for the variable Type in the column dimension.

2. The second TABLE statement produces a two-dimensional summary table with the values for the variable Type in the row dimension and the variable Units in the column dimension.

3. The third TABLE statement produces a three-dimensional summary table with:
   - the values for the variable SalesRep in the page dimension
   - the values for the variable Type in the row dimension
   - the variable AmountSold in the column dimension

The following summary table displays the results of this program:

*Figure 28.5   Multiple Tables Produced by a Single PROC TABULATE Step*

**TruBlend Coffee Makers, Inc.**
**Sales of Deluxe Model Versus Standard Model**

|        | Type   |          |
|--------|--------|----------|
|        | Deluxe | Standard |
|        | N      | N        |
|        | 16     | 94       |

**TruBlend Coffee Makers, Inc.**
**Sales of Deluxe Model Versus Standard Model**

|          | Units  |
|----------|--------|
|          | Sum    |
| Type     |        |
| Deluxe   | 2,525  |
| Standard | 38,464 |

**TruBlend Coffee Makers, Inc.**
**Sales of Deluxe Model Versus Standard Model**

SalesRep Garcia

|          | AmountSold |
|----------|------------|
|          | Sum        |
| Type     |            |
| Deluxe   | 46,778     |
| Standard | 465,293    |

**TruBlend Coffee Makers, Inc.**
**Sales of Deluxe Model Versus Standard Model**

SalesRep Hollingsworth

|          | AmountSold |
|----------|------------|
|          | Sum        |
| Type     |            |
| Deluxe   | 37,620     |
| Standard | 309,626    |

**TruBlend Coffee Makers, Inc.**
**Sales of Deluxe Model Versus Standard Model**

SalesRep Jensen

|          | AmountSold |
|----------|------------|
|          | Sum        |
| Type     |            |
| Deluxe   | 40,590     |
| Standard | 420,573    |

# Creating More Sophisticated Summary Tables

## Creating Hierarchical Tables to Report on Subgroups

You can create a hierarchical table to report on subgroups of your data by crossing elements within a dimension. Crossing elements is the operation that combines two or more elements, such as class variables, analysis variables, format modifiers, statistics, or styles. Dimensions are automatically crossed. When you cross variables in a single dimension expression, values for one variable are placed within the values for the other variable in the same dimension. This forms a hierarchy of variables and, therefore, a hierarchical table. The order in which variables are listed when they are crossed determines the order of the headings in the table. In the column dimension, variables are stacked top to bottom; in the row dimension, left to right; and in the page dimension, front to back. You cross elements in a dimension expression by putting an asterisk between them. Note that two analysis variables cannot be crossed. Also, because dimensions are automatically crossed, all analysis variables must occur in one dimension.

The PROC TABULATE step that follows creates a two-dimensional summary table that crosses two variables and that answers the question, "What was the amount sold of each type of coffee maker by each sales representative?"

```
options linesize=84 pageno=1 nodate;

proc tabulate data=year_sales format=comma10.;
   title1 'TruBlend Coffee Makers, Inc.';
   title2 'Amount Sold Per Item by Each Sales Representative';
   class SalesRep Type;
   var AmountSold;
   table SalesRep*Type,
         AmountSold;
run;
```

The expression **SalesRep*Type** in the row dimension uses the asterisk operator to cross the values of the variable SalesRep with the values of the variable Type. Because SalesRep is listed before Type when crossed, and because the elements are crossed in the row dimension, values for Type are listed to the right of values of SalesRep. Values for Type are repeated for each value of SalesRep.

The following summary table displays the results:

*Figure 28.6* Crossing Variables

```
                 TruBlend Coffee Makers, Inc.
           Amount Sold Per Item by Each Sales Representative

                                      AmountSold
                                         Sum
           SalesRep       Type
           Garcia         Deluxe         46,778
                          Standard      465,293
           Hollingsworth  Deluxe         37,620
                          Standard      309,626
           Jensen         Deluxe         40,590
                          Standard      420,573
```

Notice the hierarchy of values that are created when the values for Type are repeated to the right of each value of SalesRep.

## Formatting Output

You can override formats in summary table output by crossing variables with format modifiers. You cross a variable with a format modifier by putting an asterisk between them.

The PROC TABULATE step that follows creates a two-dimensional summary table that crosses a variable with a format modifier and that answers the question, "What was the amount sold of each type of coffee maker by each sales representative?"

```
options linesize=84 pageno=1 nodate;

proc tabulate data=year_sales format=comma10.;
   title1 'TruBlend Coffee Makers, Inc.';
   title2 'Amount Sold Per Item by Each Sales Representative';
   class SalesRep Type;
   var AmountSold;
   table SalesRep*Type,
         AmountSold*f=dollar16.2;
run;
```

The expression **AmountSold*f=dollar16.2** in the column dimension uses the asterisk operator to cross the values of the variable **AmountSold** with the SAS format modifier **f=dollar16.2**. The values for AmountSold are now displayed using the DOLLAR16.2 format. The DOLLAR16.2 format is better suited for dollar figures than the COMMA10. format, which is specified as the default in the PROC TABULATE statement.

The following summary table displays the results:

**Figure 28.7** Crossing Variables with Format Modifiers

TruBlend Coffee Makers, Inc.
Amount Sold Per Item by Each Sales Representative

| SalesRep | Type | AmountSold Sum |
|---|---|---|
| Garcia | Deluxe | $46,777.50 |
|  | Standard | $465,293.28 |
| Hollingsworth | Deluxe | $37,620.00 |
|  | Standard | $309,626.10 |
| Jensen | Deluxe | $40,590.00 |
|  | Standard | $420,572.60 |

## Calculating Descriptive Statistics

You can request descriptive statistics for a variable by crossing that variable with the appropriate statistic keyword. Crossing either a class variable or an analysis variable with a statistic tells PROC TABULATE what type of calculations to perform. Note that two statistics cannot be crossed. Also, because dimensions are automatically crossed, all statistics must occur in one dimension.

The default statistic crossed with a class variable is the N statistic or frequency. Class variables can be crossed only with frequency and percent frequency statistics. The default statistic crossed with an analysis variable is the SUM statistic. Analysis variables can be crossed with any of the many descriptive statistics that are available with PROC TABULATE including commonly used statistics like MIN, MAX, MEAN, STD, and MEDIAN. For a complete list of statistics available for use with analysis variables, see "Statistics Available in PROC TABULATE" in the *Base SAS Procedures Guide*.

The PROC TABULATE step that follows creates a two-dimensional summary table that crosses elements with a statistic and that answers the question, "What was the average amount per sale of each type of coffee maker by each sales representative?"

```
options linesize=84 pageno=1 nodate;

proc tabulate data=year_sales format=comma10.;
   title1 'TruBlend Coffee Makers, Inc.';
   title2 'Average Amount Sold Per Item by Each Sales Representative';
   class SalesRep Type;
   var AmountSold;
   table SalesRep*Type,
         AmountSold*mean*f=dollar16.2;
run;
```

In this program, the column dimension crosses the variable `AmountSold` with the statistic `mean` and with the format modifier `f=dollar16.2`. The MEAN statistic provides the arithmetic mean for AmountSold.

The following summary table displays the results:

*Figure 28.8* Crossing a Variable with a Statistic

TruBlend Coffee Makers, Inc.
Average Amount Sold Per Item by Each Sales Representative

|  |  | AmountSold |
|---|---|---|
|  |  | Mean |
| SalesRep | Type |  |
| Garcia | Deluxe | $11,694.38 |
|  | Standard | $12,924.81 |
| Hollingsworth | Deluxe | $4,702.50 |
|  | Standard | $12,901.09 |
| Jensen | Deluxe | $10,147.50 |
|  | Standard | $12,369.78 |

## Reporting on Multiple Statistics

You can create summary tables that report on two or more statistics by concatenating variables. *Concatenating* is the operation that joins the information of two or more elements, such as class variables, analysis variables, or statistics, by placing the output of the second and subsequent elements immediately after the output of the first element.

You concatenate elements in a dimension expression by putting a blank space between them.

The PROC TABULATE step that follows creates a two-dimensional summary table that uses concatenation and that answers the question, "How many sales were made, and what was the total sales figure for each type of coffee maker sold by each sales representative?"

```
options linesize=84 pageno=1 nodate;

proc tabulate data=year_sales format=comma10.;
   title1 'TruBlend Coffee Makers, Inc.';
   title2 'Sales Summary by Representative and Product';
   class SalesRep Type;
   var AmountSold;
   table SalesRep*Type,
         AmountSold*n AmountSold*f=dollar16.2;
run;
```

In this program, because the expressions **AmountSold*n** and **AmountSold*f=dollar16.2** in the column dimension are separated by a blank space, their output is concatenated.

The following summary table displays the results. In this summary table the frequency (N) of AmountSold is shown in the same table as the SUM of AmountSold.

*Figure 28.9  Concatenating Variables*

TruBlend Coffee Makers, Inc.
Sales Summary by Representative and Product

|  |  | AmountSold N | AmountSold Sum |
|---|---|---|---|
| SalesRep | Type |  |  |
| Garcia | Deluxe | 4 | $46,777.50 |
|  | Standard | 36 | $465,293.28 |
| Hollingsworth | Deluxe | 8 | $37,620.00 |
|  | Standard | 24 | $309,626.10 |
| Jensen | Deluxe | 4 | $40,590.00 |
|  | Standard | 34 | $420,572.60 |

## Reducing Code and Applying a Single Label to Multiple Elements

You can use parentheses to group concatenated elements (variables, formats, statistics, and so on) that are concatenated or crossed with a common element. This can reduce the amount of code used and can change how labels are displayed. The PROC TABULATE step that follows uses parentheses to group elements that are crossed with AmountSold and answers the question, "How many sales were made, and what was the total sales figure for each type of coffee maker sold by each sales representative?"

```
options linesize=84 pageno=1 nodate;
```

```
proc tabulate data=year_sales format=comma10.;
   title1 'TruBlend Coffee Makers, Inc.';
   title2 'Sales Summary by Representative and Product';
   class SalesRep Type;
   var AmountSold;
   table SalesRep*Type,
         AmountSold*(n sum*f=dollar16.2);
run;
```

In this program, `AmountSold*(n sum*f=dollar16.2)` takes the place of `AmountSold*n AmountSold*f=dollar16.2`. Notice the default statistic SUM from `AmountSold*f=dollar16.2` must now be included in the expression. This is because the format modifier must be crossed with a variable or a statistic. It cannot be in the expression by itself.

The following summary table displays the results:

*Figure 28.10* Using Parentheses to Group Elements

TruBlend Coffee Makers, Inc.
Sales Summary by Representative and Product

| SalesRep | Type | N | Sum |
|---|---|---|---|
| Garcia | Deluxe | 4 | $46,777.50 |
|  | Standard | 36 | $465,293.28 |
| Hollingsworth | Deluxe | 8 | $37,620.00 |
|  | Standard | 24 | $309,626.10 |
| Jensen | Deluxe | 4 | $40,590.00 |
|  | Standard | 34 | $420,572.60 |

(AmountSold spans N and Sum columns)

Note that the label, AmountSold, spans multiple columns rather than appearing twice in the summary table, as it does in Figure 28.9 on page 489.

### Getting Summaries for All Variables

You can summarize all of the class variables in a dimension with the universal class variable ALL. ALL can be concatenated with each of the three dimensions of the TABLE statement and within groups of elements delimited by parentheses. The PROC TABULATE step that follows creates a two-dimensional summary table with the universal class variable ALL, and answers the question, "For each sales representative and for all of the sales representatives as a group, how many sales were made, what was the average amount per sale, and what was the amount sold?"

```
options linesize=84 pageno=1 nodate;

proc tabulate data=year_sales format=comma10.;
   title1 'TruBlend Coffee Makers, Inc.';
   title2 'Sales Report';
```

```
       class SalesRep Type;
       var AmountSold;
       table SalesRep*Type all,
             AmountSold*(n (mean sum)*f=dollar16.2);
run;
```

In this program, the TABLE statement now includes the universal class variable ALL in the row dimension. SalesRep and Type are summarized.

The following summary table displays the results. This summary table reports the frequency (N), the MEAN, and the SUM of AmountSold for each category of SalesRep and Type. This data has been summarized for all categories of SalesRep and Type in the row labeled All.

**Figure 28.11**  *Crossing with the Universal Class Variable ALL*

TruBlend Coffee Makers, Inc.
Sales Report

|  |  | \multicolumn{3}{c}{AmountSold} |
|---|---|---|---|---|
|  |  | N | Mean | Sum |
| SalesRep | Type |  |  |  |
| Garcia | Deluxe | 4 | $11,694.38 | $46,777.50 |
|  | Standard | 36 | $12,924.81 | $465,293.28 |
| Hollingsworth | Deluxe | 8 | $4,702.50 | $37,620.00 |
|  | Standard | 24 | $12,901.09 | $309,626.10 |
| Jensen | Deluxe | 4 | $10,147.50 | $40,590.00 |
|  | Standard | 34 | $12,369.78 | $420,572.60 |
| All |  | 110 | $12,004.36 | $1,320,479.48 |

## *Defining Labels*

You can add your own labels to a summary table or remove headings from a summary table by assigning labels to variables in the TABLE statement. Simply follow the variable with an equal sign (=) followed by either the desired label or by a blank space in quotation marks. A blank space in quotation marks removes the heading from the summary table. The PROC TABULATE step that follows creates a two-dimensional summary table that uses labels in the TABLE statement and that answers the question, "What is the percent of total sales and average amount sold by each sales representative of each type of coffee maker and all coffee makers?"

```
options linesize=84 pageno=1 nodate;

proc tabulate data=year_sales format=comma10.;
   title1 'TruBlend Coffee Makers, Inc.';
   title2 'Sales Performance';
   class SalesRep Type;
   var AmountSold;
   table SalesRep='Sales Representative' ❶ *
         (Type='Type of Coffee Maker' ❶ all) all,
         AmountSold=' ' ❹ *
```

```
              (N='Sales' 2
              SUM='Amount' 2 *f=dollar16.2
              colpctsum='% Sales' 3
              mean='Average Sale' 2 *f=dollar16.2);
run;
```

The numbered items in the previous program correspond to the following:

1  The variables SalesRep and Type are assigned labels.

2  The frequency statistic N, the statistic SUM, and the statistic MEAN are assigned labels.

3  The statistic COLPCTSUM is used to calculate the percentage of the value in a single table cell in relation to the total of the values in the column and is assigned the label `% Sales'.

4  The variable AmountSold is assigned a blank label. As a result, the heading for AmountSold does not appear in the summary table.

The following summary table displays the results. In this table, no heading for the variable AmountSold is displayed. The labels `Sales', `Amount', `% Sales', and `Average Sale' replace the frequency (N), SUM, COLPCTSUM, and MEAN respectively. Labels replace the variables SalesRep and Type.

**Figure 28.12**  *Using Labels to Customize Summary Tables*

TruBlend Coffee Makers, Inc.
Sales Performance

|                      |                     | Sales | Amount        | % Sales | Average Sale |
|----------------------|---------------------|-------|---------------|---------|--------------|
| Sales Representative | Type of Coffee Maker |       |               |         |              |
| Garcia               | Deluxe              | 4     | $46,777.50    | 4       | $11,694.38   |
|                      | Standard            | 36    | $465,293.28   | 35      | $12,924.81   |
|                      | All                 | 40    | $512,070.78   | 39      | $12,801.77   |
| Hollingsworth        | Type of Coffee Maker |       |               |         |              |
|                      | Deluxe              | 8     | $37,620.00    | 3       | $4,702.50    |
|                      | Standard            | 24    | $309,626.10   | 23      | $12,901.09   |
|                      | All                 | 32    | $347,246.10   | 26      | $10,851.44   |
| Jensen               | Type of Coffee Maker |       |               |         |              |
|                      | Deluxe              | 4     | $40,590.00    | 3       | $10,147.50   |
|                      | Standard            | 34    | $420,572.60   | 32      | $12,369.78   |
|                      | All                 | 38    | $461,162.60   | 35      | $12,135.86   |
| All                  |                     | 110   | $1,320,479.48 | 100     | $12,004.36   |

## Using Styles and the Output Delivery System

If you use the Output Delivery System to create output from PROC TABULATE, for any destination other than Listing or Output destinations, you can do the following:

- Set certain style elements (such as font style, font weight, and color) that the procedure uses for various parts of the table.

*Creating More Sophisticated Summary Tables* **493**

- Specify style elements for the labels for variables by adding the option to the CLASS statement.
- Specify style elements for cells in the summary table by crossing the STYLE= option with an element of a dimension expression.

When it is used in a dimension expression, the STYLE= option must be enclosed within square brackets ([ and ]) or braces ({ and }). The PROC TABULATE step that follows creates a two-dimensional summary table that uses the STYLE= option in a CLASS statement and in the TABLE statement and that answers the question, "What is the percent of total sales and average amount sold by each sales representative of each type of coffee maker and all coffee makers?"

```
options linesize=84 pageno=1 nodate;

ods html file='summary-table.htm'; ❶
ods pdf file='summary-table.pdf'; ❷

proc tabulate data=year_sales format=comma10.;
    title1 'TruBlend Coffee Makers, Inc.';
    title2 'Sales Performance';
    class SalesRep;
    class Type / style=[font_style=italic] ❸;
    var AmountSold;
    table SalesRep='Sales Representative'*(Type='Type of Coffee Maker'
        all*[style=[background=yellow font_weight=bold]] ❹)
        all*[style=[font_weight=bold]] ❺,
        AmountSold=' '*(colpctsum='% Sales' mean='Average Sale'*
        f=dollar16.2);
run;

ods html close; ❻
ods pdf close; ❼
```

The numbered items in the previous program correspond to the following:

1. The HTML destination is open by default. You can use the ODS HTML to specify the name of your html file. The FILE= opion identifies the file that contains the HTML output. Some browsers require an extension of HTM or HTML on the filename.

2. The ODS PDF statement opens the PDF destination and creates PDF output. The FILE= option identifies the file that contains the PDF output.

3. The STYLE= option is specified in the second CLASS statement, which sets the font style of the label for Type to italic. The label for SalesRep is not affected by the STYLE= option because it is in a separate CLASS statement.

4. The universal class variable ALL is crossed with the STYLE= option, which sets the background for the table cells to yellow and the font weight for these cells to bold.

5. The universal class variable ALL is crossed with the STYLE= option, which sets the font weight for the table cells to bold.

6. The ODS HTML CLOSE statement closes the HTML destination and all of the files that are associated with it.

7. The ODS PDF CLOSE statement closes the PDF destination. You must close the PDF destination before you view the PDF output.

The following summary table displays the HTML results:

**Figure 28.13** *Using Style Attributes and the ODS HTML Statement*

TruBlend Coffee Makers, Inc.
Sales Performance

| Sales Representative | Type of Coffee Maker | % Sales | Average Sale |
|---|---|---|---|
| Garcia | Deluxe | 4 | $11,694.38 |
|  | Standard | 35 | $12,924.81 |
|  | All | **39** | **$12,801.77** |
| Hollingsworth | *Type of Coffee Maker* |  |  |
|  | Deluxe | 3 | $4,702.50 |
|  | Standard | 23 | $12,901.09 |
|  | All | **26** | **$10,851.44** |
| Jensen | *Type of Coffee Maker* |  |  |
|  | Deluxe | 3 | $10,147.50 |
|  | Standard | 32 | $12,369.78 |
|  | All | **35** | **$12,135.86** |
| All |  | **100** | **$12,004.36** |

This summary table shows the effects of the three uses of the STYLE= option with the ODS HTML statement in the previous SAS program:

- The repeated label, Type of Coffee Maker, is in italics.
- The subtotals for each value of sales representative are highlighted in a lighter color (yellow) and are bold.
- The totals for all sales representatives are bold.

The following summary table displays the PDF results:

**Figure 28.14** *Using Style Attributes and the ODS PDF Statement*

[Screenshot of a PDF viewer displaying a summary table titled "TruBlend Coffee Makers, Inc. Sales Performance" with columns for Sales Representative, Type of Coffee Maker, % Sales, and Average Sale. Data shown:

- Garcia: Deluxe 4 $11,694.38; Standard 35 $12,924.81; All 39 $12,801.77
- Hollingsworth: Deluxe 3 $4,702.50; Standard 23 $12,901.09; All 26 $10,851.44
- Jensen: Deluxe 3 $10,147.50; Standard 32 $12,369.78; All 35 $12,135.86
- All: 100 $12,004.36

Bookmarks panel shows: The Tabulate Procedure > Cross-tabular summary report > Table 1]

This summary table shows the effects of the three uses of the STYLE= option with the ODS PDF statement in the previous SAS program:

- The repeated label, Type of Coffee Maker, is in italics.
- The subtotals for each value of sales representative are highlighted and are bold.
- The totals for all sales representatives are bold.

## Ordering Class Variables

You can control the order in which class variable values and their headings are displayed in a summary table with the ORDER= option. You can use the ORDER= option with the PROC TABULATE statement and with individual CLASS statements. The syntax is `ORDER=sort-order`. The four possible sort orders (DATA, FORMATTED, FREQ, and UNFORMATTED) are defined in "Summary" on page 496. The PROC TABULATE step that follows creates a two-dimensional summary table that uses the ORDER= option with the PROC TABULATE statement to order all class variables by frequency, and that answers the question, "Which quarter produced the greatest number of sales, and which sales representative made the most sales overall?"

```
options linesize=84 pageno=1 nodate;

proc tabulate data=year_sales format=comma10. order=freq;
   title1 'TruBlend Coffee Makers, Inc.';
```

```
            title2 'Quarterly Sales and Representative Sales by Frequency';
            class SalesRep Quarter;
            table SalesRep all,
                  Quarter all;
run;
```

The following summary table displays the results of this program. The order of the values of the class variable Quarter shows that most sales occurred in quarter 3 followed by quarters 1, 2, and then 4. The order of the values of the class variable SalesRep shows that Garcia made the most sales overall, followed by Jensen and then Hollingsworth. The universal class variable ALL is included in both dimensions of this example to show the frequency data that SAS used to order the data when creating the summary table.

**Figure 28.15** *Ordering Class Variables*

**TruBlend Coffee Makers, Inc.**
**Quarterly Sales and Representative Sales by Frequency**

|  | Quarter |  |  |  | All |
|---|---|---|---|---|---|
|  | 3 | 1 | 2 | 4 |  |
|  | N | N | N | N | N |
| SalesRep |  |  |  |  |  |
| Garcia | 21 | 8 | 6 | 5 | 40 |
| Jensen | 21 | 5 | 6 | 6 | 38 |
| Hollingsworth | 15 | 5 | 6 | 6 | 32 |
| All | 57 | 18 | 18 | 17 | 110 |

# Summary

## Global Statement

TITLE<n> <'title'>;
: specifies a title. The argument *n* is a number from 1 to 10 that immediately follows the word TITLE, with no intervening blank, and specifies the level of the TITLE. The text of each *title* can be up to 132 characters long (256 characters long in some operating environments) and must be enclosed in single or double quotation marks.

## TABULATE Procedure Statements

**PROC TABULATE** <*option(s)*>;

**CLASS** *variable(s)</option(s)>*;

**VAR** *analysis-variable(s)*;

**TABLE** <<*page-expression,*> *row-expression,*> *column-expression*;

PROC TABULATE <option(s)>;
   starts the procedure.

   You can specify the following *options* in the PROC TABULATE statement:

   DATA=SAS-data-set
      specifies the *SAS-data-set* to be used by PROC TABULATE. If you omit the DATA= option, then the TABULATE procedure uses the SAS data set that was created most recently in the current job or session.

   FORMAT=format-name
      specifies a default format for formatting the value in each cell in the table. You can specify any valid SAS numeric format or user-defined format.

   MISSING
      considers missing values as valid values to create the combinations of class variables. A heading for each missing value appears in the table.

   ORDER=DATA | FORMATTED | FREQ | UNFORMATTED
      specifies the sort order that is used to create the unique combinations of the values of the class variables, which form the headings of the table. A brief description of each sort order follows:

   DATA
      orders values according to their order in the input data set.

   FORMATTED
      orders values by their ascending formatted values. This order depends on your operating environment.

   FREQ
      orders values by descending frequency count.

   ORDER=
      used in a CLASS statement overrides ORDER= used in the PROC TABULATE statement.

   UNFORMATTED
      orders values by their unformatted values, which yields the same order as PROC SORT. This order depends on your operating environment. This sort sequence is particularly useful for displaying dates chronologically.

CLASS variable(s)/option(s);
   identifies class variables for the table. Class variables determine the categories that PROC TABULATE uses to calculate statistics.

   MISSING
      considers missing values as valid values to create the combinations of class variables. A heading for each missing value appears in the table. If MISSING should apply only to a subset of the class variables, then specify MISSING in a separate CLASS statement with the subset of the class variables.

   ORDER=DATA | FORMATTED | FREQ | UNFORMATTED
      specifies the sort order used to create the unique combinations of the values of the class variables, which form the headings of the table. If ORDER= should apply only to a subset of the class variables, then specify ORDER= in a separate CLASS statement with the subset of the class variables. In this way, a separate sort order can be specified for each class variable. A brief description of each sort order follows:

   DATA
      orders values according to their order in the input data set.

FORMATTED
: orders values by their ascending formatted values. This order depends on your operating environment.

FREQ
: orders values by descending frequency count.

ORDER=
: used in a CLASS statement overrides ORDER= used in the PROC TABULATE statement.

UNFORMATTED
: orders values by their unformatted values, which yields the same order as PROC SORT. This order depends on your operating environment. This sort sequence is particularly useful for displaying dates chronologically.

VAR analysis-variable(s);
: identifies analysis variables for the table. Analysis variables contain values for which you want to compute statistics.

TABLE <<page-expression, >row-expression,> column-expression;
: defines the table that you want PROC TABULATE to produce. You must specify at least one TABLE statement. In the TABLE statement you specify page-expressions, row-expressions, and column-expressions, all of which are constructed in the same way and are referred to collectively as dimension expressions. Use commas to separate dimension expressions from one another. You define relationships among variables, statistics, and other elements within a dimension by combining them with one or more operators. Operators are symbols that tell PROC TABULATE what actions to perform on the variables, statistics, and other elements. The table that follows lists the common operators and the actions that they symbolize:

*Table 28.1* TABLE Statement Operators

| Operator | Action |
| --- | --- |
| , comma | Separates dimensions of the table |
| * asterisk | Crosses elements within a dimension |
| blank space | Concatenates elements within a dimension |
| = equal | Overrides default cell format or assigns label to an element |
| ( ) parentheses | Groups elements and associates an operator with each concatenated element in the group |
| [ ] square brackets | Groups the STYLE= option for crossing, and groups style attribute specifications within the STYLE= option |
| { } braces | Groups the STYLE= option for crossing, and groups style attribute specifications within the STYLE= option |

# Learning More

Locating procedure output
> See Chapter 33, "Understanding and Customizing SAS Output: The Basics," on page 613.

Missing values
> For a discussion about missing values, see *SAS Language Reference: Concepts*. Information about handling missing values is also in *PROC TABULATE by Example*.

ODS
> For complete documentation on how to use the Output Delivery System, see *SAS Output Delivery System: User's Guide*.

PROC TABULATE
> See the TABULATE procedure in the *Base SAS Procedures Guide*.
>
> For a detailed discussion and comprehensive examples of the TABULATE procedure, see *PROC TABULATE by Example*.

SAS formats
> See *SAS Formats and Informats: Reference*. Many formats are available with SAS, such as fractions, hexadecimal values, roman numerals, Social Security numbers, date and time values, and numbers written as words.

Statistics
> For a list of the statistics available in the TABULATE procedure, see the discussion of concepts in the TABULATE procedure in the *Base SAS Procedures Guide*. For more information about the listed statistics, see the discussion of elementary statistics in the appendix of the *Base SAS Procedures Guide*.

Style attributes
> For information about style attributes that can be set for a style element by using the Output Delivery System, see *Base SAS Procedures Guide* or *SAS Output Delivery System: User's Guide*.

Summary tables
> For additional examples of how to produce a variety of summary tables, see *SAS Guide to Report Writing Examples*.
>
> For a discussion of how to use the REPORT procedure to create summary tables, see the REPORT procedure in *Base SAS Procedures Guide*.

Tabular reports
> For interactive online examples and discussion, see lessons related to creating tabular reports in SAS Online Tutor for Version 8: SAS Programming.

Title statement
> See Chapter 27, "Producing Detail Reports with the PRINT Procedure," on page 433.

# Chapter 29
# Creating Detail and Summary Reports with the REPORT Procedure

**Introduction to Creating Detail and Summary Reports with the REPORT Procedure** .................................................. **501**
  Purpose .................................................. 501
  Prerequisites .................................................. 502

**Understanding How to Construct a Report** .................................................. **502**
  Using the Report Writing Tools .................................................. 502
  Types of Reports .................................................. 502
  Laying Out a Report .................................................. 503

**Input File and SAS Data Set for Examples** .................................................. **504**

**Creating Simple Reports** .................................................. **505**
  Displaying All the Variables .................................................. 505
  Specifying and Ordering the Columns .................................................. 507
  Ordering the Rows .................................................. 508
  Consolidating Several Observations into a Single Row .................................................. 510
  Changing the Default Order of the Rows .................................................. 511

**Creating More Sophisticated Reports** .................................................. **514**
  Adjusting the Column Layout .................................................. 514
  Customizing Column Headings .................................................. 515
  Specifying Formats .................................................. 517
  Using Variable Values as Column Headings .................................................. 518
  Summarizing Groups of Observations .................................................. 520

**Summary** .................................................. **523**
  PROC REPORT Statements .................................................. 523

**Learning More** .................................................. **527**

## Introduction to Creating Detail and Summary Reports with the REPORT Procedure

### Purpose

SAS provides a variety of report writing tools that produce detail and summary reports. The reports enable you to communicate information about your data in an organized, concise manner. The REPORT procedure enables you to create detail and summary reports in a single report writing tool.

**502** Chapter 29 • *Creating Detail and Summary Reports with the REPORT Procedure*

In this section, you learn how to use PROC REPORT to do the following:

- produce simple detail reports
- produce simple summary reports
- produce enhanced reports by adding additional statements that order and group observations, sum columns, and compute overall totals
- customize the appearance of reports by adding column spacing, column labels, line separators, and formats

## Prerequisites

To understand the examples in this section, you should be familiar with the following features and concepts:

- data set options
- the TITLE statement
- the LABEL statement
- WHERE processing
- creating and assigning SAS formats

# Understanding How to Construct a Report

## Using the Report Writing Tools

The REPORT procedure combines the features of PROC MEANS, PROC PRINT, and PROC TABULATE along with features of the DATA step report writing into a powerful report writing tool. PROC REPORT enables you to do the following:

- Create customized, presentation-quality reports.
- Develop and store report definitions that control the structure and layout.
- View previously defined reports.
- Generate multiple reports from one report definition.

There are three different ways that you can use PROC REPORT to construct reports:

- in a nonwindowing environment where you use PROC REPORT to submit a series of statements. This is the default environment.
- in a windowing environment with a prompting facility.
- in a windowing environment without a prompting facility.

The windowing environment requires minimal SAS programming skills and allows immediate, visual feedback as you develop the report. This section explains how you use the nonwindowing environment to create summary and detail reports.

## Types of Reports

The REPORT procedure enables you to construct two types of reports:

detail report
: contains one row for every observation that is selected for the report. For more information, see Output 29.1 on page 506. Each of these rows is a detail row.

summary report
: consolidates data so that each row represents multiple observations. For more information, see Output 29.5 on page 511. Each of these rows is also called a detail row.

Both detail and summary reports can contain summary lines as well as detail rows. A summary line summarizes numerical data for a set of detail rows or for all detail rows. You can use PROC REPORT to provide both default summaries and customized summaries.

## Laying Out a Report

### Establishing the Layout

If you first decide on the layout of the report, then creating the report is easier. You need to determine the following:

- which columns to display in the report
- the order of the columns and rows
- how to label the rows and columns
- which statistics to display
- whether to display a column for each value of a particular variable
- whether to display a row for every observation, or to consolidate multiple observations in a single row

Once you establish the layout of the report, use the COLUMN statement and DEFINE statement in the PROC REPORT step to construct the layout.

### Constructing the Layout

The COLUMN statement lists the report items to include as columns of the report, describes the arrangement of the columns, and defines headers that span multiple columns. A report item is a data set variable, a calculated statistic, or a variable that you compute based on other items in the report.

The DEFINE statement defines the characteristics of an item in the report. These characteristics include how PROC REPORT uses an item in the report, the text of the column heading, and the format to display the values.

You control much of a report's layout by the usages that you specify for variables in the DEFINE statements. The types of variable usages are:

ACROSS
: creates a column for each value of an ACROSS variable.

ANALYSIS
: computes a statistic from a numeric variable for all the observations represented by a cell of the report. The value of the variable depends on where it appears in the report. By default, PROC REPORT treats all numeric variables as ANALYSIS variables and computes the sum.

COMPUTED
: computes a report item from variables that you define for the report. They are not in the input data set, and PROC REPORT does not add them to the input data set.

DISPLAY
> displays a row for every observation in the input data set. By default, PROC REPORT treats all character variables as DISPLAY variables.

GROUP
> consolidates into one row all of the observations from the data set that have a unique combination of the formatted values for all GROUP variables.

ORDER
> specifies to order the rows for every observation in the input data set according to the ascending, formatted values of the ORDER variable.

The position and usage of each variable in the report determine the report's structure and content. For example, PROC REPORT orders the detail rows of the report according to the values of ORDER and GROUP variables (from left to right). Similarly, PROC REPORT orders columns for an ACROSS variable from top to bottom, according to the values of the variable. For a complete discussion of how PROC REPORT determines the layout of a report, see the *Base SAS Procedures Guide*.

## Input File and SAS Data Set for Examples

The examples in this section use one input file and one SAS data set. For more information, see "The YEAR_SALES Data Set" on page 816.

The input file contains sales records for a company, TruBlend Coffee Makers, that distributes the coffee machines. The file has the following structure:

```
01    1    Hollingsworth  Deluxe      260    49.50
01    1    Garcia         Standard     41    30.97
01    1    Hollingsworth  Deluxe      330    49.50
01    1    Jensen         Standard   1110    30.97
01    1    Garcia         Standard    715    30.97
01    1    Jensen         Deluxe      675    49.50
02    1    Jensen         Standard     45    30.97
02    1    Garcia         Deluxe       10    49.50

...more data lines...

12    4    Hollingsworth  Deluxe      125    49.50
12    4    Jensen         Standard   1254    30.97
12    4    Hollingsworth  Deluxe      175    49.50
```

The input file contains the following values from left to right:

- the month in which a sale was made
- the quarter of the year in which a sale was made
- the name of the sales representative
- the type of coffee maker sold (standard or deluxe)
- the number of units sold
- the price of each unit in US dollars

The SAS data set is named YEAR_SALES. This data set contains all the sales data from the input file and a new variable named AmountSold, which is created by multiplying Units by Price.

The following program creates the SAS data set that this section uses:

```
data year_sales;
   infile 'your-input-file';
   input Month $ Quarter $ SalesRep $14. Type $ Units Price;
   AmountSold = Units * Price;
run;
```

## Creating Simple Reports

### Displaying All the Variables

By default, PROC REPORT uses all of the variables in the data set. The layout of the report depends on the type of variables in the data set. If the data set contains any character variables, then PROC REPORT generates a simple detail report that lists the values of all the variables and the observations in the data set. If the data set contains only numeric variables, then PROC REPORT sums the value of each variable over all observations in the data set and produces a one-line summary of the sums. To produce a detail report for a data set with only numeric values, you have to define the columns in the report.

By default, PROC REPORT sends your results to the SAS procedure output. The NOWINDOWS (NOWD) option does not have to be specified. To request that PROC REPORT open the REPORT window, specify the WINDOWS option. The REPORT window enables you to modify a report repeatedly and see the modifications immediately.

The following PROC REPORT step creates the default detail report for the first quarter sales:

```
proc report data=year_sales;
   where quarter='1';
   title1 'TruBlend Coffee Makers, Inc.';
   title2 'First Quarter Sales Report';
run;
```

The WHERE statement specifies a condition that SAS uses to select observations from the YEAR_SALES data set. Before PROC REPORT builds the report, SAS selectively processes observations so that the report contains only data for the observations from the first quarter. For more information about WHERE processing, see "Selecting Observations" on page 444.

The following detail report shows all the variable values for those observations in YEAR_SALES that contains first quarter sales data:

*Output 29.1*  *The Default Report When the Data Set Contains Character Values*

TruBlend Coffee Makers, Inc. 2
First Quarter Sales Report

1

| Month | Quarter | SalesRep | Type | Units | Price | AmountSold |
|---|---|---|---|---|---|---|
| 01 | 1 | Hollingsworth | Deluxe | 260 | 49.5 | 12870 |
| 01 | 1 | Garcia | Standard | 41 | 30.97 | 1269.77 |
| 01 | 1 | Hollingsworth | Standard | 330 | 30.97 | 10220.1 |
| 01 | 1 | Jensen | Standard | 110 | 30.97 | 3406.7 |
| 01 | 1 | Garcia | Deluxe | 715 | 49.5 | 35392.5 |
| 01 | 1 | Jensen | Standard | 675 | 30.97 | 20904.75 |
| 02 | 1 | Garcia | Standard | 2045 | 30.97 | 63333.65 |
| 02 | 1 | Garcia | Deluxe | 10 | 49.5 | 495 |
| 02 | 1 | Garcia | Standard | 40 | 30.97 | 1238.8 |
| 02 | 1 | Hollingsworth | Standard | 1030 | 30.97 | 31899.1 |
| 02 | 1 | Jensen | Standard | 153 | 30.97 | 4738.41 |
| 02 | 1 | Garcia | Standard | 98 | 30.97 | 3035.06 |
| 03 | 1 | Hollingsworth | Standard | 125 | 30.97 | 3871.25 |
| 03 | 1 | Jensen | Standard | 154 | 30.97 | 4769.38 |
| 03 | 1 | Garcia | Standard | 118 | 30.97 | 3654.46 |
| 03 | 1 | Hollingsworth | Standard | 25 | 30.97 | 774.25 |
| 03 | 1 | Jensen | Standard | 525 | 30.97 | 16259.25 |
| 03 | 1 | Garcia | Standard | 310 | 30.97 | 9600.7 |

The following list corresponds to the numbered items in the preceding report:

1 The order of the columns corresponds to the position of the variables in the data set.

2 The top of the report has a title, produced by the TITLE statement.

The following PROC REPORT step produces the default summary report when the YEAR_SALES data set contains only numeric values:

```
proc report data=year_sales (keep=Units AmountSold);
   title1 'TruBlend Coffee Makers, Inc.';
   title2 'Total Yearly Sales';
run;
```

The KEEP= data set option specifies to process only the numeric variables Units and AmountSold. PROC REPORT uses these variables to create the report.

The following report displays a one-line summary for the two numeric variables:

*Output 29.2* *The Default Report When the Data Set Contains Only Numeric Values*

```
TruBlend Coffee Makers, Inc.
      Total Yearly Sales

       Units    AmountSold

       40989    1320479.5
```

PROC REPORT computed the one-line summary for Units and AmountSold by summing the value of each variable for all the observations in the data set.

## Specifying and Ordering the Columns

The first step in constructing a report is to select the columns that you want to appear in the report. By default, the report contains a column for each variable and the order of the columns corresponds to the order of the variables in the data set.

You use the COLUMN statement to specify the variables to use in the report and the arrangement of the columns. In the COLUMN statement, you can list data set variables, statistics that are calculated by PROC REPORT, or variables that are computed from other items in the report.

The following program creates a four column sales report for the first quarter:

```
proc report data=year_sales;
   where Quarter='1';
   column SalesRep Month Type Units;
   title1 'TruBlend Coffee Makers, Inc.';
   title2 'First Quarter Sales Report';
run;
```

The COLUMN statement specifies the order of the items in the report. The first column lists the values in SalesRep, the second column lists the values in Month, and so on.

The following output displays the report:

*Output 29.3  Displaying Selected Columns*

**TruBlend Coffee Makers, Inc.**
**First Quarter Sales Report**

| SalesRep | Month | Type | Units |
|---|---|---|---|
| Hollingsworth | 01 | Deluxe | 260 |
| Garcia | 01 | Standard | 41 |
| Hollingsworth | 01 | Standard | 330 |
| Jensen | 01 | Standard | 110 |
| Garcia | 01 | Deluxe | 715 |
| Jensen | 01 | Standard | 675 |
| Garcia | 02 | Standard | 2045 |
| Garcia | 02 | Deluxe | 10 |
| Garcia | 02 | Standard | 40 |
| Hollingsworth | 02 | Standard | 1030 |
| Jensen | 02 | Standard | 153 |
| Garcia | 02 | Standard | 98 |
| Hollingsworth | 03 | Standard | 125 |
| Jensen | 03 | Standard | 154 |
| Garcia | 03 | Standard | 118 |
| Hollingsworth | 03 | Standard | 25 |
| Jensen | 03 | Standard | 525 |
| Garcia | 03 | Standard | 310 |

### Ordering the Rows

You control much of the layout of a report by deciding how you use the variables. You tell PROC REPORT how to use a variable by specifying a usage option in the DEFINE statement for the variable.

To specify the order of the rows in the report, you can use the ORDER option in one or more DEFINE statements. PROC REPORT orders the rows of the report according to the values of the ORDER variables. If the report contains multiple ORDER variables, then PROC REPORT first orders rows according to the values of the first ORDER variable in the COLUMN statement.[1] Within each value of the first ORDER variable, the procedure orders rows according to the values of the second ORDER variable in the COLUMN statement, and so on.

---

[1] If you omit the COLUMN statement, then PROC REPORT processes the ORDER variables according to their position in the input data set.

The following program creates a detail report of sales for the first quarter that is ordered by the sales representatives and month:

```
proc report data=year_sales nowindows;
   where Quarter='1';
   column SalesRep Month Type Units;
   define SalesRep / order;
   define Month / order;
   title1 'TruBlend Coffee Makers, Inc.';
   title2 'First Quarter Sales Report';
run;
```

The DEFINE statements specify that SalesRep and Month are the ORDER variables. The COLUMN statement specifies the order of the columns. By default, the rows are ordered by the ascending formatted values of SalesRep. The rows for each sales representative are ordered by the values of Month.

The following output displays the report:

**Output 29.4** *Ordering the Rows*

### TruBlend Coffee Makers, Inc.
### First Quarter Sales Report

| SalesRep | Month | Type | Units |
|---|---|---|---|
| Garcia | 01 | Standard | 41 |
|  |  | Deluxe | 715 |
|  | 02 | Standard | 2045 |
|  |  | Deluxe | 10 |
|  |  | Standard | 40 |
|  |  | Standard | 98 |
|  | 03 | Standard | 118 |
|  |  | Standard | 310 |
| Hollingsworth | 01 | Deluxe | 260 |
|  |  | Standard | 330 |
|  | 02 | Standard | 1030 |
|  | 03 | Standard | 125 |
|  |  | Standard | 25 |
| Jensen | 01 | Standard | 110 |
|  |  | Standard | 675 |
|  | 02 | Standard | 153 |
|  | 03 | Standard | 154 |
|  |  | Standard | 525 |

PROC REPORT does not repeat the values of the ORDER variables from one row to the next when the values are the same.

## Consolidating Several Observations into a Single Row

You can create summary reports with PROC REPORT by defining one or more GROUP variables. A group is a set of observations that has a unique combination of values for all GROUP variables. PROC REPORT tries to consolidate, or summarize, each group into one row of the report.

To consolidate all columns across a row, you must define all variables in the report as either GROUP, ANALYSIS, COMPUTED, or ACROSS. The GROUP option in one or more DEFINE statements identifies the variables that PROC REPORT uses to form groups. You can define more than one variable as a GROUP variable, but GROUP variables must precede variables of the other types of usage. PROC REPORT determines the nesting by the order of the variables in the COLUMN statement. For more information about defining the usage of a variable, see "Constructing the Layout" on page 503.

The value of an ANALYSIS variable for a group is the value of the statistic that PROC REPORT computes for all observations in a group. For each ANALYSIS variable, you can specify the statistic in the DEFINE statement. By default, PROC REPORT uses all numeric variables as the ANALYSIS variables and computes the SUM statistic. The statistics that you can request in the DEFINE statement are as follows:

*Table 29.1* Descriptive Statistics

Descriptive statistic keywords

| | |
|---|---|
| CSS | PCTSUM |
| CV | RANGE |
| MAX | STD |
| MEAN | STDERR |
| MIN | SUM |
| N | SUMWGT |
| NMISS | USS |
| PCTN | VAR |

Quantile statistic keywords

| | |
|---|---|
| MEDIAN \| P50 | Q3 \| P75 |
| P1 | P90 |
| P5 | P95 |
| P10 | P99 |
| Q1 \| P25 | QRANGE |

Hypothesis testing keyword

PRT | PROBT                T

For definitions and discussion of these statistics, see "SAS Elementary Statistics Procedures" in *Base SAS Procedures Guide*.

The following program creates a summary report that shows the total yearly sales for each sales representative:

```
proc report data=year_sales;
   column SalesRep Units AmountSold;
   define SalesRep /group; 1
   define Units / analysis sum; 2
    define AmountSold/ analysis sum; 3
   title1 'TruBlend Coffee Makers Sales Report';
   title2 'Total Yearly Sales';
run;
```

The following list corresponds to the numbered items in the preceding program:

- The DEFINE statement specifies that SalesRep is the GROUP variable.
- The DEFINE statement specifies that Units is an ANALYSIS variable and specifies that PROC REPORT computes the SUM statistic.
- The DEFINE statement specifies that AmountSold is an ANALYSIS variable and specifies that PROC REPORT computes the SUM statistic.

The following output displays the report:

*Output 29.5   Grouping Multiple Observations in a Summary Report*

**TruBlend Coffee Makers Sales Report**
**Total Yearly Sales**

| SalesRep | Units | AmountSold |
| --- | --- | --- |
| Garcia | 15969 | 512070.78 |
| Hollingsworth | 10620 | 347246.1 |
| Jensen | 14400 | 461162.6 |

Each row of the report represents one group and summarizes all observations that have a unique value for SalesRep. PROC REPORT orders these rows in ascending order of the GROUP variable, which in this example is the sales representative ordered alphabetically. The values of the ANALYSIS variables are the sum of Units and AmountSold for all observations in a group, which in this case is the total units and amount sold by each sales representative.

## Changing the Default Order of the Rows

You can modify the default ordering sequence for the rows of a report by using the ORDER= or DESCENDING option in the DEFINE statement. The ORDER= option specifies the sort order for a variable. You can order the rows by:

DATA
: the order of the data in the input data set.

FORMATTED
: ascending formatted values.

FREQ
: ascending frequency count.

INTERNAL
: ascending unformatted or internally stored values.

By default, PROC REPORT uses the formatted values of a variable to order the rows. The DESCENDING option reverses the sort sequence so that PROC REPORT uses descending values to order the rows.

The following program creates a detail report of the first quarter sales that is ordered by number of sales:

```
proc report data=year_sales;
   where Quarter='1';
   column SalesRep Type Units Month;
   define SalesRep / order order=freq;
   define Units / order descending;
   define Type / order;
   title1 'TruBlend Coffee Makers, Inc.';
   title2 'First Quarter Sales Report';
run;
```

The following list corresponds to the numbered items in the preceding program:

- The DEFINE statements specify that SalesRep, Units, and Type are ORDER variables that correspond to the number of sales each sales representative made.

- The ORDER=FREQ option orders the rows of the report by the frequency of SalesRep.

- The DESCENDING option orders the rows for UNITS from the largest to the smallest value.

The following output displays the report:

**Output 29.6** *Changing the Order Sequence of the Rows*

### TruBlend Coffee Makers, Inc.
### First Quarter Sales Report

| ② SalesRep | ③ Type | Units | ① Month |
|---|---|---|---|
| Hollingsworth | Deluxe | 260 | 01 |
|  | Standard | 1030 | 02 |
|  |  | 330 | 01 |
|  |  | 125 | 03 |
|  |  | 25 | 03 |
| Jensen | Standard | ④ 675 | 01 |
|  |  | 525 | 03 |
|  |  | 154 | 03 |
|  |  | 153 | 02 |
|  |  | 110 | 01 |
| Garcia | Deluxe | 715 | 01 |
|  |  | 10 | 02 |
|  | Standard | 2045 | 02 |
|  |  | 310 | 03 |
|  |  | 118 | 03 |
|  |  | 98 | 02 |
|  |  | 41 | 01 |
|  |  | 40 | 02 |

The following list corresponds to the numbered items in the preceding report:

1. The order of the columns corresponds to the order in which the variables are specified in the COLUMN statement. The order of the DEFINE statements does not affect the order of the columns.

2. The order of the rows is by ascending frequency of SalesRep so that the sales representative with the least number of sales (observations) appears first while the sales representative with the greatest number of sales appears last.

3. The order of the rows within SalesRep is by ascending formatted values of Type so that sales information about the deluxe coffee maker occurs before the standard coffee maker.

4. The order of the rows within Type is by descending formatted values of Units so that the observation with the highest number of units sold appears first.

## Creating More Sophisticated Reports

### Adjusting the Column Layout

#### Understanding Column Width and Spacing for ODS LISTING Output

If you are working with the ODS LISTING destination, you can modify the column spacing (SPACING=) and the column width (COLWIDTH=) by specifying options in either the PROC REPORT statement or the DEFINE statement.

To control the spacing between columns, you can use the SPACING= option in the following statements:

- PROC REPORT statement to specify the default number of blank characters between all columns

- DEFINE statement to override the default value and to specify the number of blank characters to the left of a particular column

In the LISTING Output, PROC REPORT inserts two blank spaces between the columns. To remove space between columns, specify SPACING=0. The maximum space that PROC REPORT allows between columns depends on the number of columns in the report. The sum of all column widths plus the blank characters to left of each column cannot exceed the line size.

To specify the column widths, you can use the following options:

- the COLWIDTH= option in the PROC REPORT statement to specify the default number of characters for columns that contain computed variables or numeric data set variables

- the WIDTH= option in the DEFINE statement to specify the width of the column that PROC REPORT uses to display a report item

By default, the column width is nine characters for numeric values. You can specify the column width as small as one character and as large as the line size. PROC REPORT sets the width of a column by first looking at the WIDTH= option in the DEFINE statement. If you omit WIDTH=, then PROC REPORT uses a column width large enough to accommodate the format for a report item. If you do not assign a format, then the column width is either the length of the character variable or the value of the COLWIDTH= option.

You can adjust the column layout by specifying how to align the formatted values of a report item and the column heading with the column width. The following options in the DEFINE statement align the columns:

CENTER
   centers the column values and column heading.

LEFT
   left-aligns the column values and column heading

RIGHT
   right-aligns the column values and column heading.

## Modifying the Column Width and Spacing

The following program modifies column spacing in a summary report that shows the total yearly sales for each sales representative:

```
options linesize=80 pageno=1 nodate;

proc report data=year_sales spacing=3;
   column SalesRep Units AmountSold;
   define SalesRep /group right;
   define Units / analysis sum width=5;
   define AmountSold/ analysis sum width=10;
   title1 'TruBlend Coffee Makers Sales Report';
   title2 'Total Yearly Sales';
run;
```

The following list corresponds to the numbered items in the preceding program:

- The SPACING= option in the PROC REPORT statement inserts three blank characters between all the columns.

- The RIGHT option in the DEFINE statement right-aligns the name of the sales representative and the column heading in the column.

- The WIDTH= options in the DEFINE statements specify enough space to accommodate column headings on one line.

The following output displays the report:

**Output 29.7** *Adjusting Column Width and Spacing for LISTING Output*

```
               TruBlend Coffee Makers Sales Report                   1
                        Total Yearly Sales

              SalesRep   Units   AmountSold
                Garcia   15969    512070.78
         Hollingsworth   10620     347246.1
                Jensen   14400     461162.6
```

The column width for SalesRep is 14 characters wide, which is the length of the variable.

## Customizing Column Headings

### Understanding the Structure of Column Headings

In ODS LISTING output, PROC REPORT does not insert a vertical space beneath column headings to visually separate the detail rows from the headers. To improve the appearance of a report generated using the ODS LISTING destination, you can underline the column headings, insert a blank line beneath column headings, and specify your own column headings. The HEADLINE and HEADSKIP options in the PROC REPORT statement enable you to underline the column headings and insert a blank line after the column headings, respectively.

By default, SAS uses the variable name or the variable label, if the data set variable was previously assigned a label, for the column heading. To specify a different column heading, place text between single or double quotation marks in the DEFINE statement for the report item.

By default, PROC REPORT produces line breaks in the column heading based on the width of the column. When you use multiple sets of quotation marks in the label, each set defines a separate line of the header. If you include split characters in the label, then PROC REPORT breaks the header when it reaches the split character and continues the header on the next line. By default, the split character is the slash (/). Use the SPLIT= option in the PROC REPORT statement to specify an alternative split character.

### Modifying the Column Headings in the LISTING Output

The following program creates a summary report with multiple-line column headings for the variables SalesRep, Units, and AmountSold:

```
ods listing;
ods html close;
options linesize=80 pageno=1 nodate;

proc report data=year_sales nowindows spacing=3 headskip;
   column SalesRep Units AmountSold;
   define SalesRep /group 'Sales/Representative';
   define Units / analysis sum 'Units Sold' width=5;
   define AmountSold/ analysis sum 'Amount' 'Sold';
   title1 'TruBlend Coffee Makers Sales Report';
   title2 'Total Yearly Sales';
run;
ods listing close;
ods html;
```

The following list corresponds to the numbered items in the preceding program:

- The HEADSKIP option inserts a blank line after the column headings.

- The text in quotation marks specifies the column headings.

The following output displays the report:

**Output 29.8** *Modifying the Column Headings for LISTING Output*

```
                  TruBlend Coffee Makers Sales Report                    1
                           Total Yearly Sales

                  Sales           Units       Amount
                  Representative  Sold        Sold

                  Garcia          15969       512070.78
                  Hollingsworth   10620       347246.1
                  Jensen          14400       461162.6
```

The label Units Sold is split between two lines because the column width for this report item is 5 characters wide. The SPLIT= option in the PROC REPORT statement identifies where the label for SalesRep, Sales Representative, is split. In contrast, the label for AmountSold identifies where to split the label by using multiple sets of quotation marks.

## Specifying Formats

### Using SAS Formats

A simple and effective way to enhance the readability of your reports is to specify a format for the report items. To assign a format to a column, you can use the FORMAT statement or the FORMAT= option in the DEFINE statement. The FORMAT statement only works for data set variables. The FORMAT= option assigns a SAS format or a user-defined format to any report item.

PROC REPORT determines how to format a report item by searching for the format to use in these places and in this order:

1. the FORMAT= option in the DEFINE statement
2. the FORMAT statement
3. the data set

PROC REPORT uses the first format that it finds. If you have not assigned a format, then PROC REPORT uses the BEST9. format for numeric variables and the $w. format for character variables.

### Applying Formats to Report Items

The following program illustrates how to apply formats to the columns of a summary report of total yearly sales for each sales representative:

```
proc report data=year_sales;
   column SalesRep Units AmountSold;
   define SalesRep / group 'Sales/Representative';
   define Units / analysis sum 'Units Sold' format=comma7.;
   define AmountSold / analysis sum 'Amount' 'Sold' format=dollar14.2;
   title1 'TruBlend Coffee Makers Sales Report';
   title2 'Total Yearly Sales';
run;
```

PROC REPORT applies the COMMA7. format to the values of the variable Units and the DOLLAR14.2 format to the values of the variable AmountSold.

The following output displays the report:

**Output 29.9** *Formatting the Numeric Columns*

### TruBlend Coffee Makers Sales Report
### Total Yearly Sales

| Sales Representative | Units Sold | Amount Sold |
|---|---|---|
| Garcia | 15,969 | $512,070.78 |
| Hollingsworth | 10,620 | $347,246.10 |
| Jensen | 14,400 | $461,162.60 |

The following list corresponds to the numbered items in the preceding report:

1  The variable AmountSold uses the DOLLAR14.2 format for a maximum column width of 14 spaces. Two spaces are reserved for the decimal part of a value. The remaining 12 spaces include the decimal point, whole numbers, the dollar sign, commas, and a minus sign if a value is negative.

2  The variable Units uses the COMMA7. format for a maximum column width of seven spaces. The column width includes the numeric value, commas, and a minus sign if a value is negative.

These formats do not affect the actual data values that are stored in the SAS data set. That is, the formats only affect how values appear in a report.

### Using Variable Values as Column Headings

#### Creating the Column Headings

To create column headings from the values of the data set variables and produce crosstabulations, you can use the ACROSS option in a DEFINE statement. When you define an ACROSS variable, PROC REPORT creates a column for each value of the ACROSS variable.

Columns created by an ACROSS variable contain statistics or computed values. If nothing is above or below an ACROSS variable, then PROC REPORT displays the number of observations in the input data set that belong to a cell of the report (N statistic). A cell is a single unit of a report, formed by the intersection of a row and a column.

The examples in this section show you how to display frequency counts (the N statistic) and statistics that are computed for ANALYSIS variables. For information about placing computed variables in the cells of the report, see "REPORT Procedure" in *Base SAS Procedures Guide*.

#### Creating Frequency Counts

The following program creates a report that tabulates the number of sales for each sales representative:

```
proc report data=year_sales colwidth=5;
   column SalesRep Type n;
   define SalesRep /group 'Sales Representative';
   define Type / across 'Type of Coffee Maker';
   define n / 'Total';
   title1 'TruBlend Coffee Makers Yearly Sales Report';
   title2 'Number of Sales';
 run;
```

The following list corresponds to the numbered items in the preceding program:

- The COLUMN statement specifies that the report contain two data set variables and a calculated statistic, N. The N statistic causes PROC REPORT to add a third column that displays the number of observations for each sales representative.

- The DEFINE statement specifies that Type is an ACROSS variable.

The following output displays the report:

**Output 29.10** *Showing Frequency Counts*

### TruBlend Coffee Makers Yearly Sales Report
### Number of Sales

|  | Coffee Maker 1 |  |  |
| --- | --- | --- | --- |
| Sales Representative | Deluxe | Standard | Total 2 |
| Garcia | 4 | 36 | 40 |
| Hollingsworth | 8 | 24 | 32 |
| Jensen | 4 | 34 | 38 |

The following list corresponds to the numbered items in the preceding report:

1. Type is an ACROSS variable with nothing above or below it. Therefore, the report shows how many observations the input data set contains for each sales representative and coffee maker type.

2. The column for N statistic is labeled Total and contains the total number of observations for each sales representative.

By default, PROC REPORT ordered the columns of the ACROSS variable according to its formatted values. You can use the ORDER= option in the DEFINE statement to alter the sort order for an ACROSS variable. See for more information.

### Sharing a Column with Multiple Analysis Variables

You can create sophisticated crosstabulation by having the value of ANALYSIS variables appear in columns that the ACROSS variable creates. When an ACROSS variable shares columns with one or more ANALYSIS variables, PROC REPORT stacks the columns. For example, you can share the columns of the ACROSS variable Type with the ANALYSIS variable Units so that each column contains the number of units sold for a type of coffee maker.

To stack the value of an ANALYSIS variable in the columns created by the ACROSS variable, place that variable next to the ACROSS variable in the COLUMN statement:

```
column SalesRep Type, Unit;
```

The comma separates the ACROSS variable from the ANALYSIS variable. To specify multiple ANALYSIS variables, list their names in parentheses next to the ACROSS variable in the COLUMN statement:

```
column SalesRep Type,(Unit AmountSold);
```

If you place the ACROSS variable before the ANALYSIS variable, then the name and values of the ACROSS variable are above the name of the ANALYSIS variable in the report. If you place the ACROSS variable after the ANALYSIS variable, then the name and the values of the ACROSS variable are below the name of the ANALYSIS variable.

By default, PROC REPORT calculates the SUM statistic for the ANALYSIS variables. To display another statistic for the column, use the DEFINE statement to specify the statistic that you want computed for the ANALYSIS variable. See Table 29.1 on page 510 for a list of the available statistics.

The following program creates a report that tabulates the number of coffee makers sold and the average sale in dollars for each sales representative:

```
proc report data=year_sales;
   column SalesRep Type,(Units Amountsold);
   define SalesRep /group 'Sales Representative';
   define Type / across '';
   define units / analysis sum 'Units Sold' format=comma7.;
   define AmountSold / analysis mean 'Average/Sale' format=dollar12.2;
   title1 'TruBlend Coffee Makers Yearly Sales Report';
run;
```

The following list corresponds to the numbered items in the preceding program:

- The COLUMN statement creates columns for SalesRep and Type. The ACROSS variable Type shares its columns with the ANALYSIS variables Units and AmountSold.

- The DEFINE statement uses a blank as the label of Type in the column heading.

- The DEFINE statement uses the ANALYSIS variable Units to compute a SUM statistic.

- The DEFINE statement uses the ANALYSIS variable AmountSold to compute a MEAN statistic.

The following output displays the report:

**Output 29.11** *Sharing a Column with Multiple Analysis Variables*

### TruBlend Coffee Makers Yearly Sales Report

|  | Deluxe | | Standard | |
| --- | --- | --- | --- | --- |
| Sales Representative | Units Sold | Average Sale | Units Sold | Average Sale |
| Garcia | 945 | $11,694.38 | 15,024 | $12,924.81 |
| Hollingsworth | 760 | $4,702.50 | 9,860 | $12,901.09 |
| Jensen | 820 | $10,147.50 | 13,580 | $12,369.78 |

The values in the columns for a particular type of coffee maker are the total units sold and the average dollar sale for each sales representative.

## Summarizing Groups of Observations

### Using Group Summaries

For some reports, you might want to summarize information about a group of observations and visually separate each group. To do so, you can create a break in the report before or after each group.

To visually separate each group, you insert lines of text, called break lines, at a break. Break lines can occur at the beginning or end of a report, at the top or bottom of each page, and whenever the value of a group or order variable changes. The break line can contain the following items:

- text (including blanks)
- summaries of statistics
- report variables
- computed variables

To create group summaries, use the BREAK statement. A BREAK statement must include (in this order) the following:

- the keyword BREAK
- the location of the break (BEFORE or AFTER)
- the name of a GROUP variable that is called the break value

PROC REPORT creates a break each time the value of the break variable changes. If you want summaries to appear before the first row of each group, then use the BEFORE argument. If you want the summaries to appear after the last row of each group, then use the AFTER argument.

To create summary information for the whole report, use the RBREAK statement. An RBREAK statement must include (in this order) the following:

- the keyword RBREAK
- the location of the break (BEFORE or AFTER)

When you use the RBREAK statement, PROC REPORT inserts text, summary statistics for the entire report, or computed variables at the beginning or end of the detail rows of a report. If you want the summary to appear before the first row of the report, then use the BEFORE argument. If you want the summaries to appear after the last row of each group, then use the AFTER argument.

Both the BREAK and RBREAK statements support options that control the appearance of the group and the report summaries. You can use any combination of options in the statement in any order. For a list of available options, see "REPORT Procedure" in *Base SAS Procedures Guide*.

### *Creating Group Summaries*
The following program creates a summary report that uses break lines to display subtotals with yearly sales for each sales representative, and a yearly grand total for all sales representatives:

```
ods listing;
options linesize=80 pageno=1 nodate linesize=84;

proc report data=year_sales headskip;
    column Salesrep Quarter Units AmountSold;
    define SalesRep / group 'Sales Representative';
    define Quarter / group center;
    define Units / analysis sum 'Units Sold' format=comma7.;
    define AmountSold / analysis sum 'Amount/Sold' format=dollar14.2;
    break after SalesRep / summarize skip ol suppress;
    rbreak after / summarize skip dol;
    title1 'TruBlend Coffee Makers Sales Report';
```

```
        title2 'Total Yearly Sales';
run;
ods listing close;
```

The following list corresponds to the numbered items in the preceding program:

- The CENTER option (LISTING output only) in the DEFINE statement centers the values of the variable Quarter and the label of the column heading.

- The BREAK statement adds break lines after a change in the value of the GROUP variable SalesRep. The SUMMARIZE option writes a summary line to summarize the statistics for each group of break lines. The SKIP option (LISTING output only) inserts a blank line after each group of break lines. The OL option (LISTING output only) writes a line of hyphens (-) above each value in the summary line. The SUPPRESS option suppresses printing the value of the break variable and the overlines in the break variable column.

- The RBREAK statement adds a break line at the end of the report. The SUMMARIZE option writes a summary line that summarizes the SUM statistics for the ANALYSIS variables Units and AmountSold. The SKIP option (LISTING output only) inserts a blank line before the break line. The DOL option (LISTING output only) writes a line of equal signs (=) above each value in the summary line.

The following displays the report for the LISTING destination:

*Output 29.12  Creating Group Summaries*

```
                    TruBlend Coffee Makers Sales Report
 1
                           Total Yearly Sales

               Sales                      Units            Amount
               Representative  Quarter    Sold             Sold

               Garcia          1          3,377            $118,019.94
                               2          3,515            $108,859.55
                               3          7,144            $225,326.28
                               4          1,933            $59,865.01
                                          -------          --------------
                                          15,969 [1]
 $512,070.78 [1]

               Hollingsworth   1          1,770            $59,634.70
                               2          3,090            $96,160.55
                               3          3,285            $109,704.35
                               4          2,475            $81,746.50
                                          -------          --------------
                                          10,620           $347,246.10

               Jensen          1          1,617            $50,078.49
                               2          2,413            $74,730.61
                               3          6,687            $222,290.99
                               4          3,683            $114,062.51
                                          -------          --------------
                                          14,400           $461,162.60

                                          =======          ==============
                                          40,989 [2]
 $1,320,479.48 [2]
```

The following list corresponds to the numbered items in the preceding LISTING output:

1 The values of the ANALYSIS variables Units and AmountSold in the group summary lines are sums for all rows in the group (subtotals).

2 The values of the ANALYSIS variables Units and AmountSold in the report summary line are sums for all rows in the report (grand totals).

In this report, Units and AmountSold are ANALYSIS variables that are used to calculate the SUM statistic. If these variables were defined to calculate a different statistic, then the values in the summary lines would be the value of that statistic for all rows in the group and all rows in the report.

# Summary

## PROC REPORT Statements

**PROC REPORT** <DATA=*SAS-data-set*> <*option(s)*>;
**BREAK** *location break-variable* </*option(s)*>;
**COLUMN** *column-specification(s)*;
**DEFINE** *report-item* /<*usage*> <*option(s)*>;
**RBREAK** *location*</*option(s)*>;
**TITLE**<*n*> <*'title'*>;
**WHERE** *where-expression*;

PROC REPORT <DATA=SAS-data-set> <option(s)>;
   starts the procedure. If no other statements are used, then SAS shows all variables in the SAS-data-set in a detail report in the REPORT window. If the data set contains only numeric data, then PROC REPORT shows all variables in a summary report. Other statements, listed below, enable you to control the structure of the report.

   You can specify the following options in the PROC REPORT statement:

COLWIDTH=column-width
   specifies the default number of characters for columns that contain computed variables or numeric data set variables. This option only affects the LISTING output.

DATA=SAS-data-set
   names the SAS data set that PROC REPORT uses. If you omit DATA=, then PROC REPORT uses the most recently created data set.

HEADLINE
   inserts a line of hyphens (-) under the column headings at the top of each page of the report. This option only affects the LISTING output.

HEADSKIP
   inserts a blank line beneath all column headings (or beneath the line that the HEADLINE option inserts) at the top of each page of the report. This option only affects the LISTING output.

SPACING=space-between-columns
   specifies the number of blank characters between columns. For each column, the sum of its width and the blank characters between it and the column to its left cannot exceed the line size. This option only affects the LISTING output.

**SPLIT='character'**
  specifies the split character. PROC REPORT breaks a column heading when it reaches that character and continues the header on the next line. The split character itself is not part of the column heading, although each occurrence of the split character is counted toward the 256-character maximum for a label. In the LISTING output, the split character works in header rows and data values. In all other destinations, the split character only works in header row.

**WINDOWS | NOWINDOWS**
  selects a windowing or nonwindowing environment.

  When you use NOWINDOWS, PROC REPORT runs without the REPORT window and sends its results to the SAS procedure output. NOWINDOWS (alias NOWD) is the default. When you use WINDOWS, SAS opens the REPORT window, which enables you to modify a report repeatedly and to see the modifications immediately.

**BREAK location break-variable </option(s)>;**
  produces a default summary at a break (a change in the value of a GROUP or ORDER variable). The information in a summary applies to a set of observations. The observations share a unique combination of values for the break variable and all other GROUP or ORDER variables to the left of the break variable in the report.

  You must specify the following arguments in the BREAK statement:

location
  controls the placement of the break lines, where location is

  **AFTER**
    places the break lines immediately after the last row of each set of rows that have the same value for the break variable.

  **BEFORE**
    places the break lines immediately before the first row of each set of rows that have the same value for the break variable.

break-variable
  is a GROUP or ORDER variable. PROC REPORT writes break lines each time the value of this variable changes. You can specify the following options in the BREAK statement:

  **OL**
    inserts a line of hyphens (-) above each value that appears in the summary line. This option only affects the LISTING output.

  **SKIP**
    writes a blank line for the last break line. This option only affects the LISTING output.

  **SUMMARIZE**
    writes a summary line in each group of break lines.

  **SUPPRESS**
    suppresses the printing of the value of the break variable in the summary line, and of any underlining or overlining in the break lines.

**COLUMN <column-specification(s)>;**
  identifies items that form columns in the report and describes the arrangement of all columns. You can specify the following column-specification(s) in the COLUMN statement:

  - *report-item(s)*

- *report-item-1, report-item-2 <. . . , report-item-n>*

  where *report-item* identifies items that form columns in the report. A report-item is either the name of a data set variable, a computed variable, or a statistic.

  report-item-1, report-item-2 <. . . , report-item-n>
  : identifies report items that collectively determine the contents of the column or columns. These items are said to be stacked in the report because each item generates a header, and the headers are stacked one above the other. The header for the leftmost item is on top. If one of the items is an ANALYSIS variable, then a computed variable, or a statistic, its values fill the cells in that part of the report. Otherwise, PROC REPORT fills the cells with frequency counts.

DEFINE report-item / <usage> <option(s)>;
: describes how to use and display a report item. A report item is either the name or alias (established in the COLUMN statement) of a data set variable, a computed variable, or a statistic. The usage of the report item is

  - ACROSS
  - ANALYSIS
  - COMPUTED
  - DISPLAY
  - GROUP
  - ORDER

  You can specify the following options in the DEFINE statement:

CENTER
: centers the formatted values of the report item within the column width, and centers the column heading over the values.

column-header
: defines the column heading for the report item. Enclose each header in single or double quotation marks. When you specify multiple column headings, PROC REPORT uses a separate line for each one. The split character also splits a column heading over multiple lines.

DESCENDING
: reverses the order in which PROC REPORT displays rows or values of a GROUP, ORDER, or ACROSS variable.

FORMAT=format
: assigns a SAS format or a user-defined format to the report item. This format applies to *report-item* as PROC REPORT displays it; the format does not alter the format associated with a variable in the data set.

ORDER=DATA | FORMATTED | FREQ | INTERNAL
: orders the values of a GROUP, ORDER, or ACROSS variable according to the specified order, where

DATA
: orders values according to their order in the input data set.

FORMATTED
: orders values by their formatted (external) values. By default, the order is ascending.

FREQ
: orders values by ascending frequency count.

INTERNAL
: orders values by their unformatted values, which yields the same order that PROC SORT would yield. This order is operating environment dependent. This sort sequence is particularly useful for displaying dates chronologically.

RIGHT
: right-justifies the formatted values of the specified report item within the column width and right-justifies the column headings over the values. If the format width is the same as the width of the column, then RIGHT has no affect on the placement of values. This option only affects the LISTING output.

SPACING=horizontal-positions
: defines the number of blank characters to leave between the column that is being defined and the column immediately to its left. For each column, the sum of its width and the blank characters between it and the column to its left cannot exceed the line size. This option only affects the LISTING output.

statistic
: associates a statistic with an ANALYSIS variable. PROC REPORT uses this statistic to calculate values for the ANALYSIS variable for the observations represented by each cell of the report. If you do not associate a statistic with the variable, then PROC REPORT calculates the SUM statistic. You cannot use *statistic* in the definition of any other type of variable.

WIDTH=column-width
: defines the width of the column in which PROC REPORT displays *report-item*. This option only affects the LISTING output.

RBREAK location </option(s)>;
: produces a default summary at the beginning or end of a report.

You must specify the following argument in the RBREAK statement:

location
: controls the placement of the break lines and is either

AFTER
: places the break lines at the end of the report.

BEFORE
: places the break lines at the beginning of the report.

You can specify the following options in the RBREAK statement:

DOL
: specifies to double overline each value that appears in the summary line. This option only affects the LISTING output.

SKIP
: writes a blank line after the last break line of a break located at the beginning of the report. This option only affects the LISTING output.

SUMMARIZE
: includes a summary line as one of the break lines. A summary line at the beginning or end of a report contains values for statistics, ANALYSIS variables, or computed variables.

TITLE<n> <'title'>;
: specifies a title. The argument *n* is a number from 1 to 10 that immediately follows the word TITLE, with no intervening blank, and it specifies the level of the TITLE. The text of each *title* must be enclosed in single or double quotation marks. The maximum title length depends on your operating environment and the value of the

LINESIZE= system option. Refer to the SAS documentation for your operating environment for more information.

WHERE where-expression;
: subsets the input data set by identifying certain conditions that each observation must meet before an observation is available for processing. *Where-expression* defines the condition. The condition is a valid arithmetic or logical expression that generally consists of a sequence of operands and operators.

# Learning More

KEEP= data set option
: For an additional example, see "Reading Selected Variables" on page 96. For complete documentation about the KEEP= data set option, see "KEEP= Data Set Option" in *SAS Data Set Options: Reference*.

PROC PRINT
: For a discussion of how to create several types of detail reports, see Chapter 27, "Producing Detail Reports with the PRINT Procedure," on page 433.

PROC REPORT
: For complete documentation, see "REPORT Procedure" in *Base SAS Procedures Guide*.

PROC TABULATE
: For a discussion of how to create several types of summary reports, see Chapter 28, "Creating Summary Tables with the TABULATE Procedure," on page 473.

SAS formats
: For complete documentation, see *SAS Formats and Informats: Reference*. Many formats are available with the SAS software, such as fractions, hexadecimal values, roman numerals, Social Security numbers, date and time values, and numbers written as words.

WHERE statement
: For a discussion, see "Understanding the WHERE Statement" on page 444. For complete reference documentation about the WHERE statement, see *SAS Statements: Reference*. For a complete discussion of WHERE processing, see *SAS Language Reference: Concepts*.

# Part 7

# Producing Plots and Charts

*Chapter 30*
**Plotting the Relationship between Variables** .................... *531*

*Chapter 31*
**Producing Charts to Summarize Variables** ...................... *551*

# Chapter 30
# Plotting the Relationship between Variables

Introduction to Plotting the Relationship between Variables .................. **531**
    Overview ....................................................................... 531
    Prerequisites .................................................................. 532

Input File and SAS Data Set for Examples ............................... **532**

Plotting One Set of Variables .............................................. **535**
    Understanding the PLOT Statement ................................... 535
    Example ........................................................................ 536

Enhancing the Plot ............................................................ **537**
    Specifying the Axes Labels .............................................. 537
    Specifying the Tick Marks Values ..................................... 538
    Specifying Plotting Symbols ............................................ 539
    Removing the Legend ..................................................... 540

Plotting Multiple Sets of Variables ...................................... **541**
    Creating Multiple Plots on Separate Pages ........................ 541
    Creating Multiple Plots on the Same Page ........................ 544
    Plotting Multiple Sets of Variables on the Same Axes ........ 546

Summary ......................................................................... **547**
    PROC PLOT Statements ................................................. 547

Learning More ................................................................. **549**

## Introduction to Plotting the Relationship between Variables

### Overview

An effective way to examine the relationship between variables is to plot their values. You can use the PLOT procedure to display relationships and patterns in the data.

This section covers the following topics:

- plot one set of variables
- enhance the appearance of a plot
- create multiple plots on separate pages

- create multiple plots on the same page
- plot multiple sets of variables on the same pair of axes

## Prerequisites

To understand the examples in this section, you should be familiar with the following features and concepts:

- the LOG function
- the FORMAT statement
- the LABEL statement
- the TITLE statement

# Input File and SAS Data Set for Examples

The examples in this section use one input file and one SAS data set. The input file contains information about the high and low values of the Dow Jones Industrial Average from 1968 to 2008. The input file has the following structure:

```
1968   03DEC1968    985.21  21MAR1968    825.13
1969   14MAY1969    968.85  17DEC1969    769.93
1970   29DEC1970    842.00  06MAY1970    631.16
1971   28APR1971    950.82  23NOV1971    797.97
1972   11DEC1972   1036.27  26JAN1972    889.15
...more data lines...
2005   04MAR2005  10940.55  20APR2005  10012.36
2006   27DEC2006  12510.57  20JAN2006  10667.39
2007   09OCT2007  14164.53  05MAR2007  12050.41
2008   02MAY2008  13058.20  10OCT2008   8451.19
```

The input file contains the following values from left to right:

- the year that the observation describes
- the date of the yearly high for the Dow Jones Industrial Average
- the yearly high value for the Dow Jones Industrial Average
- the date of the yearly low for the Dow Jones Industrial Average
- the yearly low value for the Dow Jones Industrial Average

The following program creates the SAS data set HIGHLOW:

```
data highlow;
   infile 'your-input-file';
   input Year @7 DateOfHigh:date9. DowJonesHigh @26 DateOfLow:date9. DowJonesLow;
   format LogDowHigh LogDowLow 5.2 DateOfHigh DateOfLow date9.;
   LogDowHigh=log(DowJonesHigh);
   LogDowLow=log(DowJonesLow);
run;
```

The computed variables LogDowHigh and LogDowLow contain the log transformation of the yearly high and low values for the Dow Jones Industrial Average.

```
proc print data=highlow;
   title 'Dow Jones Industrial Average Yearly High and Low Values';
run;


data highlow;
   input Year @7 DateOfHigh:date9. DowJonesHigh @26 DateOfLow:date9. DowJonesLow;
   format LogDowHigh LogDowLow 5.2 DateOfHigh DateOfLow date9.;
   LogDowHigh=log(DowJonesHigh);
   LogDowLow=log(DowJonesLow);
datalines;
1968   03DEC1968  985.21   21MAR1968  825.13
1969   14MAY1969  968.85   17DEC1969  769.93
1970   29DEC1970  842.00   06MAY1970  631.16
1971   28APR1971  950.82   23NOV1971  797.97
1972   11DEC1972 1036.27   26JAN1972  889.15
1973   11JAN1973 1051.70   05DEC1973  788.31
1974   13MAR1974  891.66   06DEC1974  577.60
1975   15JUL1975  881.81   02JAN1975  632.04
1976   21SEP1976 1014.79   02JAN1976  858.71
1977   03JAN1977  999.75   02NOV1977  800.85
1978   08SEP1978  907.74   28FEB1978  742.12
1979   05OCT1979  897.61   07NOV1979  796.67
1980   20NOV1980 1000.17   21APR1980  759.13
1981   27APR1981 1024.05   25SEP1981  824.01
1982   27DEC1982 1070.55   12AUG1982  776.92
1983   29NOV1983 1287.20   03JAN1983 1027.04
1984   06JAN1984 1286.64   24JUL1984 1086.57
1985   16DEC1985 1553.10   04JAN1985 1184.96
1986   02DEC1986 1955.57   22JAN1986 1502.29
1987   25AUG1987 2722.42   19OCT1987 1738.74
1988   21OCT1988 2183.50   20JAN1988 1879.14
1989   09OCT1989 2791.41   03JAN1989 2144.64
1990   16JUL1990 2999.75   11OCT1990 2365.10
1991   31DEC1991 3168.83   09JAN1991 2470.30
1992   01JUN1992 3413.21   09OCT1992 3136.58
1993   29DEC1993 3794.33   20JAN1993 3241.95
1994   31JAN1994 3978.36   04APR1994 3593.35
1995   13DEC1995 5216.47   30JAN1995 3832.08
1996   27DEC1996 6560.91   10JAN1996 5032.94
1997   06AUG1997 8259.31   11APR1997 6391.69
1998   23NOV1998 9374.27   31AUG1998 7539.07
1999   31DEC1999 11497.12  22JAN1999 9120.67
2000   14JAN2000 11722.98  07MAR2000 9796.04
2001   21MAY2001 11337.92  21SEP2001 8235.81
2002   19MAR2002 10635.25  09OCT2002 7286.27
2003   31DEC2003 10453.92  11MAR2003 7524.06
2004   28DEC2004 10854.54  25OCT2004 9749.99
2005   04MAR2005 10940.55  20APR2005 10012.36
2006   27DEC2006 12510.57  20JAN2006 10667.39
2007   09OCT2007 14164.53  05MAR2007 12050.41
2008   02MAY2008 13058.20  10OCT2008 8451.19
;
run;
```

**534** Chapter 30 • *Plotting the Relationship between Variables*

*Figure 30.1* SAS Output for the HIGHLOW Data Set

Dow Jones Industrial Average Yearly High and Low Values

| Obs | Year | DateOfHigh | DowJonesHigh | DateOfLow | DowJonesLow | LogDowHigh | LogDowLow |
|---|---|---|---|---|---|---|---|
| 1 | 1968 | 03DEC1968 | 985.21 | 21MAR1968 | 825.13 | 6.89 | 6.72 |
| 2 | 1969 | 14MAY1969 | 968.85 | 17DEC1969 | 769.93 | 6.88 | 6.65 |
| 3 | 1970 | 29DEC1970 | 842.00 | 06MAY1970 | 631.16 | 6.74 | 6.45 |
| 4 | 1971 | 28APR1971 | 950.82 | 23NOV1971 | 797.97 | 6.86 | 6.68 |
| 5 | 1972 | 11DEC1972 | 1036.27 | 26JAN1972 | 889.15 | 6.94 | 6.79 |
| 6 | 1973 | 11JAN1973 | 1051.70 | 05DEC1973 | 788.31 | 6.96 | 6.67 |
| 7 | 1974 | 13MAR1974 | 891.66 | 06DEC1974 | 577.60 | 6.79 | 6.36 |
| 8 | 1975 | 15JUL1975 | 881.81 | 02JAN1975 | 632.04 | 6.78 | 6.45 |
| 9 | 1976 | 21SEP1976 | 1014.79 | 02JAN1976 | 858.71 | 6.92 | 6.76 |
| 10 | 1977 | 03JAN1977 | 999.75 | 02NOV1977 | 800.85 | 6.91 | 6.69 |
| 11 | 1978 | 08SEP1978 | 907.74 | 28FEB1978 | 742.12 | 6.81 | 6.61 |
| 12 | 1979 | 05OCT1979 | 897.61 | 07NOV1979 | 796.67 | 6.80 | 6.68 |
| 13 | 1980 | 20NOV1980 | 1000.17 | 21APR1980 | 759.13 | 6.91 | 6.63 |
| 14 | 1981 | 27APR1981 | 1024.05 | 25SEP1981 | 824.01 | 6.93 | 6.71 |
| 15 | 1982 | 27DEC1982 | 1070.55 | 12AUG1982 | 776.92 | 6.98 | 6.66 |
| 16 | 1983 | 29NOV1983 | 1287.20 | 03JAN1983 | 1027.04 | 7.16 | 6.93 |
| 17 | 1984 | 06JAN1984 | 1286.64 | 24JUL1984 | 1086.57 | 7.16 | 6.99 |
| 18 | 1985 | 16DEC1985 | 1553.10 | 04JAN1985 | 1184.96 | 7.35 | 7.08 |
| 19 | 1986 | 02DEC1986 | 1955.57 | 22JAN1986 | 1502.29 | 7.58 | 7.31 |
| 20 | 1987 | 25AUG1987 | 2722.42 | 19OCT1987 | 1738.74 | 7.91 | 7.46 |
| 21 | 1988 | 21OCT1988 | 2183.50 | 20JAN1988 | 1879.14 | 7.69 | 7.54 |
| 22 | 1989 | 09OCT1989 | 2791.41 | 03JAN1989 | 2144.64 | 7.93 | 7.67 |
| 23 | 1990 | 16JUL1990 | 2999.75 | 11OCT1990 | 2365.10 | 8.01 | 7.77 |
| 24 | 1991 | 31DEC1991 | 3168.83 | 09JAN1991 | 2470.30 | 8.06 | 7.81 |
| 25 | 1992 | 01JUN1992 | 3413.21 | 09OCT1992 | 3136.58 | 8.14 | 8.05 |
| 26 | 1993 | 29DEC1993 | 3794.33 | 20JAN1993 | 3241.95 | 8.24 | 8.08 |
| 27 | 1994 | 31JAN1994 | 3978.36 | 04APR1994 | 3593.35 | 8.29 | 8.19 |
| 28 | 1995 | 13DEC1995 | 5216.47 | 30JAN1995 | 3832.08 | 8.56 | 8.25 |
| 29 | 1996 | 27DEC1996 | 6560.91 | 10JAN1996 | 5032.94 | 8.79 | 8.52 |
| 30 | 1997 | 06AUG1997 | 8259.31 | 11APR1997 | 6391.69 | 9.02 | 8.76 |
| 31 | 1998 | 23NOV1998 | 9374.27 | 31AUG1998 | 7539.07 | 9.15 | 8.93 |
| 32 | 1999 | 31DEC1999 | 11497.12 | 22JAN1999 | 9120.67 | 9.35 | 9.12 |
| 33 | 2000 | 14JAN2000 | 11722.98 | 07MAR2000 | 9796.04 | 9.37 | 9.19 |
| 34 | 2001 | 21MAY2001 | 11337.92 | 21SEP2001 | 8235.81 | 9.34 | 9.02 |
| 35 | 2002 | 19MAR2002 | 10635.25 | 09OCT2002 | 7286.27 | 9.27 | 8.89 |
| 36 | 2003 | 31DEC2003 | 10453.92 | 11MAR2003 | 7524.06 | 9.25 | 8.93 |
| 37 | 2004 | 28DEC2004 | 10854.54 | 25OCT2004 | 9749.99 | 9.29 | 9.19 |
| 38 | 2005 | 04MAR2005 | 10940.55 | 20APR2005 | 10012.36 | 9.30 | 9.21 |
| 39 | 2006 | 27DEC2006 | 12510.57 | 20JAN2006 | 10667.39 | 9.43 | 9.27 |
| 40 | 2007 | 09OCT2007 | 14164.53 | 05MAR2007 | 12050.41 | 9.56 | 9.40 |
| 41 | 2008 | 02MAY2008 | 13058.20 | 10OCT2008 | 8451.19 | 9.48 | 9.04 |

# Plotting One Set of Variables

## Understanding the PLOT Statement

The PLOT procedure produces two-dimensional graphs that plot one variable against another within a set of coordinate axes. The coordinates of each point on the plot correspond to the values of two variables. Graphs are automatically scaled to the values of your data, although you can control the scale by specifying the coordinate axes.

You can create a simple two-dimensional plot for one set of measures by using the following PLOT statement:

**PROC PLOT** <DATA=*SAS-data-set*>;

   **PLOT** *vertical*horizontal*;

where *vertical* is the name of the variable to plot on the vertical axis and *horizontal* is the name of the variable to plot on the horizontal axis.

By default, PROC PLOT selects plotting symbols. The data determines the labels for the axes, the values of the axes, and the values of the tick marks. The plot displays the following:

- the name of the vertical variable that is next to the vertical axis and the name of the horizontal variable that is beneath the horizontal axis

- the axes and the tick marks that are based on evenly spaced intervals

- the letter A as the plotting symbol to indicate one observation; the letter B as the plotting symbol if two observations coincide; the letter C if three coincide, and so on

- a legend with the name of the variables in the plot and meaning of the plotting symbols

The following display shows the axes, values, and tick marks on a plot.

**Figure 30.2** *Diagram of Axes, Values, and Tick Marks*

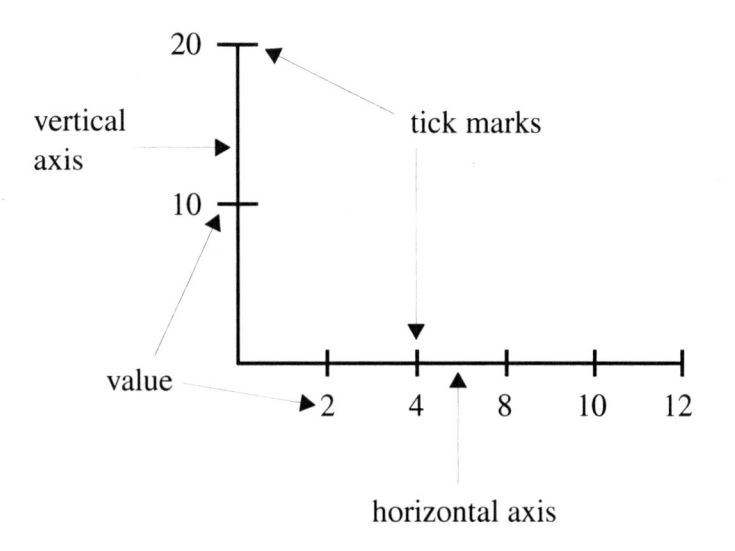

*Note:* PROC PLOT is an interactive procedure. After you issue the PROC PLOT statement, you can continue to submit any statements that are valid with the procedure without resubmitting the PROC statement. Therefore, you can easily and quickly experiment with changing labels, values for tick marks, and so on.

## Example

The following program uses the PLOT statement to create a simple plot that shows the trend in high Dow Jones values from 1968 to 2008:

```
proc plot data=highlow;
   plot DowJonesHigh*Year;
   title 'Dow Jones Industrial Average Yearly High';
run;
```

The following output shows the plot:

**Figure 30.3** *Using a Simple Plot to Show Data Trends*

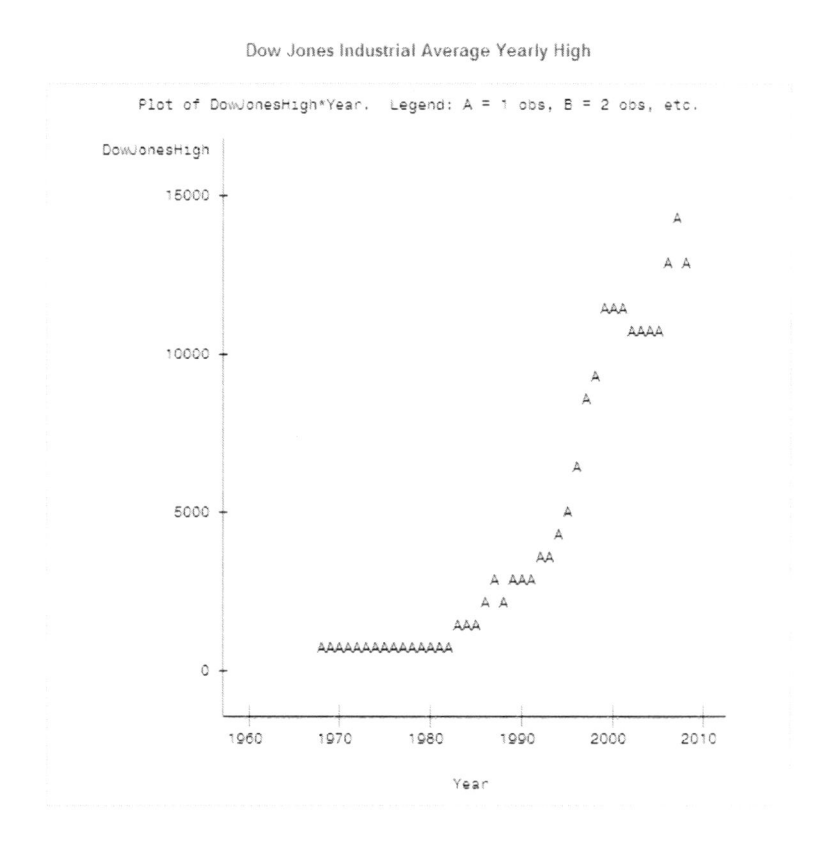

```
                    Dow Jones Industrial Average Yearly High

       Plot of DowJonesHigh*Year.   Legend: A = 1 obs, B = 2 obs, etc.

  DowJonesHigh |
               |
         15000 +
               |                                                A
               |
               |                                              A A
               |                                          AAA
               |                                          AAAA
         10000 +                                        A
               |                                        A
               |
               |                                      A
               |
          5000 +                                    A
               |                                    A
               |                                  AA
               |                              A AAA
               |                              A A
               |                            AAA
             0 +            AAAAAAAAAAAAAA
               |
               +-------+-------+-------+-------+-------+-------+
                     1960    1970    1980    1990    2000    2010
                                        Year
```

The plot graphically depicts the exponential trend in the high value of the Dow Jones Industrial Average over the past 50 years. The greatest growth has occurred in the past 10 years, increasing by almost 6,000 points.

# Enhancing the Plot

## *Specifying the Axes Labels*

Sometimes you might want to supply additional information about the axes. You can enhance the plot by specifying the labels for the vertical and horizontal axes.

The following program plots the log transformation of DowJonesHigh for each year and uses the LABEL statement to change the axes labels:

```
proc plot data=highlow;
   plot LogDowHigh*Year;
   label LogDowHigh='Log of Highest Value'
         Year='Year Occurred';
   title 'Dow Jones Industrial Average Yearly High';
run;
```

The following output shows the plot:

**Figure 30.4** *Specifying the Labels for the Axes*

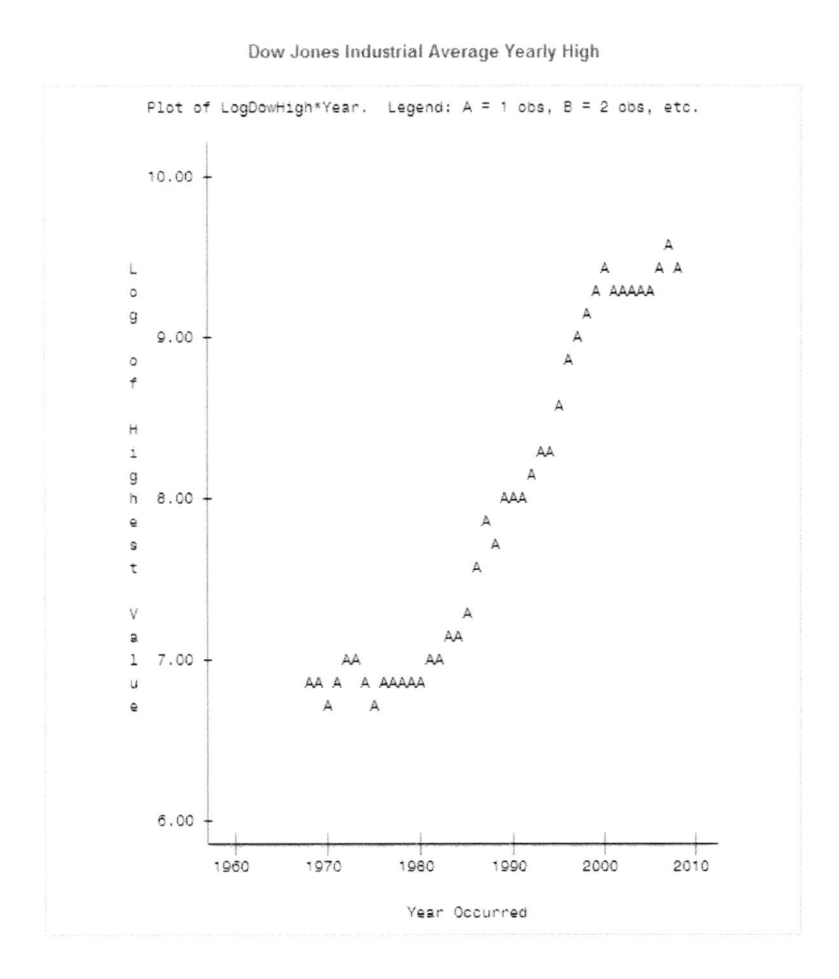

Plotting the log transformation of DowJonesHigh changes the exponential trend to a linear trend. The label for each variable is centered parallel to its axis.

## Specifying the Tick Marks Values

In the previous plots, the range on the horizontal axis is from 1960 to 2010. Tick marks and labels representing the years are spaced at intervals of 10. You can control the selection of the range and the interval on the horizontal axis with the HAXIS= option in the PLOT statement. A corresponding PLOT statement option, VAXIS=, controls the values of the tick mark on the vertical axis.

The forms of the HAXIS= and VAXIS= options follow. You must precede the first option in a PLOT statement with a slash.

**PLOT** *vertical\*horizontal* / HAXIS=*tick-value-list*;

**PLOT** *vertical\*horizontal* / VAXIS=*tick-value-list*;

where *tick-value-list* is a list of all values to assign to tick marks.

For example, to specify tick marks every five years from 1969 to 2010, use the following option:

```
haxis=1960 1965 1970 1975 1980 1985 1990 1995 2000 2005 2010
```

Or, you can abbreviate this list of tick marks:

```
haxis=1960 to 2010 by 5
```

The following program uses the HAXIS= option to specify the tick mark values for the horizontal axis:

```
proc plot data=highlow;
   plot LogDowHigh*Year / haxis=1968 to 2008 by 4;
   label LogDowHigh='Log of Highest Value'
         Year='Year Occurred';
   title 'Dow Jones Industrial Average Yearly High';
run;
```

The following output shows the plot:

*Figure 30.5* Specifying the Range and the Intervals of the Horizontal Axis

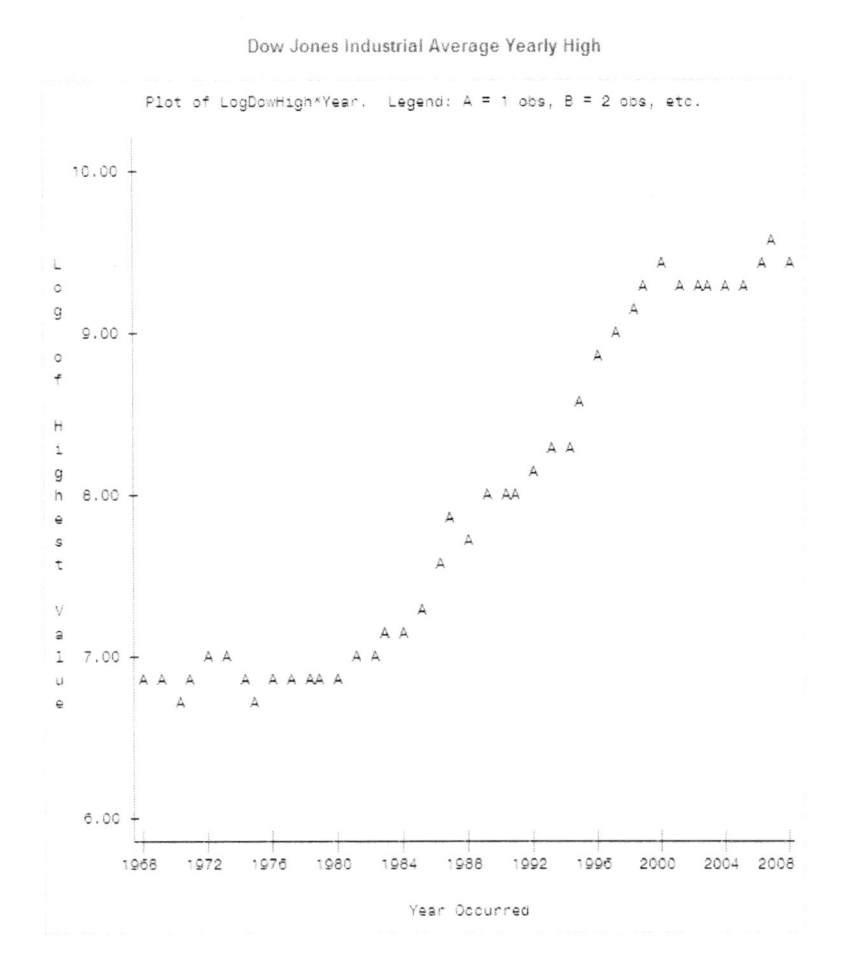

```
                      Dow Jones Industrial Average Yearly High

           Plot of LogDowHigh*Year.    Legend: A = 1 obs, B = 2 obs, etc.

     10.00 +
                                                                           A
  L                                                                A     A A
  o                                                          A  A AA A A
  g                                                       A
      9.00 +                                            A
  o                                                   A
  f
                                                  A
  H                                             A A
  i                                           A
  g                                       A AA
  h   8.00 +
  e                                     A
  s                                   A
  t
                                  A
  V                             A A
  a                         A A
  l   7.00 +       A A
  u           A A A    A  A A AA A
  e             A        A

      6.00 +
           +-----+-----+-----+-----+-----+-----+-----+-----+-----+-----+
          1968  1972  1976  1980  1984  1988  1992  1996  2000  2004  2008

                                     Year Occurred
```

The range of the horizontal axis is from 1968 to 2008, and the tick marks are now arranged at four-year intervals.

## Specifying Plotting Symbols

By default, PROC PLOT uses the letter A as the plotting symbol to indicate one observation, the letter B as the plotting symbol if two observations coincide, the letter C if three coincide, and so on. The letter Z represents 26 or more coinciding observations.

If you are plotting two sets of data on the same pair of axes, you can use the following form of the PLOT statement to specify your own plotting symbols:

**PLOT** *vertical\*horizontal='character'*;

where *character* is a plotting symbol to mark each point on the plot. PROC PLOT uses this character to represent values from one or more observations.

The following program uses the plus sign (+) as the plotting symbol for the plot:

```
proc plot data=highlow;
   plot LogDowHigh*Year='+' / haxis=1968 to 2008 by 4;
   label LogDowHigh='Log of Highest Value'
         Year='Year Occurred';
   title 'Dow Jones Industrial Average Yearly High';
```

run;

The plotting symbol must be enclosed in either single or double quotation marks.

The following output shows the plot:

**Figure 30.6** *Specifying a Plotting Symbol*

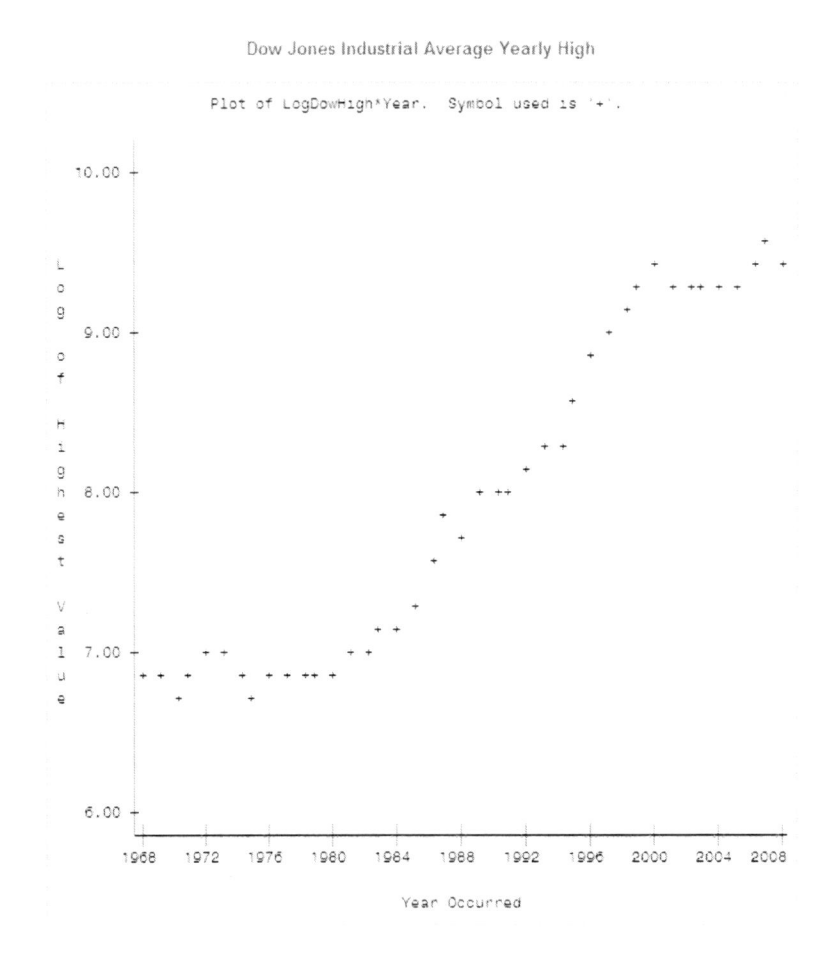

*Note:* When a plotting symbol is specified, PROC PLOT uses that symbol for all points on the plot regardless of how many observations might coincide. If observations coincide, then a message appears at the bottom of the plot telling how many observations are hidden.

## Removing the Legend

Often, a few simple changes to a plot can improve its appearance. You can draw a frame around the entire plot, rather than just on the left side and bottom. This makes it easier to determine the values that the plotting symbols represent on the left side of the plot. Also, you can suppress the legend when the labels clearly identify the variables in the plot or when the association between the plotting symbols and the variables is clear.

The following program uses the NOLEGEND option to suppress the legend and the BOX option to box the entire plot:

```
proc plot data=highlow nolegend;
   plot LogDowHigh*Year='+' / haxis=1968 to 2008 by 4
                              box;
   label LogDowHigh='Log of Highest Value'
```

```
            Year='Year Occurred';
   title 'Dow Jones Industrial Average Yearly High';
run;
```

The following output shows the plot:

**Figure 30.7** Removing the Legend

*Dow Jones Industrial Average Yearly High*

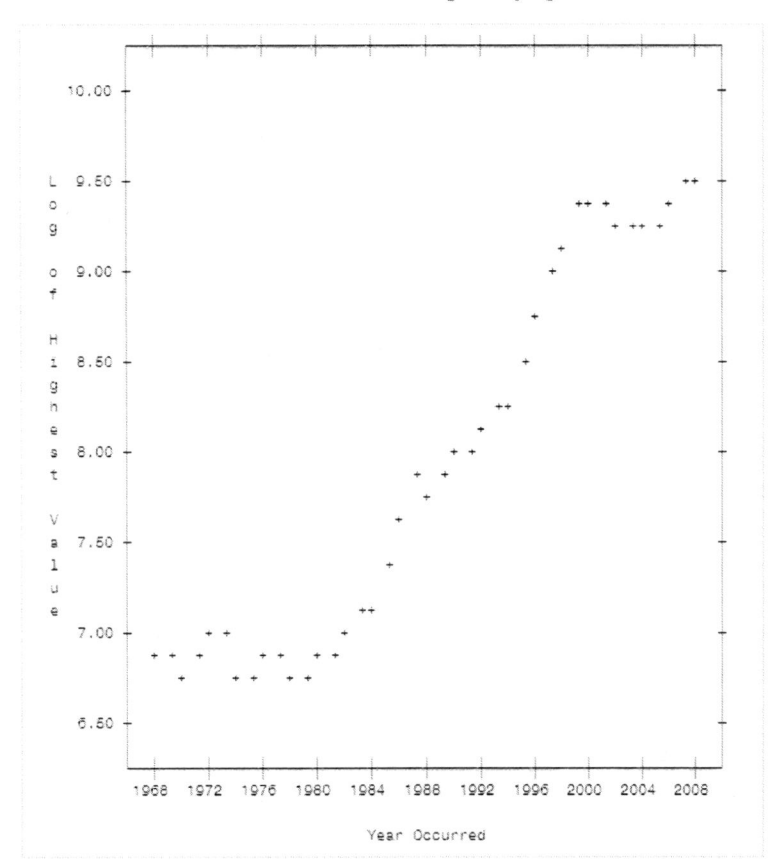

[scatter plot showing yearly high values from 1968 to 2008, ranging from approximately 6.50 to 9.50+ on the Log of Highest Value axis]

# Plotting Multiple Sets of Variables

## Creating Multiple Plots on Separate Pages

You can compare trends for different sets of measures by creating multiple plots. To request more than one plot from the same SAS data set, simply specify additional sets of variables in the PLOT statement. The form of the statement is

**PLOT** *vertical-1\*horizontal-1 vertical-2\*horizontal-2*;

All the options that you list in a PLOT statement apply to all of the plots that the statement produces.

The following program uses the PLOT statement to produce separate plots of the highest and lowest values of the Dow Jones Industrial Average from 1968 to 2008:

```
proc plot data=highlow;
```

```
      plot LogDowHigh*Year='+' LogDowLow*Year='o'
                          / haxis=1968 to 2008 by 4 box;
      label LogDowHigh='Log of Highest Value'
            LogDowLow='Log of Lowest Value'
            Year='Year Occurred';
      title 'Dow Jones Industrial Average Yearly High';
run;
```

The following output shows the plots:

**Figure 30.8** *Creating Multiple Plots on Separate Pages*

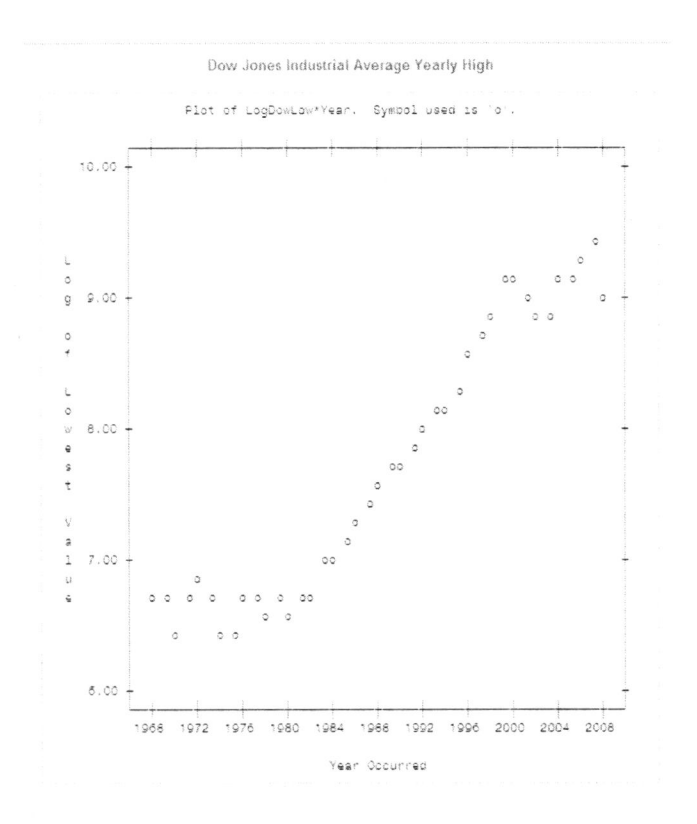

The plots appear on separate pages and use different vertical axes. Different plotting symbols represent the high and low values of the Dow Jones Industrial Average.

### Creating Multiple Plots on the Same Page

You can more easily compare the trends in different sets of measures when the plots appear on the same page. PROC PLOT provides two options that display multiple plots on the same page:

- the VPERCENT= option
- the HPERCENT= option

You can specify these options in the PROC PLOT statement by using one of the following forms:

**PROC PLOT** <DATA=*SAS-data-set*> VPERCENT=*number*;

**PROC PLOT** <DATA=*SAS-data-set*> HPERCENT=*number*;

where *number* is the percent of the vertical or the horizontal space given to each plot. You can substitute the aliases VPCT= and HPCT= for these options.

To fit two plots on a page, one beneath the other, as in Figure 30.9 on page 544, use VPERCENT=50; to fit three plots, use VPERCENT=33; and so on. To fit two plots on a page, side by side, use HPERCENT=50; to fit three plots, as in Figure 30.10 on page 545, use HPERCENT=33; and so on. Figure 30.11 on page 545 combines both of these options in the same PLOT statement to create a matrix of plots. Because the VPERCENT= option and the HPERCENT= option appear in the PROC PLOT statement, they affect all plots that are created in the PROC PLOT step.

The following examples show the position of the plots:

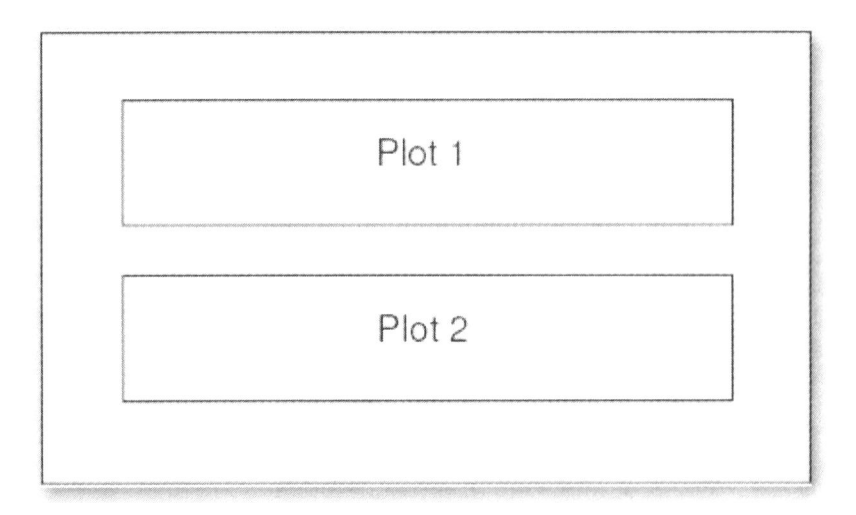

**Figure 30.9** Plots Produced with VPERCENT=50

*Figure 30.10* Plots Produced with HPERCENT=33

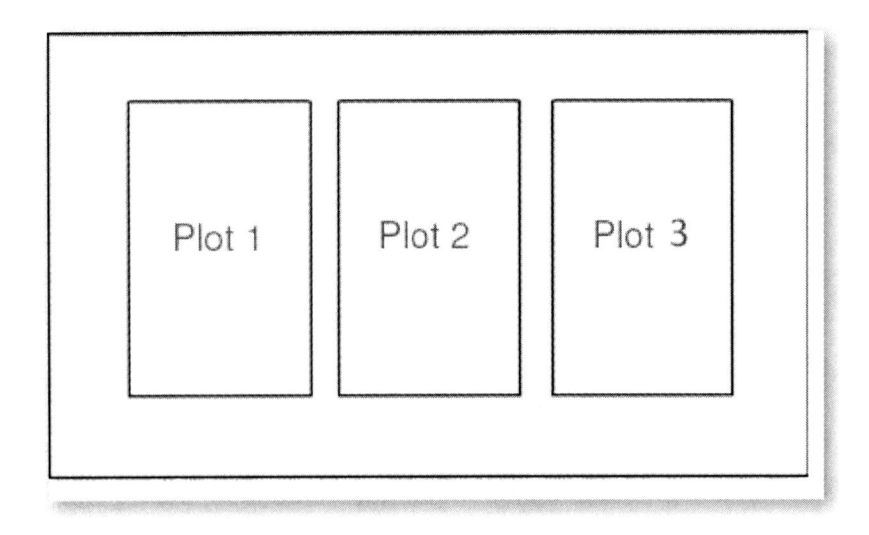

*Figure 30.11* Plots Produced with VPERCENT=50 and HPERCENT=33

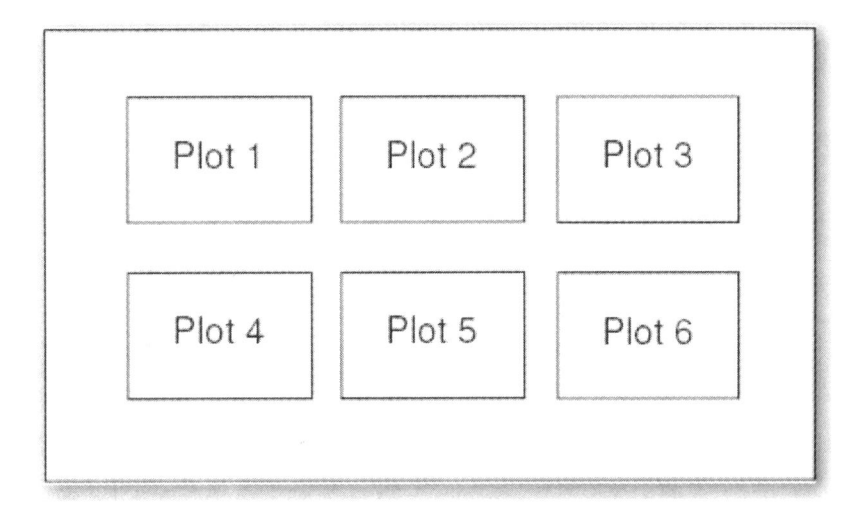

The following program uses the VPERCENT= option to display two plots on the same page so that you can more easily compare the high and the low Dow Jones values:

```
proc plot data=highlow vpercent=50;
   plot LogDowHigh*Year='+' LogDowLow*Year='o'
                          / haxis=1968 to 2008 by 4 box;
   label LogDowHigh='Log of High'
         LogDowLow='Log of Low'
         Year='Year Occurred';
   title 'Dow Jones Industrial Average Yearly High';
run;
```

In the output, PROC PLOT uses 50% of the vertical space on the page to display each plot.

**546** Chapter 30 • *Plotting the Relationship between Variables*

The following output displays the plots:

**Figure 30.12** *Creating Multiple Plots on the Same Page*

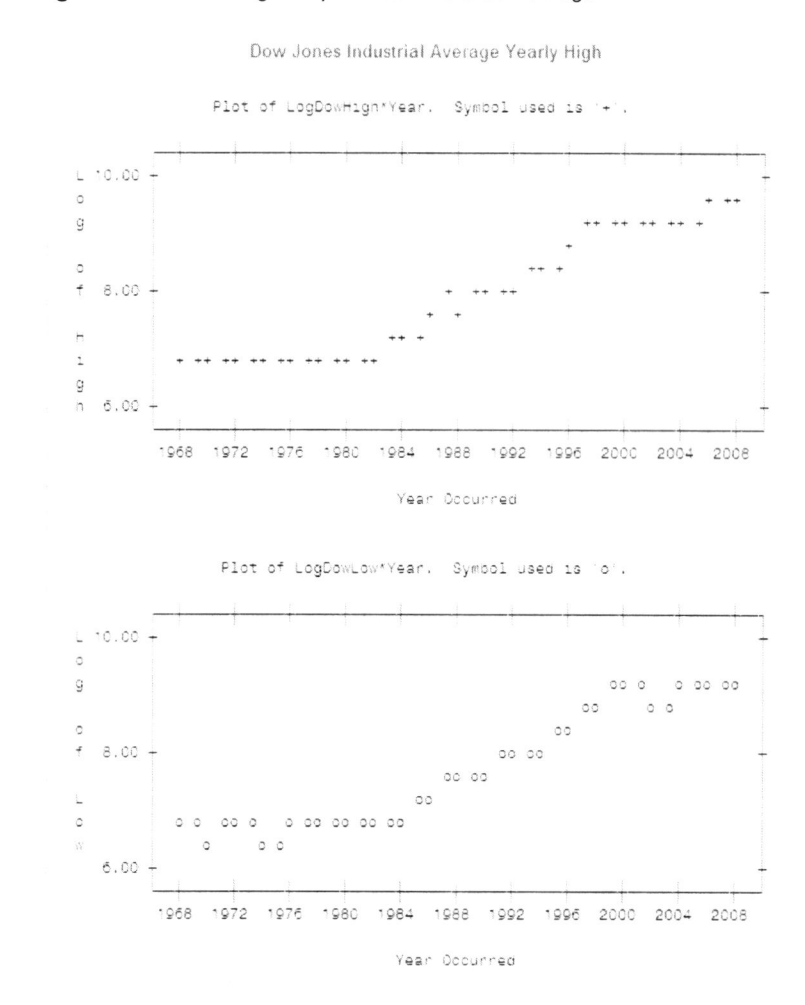

[Figure: Dow Jones Industrial Average Yearly High - two plots, one of LogDowHigh*Year using symbol '+' and one of LogDowLow*Year using symbol 'o', both plotted against Year Occurred from 1968 to 2008]

The two plots appear on the same page, one beneath the other.

## Plotting Multiple Sets of Variables on the Same Axes

The easiest way to compare trends in multiple sets of measures is to superimpose the plots on one set of axes by using the OVERLAY option in the PLOT statement. The variable names, or variable labels if they exist, from the first plot become the axes labels. Unless you use the HAXIS= option or the VAXIS= option, PROC PLOT automatically scales the axes to best fit all the variables.

The following program uses the OVERLAY option to plot the high and the low Dow Jones Industrial Average values on the same pair of axes:

```
proc plot data=highlow;
   plot LogDowHigh*Year='+' LogDowLow*Year='o'
                   / haxis=1968 to 2008 by 4
                     overlay box;
   label LogDowHigh='Log of High or Low'
         Year='Year Occurred';
   title 'Dow Jones Industrial Average';
run;
```

A new label for the variable LogDowHigh is specified because PROC PLOT uses only this variable to label the vertical axis.

The following output displays the plot:

**Figure 30.13** *Overlaying Two Plots*

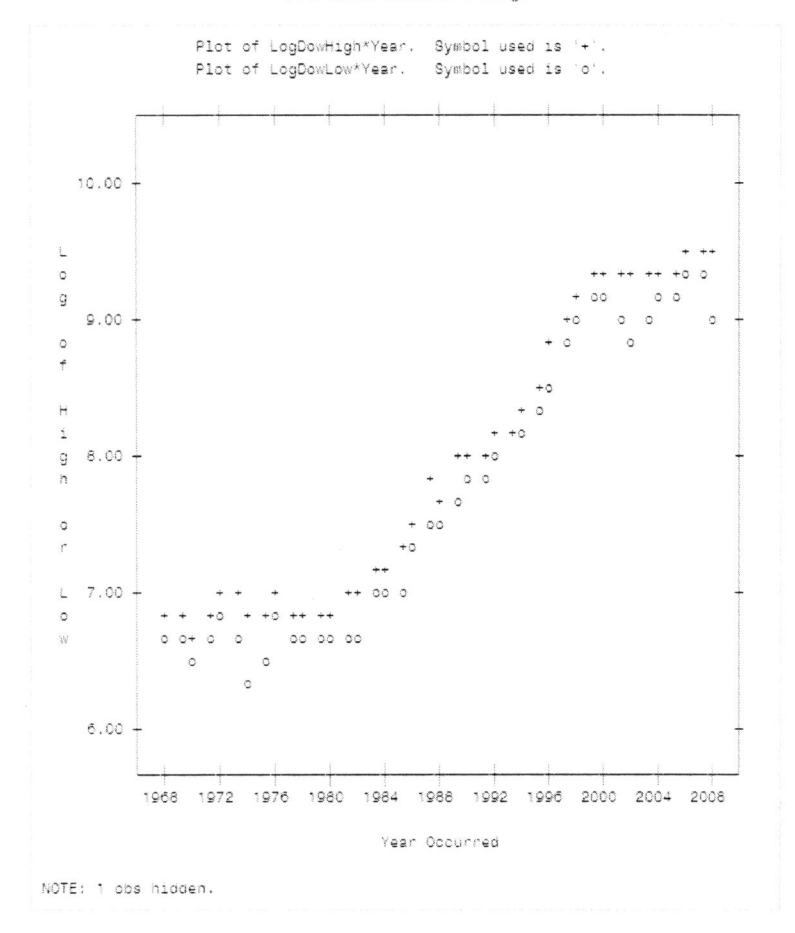

```
                      Dow Jones Industrial Average

         Plot of LogDowHigh*Year.   Symbol used is '+'.
         Plot of LogDowLow*Year.    Symbol used is 'o'.

        +----+----+----+----+----+----+----+----+----+----+----+
  10.00 +                                                      +
 L      |
 o      |                                                +  ++
 g      |                                        ++ ++ ++ +o o
        |                                      + oo       o  o
   9.00 +                                    +o   o o         o
 o      |                                   + o o     o
 f      |
        |                                 +o
 H      |                                + o
 i      |                              +  +o
 g 8.00 +                           ++ +o
 h      |                           +  o o
        |                          + o
 o      |                        + oo
 r      |                       +o
        |                      ++
 L 7.00 +         + +  +      ++ oo o                           +
 o      |      + + +o  + +o ++ ++
 w      |       o o+ o  o    oo oo oo
        |            o    o
        |               o
   6.00 +                                                      +
        +----+----+----+----+----+----+----+----+----+----+----+
          1968 1972 1976 1980 1984 1988 1992 1996 2000 2004 2008

                              Year Occurred

NOTE: 1 obs hidden.
```

The linear trends in the high and low Dow Jones values over the years from 1968 to 2008 are easily noticed.

*Note:* When the SAS system option OVP is in effect and overprinting is allowed, the plots are superimposed. Otherwise, when NOOVP is in effect, PROC PLOT uses the plotting symbol from the first plot to represent points that appear in more than one plot. In such a case, the output includes a message telling you how many observations are hidden.

# Summary

## PROC PLOT Statements

**PROC PLOT** <DATA=*SAS-data-set*> <*options*>;
   **LABEL** *variable='label'*;

**PLOT** *request-list* </*option(s)*>;

**TITLE**<*n*> <'*title*'>;

PROC PLOT <DATA=*SAS-data-set*> <*option(s)*> ;
: starts the PLOT procedure. You can specify the following *option(s)* in the PROC PLOT statement:

DATA=*SAS-data-set*
: names the SAS data set that PROC PLOT uses. If you omit DATA=, then PROC PLOT uses the most recently created data set.

HPERCENT=*percent(s)*
: specifies one or more percentages of the available horizontal space to use for each plot. HPERCENT= enables you to put multiple plots on one page. PROC PLOT tries to fit as many plots as possible on a page. After using each of the *percent(s)*, PROC PLOT cycles back to the beginning of the list. A zero in the list forces PROC PLOT to go to a new page even though it could fit the next plot on the same page.

NOLEGEND
: suppresses the default legend. The legend lists the names of the variables being plotted and the plotting symbols that are used in the plot.

VPERCENT=*percent(s)*
: specifies one or more percentages of the available vertical space to use for each plot. If you use a percentage greater than 100, then PROC PLOT prints sections of the plot on successive pages.

LABEL *variable*='*label*';
: specifies to use labels for the axes. *Variable* names the variable to label and *label* specifies a string of up to 256 characters, which includes blanks. The *label* must be enclosed in single or double quotation marks.

PLOT *request-list* </*option(s)*>;
: enables you to request individual plots in the *request-list* in the PLOT statement. Each element in the list has the following form:

: *vertical*\**horizontal*<='*symbol*'>

: where *vertical* and *horizontal* are the names of the variables that appear on the axes and *symbol* is the character to use for all points on the plot.

: You can request any number of plot statements in one PROC PLOT step. A list of options pertains to a single plot statement.

BOX
: draws a box around the entire plot, rather than only on the left side and bottom.

HAXIS=<*tick-value-list*>
: specifies the tick mark values for the horizontal axis. The *tick-value-list* consists of a list of values to use for tick marks.

OVERLAY
: superimposes all of the plots that are requested in the PLOT statement on one set of axes. The variable names, or variable labels if they exist, from the first plot are used to label the axes. Unless you use the HAXIS= or the VAXIS= option, PROC PLOT automatically scales the axes in the way that best fits all the variables.

VAXIS=<*tick-value-list*>
: specifies tick mark values for the vertical axis. The *tick-value-list* consists of a list of values to use for tick marks.

TITLE<*n*> <'*title*'>;
> specifies a title. The argument *n* is a number from 1 to 10 that immediately follows the word TITLE, with no intervening blank, and specifies the level of the TITLE. The text of each *title* must be enclosed in single or double quotation marks. The maximum title length that is allowed depends on your operating environment and the value of the LINESIZE= system option. Refer to the SAS documentation for your operating environment for more information.

# Learning More

PROC CHART and PROC UNIVARIATE
> When you are preparing graphics presentations, some data lends itself to charts, whereas other data is better suited for plots. For a discussion about how to make a variety of charts, see Chapter 31, "Producing Charts to Summarize Variables," on page 551.

PROC PLOT
> You can also use PROC PLOT to create contour plots, to draw a reference line at a particular value on a plot, and to change the borders of the plot. For complete documentation, see *Base SAS Procedures Guide*.

SAS functions
> SAS provides a wide array of numeric functions that include arithmetic and algebraic expressions, trigonometric and hyperbolic expressions, probability distributions, simple statistics, and random number generation. For complete documentation, see *SAS Functions and CALL Routines: Reference*.

# Chapter 31
# Producing Charts to Summarize Variables

| | |
|---|---|
| **Introduction to Producing Charts to Summarize Variables** | 552 |
| Purpose | 552 |
| Prerequisites | 552 |
| **Understanding the Charting Tools** | 552 |
| **Input File and SAS Data Set for Examples** | 553 |
| **Charting Frequencies with the CHART Procedure** | 555 |
| Types of Frequency Charts | 555 |
| Creating Vertical Bar Charts | 556 |
| Creating a Horizontal Bar Chart | 558 |
| Creating Block Charts | 560 |
| Creating Pie Charts | 561 |
| **Customizing Frequency Charts** | 563 |
| Changing the Number of Ranges | 563 |
| Specifying Midpoints for a Numeric Variable | 564 |
| Specifying the Number of Midpoints in a Chart | 566 |
| Charting Every Value | 567 |
| Charting the Frequency of a Character Variable | 569 |
| Charting Mean Values | 572 |
| Creating a Three-Dimensional Chart | 573 |
| **Creating High-Resolution Histograms** | 574 |
| Understanding How to Use the HISTOGRAM Statement | 574 |
| Understanding How to Use SAS/GRAPH to Create Histograms | 575 |
| Creating a Simple Histogram | 575 |
| Changing the Axes of a Histogram | 577 |
| Displaying Summary Statistics in a Histogram | 582 |
| Creating a Comparative Histogram | 584 |
| **Summary** | 586 |
| PROC CHART Statements | 586 |
| PROC UNIVARIATE Statements | 588 |
| GOPTIONS Statement | 590 |
| FORMAT Statement | 590 |
| **Learning More** | 590 |

# Introduction to Producing Charts to Summarize Variables

## Purpose

Charts, like plots, provide a technique to summarize data graphically. You can use a chart to show the values of a single variable or several variables. A bar chart also enables you to graphically examine the distribution of the values of a variable.

In this section, you will learn how to create the following:

- vertical bar charts
- horizontal bar charts
- pie charts
- block charts
- high-resolution histograms and comparative histograms

The examples range in complexity from simple frequency bar charts to more complex charts that group variables and include summary statistics.

## Prerequisites

To understand the examples in this section, you should be familiar with the following features and concepts:

- the LABEL statement
- the TITLE statement
- SAS system options
- creating and assigning SAS formats

# Understanding the Charting Tools

Base SAS software provides two procedures that produce charts:

- PROC CHART
- PROC UNIVARIATE

PROC CHART produces a variety of charts for character or numeric variables. The charts include vertical and horizontal bar charts, block charts, pie charts, and star charts. These types of charts graphically display the values of a variable or a statistic that are associated with those values. PROC UNIVARIATE produces histograms for continuous numeric variables that enable you to visualize the distribution of your data.

PROC CHART is a useful tool to visualize data quickly. However, you can use PROC GCHART[1] to produce high-resolution, publication-quality bar charts that include color

---

[1] PROC GCHART and PROC CHART produce identical charts.

and various fonts when your site licenses SAS/GRAPH software. You can use PROC UNIVARIATE to customize the histograms by adding tables with summary statistics directly on the graphical display. PROC UNIVARIATE also enables you to overlay the histogram with fitted density curves or kernel density estimates so that you can examine the underlying distribution of your data.

# Input File and SAS Data Set for Examples

The examples in this section use one input file and one SAS data set. For a complete listing of the input data, see "The YEAR_SALES Data Set" on page 816. The input file contains the enrollment and exam grades for an introductory chemistry course. The 50 students enrolled in the course attend several lectures, and a discussion section one day a week. The input file has the following structure:

```
Abdallah        F Mon   46 Anderson        M Wed   75
Aziz            F Wed   67 Bayer           M Wed   77
Bhatt           M Fri   79 Blair           F Fri   70
Bledsoe         F Mon   63 Boone           M Wed   58
Burke           F Mon   63 Chung           M Wed   85
Cohen           F Fri   89 Drew            F Mon   49
Dubos           M Mon   41 Elliott         F Wed   85
...more data lines...
Simonson        M Wed   62 Smith N         M Wed   71
Smith R         M Mon   79 Sullivan        M Fri   77
Swift           M Wed   63 Wolfson         F Fri   79
Wong            F Fri   89 Zabriski        M Fri   89
```

The input file contains the following values from left to right:

- the student's last name (and first initial if necessary)
- the student's gender (F or M)
- the day of the week for the student's discussion section (Mon, Wed, or Fri)
- the student's first exam grade

The following program creates the GRADES data set that this section uses. This example shows the first fifteen observations:

```
options pagesize=60 linesize=80 pageno=1 nodate;

data grades;
   infile 'your-input-file';
   input Name & $14. Gender : $2. Section : $3. ExamGrade1 @@;
run;

proc print data=grades;
   title 'Introductory Chemistry Exam Scores';
run;

options obs=15;
data grades;
   input Name &$14. Gender :$2. Section :$3. ExamGrade1 @@;
   datalines;
Abdallah        F Mon   46 Anderson        M Wed   75
Aziz            F Wed   67 Bayer           M Wed   77
Bhatt           M Fri   79 Blair           F Fri   70
```

```
   Bledsoe      F Mon  63  Boone        M Wed  58
   Burke        F Mon  63  Chung        M Wed  85
   Cohen        F Fri  89  Drew         F Mon  49
   Dubos L      M Mon  41  Elliott      F Wed  85
   Farmer       F Wed  58  Franklin     F Wed  59
   Freeman      F Mon  79  Friedman     M Mon  58
   Gabriel      M Fri  75  Garcia       M Mon  79
   Harding      M Mon  49  Hazelton     M Mon  55
   Hinton       M Fri  85  Hung         F Fri  98
   Jacob        F Wed  64  Janeway      F Wed  51
   Jones        F Mon  39  Jorgensen    M Mon  63
   Judson       F Fri  89  Kuhn         F Mon  89
   LeBlanc      F Fri  70  Lee          M Fri  48
   Litowski     M Fri  85  Malloy       M Wed  79
   Meyer        F Fri  85  Nichols      M Mon  58
   Oliver       F Mon  41  Park         F Mon  77
   Patel        M Wed  73  Randleman    F Wed  46
   Robinson     M Fri  64  Shien        M Wed  55
   Simonson     M Wed  62  Smith N      M Wed  71
   Smith R      M Mon  79  Sullivan     M Fri  77
   Swift        M Wed  63  Wolfson      F Fri  79
   Wong         F Fri  89  Zabriski     M Fri  89
   ;
```

*Note:* Most output in this section uses an OPTIONS statement that specifies PAGESIZE=40 and LINESIZE=80. Other examples use an OPTIONS statement with a different line size or page size to make a chart more readable. When the PAGESIZE= and LINESIZE= options are set, they remain in effect until you reset the options with another OPTIONS statement, or you end the SAS session.

The following output displays the first 15 observations:

**Figure 31.1** Introductory Chemistry Exam Scores

### Introductory Chemistry Exam Scores

| Obs | Name | Gender | Section | ExamGrade1 |
|---|---|---|---|---|
| 1 | Abdallah | F | Mon | 46 |
| 2 | Anderson | M | Wed | 75 |
| 3 | Aziz | F | Wed | 67 |
| 4 | Bayer | M | Wed | 77 |
| 5 | Bhatt | M | Fri | 79 |
| 6 | Blair | F | Fri | 70 |
| 7 | Bledsoe | F | Mon | 63 |
| 8 | Boone | M | Wed | 58 |
| 9 | Burke | F | Mon | 63 |
| 10 | Chung | M | Wed | 85 |
| 11 | Cohen | F | Fri | 89 |
| 12 | Drew | F | Mon | 49 |
| 13 | Dubos | M | Mon | 41 |
| 14 | Elliott | F | Wed | 85 |
| 15 | Farmer | F | Wed | 58 |

You can create bar charts with this data set to do the following:

- Examine the distribution of grades.
- Determine a letter grade for each student.
- Compare the number of students in each section.
- Compare the number of males and females in each section.
- Compare the performance of the students in different sections.

# Charting Frequencies with the CHART Procedure

## Types of Frequency Charts

By default, PROC CHART creates a frequency chart in which each bar, section, or block in the chart represents a range of values. By default, PROC CHART selects ranges based on the values of the chart variable. At the center of each range is a midpoint. A midpoint does not always correspond to an actual value of the chart variable. The size of each bar, block, or section represents the number of observations that fall in that range.

PROC CHART makes several types of charts:

vertical and horizontal bar charts
: display the magnitude of data with the length or height of bars.

block charts
: display the relative magnitude of data with blocks of varying size.

pie charts
: display data as wedge-shaped sections of a circle that represent the relative contribution of each section to the whole circle.

star charts
: display data as bars that radiate from a center point, like spokes in a wheel.

The shape of each type of chart emphasizes a certain aspect of the data. The chart that you choose depends on the nature of your data and the aspect that you want to emphasize.

## Creating Vertical Bar Charts

### Understanding Vertical Bar Charts

A vertical bar chart emphasizes individual ranges. The horizontal, or midpoint, axis shows the values of the variable divided into ranges. By default, the vertical axis shows the frequency of values for a given range. The differences in bar heights enable you to quickly determine which ranges contain many observations and which contain few observations.

The VBAR statement in a PROC CHART step produces vertical bar charts. If you use the VBAR statement without any options, then PROC CHART automatically does the following:

- scales the vertical axis
- determines the bar width
- selects the spacing between bars
- labels the axes

For continuous numeric data, PROC CHART determines the number of bars and the midpoint for each bar from the minimum and maximum value of the chart variable. For character variables or discrete numeric variables, PROC CHART creates a bar for each value of the chart variable. However, you can change how PROC CHART determines the axes by using options.

*Note:* If the number of characters per line (LINESIZE=) is not sufficient to display vertical bars, then PROC CHART automatically produces a horizontal bar chart.

### The Program

The following program uses the VBAR statement to create a vertical bar chart of frequencies for the numeric variable ExamGrade1:

```
options pagesize=40 linesize=80 pageno=1 nodate;

proc chart data=grades;
   vbar ExamGrade1;
   title 'Grades for First Chemistry Exam';
```

```
run;
```

The following output displays the bar chart:

**Figure 31.2** *Using a Vertical Bar Chart to Show Frequencies*

```
                    Grades for First Chemistry Exam

Frequency

14 ┤                     xxxxx
                         xxxxx
13 ┤                     xxxxx
                         xxxxx
12 ┤                     xxxxx
                         xxxxx
11 ┤                     xxxxx              xxxxx
                         xxxxx              xxxxx
10 ┤                     xxxxx              xxxxx     xxxxx
                         xxxxx              xxxxx     xxxxx
 9 ┤                     xxxxx              xxxxx     xxxxx
                         xxxxx              xxxxx     xxxxx
 8 ┤                     xxxxx              xxxxx     xxxxx
                         xxxxx              xxxxx     xxxxx
 7 ┤                     xxxxx              xxxxx     xxxxx
                         xxxxx              xxxxx     xxxxx
 6 ┤         xxxxx       xxxxx              xxxxx     xxxxx
             xxxxx       xxxxx              xxxxx     xxxxx
 5 ┤         xxxxx       xxxxx    xxxxx     xxxxx     xxxxx
             xxxxx       xxxxx    xxxxx     xxxxx     xxxxx
 4 ┤         xxxxx       xxxxx    xxxxx     xxxxx     xxxxx
             xxxxx       xxxxx    xxxxx     xxxxx     xxxxx
 3 ┤ xxxxx   xxxxx       xxxxx    xxxxx     xxxxx     xxxxx
     xxxxx   xxxxx       xxxxx    xxxxx     xxxxx     xxxxx
 2 ┤ xxxxx   xxxxx       xxxxx    xxxxx     xxxxx     xxxxx
     xxxxx   xxxxx       xxxxx    xxxxx     xxxxx     xxxxx
 1 ┤ xxxxx   xxxxx       xxxxx    xxxxx     xxxxx     xxxxx     xxxxx
     xxxxx   xxxxx       xxxxx    xxxxx     xxxxx     xxxxx     xxxxx
     ─────────────────────────────────────────────────────────────────
       40      50          60       70        80        90       100

                           ExamGrade1 Midpoint
```

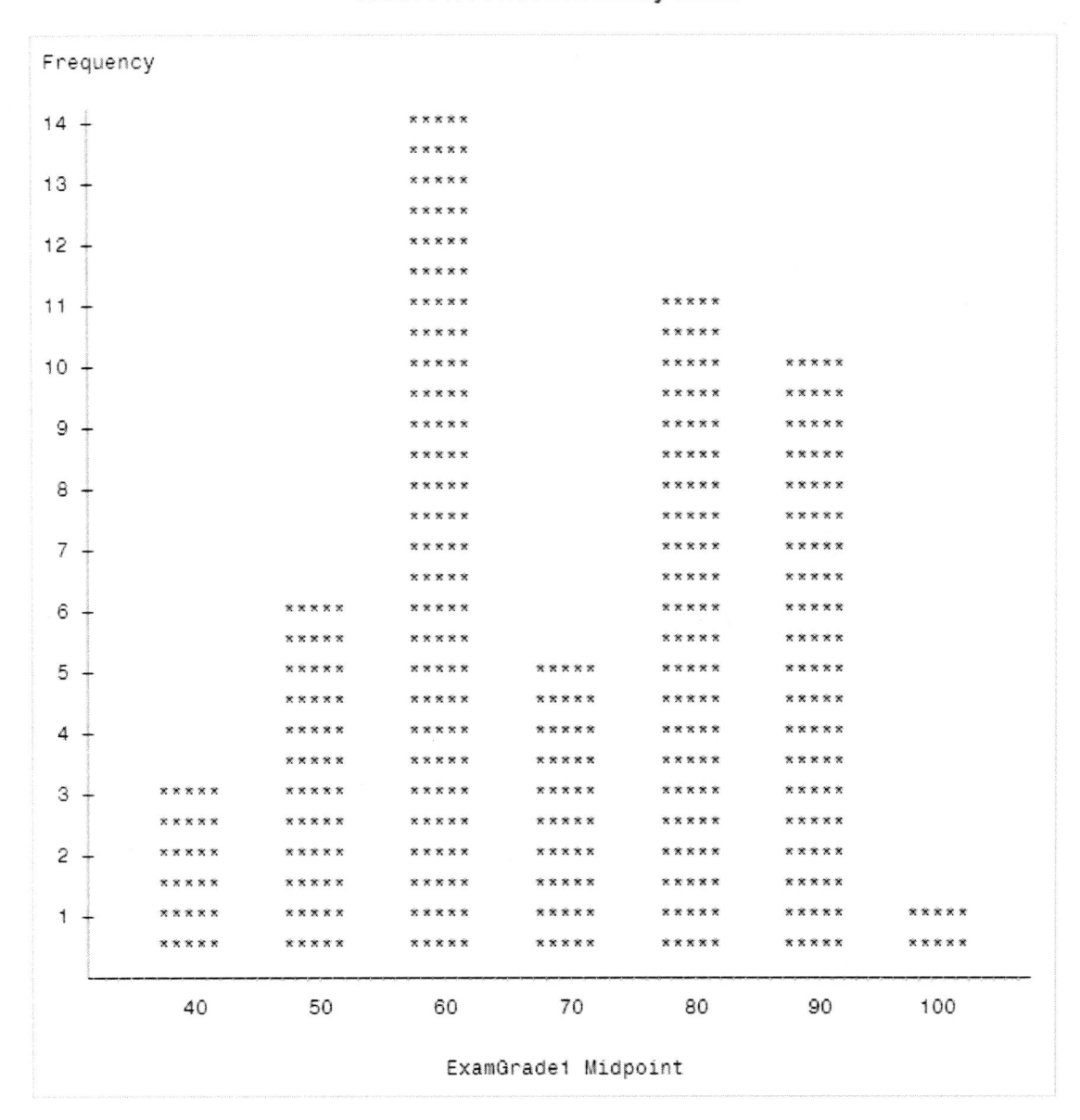

The midpoint axis for the above chart ranges from 40 to 100 and is incremented in intervals of 10. The following table shows the values and frequency of each bar:

**Table 31.1** *Values and Frequency*

| Range    | Midpoint | Frequency |
|----------|----------|-----------|
| 35 to 44 | 40       | 3         |
| 45 to 54 | 50       | 6         |

| Range | Midpoint | Frequency |
|---|---|---|
| 55 to 64 | 60 | 14 |
| 65 to 74 | 70 | 5 |
| 75 to 84 | 80 | 11 |
| 85 to 94 | 90 | 10 |
| 95 to 104 | 10 | 1 |

*Note:* Because PROC CHART selects the size of the ranges and the location of their midpoints based on all values of the numeric variable, the highest and lowest ranges can extend beyond the values in the data. In this example the lowest grade is 39 while the lowest range extends from 35 to 44. Similarly, the highest grade is 98 while the highest range extends from 95 to 104.

## Creating a Horizontal Bar Chart

### Understanding Horizontal Bar Charts

A horizontal bar chart has essentially the same characteristics as a vertical bar chart. Both charts emphasize individual ranges. However, a horizontal bar chart rotates the bars so that the horizontal axis shows frequency and the vertical axis shows the values of the chart variable. To the right of the horizontal bars, PROC CHART displays a table of statistics that summarizes the data.

The HBAR statement in a PROC CHART step produces horizontal bar charts. By default, the table of statistics includes frequency, cumulative frequency, percentage, and cumulative percentage. You can request specific statistics so that the table contains only these statistics and the frequency.

### Understanding HBAR Statistics

The default horizontal bar chart uses less space than charts of other shapes. PROC CHART takes advantage of the small size of horizontal bar charts and displays statistics to the right of the chart. The statistics include

Frequency
    is the number of observations in a given range.

Cumulative Frequency
    is the number of observations in all ranges up to and including a given range. The cumulative frequency for the last range is equal to the number of observations in the data set.

Percent
    is the percentage of observations in a given range.

Cumulative Percent
    is the percentage of observations in all ranges up to and including a given range. The cumulative percentage for the last range is always 100.

Various options enable you to control the statistics that appear in the table. You can select the statistics by using the following options: FREQ, CFREQ, PERCENT, and CPERCENT. To suppress the table of statistics, use the NOSTAT option.

## The Programs

The following program uses the HBAR statement to create a horizontal bar chart of the frequency for the variable ExamGrade1:

```
options pagesize=40 linesize=80 pageno=1 nodate;

proc chart data=grades;
   hbar Examgrade1;
   title 'Grades for First Chemistry Exam';
run;
```

The following output displays the bar chart:

*Figure 31.3* Using a Horizontal Bar Chart to Show Frequencies

```
                       Grades for First Chemistry Exam

ExamGrade1                                           Cum.            Cum.
 Midpoint                                   Freq     Freq  Percent  Percent

    40     ******                              3        3    6.00     6.00

    50     ************                        6        9   12.00    18.00

    60     ******************************     14       23   28.00    46.00

    70     **********                          5       28   10.00    56.00

    80     ***********************            11       39   22.00    78.00

    90     ********************               10       49   20.00    98.00

   100     **                                  1       50    2.00   100.00
           +----+----+----+----+----+----+----+
                2    4    6    8   10   12   14
                           Frequency
```

The cumulative percent shows that the median grade for the exam (the grade that 50% of observations lie above and 50% below) lies within the midpoint of 70.

The next example produces the same horizontal bar chart as above, but the program uses the NOSTAT option to eliminate the table of statistics.

```
options pagesize=40 linesize=80 pageno=1 nodate;

proc chart data=grades;
   hbar Examgrade1 / nostat;
   title 'Grades for First Chemistry Exam';
run;
```

The following output displays the bar chart:

**Figure 31.4** *Removing Statistics from a Horizontal Bar Chart*

```
                      Grades for First Chemistry Exam

   ExamGrade1
    Midpoint

       40      xxxxxxxxxxx

       50      xxxxxxxxxxxxxxxxxxxxxxxx

       60      xxxxxxxxxxxxxxxxxxxxxxxxxxxxxxxxxxxxxxxxxxxxxxxxxxxxxxxx

       70      xxxxxxxxxxxxxxxxxxx

       80      xxxxxxxxxxxxxxxxxxxxxxxxxxxxxxxxxxxxxxxxxxxx

       90      xxxxxxxxxxxxxxxxxxxxxxxxxxxxxxxxxxxxxx

      100      xxxx

               +----+----+----+----+----+----+----+----+----+----+----+----+----+----+
               1    2    3    4    5    6    7    8    9   10   11   12   13   14
                                             Frequency
```

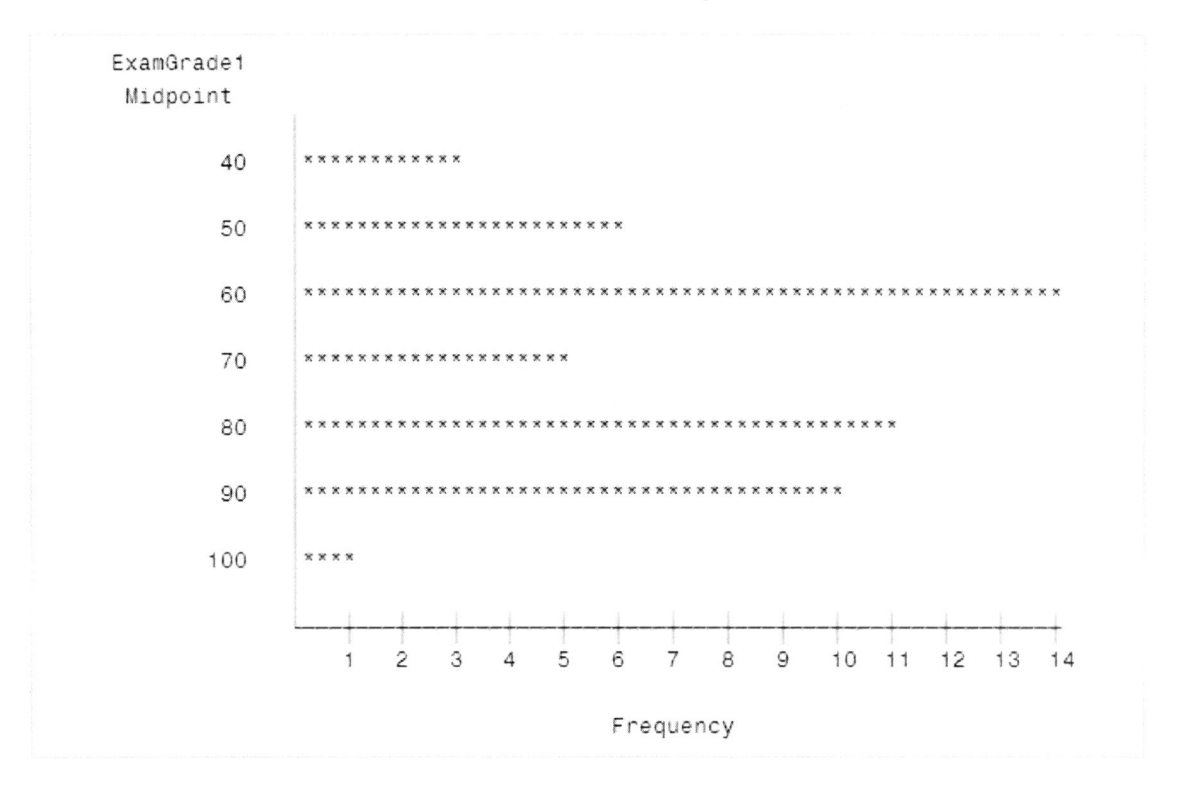

## Creating Block Charts

### Understanding Block Charts

A block chart shows the relative magnitude of data by using blocks of varying height. Each block in a square represents a category of data. A block chart is similar to a vertical bar chart. It uses a more sophisticated presentation of the data to emphasize the individual ranges. However, a block chart is less precise than a bar chart because the maximum height of a block is 10 lines.

The BLOCK statement in a PROC CHART step produces a block chart. You can also use the BLOCK statement to create three-dimensional frequency charts. For an example, see "Creating a Three-Dimensional Chart" on page 573. If you create block charts with a large number of charted values, then you might have to adjust the SAS system options LINESIZE= and PAGESIZE= so that the block chart fits on one page.

*Note:* If the line size or page size is not sufficient to display all the bars, then PROC CHART automatically produces a horizontal bar chart.

### The Program

The following program uses the BLOCK statement to create a block frequency chart for the numeric variable ExamGrade1:

```
options linesize=120 pagesize=40 pageno=1 nodate;
```

```
proc chart data=grades;
   block Examgrade1;
   title 'Grades for First Chemistry Exam';
run;
```

The OPTIONS statement increases the line size to 120.

The following output displays the block chart:

*Figure 31.5  Using a Block Chart to Show Frequencies*

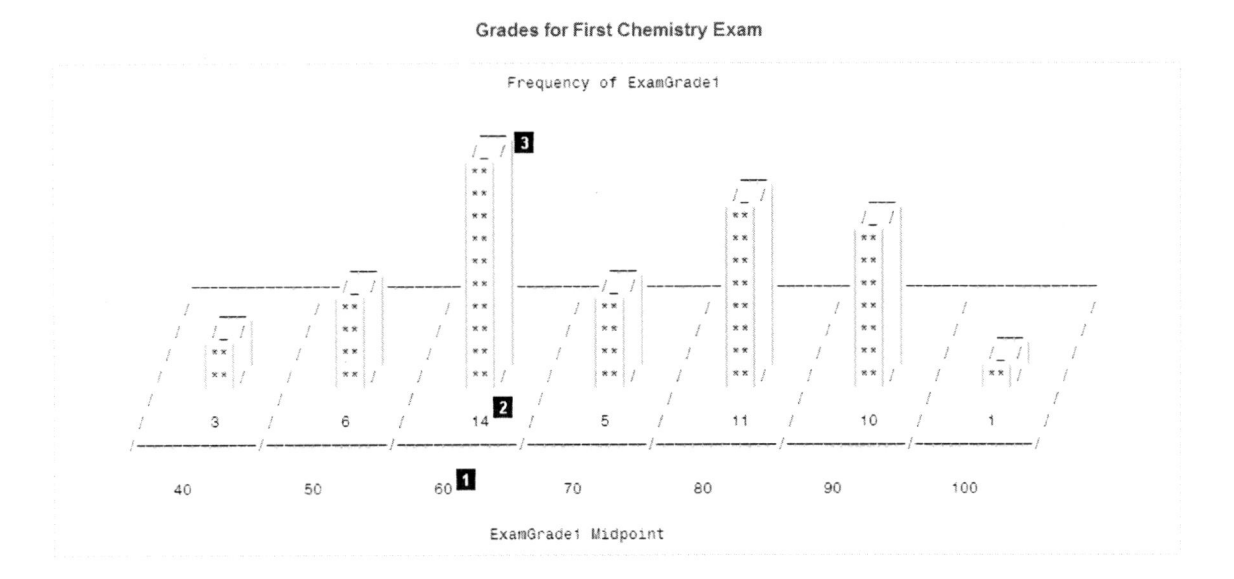

```
                        Grades for First Chemistry Exam
                            Frequency of ExamGrade1

                                    /_/  3
                                    **
                                    **
                                    **
                                    **
                                    **
                                    **         /_/       /_/
                                    **         **        **
          ————————/_/————————————— ** ———/_/— ** ——————— ** ————————
          /    /   /   **  /   **  /  **  /  **  /   **  /   /_/  /
          / /_/ /   /  **  /   **  /  **  /  **  /   **  /   **   /
          / **  /   /  **  /   **  /  **  /  **  /   **  /   **   /
          / **  /   /  **  /   **  /  **  /  **  /   **  /   **   /
          /     /   /      /       /      /      /       /        /
               3       6        14 2     5       11     10      1

         /————————/————————/————————/————————/————————/————————/
         40       50       60 1     70       80       90      100
                                ExamGrade1 Midpoint
```

The chart shows the effects of using the BLOCK statement.

1. PROC CHART uses the same midpoints for both the bar chart and block chart. The midpoints appear beneath the chart.
2. The number of observations represented by each block appear beneath the block.
3. The height of a block is proportional to the number of observations in a block.

## Creating Pie Charts

### Understanding Pie Charts

A pie chart emphasizes the relative contribution of parts (a range of values) to the whole. Graphing the distribution of grades as a pie chart shows you the size of each range relative to the others just as the vertical bar chart does. However, the pie chart also enables you to visually compare the number of grades in a range to the total number of grades.

The PIE statement in a PROC CHART step produces a pie chart. PROC CHART determines the number of sections for the pie chart the same way it determines the number of bars for a vertical chart, with one exception: if any slices of the pie account for fewer than three print positions, then PROC CHART groups them into a category called "Other."

PROC CHART displays the values of the midpoints around the perimeter of the pie chart. Inside each section of the chart, PROC CHART displays the number of observations in the range and the percentage of observations that the number represents.

The SAS system options LINESIZE= and PAGESIZE= determine the size of the pie. If your printer does not print 6 lines per inch and 10 columns per inch, then the pie looks elliptical. To make a circular pie chart, you must use the LPI= option in the PROC CHART statement. For more information, see the CHART procedure in the *Base SAS Procedures Guide*.

### The Program

The following program uses the PIE statement to create a pie chart of frequencies for the numeric variable ExamGrade1:

```
options pagesize=40 linesize=80 pageno=1 nodate;

proc chart data=grades;
   pie ExamGrade1;
   title 'Grades for First Chemistry Exam';
run;
```

The following output displays the pie chart:

*Figure 31.6* Using a Pie Chart to Show Frequencies

```
                    Grades for First Chemistry Exam

                           Frequency of ExamGrade1


                    60    ************
                       ****              ****
                      ***                  ***
                     **                      **
                    **                        ** 50
                   *                            *
                   *         14                 *
                  **        28.00%      6      **
                  *                    12.00%   *
                 *                               *
                **                              **
                *                         *      40
                *                    3         *
                *                    6.00%     *
                *    5         +         1    *
           70   *   10.00%              2.00%  *  Other
                *                              *
                *                              *
                **                            **
                 *                10           *
                  *              20.00%       *
                  **   11                    **
                   *  22.00%                 *
                   *                         *
                    **                  ** 90
                     **                **
                      ***             ***
                    80  ****         ****
                          ************
```

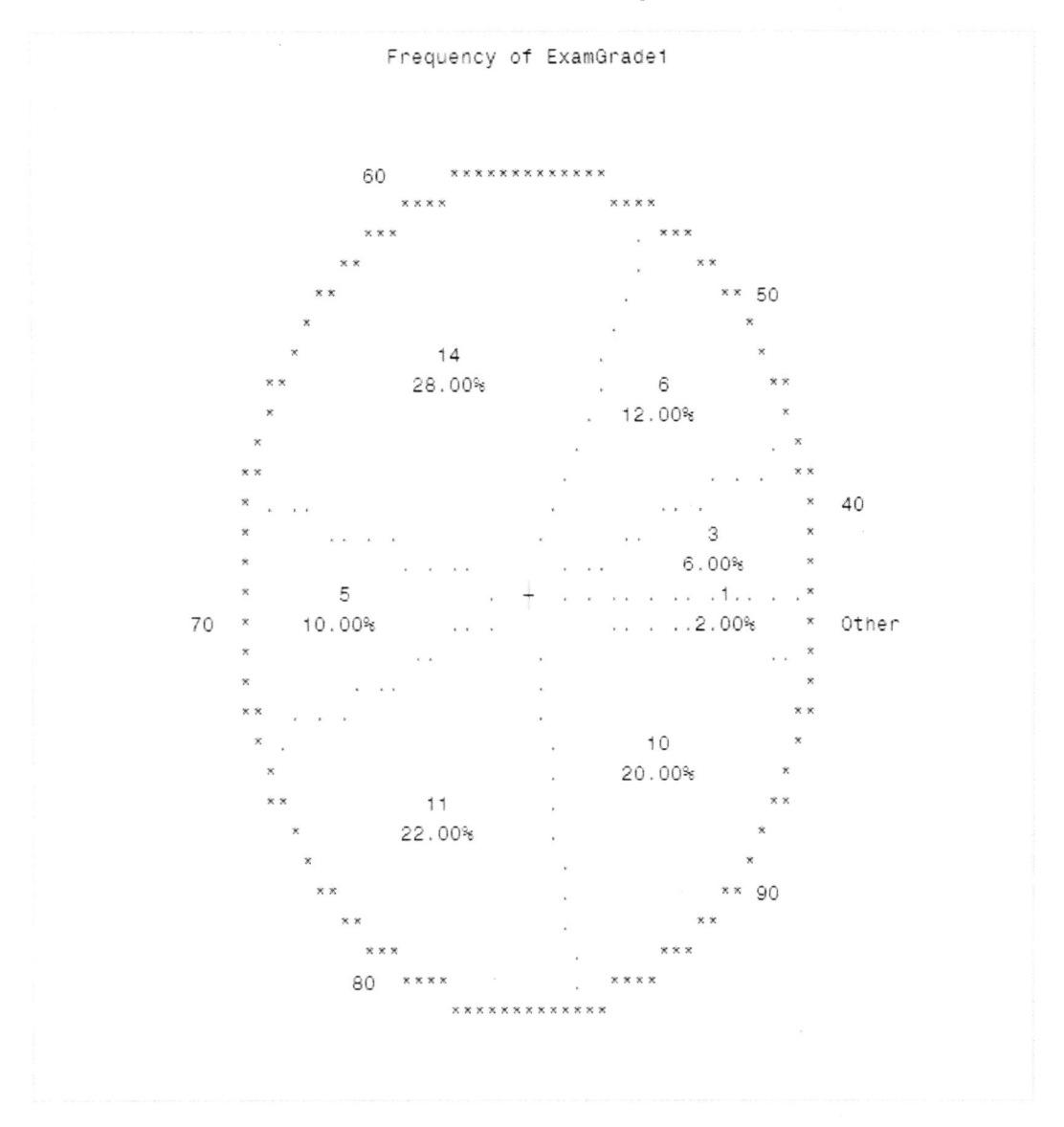

In this pie chart the **Other** section represents the one grade in the range with a midpoint of 100. The size of a section corresponds to the number of observations that fall in its range.

# Customizing Frequency Charts

## *Changing the Number of Ranges*

You can change the appearance of the charts in the following ways:

**Table 31.2** Chart Appearance

| Action | Option |
|---|---|
| Specify midpoints that define the range of values that each bar, block, or section represents | MIDPOINTS= option |
| Specify the number of bars on the chart and let PROC CHART compute the midpoints | LEVELS= option |
| Specify a variable that contains discrete numeric values. PROC CHART produces a bar chart with a bar for each distinct value | DISCRETE option |

*Note:* Most examples in this section use vertical bar charts. However, unless documented otherwise, you can use any of the options in the PIE, BLOCK, or HBAR statements.

### Specifying Midpoints for a Numeric Variable

You can specify midpoints for a continuous numeric variable by using the MIDPOINTS= option in the VBAR statement. The form of this option is

**VBAR** *variable* / MIDPOINTS=*midpoints-list*;

where midpoints-list is a list of the numbers to use as midpoints.

For example, to specify the traditional grading ranges with midpoints from 55 to 95, use the following option:

```
midpoints=55 65 75 85 95
```

Or, you can abbreviate the list of midpoints:

```
midpoints=55 to 95 by 10
```

The corresponding ranges are as follows:

```
50 to 59
60 to 69
70 to 79
80 to 89
90 to 99
```

The following program uses the MIDPOINTS= option to create a bar chart for ExamGrade1:

```
options pagesize=40 linesize=80 pageno=1 nodate;

proc chart data=grades;
   vbar Examgrade1 / midpoints=55 to 95 by 10;
   title 'Assigning Grades for First Chemistry Exam';
run;
```

The MIDPOINTS= option forces PROC CHART to center the five bars around the traditional midpoints for exam grades.

The following output displays the bar chart:

**Figure 31.7** *Specifying the Midpoints for a Vertical Bar Chart*

```
                Assigning Grades for First Chemistry Exam

  Frequency

  16 -       xxxxx
             xxxxx
  15 -       xxxxx              xxxxx
             xxxxx              xxxxx
  14 -       xxxxx              xxxxx
             xxxxx              xxxxx
  13 -       xxxxx              xxxxx
             xxxxx              xxxxx
  12 -       xxxxx              xxxxx
             xxxxx              xxxxx
  11 -       xxxxx              xxxxx
             xxxxx              xxxxx
  10 -       xxxxx              xxxxx     xxxxx
             xxxxx              xxxxx     xxxxx
   9 -       xxxxx              xxxxx     xxxxx
             xxxxx              xxxxx     xxxxx
   8 -       xxxxx     xxxxx    xxxxx     xxxxx
             xxxxx     xxxxx    xxxxx     xxxxx
   7 -       xxxxx     xxxxx    xxxxx     xxxxx
             xxxxx     xxxxx    xxxxx     xxxxx
   6 -       xxxxx     xxxxx    xxxxx     xxxxx
             xxxxx     xxxxx    xxxxx     xxxxx
   5 -       xxxxx     xxxxx    xxxxx     xxxxx
             xxxxx     xxxxx    xxxxx     xxxxx
   4 -       xxxxx     xxxxx    xxxxx     xxxxx
             xxxxx     xxxxx    xxxxx     xxxxx
   3 -       xxxxx     xxxxx    xxxxx     xxxxx
             xxxxx     xxxxx    xxxxx     xxxxx
   2 -       xxxxx     xxxxx    xxxxx     xxxxx
             xxxxx     xxxxx    xxxxx     xxxxx
   1 -       xxxxx     xxxxx    xxxxx     xxxxx    xxxxx
             xxxxx     xxxxx    xxxxx     xxxxx    xxxxx
             ------------------------------------------------
               55        65       75        85       95

                       ExamGrade1 Midpoint
```

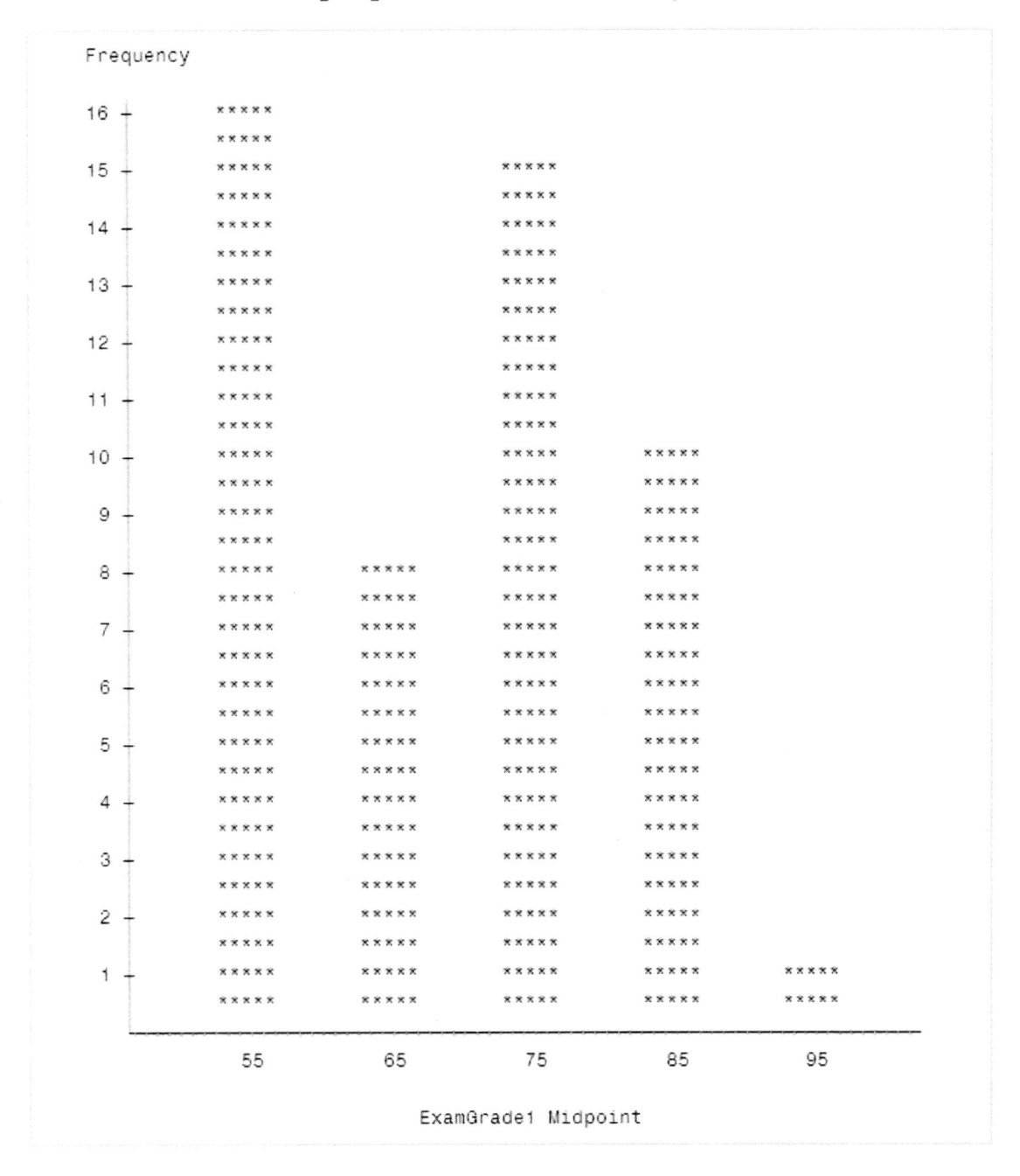

A traditional method to assign grades assumes that the data is normally distributed. However, the bars do not appear as a normal (bell-shaped) curve. If grades are assigned based on these midpoints and the traditional pass or fail boundary of 60, then a substantial portion of the class fails the exam because more observations fall in the bar around the midpoint of 55 than in any other bar.

### Specifying the Number of Midpoints in a Chart

You can specify the number of midpoints in the chart rather than the values of the midpoints by using the LEVELS= option. The procedure selects the midpoints.

The form of the option is

**VBAR** *variable* / LEVELS=*number-of-midpoints*;

where number-of-midpoints specifies the number of midpoints.

The following program uses the LEVELS= option to create a bar chart with five bars:[1]

```
options pagesize=40 linesize=80 pageno=1 nodate;

proc chart data=grades;
   vbar Examgrade1 / levels=5;
   title 'Assigning Grades for First Chemistry Exam';
run;
```

The LEVELS= option forces PROC CHART to compute only five midpoints.

---

[1] You can use SAS to normalize the data before the chart is created.

The following output displays the bar chart:

*Figure 31.8* Specifying Five Midpoints for a Vertical Bar

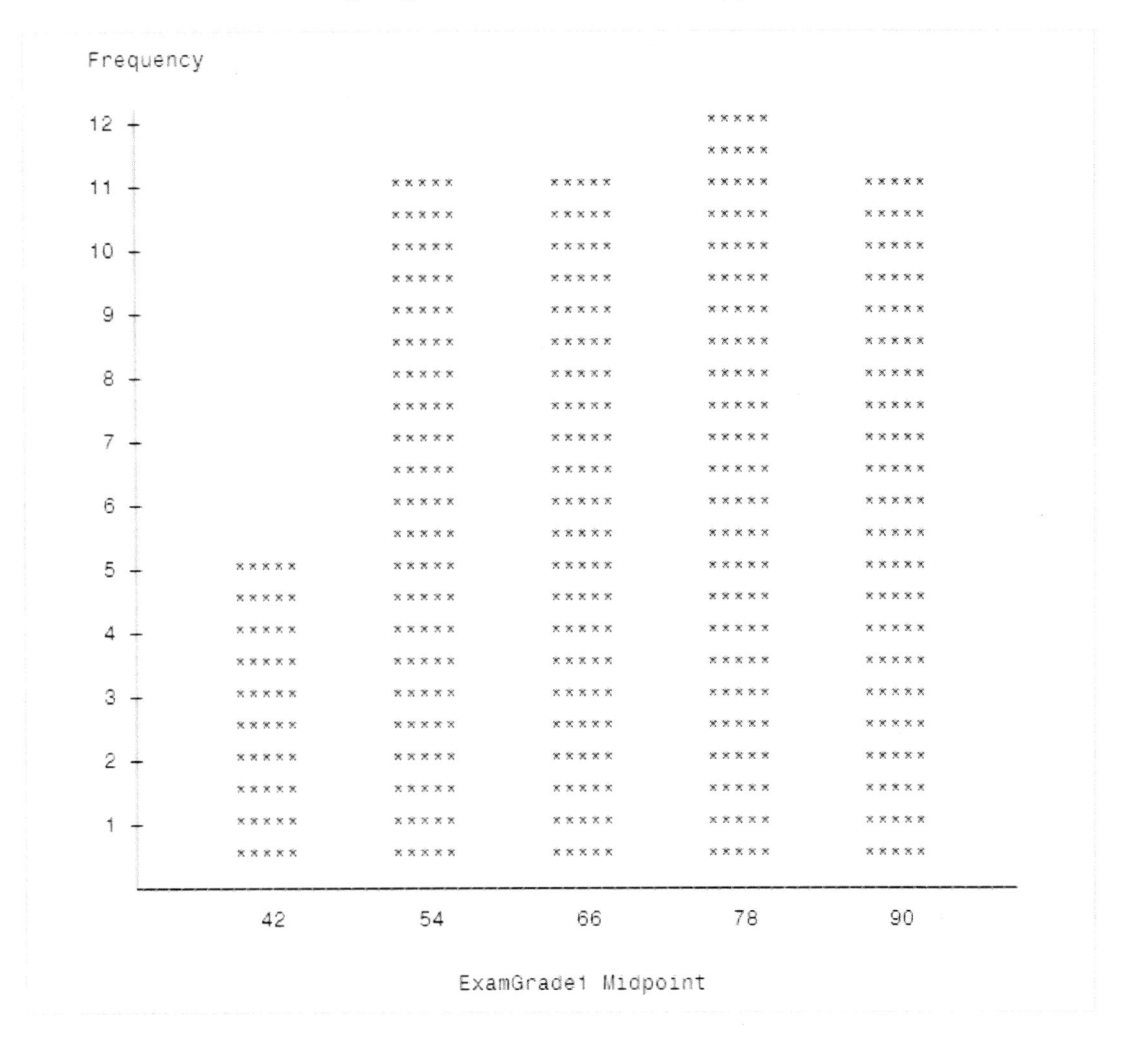

```
                      Assigning Grades for First Chemistry Exam

Frequency

  12 +                                              xxxxx
                                                    xxxxx
  11 +                    xxxxx       xxxxx         xxxxx         xxxxx
                          xxxxx       xxxxx         xxxxx         xxxxx
  10 +                    xxxxx       xxxxx         xxxxx         xxxxx
                          xxxxx       xxxxx         xxxxx         xxxxx
   9 +                    xxxxx       xxxxx         xxxxx         xxxxx
                          xxxxx       xxxxx         xxxxx         xxxxx
   8 +                    xxxxx       xxxxx         xxxxx         xxxxx
                          xxxxx       xxxxx         xxxxx         xxxxx
   7 +                    xxxxx       xxxxx         xxxxx         xxxxx
                          xxxxx       xxxxx         xxxxx         xxxxx
   6 +                    xxxxx       xxxxx         xxxxx         xxxxx
                          xxxxx       xxxxx         xxxxx         xxxxx
   5 +      xxxxx         xxxxx       xxxxx         xxxxx         xxxxx
            xxxxx         xxxxx       xxxxx         xxxxx         xxxxx
   4 +      xxxxx         xxxxx       xxxxx         xxxxx         xxxxx
            xxxxx         xxxxx       xxxxx         xxxxx         xxxxx
   3 +      xxxxx         xxxxx       xxxxx         xxxxx         xxxxx
            xxxxx         xxxxx       xxxxx         xxxxx         xxxxx
   2 +      xxxxx         xxxxx       xxxxx         xxxxx         xxxxx
            xxxxx         xxxxx       xxxxx         xxxxx         xxxxx
   1 +      xxxxx         xxxxx       xxxxx         xxxxx         xxxxx
            xxxxx         xxxxx       xxxxx         xxxxx         xxxxx
       ------------------------------------------------------------------
             42            54          66            78            90

                                ExamGrade1 Midpoint
```

Assigning grades for these midpoints results in three students with exam grades in the lowest range.

## Charting Every Value

By default, PROC CHART assumes that all numeric variables are continuous and automatically chooses intervals for them unless you use MIDPOINTS= or LEVELS=. You can specify that a numeric variable is discrete rather than continuous by using the DISCRETE option. PROC CHART creates a frequency chart with bars for each distinct value of the discrete numeric variable.

The following program uses the DISCRETE option to create a bar chart with a bar for each value of ExamGrade1:

```
options pagesize=40 linesize=80 pageno=1 nodate;

proc chart data=grades;
```

**568** *Chapter 31 • Producing Charts to Summarize Variables*

```
      vbar Examgrade1 / discrete;
      title 'Grades for First Chemistry Exam';
run;
```

The following output displays the bar chart:

*Figure 31.9* Specifying a Bar for Each Exam Grade

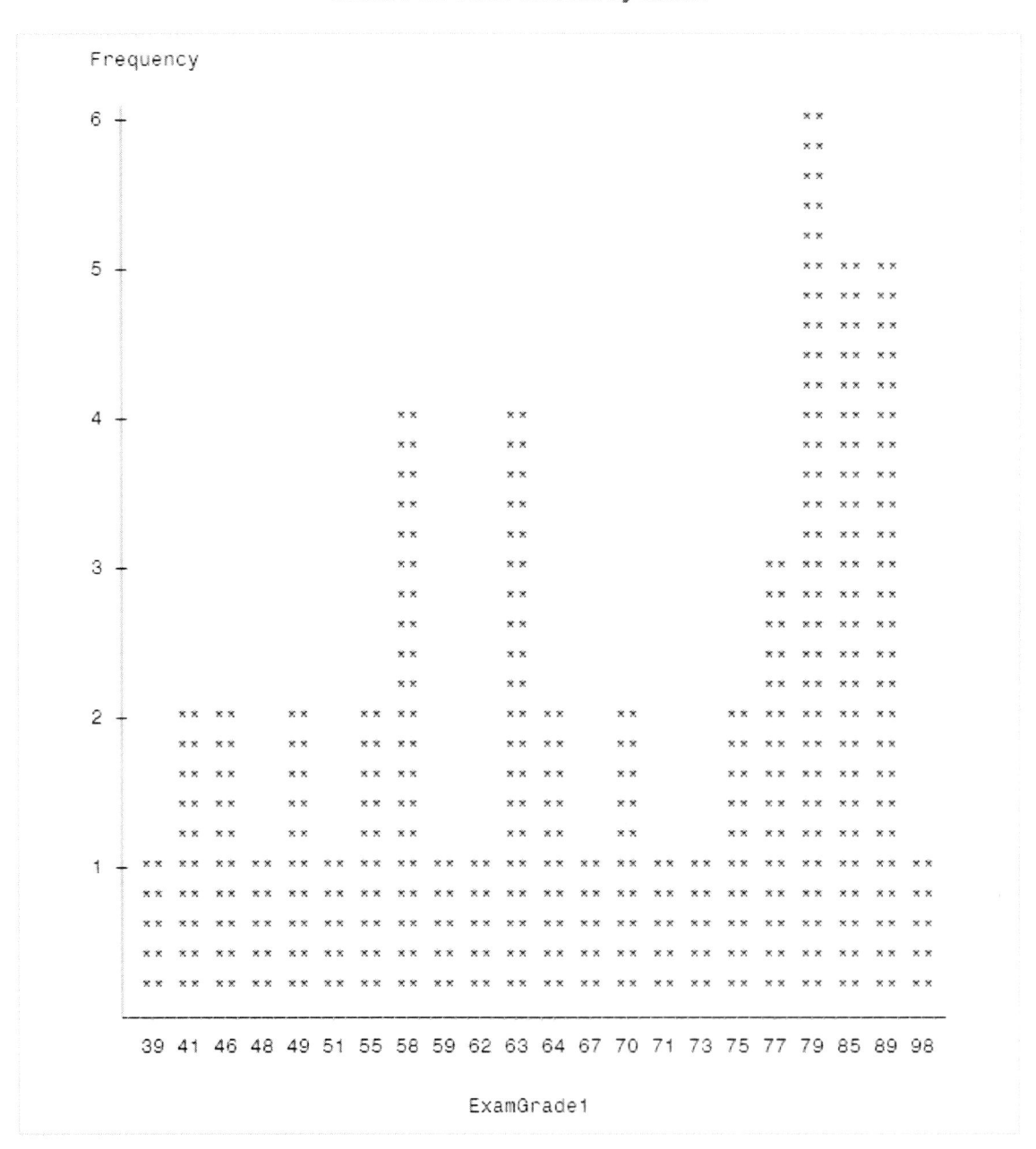

```
                    Grades for First Chemistry Exam

  Frequency

    6 +                                                      x x
                                                             x x
                                                             x x
                                                             x x
                                                             x x
    5 +                                                      x x   x x   x x
                                                             x x   x x   x x
                                                             x x   x x   x x
                                                             x x   x x   x x
                                                             x x   x x   x x
    4 +                    x x         x x                   x x   x x   x x
                           x x         x x                   x x   x x   x x
                           x x         x x                   x x   x x   x x
                           x x         x x                   x x   x x   x x
                           x x         x x                   x x   x x   x x
    3 +                    x x         x x             x x   x x   x x   x x
                           x x         x x             x x   x x   x x   x x
                           x x         x x             x x   x x   x x   x x
                           x x         x x             x x   x x   x x   x x
                           x x         x x             x x   x x   x x   x x
    2 +  x x  x x    x x   x x  x x    x x  x x   x x  x x   x x   x x   x x
         x x  x x    x x   x x  x x    x x  x x   x x  x x   x x   x x   x x
         x x  x x    x x   x x  x x    x x  x x   x x  x x   x x   x x   x x
         x x  x x    x x   x x  x x    x x  x x   x x  x x   x x   x x   x x
         x x  x x    x x   x x  x x    x x  x x   x x  x x   x x   x x   x x
    1 +  x x x x x x x x x x x x x x x x x x x x x x x x x x x x x x x x x x
         x x x x x x x x x x x x x x x x x x x x x x x x x x x x x x x x x x
         x x x x x x x x x x x x x x x x x x x x x x x x x x x x x x x x x x
         x x x x x x x x x x x x x x x x x x x x x x x x x x x x x x x x x x
         x x x x x x x x x x x x x x x x x x x x x x x x x x x x x x x x x x
         -----------------------------------------------------------------
         39 41 46 48 49 51 55 58 59 62 63 64 67 70 71 73 75 77 79 85 89 98

                                   ExamGrade1
```

The chart shows that in most cases only one or two students earned a given grade. However, clusters of three or more students earned grades of 58, 63, 77, 79, 85, and 89. The mode for this exam (most frequently earned exam grade) is 79.

*Note:* PROC CHART does not proportionally space the values of a discrete numeric variable on the horizontal axis.

### Charting the Frequency of a Character Variable

#### Overview
You can create charts of a character variable as well as a numeric variable. For example, to compare enrollment among sections, PROC CHART creates a chart that shows the number of students in each section.

Creating a frequency chart of a character variable is the same as creating a frequency chart of a numeric variable. However, the main difference between charting a numeric variable and charting a character variable is how PROC CHART selects the midpoints. By default, PROC CHART uses each value of a character variable as a midpoint, as if the DISCRETE option were in effect. You can limit the selection of midpoints to a subset of the variable's values. If you do not define a format for the chart variable, then a single bar, block, or section represents a single value of the variable.

#### Specifying Midpoints for a Character Variable
By default, the midpoints that PROC CHART uses for character variables are in alphabetical order. However, you can easily rearrange the order of the midpoints with the MIDPOINTS= option. When you use the MIDPOINTS= option for character variables, you must enclose the value of each midpoint in single or double quotation marks. Also, the values must correspond to values in the data set. For example,

```
midpoints='Mon' 'Wed' 'Fri'
```

uses the three days the class sections meet as midpoints.

The following program uses the MIDPOINTS= option to create a bar chart that shows the number of students enrolled in each section:

```
options pagesize=40 linesize=80 pageno=1 nodate;

proc chart data=grades;
   vbar Section / midpoints='Mon' 'Wed' 'Fri';
   title 'Enrollment for an Introductory Chemistry Course';
run;
```

The MIDPOINTS= option alters the chart so that the days of the week appear in chronological rather than alphabetical order.

The following output displays the bar chart:

**Figure 31.10** *Ordering Character Midpoints Chronologically*

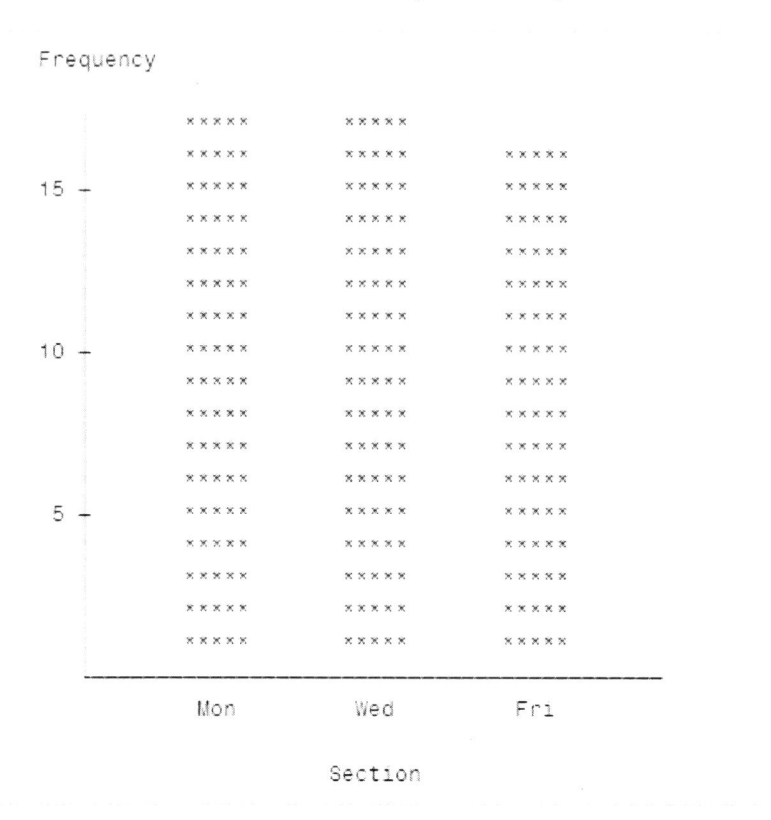

```
           Enrollment for an Introductory Chemistry Course

     Frequency

                  x x x x x      x x x x x
                  x x x x x      x x x x x      x x x x x
      15 -        x x x x x      x x x x x      x x x x x
                  x x x x x      x x x x x      x x x x x
                  x x x x x      x x x x x      x x x x x
                  x x x x x      x x x x x      x x x x x
                  x x x x x      x x x x x      x x x x x
      10 -        x x x x x      x x x x x      x x x x x
                  x x x x x      x x x x x      x x x x x
                  x x x x x      x x x x x      x x x x x
                  x x x x x      x x x x x      x x x x x
                  x x x x x      x x x x x      x x x x x
       5 -        x x x x x      x x x x x      x x x x x
                  x x x x x      x x x x x      x x x x x
                  x x x x x      x x x x x      x x x x x
                  x x x x x      x x x x x      x x x x x
                  x x x x x      x x x x x      x x x x x
              ---------------------------------------------
                    Mon            Wed            Fri

                              Section
```

The chart shows that the Monday and Wednesday sections have the same number of students; the Friday section has one less student.

## Creating Subgroups within a Range

You can show how a subgroup contributes to each bar or block by using the SUBGROUP= option in the BLOCK statement, HBAR statement, or VBAR statement. For example, you can use the SUBGROUP= option to explore patterns within a population (gender differences).

The SUBGROUP= option defines a variable called the subgroup variable. PROC CHART uses the first character of each value to fill in the portion of the bar or block that corresponds to that value, unless more than one value begins with the same first character. In that case, PROC CHART uses the letters A, B, C, and so on, to fill in the bars or blocks.

If you assign a format to the variable, then PROC CHART uses the first character of the formatted value. The characters that PROC CHART uses in the chart and the values that they represent are shown in a legend at the bottom of the chart.

PROC CHART orders the subgroup symbols as A through Z, and as 0 through 9, with the characters in ascending order. PROC CHART calculates the height of a bar or block for each subgroup individually and rounds the percentage of the total bar up or down. So the total height of the bar might be greater or less than the height of the same bar without the SUBGROUP= option.

# Customizing Frequency Charts

The following program uses GENDER as the subgroup variable to show how many members in each section are male and female:

```
options pagesize=40 linesize=80 pageno=1 nodate;

proc chart data=grades;
   vbar Section / midpoints='Mon' 'Wed' 'Fri'
               subgroup=Gender;
   title 'Enrollment for an Introductory Chemistry Course';
run;
```

The following output displays the bar chart:

**Figure 31.11**  Using Gender to Form Subgroups

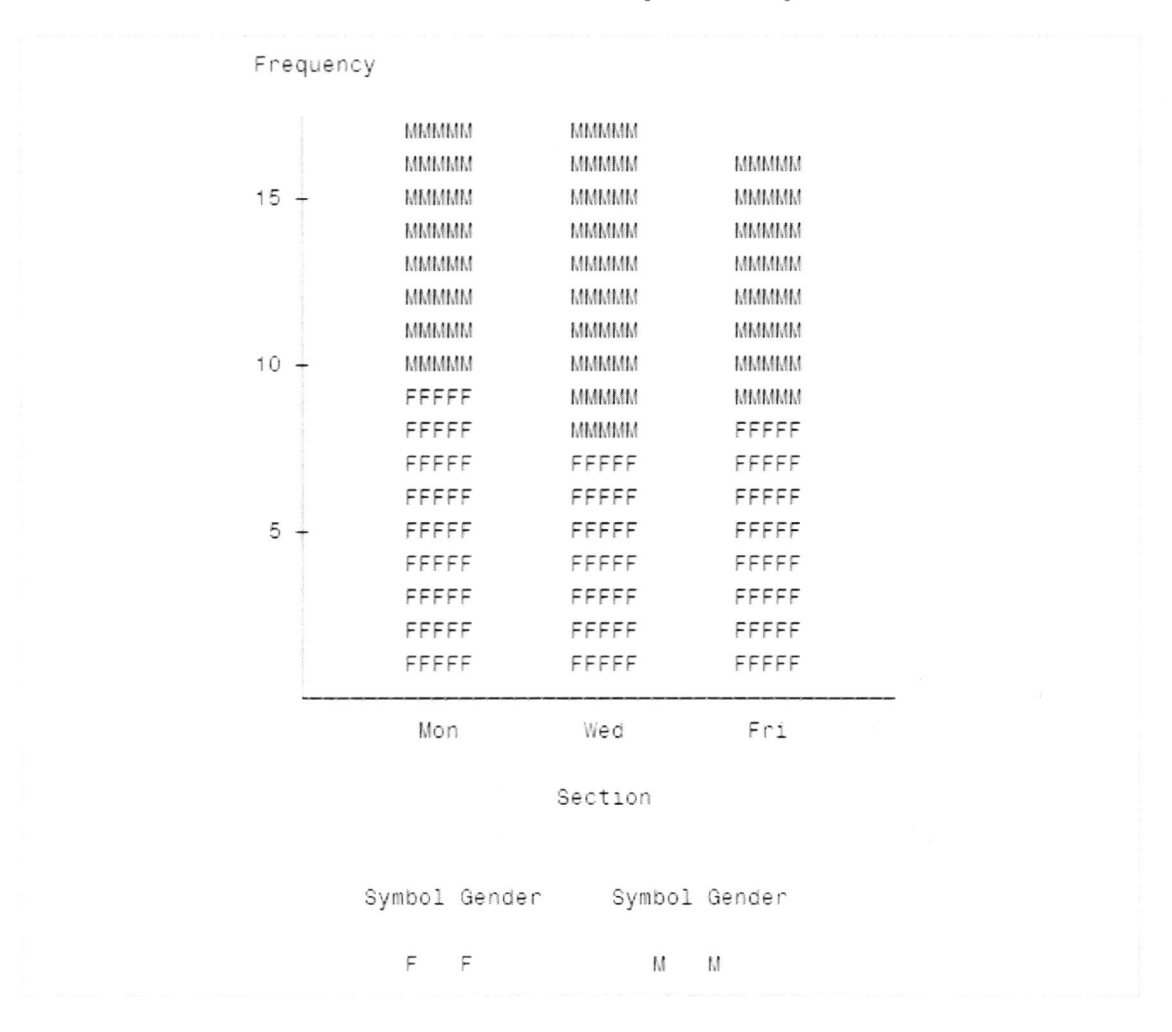

```
                Enrollment for an Introductory Chemistry Course

    Frequency
     |
     |          MMMMM    MMMMM
     |          MMMMM    MMMMM    MMMMM
  15 +          MMMMM    MMMMM    MMMMM
     |          MMMMM    MMMMM    MMMMM
     |          MMMMM    MMMMM    MMMMM
     |          MMMMM    MMMMM    MMMMM
     |          MMMMM    MMMMM    MMMMM
  10 +          MMMMM    MMMMM    MMMMM
     |          FFFFF    MMMMM    MMMMM
     |          FFFFF    MMMMM    FFFFF
     |          FFFFF    FFFFF    FFFFF
     |          FFFFF    FFFFF    FFFFF
   5 +          FFFFF    FFFFF    FFFFF
     |          FFFFF    FFFFF    FFFFF
     |          FFFFF    FFFFF    FFFFF
     |          FFFFF    FFFFF    FFFFF
     |          FFFFF    FFFFF    FFFFF
     -------------------------------------------
                 Mon      Wed      Fri

                         Section


              Symbol Gender     Symbol Gender

                F    F             M    M
```

PROC CHART fills each bar in the chart with the characters that represent the value of the variable GENDER. The portion of the bar that is filled with Fs represents the number of observations that correspond to females; the portion that is filled with Ms represents the number of observations that correspond to males. Because the value of Gender contains a single character (F or M), the symbol that PROC CHART uses as the fill character is identical to the value of the variable.

## Charting Mean Values

PROC CHART enables you to specify what the bars or sections in the chart represent. By default, each bar, block, or section represents the frequency of the chart variable. You can also identify a variable whose values determine the sizes of the bars, blocks, or sections in the chart.

You define a variable called the sumvar variable by using the SUMVAR= option. With the SUMVAR= option, you can also use the TYPE= option to specify whether the sum of the Sumvar variable or the mean of the Sumvar variable determines the size of the bars or sections. The available types are

SUM
> sums the values of the Sumvar variable in each range. Then PROC CHART uses the sums to determine the size of each bar, block, or section. SUM is the default type.

MEAN
> determines the mean value of the Sumvar variable in each range. Then PROC CHART uses the means to determine the size of each bar, block, or section.

The following program creates a bar chart grouped by gender to compare the mean value of all grades in each section:

```
options pagesize=40 linesize=80 pageno=1 nodate;

proc chart data=grades;
   vbar Section / midpoints='Mon' 'Wed' 'Fri' group=Gender
               sumvar=Examgrade1 type=mean;
   title 'Mean Exam Grade for Introductory Chemistry Sections';
run;
```

The SUMVAR= option specifies that the values of ExamGrade1 determine the size of the bars. The TYPE=MEAN option specifies to compare the mean grade for each group.

The following output displays the bar chart:

*Figure 31.12* Using the SUMVAR= Option to Compare Mean Values

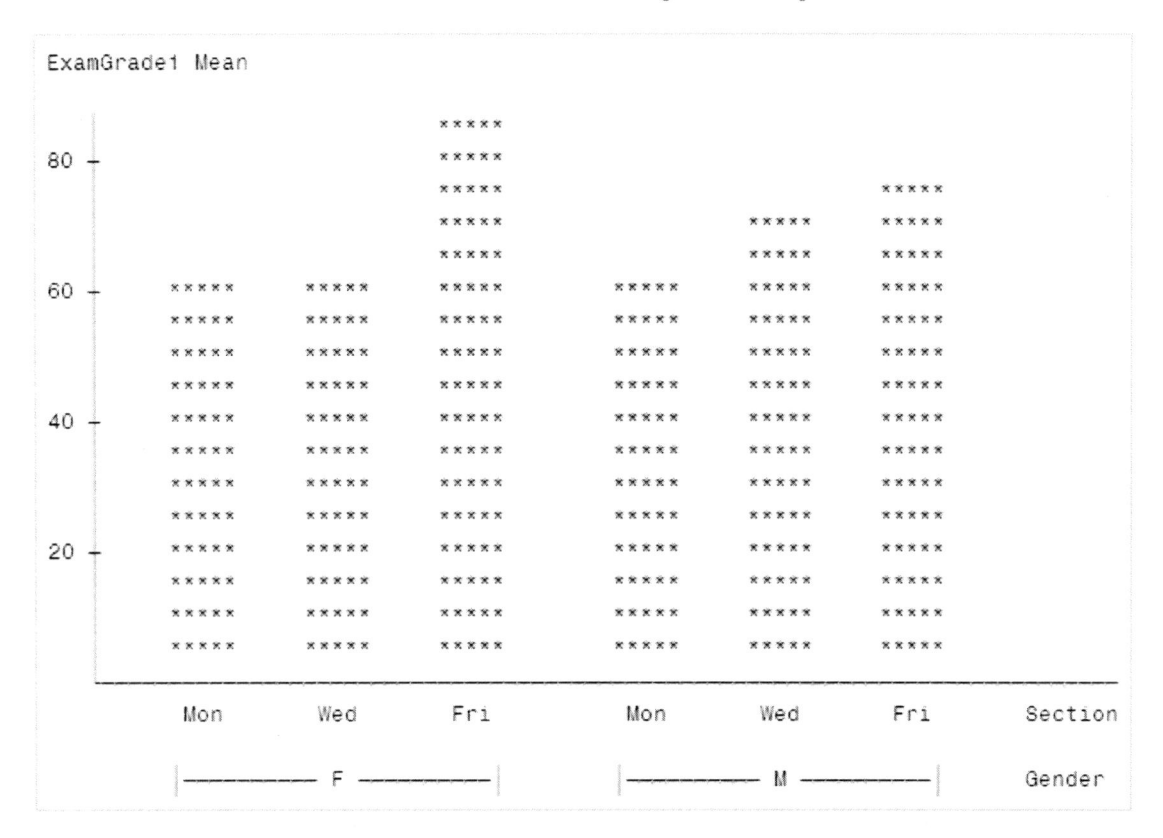

```
                      Mean Exam Grade for Introductory Chemistry Sections

 ExamGrade1 Mean
                              xxxxx
    80 -                      xxxxx
                              xxxxx                        xxxxx
                              xxxxx            xxxxx       xxxxx
                              xxxxx            xxxxx       xxxxx
    60 -   xxxxx    xxxxx     xxxxx    xxxxx   xxxxx       xxxxx
           xxxxx    xxxxx     xxxxx    xxxxx   xxxxx       xxxxx
           xxxxx    xxxxx     xxxxx    xxxxx   xxxxx       xxxxx
           xxxxx    xxxxx     xxxxx    xxxxx   xxxxx       xxxxx
    40 -   xxxxx    xxxxx     xxxxx    xxxxx   xxxxx       xxxxx
           xxxxx    xxxxx     xxxxx    xxxxx   xxxxx       xxxxx
           xxxxx    xxxxx     xxxxx    xxxxx   xxxxx       xxxxx
           xxxxx    xxxxx     xxxxx    xxxxx   xxxxx       xxxxx
    20 -   xxxxx    xxxxx     xxxxx    xxxxx   xxxxx       xxxxx
           xxxxx    xxxxx     xxxxx    xxxxx   xxxxx       xxxxx
           xxxxx    xxxxx     xxxxx    xxxxx   xxxxx       xxxxx
           xxxxx    xxxxx     xxxxx    xxxxx   xxxxx       xxxxx
         ----------------------------------------------------------
            Mon      Wed       Fri      Mon     Wed         Fri        Section

           |------------ F ------------|   |------------ M ------------|   Gender
```

The chart shows that the females in the Friday section achieved the highest mean grade, followed by the males in the same section.

## Creating a Three-Dimensional Chart

Complicated relationships such as the ones charted with the GROUP= option might be easier to understand if you present them as three-dimensional block charts. The following program uses the BLOCK statement to create a block chart for the numeric variable ExamGrade1:

```
options linesize=120 pagesize=40 pageno=1 nodate;
proc chart data=grades;
   block Section / midpoints='Mon' 'Wed' 'Fri'
                   sumvar=Examgrade1 type=mean
                   group=Gender;
   format Examgrade1 4.1;
   title 'Mean Exam Grade for Introductory Chemistry Sections';
run;
```

The FORMAT statement specifies the number of decimals that PROC CHART uses to report the mean value of ExamGrade1 beneath each block.

*Note:* If the line size or page size is not sufficient to display all the bars, then PROC CHART produces a horizontal bar chart.

The following output displays the block chart:

**Figure 31.13** *Using a Block Chart to Compare Group Means*

```
                    Mean Exam Grade for Introductory Chemistry Sections

                    Mean of ExamGrade1 by Section grouped by Gender
```

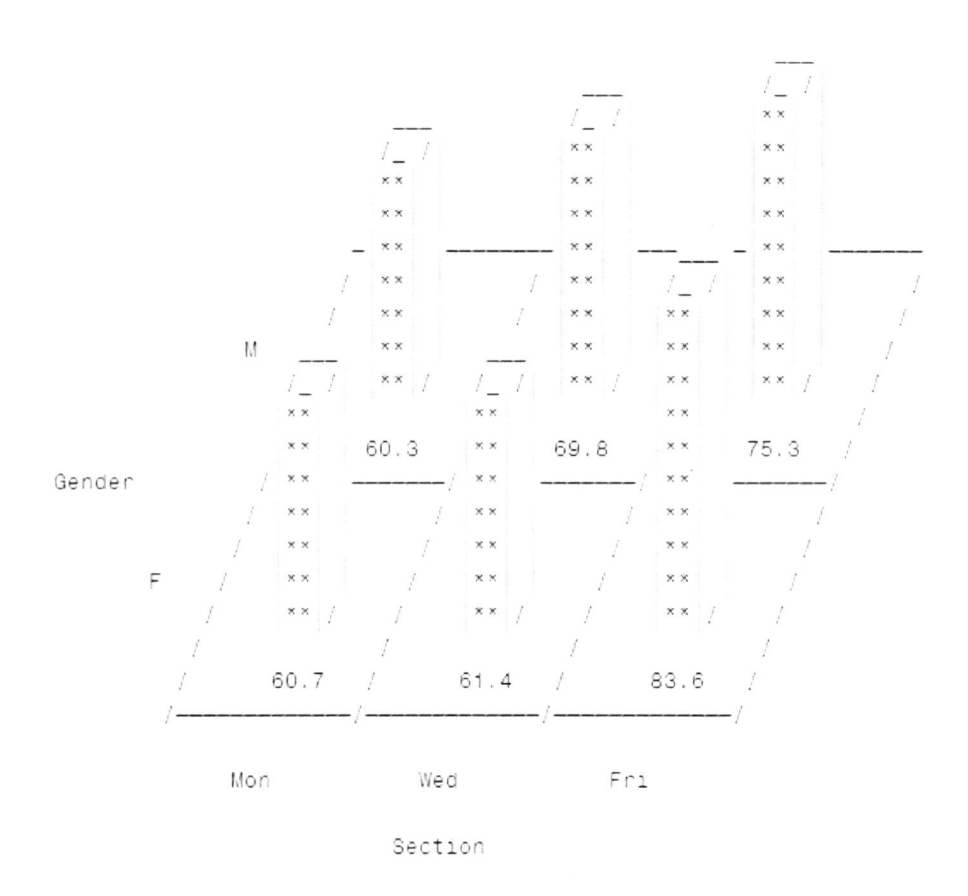

The value that is shown beneath each block is the mean of ExamGrade1 for that combination of Section and Gender. You can easily see that both females and males in the Friday section earned higher grades than their counterparts in the other sections.

# Creating High-Resolution Histograms

## Understanding How to Use the HISTOGRAM Statement

A histogram is similar to a vertical bar chart. This type of bar chart emphasizes the individual ranges of continuous numeric variables and enables you to examine the distribution of your data.

The HISTOGRAM statement in a PROC UNIVARIATE step produces histograms and comparative histograms. PROC UNIVARIATE creates a histogram by dividing the data into intervals of equal length, counting the number of observations in each interval, and plotting the counts as vertical bars that are centered around the midpoint of each interval.

If you use the HISTOGRAM statement without any options, then PROC UNIVARIATE automatically does the following:

- scales the vertical axis to show the percentage of observations in an interval
- determines the bar width based on the method of Terrell and Scott (1985)
- labels the axes

The HISTOGRAM statement provides various options that enable you to control the layout of the histogram and enhance the graph. You can also fit families of density curves and superimpose kernel density estimates on the histograms, which can be useful in examining the data distribution. For additional information about the density curves that SAS computes, see the UNIVARIATE procedure in the *Base SAS Procedures Guide*.

## Understanding How to Use SAS/GRAPH to Create Histograms

If your site licenses SAS/GRAPH software, then you can use the HISTOGRAM statement to create high-resolution graphs. When you create charts with a graphics device, you can also use the AXIS, LEGEND, PATTERN, and SYMBOL statements to enhance your plots.

To control the appearance of a high-resolution graph, you can specify a GOPTIONS statement before the PROC step that creates the graph. The GOPTIONS statement changes the values of the graphics options that SAS uses when graphics output is created. Graphics options affect the characteristics of a graph, such as size, colors, type fonts, fill patterns, and line thickness. In addition, they affect the settings of device parameters such as the appearance of the display, the type of output that is produced, and the destination of the output.

Most of the examples in this section use the following GOPTIONS statement:

```
goptions reset=global
         gunit=pct
         hsize= 5.625 in
         vsize= 3.5 in
         htitle=4
         htext=3
         vorigin=0 in
         horigin= 0 in
         cback=white border
         ctext=black
         colors=(black blue green red yellow)
         ftext=swiss
         lfactor=3;
```

For additional information about how to modify the appearance of your graphics output, see *SAS/GRAPH: Reference*.

## Creating a Simple Histogram

The following program uses the HISTOGRAM statement to create a histogram for the numeric variable ExamGrade1:

```
proc univariate data=grades noprint;
   histogram ExamGrade1;
   title 'Grades for First Chemistry Exam';
```

```
run;
```

The NOPRINT option suppresses the tables of statistics that the PROC UNIVARIATE statement creates.

The following figure displays the histogram:

*Figure 31.14* Using a Histogram to Show Percentages

### Grades for First Chemistry Exam

The UNIVARIATE Procedure

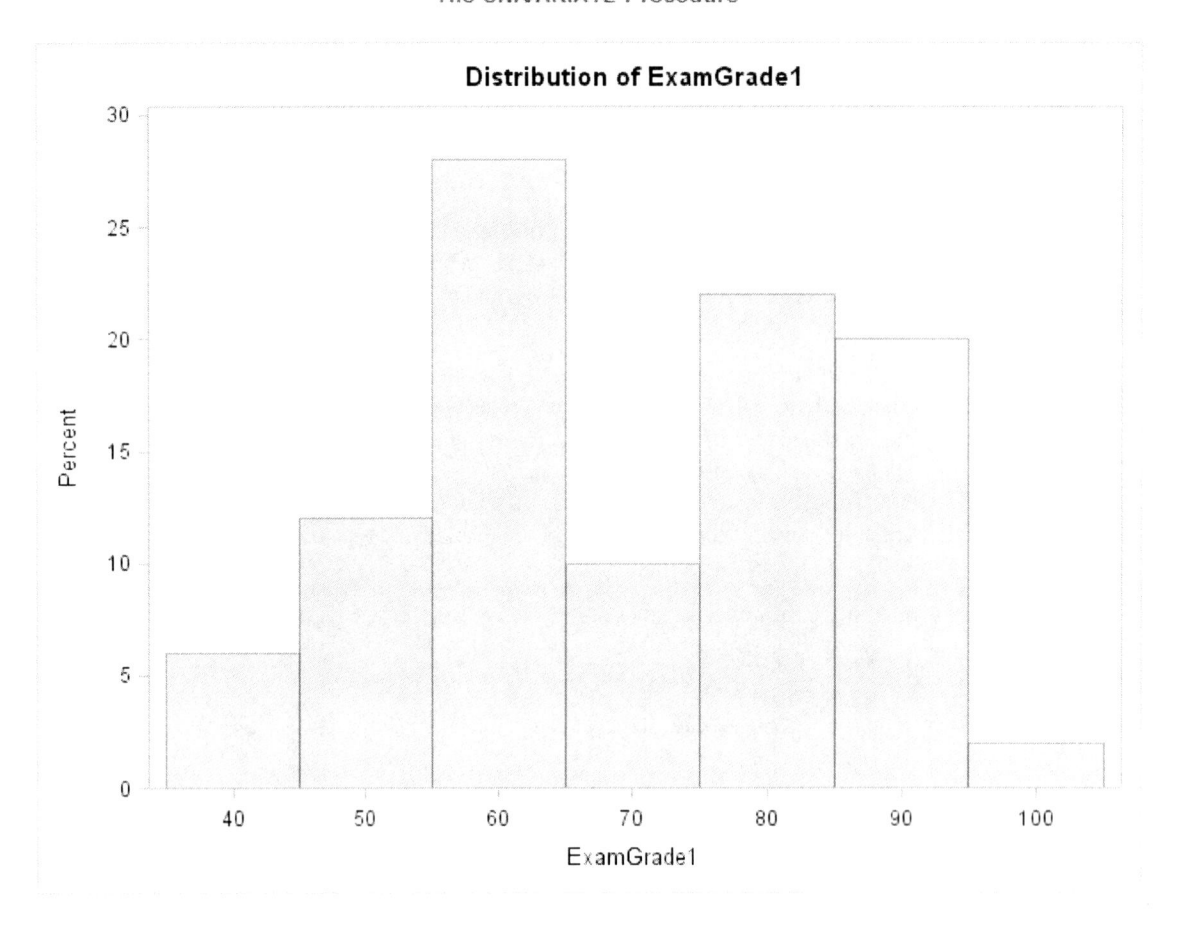

**Distribution of ExamGrade1**

The midpoint axis for the above histogram goes from 40 to 100 and is incremented in intervals of 10. The following table shows the values:

*Table 31.3* Histogram Values

| Interval | Midpoint |
| --- | --- |
| 35 to 44 | 40 |
| 45 to 54 | 50 |
| 55 to 64 | 60 |

| Interval | Midpoint |
|---|---|
| 65 to 74 | 70 |
| 75 to 84 | 80 |
| 85 to 94 | 90 |
| 95 to 104 | 10 |

*Note:* Because PROC UNIVARIATE selects the size of the intervals and the location of their midpoints based on all values of the numeric variable, the highest and lowest intervals can extend beyond the values in the data. In this example the lowest grade is 39 while the lowest interval extends from 35 to 44. Similarly, the highest grade is 98 while the highest interval extends from 95 to 104.

## Changing the Axes of a Histogram

### Enhancing the Vertical Axis

The exact value of a histogram bar is sometimes difficult to determine. By default, PROC UNIVARIATE does not provide minor tick marks between the vertical axis values (major tick marks). You can specify the number of minor tick marks between major tick marks with the VMINOR= option.

To make it easier to see the location of major tick marks, you can use the GRID option to add grid lines on the histogram. Grid lines are horizontal lines that are positioned at major tick marks on the vertical axis. PROC UNIVARIATE provides two options to change the appearance of the grid line:

CGRID=   sets the color of the grid lines.

LGRID=   sets the line type of the grid lines.

By default, PROC UNIVARIATE draws a solid line using the first color in the device color list. For a list of the available line types, see *SAS/GRAPH: Reference*.

The following program creates a histogram that shows minor tick marks and grid lines for the numeric variable ExamGrade1:

```
proc univariate data=grades noprint;
   histogram Examgrade1 / vminor=4 grid lgrid=34;
   title 'Grades for First Chemistry Exam';
run;
```

Four minor tick marks are inserted between each major tick mark. Closely spaced dots are used to draw the grid lines.

**578** Chapter 31 • *Producing Charts to Summarize Variables*

The following figure displays the histogram:

***Figure 31.15*** *Specifying Grid Lines for a Histogram*

Grades for First Chemistry Exam

The UNIVARIATE Procedure

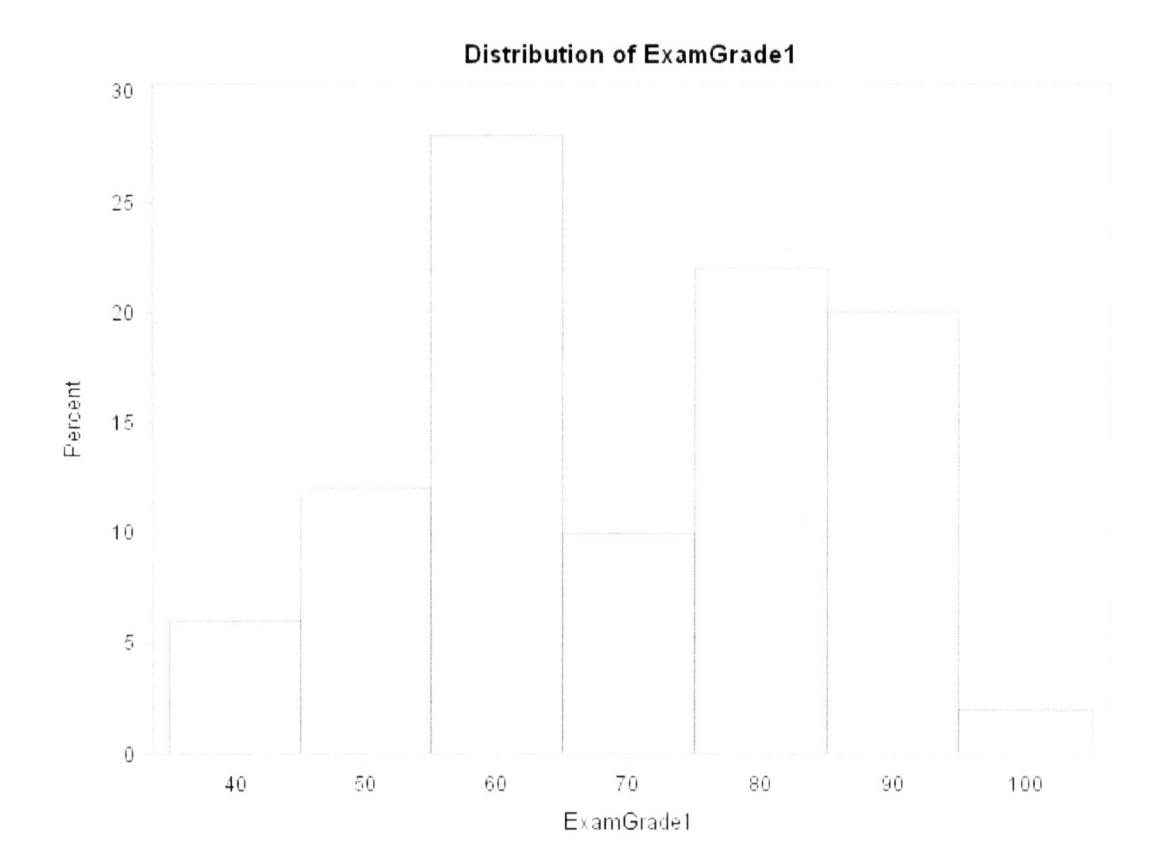

**Distribution of ExamGrade1**

Now, the height of each histogram bar is easily determined from the chart. The following table shows the percentage each interval represents:

***Table 31.4*** *Percentage of Each Interval*

| Interval | Percent |
|---|---|
| 35 to 44 | 6 |
| 45 to 54 | 12 |
| 55 to 64 | 28 |
| 65 to 74 | 10 |
| 75 to 84 | 22 |
| 85 to 94 | 20 |

| Interval | Percent |
|---|---|
| 95 to 104 | 2 |

### *Specifying the Vertical Axis Values*

PROC UNIVARIATE enables you to specify what the bars in the histogram represent, and the values of the vertical axis. By default, each bar represents the percentage of observations that fall into the given interval.

The VSCALE= option enables you to specify the following scales for the vertical axis:

- COUNT
- PERCENT
- PROPORTION

The VAXIS= option enables you to specify evenly spaced tick mark values for the vertical axis. The form of this option is

**HISTOGRAM** *variable* / VAXIS=*value-list*;

where value-list is a list of numbers to use as major tick mark values. The first value is always equal to zero and the last value is always greater than or equal to the height of the largest bar.

The following program creates a histogram that shows counts on the vertical axis for the numeric variable ExamGrade1:

```
proc univariate data=grades noprint;
   histogram Examgrade1 / vscale=count vaxis=0 to 16 by 2 vminor=1;
   title 'Grades for First Chemistry Exam';
run;
```

The values of the vertical axis range from 0 to 16 in increments of two. One minor tick mark is inserted between each major tick mark.

The following figure displays the histogram:

*Figure 31.16*  *Using a Histogram to Show Counts*

**Grades for First Chemistry Exam**

The UNIVARIATE Procedure

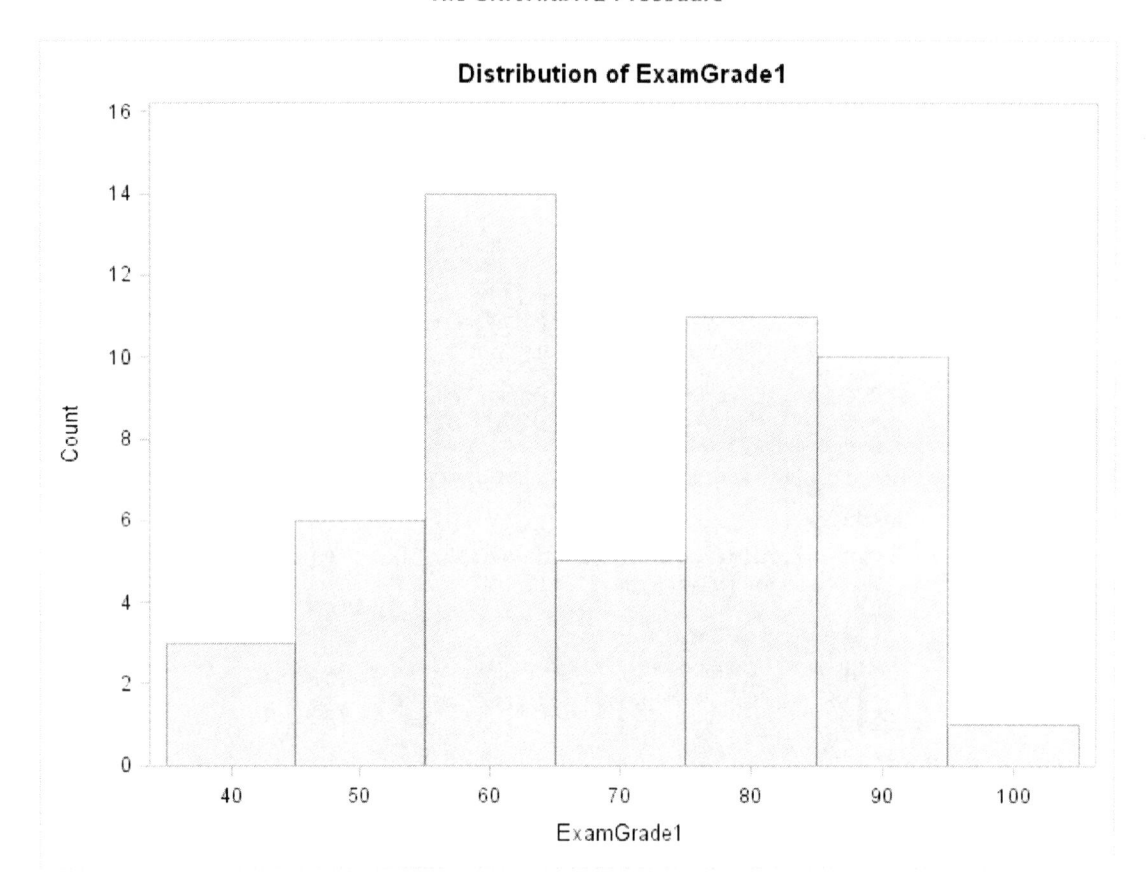

**Distribution of ExamGrade1**

### Specifying the Midpoints of a Histogram

You can control the width of the histogram bars by using the MIDPOINTS= option. PROC UNIVARIATE uses the value of the midpoints to determine the width of the histogram bars. The difference between consecutive midpoints is the bar width.

To specify midpoints, use the MIDPOINTS= option in the HISTOGRAM statement. The form of the MIDPOINTS= option is

**HISTOGRAM** *variable* / MIDPOINTS=*midpoint-list*;

where midpoint-list is a list of numbers to use as midpoints. You must use evenly spaced midpoints that are listed in increasing order.

For example, to specify the traditional grading ranges with midpoints from 55 to 95, use the following option:

```
midpoints=55 65 75 85 95
```

Or, you can abbreviate this list of midpoints:

```
midpoints=55 to 95 by 10
```

The following program uses the MIDPOINTS= option to create a histogram for the numeric variable ExamGrade1:

```
proc univariate data=grades
   (where=(examgrade1 ge 55 and examgrade1 le 100 )) noprint;
   histogram Examgrade1 / vscale=count vaxis=0 to 16 by 2 vminor=1
                          midpoints=55 65 75 85 95 hoffset=10
                          vaxislabel='Frequency';
   title 'Grades for First Chemistry Exam';
run;
```

The following list corresponds to the items in the preceding program:

- The MIDPOINTS= option forces PROC UNIVARIATE to center the five bars around the traditional midpoints for exam grades.

- The HOFFSET= option uses a 10% offset at both ends of the horizontal axis.

- The VAXISLABEL= option uses Frequency as the label for the vertical axis. The default label is Count.

The following figure displays the histogram:

*Figure 31.17*    *Specifying Five Midpoints for a Histogram*

**Grades for First Chemistry Exam**

The UNIVARIATE Procedure

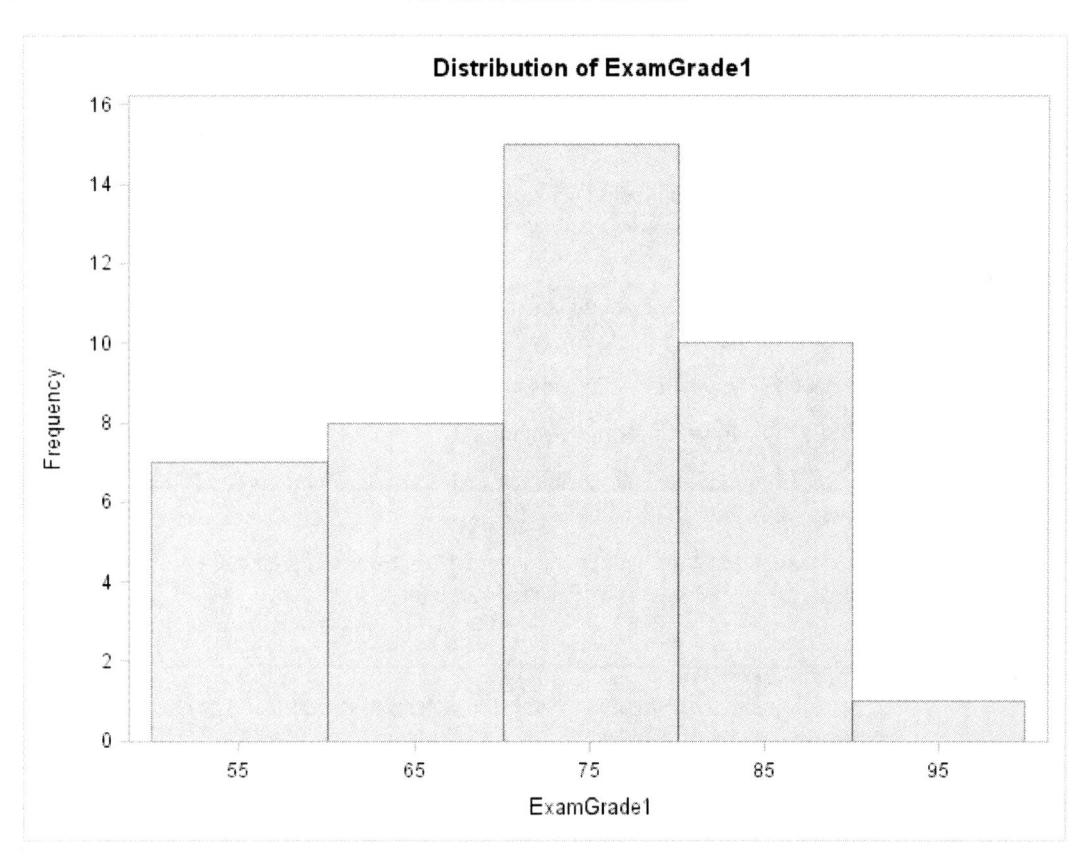

The midpoint axis for the above histogram goes from 55 to 95 and is incremented in intervals of 10. The histogram excludes any exam scores that are below 50.

## Displaying Summary Statistics in a Histogram

### Understanding How to Use the INSET Statement

PROC UNIVARIATE enables you to add a box or table of summary statistics, called an inset, directly in the histogram. Typically, an inset shows statistics that PROC UNIVARIATE has calculated, but an inset can also display values that you provide in a SAS data set.

To add a table of summary statistics, use the INSET statement. You can use multiple INSET statements in the UNIVARIATE procedure to add more than one table to a histogram. The INSET statements must follow the HISTOGRAM statement that creates the plot that you want augmented. The inset appears in all the graphs that the preceding HISTOGRAM statement produces.

The form of the INSET statement is as follows:

**INSET** *<keyword(s)> </ option(s)>*

You specify the keywords for inset statistics (such as N, MIN, MAX, MEAN, and STD) immediately after the word INSET. You can also specify the keyword DATA= followed by the name of a SAS data set to display customized statistics that are stored in a SAS data set. The statistics appear in the order in which you specify the keywords.

By default, PROC UNIVARIATE uses appropriate labels and appropriate formats to display the statistics in the inset. To customize a label, specify the keyword followed by an equal sign (=) and the desired label in quotation marks. To customize the format, specify a numeric format in parentheses after the keyword. You can assign labels that are up to 24 characters. If you specify both a label and a format for a keyword, then the label must appear before the format. For example,

```
inset n='Sample Size' std='Std Dev' (5.2);
```

requests customized labels for two statistics (sample size and standard deviation). The standard deviation is also assigned a format that has a field width of five and includes two decimal places.

Various options enable you to customize the appearance of the inset. For example, you can do the following:

- Specify the position of the inset.
- Specify a heading for the inset table.
- Specify graphical enhancements, such as background colors, text colors, text height, text font, and drop shadows.

For a complete list of the keywords and the options that you can use in the INSET statement, see the *Base SAS Procedures Guide*.

### The Program

The following program uses the INSET statement to add summary statistics for the numeric variable ExamGrade1 to the histogram:

```
proc univariate data=grades noprint;
    histogram Examgrade1 /vscale=count vaxis=0 to 16 by 2 vminor=1 hoffset=10
                    midpoints=55 65 75 85 95 vaxislabel='Frequency';
    inset n='No. Students' mean='Mean Grade' min='Lowest Grade' ❶
        max='Highest Grade' / header='Summary Statistics' position=ne ❷ ❸
                    format=3. ❹;
```

```
   title 'Grade Distribution for the First Chemistry Exam';
run;
```

The following list corresponds to the numbered items in the preceding program:

1   The statistical keywords N, MEAN, MIN, and MAX specify that the number of observations, the mean exam grade, the minimum exam grade, and the maximum exam grade appear in the inset. Each keyword is assigned a customized label to identify the statistic in the inset.

2   The HEADER= option specifies the heading text that appears at the top of the inset.

3   The POSITION= option uses a compass point to position the inset. The table appears at the northeast corner of the histogram.

4   The FORMAT= option requests a format with a field width of three for all the statistics in the inset.

*Figure 31.18*   *Adding an Inset to a Histogram*

Grade Distribution for the First Chemistry Exam

The UNIVARIATE Procedure

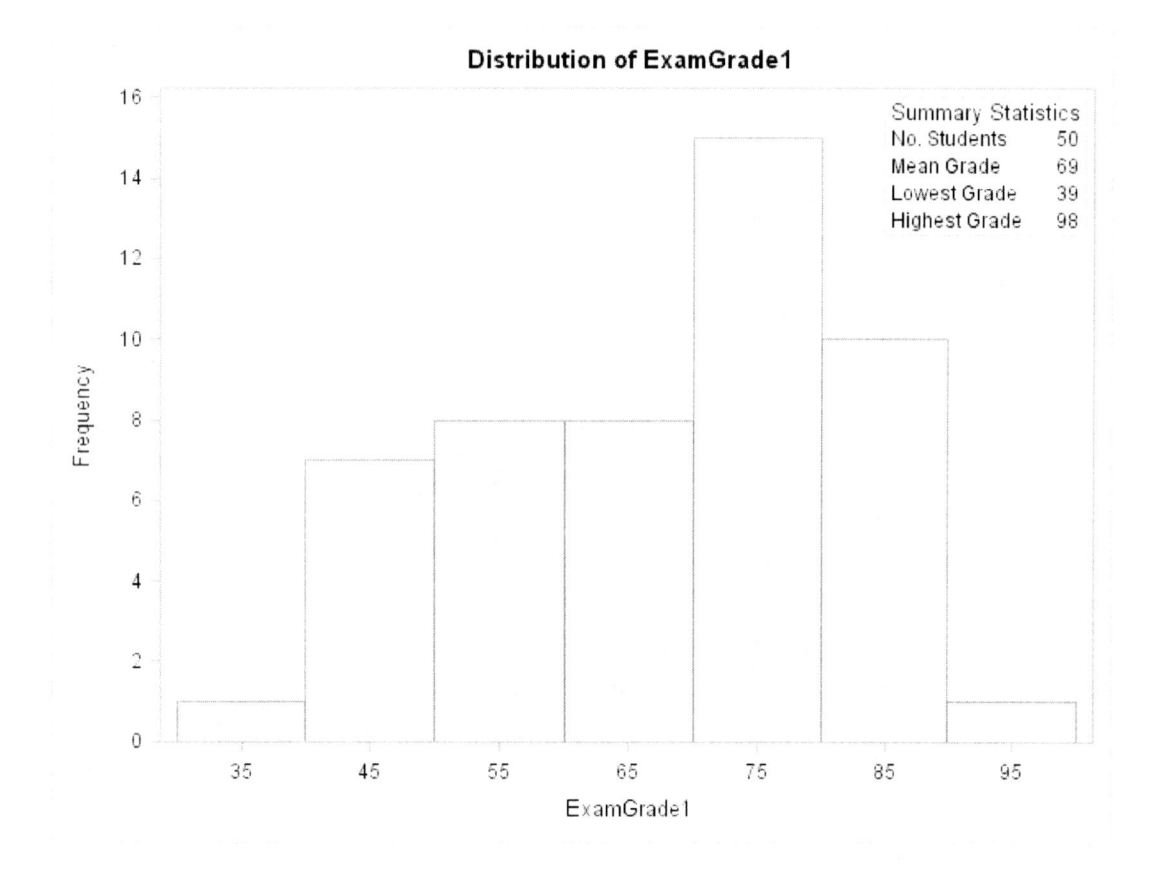

**Distribution of ExamGrade1**

| Summary Statistics | |
|---|---|
| No. Students | 50 |
| Mean Grade | 69 |
| Lowest Grade | 39 |
| Highest Grade | 98 |

The histogram shows the data distribution. The table of summary statistics in the upper right corner of the histogram provides information about the sample size, the mean grade, the lowest value, and the highest value.

### Creating a Comparative Histogram

#### Understanding Comparative Histograms

A comparative histogram is a series of component histograms that are arranged as an array or a matrix. PROC UNIVARIATE uses uniform horizontal and vertical axes to display the component histograms. This enables you to use the comparative histogram to visually compare the distribution of a numeric variable across the levels of up to two classification variables.

You use the CLASS statement with a HISTOGRAM statement to create either a one-way or a two-way comparative histogram. The form of the CLASS statement is as follows:

**CLASS** *variable-1 <(variable-option(s))> <variable-2 <(variable-option(s))>> </ options>*;

Class variables can be numeric or character. Class variables can have continuous values, but they typically have a few discrete values that define levels of the variable. You can reduce the number of classification levels by using a FORMAT statement to combine the values of a class variable.

When you specify one class variable, PROC UNIVARIATE displays an array of component histograms (stacked or side-by-side). To create the one-way comparative histogram, PROC UNIVARIATE categorizes the values of the analysis variable by the formatted values (levels) of the class variable. Each classification level generates a separate histogram.

When you specify two class variables, PROC UNIVARIATE displays a matrix of component plots. To create the two-way comparative histogram, PROC UNIVARIATE categorizes the values of the analysis variable by the cross-classified values (levels) of the class variables. Each combination of the cross-classified levels generates a separate histogram. The levels of class variable-1 are the labels for the rows of the matrix, and the levels of class variable-2 are the labels for the columns of the matrix.

You can specify options in the HISTOGRAM statement to customize the appearance of the comparative histogram. For example, you can do the following:

- Specify the number of rows for the comparative histogram.
- Specify the number of columns for the comparative histogram.
- Specify graphical enhancements, such as background colors and text colors for the labels.

For a complete list of the keywords and the options that you can use in the HISTOGRAM statement, see the *Base SAS Procedures Guide*.

#### The Program

The following program uses the CLASS statement to create a comparative histogram by gender and section for the numeric variable ExamGrade1:

```
proc format;
    value $gendfmt 'M'='Male'
                   'F'='Female'; 1
run;

proc univariate data=grades noprint;
    class Gender 2 Section(order=data); 3
    histogram Examgrade1 / midpoints=45 to 95 by 10 vscale=count vaxis=0 to 6 by 2
```

```
                           vaxislabel='Frequency' turnvlabels ❹nrows=2 ncols=3 ❺
                           cframe=ligr ❻cframeside=gwh cframetop=gwh cfill=gwh; ❼
   inset mean(4.1) n / noframe ❽position=(2,65); ❾
   format Gender $gendfmt.; ❶
   title 'Grade Distribution for the First Chemistry Exam';
run;
```

The following list corresponds to the numbered items in the preceding program:

1  PROC FORMAT creates a user-written format that labels Gender with a character string. The FORMAT statement assigns the format to Gender.

2  The CLASS statement creates a two-way comparative histogram that uses Gender and Section as the classification variables. PROC UNIVARIATE produces a component histogram for each level (a distinct combination of values) of these variables.

3  The ORDER= option positions the values of Section according to their order in the input data set. The comparative histogram displays the levels of Section according to the days of the week (Mon, Wed, and Fri). The default order of the levels is determined by sorting the internal values of Section (Fri, Mon, and Wed).

4  The TURNVLABELS option turns the characters in the vertical axis labels so that they are displayed vertically instead of horizontally.

5  The NROWS= option and the NCOLS= option specify a 2 × 3 arrangement for the component histograms.

6  The CFRAME= option specifies the color that fills the area of each component histogram that is enclosed by the axes and the frame. The CFRAMESIDE= option and the CFRAMETOP= option specify the color to fill the frame area for the column labels and the row labels that appear down the side and across the top of the comparative histogram. By default, these areas are not filled.

7  The CFILL= option specifies the color to fill the bars of each component histogram. By default, the bars are not filled.

8  The NOFRAME option suppresses the frame around the inset table.

9  The POSITION= option uses axis percentage coordinates to position the inset. The position of the bottom left corner of the inset is 2% of the way across the horizontal axis and 65% of the way up the vertical axis.

**586** Chapter 31 • *Producing Charts to Summarize Variables*

The following figure displays the comparative histogram:

***Figure 31.19*** *Using a Comparative Histogram to Examine Exam Grades by Gender and Section*

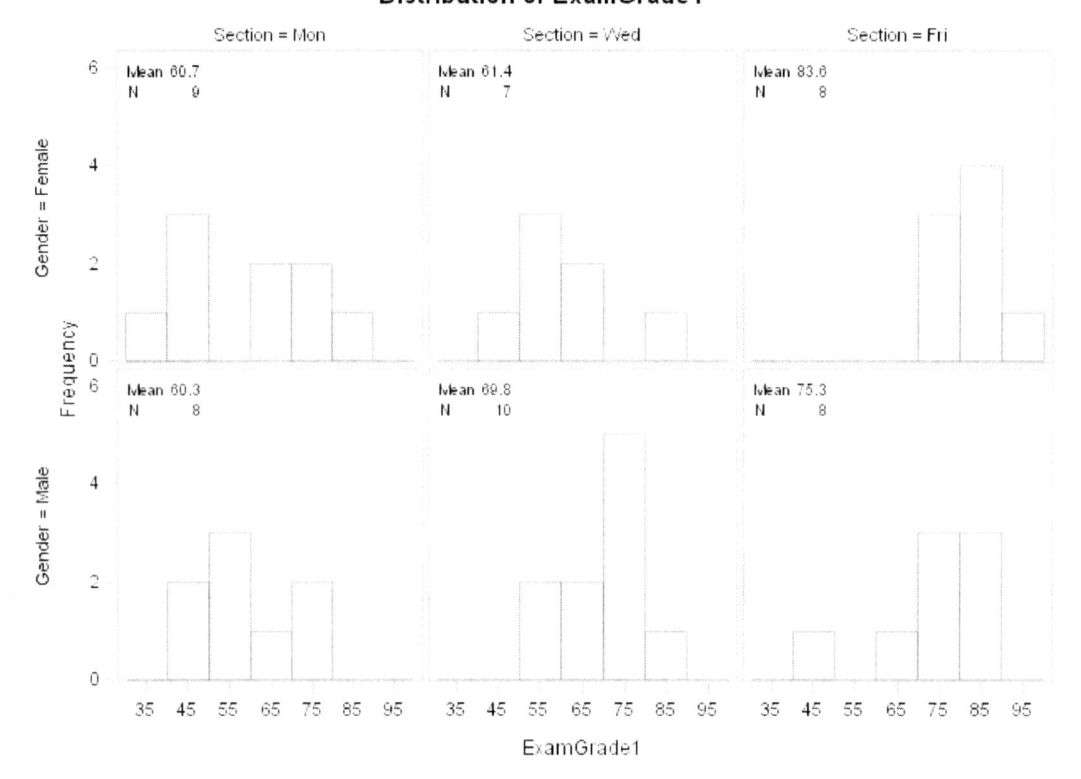

The comparative histogram is a 2 × 3 matrix of component histograms for each combination of Section and Gender. Each component histogram displays a table of statistics that reports the mean of ExamGrade1 and the number of students. You can easily see that both females and males in the Friday section earned higher grades than their counterparts in the other sections.

# Summary

## PROC CHART Statements

**PROC CHART** <DATA=*SAS-data-set*> <*options*>;
  *chart-type variable(s)* </*options*>;

PROC CHART <DATA=SAS-data-set> <options> ;
  starts the CHART procedure. You can specify the following options in the PROC CHART statement:

DATA=SAS-data-set
: names the SAS data set that PROC CHART uses. If you omit DATA=, then PROC CHART uses the most recently created data set.

LPI=value
: specifies the proportions of PIE and STAR charts.

chart-type variable(s) < /options>;
: is a chart statement where

chart-type
: specifies the type of chart and can be any of the following:

- BLOCK
- HBAR
- PIE
- VBAR

You can use any number of chart statements in one PROC CHART step. A list of options pertains to a single chart statement.

variable(s)
: identifies the variables to chart (called the chart variables).

options
: specifies a list of options. Not all types of chart support all options.

You can use the following options in the VBAR, HBAR, and BLOCK statements:

GROUP=variable
: produces a set of bars or blocks for each value of variable.

SUBGROUP=variable
: proportionally fills each block or bar with characters that represent different values of variable. You can use the following options in the VBAR, HBAR, BLOCK, and PIE statements:

DISCRETE
: creates a bar, block, or section for every value of the chart variable.

LEVELS=number-of-midpoints
: specifies the number-of-midpoints. The procedure selects the midpoints.

MIDPOINTS=midpoints-list
: specifies the values of the midpoints.

SUMVAR=variable
: specifies the variable to use to determine the size of the bars, blocks, or sections.

TYPE=SUM|MEAN
: specifies the type of chart to create, where

SUM
: sums the values of the Sumvar variable in each range. Then PROC CHART uses the sums to determine the size of each bar, block, or section.

MEAN
: determines the mean value of the Sumvar variable in each range. Then PROC CHART uses the means to determine the size of each bar, block, or section.

You can use the following options in the HBAR statement:

**NOSTAT**
   suppresses the printing of the statistics that accompany the chart by default.

**FREQ**
   requests frequency statistics.

**CFREQ**
   requests cumulative frequency statistics.

**PERCENT**
   requests percentage statistics.

**CPERCENT**
   requests cumulative percentage statistics.

## PROC UNIVARIATE Statements

**PROC UNIVARIATE** <option(s)>;

**CLASS** variable-1 <(variable-option(s))> <variable-2 <(variable-option(s))>> </option(s)>;

**HISTOGRAM** <variable(s)> </option(s)>;

**INSET** <keyword(s)> </option(s)>;

PROC UNIVARIATE option(s);
   starts the UNIVARIATE procedure. You can specify the following options in the PROC UNIVARIATE statement:

   DATA=SAS-data-set
      names the SAS data set that PROC UNIVARIATE uses. If you omit DATA=, then PROC UNIVARIATE uses the most recently created data set.

   NOPRINT
      suppresses the descriptive statistics that the PROC UNIVARIATE statement creates.

CLASS variable-1<(variable-option(s))> <variable-2<(variable-option(s))>> </ option(s)>;
   specifies up to two variables whose values determine the classification levels for the component histograms. Variables in a CLASS statement are referred to as class variables.

   You can specify the following option(s) in the CLASS statement:

   ORDER=DATA | FORMATTED | FREQ | INTERNAL
      specifies the display order for the class variable values, where

      DATA
         orders values according to their order in the input data set.

      FORMATTED
         orders values by their ascending formatted values. This order depends on your operating environment.

      FREQ
         orders values by descending frequency count so that levels with the most observations are listed first.

      INTERNAL
         orders values by their unformatted values, which yields the same order as PROC SORT. This order depends on your operating environment.

**HISTOGRAM** <variable(s)> </option(s)>;

creates histograms and comparative histograms using high-resolution graphics for the analysis variables that are specified. If you omit *variable(s)* in the HISTOGRAM statement, then the procedure creates a histogram for each variable that you list in the VAR statement, or for each numeric variable in the DATA= data set if you omit a VAR statement.

You can specify the following options in the PROC UNIVARIATE statement:

**CGRID**=color

specifies the color for grid lines when a grid is displayed on the histogram.

**GRID**

specifies to display a grid on the histogram. Grid lines are horizontal lines that are positioned at major tick marks on the vertical axis.

**HOFFSET**=value

specifies the offset in percentage screen units at both ends of the horizontal axis.

**GRID**

specifies to display a grid on the histogram. Grid lines are horizontal lines that are positioned at major tick marks on the vertical axis.

**LGRID**=linetype

specifies the line type for the grid when a grid is displayed on the histogram. The default is a solid line.

**MIDPOINTS**=value(s)

determines the width of the histogram bars as the difference between consecutive midpoints. PROC UNIVARIATE uses the same *value(s)* for all variables. You must use evenly spaced midpoints that are listed in increasing order.

**VAXIS**=value(s)

specifies tick mark values for the vertical axis. Use evenly spaced values that are listed in increasing order. The first value must be zero and the last value must be greater than or equal to the height of the largest bar. You must scale the values in the same units as the bars.

**VMINOR**=n

specifies the number of minor tick marks between each major tick mark on the vertical axis. PROC UNIVARIATE does not label minor tick marks.

**VSCALE**=scale

specifies the scale of the vertical axis, where *scale* is

**COUNT**

scales the data in units of the number of observations per data unit.

**PERCENT**

scales the data in units of percentage of observations per data unit.

**PROPORTION**

scales the data in units of proportion of observations per data unit.

**INSET** <keyword(s)> </option(s)>;

places a box or table of summary statistics, called an inset, directly in the histogram.

You can specify the following options in the PROC UNIVARIATE statement:

**keyword(s)**

specifies one or more keywords that identify the information to display in the inset. PROC UNIVARIATE displays the information in the order in which you request the keywords. For a complete list of keywords, see the INSET statement in *SAS/GRAPH: Reference*.

FORMAT=format
: specifies a format for all the values in the inset. If you specify a format for a particular statistic, then this format overrides FORMAT=*format*.

HEADER=string
: specifies the heading text where *string* cannot exceed 40 characters.

NOFRAME
: suppresses the frame drawn around the text.

POSITION=position
: determines the position of the inset. The *position* is a compass point keyword, a margin keyword, or a pair of coordinates (*x*, *y*). The default position is NW, which positions the inset in the upper left (northwest) corner of the display.

## GOPTIONS Statement

GOPTIONS options-list;
: specifies values for graphics options. Graphics options control characteristics of the graph, such as size, colors, type fonts, fill patterns, and symbols. In addition, they affect the settings of device parameters, which are defined in the device entry. Device parameters control such characteristics as the appearance of the display, the type of output that is produced, and the destination of the output.

## FORMAT Statement

FORMAT variable format-name;
: enables you to display the value of a *variable* by using a special pattern that you specify as *format-name*.

# Learning More

PROC CHART
: For complete documentation, see the *Base SAS Procedures Guide*. In addition to the features that are described in this section, you can use PROC CHART to create star charts, to draw a reference line at a particular value on a bar chart, and to change the symbol that is used to draw charts. You can also create charts based, not only on frequency, sum, and mean, but also on cumulative frequency, percent, and cumulative percent.

PROC UNIVARIATE
: For complete documentation, see the *Base SAS Procedures Guide*.

PROC PLOT
: For a discussion about how to plot the relationship between variables, see "Introduction to Plotting the Relationship between Variables" on page 531. When you are preparing graphics presentations, some data lends itself to charts. Other data is better suited for plots.

SAS formats
: For complete documentation, see *SAS Formats and Informats: Reference*. Many formats are available with SAS, including fractions, hexadecimal values, roman numerals, Social Security numbers, date and time values, and numbers written as words.

PROC FORMAT
: For complete documentation about how to create your own formats, see the *Base SAS Procedures Guide*.

SAS/GRAPH software
: For complete documentation, see *SAS/GRAPH: Reference*. If your site has SAS/GRAPH software, then you can use the GCHART procedure to take advantage of the high-resolution graphics capabilities of output devices and produce charts that include color, different fonts, and text.

TITLE and FOOTNOTE statements
: For a discussion about using titles and footnotes in a report, see "Understanding Titles and Footnotes" on page 458.

## Part 8

# Designing Your Own Output

*Chapter 32*
**Writing Lines to the SAS Log or to an Output File** .............. *595*

*Chapter 33*
**Understanding and Customizing SAS Output: The Basics** ....... *613*

*Chapter 34*
**Understanding and Customizing SAS Output: The Output Delivery System (ODS)** ................................... *643*

# Chapter 32
# Writing Lines to the SAS Log or to an Output File

**Introduction to Writing Lines to the SAS Log or to an Output File** ........... **595**
    Purpose ................................................................ 595
    Prerequisites ........................................................... 596

**Understanding the PUT Statement** ........................................ **596**

**Writing Output without Creating a Data Set** ............................. **596**

**Writing Simple Text** ..................................................... **597**
    Writing a Character String ............................................. 597
    Writing Variable Values ................................................ 599
    Writing on the Same Line More Than Once .............................. 600
    Releasing a Held Line ................................................. 601

**Writing a Report** ........................................................ **603**
    Writing to an Output File ............................................. 603
    Designing the Report .................................................. 604
    Writing Data Values ................................................... 604
    Improving the Appearance of Numeric Data Values ...................... 605
    Writing a Value at the Beginning of Each BY Group ..................... 606
    Calculating Totals .................................................... 607
    Writing Headings and Footnotes for a One-Page Report .................. 608

**Summary** ................................................................ **610**
    Statements ............................................................ 610

**Learning More** .......................................................... **611**

## Introduction to Writing Lines to the SAS Log or to an Output File

### Purpose

In previous sections you learned how to store data values in a SAS data set and to use SAS procedures to produce a report that is based on these data values. In this section, you will learn how to do the following:

- design output by positioning data values and character strings in an output file
- prevent SAS from creating a data set by using the DATA _NULL_ statement

- produce reports by using the DATA step instead of using a procedure
- direct data to an output file by using a FILE statement

*Prerequisites*

Before proceeding with this section, you should be familiar with the concepts presented in the following sections:

- Chapter 1, "What is the SAS System?," on page 3
- Chapter 3, "Introduction to DATA Step Processing," on page 27

## Understanding the PUT Statement

When you create output using the DATA step, you can customize that output by using the PUT statement to write text to the SAS log or to another output file. The PUT statement has the following form:

**PUT**<*variable*<*format*>> <'*character-string*'>;

*variable*
    names the variable that you want to write.

*format*
    specifies a format to use when you write variable values.

'*character-string*'
    specifies a string of text to write. Be sure to enclose the string in quotation marks.

## Writing Output without Creating a Data Set

In many cases, when you use a DATA step to write a report, you do not need to create an additional data set. When you use the DATA _NULL_ statement, SAS processes the DATA step without writing observations to a data set. Using the DATA _NULL_ statement can increase program efficiency considerably.

The following is an example of a DATA _NULL_ statement:

```
data _null_;
```

The following program uses a PUT statement to write women's Olympic medalist information to the SAS log. Because the program uses a DATA _NULL_ statement, SAS does not create a data set.

```
data _null_;
   length medalist $ 19;
   input year 1-4  medalist $ 6-24  medal $ 26-31 country $ 33-35 result 37-41;
   put medalist country medal result year;
   datalines;
1984 Lingjuan Li        SILVER CHN 2559
1984 Jin-Ho Kim         BRONZE KOR 2555
1988 Soo-Nyung Kim      GOLD   KOR 2683
```

```
             Hee-Kyung Wang        SILVER KOR 2612
        1988 Young-Sook Yun        BRONZE KOR 2593
        1992 Youn-Jeong Cho        GOLD   KOR 113
        1992 Soo-Nyung Kim         SILVER KOR 105
        1992 Natalya Valeyeva      BRONZE URS
        1996 Kyung-Wook Kim        GOLD   KOR
        1996 Ying He               SILVER CHN
        1996 Olena Sadovnycha      BRONZE UKR
        2000 Mi-Jin Jun            GOLD   KOR 107
        2000 Nam-Soon Kim          SILVER KOR 106
        2000 Soo-Nyung Kim         BRONZE KOR 103
        ;
        run;
```

The following output shows the results:

**Output 32.1** *Writing to the SAS Log*

```
1    data _null_;
2       length medalist $ 19;
3       input year 1-4  medalist $ 6-24  medal $ 26-31 country $ 33-35 result 37-41;
4       put medalist country medal result year;
5       datalines;

Lingjuan Li CHN SILVER 2559 1984
Jin-Ho Kim KOR BRONZE 2555 1984
Soo-Nyung Kim KOR GOLD 2683 1988
Hee-Kyung Wang KOR SILVER 2612 .
Young-Sook Yun KOR BRONZE 2593 1988
Youn-Jeong Cho KOR GOLD 113 1992
Soo-Nyung Kim KOR SILVER 105 1992
Natalya Valeyeva URS BRONZE . 1992
Kyung-Wook Kim KOR GOLD . 1996
Ying He CHN SILVER . 1996
Olena Sadovnycha UKR BRONZE . 1996
Mi-Jin Jun KOR GOLD 107 2000
Nam-Soon Kim KOR SILVER 106 2000
Soo-Nyung Kim KOR BRONZE 103 2000
NOTE: DATA statement used (Total process time):
      real time           0.02 seconds
      cpu time            0.03 seconds


20   ;
21   run;
```

SAS indicates missing numeric values with a period. Note that the log contains one missing observation for the variable year and four missing observations for the variable result.

# Writing Simple Text

## Writing a Character String

In its simplest form, the PUT statement writes the character string that you specify to the SAS log, to a procedure output file, or to an external file. If you omit the destination (as

**598** Chapter 32 • *Writing Lines to the SAS Log or to an Output File*

in this example), then SAS writes the string to the log. In the following example, SAS executes the PUT statement once during each iteration of the DATA step. If SAS encounters missing values for the variables Year and Result, then the PUT statement writes a message to the log.

```
data _null_;
   length medalist $ 19;
   input year 1-4  medalist $ 6-24  medal $ 26-31 country $ 33-35 result 37-41;
   if year=. then put '*** Missing Year';
      else
   if result=. then put '*** Missing Results';

   datalines;
1984 Lingjuan Li         SILVER CHN 2559
1984 Jin-Ho Kim          BRONZE KOR 2555
1988 Soo-Nyung Kim       GOLD   KOR 2683
     Hee-Kyung Wang      SILVER KOR 2612
1988 Young-Sook Yun      BRONZE KOR 2593
1992 Youn-Jeong Cho      GOLD   KOR 113
1992 Soo-Nyung Kim       SILVER KOR 105
1992 Natalya Valeyeva    BRONZE URS
1996 Kyung-Wook Kim      GOLD   KOR
1996 Ying He             SILVER CHN
1996 Olena Sadovnycha    BRONZE UKR
2000 Mi-Jin Jun          GOLD   KOR 107
2000 Nam-Soon Kim        SILVER KOR 106
2000 Soo-Nyung Kim       BRONZE KOR 103
;
run;
```

The following output shows the results:

**Output 32.2** *Writing a Character String to the SAS Log*

```
22   data _null_;
23      length medalist $ 19;
24      input year 1-4  medalist $ 6-24  medal $ 26-31 country $ 33-35 result 37-41;
25      if year=. then put '*** Missing Year';
26         else
27      if result=. then put '*** Missing Results';
28
29      datalines;

*** Missing Year
*** Missing Results
*** Missing Results
*** Missing Results
*** Missing Results
NOTE: DATA statement used (Total process time):
      real time          0.00 seconds
      cpu time           0.01 seconds



44   ;
45   run;
```

### Writing Variable Values

The previous example shows that the value for Year is missing one observation in the data set and the variable Result is missing four observations. To identify which observations have the missing values, write the value of one or more variables along with the character string. The following program writes the value of Year and Result, as well as the character string:

```
data _null_;
   length medalist $ 19;
   input year 1-4  medalist $ 6-24  medal $ 26-31 country $ 33-35 result 37-41;
   if year=. then put '*** Missing Year' medalist country;
      else
   if result=. then put '*** Missing Results' medalist country;
   datalines;
1984 Lingjuan Li        SILVER CHN 2559
1984 Jin-Ho Kim         BRONZE KOR 2555
1988 Soo-Nyung Kim      GOLD   KOR 2683
     Hee-Kyung Wang     SILVER KOR 2612
1988 Young-Sook Yun     BRONZE KOR 2593
1992 Youn-Jeong Cho     GOLD   KOR 113
1992 Soo-Nyung Kim      SILVER KOR 105
1992 Natalya Valeyeva   BRONZE URS
1996 Kyung-Wook Kim     GOLD   KOR
1996 Ying He            SILVER CHN
1996 Olena Sadovnycha   BRONZE UKR
2000 Mi-Jin Jun         GOLD   KOR 107
2000 Nam-Soon Kim       SILVER KOR 106
2000 Soo-Nyung Kim      BRONZE KOR 103
;
run;
```

Notice that the last character in each of the strings is blank. This is an example of list output. In list output, SAS automatically moves one column to the right after writing a variable value, but not after writing a character string. The simplest way to include the required space is to include it in the character string.

SAS keeps track of its position in the output line with a pointer. Another way to describe the action in this PUT statement is to say that in list output, the pointer moves one column to the right after writing a variable value, but not after writing a character string. In later parts of this section, you will learn ways to move the pointer to control where the next piece of text is written.

The following output shows the results:

**Output 32.3** *Writing a Character String and Variable Values*

```
46     data _null_;
47        length medalist $ 19;
48        input year 1-4  medalist $ 6-24  medal $ 26-31 country $ 33-35 result 37-41;
49        if year=. then put '*** Missing Year' medalist country;
50           else
51        if result=. then put '*** Missing Results' medalist country;
52
53
54        datalines;

*** Missing YearHee-Kyung Wang KOR
*** Missing ResultsNatalya Valeyeva URS
*** Missing ResultsKyung-Wook Kim KOR
*** Missing ResultsYing He CHN
*** Missing ResultsOlena Sadovnycha UKR
NOTE: DATA statement used (Total process time):
      real time           0.00 seconds
      cpu time            0.01 seconds


69     ;
70     run;
```

## Writing on the Same Line More Than Once

By default, each PUT statement begins on a new line. However, you can write on the same line if you use more than one PUT statement and at least one trailing @ ("at" sign).

The trailing @ is a type of pointer control called a line-hold specifier. Pointer controls are one way to specify where SAS writes text. In the following example, using the trailing @ causes SAS to write the item in the second PUT statement on the same line rather than on a new line. The execution of either PUT statement holds the output line for further writing because each PUT statement has a trailing @. SAS continues to write on that line when a later PUT statement in the same iteration of the DATA step is executed, and when a PUT statement in a later iteration is executed.

```
data _null_;
   length medalist $ 19;
   input year 1-4  medalist $ 6-24  medal $ 26-31 country $ 33-35 result 37-41;
   if year=. then put '*** Missing Year' medalist country @;
      else
   if result=. then put '*** Missing Results' medalist country @;
   datalines;
1984 Lingjuan Li            SILVER CHN 2559
1984 Jin-Ho Kim             BRONZE KOR 2555
1988 Soo-Nyung Kim          GOLD   KOR 2683
     Hee-Kyung Wang         SILVER KOR 2612
1988 Young-Sook Yun         BRONZE KOR 2593
1992 Youn-Jeong Cho         GOLD   KOR 113
1992 Soo-Nyung Kim          SILVER KOR 105
1992 Natalya Valeyeva       BRONZE URS
1996 Kyung-Wook Kim         GOLD   KOR
1996 Ying He                SILVER CHN
1996 Olena Sadovnycha       BRONZE UKR
2000 Mi-Jin Jun             GOLD   KOR 107
2000 Nam-Soon Kim           SILVER KOR 106
```

```
               2000 Soo-Nyung Kim         BRONZE KOR 103
            ;
            run;
```

The following output shows the results:

**Output 32.4**  *Writing on the Same Line More Than Once*

```
71     data _null_;
72        length medalist $ 19;
73        input year 1-4  medalist $ 6-24  medal $ 26-31 country $ 33-35 result 37-41;
74        if year=. then put '*** Missing Year' medalist country @;
75           else
76        if result=. then put '*** Missing Results' medalist country @;
77
78
79        datalines;

*** Missing YearHee-Kyung Wang KOR *** Missing ResultsNatalya Valeyeva URS *** Missing Results
Kyung-Wook Kim KOR *** Missing ResultsYing He CHN *** Missing ResultsOlena Sadovnycha UKR
NOTE: DATA statement used (Total process time):
      real time           0.00 seconds
      cpu time            0.01 seconds


94     ;
95     run;
```

If the output line were long enough, then SAS would write all three messages about missing data on a single line. Because the line is not long enough, SAS continues writing on the next line. When it determines that an individual data value or character string does not fit on a line, SAS brings the entire item down to the next line. SAS does not split a data value or character string.

### Releasing a Held Line

In the following example, the input file has six missing values. One record has missing values for both the Year and Result variables. Four other records have missing values for either the Year or the Result variable.

To improve the appearance of your report, you can write all the missing variables for each observation on a separate line. When values for the two variables Year and Result are missing, two PUT statements write to the same line. When either Year or Result is missing, only one PUT statement writes to that line.

SAS determines where to write the output by the presence of the trailing @ in the PUT statement and the presence of a null PUT statement that releases the hold on the line. Executing a PUT statement with a trailing @ causes SAS to hold the current output line for further writing. It holds the line either in the current iteration of the DATA step or in a future iteration. Executing a PUT statement without a trailing @ releases the held line.

To release a line without writing a message, use a null PUT statement:

```
put;
```

A null PUT statement has the same characteristics of other PUT statements: by default, it writes output to a new line, writes what you specify in the statement (nothing in this case), and releases the line when it finishes executing. If a trailing @ is in effect, then the null PUT statement begins on the current line, writes nothing, and releases the line.

The following program shows how to write one or more items to the same line:

- If a value for the variable Year is missing, then the first PUT statement holds the line in case the variable Result is missing a value for that observation.
- If a value for the variable Result is missing, then the next PUT statement writes a message and releases the line.
- If the variable Result does not have a missing value, but if a message has been written for the variable Year (year=.), then the null PUT statement releases the line.
- If neither the variable Year nor the variable Result has missing values, then the line is not released and no PUT statement is executed.

```
data _null_;
   length medalist $ 19;
   input year 1-4  medalist $ 6-24  medal $ 26-31 country $ 33-35 result 37-41;
   if year=. then put '*** Missing Year' medalist country @;
   if result=. then put '*** Missing Results' medalist country ;
      else if year=. then put;

   datalines;
1984 Lingjuan Li          SILVER CHN 2559
1984 Jin-Ho Kim           BRONZE KOR 2555
1988 Soo-Nyung Kim        GOLD   KOR 2683
     Hee-Kyung Wang       SILVER KOR 2612
1988 Young-Sook Yun       BRONZE KOR 2593
1992 Youn-Jeong Cho       GOLD   KOR 113
1992 Soo-Nyung Kim        SILVER KOR 105
     Natalya Valeyeva     BRONZE URS
1996 Kyung-Wook Kim       GOLD   KOR
1996 Ying He              SILVER CHN
1996 Olena Sadovnycha     BRONZE UKR
2000 Mi-Jin Jun           GOLD   KOR 107
2000 Nam-Soon Kim         SILVER KOR 106
2000 Soo-Nyung Kim        BRONZE KOR 103
;
run;
```

The following output shows the results:

**Output 32.5** *Writing One or More Times to a Line and Releasing the Line*

```
1     data _null_;
2        length medalist $ 19;
3        input year 1-4  medalist $ 6-24  medal $ 26-31 country $ 33-35 result 37-41;
4        if year=. then put '*** Missing Year' medalist country @;
5        if result=. then put '*** Missing Results' medalist country ;
6           else if year=. then put;
7
8        datalines;
*** Missing YearHee-Kyung Wang KOR
*** Missing YearNatalya Valeyeva URS *** Missing ResultsNatalya Valeyeva URS
*** Missing ResultsKyung-Wook Kim KOR
*** Missing ResultsYing He CHN
*** Missing ResultsOlena Sadovnycha UKR
NOTE: DATA statement used (Total process time):
      real time           0.46 seconds
      cpu time            0.03 seconds


23    ;
24    run;
```

# Writing a Report

## Writing to an Output File

The PUT statement writes lines of text to the SAS log. However, the SAS log is not usually a good destination for a formal report because it also contains the source statements for the program and messages from SAS.

The simplest destination for a printed report is the SAS output file, which is the same place SAS writes output from procedures. SAS automatically defines various characteristics such as page numbers for the procedure output file, and you can take advantage of them instead of defining all the characteristics yourself.

To route lines to the procedure output file, use the FILE statement. The FILE statement has the following form:

**FILE** PRINT <*options*>;

PRINT is a reserved fileref that directs output that is produced by PUT statements to the same print file as the output that is produced by SAS procedures.

*Note:* Make sure that the FILE statement precedes the PUT statement in the program code.

FILE statement *options* specify options that you can use to customize output. The report that is produced in this section uses the following options:

NOTITLES
: eliminates the default title line and makes that line available for writing. By default, the procedure output file contains the title "The SAS System." Because the report creates another title that is descriptive, you can remove the default title by specifying the NOTITLES option.

**FOOTNOTES**
controls whether currently defined footnotes are written to the report.

*Note:* When you use the FILE statement to include footnotes in a report, you must use the FOOTNOTES option in the FILE statement and include a FOOTNOTE statement in your program. The FOOTNOTE statement contains the text of the footnote.

*Note:* You can also remove the default title with a null TITLE statement: `title;`. In this case, SAS writes a line that contains only the date and page number in place of the default title. The line is not available for writing other text.

## Designing the Report

After choosing a destination for your report, the next step in producing a report is to decide how you want it to look. You create the design and determine which lines and columns the text will occupy. Planning how you want your final report to look helps you write the necessary PUT statements to produce the report. The rest of the examples in this section show how to modify a program to produce a final report that resembles the one shown here.

*Output 32.6* Morning and Evening Newspaper Circulation Report

```
              Morning and Evening Newspaper Circulation

    State          Year                   Thousands of Copies
                                          Morning      Evening

    Massachusetts  1999                     798.4        984.7
                   1998                     834.2        793.6
                   1997                     750.3           .
                                          ------       ------
                   Total for each category 2382.9       1778.3
                           Combined total           4161.2

    Alabama        1999                        .         698.4
                   1998                     463.8        522.0
                   1997                     583.2        234.9
                   1996                        .         339.6
                                          ------       ------
                   Total for each category 1047.0       1794.9
                           Combined total           2841.9


                              Preliminary Report
```

## Writing Data Values

After you design your report, you can begin to write the program that will create it. The following program shows how to display the data values for the YEAR, MORNING_COPIES, and EVENING_COPIES variables in specific positions.

In a PUT statement, the @ followed by a number is a pointer control, but it is different from the trailing @ described earlier. The @*n* argument is a column-pointer control. It tells SAS to move to column *n*. In this example the pointer moves to the specified locations, and the PUT statement writes values at those points. Using pointer controls is a simple but useful way of writing data values in columns.

```
title;
```

```
data _null_;
   input state $ morning_copies evening_copies year;
   file print notitles;
   put @26 year @53 morning_copies @66 evening_copies;
   datalines;
Massachusetts 798.4 984.7 1999
Massachusetts 834.2 793.6 1998
Massachusetts 750.3 .     1997
Alabama       .     698.4 1999
Alabama       463.8 522.0 1998
Alabama       583.2 234.9 1997
Alabama       .     339.6 1996
;
run;
```

The following output shows the results:

*Output 32.7*  Data Values in Specific Locations in the Output

```
                         1999          798.4         984.7
                         1998          834.2         793.6
                         1997          750.3            .
                         1999            .           698.4
                         1998          463.8         522
                         1997          583.2         234.9
                         1996            .           339.6
```

## *Improving the Appearance of Numeric Data Values*

In the design for your report, all numeric values are aligned on the decimal point. To achieve this result, you have to alter the appearance of the numeric data values by using SAS formats. In the input data all values for MORNING_COPIES and EVENING_COPIES contain one decimal place, except in one case where the decimal value is 0. In list output SAS writes values in the simplest way, that is, by omitting the 0s in the decimal portion of a value. In formatted output, you can show one decimal place for every value by associating a format with a variable in the PUT statement. Using a format can also align your output values.

The format that is used in the program is called the *w.d* format. The *w.d* format specifies the number of columns to be used for writing the entire value, including the decimal point. It also specifies the number of columns to be used for writing the decimal portion of each value. In this example the format 5.1 causes SAS to use five columns, including one decimal place, for writing each value. Therefore, SAS prints the 0s in the decimal portion as necessary. The format also aligns the periods that SAS uses to indicate missing values with the decimal points.

```
title;
data _null_;
   input state $ morning_copies evening_copies year;
   file print notitles;
   put @26 year @53 morning_copies 5.1 @66 evening_copies 5.1;
   datalines;
Massachusetts 798.4 984.7 1999
Massachusetts 834.2 793.6 1998
Massachusetts 750.3 .     1997
Alabama       .     698.4 1999
```

```
          Alabama          463.8 522.0 1998
          Alabama          583.2 234.9 1997
          Alabama            .   339.6 1996
          ;
          run;
```

The following output shows the results:

**Output 32.8**  *Formatted Numeric Output*

```
           1999              798.4        984.7
           1998              834.2        793.6
           1997              750.3          .
           1999                .          698.4
           1998              463.8        522.0
           1997              583.2        234.9
           1996                .          339.6
```

## *Writing a Value at the Beginning of Each BY Group*

The next step in creating your report is to add the name of the state to your output. If you include the name of the state in the PUT statement with other data values, then the state will appear on every line. However, remembering what you want your final report to look like, you need to write the name of the state only for the first observation of a particular state. Performing a task once for a group of observations requires the use of the BY statement for BY-group processing. The BY statement has the following form:

**BY** *by-variable(s)*<NOTSORTED>;

The *by-variable* names the variable by which the data set is sorted. The optional NOTSORTED option specifies that observations with the same BY value are grouped together but are not necessarily sorted in alphabetical or numerical order.

For BY-group processing,

- ensure that observations come from a SAS data set, not an external file.
- when the data is grouped in BY groups but the groups are not necessarily in alphabetical order, use the NOTSORTED option in the BY statement. For example, use

```
by state notsorted;
```

The following program creates a permanent SAS data set named NEWS.CIRCULATION, and writes the name of the state on the first line of the report for each BY group.

```
title;
libname news 'SAS-data-library';
data news.circulation;
   length state $ 15;
   input state $ morning_copies evening_copies year;
   datalines;
Massachusetts 798.4 984.7 1999
Massachusetts 834.2 793.6 1998
Massachusetts 750.3  .    1997
Alabama         .   698.4 1999
Alabama       463.8 522.0 1998
Alabama       583.2 234.9 1997
```

```
              Alabama         .     339.6 1996
;

data _null_;
   set news.circulation;
   by state notsorted;
   file print notitles;
   if first.state then put / @7 state @;
   put @26 year @53 morning_copies 5.1 @66 evening_copies 5.1;
run;
```

During the first observation for a given state, a PUT statement writes the name of the state and holds the line for further writing (the year and circulation figures). The next PUT statement writes the year and circulation figures and releases the held line. In observations after the first, only the second PUT statement is processed. It writes the year and circulation figures and releases the line as usual.

The first PUT statement contains a slash (/), a pointer control that moves the pointer to the beginning of the next line. In this example, the PUT statement prepares to write on a new line (the default action). Then the slash moves the pointer to the beginning of the next line. As a result, SAS skips a line before writing the value of STATE. In the output, a blank line separates the data for Massachusetts from the data for Alabama. The output for Massachusetts also begins one line farther down the page than it would have otherwise. (That blank line is used later in the development of the report.)

The following output shows the results:

*Output 32.9  Effect of BY-Group Processing*

```
     Massachusetts    1999          798.4        984.7
                      1998          834.2        793.6
                      1997          750.3           .

     Alabama          1999             .         698.4
                      1998          463.8        522.0
                      1997          583.2        234.9
                      1996             .         339.6
```

## Calculating Totals

The next step is to calculate the total morning circulation figures, total evening circulation figures, and total overall circulation figures for each state. Sum statements accumulate the totals, and assignment statements start the accumulation at 0 for each state. When the last observation for a given state is being processed, an assignment statement calculates the overall total, and a PUT statement writes the totals and additional descriptive text.

```
libname news 'SAS-data-library';
title;
data _null_;
   set news.circulation;
   by state notsorted;
   file print notitles;
      /* Set values of accumulator variables to 0 */
      /* at beginning of each BY group.           */
      if first.state then
         do;
```

```
                          morning_total=0;
                          evening_total=0;
                          put / @7 state @;
                     end;
                put @26 year @53 morning_copies 5.1 @66 evening_copies 5.1;

                /* Accumulate separate totals for morning and  */
                /* evening circulations.                        */
                morning_total+morning_copies;
                evening_total+evening_copies;

                /* Calculate total circulation at the end of    */
                /* each BY group.                               */

                if last.state then
                     do;
                          all_totals=morning_total+evening_total;
                          put @52 '------' @65 '------' /
                              @26 'Total for each category'
                              @52 morning_total 6.1 @65 evening_total 6.1 /
                              @35 'Combined total' @59 all_totals 6.1;
                     end;
           run;
```

The following output shows the results:

***Output 32.10*** *Calculating and Writing Totals for Each BY Group*

```
     Massachusetts    1999                      798.4       984.7
                      1998                      834.2       793.6
                      1997                      750.3           .
                                               ------      ------
                      Total for each category  2382.9      1778.3
                            Combined total           4161.2

     Alabama          1999                          .       698.4
                      1998                      463.8       522.0
                      1997                      583.2       234.9
                      1996                          .       339.6
                                               ------      ------
                      Total for each category  1047.0      1794.9
                            Combined total           2841.9
```

Notice that Sum statements ignore missing values when they accumulate totals. Also, by default, Sum statements assign the accumulator variables (in this case, MORNING_TOTAL, and EVENING_TOTAL) an initial value of 0. Therefore, although the assignment statements in the DO group are executed for the first observation for both states, you need them only for the second state.

### Writing Headings and Footnotes for a One-Page Report

The report is complete except for the title lines, column headings, and footnote. Because this is a simple, one-page report, you can write the heading with a PUT statement that is executed only during the first iteration of the DATA step. The automatic variable _N_ counts the number of times the DATA step has iterated or looped, and the PUT statement is executed when the value of _N_ is 1.

The FOOTNOTES option in the FILE statement and the FOOTNOTE statement create the footnote. The following program is complete:

```
libname news 'SAS-data-library';
title;
data _null_;
   set news.circulation;
   by state notsorted;
   file print notitles footnotes;
   if _n_=1 then put @16 'Morning and Evening Newspaper Circulation' //
                     @7  'State' @26 'Year' @51 'Thousands of Copies' /
                     @51 'Morning      Evening';
   if first.state then
      do;
         morning_total=0;
         evening_total=0;
         put / @7 state @;
      end;
   put @26 year @53 morning_copies 5.1 @66 evening_copies 5.1;
   morning_total+morning_copies;
   evening_total+evening_copies;
   if last.state then
      do;
         all_totals=morning_total+evening_total;
         put @52 '------' @65 '------' /
             @26 'Total for each category'
             @52 morning_total 6.1 @65 evening_total 6.1 /
             @35 'Combined total' @59 all_totals 6.1;
      end;
   footnote 'Preliminary Report';
run;
```

The following output shows the results:

*Output 32.11  The Final Report*

```
          Morning and Evening Newspaper Circulation
     State              Year             Thousands of Copies
                                         Morning      Evening

     Massachusetts      1999              798.4         984.7
                        1998              834.2         793.6
                        1997              750.3           .
                                         ------        ------
                        Total for each category  2382.9  1778.3
                              Combined total        4161.2

     Alabama            1999                .           698.4
                        1998              463.8         522.0
                        1997              583.2         234.9
                        1996                .           339.6
                                         ------        ------
                        Total for each category  1047.0  1794.9
                              Combined total        2841.9




                            Preliminary Report
```

Notice that a blank line appears between the last line of the heading and the first data for Massachusetts, although the PUT statement for the heading does not write a blank line. The line comes from the slash (/) in the PUT statement that writes the value of STATE in the first observation of each BY group.

Executing a PUT statement during the first iteration of the DATA step is a simple way to produce headings, especially when a report is only one page long.

## Summary

### Statements

BY *variable-1* <... *variable-n* > <NOTSORTED>;
: indicates that all observations with common values of the BY variables are grouped together. The NOTSORTED option indicates that the variables are grouped but that the groups are not necessarily in alphabetical or numerical order.

DATA _NULL_;
: specifies that SAS will not create an output data set.

FILE PRINT <NOTITLES> <FOOTNOTES>;
: directs output to the SAS procedure output file. Place the FILE statement before the PUT statements that write to that file. The NOTITLES option suppresses titles that are currently in effect, and makes the lines unavailable for writing other text. The FOOTNOTES option, along with the FOOTNOTE statement, writes a footnote to the file.

PUT;
: by default, begins a new line and releases a previously held line. A PUT statement that does not write any text is known as a null PUT statement.

PUT <*variable* <*format*>> <*character string*>;
: writes lines to the destination that is specified in the FILE statement. If no FILE statement is present, then the PUT statement writes to the SAS log. By default, each PUT statement begins on a new line, writes what is specified, and releases the line. A DATA step can contain any number of PUT statements.

By default, SAS writes a variable or character-string at the current position in the line. SAS automatically moves the pointer one column to the right after writing a variable value but not after writing a character string. That is, SAS places a blank after a variable value but not after a character string. This form of output is called list output. If you place a format after a variable name, then SAS writes the value of the variable beginning at its current position in the line. SAS also uses the format that you specify. The position of the pointer after a formatted value is the following column. That is, SAS does not automatically skip a column. Using a format in a PUT statement is called formatted output. You can combine list and formatted output in a single PUT statement.

PUT<@*n*> <*variable* <*format*>> <*character-string*> </> <@>;
: writes lines to the destination that is specified in the FILE statement. If no FILE statement is present, then the PUT statement writes to the SAS log. The @n pointer control moves the pointer to column n in the current line. The / moves the pointer to the beginning of a new line. (You can use slashes anywhere in the PUT statement to skip lines.) Multiple slashes skip multiple lines. The trailing @, if present, must be the last item in the PUT statement. Executing a PUT statement with a trailing @ holds the current line for use by a later PUT statement either in the same iteration of

the DATA step or a later iteration. Executing a PUT statement without a trailing @ releases a held line.

TITLE;
specifies title lines for SAS output.

## Learning More

Pointer controls
For more information about pointer controls, see the PUT statement in *SAS Statements: Reference*.

Statements
For more information about the statements that are described in this section, see *SAS Statements: Reference*.

ODS
For information about using ODS with the PUT statement and the DATA step, see *SAS Output Delivery System: User's Guide*.

## Chapter 33
# Understanding and Customizing SAS Output: The Basics

**Introduction to the Basics of Understanding and Customizing SAS Output** .... 614
    Purpose ................................................................. 614
    Prerequisites ............................................................ 614

**Understanding Output** ..................................................... 614
    Output from Procedures ................................................. 614
    Output from DATA Step Applications ..................................... 615
    Output from the Output Delivery System (ODS) ........................... 615

**Input SAS Data Set for Examples** ........................................... 616

**Locating Procedure Output** ................................................. 617

**Making Output Informative** ................................................. 618
    Adding Titles ........................................................... 618
    Adding Footnotes ....................................................... 620
    Labeling Variables ...................................................... 621
    Developing Descriptive Output .......................................... 623

**Controlling Output Appearance of Listing Output** ........................... 624
    Specifying SAS System Options .......................................... 624
    Numbering Pages ....................................................... 625
    Centering Output ....................................................... 625
    Specifying Page and Line Size ........................................... 625
    Writing Date and Time Values ........................................... 626
    Choosing Options Selectively ............................................ 626

**Controlling the Appearance of Pages** ....................................... 627
    Input Data Set for Examples of Multiple-page Reports .................... 627
    Writing Centered Title and Column Headings ............................ 629
    Writing Titles and Column Headings in Specific Columns ................. 631
    Changing a Portion of a Heading ........................................ 633
    Controlling Page Divisions .............................................. 635

**Representing Missing Values** ............................................... 638
    Recognizing Default Values ............................................. 638
    Customizing Output of Missing Values By Using a System Option ......... 639
    Customizing Output of Missing Values By Using a Procedure ............. 640

**Summary** .................................................................. 641
    Statements ............................................................. 641
    SAS System Options .................................................... 641

**Learning More** ............................................................. 642

# Introduction to the Basics of Understanding and Customizing SAS Output

## Purpose

In this section, you will learn to understand your output so that you can enhance its appearance and make it more informative. It discusses DATA step and PROC step output.

This section describes how to enhance the appearance of your output by doing the following:

- adding titles, column headings, footnotes, and labels
- customizing headings
- changing a portion of a heading
- numbering pages and controlling page divisions
- printing date and time values
- representing missing numeric values with a character

## Prerequisites

Before proceeding with this section, you should understand the concepts that are presented in the following sections:

- Chapter 3, "Introduction to DATA Step Processing," on page 27
- Chapter 32, "Writing Lines to the SAS Log or to an Output File," on page 595

# Understanding Output

## Output from Procedures

When you invoke a SAS procedure, SAS analyzes or processes your data. You can read a SAS data set, compute statistics, print results, or create a new data set. One of the results of executing a SAS procedure is creating procedure output. The location of procedure output varies with the method of running SAS, the operating environment, and the options that you use. The form and content of the output varies with each procedure. Some procedures, such as the SORT procedure, do not produce printed output.

SAS has numerous procedures that you can use to process your data. For example, you can use the PRINT procedure to print a report that lists the values of each variable in your SAS data set. You can use the MEANS procedure to compute descriptive statistics for variables across all observations and within groups of observations. You can use the UNIVARIATE procedure to produce information about the distribution of numeric variables. For a graphic representation of your data, you can use the CHART procedure. Many other procedures are available through SAS.

## Output from DATA Step Applications

Although output is usually generated by a procedure, you can also generate output by using a DATA step application. Using the DATA step, you can do the following:

- create a SAS data set
- write to an external file
- produce a report

To generate output, you can use the FILE and PUT statements together within the DATA step. Use the FILE statement to identify your current output file. Then use the PUT statement to write lines that contain variable values or text strings to the output file. You can write the values in column, list, or formatted style.

You can use the FILE and PUT statements to target a subset of data. If you have a large data set that includes unnecessary information, this type of DATA step processing can save time and computer resources. Write your code so that the FILE statement executes before a PUT statement in the current execution of a DATA step. Otherwise, your data is written to the SAS log.

If you have a SAS data set, you can use the FILE and PUT statements to create an external file that another computer language can process. For example, you can create a SAS data set that lists the test scores for high school students. You can then use this file as input to a Fortran program that analyzes test scores. The following table lists the variables and the column positions that an existing Fortran program expects to find in the input SAS data set:

*Table 33.1  Variables and Column Positions*

| Variable | Column location |
| --- | --- |
| YEAR | 10-13 |
| TEST | 15-25 |
| GENDER | 30 |
| SCORE | 35-37 |

You can use the FILE and PUT statements in the DATA step to create the data set that the Fortran program reads:

```
data _null_;
   set out.sats1;
   file 'your-output-file';
   put @10 year @15 test
       @30 gender @35 score;
run;
```

## Output from the Output Delivery System (ODS)

Beginning with Version 7, procedure output is much more flexible because of the Output Delivery System (ODS). ODS is a method of delivering output in a variety of formats and of making the formatted output easy to access. Important features of ODS include the following:

- ODS combines raw data with one or more table definitions to produce one or more output objects. When you send these objects to any or all ODS destinations, your output is formatted according to the instructions in the table definition. ODS destinations can produce an output data set, traditional monospace output, output that is formatted for a high-resolution printer, output that is formatted in HyperText Markup Language (HTML), and so on.

- ODS provides table definitions that define the structure of the output from procedures and from the DATA step. You can customize the output by modifying these definitions or by creating your own definitions.

- ODS provides a way for you to choose individual output objects to send to ODS destinations. For example, PROC UNIVARIATE produces five output objects. You can easily create HTML output, an output data set, traditional Listing output, or Printer output from any or all of these output objects. You can send different output objects to different destinations.

- ODS stores a link to each output object in the Results folder in the Results window.

In addition, ODS removes responsibility for formatting output from individual procedures and from the DATA step. The procedure or DATA step supplies raw data and the name of the table definition that contains the formatting instructions. Then ODS formats the output. Because formatting is now centralized in ODS, the addition of a new ODS destination does not affect any procedures or the DATA step. As future destinations are added to ODS, they will automatically become available to the DATA step and to all procedures that support ODS.

For more information and examples, see Chapter 34, "Understanding and Customizing SAS Output: The Output Delivery System (ODS)," on page 643.

## Input SAS Data Set for Examples

The following program creates a SAS data set that contains Scholastic Aptitude Test (SAT) information for university-bound high school seniors from 1972 through 1998. (To view the entire DATA step, see "The UNIVERSITY_TEST_SCORES Data Set" on page 815.) The data set in this example is stored in a SAS library that is referenced by the libref ADMIN. For selected years between 1972 and 1998, the data set shows estimated scores that are based on the total number of students nationwide taking the test. Scores are estimated for male (m) and female (f) students, for both the verbal and math portions of the test.

```
options pagesize=60 linesize=80 pageno=1 nodate;
libname admin 'your-data-library';

data admin.sat_scores;
   input Test $ Gender $ Year SATscore @@;
   datalines;
Verbal m 1972 531   Verbal f 1972 529
Verbal m 1973 523   Verbal f 1973 521
Verbal m 1974 524   Verbal f 1974 520
   ...more SAS data lines...
Math   m 1996 527   Math   f 1996 492
Math   m 1997 530   Math   f 1997 494
Math   m 1998 531   Math   f 1998 496
```

```
;
proc print data=admin.sat_scores;
run;
```

The following output displays a partial list of the results:

**Output 33.1** *The ADMIN.SAT_SCORES Data Set: Partial HTML Output*

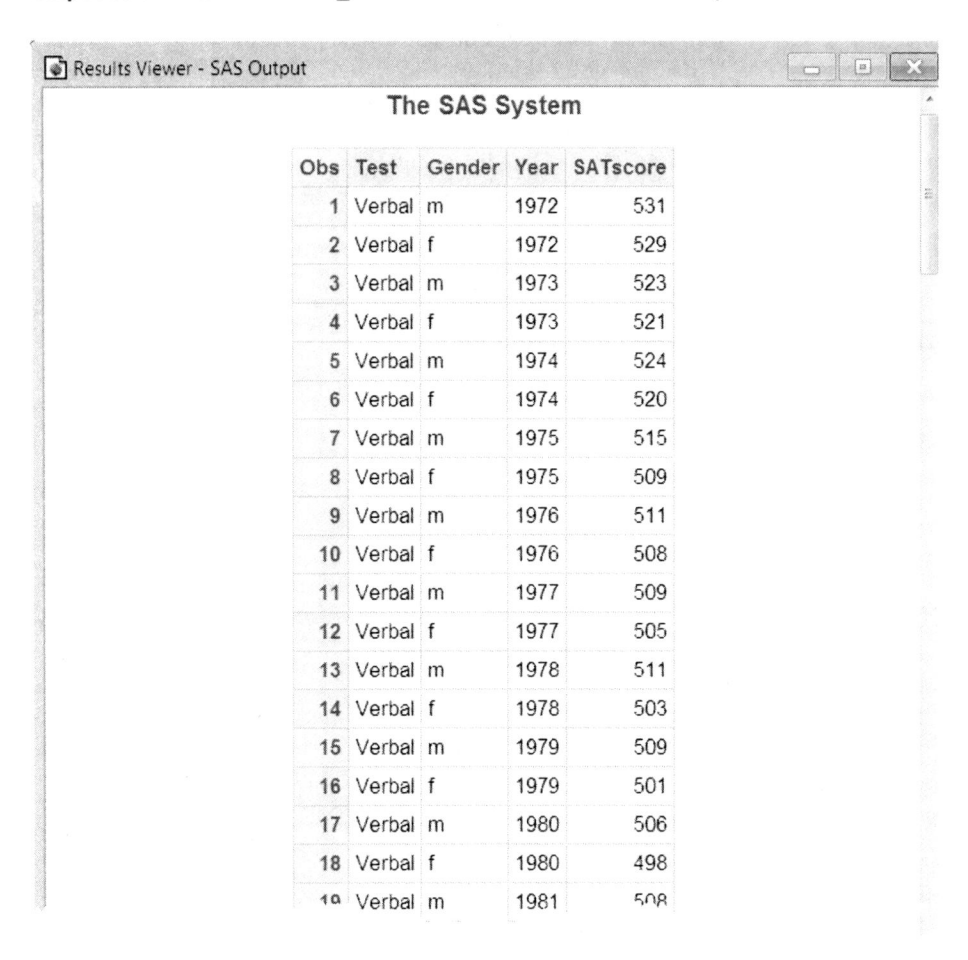

| Obs | Test | Gender | Year | SATscore |
|---|---|---|---|---|
| 1 | Verbal | m | 1972 | 531 |
| 2 | Verbal | f | 1972 | 529 |
| 3 | Verbal | m | 1973 | 523 |
| 4 | Verbal | f | 1973 | 521 |
| 5 | Verbal | m | 1974 | 524 |
| 6 | Verbal | f | 1974 | 520 |
| 7 | Verbal | m | 1975 | 515 |
| 8 | Verbal | f | 1975 | 509 |
| 9 | Verbal | m | 1976 | 511 |
| 10 | Verbal | f | 1976 | 508 |
| 11 | Verbal | m | 1977 | 509 |
| 12 | Verbal | f | 1977 | 505 |
| 13 | Verbal | m | 1978 | 511 |
| 14 | Verbal | f | 1978 | 503 |
| 15 | Verbal | m | 1979 | 509 |
| 16 | Verbal | f | 1979 | 501 |
| 17 | Verbal | m | 1980 | 506 |
| 18 | Verbal | f | 1980 | 498 |
| 19 | Verbal | m | 1981 | 508 |

# Locating Procedure Output

The location of your procedure output depends on the method that you use to start, run, and exit SAS. It also depends on your operating environment and on the settings of SAS system options. The following table shows the default location for each method of operation.

**Table 33.2** *Default Locations for Procedure Output*

| Method of Operation | Location of Procedure Output |
|---|---|
| Windowing environment | Results Viewer and Results windows |
| Interactive line mode | On the terminal display, as each step executes |

| Method of Operation | Location of Procedure Output |
|---|---|
| Noninteractive SAS programs | Depends on the operating environment |
| Batch jobs | Line printer or disk file |

By default, SAS stores output in your Work directory. In the SAS windowing environment for Windows and UNIX, after you have opened and closed the HTML destination, your output goes to your current working directory. You can use the ODS PREFERENCE statement anytime during your SAS session to return to the default behavior. This action is helpful when you are creating multiple graphics and do not want them to accumulate in your current working directory.

# Making Output Informative

## Adding Titles

At the top of each page of output, SAS automatically writes the following title:

```
The SAS System
```

You can make output more informative by using the TITLE statement to specify your own title. A TITLE statement writes the title that you specify at the top of every page. The form of the TITLE statement is:

**TITLE**<*n*> <*'text'*>;

where *n* specifies the relative line that contains the title, and *text* specifies the text of the title. The value of *n* can be 1 to 10. If you omit *n*, SAS assumes a value of 1. Therefore, you can specify TITLE or TITLE1 for the first title line. By default, SAS centers a title.

To add the title 'SAT Scores by Year, 1972-1998' to your output, use the following TITLE statement:

```
title 'SAT Scores by Year, 1972-1998';
```

The TITLE statement is a global statement. This means that within a SAS session, SAS continues to use the most recently created title until you change or eliminate it, even if you generate different output later. You can use the TITLE statement anywhere in your program.

You can specify up to ten titles per page by numbering them in ascending order. If you want to add a subtitle to your previous title, then number your titles by the order in which you want them to appear. For example, use the TITLE3 statement to create the subtitle 'Separate Statistics by Test Type'. To add a blank line between titles, skip a number as you number your TITLE statements. Your TITLE statements now become

```
title1 'SAT Scores by Year, 1972-1998';
title3 'Separate Statistics by Test Type';
```

To modify a title line, you change the text in the title and resubmit your program, including all of the TITLE statements. Be aware that a TITLE statement for a given line cancels the previous TITLE statement for that line and for all lines with higher-numbered titles.

To eliminate all titles including the default title, specify

```
title;
```
or
```
title1;
```

The following example shows how to use multiple TITLE statements.

```
libname admin 'SAS-data-library';

data report;
   set admin.sat_scores;
   if year ge 1995 then output;
   title1 'SAT Scores by Year, 1995-1998';
   title3 'Separate Statistics by Test Type';
run;

proc print data=report;
run;
```

The following output displays the results:

*Output 33.2* *Report Showing Multiple TITLE Statements*

**SAT Scores by Year, 1995-1998**

**Separate Statistics by Test Type**

| Obs | Test | Gender | Year | SATscore |
|---|---|---|---|---|
| 1 | Verbal | m | 1995 | 505 |
| 2 | Verbal | f | 1995 | 502 |
| 3 | Verbal | m | 1996 | 507 |
| 4 | Verbal | f | 1996 | 503 |
| 5 | Verbal | m | 1997 | 507 |
| 6 | Verbal | f | 1997 | 503 |
| 7 | Verbal | m | 1998 | 509 |
| 8 | Verbal | f | 1998 | 502 |
| 9 | Math | m | 1995 | 525 |
| 10 | Math | f | 1995 | 490 |
| 11 | Math | m | 1996 | 527 |
| 12 | Math | f | 1996 | 492 |
| 13 | Math | m | 1997 | 530 |
| 14 | Math | f | 1997 | 494 |
| 15 | Math | m | 1998 | 531 |
| 16 | Math | f | 1998 | 496 |

Although the TITLE statement can appear anywhere in your program, you can associate the TITLE statement with a particular procedure step by positioning it in one of the following locations:

- before the step that produces the output

- after the procedure statement but before the next DATA or RUN statement, or the next procedure

Remember that the TITLE statement applies globally until you change or eliminate it.

## Adding Footnotes

The FOOTNOTE statement follows the same guidelines as the TITLE statement. The FOOTNOTE statement is a global statement. This means that within a SAS session, SAS continues to use the most recently created footnote until you change or eliminate it, even if you generate different output later. You can use the FOOTNOTE statement anywhere in your program.

A footnote writes up to ten lines of text at the bottom of the procedure output or DATA step output. The form of the FOOTNOTE statement is:

**FOOTNOTE**<*n*> <*'text'*>;

where *n* specifies the relative line to be occupied by the footnote, and *text* specifies the text of the footnote. The value of *n* can be 1 to 10. If you omit *n*, SAS assumes a value of 1.

To add the footnote '1967 and 1970 SAT scores estimated based on total number of people taking the SAT,' specify the following statements anywhere in your program:

```
footnote1 '1967 and 1970 SAT scores estimated based on total number';
footnote2 'of people taking the SAT';
```

You can specify up to ten lines of footnotes per page by numbering them in ascending order. When you alter the text of one footnote in a series and execute your program again, SAS changes the text of that footnote. However, if you execute your program with numbered FOOTNOTE statements, SAS eliminates all higher-numbered footnotes.

```
footnote;
```

or

```
footnote1;
```

The following example shows how to use multiple FOOTNOTE statements.

```
libname admin 'SAS-data-library';

data report;
   set admin.sat_scores;
   if year ge 1996 then output;
   title1 'SAT Scores by Year, 1996-1998';
   title3 'Separate Statistics by Test Type';
   footnote1 '1996 through 1998 SAT scores estimated based on total number';
   footnote2 'of people taking the SAT';
run;

proc print data=report;
run;
```

The following output displays the results:

*Output 33.3   Report Showing a Footnote*

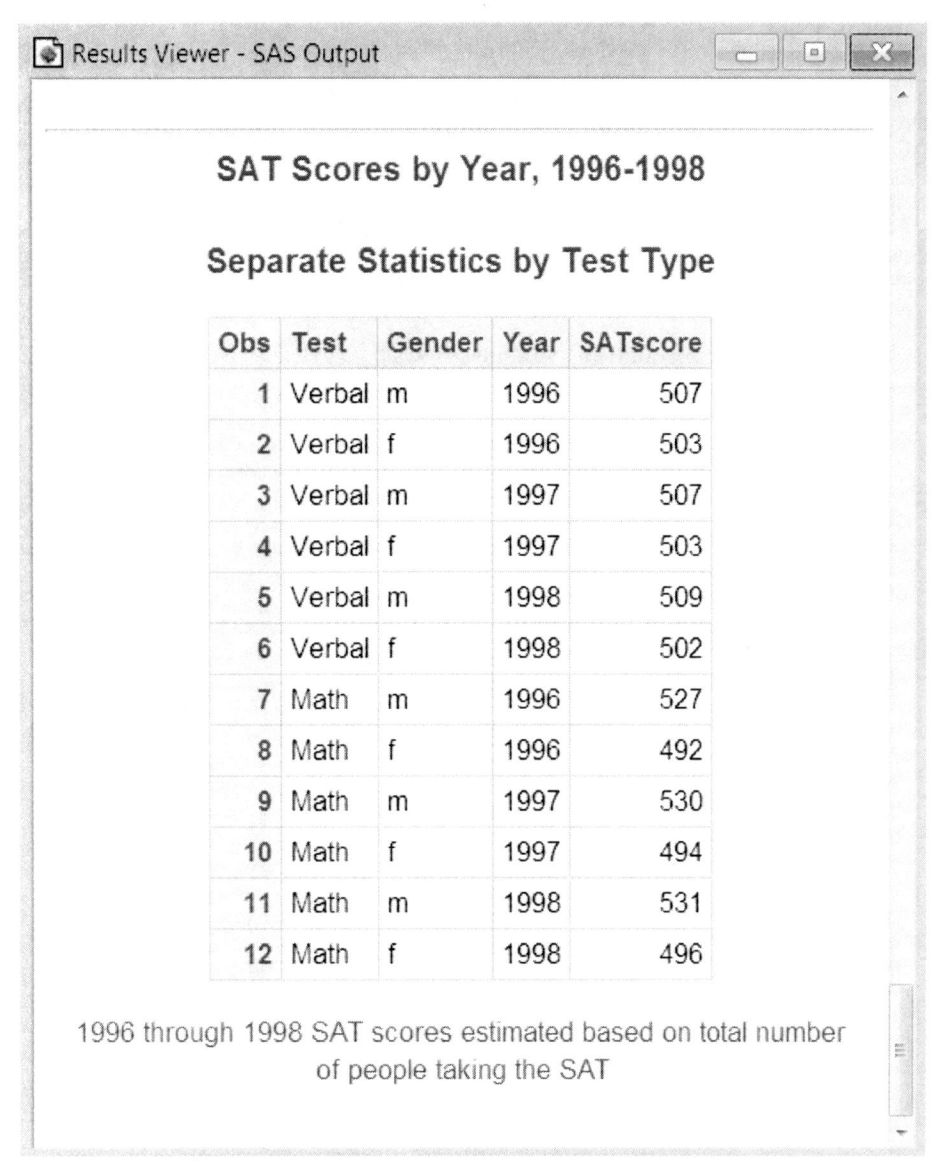

### SAT Scores by Year, 1996-1998

### Separate Statistics by Test Type

| Obs | Test | Gender | Year | SATscore |
|---|---|---|---|---|
| 1 | Verbal | m | 1996 | 507 |
| 2 | Verbal | f | 1996 | 503 |
| 3 | Verbal | m | 1997 | 507 |
| 4 | Verbal | f | 1997 | 503 |
| 5 | Verbal | m | 1998 | 509 |
| 6 | Verbal | f | 1998 | 502 |
| 7 | Math | m | 1996 | 527 |
| 8 | Math | f | 1996 | 492 |
| 9 | Math | m | 1997 | 530 |
| 10 | Math | f | 1997 | 494 |
| 11 | Math | m | 1998 | 531 |
| 12 | Math | f | 1998 | 496 |

1996 through 1998 SAT scores estimated based on total number of people taking the SAT

Although the FOOTNOTE statement can appear anywhere in your program, you can associate the FOOTNOTE statement with a particular procedure step by positioning it at one of the following locations:

- after the RUN statement for the previous step
- after the procedure statement but before the next DATA or RUN statement, or before the next procedure

Remember that the FOOTNOTE statement applies globally until you change or eliminate it.

## Labeling Variables

In procedure output, SAS automatically writes the variables with the names that you specify. However, you can designate a label for some or all of your variables by specifying a LABEL statement either in the DATA step or, with some procedures, in the PROC step of your program. Your label can be up to 256 characters long, including blanks.

For example, to describe the variable SATscore with the phrase 'SAT Score,' specify

```
label SATscore ='SAT Score';
```

If you specify the LABEL statement in the DATA step, the label is permanently stored in the data set. If you specify the LABEL statement in the PROC step, the label is associated with the variable only for the duration of the PROC step. In either case, when a label is assigned, it is written with almost all SAS procedures. The exception is the PRINT procedure. Whether you put the LABEL statement in the DATA step or in the PROC step, with the PRINT procedure, you must specify the LABEL option as follows:

```
proc print data=report label;
run;
```

The following example shows how to use a label statement.

```
libname admin 'SAS-data-library';

data report;
   set admin.sat_scores;
   if year ge 1996 then output;
   label Test='Test Type'
         SATscore='SAT Score';
   title1 'SAT Scores by Year, 1996-1998';
   title3 'Separate Statistics by Test Type';
   footnote1 '1967 and 1970 SAT scores estimated based on total number';
   footnote2 'of people taking the SAT';
run;

proc print data=report label;
run;
```

The following output displays the results:

*Output 33.4   Variable Labels in SAS Output*

### SAT Scores by Year, 1996-1998

### Separate Statistics by Test Type

| Obs | Test Type | Gender | Year | SAT Score |
|-----|-----------|--------|------|-----------|
| 1   | Verbal    | m      | 1996 | 507       |
| 2   | Verbal    | f      | 1996 | 503       |
| 3   | Verbal    | m      | 1997 | 507       |
| 4   | Verbal    | f      | 1997 | 503       |
| 5   | Verbal    | m      | 1998 | 509       |
| 6   | Verbal    | f      | 1998 | 502       |
| 7   | Math      | m      | 1996 | 527       |
| 8   | Math      | f      | 1996 | 492       |
| 9   | Math      | m      | 1997 | 530       |
| 10  | Math      | f      | 1997 | 494       |
| 11  | Math      | m      | 1998 | 531       |
| 12  | Math      | f      | 1998 | 496       |

1996 through 1998 SAT scores estimated based on total number of people taking the SAT

## Developing Descriptive Output

The following example incorporates the TITLE, LABEL, and FOOTNOTE statements, and produces output.

```
libname admin 'SAS-data-library';

proc sort data=admin.sat_scores;
   by gender;
run;

proc means data=admin.sat_scores maxdec=2 fw=8;
   by gender;
   label SATscore='SAT score';
   title1 'SAT Scores by Year, 1967-1976';
   title3 'Separate Statistics by Test Type';
   footnote1 '1972 and 1976 SAT scores estimated based on the';
   footnote2 'total number of people taking the SAT';
run;
```

The following output displays the results:

**Output 33.5** *Titles, Labels, and Footnotes in SAS Output*

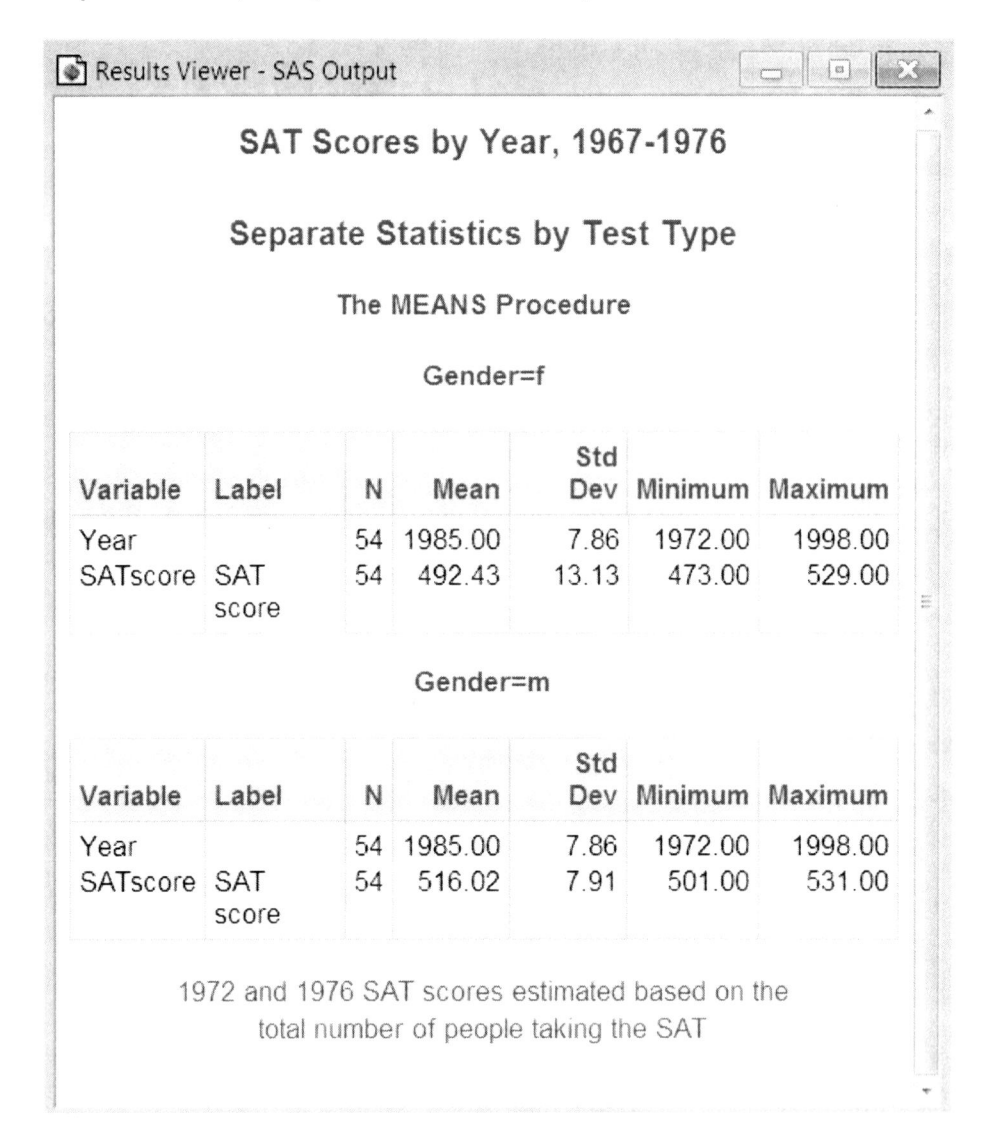

```
SAT Scores by Year, 1967-1976

Separate Statistics by Test Type

The MEANS Procedure

Gender=f
```

| Variable | Label | N | Mean | Std Dev | Minimum | Maximum |
|---|---|---|---|---|---|---|
| Year |  | 54 | 1985.00 | 7.86 | 1972.00 | 1998.00 |
| SATscore | SAT score | 54 | 492.43 | 13.13 | 473.00 | 529.00 |

```
Gender=m
```

| Variable | Label | N | Mean | Std Dev | Minimum | Maximum |
|---|---|---|---|---|---|---|
| Year |  | 54 | 1985.00 | 7.86 | 1972.00 | 1998.00 |
| SATscore | SAT score | 54 | 516.02 | 7.91 | 501.00 | 531.00 |

1972 and 1976 SAT scores estimated based on the total number of people taking the SAT

# Controlling Output Appearance of Listing Output

## Specifying SAS System Options

You can enhance the appearance of your Listing output by specifying SAS system options in the OPTIONS statement. The changes that result from specifying system options remain in effect for the rest of the job, session, or SAS process, or until you issue another OPTIONS statement to change the options.

You can specify SAS system options through the OPTIONS statement, through the OPTIONS window, at SAS invocation, at the initiation of a SAS process, and in a configuration file. Default option settings can vary among sites. To determine the settings at your site, execute the OPTIONS procedure or browse the OPTIONS window.

The OPTIONS statement has the following form:

**OPTIONS** *option(s)*;

where *option* specifies one or more SAS options that you want to change.

*Note:* An OPTIONS statement can appear at any place in a SAS program, except within data lines.

## Numbering Pages

By default, SAS numbers pages of output starting with page 1. However, you can suppress page numbers with the NONUMBER system option. To suppress page numbers, specify the following OPTIONS statement:

```
options nonumber;
```

This option, like all SAS system options, remains in effect for the duration of your session or until you change it. Change the option by specifying

```
options number;
```

You can use the PAGENO= system option to specify a beginning page number for the next page of output that SAS writes. The PAGENO= option enables you to reset page numbering in the middle of a SAS session. For example, the following OPTIONS statement resets the next output page number to 5:

```
options pageno=5;
```

## Centering Output

By default, SAS centers both the output and output titles. However, you can left-align your output by specifying the following OPTIONS statement:

```
options nocenter;
```

The NOCENTER option remains in effect for the duration of your SAS session or until you change it. Change the option by specifying

```
options center;
```

## Specifying Page and Line Size

Procedure output is scaled automatically to fit the size of the page and line. The number of lines per page and the number of characters per line of printed output are determined by the settings of the PAGESIZE= and LINESIZE= system options. The default settings vary from site to site and are further affected by the machine, operating environment, and method of running SAS. For example, when SAS runs in interactive mode, the PAGESIZE= option by default assumes the size of the device that you specify. You can adjust both your page size and line size by resetting the PAGESIZE= and LINESIZE= options.

For example, you can specify the following OPTIONS statement:

```
options pagesize=40 linesize=64;
```

The PAGESIZE= and LINESIZE= options remain in effect for the duration of your SAS session or until you change them.

### Writing Date and Time Values

By default, SAS writes at the top of your output the beginning date and time of the SAS session during which your job executed. This automatic record is especially useful when you execute a program many times. However, you can use the NODATE system option to specify that these values not appear. To do this, specify the following OPTIONS statement:

```
options nodate;
```

The NODATE option remains in effect for the duration of your SAS session or until you change it.

### Choosing Options Selectively

Choose the system options that you need to meet your specifications. The following program, which uses the conditional IF-THEN/ELSE statement to subset the data set, includes a number of SAS options. The OPTIONS statement specifies a line size of 64, left-aligns the output, numbers the output pages and supplies the date on which the SAS session was started. Because HTML output is created by default, you should first close the ODS HTML destination with the ODS HTML CLOSE statement. If you do not want to create HTML output, closing the HTML destination saves system resources. The ODS LISTING destination is closed by default. You must specify the ODS LISTING statement to open the LISTING destination and create Listing output.

```
ods html close;
ods listing;
options linesize=64 nocenter number date;

libname admin 'SAS-data-library';

data high_scores;
   set admin.sat_scores;
   if SATscore < 525 then delete;
run;

proc print data=high_scores;
   title 'SAT Scores: 525 and Above';
run;
ods listing close;
```

The following output displays the results:

*Output 33.6  Effect of System Options on Listing Output*

```
SAT Scores: 525 and Above                                              1
                            14:30 Thursday, May 2, 2013


Obs     Test      Gender     Year     SATscore

 1      Verbal       m       1972       531
 2      Verbal       f       1972       529
 3      Math         m       1972       527
 4      Math         m       1973       525
 5      Math         m       1995       525
 6      Math         m       1996       527
 7      Math         m       1997       530
 8      Math         m       1998       531
```

# Controlling the Appearance of Pages

## Input Data Set for Examples of Multiple-page Reports

In the sections that follow, you learn how to customize multiple-page reports.

The following program creates and prints a SAS data set that contains newspaper circulation figures for morning and evening editions. Each record lists the state, morning circulation figures (in thousands), evening circulation figures (in thousands), and year that the data represents. Because HTML output is created by default, you should first close the ODS HTML destination with the ODS HTML CLOSE statement. If you do not want to create HTML output, closing the HTML destination saves system resources. The ODS LISTING destination is closed by default. You must specify the ODS LISTING statement to open the LISTING destination and create Listing output.

```
data circulation_figures;
   length state $ 15;
   input state $ morning_copies evening_copies year;
   datalines;
Colorado     738.6  210.2  1984
Colorado     742.2  212.3  1985
Colorado     731.7  209.7  1986
Colorado     789.2  155.9  1987
Vermont      623.4  566.1  1984
Vermont      533.1  455.9  1985
Vermont      544.2  566.7  1986
Vermont      322.3  423.8  1987
Alaska        51.0   80.7  1984
Alaska        58.7   78.3  1985
Alaska        59.8   70.9  1986
Alaska        64.3   64.6  1987
Alabama      256.3  480.5  1984
Alabama      291.5  454.3  1985
Alabama      303.6  454.7  1986
Alabama        .    454.5  1987
Maine          .      .    1984
```

```
            Maine       .    68.0 1985
            Maine     222.7  68.6 1986
            Maine     224.1  66.7 1987
            Hawaii    433.5 122.3 1984
            Hawaii    455.6 245.1 1985
            Hawaii    499.3 355.2 1986
            Hawaii    503.2 488.6 1987
            ;
            run;


            ods html close;
            ods listing;
            proc print data=circulation_figures;

            ods listing close;
```

The following output displays the results:

**Output 33.7**  *SAS Data Set CIRCULATION_FIGURES*

```
                        The SAS System                              1

                       morning_  evening_
            Obs   state  copies    copies    year

             1   Colorado  738.6    210.2    1984
             2   Colorado  742.2    212.3    1985
             3   Colorado  731.7    209.7    1986
             4   Colorado  789.2    155.9    1987
             5   Vermont   623.4    566.1    1984
             6   Vermont   533.1    455.9    1985
             7   Vermont   544.2    566.7    1986
             8   Vermont   322.3    423.8    1987
             9   Alaska     51.0     80.7    1984
            10   Alaska     58.7     78.3    1985
            11   Alaska     59.8     70.9    1986
            12   Alaska     64.3     64.6    1987
            13   Alabama   256.3    480.5    1984
            14   Alabama   291.5    454.3    1985
            15   Alabama   303.6    454.7    1986
```

```
                        The SAS System                              2

                       morning_  evening_
            Obs   state  copies    copies    year

            16   Alabama     .      454.5    1987
            17   Maine       .        .      1984
            18   Maine       .       68.0    1985
            19   Maine     222.7     68.6    1986
            20   Maine     224.1     66.7    1987
            21   Hawaii    433.5    122.3    1984
            22   Hawaii    455.6    245.1    1985
            23   Hawaii    499.3    355.2    1986
            24   Hawaii    503.2    488.6    1987
```

## Writing Centered Title and Column Headings

Producing centered titles with TITLE statements is easy, because centering is the default for the TITLE statement. Producing column headings is not so easy. You must insert the correct number of blanks in the TITLE statements so that the entire title, when centered, causes the text to fall in the correct columns. The following example shows how to write centered lines and column headings. The titles and column headings appear at the top of every page of output.

```
ods html close;
ods listing;

options linesize=80 pagesize=20 nodate;

data report1;
   infile 'your-data-file';
   input state $ morning_copies evening_copies year;
run;

title 'Morning and Evening Newspaper Circulation';
title2;
title3
       'State            Year                  Thousands of Copies';
title4
       '                                          Morning
Evening';

data _null_;
   set report1;
   by state notsorted;
   file print;
   if first.state then
      do;
         morning_total=0;
         evening_total=0;
         put / @7 state @;
      end;
   put @26 year @53 morning_copies 5.1 @66 evening_copies 5.1;
   morning_total+morning_copies;
   evening_total+evening_copies;
   if last.state then
      do;
         all_totals=morning_total+evening_total;
         put @52 '------' @65 '------' /
             @26 'Total for each category'
             @52 morning_total 6.1 @65 evening_total 6.1
 /
             @35 'Combined total' @59 all_totals 6.1;
      end;
run;
ods listing close;
```

The following output displays the results:

**Output 33.8** Centered Lines and Column Headings in SAS Output

```
              Morning and Evening Newspaper Circulation                      1

     State        Year              Thousands of Copies
                                     Morning      Evening

   Colorado       1984                 738.6       210.2
                  1985                 742.2       212.3
                  1986                 731.7       209.7
                  1987                 789.2       155.9
                                       ------      ------
                  Total for each category  3001.7   788.1
                         Combined total         3789.8

   Vermont        1984                 623.4       566.1
                  1985                 533.1       455.9
                  1986                 544.2       566.7
                  1987                 322.3       423.8
                                       ------      ------
                  Total for each category  2023.0  2012.5
                         Combined total         4035.5
```

```
              Morning and Evening Newspaper Circulation                      2

     State        Year              Thousands of Copies
                                     Morning      Evening

   Alaska         1984                  51.0        80.7
                  1985                  58.7        78.3
                  1986                  59.8        70.9
                  1987                  64.3        64.6
                                       ------      ------
                  Total for each category   233.8    294.5
                         Combined total          528.3

   Alabama        1984                 256.3       480.5
                  1985                 291.5       454.3
                  1986                 303.6       454.7
                  1987                    .        454.5
                                       ------      ------
                  Total for each category   851.4   1844.0
                         Combined total         2695.4
```

```
              Morning and Evening Newspaper Circulation                    3
    State          Year                Thousands of Copies
                                         Morning      Evening

    Maine          1984                    .              .
                   1985                    .            68.0
                   1986                  222.7          68.6
                   1987                  224.1          66.7
                                         ------         ------
                   Total for each category 446.8        203.3
                              Combined total       650.1

    Hawaii         1984                  433.5         122.3
                   1985                  455.6         245.1
                   1986                  499.3         355.2
                   1987                  503.2         488.6
                                         ------        ------
                   Total for each category 1891.6      1211.2
                              Combined total      3102.8
```

When you create titles and column headings with TITLE statements, consider the following:

- SAS writes page numbers on title lines by default. Therefore, page numbers appear in this report. If you do not want page numbers, specify the NONUMBER system option.

- The PUT statement pointer begins on the first line after the last TITLE statement. SAS does not skip a line before beginning the text as it does with procedure output. In this example, the blank line between the TITLE4 statement and the first line of data for each state is produced by the slash (/) in the PUT statement in the FIRST.STATE group.

## *Writing Titles and Column Headings in Specific Columns*

The easiest way to program headings in specific columns is to use a PUT statement. Instead of calculating the exact number of blanks that are required to make text fall in particular columns, you move the pointer to the appropriate column with pointer controls and write the text. To write headings with a PUT statement, you must execute the PUT statement at the beginning of each page, regardless of the observation that is being processed or the iteration of the DATA step. The FILE statement with the HEADER= option specifies the headings that you want to write.

Use the following form of the FILE statement to specify column headings.

**FILE** PRINT HEADER=*label*;

PRINT is a reserved fileref that directs output that is produced by any PUT statements to the same print file as the output that is produced by SAS procedures. The *label* variable defines a statement label that identifies a group of SAS statements that execute each time SAS begins a new output page.

The following program uses the HEADER= option of the FILE statement to add a header routine to the DATA step. The routine uses pointer controls in the PUT statement to write the title, skip two lines, and then write column headings in specific locations.

```
   ods html close;
   ods listing;
```

```
    options linesize=80 pagesize=24;

data _null_;
   set circulation_figures;
   by state notsorted;
   file print notitles header=pagetop;
❶
   if first.state then
        do;
            morning_total=0;
            evening_total=0;
            put / @7 state @;
        end;
   put @26 year @53 morning_copies 5.1 @66 evening_copies 5.1;
   morning_total+morning_copies;
   evening_total+evening_copies;
   if last.state then
      do;
         all_totals=morning_total+evening_total;
         put @52 '------' @65 '------' /
             @26 'Total for each category'
             @52 morning_total 6.1 @65 evening_total 6.1 /
             @35 'Combined total' @59 all_totals 6.1;
      end;
   return; ❷
   pagetop: ❸
      put @16 'Morning and Evening Newspaper Circulation' //
          @7 'State' @26 'Year' @51 'Thousands of Copies'/
          @51 'Morning      Evening';
   return; ❹
   run;
ods listing close;
```

The following list corresponds to the numbered items in the preceding program:

1   The PRINT fileref in the FILE statement creates Listing output. The NOTITLES option eliminates title lines so that the lines can be used by the PUT statement. The HEADER= option defines a statement label that points to a group of SAS statements that executes each time SAS begins a new output page. (You can use the HEADER= option only for creating print files.)

2   The RETURN statement that is located before the header routine marks the end of the main part of the DATA step. It causes execution to return to the beginning of the step for another iteration. Without this return statement, the statements in the header routine would be executed during each iteration of the DATA step, as well as at the beginning of each page.

3   The pagetop: label identifies the header routine. Each time SAS begins a new page, execution moves from its current position to the label pagetop: and continues until SAS encounters the RETURN statement. When execution reaches the RETURN statement at the end of the header routine, execution returns to the statement that was being executed when SAS began a new page.

4   The RETURN statement ends the header routine. Execution returns to the statement that was being executed when SAS began a new page.

The following output displays the results:

**Output 33.9**  *Title and Column Headings in Specific Locations*

```
              Morning and Evening Newspaper Circulation
   State         Year              Thousands of Copies
                                   Morning      Evening

   Colorado      1984                738.6       210.2
                 1985                742.2       212.3
                 1986                731.7       209.7
                 1987                789.2       155.9
                                    ------      ------
                 Total for each category 3001.7   788.1
                      Combined total        3789.8

   Vermont       1984                623.4       566.1
                 1985                533.1       455.9
                 1986                544.2       566.7
                 1987                322.3       423.8
                                    ------      ------
                 Total for each category 2023.0   2012.5
                      Combined total        4035.5

   Alaska        1984                 51.0        80.7
                 1985                 58.7        78.3
                 1986                 59.8
70.9
```

```
              Morning and Evening Newspaper Circulation
   State         Year              Thousands of Copies
                                   Morning      Evening
                 1987                 64.3        64.6
                                    ------      ------
                 Total for each category  233.8    294.5
                      Combined total         528.3

   Alabama       1984                256.3       480.5
                 1985                291.5       454.3
                 1986                303.6       454.7
                 1987                   .        454.5
                                    ------      ------
                 Total for each category  851.4   1844.0
                      Combined total        2695.4

   Maine         1984                   .           .
                 1985                   .         68.0
                 1986                222.7        68.6
                 1987                224.1        66.7
                                    ------      ------
                 Total for each category  446.8    203.3
                      Combined total         650.1
```

## Changing a Portion of a Heading

You can use variable values to create headings that change on every page. For example, if you eliminate the default page numbers in the procedure output file, you can create your own page numbers as part of the heading. You can also write the numbers differently from the default method. For example, you can write "Page 1" rather than "1." Page numbers are an example of a heading that changes with each new page.

**634** *Chapter 33* • *Understanding and Customizing SAS Output: The Basics*

The following program creates page numbers using a Sum statement and writes the numbers as part of the header routine.

```
ods html close;
ods listing;
options linesize=80 pagesize=24;

data _null_;
   set circulation_figures;
   by state notsorted;
   file print notitles header=pagetop;
   if first.state then
         do;
            morning_total=0;
            evening_total=0;
            put / @7 state @;
         end;
   put @26 year @53 morning_copies 5.1 @66 evening_copies 5.1;
   morning_total+morning_copies;
   evening_total+evening_copies;
   if last.state then
      do;
         all_totals=morning_total+evening_total;
         put @52 '------' @65 '------' /
             @26 'Total for each category'
             @52 morning_total 6.1 @65 evening_total 6.1 /
             @35 'Combined total' @59 all_totals 6.1;
      end;
   return;

   pagetop:
      pagenum+1;  ❶
      put @16 'Morning and Evening Newspaper Circulation'
          @67 'Page ' pagenum //
      ❷
          @7 'State' @26 'Year' @51 'Thousands of Copies'/
          @51 'Morning     Evening';
   return;
run;
ods listing close;
```

The following list corresponds to the numbered items in the preceding program:

**1** In this Sum statement, SAS adds the value 1 to the accumulator variable PAGENUM each time a new page begins.

**2** The literal Page and the current page number are printed at the top of each new page.

The following output displays the results:

*Output 33.10* Changing a Portion of a Heading

```
              Morning and Evening Newspaper Circulation        Page 1
       State          Year              Thousands of Copies
                                        Morning        Evening

       Colorado       1984                738.6          210.2
                      1985                742.2          212.3
                      1986                731.7          209.7
                      1987                789.2          155.9
                                         ------         ------
                      Total for each category 3001.7      788.1
                             Combined total          3789.8

       Vermont        1984                623.4          566.1
                      1985                533.1          455.9
                      1986                544.2          566.7
                      1987                322.3          423.8
                                         ------         ------
                      Total for each category 2023.0     2012.5
                             Combined total          4035.5

       Alaska         1984                 51.0           80.7
                      1985                 58.7           78.3
                      1986                 59.8
70.9
```

```
              Morning and Evening Newspaper Circulation        Page 2
       State          Year              Thousands of Copies
                                        Morning        Evening
                      1987                 64.3           64.6
                                         ------         ------
                      Total for each category  233.8      294.5
                             Combined total           528.3

       Alabama        1984                256.3          480.5
                      1985                291.5          454.3
                      1986                303.6          454.7
                      1987                   .           454.5
                                         ------         ------
                      Total for each category  851.4     1844.0
                             Combined total          2695.4

       Maine          1984                   .              .
                      1985                   .             68.0
                      1986                222.7           68.6
                      1987                224.1           66.7
                                         ------         ------
                      Total for each category  446.8      203.3
                             Combined total           650.1
```

## Controlling Page Divisions

The report in Output 33.10 on page 635 automatically splits the data for Alaska over two pages. To make attractive page divisions, you need to know that there is sufficient space on a page to print all the data for a particular state before you print any data for it.

First, you must know how many lines are needed to print a group of data. Then you use the LINESLEFT= option in the FILE statement to create a variable whose value is the number of lines remaining on the current page. Before you begin writing a group of data, compare the number of lines that you need to the value of that variable. If more lines are required than are available, use the _PAGE_ pointer control to advance the pointer to the first line of a new page.

In your report, the maximum number of lines that you need for any state is eight (four years of circulation data for each state plus four lines for the underline, the totals, and the blank line between states). The following program creates a variable named CKLINES and compares its value to eight at the beginning of each BY group. If the value is less than eight, SAS begins a new page before writing that state.

```
ods html close;
ods listing;
options pagesize=24;

data _null_;
   set circulation_figures;
   by state notsorted;
   file print notitles header=pagetop linesleft=cklines;
   if first.state then
        do;
            morning_total=0;
            evening_total=0;
            if cklines<8 then put _page_;
            put / @7 state @;
        end;
   put @26 year @53 morning_copies 5.1 @66 evening_copies 5.1;
   morning_total+morning_copies;
   evening_total+evening_copies;
   if last.state then
      do;
         all_totals=morning_total+evening_total;
         put @52 '------' @65 '------' /
             @26 'Total for each category'
             @52 morning_total 6.1 @65 evening_total 6.1 /
             @35 'Combined total' @59 all_totals 6.1;
      end;
   return;

   pagetop:
      pagenum+1;
      put @16 'Morning and Evening Newspaper Circulation'
              @67 'Page ' pagenum //
          @7 'State' @26 'Year' @51 'Thousands of Copies'/
          @51 'Morning     Evening';
      return;
run;
ods listing close;
```

The following output displays the results:

**Output 33.11**  Output with Specific Page Divisions

```
          Morning and Evening Newspaper Circulation         Page 1

  State         Year                Thousands of Copies
                                    Morning      Evening

  Colorado      1984                  738.6        210.2
                1985                  742.2        212.3
                1986                  731.7        209.7
                1987                  789.2        155.9
                                     ------       ------
                Total for each category 3001.7      788.1
                      Combined total          3789.8

  Vermont       1984                  623.4        566.1
                1985                  533.1        455.9
                1986                  544.2        566.7
                1987                  322.3        423.8
                                     ------       ------
                Total for each category 2023.0     2012.5
                      Combined total          4035.5
```

```
          Morning and Evening Newspaper Circulation         Page 2

  State         Year                Thousands of Copies
                                    Morning      Evening

  Alaska        1984                   51.0         80.7
                1985                   58.7         78.3
                1986                   59.8         70.9
                1987                   64.3         64.6
                                     ------       ------
                Total for each category  233.8      294.5
                      Combined total           528.3

  Alabama       1984                  256.3        480.5
                1985                  291.5        454.3
                1986                  303.6        454.7
                1987                    .          454.5
                                     ------       ------
                Total for each category  851.4     1844.0
                      Combined total          2695.4
```

```
          Morning and Evening Newspaper Circulation         Page 3

  State         Year                Thousands of Copies
                                    Morning      Evening

  Maine         1984                    .            .
                1985                    .           68.0
                1986                  222.7         68.6
                1987                  224.1         66.7
                                     ------       ------
                Total for each category  446.8      203.3
                      Combined total           650.1
```

## Representing Missing Values

### Recognizing Default Values

In the following example, numeric data for male verbal and math scores is missing for 1972. Character data for gender is missing for math scores in 1975. By default, SAS replaces a missing numeric value with a period, and a missing character value with a blank when it creates the data set.

```
libname admin 'SAS-data-library';
data admin.sat_scores2;
   input Test $ 1-8 Gender $ 10 Year 12-15 SATscore 17-19;
   datalines;
verbal    m 1972 .
verbal    f 1972 529
verbal    m 1975 515
verbal    f 1975 509
math      m 1972 .
math      f 1972 489
math        1975 518
math        1975 479
;
run;


ods html close;
ods listing;

options pagesize=60 linesize=80 pageno=1 nodate;

proc print data=admin.sat_scores2;
   title 'SAT Scores for Years 1972 and 1975';
run;

ods listing close;
```

The following output displays the results:

**Output 33.12** *Default Display of Missing Values*

```
             SAT Scores for Years 1972 and 1975                    1

        Obs    Test      Gender    Year     SATscore

         1     verbal      m       1972         .
         2     verbal      f       1972        529
         3     verbal      m       1975        515
         4     verbal      f       1975        509
         5     math        m       1972         .
         6     math        f       1972        489
         7     math                1975        518
         8     math                1975        479
```

## Customizing Output of Missing Values By Using a System Option

If your data set contains missing numeric values, you can use the MISSING= system option to display the missing values as a single character rather than as the default period. You specify the character that you want to use as the value of the MISSING= option. You can specify any single character.

In the following program, the MISSING= option in the OPTIONS statement causes the PRINT procedure to display the letter M, rather than a period, for each numeric missing value.

```
options missing='M' pageno=1;

libname admin 'SAS-data-library';
data admin.sat_scores2;
    input Test $ 1-8 Gender $ 10 Year 12-15 SATscore 17-19;
    datalines;
verbal    m 1972
verbal    f 1972 529
verbal    m 1975 515
verbal    f 1975 509
math      m 1972
math      f 1972 489
math        1975 518
math        1975 479
;

ods html close;
ods listing;

proc print data=admin.sat_scores2;
    title 'SAT Scores for Years 1972 and 1975';
run;
ods listing close;
```

The following output displays the results:

**Output 33.13** *Customized Output of Missing Numeric Values*

```
              SAT Scores for Years 1972 and 1975                  1

          Obs    Test      Gender    Year    SATscore

           1     verbal       m      1972       M
           2     verbal       f      1972      529
           3     verbal       m      1975      515
           4     verbal       f      1975      509
           5     math         m      1972       M
           6     math         f      1972      489
           7     math                1975      518
           8     math                1975      479
```

### Customizing Output of Missing Values By Using a Procedure

Using the FORMAT procedure is another way to represent missing numeric values. It enables you to customize missing values by formatting them. You first use the FORMAT procedure to define a format, and then use a FORMAT statement in a PROC or DATA step to associate the format with a variable.

The following program uses the FORMAT procedure to define a format, and then uses a FORMAT statement in the PROC step to associate the format with the variable SCORE. Note that you do not follow the format name with a period in the VALUE statement but a period always accompanies the format when you use it in a FORMAT statement.

```
ods html close;
ods listing;

options pageno=1;
libname admin 'SAS-data-library';

proc format;
   value xscore .='score unavailable';
run;

proc print data=admin.sat_scores2;
   format SATscore xscore.;
   title 'SAT Scores for Years 1972 and 1975';
run;

ods listing close;
```

The following output displays the results:

**Output 33.14** *Numeric Missing Values Replaced by a Format*

```
              SAT Scores for Years 1972 and 1975                 1

     Obs     Test      Gender      Year         SATscore

      1      verbal       m        1972      score unavailable
      2      verbal       f        1972                    529
      3      verbal       m        1975                    515
      4      verbal       f        1975                    509
      5      math         m        1972      score unavailable
      6      math         f        1972                    489
      7      math                  1975                    518
      8      math                  1975                    479
```

# Summary

## Statements

FILE *file-specification*;
: identifies an external file that the DATA step uses to write output from a PUT statement.

FILE PRINT <HEADER=*label*> <LINESLEFT=*number-of-lines*>;
: directs the output that is produced by any PUT statements to the same print file as the output that is produced by SAS procedures. The HEADER option defines a statement label that identifies a group of SAS statements that you want to execute each time SAS begins a new output page. The LINESLEFT= option defines a variable whose value is the number of lines left on the current page.

FOOTNOTE <*n*> <'*text*'>;
: specifies up to ten footnote lines to be printed at the bottom of a page of output. The variable *n* specifies the relative line to be occupied by the footnote, and *text* specifies the text of the footnote.

LABEL variable='*label*';
: associates the variable that you specify with the descriptive text that you specify as the label. Your label can be up to 256 characters long, including blanks. You can use the LABEL statement in either the DATA step or the PROC step.

ODS LISTING <*option(s)*>
: opens, manages, or closes the LISTING destination.

ODS HTML <*option(s)*>
: opens, manages, or closes the HTML destination.

OPTIONS *option(s)*;
: changes the value of one or more SAS system options.

TITLE <*n*> <'*text*'>;
: specifies up to ten title lines to be printed on each page of the procedure output file and other SAS output. The variable *n* specifies the relative line that contains the title line, and *text* specifies the text of the title.

## SAS System Options

NUMBER | NONUMBER
: controls whether the page number is printed on the first title line of each page of output.

PAGENO=n
: resets the page number for the next page of output.

CENTER | NOCENTER
: controls whether SAS procedure output is centered.

PAGESIZE=n
: specifies the number of lines that can be printed per page of output.

LINESIZE=n
: specifies the printer line width for the SAS log and the standard procedure output file used by the DATA step and procedures.

DATE | NODATE
: controls whether the date and time are printed at the top of each page of the SAS log, the standard print file, or any file with the PRINT attribute.

MISSING='character'
: specifies the character to be printed for missing numeric variable values.

## Learning More

ODS
- "Dictionary of ODS Language Statements" in *SAS Output Delivery System: User's Guide*

SAS output
- Chapter 32, "Writing Lines to the SAS Log or to an Output File," on page 595
- Chapter 34, "Understanding and Customizing SAS Output: The Output Delivery System (ODS)," on page 643

*Chapter 34*
# Understanding and Customizing SAS Output: The Output Delivery System (ODS)

**Introduction to Customizing SAS Output By Using the Output Delivery System** .................................................. **644**
    Purpose .................................................................. 644
    Prerequisites ............................................................ 644

**Input Data Set for Examples** ................................................ **644**

**Understanding ODS Output Formats and Destinations** ...................... **645**

**Selecting an Output Format** ................................................. **647**

**Creating Formatted Output** .................................................. **648**
    Creating HTML Output for a Web Browser ............................... 648
    Creating PDF Output for Adobe Acrobat and Other Applications ........... 656
    Creating RTF and PowerPoint Output ................................... 658

**Selecting the Output That You Want to Format** ............................. **661**
    Identifying Output ...................................................... 661
    Selecting and Excluding Program Output ................................ 663
    Creating a SAS Data Set ................................................. 665

**Customizing ODS Output** ................................................... **667**
    Customizing ODS Output at the Level of a SAS Job ...................... 667
    Customizing ODS Output By Using a Template .......................... 670

**Storing Links to ODS Output** ............................................... **678**

**Summary** ................................................................... **680**
    ODS Statements ......................................................... 680
    PROC SORT Statements ................................................. 682
    PROC TABULATE Statements ........................................... 682
    PROC TEMPLATE Statements ........................................... 682
    PROC UNIVARIATE Statements ......................................... 683

**Learning More** ............................................................. **683**

# Introduction to Customizing SAS Output By Using the Output Delivery System

## Purpose

The Output Delivery System (ODS) enables you to produce output in a variety of formats, such as:

- HTML files
- PDF files
- RTF files (for use with Microsoft Word)
- PowerPoint slides
- output data sets

In this chapter, you will learn how to create ODS output for the formats that are listed above.

## Prerequisites

Before using this chapter, you should be familiar with the concepts presented in:

- Chapter 1, "What is the SAS System?," on page 3
- Chapter 24, "Directing SAS Output and the SAS Log," on page 391

You should also be familiar with DATA step processing, and creating procedure output.

# Input Data Set for Examples

The following program creates the original PROC TABULATE, PROC UNIVARIATE, and PROC SGPANEL output. These procedures produce summary statistics for the sales of office furniture in various countries. When you run this example program, you are creating ODS output. By default, HTML is created when you run code in the SAS windowing environment for Windows or UNIX. Your output (including your graphics) is sent to your Work directory. This output is viewable in the Results Viewer.

*Note:* By default, SAS stores output created by ODS in your Work directory. In the SAS windowing environment for Windows and UNIX, after you have opened and closed the HTML destination, your output goes to your current working directory. You can use the ODS PREFERENCE statement anytime during your SAS session to return to the default behavior. This action is helpful when you are creating multiple graphics and do not want them to accumulate in your current working directory.

The program creates the following:

- one output object for PROC TABULATE
- fifteen output objects for PROC UNIVARIATE
- one output object for PROC SGPANEL

The following program creates the output that this chapter uses.

```
options nodate nonumber;
proc sort data=sashelp.prdsale out=prdsale;
    by Country;
run;

title;

proc tabulate data=prdsale;
    class region division prodtype;
    classlev region division prodtype;
    var actual;
    keyword all sum;
    keylabel all='Total';
    table (region all)*(division all),
          (prodtype all)*(actual*f=dollar10.) /
          misstext=[label='Missing']
          box=[label='Region by Division and Type'];
run;

title 'Actual Product Sales';
title2 '(millions of dollars)';

proc univariate data=prdsale;
    by Country;
    var actual;
run;

title 'Sales Figures for First Quarter by Product';

proc sgpanel data=prdsale;
    where quarter=1;
    panelby product / novarname;
    vbar region / response=predict;
    vline region / response=actual lineattrs=GraphFit;
    colaxis fitpolicy=thin;
    rowaxis label='Sales';
run;
```

*Note:* The examples use filenames that might not be valid in all operating environments. For information about how your operating environment uses file specifications, see the documentation for your operating environment.

# Understanding ODS Output Formats and Destinations

The Output Delivery System (ODS) enables you to produce output in a variety of formats that you can easily access. ODS creates various types of tabular output by combining raw data with one or more table templates to produce one or more output objects. The basic component of ODS functionality is the output object. The PROC or DATA step that you run provides the data component (raw data) and the name of the table template that contains the formatting instructions. The data component and table template together form the output object. These objects can be sent to any or all ODS destinations, such as PDF, HTML, RTF, or LISTING. By default, in the SAS windowing environment for Windows and UNIX, SAS uses ODS to produce HTML output. By default, in batch mode, SAS produces LISTING output. By specifying an ODS destination, you control the type of output that SAS creates.

The following figure illustrates the concept of output. The data and the table template form an output object, which creates the type of ODS output that you specified in the ODS template.

*Figure 34.1* Model of the Production of ODS Output

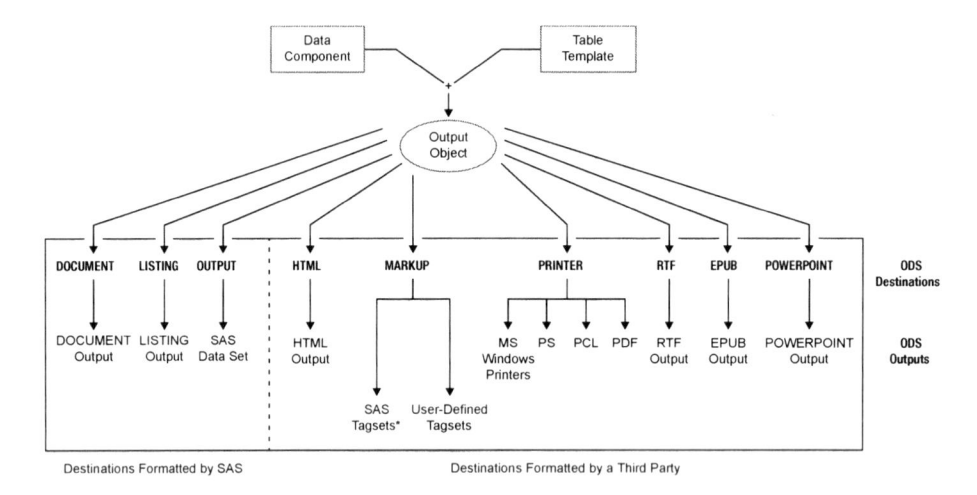

The following definitions describe the terms in the preceding figure:

data component
: Each procedure that supports ODS and each DATA step produces data, which contains the results (numbers and characters) of the step in a form similar to a SAS data set.

table template
: The table template is a set of instructions that describes how to format the data. This description includes but is not limited to the following items:

    - the order of the columns
    - text and order of column headings
    - formats for data
    - font sizes and font faces

output object
: ODS combines formatting instructions with the data to produce an output object. The output object, therefore, contains both the results of the procedure or DATA step and information about how to format the results. An output object has a name, a label, and a path.

    *Note:* Although many output objects include formatting instructions, not all of them do. In some cases the output object consists of only the data.

ODS destinations
: An ODS destination specifies a specific type of output. ODS supports a number of destinations, including but not limited to the following:

    EPUB
    : produces output with the .epub extension. E-books that use the .epub format can be read by a wide variety of e-book readers.

HTML
: produces HTML 4.0 output that contains embedded style sheets. You can access the output on the web with your web browser.

OUTPUT
: produces a SAS data set.

POWERPOINT
: produces PowerPoint output for Microsoft PowerPoint.

LISTING
: produces traditional SAS output (monospace format).

PDF
: produces PDF output, a form of output that is read by Adobe Acrobat and other applications.

PS
: produces PDF output, a form of output that is read by Adobe Acrobat and other applications.

PRINTER
: produces printable output.

RTF
: produces output written in Rich Text Format for use with Microsoft Word 2002.

ODS output
: ODS output consists of formatted output from any of the ODS destinations.

For detailed information about ODS, see *SAS Output Delivery System: User's Guide*. For information about valid ODS destinations, see "Dictionary of ODS Language Statements" in *SAS Output Delivery System: User's Guide*.

## Selecting an Output Format

You select the format for your output by opening and closing ODS destinations in your program. When one or more destinations are open, ODS can send output objects to them and produce formatted output. When a destination is closed, ODS does not send an output object to it and no output is produced.

By default, all programs automatically produce HTML output along with output for other destinations that you specifically open. Therefore, by default, the HTML destination is open, and all other destinations are closed.

To create formatted output, open one or more destinations by using the following statements:

**ODS** *destination file-specification(s)*;

The *destination* is one of the valid ODS destinations. The argument *file-specification* opens the destination and specifies one or more files to write to.

To view or print the ODS output that you have selected, you need to close all the destinations that you opened, except for the HTML destination. You can use separate statements to close individual destinations, or use one statement to close all destinations (including the HTML destination). To close ODS destinations, use the following statements:

**ODS** *destination* CLOSE;

**ODS _ALL_ CLOSE**;

Before you can view ODS output (other than the default destination), the ODS destination must be closed. If you run an example without closing the destination, the SAS log lists the file as being created, but you cannot access the file or see the file when you issue the ls command. To access the file, execute the following statement in your SAS session: ODS *destination* CLOSE;

*Note:* The ODS _ALL_ CLOSE statement, which closes all open destinations, is available with SAS Release 8.2 and higher.

In some cases you might not want to create HTML output. Use the `ODS HTML CLOSE;` statement at the beginning of your program to close the HTML destination and prevent SAS from producing HTML output. Closing unnecessary destinations conserves system resources.

*Note:* Because ODS statements are global statements, it is good practice to open the HTML destination at the end of your program. If you execute other programs in your current SAS session, HTML output is then available. To open the HTML destination, use the `ODS HTML;` statement at the end of your program.

# Creating Formatted Output

## Creating HTML Output for a Web Browser

### Understanding the Four Types of HTML Output Files

HTML output is created b default. However, if you want to use ODS HTML statement options, or create a table file, page file, or frame file, you must use the ODS HTML statement. When you use the ODS HTML statement, you can create output that is formatted in HTML. You can browse the output files with Internet Explorer, Netscape, or any other browser that fully supports the HTML 4.

The ODS HTML statement can create four types of HTML files:

- a body file that contains the results of the DATA step or procedure
- a table of contents that links to items in the body file
- a table of pages that links to items in the body file
- a frame file that displays the results of the procedure or DATA step, the table of contents, and the table of pages

The body file is created by default with all ODS HTML output. If you do not want to link to your output, then creating a table of contents, a table of pages, and a frame file is not necessary.

### Creating HTML Output: The Simplest Case

To produce the simplest type of HTML output, the only file that you need to create is a body file. The following example executes the SORT, MEANS, TABULATE, UNIVARIATE, and SGPANEL procedures. These files contain summary statistics for the sales of office furniture in various countries. Notice that no ODS statement is needed, because the HTML destination is open and SAS automatically creates the HTML body file.

```
options nodate nonumber;
proc sort data=sashelp.prdsale out=prdsale;
    by Country;
```

```
run;

proc tabulate data=prdsale;
   class region division prodtype;
   classlev region division prodtype;
   var actual;
   keyword all sum;
   keylabel all='Total';
   table (region all)*(division all),
         (prodtype all)*(actual*f=dollar10.) /
         misstext=[label='Missing']
         box=[label='Region by Division and Type'];
 run;

title 'Actual Product Sales';
title2 '(millions of dollars)';

proc univariate data=prdsale;
   by Country;
   var actual;
run;

title 'Sales Figures for First Quarter by Product';

proc sgpanel data=prdsale;
    where quarter=1;
    panelby product / novarname;
    vbar region / response=predict;
    vline region / response=actual lineattrs=GraphFit;
    colaxis fitpolicy=thin;
    rowaxis label='Sales';
run;
```

You can view the output in the Results Viewer window, and browse the output objects in the Results window. You can view information about each output object such as name, type, description, and template in the Table Properties window.

*Figure 34.2* Output Viewed in the Results Window

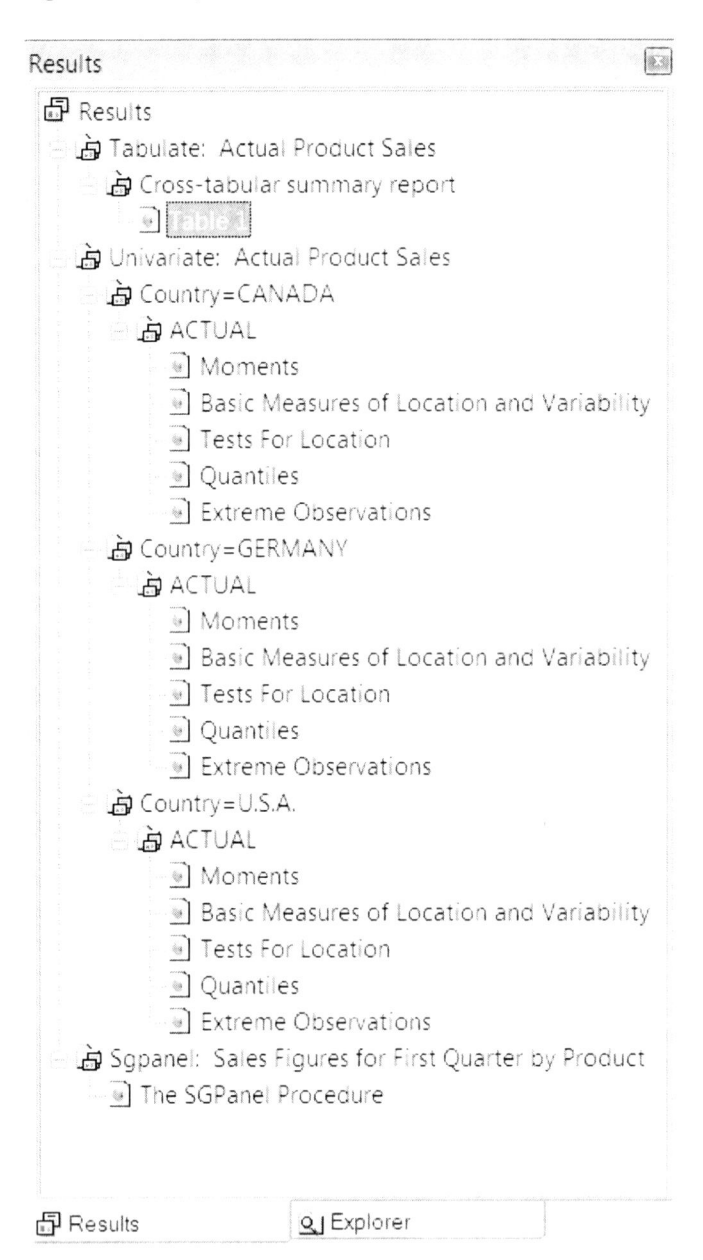

In the Results window, the following output corresponds to the output object Table 1 under **Tabulate: Actual Product Sales ⇨ Cross-tabular summary report**. The name of this output object is Table 1.

*Figure 34.3* PROC TABULATE Output

**Actual Product Sales**
**(millions of dollars)**

| Region by Division and Type | | Product type | | Total |
|---|---|---|---|---|
| | | FURNITURE | OFFICE | |
| | | Actual Sales | Actual Sales | Actual Sales |
| | | Sum | Sum | Sum |
| Region | Division | | | |
| EAST | CONSUMER | $72,570 | $108,686 | $181,256 |
| | EDUCATION | $73,901 | $115,104 | $189,005 |
| | Total | $146,471 | $223,790 | $370,261 |
| WEST | Division | | | |
| | CONSUMER | $76,209 | $105,020 | $181,229 |
| | EDUCATION | $67,945 | $110,902 | $178,847 |
| | Total | $144,154 | $215,922 | $360,076 |
| Total | Division | | | |
| | CONSUMER | $148,779 | $213,706 | $362,485 |
| | EDUCATION | $141,846 | $226,006 | $367,852 |
| | Total | $290,625 | $439,712 | $730,337 |

You can view the properties of each output object in the Table Properties window by right-clicking on the highlight output object and selecting **Properties** from the drop down list.

*Figure 34.4* PROC TABULATE: Properties of Table 1

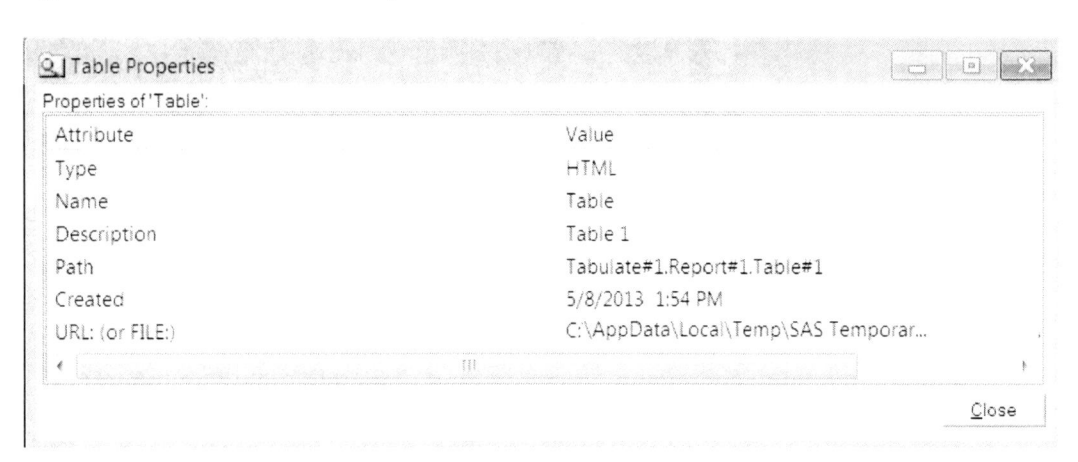

**Table Properties**
Properties of 'Table':

| Attribute | Value |
|---|---|
| Type | HTML |
| Name | Table |
| Description | Table 1 |
| Path | Tabulate#1.Report#1.Table#1 |
| Created | 5/8/2013 1:54 PM |
| URL: (or FILE:) | C:\AppData\Local\Temp\SAS Temporar... |

In the Results window, the following output corresponds to the output object Moments under: **Univariate: Actual Product Sales** ⇨ **Country=CANADA** ⇨ **ACTUAL**. The name of this output object is Moments.

*Figure 34.5* PROC UNIVARIATE Output

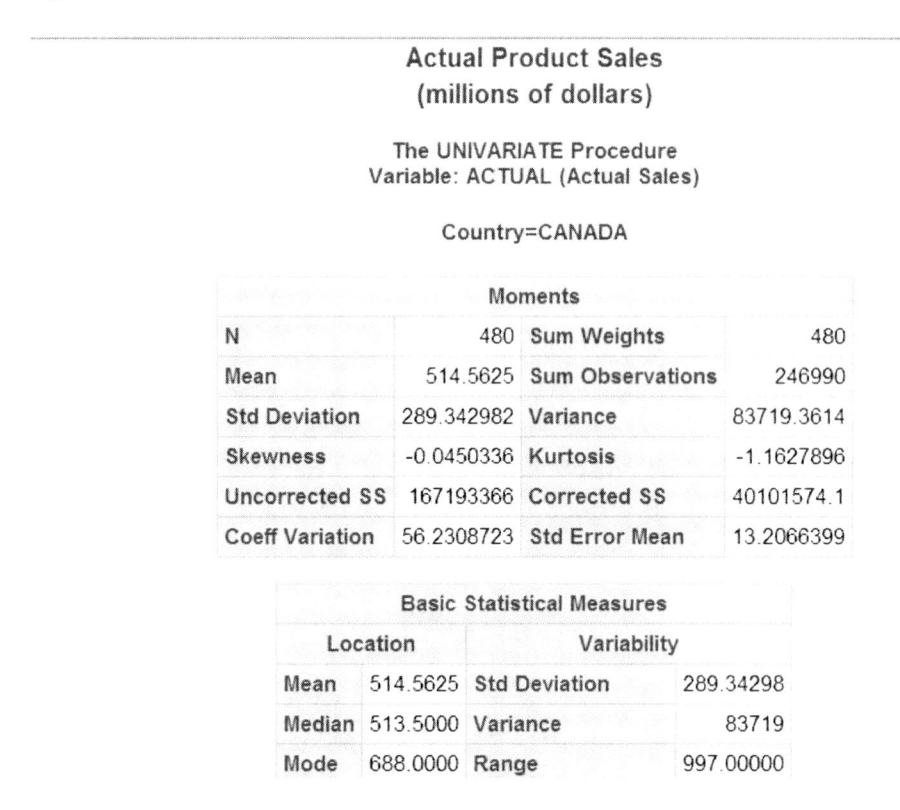

**Actual Product Sales**
**(millions of dollars)**

The UNIVARIATE Procedure
Variable: ACTUAL (Actual Sales)

Country=CANADA

| Moments |  |  |  |
|---|---|---|---|
| N | 480 | Sum Weights | 480 |
| Mean | 514.5625 | Sum Observations | 246990 |
| Std Deviation | 289.342982 | Variance | 83719.3614 |
| Skewness | -0.0450336 | Kurtosis | -1.1627896 |
| Uncorrected SS | 167193366 | Corrected SS | 40101574.1 |
| Coeff Variation | 56.2308723 | Std Error Mean | 13.2066399 |

| Basic Statistical Measures |  |  |  |
|---|---|---|---|
| Location |  | Variability |  |
| Mean | 514.5625 | Std Deviation | 289.34298 |
| Median | 513.5000 | Variance | 83719 |
| Mode | 688.0000 | Range | 997.00000 |

You can view the properties of each output object in the Table Properties window by right-clicking on the highlight output object and selecting **Properties** from the drop down list.

*Figure 34.6* PROC UNIVARIATE: Properties of Moments

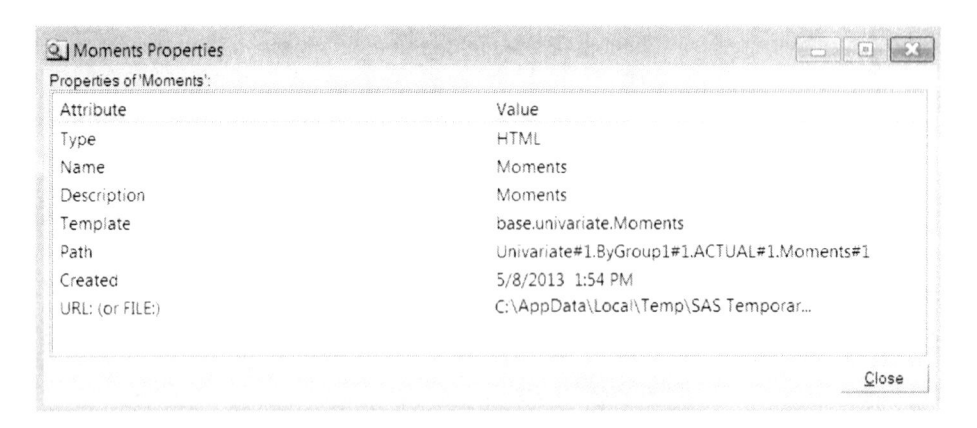

**Moments Properties**

Properties of 'Moments':

| Attribute | Value |
|---|---|
| Type | HTML |
| Name | Moments |
| Description | Moments |
| Template | base.univariate.Moments |
| Path | Univariate#1.ByGroup1#1.ACTUAL#1.Moments#1 |
| Created | 5/8/2013 1:54 PM |
| URL: (or FILE:) | C:\AppData\Local\Temp\SAS Temporar... |

Close

In the Results window, the following output corresponds to the output object **The SGPanel Procedure** under: **Sgpanel: Sales Figures for First Quarter by Product**. The name of this output object is SGPanel.

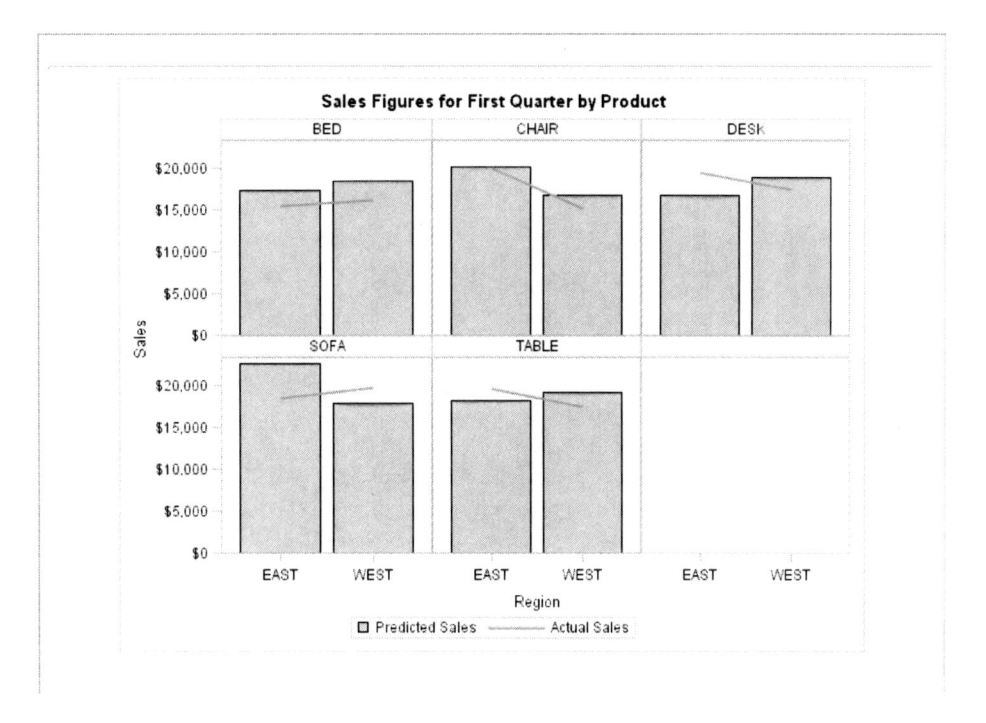

*Figure 34.7   PROC SGPANEL Output*

You can view the properties of each output object in the Table Properties window by right-clicking on the highlight output object and selecting **Properties** from the drop down list.

*Figure 34.8   PROC SGPANEL: Properties of SGPanel*

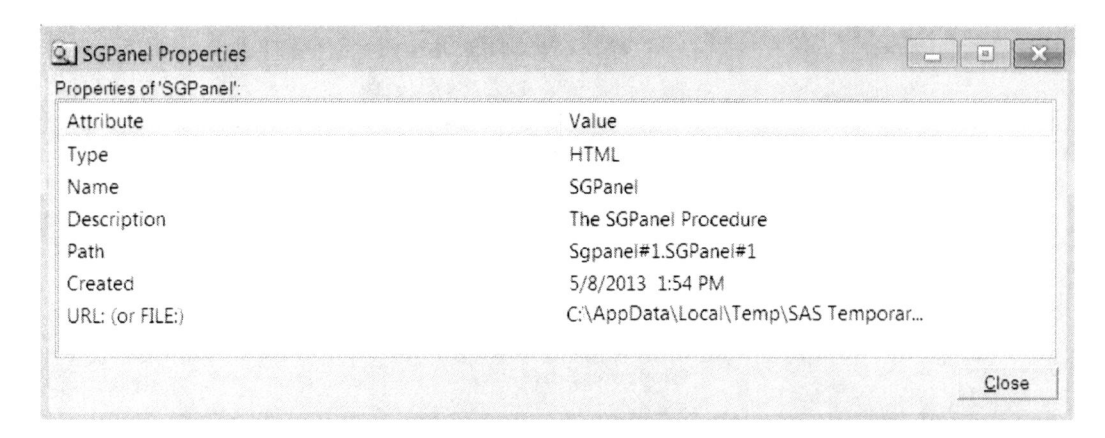

## Creating HTML Output: Linking Results with a Table of Contents

The ODS HTML destination enables you to link to your results from a table of contents and a table of pages. To do this, you need to create the following HTML files: a body file, a frame file, a table of contents, and a table of pages. When you view the frame file and select a link in the table of contents or the table of pages, the HTML table that contains the selected part of the procedure results appears at the top of your browser.

The following example creates multiple pages of output from the UNIVARIATE procedure. You can access specific output results (tables) from links in the table of contents or the table of pages. The results contain statistics for the average SAT scores of entering first-year college classes. The output is grouped by the value of Gender in the CLASS statement and by the value of Test in the BY statement.

```
options nodate nonumber;
proc sort data=sashelp.prdsale out=prdsale;
    by Country;
run;
```

**❶** ```
ods html file='SalesFig-body.htm'
        contents='SalesFig-contents.htm'
        page='SalesFig-page.htm'
        frame='SalesFig-frame.htm';
```

**❷** ```
proc tabulate data=prdsale;
   class region division prodtype;
   classlev region division prodtype;
   var actual;
   keyword all sum;
   keylabel all='Total';
   table (region all)*(division all),
         (prodtype all)*(actual*f=dollar10.) /
         misstext=[label='Missing']
         box=[label='Region by Division and Type'];
run;

title 'Actual Product Sales';
title2 '(millions of dollars)';
```

**❸** ```
proc univariate data=prdsale;
   by Country;
   var actual;
run;

title 'Sales Figures for First Quarter by Product';
```

**❹** ```
proc sgpanel data=prdsale;
   where quarter=1;
   panelby product / novarname;
   vbar region / response=predict;
   vline region / response=actual lineattrs=GraphFit;
   colaxis fitpolicy=thin;
   rowaxis label='Sales';
run;
```

**❺** ```ods html close;```
**❻** ```ods html;```

1  The ODS HTML statement opens the HTML destination and creates four types of files:

   - the body file (created with the FILE= option), which contains the formatted data
   - the contents file, which is a table of contents with links to items in the body file
   - the page file, which is a table of pages with links to items in the body file
   - the frame file, which displays the table of contents, the table of pages, and the body file

**2** The TABULATE procedure creates a summary report for actual product sales.

**3** The UNIVARIATE procedure creates moments, basic measures, quantiles, and extreme observations tables for the actual sales for each country.

**4** The SGPANEL procedure creates a graph of the sales figures for the first quarter by product.

**5** The ODS HTML CLOSE statement closes the HTML destination to make output available for viewing.

**6** The ODS HTML statement reopens the HTML destination so that the next program that you run can produce HTML output.

The following SAS log shows that four HTML files are created with the ODS HTML statement:

*Log 34.1    Partial SAS Log: HTML File Creation*

```
1      options nodate nonumber;
2      proc sort data=sashelp.prdsale out=prdsale;
3          by Country;
4      run;

NOTE: There were 1440 observations read from the data set SASHELP.PRDSALE.
NOTE: The data set WORK.PRDSALE has 1440 observations and 10 variables.
NOTE: PROCEDURE SORT used (Total process time):
      real time           0.03 seconds
      cpu time            0.03 seconds


5
6      ods html file='SalesFig-body.htm'
7               contents='SalesFig-contents.htm'
8               page='SalesFig-page.htm'
9               frame='SalesFig-frame.htm';
NOTE: Writing HTML Body file: SalesFig-body.htm
NOTE: Writing HTML Contents file: SalesFig-contents.htm
NOTE: Writing HTML Pages file: SalesFig-page.htm
NOTE: Writing HTML Frame file: SalesFig-frame.htm
10
11
12     proc tabulate data=prdsale;
13     class region division prodtype;
```

The following output displays the frame file, name SalesFig-frame.htm, which displays the table of contents (upper left side), the table of pages (lower left side), and the body file (right side). Both the Table of Contents and the Table of Pages contain links to the results in the body file. If you click on a link in the Table of Contents or the Table of Pages, SAS displays the corresponding results at the top of the browser.

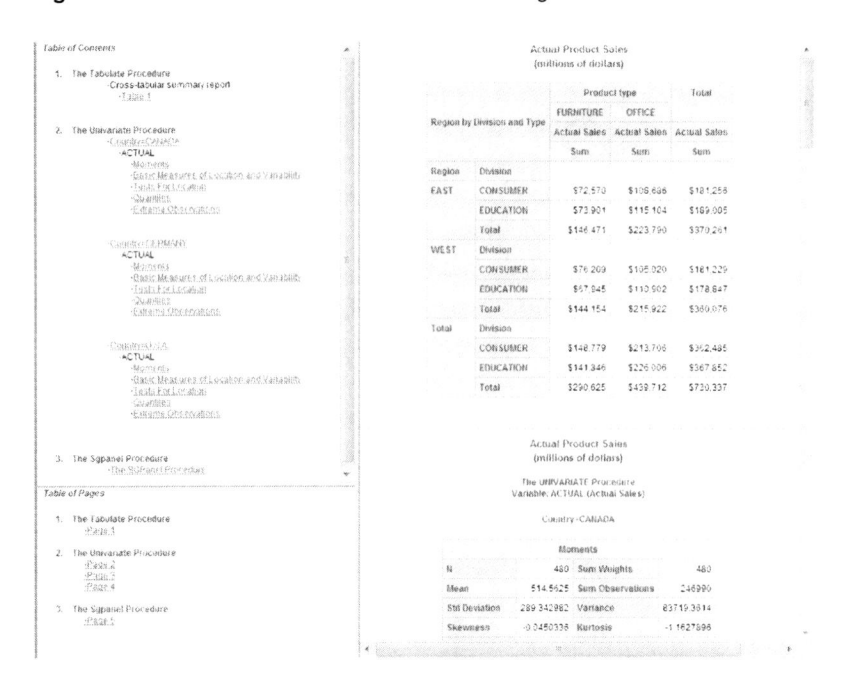

**Figure 34.9** View of the HTML Frame File SalesFig-frame.htm

## Creating PDF Output for Adobe Acrobat and Other Applications

You can create output that is formatted for Adobe Acrobat and other applications. Before you can access the file, however, you must close the PDF destination.

The following example executes the SORT, MEANS, TABULATE, UNIVARIATE, and SGPANEL procedures. These files contain summary statistics for the sales of office furniture in various countries.

```
options nodate nonumber;
proc sort data=sashelp.prdsale out=prdsale;
    by Country;
run;

❶ ods html close;
❷ ods pdf file='odspdf_output.pdf';

❸ proc tabulate data=prdsale;
    class region division prodtype;
    classlev region division prodtype;
    var actual;
    keyword all sum;
    keylabel all='Total';
    table (region all)*(division all),
          (prodtype all)*(actual*f=dollar10.) /
           misstext=[label='Missing']
           box=[label='Region by Division and Type'];
run;

title 'Actual Product Sales';
title2 '(millions of dollars)';

❹ proc univariate data=prdsale;
```

```
      by Country;
      var actual;
run;

title 'Sales Figures for First Quarter by Product';

5 proc sgpanel data=prdsale;
      where quarter=1;
      panelby product / novarname;
      vbar region / response=predict;
      vline region / response=actual lineattrs=GraphFit;
      colaxis fitpolicy=thin;
      rowaxis label='Sales';
run;

6 ods pdf close;
7 ods html;
```

The following list corresponds to the numbered items in the preceding program:

1 By default, the HTML destination is open. To conserve resources, the program uses the ODS HTML CLOSE statement to close this destination.

2 The ODS PDF statement opens the PDF destination and specifies the file to write to.

3 The TABULATE procedure creates a summary report for actual product sales.

4 The UNIVARIATE procedure creates moments, basic measures, quantiles, and extreme observations tables for the actual sales for each country.

5 The SGPANEL procedure creates a graph of the sales figures for the first quarter by product.

6 The ODS PDF CLOSE statement closes the PDF destination to make output available for viewing.

7 The ODS HTML statement reopens the HTML destination so that the next program that you run can produce HTML output.

**658** Chapter 34 • *Understanding and Customizing SAS Output: The Output Delivery System (ODS)*

*Figure 34.10* ODS Output: PDF Format

## Creating RTF and PowerPoint Output

You can send output to multiple destinations at the same time. The ODS RTF statement create output that is formatted for use with Microsoft Word. The ODS POWERPOINT statement creates output that is formatted for Microsoft PowerPoint. Before you can access the files, you must close the destinations. You can close all open destinations with the ODS _ALL_ CLOSE statement.

The following example executes the SORT, MEANS, TABULATE, UNIVARIATE, and SGPANEL procedures. These files contain summary statistics for the sales of office furniture in various countries.

```
options nodate nonumber;
proc sort data=sashelp.prdsale out=prdsale;
    by Country;
run;

❶ods html close;
❷ods rtf file='odsrtf_output.rtf';
❸ods powerpoint file='odspp_output.ppt';

❹proc tabulate data=prdsale;
   class region division prodtype;
   classlev region division prodtype;
   var actual;
   keyword all sum;
   keylabel all='Total';
   table (region all)*(division all),
         (prodtype all)*(actual*f=dollar10.) /
         misstext=[label='Missing']
         box=[label='Region by Division and Type'];
run;
```

```
   title 'Actual Product Sales';
   title2 '(millions of dollars)';

5  proc univariate data=prdsale;
      by Country;
      var actual;
   run;

   title 'Sales Figures for First Quarter by Product';

6  proc sgpanel data=prdsale;
      where quarter=1;
      panelby product / novarname;
      vbar region / response=predict;
      vline region / response=actual lineattrs=GraphFit;
      colaxis fitpolicy=thin;
      rowaxis label='Sales';
   run;

7  ods _all_ close;
8  ods html;
```

The following list corresponds to the numbered items in the preceding program:

1  By default, the HTML destination is open. To conserve resources, the program uses the ODS HTML CLOSE statement to close this destination.

2  The ODS RTF statement opens the RTF destination and specifies the file to write to.

3  The ODS POWERPOINT statement opens the POWERPOINT destination and specifies the file to write to.

4  The TABULATE procedure creates a summary report for actual product sales.

5  The UNIVARIATE procedure creates moments, basic measures, quantiles, and extreme observations tables for the actual sales for each country.

6  The SGPANEL procedure creates a graph of the sales figures for the first quarter by product.

7  The ODS _ALL_ CLOSE statement closes all open destinations to make output available for viewing.

8  The ODS HTML statement reopens the HTML destination so that the next program that you run can produce HTML output.

**660** Chapter 34 • Understanding and Customizing SAS Output: The Output Delivery System (ODS)

The following output displays the first page of the RTF output:

*Figure 34.11* ODS Output: RTF Format

*Actual Product Sales*
*(millions of dollars)*

|  |  | Product type | | |
|---|---|---|---|---|
| Region by Division and Type | | FURNITURE | OFFICE | Total |
| | | Actual Sales | Actual Sales | Actual Sales |
| | | Sum | Sum | Sum |
| Region | Division | | | |
| EAST | CONSUMER | $72,570 | $108,686 | $181,256 |
|  | EDUCATION | $73,901 | $115,104 | $189,005 |
|  | Total | $146,471 | $223,790 | $370,261 |
| WEST | Division | | | |
|  | CONSUMER | $76,209 | $105,020 | $181,229 |
|  | EDUCATION | $67,945 | $110,902 | $178,847 |
|  | Total | $144,154 | $215,922 | $360,076 |
| Total | Division | | | |
|  | CONSUMER | $148,779 | $213,706 | $362,485 |
|  | EDUCATION | $141,846 | $226,006 | $367,852 |
|  | Total | $290,625 | $439,712 | $730,337 |

*Figure 34.12* ODS Output: PowerPoint Format

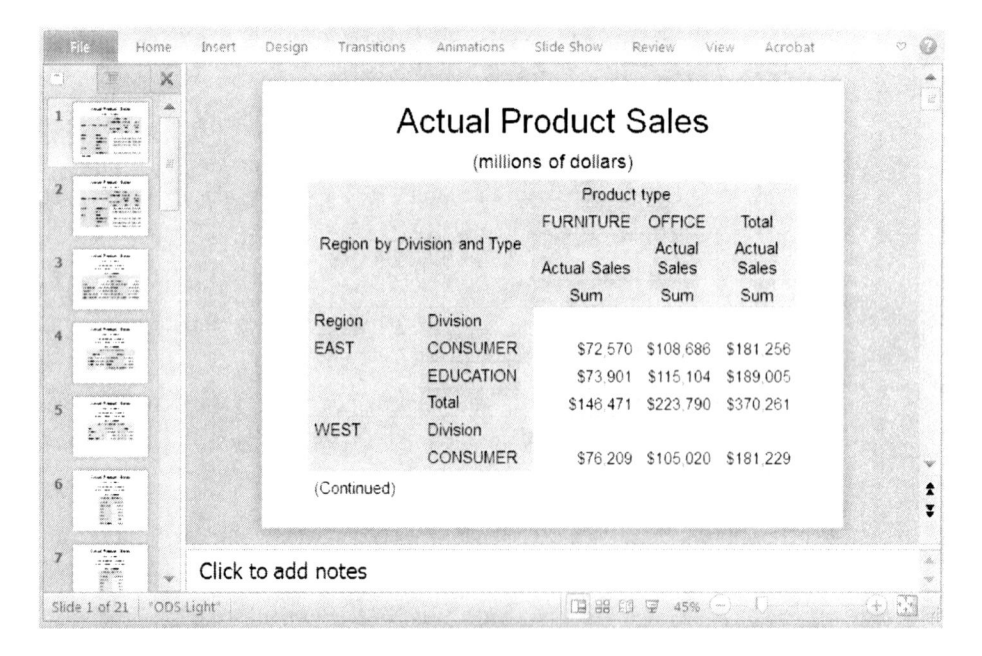

## Selecting the Output That You Want to Format

### *Identifying Output*

Program output, in the form of output objects, contain both the results of a procedure or DATA step and information about how to format the results. To select an output object for formatting, you need to know which output objects your program creates. To identify the output objects, you can use the ODS TRACE statement. The simplest form of the ODS TRACE statement is as follows:

**ODS TRACE** ON | OFF;

ODS TRACE determines whether to write to the SAS log a record of each output object that a program creates. The ON option writes the trace record to the log, and the OFF option suppresses the writing of the trace record.

The trace record has the following components:

Name
    is the name of the output object.

Label
    is the label that briefly describes the contents of the output object.

Template
    is the name of the table template that ODS used to format the output object.

Path
    shows the location of the output object.

In the ODS SELECT statement in your program, you can refer to an output object by name, label, or path.

The following program executes the TABULATE, UNIVARIATE, and SGPANEL procedures and writes a trace record to the SAS log.

```
ods trace on;

options nodate nonumber;
proc sort data=sashelp.prdsale out=prdsale;
    by Country;
run;

proc tabulate data=prdsale;
   class region division prodtype;
   classlev region division prodtype;
   var actual;
   keyword all sum;
   keylabel all='Total';
   table (region all)*(division all),
         (prodtype all)*(actual*f=dollar10.) /
         misstext=[label='Missing']
         box=[label='Region by Division and Type'];
run;

title 'Actual Product Sales';
title2 '(millions of dollars)';
```

```
proc univariate data=prdsale;
   by Country;
   var actual;
run;

title 'Sales Figures for First Quarter by Product';

proc sgpanel data=prdsale;
   where quarter=1;
   panelby product / novarname;
   vbar region / response=predict;
   vline region / response=actual lineattrs=GraphFit;
   colaxis fitpolicy=thin;
   rowaxis label='Sales';
run;

ods trace off;
```

The following output displays the results of ODS TRACE.

**Log 34.2** *Partial ODS TRACE Output in the Log*

```
22   proc univariate data=prdsale;
23   by Country;
24   var actual;
25   run;


Output Added:
-------------
Name:       Moments
Label:      Moments
Template:   base.univariate.Moments
Path:       Univariate.ByGroup1.ACTUAL.Moments
-------------

......


26   title 'Sales Figures for First Quarter by Product';
27   proc sgpanel data=prdsale;
28       where quarter=1;
29       panelby product / novarname;
30       vbar region / response=predict;
31       vline region / response=actual lineattrs=GraphFit;
32       colaxis fitpolicy=thin;
33       rowaxis label='Sales';
34   run;

NOTE: PROCEDURE SGPANEL used (Total process time):
      real time           2.00 seconds
      cpu time            0.21 seconds


Output Added:
-------------
Name:     SGPanel
Label:    The SGPanel Procedure
Path:     Sgpanel.SGPanel
-------------
NOTE: There were 360 observations read from the data set WORK.PRDSALE.
      WHERE quarter=1;

35
36   ods trace off;
```

## Selecting and Excluding Program Output

For each destination, ODS maintains a selection list or an exclusion list. The selection list is a list of output objects that produce formatted output. The exclusion list is a list of output objects for which no output is produced.

You can select and exclude output objects by specifying the destination in an ODS SELECT or ODS EXCLUDE statement. If you do not specify a destination, ODS sends output to all open destinations.

Selection and exclusion lists can be modified and reset at different points in a SAS session, such as at procedure boundaries. If you end each procedure with an explicit QUIT statement, rather than waiting for the next PROC or DATA step to end it for you, the QUIT statement resets the selection list.

To choose one or more output objects and send them to open ODS destinations, use the ODS SELECT statement. The simplest form of the ODS SELECT statement is as follows:

**ODS SELECT** <ODS-destination> output-object(s);

The argument *ODS-destination* identifies the output format, and *output-object* specifies one or more output objects to add to a selection list.

To exclude one or more output objects from being sent to open destinations, use the ODS EXCLUDE statement. The simplest form of the ODS EXCLUDE statement is as follows:

**ODS EXCLUDE** <ODS-destination> output-object(s);

The argument *ODS-destination* identifies the output format, and *output-object* specifies one or more output objects to add to an exclusion list.

The following example executes the TABULATE, UNIVARIATE, and SGPANEL procedures. The ODS SELECT statement uses the name component in the trace records to select only the ExtremeObs, Quantiles, and Moments output objects for PROC UNIVARIATE. Because the ODS SELECT and ODS EXCLUDE statements reset at procedure boundaries, you must specify the ODS EXCLUDE statement before the PROC TABULATE and PROC SGPANEL steps to exclude those output objects.

You can use the ODS EXCLUDE statement before the PROC UNIVARIATE step instead of the ODS SELECT statement. The following ODS EXCLUDE statement gives you the same results:

```
ods exclude BasicMeasures TestsForLocation;

    ods html close;
    options nodate nonumber;
    proc sort data=sashelp.prdsale out=prdsale;
        by Country;
    run;

    ods pdf file='PDFPrdsale.pdf';

    ods exclude table;

    title 'Actual Product Sales';
    title2 '(millions of dollars)';

    proc tabulate data=prdsale;
        class region division prodtype;
        classlev region division prodtype;
        var actual;
        keyword all sum;
        keylabel all='Total';
        table (region all)*(division all),
            (prodtype all)*(actual*f=dollar10.) /
            misstext=[label='Missing']
            box=[label='Region by Division and Type'];
    run;

    title;
    title2;

    ods select ExtremeObs Quantiles Moments;
```

## Selecting the Output That You Want to Format

```
proc univariate data=prdsale;
    by Country;
    var actual;
run;

title 'Sales Figures for First Quarter by Product';

ods exclude sgpanel;

proc sgpanel data=prdsale;
    where quarter=1;
    panelby product / novarname;
    vbar region / response=predict;
    vline region / response=actual lineattrs=GraphFit;
    colaxis fitpolicy=thin;
    rowaxis label='Sales';
run;

ods pdf close;
ods html open;
```

The following two displays show the results in PDF format. They show the ExtremeObs, Quantiles, and Moments tables for PROC UNIVARIATE.

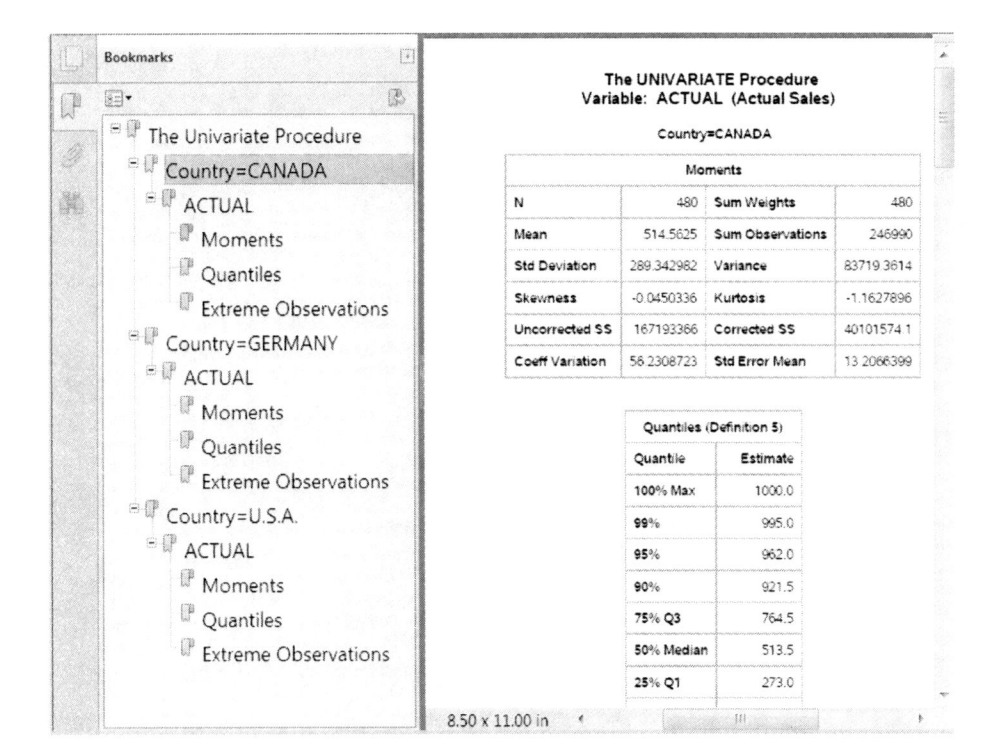

**Figure 34.13** ODS SELECT Statement: PDF Format

### Creating a SAS Data Set

ODS enables you to create a SAS data set from an output object. To create a single output data set, use the following form of the ODS OUTPUT statement:

**ODS OUTPUT** *output-object(s)=SAS-data-set*;

The argument *output-object* specifies one or more output objects to turn into a SAS data set, and *SAS-data-set* specifies the data set that you want to create.

In the following program, ODS opens the Output destination and creates the SAS data set MYFILE.MEASURES from the output object BasicMeasures. ODS then closes the Output destination.

```
data sat_scores;
   input Test $ Gender $ Year SATscore @@;
   datalines;
Verbal m 1972 531   Verbal f 1972 529
Verbal m 1973 523   Verbal f 1973 521
Verbal m 1974 524   Verbal f 1974 520
Verbal m 1975 515   Verbal f 1975 509
Verbal m 1976 511   Verbal f 1976 508
Verbal m 1977 509   Verbal f 1977 505
Verbal m 1978 511   Verbal f 1978 503
Verbal m 1979 509   Verbal f 1979 501
Verbal m 1980 506   Verbal f 1980 498
Verbal m 1981 508   Verbal f 1981 496
Math   m 1985 522   Math   f 1985 480
Math   m 1986 523   Math   f 1986 479
Math   m 1987 523   Math   f 1987 481
Math   m 1988 521   Math   f 1988 483
Math   m 1989 523   Math   f 1989 482
Math   m 1990 521   Math   f 1990 483
Math   m 1991 520   Math   f 1991 482
Math   m 1992 521   Math   f 1992 484
Math   m 1993 524   Math   f 1993 484
Math   m 1994 523   Math   f 1994 487
Math   m 1995 525   Math   f 1995 490
Math   m 1996 527   Math   f 1996 492
Math   m 1997 530   Math   f 1997 494
Math   m 1998 531   Math   f 1998 496
;

proc sort data=sat_scores out=sorted_scores;
   by Test;
run;

   ods html close;  [1]
   ods output BasicMeasures=measures;
   [2]

   proc univariate data=sat_scores;
   [3]
      var SATscore;
      class Gender;
   run;

   ods output close;  [4]
   ods html;  [5]
```

The following list corresponds to the numbered items in the preceding program:

1  By default, the HTML destination is open. To conserve resources, the ODS HTML CLOSE statement closes this destination.

2  The ODS OUTPUT statement opens the Output destination and specifies the permanent data set to create from the output object BasicMeasures.

3  The UNIVARIATE procedure produces summary statistics for the average SAT scores of entering first-year college students. The output is grouped by the CLASS variable Gender.

4  The ODS OUTPUT CLOSE statement closes the Output destination.

5  The ODS HTML statement reopens the default HTML destination so that the next program that you run can produce HTML output.

The following SAS log shows that the WORK.MEASURES data set was created with the ODS OUTPUT statement:

**Log 34.3**  Partial SAS Log: SAS Data Set Creation

```
166
167
168  ods html close;
169  ods output BasicMeasures=measures;
170
171
172  proc univariate data=sat_scores;
173
174     var SATscore;
175     class Gender;
176  run;

NOTE: The data set WORK.MEASURES has 8 observations and 6 variables.
NOTE: PROCEDURE UNIVARIATE used (Total process time):
      real time           0.01 seconds
      cpu time            0.01 seconds


177
178  ods output close;
179  ods html;
NOTE: Writing HTML Body file: sashtml3.htm
```

# Customizing ODS Output

## Customizing ODS Output at the Level of a SAS Job

ODS provides a way for you to customize output at the level of the SAS job. To do this, you use a style template, which describes how to show such items as color, font face, font size, and so on. The style template determines the appearance of the output. The FancyPrinter style template is one of several that is available with SAS.

To view the available ODS style templates, run the following code:

```
proc template;
   list styles;
```

```
run;
```

**Output 34.1** *Sample Listing of ODS Styles*

The SAS System

| | Listing of: SASHELP.TMPLMST | |
|---|---|---|
| | Path Filter is: Styles | |
| | Sort by: PATH/ASCENDING | |
| Obs | Path | Type |
| 1 | Styles | Dir |
| 2 | Styles.Analysis | Style |
| 3 | Styles.BarrettsBlue | Style |
| 4 | Styles.BlockPrint | Style |
| 5 | Styles.Daisy | Style |
| 6 | Styles.Default | Style |
| 7 | Styles.Dove | Style |
| 8 | Styles.Dtree | Style |
| 9 | Styles.EGDefault | Style |
| 10 | Styles.FancyPrinter | Style |
| 11 | Styles.Festival | Style |
| 12 | Styles.FestivalPrinter | Style |
| 13 | Styles.Gantt | Style |
| 14 | Styles.GrayscalePrinter | Style |
| 15 | Styles.HTMLBlue | Style |
| 16 | Styles.Harvest | Style |
| 17 | Styles.HighContrast | Style |
| 18 | Styles.Journal | Style |
| 19 | Styles.Journal1a | Style |
| 20 | Styles.Journal2 | Style |

The following example uses the FancyPrinter style template to customize program output. The STYLE= option in the ODS PRINTER statement specifies that the program use the FancyPrinter style. This style causes ODS to write the titles, footnote, and variable names of the printer output in italics. The PDFTOC= option controls the level of the expansion of the table of contents in PDF documents. The PDFTOC=1 option specifies that the TOC has two levels of expansion.

```
options nodate nonumber;
proc sort data=sashelp.prdsale out=prdsale;
    by Country;
run;

ods html close;
ods pdf file='odspdf_output_custom.pdf' pdftoc=2 style=fancyprinter;
title;
```

```
proc tabulate data=prdsale;
   class region division prodtype;
   classlev region division prodtype;
   var actual;
   keyword all sum;
   keylabel all='Total';
   table (region all)*(division all),
         (prodtype all)*(actual*f=dollar10.) /
         misstext=[label='Missing']
         box=[label='Region by Division and Type'];
 run;

title 'Actual Product Sales';
title2 '(millions of dollars)';

proc univariate data=prdsale;
   by Country;
   var actual;
run;

title 'Sales Figures for First Quarter by Product';

proc sgpanel data=prdsale;
   where quarter=1;
   panelby product / novarname;
   vbar region / response=predict;
   vline region / response=actual lineattrs=GraphFit;
   colaxis fitpolicy=thin;
   rowaxis label='Sales';
run;

ods pdf close;
ods html;
```

The following output displays the results:

**Figure 34.14** *Printer Output: Titles, Footnote, and Variables Printed in Italics*

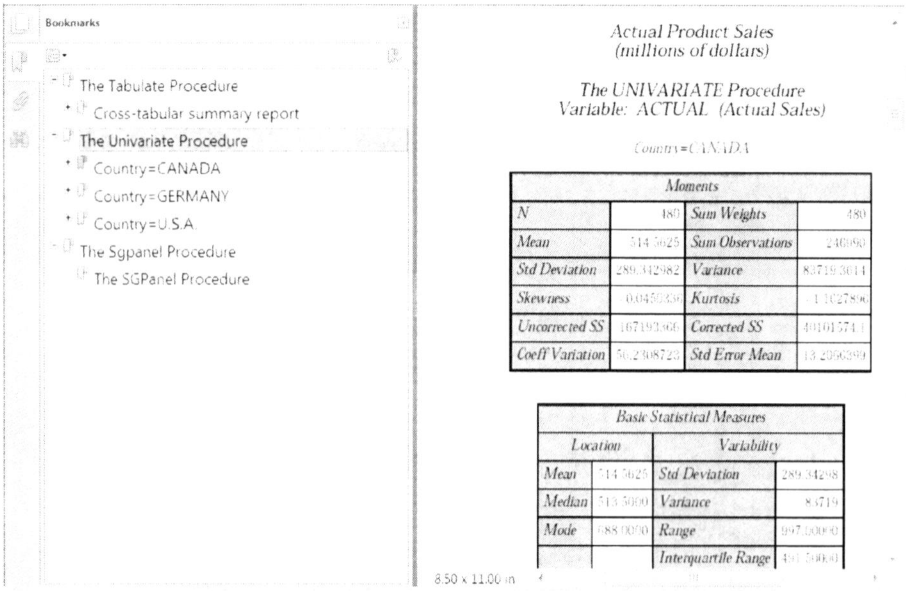

For more information about ODS styles, see "Style Templates" in *SAS Output Delivery System: User's Guide*.

## Customizing ODS Output By Using a Template

Another way to customize ODS output is by using a table template. A table template describes how to format tabular output. It can determine the order of table headings and footnotes, the order of columns, and the appearance of the output. A table template can contain one or more columns, headings, or footnotes. You can create your own table template, or modify an existing one. Many procedures that fully support ODS provide table templates that you can customize.

In the SAS Windowing Environment, to view the SAS style templates that are supplied by SAS, do the following:

1. In the Results window, select the **Results** folder. Right-click and select **Templates** to open the Templates window.

2. Expand a directory, such as **Sashelp.Tmplmst** or **Sashelp.Tmplbase** to view the contents of that directory.

3. Double-click a directory, such as **Base** to view the contents of that directory.

In this example, we are changing a template that PROC UNIVARIATE creates. To view the template names for PROC UNIVARIATE, in the Templates window, select

**Sashelp.Tmplbase** ⇨ **Base** ⇨ **Univariate**. We are using the Measures table template as a template to make changes to the output.

*Output 34.2* PROC UNIVARIATE Templates That Are Supplied by SAS

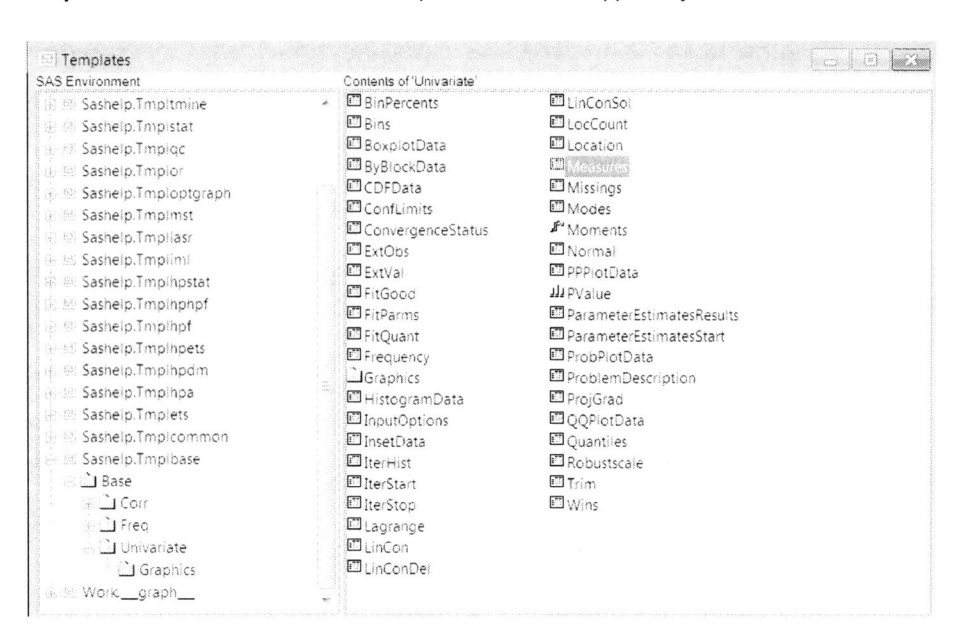

To view the Measures code, double-click **Measures**. The **Measures** template opens in the Template Browser window. You can then copy and paste the code into the Editor window for customization.

*Output 34.3* Measure Templates That Are Supplied by SAS

```
proc template;
    define table Base.Univariate.Measures / store = SASHELP.TMPLBASE;
        notes "Basic measures of location and variability";
        dynamic MeasHdr varname varlabel;
        column LocMeasure LocValue VarMeasure VarValue;
        header h1 h2 h3;
        translate _val_=._ into "";

        define h1;
            text MeasHdr;
            space = 1;
            spill_margin;
        end;

        define h2;
            text "Location";
            end = LocValue;
            start = LocMeasure;
        end;

        define h3;
            text "Variability";
            end = VarValue;
            start = VarMeasure;
        end;

        define LocMeasure;
            space = 3;
            glue = 2;
            style = Rowheader;
            print_headers = OFF;
        end;

        define LocValue;
            space = 5;
            print_headers = OFF;
        end;

        define VarMeasure;
            space = 3;
            glue = 2;
            style = Rowheader;
            print_headers = OFF;
        end;

        define VarValue;
            format = D10.;
            print_headers = OFF;
        end;
    end;
run;
*** END OF TEXT ***
```

You can create your own table template or modify an existing one by using the TEMPLATE procedure. The following is a simplified form of the TEMPLATE procedure:

**PROC TEMPLATE;DEFINE** *table-definition*;**HEADER** *header(s)*;**COLUMN** *column(s)*;**END**;

The DEFINE statement creates the table template that serves as the template for writing the output. The HEADER statement specifies the order of the headings, and the COLUMN statement specifies the order of the columns. The arguments in each of these statements point to routines in the program that format the output. The END statement ends the table template.

The following example shows how to use PROC TEMPLATE to create customized HTML and PDF output. You can first copy the SAS template code that is supplied by SAS using the method described in Output 34.3 on page 672. In this example, the SAS

*Customizing ODS Output* **673**

program creates a customized table template for the Basic Measures output table from PROC UNIVARIATE. The following customized version shows that

- the "Measures of Variability" section precedes the "Measures of Location" section
- column headings are modified
- font colors are modifies
- statistics are displayed in a bold, italic font with a 7.3 format, and a specific font color.

Create the new template.

```
❶ options nodate nonumber;

❷ proc template;
   ❸ define table base.univariate.Measures;
   ❹ header h1 h2 h3;
   ❺ column VarMeasure VarValue LocMeasure LocValue;


   ❻ define h1;
      text "Basic Statistical Measures";
      style=data{color=orange fontstyle=italic};
      spill_margin=on;
      space=1;
   end;

   ❻ define h2;
      text "Measures of Variability";
      start=VarMeasure;
      end=VarValue;
   end;

   ❻ define h3;
      text "Measures of Location";
      start=LocMeasure;
      end=LocValue;
   end;

   ❼ define LocMeasure;
      print_headers=off;
      glue=2;
      space=3;
      style=rowheader;
   end;

   ❼ define LocValue;
      print_headers=off;
      space=5;
      format=7.3;
      style=data{font_style=italic font_weight=bold color=red};
   end;

   ❼ define VarMeasure;
      print_headers=off;
      glue=2;
      space=3;
```

```
            style=rowheader;
            style=data{font_style=italic font_weight=bold color=purple};
         end;

      ⁷define VarValue;
         print_headers=off;
         format=7.3;
         style=data{font_style=italic font_weight=bold color=blue};
      end;
   ⁸end;
⁹run;
```

The following list corresponds to the numbered items in the preceding program:

1  The NODATE and NONUMBER options affect the Printer output. None of the options affects the HTML output.

2  PROC TEMPLATE begins the procedure for creating a table.

3  The DEFINE statement creates the table template base.univariate.Measures in SASUSER.

   The base.univariate.Measures table template that SAS provides is stored in a template store in the SASHELP library. (See Output 34.2 on page 670.)

4  The HEADER statement determines the order in which the table template uses the headings, which are defined later in the program.

5  The COLUMN statement determines the order in which the variables appear. PROC UNIVARIATE names the variables.

6  These DEFINE blocks define the three headings and specify the text to use for each heading. By default, a heading spans all columns. This is the case for H1. H2 spans the variables VarMeasure and VarValue. H3 spans LocMeasure and LocValue.

7  These DEFINE blocks specify characteristics for each of the four variables. They use FORMAT= to specify a format of 7.3 for LocValue and VarValue. They also use STYLE= to specify a bold, italic font for these two variables. The STYLE= option does not affect the LISTING output.

8  The END statement ends the table template.

9  The RUN statement executes the procedure.

To view the table template that PROC TEMPLATE created, in the Templates window, select **Sasuser.Templat** ⇨ **Base** ⇨ **Univariate**. By default, ODS searches for a table template in SASUSER before SASHELP. When PROC UNIVARIATE calls for a table definition by this name, ODS uses the one from SASUSER.

*Customizing ODS Output* **675**

**Output 34.4** *User Created PROC UNIVARIATE Table Template*

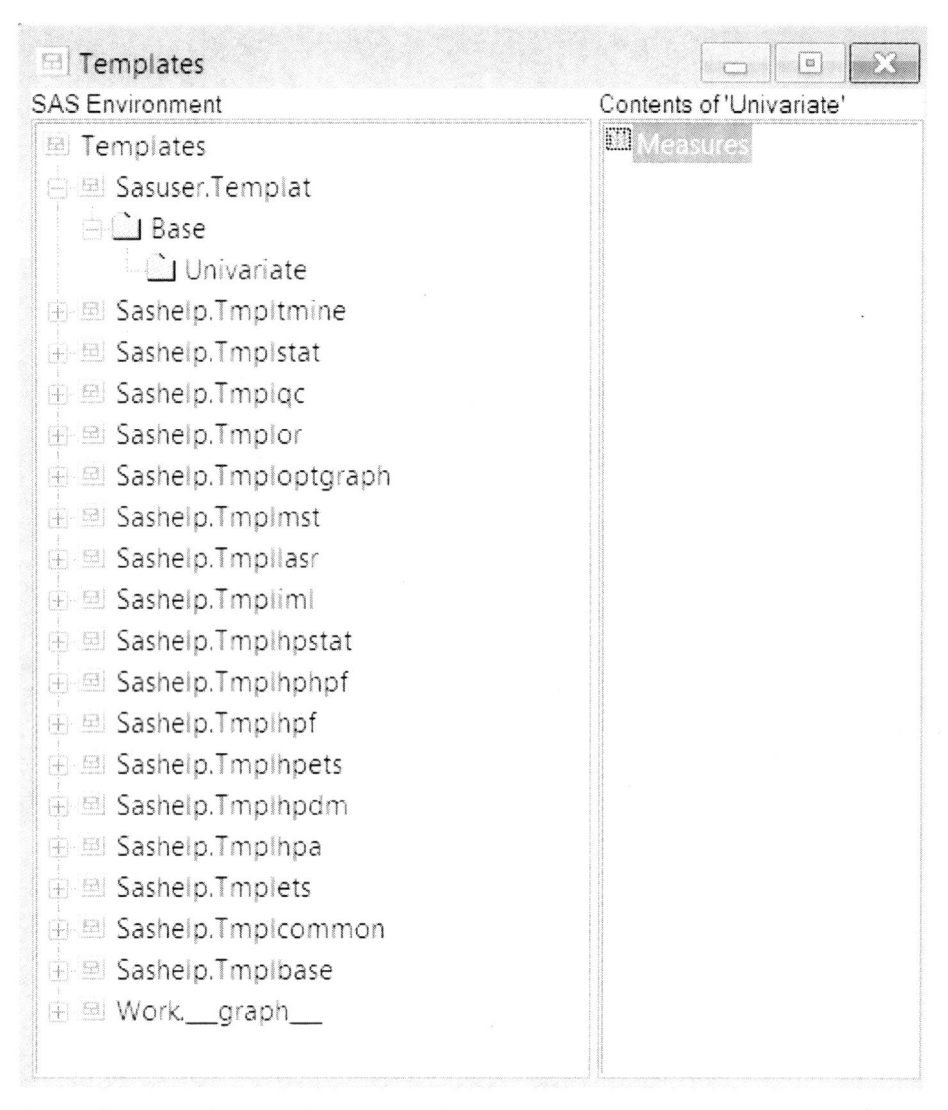

```
Templates
SAS Environment                          Contents of 'Univariate'
  Templates                                Measures
    Sasuser.Templat
      Base
        Univariate
    Sashelp.Tmpltmine
    Sashelp.Tmplstat
    Sashelp.Tmplqc
    Sashelp.Tmplor
    Sashelp.Tmploptgraph
    Sashelp.Tmplmst
    Sashelp.Tmpllasr
    Sashelp.Tmpliml
    Sashelp.Tmplhpstat
    Sashelp.Tmplhphpf
    Sashelp.Tmplhpf
    Sashelp.Tmplhpets
    Sashelp.Tmplhpdm
    Sashelp.Tmplhpa
    Sashelp.Tmplets
    Sashelp.Tmplcommon
    Sashelp.Tmplbase
    Work.__graph__
```

Create the output that uses the new template.

```
1 ods select BasicMeasures;

options nodate nonumber;
proc sort data=sashelp.prdsale out=prdsale;
    by Country;
run;

2 ods html file='SalesFig-body.htm'
        contents='SalesFig-contents.htm'
        page='SalesFig-page.htm'
        frame='SalesFig-frame.htm';

3 ods pdf file='odspdf_output_custom2.pdf' ;

title 'Actual Product Sales';
title2 '(millions of dollars)';
```

```
4 proc univariate data=prdsale;
     by Country;
     var actual;
  run;

5 ods _all_ close;
6 ods html;

7 proc template;
     delete base.univariate.Measures;
  run;
```

1 The ODS SELECT statement selects the output object that contains the basic measures.

2 The ODS HTML statement begins the program that uses the customized table template. It opens the HTML destination and identifies the files to write to.

3 The ODS PDF statement opens the PDF destination and identifies the file to write to.

4 PROC UNIVARIATE produces one object for each variable in each BY group. It uses the customized table template to format the data. By default, ODS searches for a table template in SASUSER before SASHELP. When PROC UNIVARIATE calls for a table definition by this name, ODS uses the one from SASUSER.

5 The ODS _ALL_ CLOSE statement closes all open ODS destinations.

6 The ODS HTML statement opens the HTML destination for output.

7 This PROC TEMPLATE step deletes the base.univariate.Measures template. If you do not delete it, it will be applied to all of your tabular output until you do delete it.

The following displays show the printer output:

**Output 34.5** *PDF Output with No Customization*

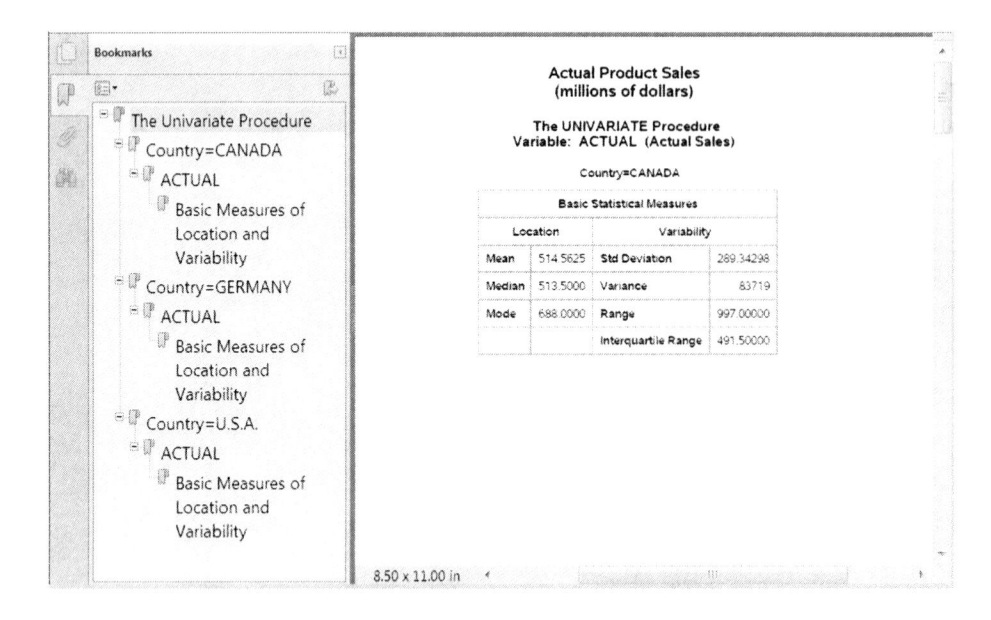

*Customizing ODS Output* **677**

***Figure 34.15*** *Customized Printer Output from the TEMPLATE Procedure*

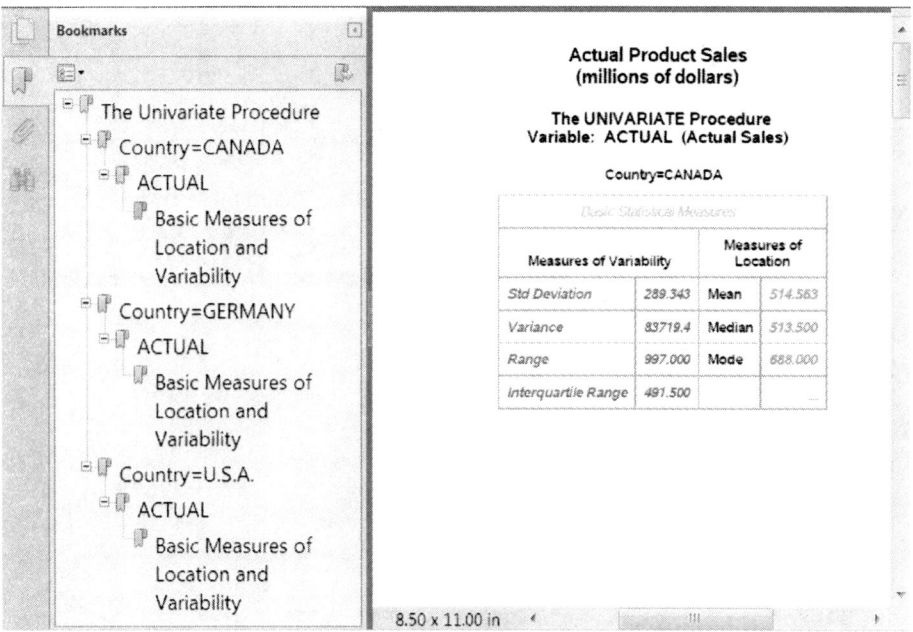

The following display shows the HTML output:

***Figure 34.16*** *Customized HTML Output from the TEMPLATE Procedure*

## Storing Links to ODS Output

When you run a procedure that supports ODS, SAS automatically stores a link to each piece of ODS output in the Results folder in the Results window. It marks the link with an icon that identifies the output destination that created the output.

In the following example, SAS executes the UNIVARIATE, TABULATE, and SGPANEL procedures and generates HTML, PowerPoint, and Rich Text Format (RTF) output.

```
options nodate nonumber;
proc sort data=sashelp.prdsale out=prdsale;
    by Country;
run;

ods html file='SalesFig-body.htm'
         contents='SalesFig-contents.htm'
         page='SalesFig-page.htm'
         frame='SalesFig-frame.htm';
ods rtf file='odsrtf_output.rtf';
ods powerpoint file='odspp_output.ppt';

proc tabulate data=prdsale;
    class region division prodtype;
    classlev region division prodtype;
    var actual;
    keyword all sum;
    keylabel all='Total';
    table (region all)*(division all),
          (prodtype all)*(actual*f=dollar10.) /
          misstext=[label='Missing']
          box=[label='Region by Division and Type'];
run;

title 'Actual Product Sales';
title2 '(millions of dollars)';

proc univariate data=prdsale;
    by Country;
    var actual;
run;

title 'Sales Figures for First Quarter by Product';

proc sgpanel data=prdsale;
    where quarter=1;
    panelby product / novarname;
    vbar region / response=predict;
    vline region / response=actual lineattrs=GraphFit;
    colaxis fitpolicy=thin;
    rowaxis label='Sales';
run;
```

```
ods _all_ close;
ods html;
```

PROC UNIVARIATE, PROC TABULATE, and PROC SGPANEL each generate a folder in the Results window. Within these folders, folder for variables, BY groups, and statistics are created. For example, PROC UNIVARIATE creates a folder name **Country** for each BY group, such as **Country=CANADA**. Within each BY group folder is a folder for each actual variable, named **ACTUAL**. Within the folder **ACTUAL** there is a folder for each statistic, such as **Moments**. The last folders, such as **Moments**, contain the output objects that the procedure creates. The output objects are represented by the appropriate icon.

Within the folder for each output object is a link to each piece of output. The icon next to the link indicates which ODS destination created the output. In this example, the Moments output was sent to the HTML, RTF. and POWEPOINT destinations.

The Results window in the display that follows shows the folders and output objects that the UNIVARIATE, TABULATE, and SGPANEL procedures creates.

**Figure 34.17** View of the Results Window

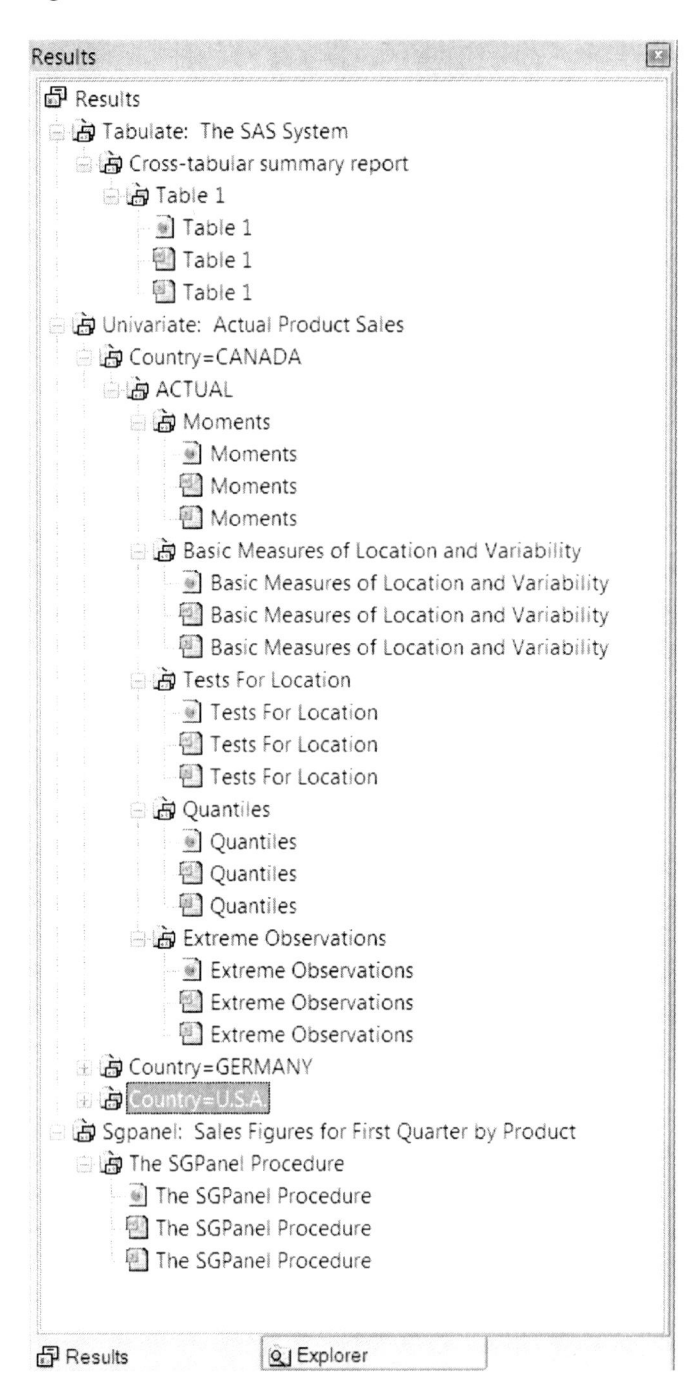

# Summary

## ODS Statements

ODS EXCLUDE <ODS-destination> *output-object(s)*;
  specifies one or more output objects to add to an exclusion list.

ODS HTML *HTML-file-specification(s)* <STYLE='style-definition'>;
: opens the HTML destination and specifies the HTML file or files to write to. After the destination is open, you can create output that is written in Hyper Text Markup Language (HTML).

    *Note:* The HTML destination is open by default.

    You can specify up to four HTML files to write to. The specifications for these files have the following form:

    BODY='*body-file-name*'
    : identifies the file that contains the HTML output.

    CONTENTS='*contents-file-name*'
    : identifies the file that contains a table of contents for the HTML output. The contents file has links to the body file.

    FRAME='*frame-file-name*'
    : identifies the file that integrates the table of contents, the page contents, and the body file. If you open the frame file, you see a table of contents, a table of pages, or both, as well as the body file. If you specify FRAME=, you must also specify CONTENTS= or PAGE= or both.

    PAGE='*page-file-name*'
    : identifies the file that contains a description of each page of the body file and links to the body file. ODS produces a new page of output whenever a procedure explicitly asks for a new page. The SAS system option PAGESIZE= has no effect on pages in HTML output.

    The STYLE= option enables you to choose HTML presentation styles. For complete descriptions of individual styles that are supplied by SAS, see "Style Templates" in *SAS Output Delivery System: User's Guide*.

ODS LISTING;
: opens the LISTING destination.

ODS LISTING CLOSE;
: closes the LISTING destination so that no LISTING output is created.

ODS OUTPUT *output-object(s)=SAS-data-set*;
: opens the Output destination and converts one or more output objects to a SAS data set.

ODS POWERPOINT *file-specification*;
: opens the POWERPOINT destination and specifies the file to write to.

ODS PDF *file-specification*;
: opens the PDF destination and specifies the file to write to.

ODS RTF *file-specification*;
: opens the RTF destination and specifies the file to write to. After the destination is open, you can create RTF output.

ODS *destination* CLOSE;
: closes the specific *destination* and enables you to view the output.

ODS _ALL_ CLOSE;
: closes all open destinations.

ODS SELECT <ODS-destination> *output-object(s)*;
: specifies one or more output objects to add to a selection list.

## PROC SORT Statements

**PROC SORT** DATA=*SAS-data-set* OUT=*SAS-data-set*;

**BY** <DESCENDING> *variable-1*<<DESCENDING> *variable-2* ...>;

PROS SORT statement
: orders SAS data set observations by the values of one or more character or numeric variables. The DATA= option specifies the input data set. The OUT= option specifies the output data set.

BY *variable(s)*;
: specifies the sorting variables.

## PROC TABULATE Statements

**PROC TABULATE** <*option(s)*>

**CLASS** *variable(s)* </ *options*>;

**CLASSLEV** *variable(s)* </ STYLE=<*style-element*>>;

**KEYLABEL** *keyword-1*='*description-1*' <*keyword-2*='*description-2*' ...>;

**KEYWORD** *keyword(s)* </ STYLE=<*style-element*>>;

**TABLE** <<*page-expression*,> *row-expression*,> *column-expression*</ *table-option(s)*>;

**VAR** *analysis-variable(s)*</ *option(s)*>;

PROC TABULATE statement
: displays descriptive statistics in tabular format

CLASS statement
: identifies class variables for the table. Class variables determine the categories that PROC TABULATE uses to calculate statistics.

CLASSLEV statement
: specifies a style element for class variable level value headings.

KEYWORD statement
: specifies a style element for keyword headings.

KEYLABEL statement
: labels a keyword for the duration of the PROC TABULATE step. PROC TABULATE uses the label anywhere that the specified keyword would

TABLE statement
: describes a table to be printed.

VAR statement
: identifies numeric variables to use as analysis variables.

## PROC TEMPLATE Statements

**PROC TEMPLATE**;

**DEFINE** <*template-type*> *template-name*</ *option(s)*>;
   *statements-and-attributes*;
      **COLUMN** *column(s)*;
      **HEADER** *header-specification(s)*;
   **END**;
   **DELETE** *item-path* / <STORE=*template-store*>
**END**;

PROC TEMPLATE statement
   begins a PROC TEMPLATE template.

DELETE statement
   deletes the specified item.

COLUMN statement
   declares a symbol as a column in the table and specifies the order of the columns.

DEFINE statement
   creates a template inside a table template. The DEFINE statement uses the COLUMN and HEADER statements to create column and table headings.

HEADER statement
   declares a symbol as a header in the table and specifies the order of the headers.

## PROC UNIVARIATE Statements

**PROC UNIVARIATE** DATA=*SAS-data-set*;
   **BY** *variable(s)*;
   **VAR** *variable(s)*;

PROC UNIVARIATE statement
   begins the UNIVARIATE procedure and generates descriptive statistics based on moments (including skewness and kurtosis), quantiles or percentiles (such as the median), frequency tables, and extreme values

CLASS statement
   specifies up to two variables whose values define the classification levels for the analysis.

VAR statement
   specifies the analysis variables and their order in the results.

# Learning More

PROC UNIVARIATE
   For detailed information about the UNIVARIATE procedure and other Base Statistical Procedures, see *Base SAS(R) 9.3 Procedures Guide: Statistical Procedures*.

ODS output
   - For information about getting started with the Output Delivery System, see *Getting Started with the SAS Output Delivery System*.
   - For detailed information about the Output Delivery System, see *SAS Output Delivery System: User's Guide*.

- For complete descriptions of individual SAS style attributes, see "Style Attributes" in *SAS Output Delivery System: User's Guide*.

- For complete descriptions of ODS styles, see "Style Templates" in *SAS Output Delivery System: User's Guide*.

- For information about valid ODS destinations, see "Dictionary of ODS Language Statements" in *SAS Output Delivery System: User's Guide*.

PROC SGPANEL

For detailed information about the SGPANEL procedure, see *SAS ODS Graphics: Procedures Guide*.

Base SAS procedures

For information about the PROC SORT and PROC TABULATE procedures, see *Base SAS Procedures Guide*.

# Part 9

# Storing and Managing Data in SAS Files

*Chapter 35*
**Understanding SAS Libraries** ................................. *687*

*Chapter 36*
**Managing SAS Libraries** ....................................... *695*

*Chapter 37*
**Getting Information about Your SAS Data Sets** ................ *701*

*Chapter 38*
**Modifying SAS Data Set Names and Variable Attributes** ........ *713*

*Chapter 39*
**Copying, Moving, and Deleting SAS Data Sets** ................. *725*

# Chapter 35
# Understanding SAS Libraries

Introduction to Understanding SAS Libraries . . . . . . . . . . . . . . . . . . . . . . . . . . . . **687**
   Purpose . . . . . . . . . . . . . . . . . . . . . . . . . . . . . . . . . . . . . . . . . . . . . . . . . . . . . . . . . . 687
   Prerequisites . . . . . . . . . . . . . . . . . . . . . . . . . . . . . . . . . . . . . . . . . . . . . . . . . . . . . . 687

**What Is a SAS Library?** . . . . . . . . . . . . . . . . . . . . . . . . . . . . . . . . . . . . . . . . . . . . . . . **688**

**Accessing a SAS Library** . . . . . . . . . . . . . . . . . . . . . . . . . . . . . . . . . . . . . . . . . . . . . **688**
   Telling SAS Where the SAS Library Is Located . . . . . . . . . . . . . . . . . . . . . . . . . 688
   Assigning a Libref . . . . . . . . . . . . . . . . . . . . . . . . . . . . . . . . . . . . . . . . . . . . . . . . . . 688
   Using Librefs for Temporary and Permanent Libraries . . . . . . . . . . . . . . . . . . . . 689

**Storing Files in a SAS Library** . . . . . . . . . . . . . . . . . . . . . . . . . . . . . . . . . . . . . . . . **690**
   What Is a SAS File? . . . . . . . . . . . . . . . . . . . . . . . . . . . . . . . . . . . . . . . . . . . . . . . . 690
   Understanding SAS Data Sets . . . . . . . . . . . . . . . . . . . . . . . . . . . . . . . . . . . . . . . . 690
   Understanding Other SAS Files . . . . . . . . . . . . . . . . . . . . . . . . . . . . . . . . . . . . . . 691

**Referencing SAS Data Sets in a SAS Library** . . . . . . . . . . . . . . . . . . . . . . . . . . . . **691**
   Understanding Data Set Names . . . . . . . . . . . . . . . . . . . . . . . . . . . . . . . . . . . . . . 691
   Using a One-Level Name . . . . . . . . . . . . . . . . . . . . . . . . . . . . . . . . . . . . . . . . . . . 692
   Using a Two-Level Name . . . . . . . . . . . . . . . . . . . . . . . . . . . . . . . . . . . . . . . . . . . 693

**Summary** . . . . . . . . . . . . . . . . . . . . . . . . . . . . . . . . . . . . . . . . . . . . . . . . . . . . . . . . . . **693**
   Statements . . . . . . . . . . . . . . . . . . . . . . . . . . . . . . . . . . . . . . . . . . . . . . . . . . . . . . . 693
   SAS Data Set Reference . . . . . . . . . . . . . . . . . . . . . . . . . . . . . . . . . . . . . . . . . . . . 693

**Learning More** . . . . . . . . . . . . . . . . . . . . . . . . . . . . . . . . . . . . . . . . . . . . . . . . . . . . . **694**

## Introduction to Understanding SAS Libraries

### Purpose

The way in which SAS handles data libraries is different from one operating environment to another. In this section, you will learn basic concepts about the SAS library and how to use libraries in SAS programs. For more detailed information, see the SAS documentation for your operating environment.

### Prerequisites

Before proceeding with this section, you should understand the concepts presented in the following sections:

- Chapter 1, "What is the SAS System?," on page 3
- Chapter 3, "Introduction to DATA Step Processing," on page 27

## What Is a SAS Library?

A SAS library is a collection of one or more SAS files that are recognized by SAS and can be referenced and stored as a unit. Each file is a member of the library. SAS libraries help to organize your work. For example, if a SAS program uses more than one SAS file, then you can keep all the files in the same library. Organizing files in libraries makes it easier to locate the files and reference them in a program.

Under most operating environments, a SAS library usually corresponds to the level of organization that the operating environment uses to organize files. For example, in directory-based operating environments, a SAS library is a group of SAS files in the same directory. The directory might contain other files, but only the SAS files are part of the SAS library.

*z/OS Specifics*
Under the z/OS operating environment, a SAS library is a specially formatted z/OS data set. This type of data set can contain only SAS files.

## Accessing a SAS Library

### Telling SAS Where the SAS Library Is Located

No matter which operating environment you are using, to access a SAS library, you must tell SAS where it is. You can do this in the following ways:

- directly specify the operating environment's physical name for the location of the SAS library. The physical name must conform to the naming conventions of your operating environment, and it must be enclosed in quotation marks. For example, in the SAS windowing environment, the following DATA statement creates a data set named MYFILE:

  ```
  data 'c:\my documents\sasfiles\myfile';
  ```

- assign a SAS *libref* (library reference), which is a SAS name that is temporarily associated with the physical location name of the SAS library.

### Assigning a Libref

After you assign a libref to the location of a SAS library, then in your SAS program, you can reference files in the library by using the libref instead of using the long physical name that the operating environment uses. The libref is a SAS name that is temporarily associated with the physical location of the SAS library. There are several ways to assign a libref:

- use the LIBNAME statement
- use the LIBNAME function
- use the New Library window from the SAS Explorer window

- for some operating environments, use operating environment commands

A common method for assigning a libref is to use the LIBNAME statement to associate a name with a SAS library. Here is the simplest form of the LIBNAME statement:

**LIBNAME** *libref* '*SAS-library*' ;

*libref*
> specifies a shortcut name to associate with the SAS library. This name must conform to the rules for SAS names. A libref cannot exceed eight characters.
>
> *z/OS Specifics*
> > Under the z/OS operating environment, the libref must also conform to the rules for operating environment names.
>
> Think of the libref as an abbreviation for the operating environment's name for the library. The libref only lasts for the duration of the SAS session. You can change the libref for any given library from session to session. That is, you do not have to use the same libref for a particular library each time you use SAS.

*SAS-library*
> specifies the physical name for the SAS library. The physical name is the name that is recognized by your operating environment. Enclose the physical name in single or double quotation marks.

Here are examples of the LIBNAME statement for different operating environments. For more examples, see the SAS documentation for your operating environment.

Windows
```
libname mydata 'c:\my documents\sasfiles';
```

UNIX
```
libname mydata '/u/myid/sasfiles';
```

z/OS
```
libname mydata 'edc.company.sasfiles';
```

When you assign a libref with the LIBNAME statement, SAS writes a note to the SAS log confirming the assignment. This note also includes the operating environment's physical name for the SAS library.

## *Using Librefs for Temporary and Permanent Libraries*

When a libref is assigned to a SAS library, you can use the libref throughout the SAS session to access the SAS files that are stored in that library. The association between a libref and a SAS library lasts only for the duration of the SAS session, or until you change the libref or discontinue it with another LIBNAME statement.

When you start a SAS session, SAS automatically assigns the libref WORK to a special SAS library. Usually, the files in the WORK library are temporary files. That is, SAS initializes the WORK library when you begin a SAS session, and deletes all files in the WORK library when you end the session. Therefore, the WORK library is a useful place to store SAS files that you do not need to save for a subsequent SAS session. The automatic deletion of the WORK library files at the end of the session prevents you from wasting disk space.

Files that are stored in any SAS library other than the WORK library are usually permanent files. That is, they are available from one SAS session to the next. Store SAS files in a permanent library if you plan to use them in multiple SAS sessions.

## Storing Files in a SAS Library

### What Is a SAS File?

You store all SAS files in a SAS library. A SAS file is a specially structured file that is created, organized, and maintained by SAS. The files reside in SAS libraries as members with specific types. Examples of SAS files are as follows:

- SAS data sets (which can be SAS data files or SAS data views)
- SAS catalogs
- SAS/ACCESS descriptor files
- stored compiled DATA step programs

*Note:* A file that contains SAS statements, even one that is created during a SAS session, is usually not considered a SAS file. For example, in directory-based operating environments, a .sas file is a text file that typically contains a program and is not considered a SAS file.

### Understanding SAS Data Sets

A SAS data set is a SAS file that is stored in a SAS library that consists of descriptor information. Descriptor information identifies the attributes of a SAS data set and its contents, and data values that are organized as a table of observations (rows) and variables (columns). A SAS data set can be either a SAS data file or a SAS data view.

If the descriptor information and the observations are in the same physical location, then the data set is a SAS data file, which has a member type DATA. A SAS data file can be associated with an index. One purpose of an index is to optimize the performance of WHERE processing. Basically, an index contains values in ascending order for a specific variable or variables. The index also includes information about the location of those values within observations in the SAS data file.

If the descriptor and the observations are stored separately, then they form a SAS data view, which has a member type VIEW. The observations in a SAS data view might be stored in a SAS data file, an external database, or an external file. The descriptor contains information about where the data is located and which observations and variables to process. You use a view like you would a SAS data file. You might use a view when you need only a subset of a large amount of data. In addition to saving storage space, views simplify maintenance because they automatically reflect any changes to the data. There are three types of SAS data views:

- DATA step views
- SAS/ACCESS views
- PROC SQL views

*Note:* SAS data views usually behave like SAS data files. Other topics in this documentation do not distinguish between the two types of SAS data sets.

### Understanding Other SAS Files

In addition to SAS data sets, a SAS library can contain the following types of SAS files:

SAS catalog
: specifies a SAS file that stores many types of information in separate units called catalog entries. Each entry is distinguished by an entry name and an entry type. Some catalog entries contain system information such as key definitions. Other catalog entries contain application information about window definitions, help windows, formats, informats, macros, or graphics output. A SAS catalog has a member type CATALOG.

SAS/ACCESS descriptor
: specifies a SAS file that contains information about the layout of an external database. SAS uses this information in order to build a SAS data view in which the observations are stored in an external database. An access descriptor has a member type ACCESS.

stored compiled DATA step program
: specifies a SAS file that contains a DATA step, which has been compiled and stored in a SAS library. A stored compiled DATA step program has a member type PROGRAM.

Complete discussion of all SAS files, except SAS data sets, is beyond the scope of this section. For more information about SAS files, see *SAS Language Reference: Concepts*.

## Referencing SAS Data Sets in a SAS Library

### Understanding Data Set Names

Every SAS data set has a two-level name of the form *libref.filename*. You can always reference a file with its two-level name. However, you can also use a one-level name (*filename*) to reference a file. By default, a one-level name references a file that uses the libref WORK for the temporary SAS library.

*Note:* This section separates the issues of permanent versus temporary files and one-level versus two-level names. Other topics in this documentation and in most SAS documentation assume typical use of the WORK libref, and refer to files that are referenced with a one-level name as temporary files. Files that are referenced with a two-level name are considered permanent files.

*Operating Environment Information*
: The documentation that is provided by the vendor for your operating environment provides information about how to create temporary and permanent files. In SAS, files in the WORK library are temporary unless you specify the NOWORKINIT and NOWORKTERM options. Files in all other SAS libraries are permanent. However, your operating environment might consider libraries and files in a different way. For example, the operating environment might enable you to create a temporary directory or a temporary z/OS data set that are deleted when you log off. Because all files in a SAS library are deleted if the underlying operating environment structure is deleted, the way the operating environment views the SAS library determines whether the library endures from one session to the next.

### Using a One-Level Name

Typically, when you reference a SAS data set with a one-level name, SAS by default uses the libref WORK for the temporary library. For example, the following program creates a temporary SAS data set named WORK.GRADES:

```
data grades;
   infile 'file-specification';
   input Name $ 1-14 Gender $ 15-20 Section $ 22-24 Grade;
run;
```

However, if you want to use a one-level name to reference a permanent SAS data set, you can assign the reserved libref USER. When USER is assigned and you reference a SAS data set with a one-level name, SAS by default uses the libref USER for a permanent SAS library. For example, the following program creates a permanent SAS data set named USER.GRADES. Note that you assign the libref USER as you do any other libref.

```
libname user 'SAS-library';
```

```
data grades;
   infile 'file-specification';
   input Name $ 1-14 Gender $ 15-20 Section $ 22-24 Grade;
run;
```

Therefore, when you reference a SAS data set with a one-level name, SAS looks for the libref USER. If it is assigned to a SAS library, then USER becomes the default libref for one-level names. If the libref USER has not been assigned, SAS uses WORK as the default libref for one-level names.

If USER is assigned, then you must use a two-level name to access a temporary data set in the WORK library. For example, if USER is assigned, then to print the data set WORK.GRADES requires a two-level name in the PROC PRINT statement:

```
proc print data=work.grades;
run;
```

If USER is assigned, then you need to make only one change in order to use the same program with files of the same name in different SAS libraries. Instead of specifying two-level names, assign USER differently in each case. For example, the following program concatenates five SAS data sets in *SAS-library-1* and puts them in a new SAS data set, WEEK, in the same library:

```
   libname user 'SAS-library-1';

   data week;
      set mon tues wed thurs fri;
   run;
```

By changing just the name of the library in the LIBNAME statement, you can combine files with the same names in another library, *SAS-library-2*:

```
libname user 'SAS-library-2';

data week;
   set mon tues wed thurs fri;
run;
```

*Note:* At your site, the libref USER might be assigned for you when you start a SAS session. Your on-site SAS support personnel knows whether the libref is assigned.

### Using a Two-Level Name

You can always reference a SAS data set with a two-level name, whether the libref that you use is WORK, USER, or some other libref that you have assigned. Usually, any two-level name with a libref other than WORK references a permanent SAS data set.

In the following program, the LIBNAME statement establishes a connection between the SAS name INTRCHEM and *SAS-library*. *SAS-library* is the physical name for the location of an existing data set or directory. The DATA step creates the SAS data set GRADES in the SAS library INTRCHEM. SAS uses the INPUT statement to construct the data set from the raw data in *file-specification*.

```
libname intrchem 'SAS-library';

data intrchem.grades;
   infile 'file-specification';
   input Name $ 1-14 Gender $ 15-20 Section $ 22-24 Grade;
run;
```

When the SAS data set INTRCHEM.GRADES is created, you can read from it by using its two-level name. The following program reads the file INTRCHEM.GRADES and creates a new SAS data set named INTRCHEM.FRIDAY, which is a subset of the original data set:

```
data intrchem.friday;
   set intrchem.grades;
      if Section='Fri';
run;
```

You can add the following PROC PRINT step to display the SAS data set INTRCHEM.FRIDAY:

```
   proc print data=intrchem.friday;
   run;
```

## Summary

### Statements

LIBNAME *libref* '*SAS-library*';
   in most operating environments, associates a *libref* with a SAS library. Enclose the name of the SAS library in single or double quotation marks.

### SAS Data Set Reference

You can reference any SAS data set with a two-level name of the form *libref.filename*. By default, if you use a one-level name to reference a SAS data set, then SAS uses the libref USER if it is assigned. If USER is not assigned, then SAS uses the libref WORK.

## Learning More

LIBNAME statement
: For more information about the LIBNAME statement, including options for the statement and information about specifying an engine other than the default engine, see "LIBNAME Statement" in *SAS Statements: Reference*.

Operating environment
: For operating environment specifics, see the SAS documentation for your operating environment.

SAS files
: Detailed information about SAS files can be found in *SAS Language Reference: Concepts*.

   For detailed information about PROC SQL views, see *SAS SQL Procedure User's Guide*.

SAS tools
: To learn about the tools that are available for managing SAS libraries, including the DATASETS procedure, see Chapter 36, "Managing SAS Libraries," on page 695.

USER libref
: For information about the USER= system option, which you can use instead of the LIBNAME statement to assign the USER libref, see "USER= System Option" in *SAS System Options: Reference*.

   *Note:* If you assign the libref both ways or if you assign it more than once with either method, then the last definition is the one that is used.

WORK library
: For information about the WORKINIT system option, see "WORKINIT System Option" in *SAS System Options: Reference*. For information about the WORKTERM system option, see "WORKTERM System Option" in *SAS System Options: Reference*. These options control when the WORK library is initialized.

# Chapter 36
# Managing SAS Libraries

Introduction to Managing SAS Libraries . . . . . . . . . . . . . . . . . . . . . . . . . . . . . . . 695
    Purpose . . . . . . . . . . . . . . . . . . . . . . . . . . . . . . . . . . . . . . . . . . . . . . . . . . . . . . . 695
    Prerequisites . . . . . . . . . . . . . . . . . . . . . . . . . . . . . . . . . . . . . . . . . . . . . . . . . . 695
Choosing Your Tools . . . . . . . . . . . . . . . . . . . . . . . . . . . . . . . . . . . . . . . . . . . . . . 695
Understanding the DATASETS Procedure . . . . . . . . . . . . . . . . . . . . . . . . . . . . 696
Looking at a PROC DATASETS Session . . . . . . . . . . . . . . . . . . . . . . . . . . . . . . 697
Summary . . . . . . . . . . . . . . . . . . . . . . . . . . . . . . . . . . . . . . . . . . . . . . . . . . . . . . . . 698
    Procedures . . . . . . . . . . . . . . . . . . . . . . . . . . . . . . . . . . . . . . . . . . . . . . . . . . . . 698
    Statements . . . . . . . . . . . . . . . . . . . . . . . . . . . . . . . . . . . . . . . . . . . . . . . . . . . . 698
Learning More . . . . . . . . . . . . . . . . . . . . . . . . . . . . . . . . . . . . . . . . . . . . . . . . . . . 698

## Introduction to Managing SAS Libraries

### Purpose

In this section, you will learn about the tools that are available for managing SAS libraries, including the DATASETS procedure. Subsequent sections describe how to use the DATASETS procedure.

### Prerequisites

Before using this section, you should understand the concepts presented in Chapter 35, "Understanding SAS Libraries," on page 687.

## Choosing Your Tools

As you accumulate more SAS files, you will need to manage the SAS libraries. Managing libraries generally involves using SAS procedures to perform routine tasks:

- listing library members and their properties
- renaming, deleting, and moving files

- renaming variables
- copying libraries and files

SAS procedures can perform any file management task for your SAS libraries with no need for operating environment commands.

Several SAS tools are available for basic file management. You can use these features alone or in combination.

SAS Explorer
: includes windows for most file management tasks, with no need for submitting SAS program statements. For example, you can create new libraries and SAS files, open existing SAS files, and perform most file management tasks such as moving, copying, and deleting files. To use SAS Explorer windows, enter `libname`, `catalog`, or `dir` in the command bar, or select the Explorer icon from the **Toolbar** menu.

CATALOG procedure
: provides catalog management utilities with the COPY and CONTENTS statements.

COPY procedure
: copies all members of a library or individual files within the library.

CONTENTS procedure
: lists library members and their properties.

DATASETS procedure
: combines all library management functions into one procedure. If you do not use SAS Explorer, or if you use SAS in batch or interactive line mode, then this procedure can save you time and resources. PROC DATASETS enables you to perform a number of management tasks such as copying, deleting, or modifying SAS files.

## Understanding the DATASETS Procedure

The DATASETS procedure is an interactive procedure. That is, the procedure remains active after a RUN statement is executed. After you start the procedure, you can continue to manipulate files within a SAS library until you have finished all the tasks that you have planned. This capability can save time and resources when you have a number of tasks to complete for one session.

Here are some important features to know about the DATASETS procedure:

- You can specify the input library in the PROC DATASETS statement.

  When you start the DATASETS procedure, you can specify the input library, which is referred to as the procedure input library. If you do not specify a library as the source of files, then SAS uses the default library, which could be the temporary library WORK or the USER library. To specify a different input library, you must start the procedure again.

- Statements execute in the order in which they are written.

  For example, you can view the contents of a SAS data set, copy a second data set from another library, and then view and compare the contents of the two data sets. To perform the tasks in the order listed above, the SAS statements that perform these tasks must be specified in the same order.

- Groups of statements can execute without a RUN statement.

For the DATASETS procedure only, SAS recognizes the following procedure statements as having an implied RUN statement, and executes them immediately when you submit them:

- APPEND statement
- CONTENTS statement
- MODIFY statement
- COPY statement
- PROC DATASETS statement

SAS reads the statements that are associated with one task until it encounters one of the statements above. SAS executes all of the preceding statements immediately and continues reading until it encounters another of the above statements. To cause the last task to execute, you must submit a RUN or QUIT statement.

*Note:* If you are running in interactive line mode, this feature enables you to receive messages that statements have already executed before you submit a RUN statement.

- The RUN statement does not stop a PROC DATASETS step.

You must submit a QUIT statement, a new PROC statement, or a DATA step to stop a PROC DATASETS step. Submitting a QUIT statement executes any statements that have not yet executed and ends the procedure.

## Looking at a PROC DATASETS Session

The following example illustrates how PROC DATASETS behaves in a typical session. In the example, a file from one SAS library is used to create a test file in another SAS library. A data set is copied and its contents are described so that the output can be visually checked to be sure that the variables are compatible with an existing file in the test library.

The following program is arranged in groups to show which statements are executed as one task. The tasks and the action by SAS are numbered in the order in which they occur in the program.

```
proc datasets library=test89;  1

   copy in=realdata out=test89;  2
   select income88;

   contents data=income88;  3
run;

   modify income88;  4
   rename Sales=Sales88;

quit;  5
```

The following list corresponds to the numbered items in the preceding program:

1  Starts the DATASETS procedure and specifies TEST89 as the procedure input library.

**698** *Chapter 36 • Managing SAS Libraries*

2. Copies the data set INCOME88 from the SAS library REALDATA to the SAS library TEST89. SAS recognizes these statements as one task. When SAS reads the CONTENTS statement, it immediately executes the COPY statement, and copies INCOME88 into the library TEST89. The CONTENTS statement acts as an implied RUN statement, which causes the COPY statement to execute. This action is more noticeable if you are running SAS in the windowing environment.

3. Describes the contents of the data set. Visually checking the output can verify that the variables are compatible with an existing SAS data set. When SAS encounters the RUN statement, it describes the content of INCOME88. Because the previous task has executed, SAS finds the data set in the procedure input library TEST89.

   After visually checking the contents, you determine that it is necessary to rename the variable Sales. Because the DATASETS procedure is still active, you can submit more statements.

4. Renames the variable Sales to Sales88.

5. Stops the DATASETS procedure. SAS executes the last two statements and ends the DATASETS procedure.

# Summary

## Procedures

PROC DATASETS <LIBRARY=*libref*>;
starts the procedure and specifies the library that the procedure processes, that is, the procedure input library. If you do not specify the LIBRARY= option, then the default is the WORK or USER library. PROC DATASETS creates a directing listing when the procedure is submitted.

## Statements

QUIT;
executes any preceding statements that have not run and stops the procedure.

RUN;
executes the preceding group of statements that have not run, without ending the procedure.

# Learning More

DATASETS procedure
To learn about using the DATASETS procedure to manage SAS libraries whose members are primarily data sets, see the following sections:

- Chapter 37, "Getting Information about Your SAS Data Sets," on page 701.
- Chapter 38, "Modifying SAS Data Set Names and Variable Attributes," on page 713.
- Chapter 39, "Copying, Moving, and Deleting SAS Data Sets," on page 725.

SAS windowing environment
> For information about managing SAS files through the SAS windowing environment, see Chapter 41, "Using the SAS Windowing Environment," on page 753.

Operating environment commands
> For information about managing SAS files using operating environment commands, see the SAS documentation for your operating environment.

# Chapter 37
# Getting Information about Your SAS Data Sets

Introduction to Getting Information about Your SAS Data Sets . . . . . . . . . . . . . . 701
    Purpose . . . . . . . . . . . . . . . . . . . . . . . . . . . . . . . . . . . . . . . . . . . . . . . . . . . . . . . . . 701
    Prerequisites . . . . . . . . . . . . . . . . . . . . . . . . . . . . . . . . . . . . . . . . . . . . . . . . . . . . 702

Input Data Library for Examples . . . . . . . . . . . . . . . . . . . . . . . . . . . . . . . . . . . . . . 702

Requesting a Directory Listing for a SAS Library . . . . . . . . . . . . . . . . . . . . . . . . 702
    Understanding a Directory Listing . . . . . . . . . . . . . . . . . . . . . . . . . . . . . . . . . . . 702
    Listing All Files in a Library . . . . . . . . . . . . . . . . . . . . . . . . . . . . . . . . . . . . . . . . 702
    Listing Files That Have the Same Member Type . . . . . . . . . . . . . . . . . . . . . . . . 704

Requesting Contents Information about SAS Data Sets . . . . . . . . . . . . . . . . . . . . 704
    Using the DATASETS Procedure for SAS Data Sets . . . . . . . . . . . . . . . . . . . . . 704
    Listing the Contents of One Data Set . . . . . . . . . . . . . . . . . . . . . . . . . . . . . . . . . 705
    Listing the Contents of All Data Sets in a Library . . . . . . . . . . . . . . . . . . . . . . . 707

Requesting Contents Information in Different Formats . . . . . . . . . . . . . . . . . . . . 708

Summary . . . . . . . . . . . . . . . . . . . . . . . . . . . . . . . . . . . . . . . . . . . . . . . . . . . . . . . . . 710
    Procedures . . . . . . . . . . . . . . . . . . . . . . . . . . . . . . . . . . . . . . . . . . . . . . . . . . . . . 710
    DATASETS Procedure Statements . . . . . . . . . . . . . . . . . . . . . . . . . . . . . . . . . . 710

Learning More . . . . . . . . . . . . . . . . . . . . . . . . . . . . . . . . . . . . . . . . . . . . . . . . . . . . 710

## Introduction to Getting Information about Your SAS Data Sets

### *Purpose*

As you create libraries of SAS data sets, SAS generates and maintains information about where the library is stored in your operating environment, how and when the data sets were created, and how their contents are defined. Using the DATASETS procedure, you can view this information without displaying the contents of the data set or referring to additional documentation.

In this section, you will learn how to get the following information about SAS libraries and SAS data sets:

- names and types of SAS files that are included in a SAS library

- names and attributes for variables in SAS data sets

- summary information about storage parameters for the operating environment
- summary information about the history and structure of SAS data sets

## Prerequisites

Before using this section, you should understand the concepts presented in the following sections:

- Chapter 35, "Understanding SAS Libraries," on page 687
- Chapter 36, "Managing SAS Libraries," on page 695

## Input Data Library for Examples

The examples in this section use a SAS library that contains information about the climate of the United States. The DATA steps that create the data sets are shown in "The USCLIM Data Sets" on page 819. Note that these DATA steps do not create catalog entries. The catalogs that you see in Figure 37.1 on page 703 have been added for illustration.

## Requesting a Directory Listing for a SAS Library

### Understanding a Directory Listing

A directory listing is a list of files in a SAS library. Each file is called a member, and each member has a member type that is assigned to it by SAS. The member type indicates the type of SAS file, such as DATA or CATALOG. When SAS processes statements, SAS not only looks for the specified file, it verifies that the file has a member type that can be processed by the statement.

The directory listing contains two main parts:

- directory information
- list of library members and their member types

### Listing All Files in a Library

To obtain a directory listing of all members in a library, you need only the PROC DATASETS statement with the LIBRARY= option. For example, the following statements send a directory listing to the Results window for a library that contains climate information. The LIBNAME statement assigns the libref USCLIM to this library:

```
libname usclim 'SAS-library';

proc datasets library=usclim;
```

The following output displays the directory listing:

*Figure 37.1* Directory Listing for the Library USCLIM

The SAS System

| Directory | |
|---|---|
| Libref | USCLIM |
| Engine | V9 |
| Physical Name | c:\Users\userid\climate |
| Filename | c:\Users\userid\climate |

| # | Name | Member Type | File Size | Last Modified |
|---|---|---|---|---|
| 1 | BASETEMP | CATALOG | 3830784 | 04/16/2013 13:22:20 |
| 2 | HIGHTEMP | DATA | 131072 | 04/16/2013 13:36:54 |
| 3 | HURRICANE | DATA | 131072 | 04/16/2013 13:36:54 |
| 4 | LOWTEMP | DATA | 131072 | 04/16/2013 13:36:54 |
| 5 | REPORT | CATALOG | 2077696 | 04/16/2013 13:22:20 |
| 6 | TEMPCHNG | DATA | 131072 | 04/16/2013 13:36:54 |

The following is a list of the items shown in the output:

Directory
: gives the physical name as well as the libref for the library. Some operating environments provide both additional and different information.

Name
: contains the second-level SAS member name that is assigned to the file. If the files are different member types, then you can have two files of the same name in one library.

Member Type
: indicates the SAS file member type. The most common member types are DATA and CATALOG. For example, the library USCLIM contains two catalogs of type CATALOG and four date sets of type DATA.

File Size
: specifies the size of the file.

Last Modified
: specifies the date on which the file was last modified.

*Note:* After you execute PROC DATASETS the first time, the DATASETS procedure continues to run. You can execute more procedure statements without executing the PROC DATASETS statement again. To end the DATASETS procedure, execute the **QUIT;** statement.

### Listing Files That Have the Same Member Type

To show only certain types of SAS files in the directory listing, use the MEMTYPE= option in the PROC DATASETS statement. The following statement produces a listing for USCLIM that contains only the information about data sets:

```
proc datasets library=usclim memtype=data;
```

The following output displays information about the data sets (member type DATA) that are stored in USCLIM:

*Figure 37.2  Directory of Data Sets Only for the Library USCLIM*

The SAS System

| Directory | |
|---|---|
| Libref | USCLIM |
| Engine | V9 |
| Physical Name | c:\Users\userid\climate |
| Filename | c:\Users\userid\climate |

| # | Name | Member Type | File Size | Last Modified |
|---|---|---|---|---|
| 1 | HIGHTEMP | DATA | 131072 | 04/16/2013 13:36:54 |
| 2 | HURRICANE | DATA | 131072 | 04/16/2013 13:36:54 |
| 3 | LOWTEMP | DATA | 131072 | 04/16/2013 13:36:54 |
| 4 | TEMPCHNG | DATA | 131072 | 04/16/2013 13:36:54 |

*Note:* Examples in this document focus on using PROC DATASETS to manage only SAS data sets. You can also list other member types by specifying MEMTYPE=. For example, MEMTYPE=CATALOG lists only SAS catalogs.

# Requesting Contents Information about SAS Data Sets

### Using the DATASETS Procedure for SAS Data Sets

To look at the contents of a SAS data set without displaying the observations, use the CONTENTS statement in the DATASETS procedure. The CONTENTS statement and its options provide descriptive information about data sets and a list of variables and their attributes.

## Listing the Contents of One Data Set

The SAS library USCLIM contains four data sets. The data set TEMPCHNG contains data for extreme changes in temperature. The following program displays the variables in the data set TEMPCHNG:

```
proc datasets library=usclim memtype=data;
   contents data=tempchng;
run;
```

The output from the CONTENTS statement produces information about the TEMPCHNG data set. The DATA= option specifies the name of the data set. The following output shows the results from the CONTENTS statement.

*Figure 37.3  Output from CONTENTS Statement for Data Set TEMPCHNG*

The SAS System

The DATASETS Procedure

| Data Set Name | USCLIM.TEMPCHNG | Observations | 5 |
|---|---|---|---|
| Member Type | DATA | Variables | 6 |
| Engine | V9 | Indexes | 0 |
| Created | 04/05/2013 09:56:16 | Observation Length | 56 |
| Last Modified | 04/05/2013 09:56:16 | Deleted Observations | 0 |
| Protection |  | Compressed | NO |
| Data Set Type |  | Sorted | NO |
| Label |  |  |  |
| Data Representation | WINDOWS_32 |  |  |
| Encoding | wlatin1 Western (Windows) |  |  |

| Engine/Host Dependent Information ||
|---|---|
| Data Set Page Size | 65536 |
| Number of Data Set Pages | 1 |
| First Data Page | 1 |
| Max Obs per Page | 1167 |
| Obs in First Data Page | 5 |
| Number of Data Set Repairs | 0 |
| ExtendObsCounter | YES |
| Filename | c:\Users\userid\climate\tempchng.sas7bdat |
| Release Created | 9.0401B0 |
| Host Created | W32_7PRO |

| Alphabetic List of Variables and Attributes |||||| 
|---|---|---|---|---|---|
| # | Variable | Type | Len | Format | Informat |
| 2 | Date | Num | 8 | DATE9. | DATE7. |
| 6 | Diff | Num | 8 |  |  |
| 4 | End_f | Num | 8 |  |  |
| 5 | Minutes | Num | 8 |  |  |
| 3 | Start_f | Num | 8 |  |  |
| 1 | State | Char | 13 |  | $CHAR13. |

Note that output from the CONTENTS statement varies for different operating environments.

The following list describes information that you might find when you use the CONTENTS statement with PROC DATASETS:

The DATASETS Procedure heading
: contains field names. Fields are empty if they do not apply to the data set. Field names are listed below:

Data Set Name
: specifies the two-level name that is assigned to the data set.

Member Type
: specifies the type of library member.

Engine
: specifies the access method that SAS uses to read from or write to the data set.

Created
: specifies the date on which the data set was created.

Last Modified
: specifies the last date that the data set was modified.

Protection
: indicates whether the data set is password protected for READ, WRITE, or ALTER operations.

Data Set Type
: applies only to files with the member type DATA. Information in this field indicates that the data set contains special observations and variables for use with SAS statistical procedures.

Label
: specifies the descriptive information that you supply in a LABEL= data set option to identify the data set.

Data Representation
: specifies the form in which data is stored in a particular operating environment.

Encoding
: specifies a mapping of a coded character set to code values. Each character in a character set maps to a unique numeric representation.

Observations
: specifies the total number of observations that are currently in the data set.

Variables
: specifies the number of variables in the data set.

Indexes
: specifies the number of indexes for the data set.

Observation Length
: specifies the length of each observation in bytes.

Deleted Observations
: specifies the number of observations that are marked for deletion, if applicable.

Compressed
: indicates whether the data has fixed-length or variable-length records. If the data set is compressed, then additional fields indicate whether new observations are added to the end of the data set or written to unused space within the data set. It

also indicates whether the data set can be randomly accessed by observation
number rather than by sequential access only.

Sorted
: indicates whether the data set has been sorted.

Engine/Host Dependent Information
: lists information about the engine, which is the mechanism for reading from and writing to files, and about how the data set is stored by the operating environment. Depending on the engine, the output in this section might differ. For more information, see the SAS documentation for your operating environment.

Alphabetical List of Variables and Attributes
: lists all the variable names in the data set in alphabetical order and describes the attributes that are assigned to the variable when it is defined. The attributes are described below:

#
: specifies the logical position of the variable in the observation. This is the number that is assigned to the variable when it is defined.

Variable
: specifies the name of the variable.

Type
: indicates whether the variable is character or numeric.

Len
: specifies the length of the variable in bytes.

Format
: specifies the format of the variable.

Informat
: specifies the informat of the variable.

In addition, if applicable, the output also displays a table that describes the following information:

- indexes for indexed variables
- any defined integrity constraints
- sort information

## *Listing the Contents of All Data Sets in a Library*

You can list the contents of all the data sets in a library by specifying the keyword _ALL_ with the DATA= option. The following statements produce a directory listing for the library and a contents listing for each data set in the directory:

```
contents data=_all_;
run;
```

To create only a directory listing, add the NODS option to the CONTENTS statement. The following statements produce a directory listing but suppress a contents listing for individual data sets. Use this form if you want the directory listing for the procedure input library:

```
contents data=_all_ nods;
run;
```

Include the libref if you want the directory listing for another library. This example specifies the library STORM:

```
contents data=storm._all_ nods;
run;
```

# Requesting Contents Information in Different Formats

For a variation of the contents listing, use the VARNUM option or the SHORT option in the CONTENTS statement. For example, the following statements produce a list of variable names in the order in which they were defined, which is their logical position in the data set:

```
contents data=tempchng varnum;
run;
```

The CONTENTS statement specifies the data set TEMPCHNG and includes the VARNUM option to list variables in order of their logical position. (By default, the CONTENTS statement lists variables alphabetically.)

The following output shows the contents in variable number order:

**Figure 37.4** *Contents of the Data Set TEMPCHNG in Variable Number Order*

The SAS System

The DATASETS Procedure

| Data Set Name | USCLIM.TEMPCHNG | Observations | 5 |
|---|---|---|---|
| Member Type | DATA | Variables | 6 |
| Engine | V9 | Indexes | 0 |
| Created | 04/05/2013 09:56:16 | Observation Length | 56 |
| Last Modified | 04/05/2013 09:56:16 | Deleted Observations | 0 |
| Protection | | Compressed | NO |
| Data Set Type | | Sorted | NO |
| Label | | | |
| Data Representation | WINDOWS_32 | | |
| Encoding | wlatin1 Western (Windows) | | |

Engine/Host Dependent Information

| Data Set Page Size | 65536 |
|---|---|
| Number of Data Set Pages | 1 |
| First Data Page | 1 |
| Max Obs per Page | 1167 |
| Obs in First Data Page | 5 |
| Number of Data Set Repairs | 0 |
| ExtendObsCounter | YES |
| Filename | c:\Users\userid\climate\tempchng.sas7bdat |
| Release Created | 9.0401B0 |
| Host Created | W32_7PRO |

Variables in Creation Order

| # | Variable | Type | Len | Format | Informat |
|---|---|---|---|---|---|
| 1 | State | Char | 13 | | $CHAR13. |
| 2 | Date | Num | 8 | DATE9. | DATE7. |
| 3 | Start_f | Num | 8 | | |
| 4 | End_f | Num | 8 | | |
| 5 | Minutes | Num | 8 | | |
| 6 | Diff | Num | 8 | | |

If you do not need all of the information in the contents listing, then you can request an abbreviated version by using the SHORT option in the CONTENTS statement. The following statements request an abbreviated version and then end the DATASETS procedure by issuing the QUIT statement:

```
   contents data=tempchng short;
run;
quit;
```

The following output lists the variable names for the TEMPCHNG data set:

*Figure 37.5* Listing Variable Names Only for the Data Set TEMPCHNG

```
                        The SAS System

                      The DATASETS Procedure

          Alphabetic List of Variables for USCLIM.TEMPCHNG

          Date Diff End_f Minutes Start_f State
```

# Summary

## Procedures

PROC DATASETS <LIBRARY=*libref* <MEMTYPE=*mtype(s)*>>;
The MEMTYPE= option restricts processing to a certain type or types of SAS files and restricts the library directory listing to SAS files of the specified member types.

## DATASETS Procedure Statements

CONTENTS <DATA=<*libref*>.*SAS-data-set*> <NODS> <SHORT> <VARNUM> ;
describes the contents of a specific SAS data set in the library. The default data set is the most recently created data set for the job or session. For the CONTENTS statement in PROC DATASETS, when you specify DATA=, the default libref is the procedure input library. However, for the CONTENTS procedure, the default libref is either WORK or USER.

Use the NODS option with the keyword _ALL_ in the DATA= option to produce only the directory listing of the library in SAS output. That is, the NODS option suppresses the contents of individual files. You cannot use the NODS option when you specify only one SAS data set in the DATA= option.

The SHORT option produces only an alphabetical list of variable names, index information, integrity constraint information, and sort information for the SAS data set.

The VARNUM option produces a list of variable names in the order in which they were defined, which is their logical position in the data set. By default, the CONTENTS statement lists variables alphabetically.

# Learning More

CATALOG procedure
You can use the CATALOG procedure to obtain contents information about catalogs. For more information, see "CATALOG" in *Base SAS Procedures Guide*.

DATASETS procedure
: The DATASETS procedure is a utility procedure that manages your SAS files. For more information about the DATASETS procedure and the CONTENTS statement, see "DATASETS" in *Base SAS Procedures Guide*.

CONTENTS procedure
: The CONTENTS procedure shows the contents of a SAS data set and writes the directory of the SAS library. For more information, see "CONTENTS" in *Base SAS Procedures Guide*.

Windowing environment
: For more information about using the windowing environment to obtain information about SAS data sets, see Chapter 41, "Using the SAS Windowing Environment," on page 753.

# Chapter 38
# Modifying SAS Data Set Names and Variable Attributes

Introduction to Modifying SAS Data Set Names and Variable Attributes . . . . . . . 713
    Purpose . . . . . . . . . . . . . . . . . . . . . . . . . . . . . . . . . . . . . . . . . . . . . . . . . . . . . . . . . 713
    Prerequisites . . . . . . . . . . . . . . . . . . . . . . . . . . . . . . . . . . . . . . . . . . . . . . . . . . . . . 714

Input Data Library for Examples . . . . . . . . . . . . . . . . . . . . . . . . . . . . . . . . . . . . . . 714

Renaming SAS Data Sets . . . . . . . . . . . . . . . . . . . . . . . . . . . . . . . . . . . . . . . . . . . . 714

Modifying Variable Attributes . . . . . . . . . . . . . . . . . . . . . . . . . . . . . . . . . . . . . . . . 716
    Understanding How to Modify Variable Attributes . . . . . . . . . . . . . . . . . . . . . . . 716
    Renaming Variables . . . . . . . . . . . . . . . . . . . . . . . . . . . . . . . . . . . . . . . . . . . . . . . 716
    Assigning, Changing, or Removing Formats . . . . . . . . . . . . . . . . . . . . . . . . . . . . 717
    Assigning, Changing, or Removing Labels . . . . . . . . . . . . . . . . . . . . . . . . . . . . . 720

Summary . . . . . . . . . . . . . . . . . . . . . . . . . . . . . . . . . . . . . . . . . . . . . . . . . . . . . . . . . 723
    DATASETS Procedure Statements . . . . . . . . . . . . . . . . . . . . . . . . . . . . . . . . . . . . 723

Learning More . . . . . . . . . . . . . . . . . . . . . . . . . . . . . . . . . . . . . . . . . . . . . . . . . . . . 724

## Introduction to Modifying SAS Data Set Names and Variable Attributes

### Purpose

SAS enables you to modify data set names and variable attributes without creating new data sets. In this section, you will learn how to use statements in the DATASETS procedure to perform the following tasks:

- rename data sets

- rename variables

- modify variable formats

- modify variable labels

This section focuses on using the DATASETS procedure to modify data sets. However, you can also use some of the illustrated statements and options to modify other types of SAS files.

*Note:* You cannot use the DATASETS procedure to change the values of observations, to create or delete variables, or to change the type or length of variables. These modifications are done with DATA step statements and functions.

## Prerequisites

Before using this section, you should understand the concepts presented in the following sections:

- Chapter 35, "Understanding SAS Libraries," on page 687
- Chapter 36, "Managing SAS Libraries," on page 695
- Chapter 37, "Getting Information about Your SAS Data Sets," on page 701

## Input Data Library for Examples

The examples in this section use a SAS library that contains information about the climate of the United States. The DATA steps that create the data sets in the SAS library are shown in "The CLIMATE, PRECIP, and STORM Data Sets" on page 820.

## Renaming SAS Data Sets

Renaming data sets is often required for effective library management. For example, you might rename a data set when you archive it or when you add new data values.

Use the CHANGE statement in the DATASETS procedure to rename one or more data sets in the same library. Here is the syntax for the CHANGE statement:

**CHANGE** *old-name=new-name*;

*old-name*
specifies the current name of the SAS data set.

*new-name*
specifies the name that you want to give the data set.

This example renames two data sets in the SAS library USCLIM, which contains information about the climate of the United States. The following program starts the DATASETS procedure. It then changes the name of the data set HIGHTEMP to USHIGH and the name of the data set LOWTEMP to USLOW:

```
libname usclim 'SAS-library';

proc datasets library=usclim;
   change hightemp=ushigh lowtemp=uslow;
run;
```

As it processes these statements, SAS sends messages to the SAS log. The messages verify that the data sets are renamed:

**Log 38.1** *Renaming Data Sets in the Library USCLIM*

```
38   proc datasets library=usclim;
39      change hightemp=ushigh lowtemp=uslow;
40   run;

NOTE: Changing the name USCLIM.HIGHTEMP to USCLIM.USHIGH (memtype=DATA).
NOTE: Changing the name USCLIM.LOWTEMP to USCLIM.USLOW (memtype=DATA).
```

The following program executes the DATASETS procedure, where you can see the results of the changes:

```
proc datasets library=usclim;
run;
```

The following output shows information about the library:

**Figure 38.1** *Renaming Data Sets in the Library USCLIM*

The SAS System

| Directory | |
|---|---|
| Libref | USCLIM |
| Engine | V9 |
| Physical Name | c:\Users\userid\climate |
| Filename | c:\Users\userid\climate |

| # | Name | Member Type | File Size | Last Modified |
|---|---|---|---|---|
| 1 | BASETEMP | CATALOG | 3830784 | 04/16/2013 13:22:20 |
| 2 | HURRICANE | DATA | 131072 | 04/16/2013 13:36:54 |
| 3 | REPORT | CATALOG | 2077696 | 04/16/2013 13:22:20 |
| 4 | TEMPCHNG | DATA | 131072 | 04/16/2013 13:36:54 |
| 5 | USHIGH | DATA | 131072 | 04/16/2013 13:36:54 |
| 6 | USLOW | DATA | 131072 | 04/16/2013 13:36:54 |

## Modifying Variable Attributes

### Understanding How to Modify Variable Attributes

Each variable in a SAS data set has attributes such as name, type, length, format, informat, label, and so on. These attributes enable you to identify a variable as well as define to SAS how the variable can be used.

By using the DATASETS procedure, you can assign, change, or remove certain attributes with the MODIFY statement and subordinate statements. For example, using MODIFY and subordinate statements enables you to perform the following tasks:

- rename variables
- assign, change, or remove a format, which changes how the values are written or displayed
- assign, change, or remove labels

*Note:* You cannot use the MODIFY statement to modify fixed attributes such as the type or length of a variable.

### Renaming Variables

You might need to rename variables, for example, before combining data sets that have one or more matching variable names. The DATASETS procedure enables you to rename one or more variables by using the MODIFY statement and its subordinate RENAME statement. Here is the syntax for the statements:

**MODIFY** *SAS-data-set*;

    **RENAME** *old-name=new-name*;

*SAS-data-set*
    specifies the name of the SAS data set that contains the variable that you want to rename.

*old-name*
    specifies the current name of the variable.

*new-name*
    specifies the name that you want to give the variable.

This example renames two variables in the data set HURRICANE, which is in the SAS library USCLIM. The following statements change the variable name State to Place and the variable name Deaths to USDeaths. The DATASETS procedure is already active, so the PROC DATASETS statement is not necessary.

```
modify hurricane;
   rename State=Place Deaths=USDeaths;
run;
```

The SAS log messages verify that the variables are renamed to Place and USDeaths as shown in the following output. All other attributes that are assigned to these variables remain unchanged.

**Log 38.2**  *Renaming Variables in the Data Set HURRICANE*

```
48         modify hurricane;
49         rename State=Place Deaths=USDeaths;
NOTE: Renaming variable State to Place.
NOTE: Renaming variable Deaths to USDeaths.
50    run;

NOTE: MODIFY was successful for USCLIM.HURRICANE.DATA.
```

## *Assigning, Changing, or Removing Formats*

SAS enables you to assign and store formats, which are used by many SAS procedures for output. Assigning, changing, or removing a format changes how the values are written or displayed. By using the DATASETS procedure, you can change a variable's format with the MODIFY statement and its subordinate FORMAT statement. You can change a variable's format either to a SAS format or to a format that you have defined and stored. You can also remove a format. Here is the syntax for these statements:

**MODIFY** *SAS-data-set*;

**FORMAT** *variable(s)* <*format*>;

*SAS-data-set*
  specifies the name of the SAS data set that contains the variable whose format you want to modify.

*variable*
  specifies the name of one or more variables whose format you want to assign, change, or remove.

*format*
  specifies the format that you want to give the variable. If you do not specify a format, then SAS removes any format that is associated with the specified variable.

When you assign or change a format, follow these rules:

- List the variable name before the format.

- List multiple variable names or use an abbreviated variable list if you want to assign the format to more than one variable.

- Do not use punctuation to separate items in the list.

The following FORMAT statement illustrates ways to include many variables and formats in the same FORMAT statement:

```
format Date1-Date5 date9. Cost1 Cost2 dollar4.2 Place $char25.;
```

The variables Date1 through Date5 are written in abbreviated list form, and the format DATE9. is assigned to all five variables. The variables Cost1 and Cost2 are listed individually before their format. The format $CHAR25. is assigned to the variable Place.

Two rules apply when you are removing formats from variables:

- List the variable names only.

- Place the variable names last in the list if you are using the same FORMAT statement to assign or change formats.

The following example shows how to change formats in the data set HURRICANE:

```
contents data=hurricane;
   modify hurricane;
      format Date monyy7. Millions;
contents data=hurricane;
run;
```

In this example, the following changes are made:

- the format for the variable Date is changed from a full spelling of the month, day, and year to an abbreviation of the month and year

- the format for the variable Millions is removed

- the contents of the data set HURRICANE is displayed before and after the changes

Note that because the FORMAT statement does not send messages to the SAS log, you must use the CONTENTS statement if you want to make sure that the changes were made.

The following output from the two CONTENTS statements displays the contents of the data set before and after the changes. The format for the variable Date is changed from WORDDATE18. to MONYY7., and the format for the variable Millions is removed.

*Figure 38.2* Modifying Variable Formats in the Data Set HURRICANE: Before Change

The SAS System

The DATASETS Procedure

| Data Set Name | USCLIM.HURRICANE | Observations | 6 |
|---|---|---|---|
| Member Type | DATA | Variables | 5 |
| Engine | V9 | Indexes | 0 |
| Created | 04/16/2013 13:36:54 | Observation Length | 48 |
| Last Modified | 04/17/2013 07:25:49 | Deleted Observations | 0 |
| Protection |  | Compressed | NO |
| Data Set Type |  | Sorted | NO |
| Label |  |  |  |
| Data Representation | WINDOWS_32 |  |  |
| Encoding | wlatin1 Western (Windows) |  |  |

Engine/Host Dependent Information

| Data Set Page Size | 65536 |
|---|---|
| Number of Data Set Pages | 1 |
| First Data Page | 1 |
| Max Obs per Page | 1361 |
| Obs in First Data Page | 6 |
| Number of Data Set Repairs | 0 |
| ExtendObsCounter | YES |
| Filename | c:\Users\users\climate\hurricane.sas7bdat |
| Release Created | 9.0401B0 |
| Host Created | W32_7PRO |

Alphabetic List of Variables and Attributes

| # | Variable | Type | Len | Format | Informat | Label |
|---|---|---|---|---|---|---|
| 2 | Date | Num | 8 | WORDDATE18. | DATE9. |  |
| 4 | Millions | Num | 8 | DOLLAR6. |  | Damage |
| 5 | Name | Char | 8 |  |  |  |
| 1 | Place | Char | 14 |  | $CHAR14. |  |
| 3 | USDeaths | Num | 8 |  |  |  |

**Figure 38.3** *Modifying Variable Formats in the Data Set HURRICANE: After Change*

```
                            The SAS System
                          The DATASETS Procedure
```

| Data Set Name | USCLIM.HURRICANE | Observations | 6 |
| --- | --- | --- | --- |
| Member Type | DATA | Variables | 5 |
| Engine | V9 | Indexes | 0 |
| Created | 04/16/2013 13:36:54 | Observation Length | 48 |
| Last Modified | 04/17/2013 07:37:02 | Deleted Observations | 0 |
| Protection |  | Compressed | NO |
| Data Set Type |  | Sorted | NO |
| Label |  |  |  |
| Data Representation | WINDOWS_32 |  |  |
| Encoding | wlatin1 Western (Windows) |  |  |

| Engine/Host Dependent Information ||
| --- | --- |
| Data Set Page Size | 65536 |
| Number of Data Set Pages | 1 |
| First Data Page | 1 |
| Max Obs per Page | 1361 |
| Obs in First Data Page | 6 |
| Number of Data Set Repairs | 0 |
| ExtendObsCounter | YES |
| Filename | c:\Users\userid\climate\hurricane.sas7bdat |
| Release Created | 9.0401B0 |
| Host Created | W32_7PRO |

| Alphabetic List of Variables and Attributes ||||||
| --- | --- | --- | --- | --- | --- |
| # | Variable | Type | Len | Format | Informat | Label |
| 2 | Date | Num | 8 | MONYY7. | DATE9. |  |
| 4 | Millions | Num | 8 |  |  | Damage |
| 5 | Name | Char | 8 |  |  |  |
| 1 | Place | Char | 14 |  | $CHAR14. |  |
| 3 | USDeaths | Num | 8 |  |  |  |

## Assigning, Changing, or Removing Labels

A label is the descriptive information that identifies variables in tables, plots, and graphs. You usually assign labels when you create a variable. If you do not assign a label, then SAS uses the variable name as the label. However, in CONTENTS output, if a label is not assigned, then the field is blank. By using the MODIFY statement and its subordinate LABEL statement, you can assign, change, or remove a label. Here is the syntax for these statements:

**MODIFY** *SAS-data-set*;

    **LABEL** *variable*=<'*label*'>;

*SAS-data-set*
    specifies the name of the SAS data set that contains the variable whose label you want to modify.

*variable*
: specifies the name of the variable whose label you want to assign, change, or remove.

*label*
: specifies the label, which can be from 1 to 256 characters, that you want to give the variable. If you do not specify a label and one exists, then SAS removes the current label.

When you use the LABEL statement, follow these rules:

- Enclose the text of the label in single or double quotation marks. If a single quotation mark appears in the label (for example, an apostrophe), then enclose the text in double quotation marks.
- Limit the label to no more than 256 characters, including blanks.
- To remove a label, use a blank as the text of the label, that is, *variable*=' '.

In the SAS data set HURRICANE, the following statements change the label for the variable Millions and assign a label for the variable Place. Because the LABEL statement does not send messages to the SAS log, the CONTENTS statement is specified to verify that the changes were made. The QUIT statement stops the DATASETS procedure.

```
   contents data=hurricane;
      modify hurricane;
         label Millions='Damage in Millions' Place='State Hardest Hit';
   contents data=hurricane;
run;
quit;
```

The following output from the two CONTENTS statements displays the contents of the data set before and after the changes:

*Figure 38.4* *Modifying Variable Labels in the Data Set HURRICANE: Before Change*

```
                          The SAS System

                        The DATASETS Procedure

Data Set Name       USCLIM.HURRICANE       Observations           6
Member Type         DATA                   Variables              5
Engine              V9                     Indexes                0
Created             04/16/2013 13:36:54    Observation Length    48
Last Modified       04/17/2013 07:37:02    Deleted Observations   0
Protection                                 Compressed            NO
Data Set Type                              Sorted                NO
Label
Data Representation WINDOWS_32
Encoding            wlatin1 Western (Windows)

                   Engine/Host Dependent Information

Data Set Page Size          65536
Number of Data Set Pages    1
First Data Page             1
Max Obs per Page            1361
Obs in First Data Page      6
Number of Data Set Repairs  0
ExtendObsCounter            YES
Filename                    c:\Users\userid\climate\hurricane.sas7bdat
Release Created             9.0401B0
Host Created                W32_7PRO

              Alphabetic List of Variables and Attributes

          #  Variable  Type  Len  Format    Informat  Label
          2  Date      Num   8    MONYY7.   DATE9.
          4  Millions  Num   8                        Damage
          5  Name      Char  8
          1  Place     Char  14             $CHAR14.
          3  USDeaths  Num   8
```

**Figure 38.5** *Modifying Variable Labels in the Data Set HURRICANE: After Change*

The SAS System

The DATASETS Procedure

| Data Set Name | USCLIM.HURRICANE | Observations | 6 |
| Member Type | DATA | Variables | 5 |
| Engine | V9 | Indexes | 0 |
| Created | 04/16/2013 13:36:54 | Observation Length | 48 |
| Last Modified | 04/17/2013 08:00:42 | Deleted Observations | 0 |
| Protection | | Compressed | NO |
| Data Set Type | | Sorted | NO |
| Label | | | |
| Data Representation | WINDOWS_32 | | |
| Encoding | wlatin1 Western (Windows) | | |

Engine/Host Dependent Information

| Data Set Page Size | 65536 |
| Number of Data Set Pages | 2 |
| First Data Page | 1 |
| Max Obs per Page | 1361 |
| Obs in First Data Page | 6 |
| Number of Data Set Repairs | 0 |
| ExtendObsCounter | YES |
| Filename | c:\Users\userid\climate\hurricane.sas7bdat |
| Release Created | 9.0401B0 |
| Host Created | W32_7PRO |

Alphabetic List of Variables and Attributes

| # | Variable | Type | Len | Format | Informat | Label |
|---|---|---|---|---|---|---|
| 2 | Date | Num | 8 | MONYY7 | DATE9 | |
| 4 | Millions | Num | 8 | | | Damage in Millions |
| 5 | Name | Char | 8 | | | |
| 1 | Place | Char | 14 | $CHAR14. | | State Hardest Hit |
| 3 | USDeaths | Num | 8 | | | |

# Summary

## DATASETS Procedure Statements

CHANGE *old-name=new-name*;
   renames the SAS data set that you specify with *old-name* to the name that you specify with *new-name*. You can rename more than one data set in the same library by using one CHANGE statement. All new names must be valid SAS names.

MODIFY *SAS-data-set*;
   identifies the SAS data set that you want to modify. These are some of the subordinate statements that you can use with the MODIFY statement:

FORMAT *variable(s)* <*format*>;
: assigns, changes, or removes the format for one or more variables. You specify the variable and the format by using the *variable* and *format* arguments. You can give more than one variable the same format by listing more than one variable before the format. Do not specify *format* if you want to remove a format.

LABEL *variable*=<'*label*'>;
: assigns, changes, or removes the label for the variable that you specify with *variable*. To remove a label, place a blank space inside the quotation marks.

RENAME *old-name=new-name*;
: changes the name of one or more variables that you specify with *old-name* to the name that you specify with *new-name*. You can rename more than one variable in the same data set by using one RENAME statement. All names must be valid SAS names.

# Learning More

Informats and formats
: Informats and formats are available for reading and displaying data. For more information, see *SAS Formats and Informats: Reference*.

LABEL statement
: The LABEL statement assigns descriptive labels to variables. For more information, see "LABEL Statement" in *SAS Statements: Reference*.

MODIFY statement
: The MODIFY statement in the DATASETS procedure has additional statements that change informats and that create and delete indexes for variables. For more information, see "DATASETS" in *Base SAS Procedures Guide*.

Renaming variables
: You can use the RENAME= data set option and the RENAME statement in the DATA step to rename variables. For more information, see "RENAME= Data Set Option" in *SAS Data Set Options: Reference* and "RENAME Statement" in *SAS Statements: Reference*.

Variables
: To learn how to create and delete variables in the DATA step, see Chapter 6, "Starting with SAS Data Sets," on page 91.

# Chapter 39
# Copying, Moving, and Deleting SAS Data Sets

Introduction to Copying, Moving, and Deleting SAS Data Sets . . . . . . . . . . . . . . . 725
    Purpose . . . . . . . . . . . . . . . . . . . . . . . . . . . . . . . . . . . . . . . . . . . . . . . . . . . . . . . . . . . 725
    Prerequisites . . . . . . . . . . . . . . . . . . . . . . . . . . . . . . . . . . . . . . . . . . . . . . . . . . . . . . . 726

**Input Data Libraries for Examples** . . . . . . . . . . . . . . . . . . . . . . . . . . . . . . . . . . . . . 726

**Copying SAS Data Sets** . . . . . . . . . . . . . . . . . . . . . . . . . . . . . . . . . . . . . . . . . . . . . . 727
    Copying from the Procedure Input Library . . . . . . . . . . . . . . . . . . . . . . . . . . . . 727
    Copying from Other Libraries . . . . . . . . . . . . . . . . . . . . . . . . . . . . . . . . . . . . . . . 729

**Copying Specific SAS Data Sets** . . . . . . . . . . . . . . . . . . . . . . . . . . . . . . . . . . . . . . 730
    Selecting Data Sets to Copy . . . . . . . . . . . . . . . . . . . . . . . . . . . . . . . . . . . . . . . . 730
    Excluding Data Sets from Copying . . . . . . . . . . . . . . . . . . . . . . . . . . . . . . . . . . 731

**Moving SAS Libraries and SAS Data Sets** . . . . . . . . . . . . . . . . . . . . . . . . . . . . . . 731
    Moving Libraries . . . . . . . . . . . . . . . . . . . . . . . . . . . . . . . . . . . . . . . . . . . . . . . . . 731
    Moving Specific Data Sets . . . . . . . . . . . . . . . . . . . . . . . . . . . . . . . . . . . . . . . . . 733

**Deleting SAS Data Sets** . . . . . . . . . . . . . . . . . . . . . . . . . . . . . . . . . . . . . . . . . . . . . 734
    Specifying Data Sets to Delete . . . . . . . . . . . . . . . . . . . . . . . . . . . . . . . . . . . . . . 734
    Specifying Data Sets to Save . . . . . . . . . . . . . . . . . . . . . . . . . . . . . . . . . . . . . . . 735

**Deleting All Files in a SAS Library** . . . . . . . . . . . . . . . . . . . . . . . . . . . . . . . . . . . 735

**Summary** . . . . . . . . . . . . . . . . . . . . . . . . . . . . . . . . . . . . . . . . . . . . . . . . . . . . . . . . . . 736
    Procedures . . . . . . . . . . . . . . . . . . . . . . . . . . . . . . . . . . . . . . . . . . . . . . . . . . . . . . 736
    DATASETS Procedure Statements . . . . . . . . . . . . . . . . . . . . . . . . . . . . . . . . . . . 736

**Learning More** . . . . . . . . . . . . . . . . . . . . . . . . . . . . . . . . . . . . . . . . . . . . . . . . . . . . . 737

## Introduction to Copying, Moving, and Deleting SAS Data Sets

### Purpose

Copying, moving, and deleting SAS data sets are the library management tasks that are performed most frequently. For example, you perform these tasks to create test files, make backups, archive files, and remove unused files. The DATASETS procedure enables you to work with all the files in a SAS library or with specific files in the library.

In this section, you will learn how to use the DATASETS procedure to perform the following tasks:

- copy an entire library
- copy specific SAS data sets
- move specific SAS data sets
- delete specific SAS data sets
- delete all files in a library

This section focuses on using the DATASETS procedure to copy, move, and delete data sets. You can also use the illustrated statements and options to copy, move, and delete other types of SAS files.

## Prerequisites

Before using this section, you should understand the concepts that are presented in the following sections:

- Chapter 35, "Understanding SAS Libraries," on page 687
- Chapter 36, "Managing SAS Libraries," on page 695
- Chapter 38, "Modifying SAS Data Set Names and Variable Attributes," on page 713

# Input Data Libraries for Examples

The examples in this section use five SAS libraries that contain sample data sets that are used to collect and store weather statistics for the United States and other countries. The libraries have the librefs PRECIP, USCLIM, CLIMATE, WEATHER, and STORM. The following LIBNAME statements assign the librefs:

```
libname precip 'SAS-library-1';
libname usclim 'SAS-library-2';
libname climate 'SAS-library-3';
libname weather 'SAS-library-4';
libname storm 'SAS-library-5';
```

*Note:* For each LIBNAME statement, *SAS-library* is a different physical name for the location of the SAS library. In order to copy all or some SAS data sets from one library to another, the input and output libraries must be in different physical locations.

The DATA steps that create the data sets in the SAS libraries CLIMATE, PRECIP, and STORM are shown in "The CLIMATE, PRECIP, and STORM Data Sets" on page 820. The DATA steps that create the data sets in the SAS library USCLIM are shown in "The USCLIM Data Sets" on page 819.

# Copying SAS Data Sets

## Copying from the Procedure Input Library

You can use the COPY statement in the DATASETS procedure to copy all or some SAS data sets from one library to another. When copying data sets, SAS duplicates the contents of each file, including the descriptor information, and updates information in the directory for each library.

*CAUTION:*
   **During processing, SAS automatically writes the data from the input library into an output data set of the same name. If there are duplicate data set names, then you do not receive a warning message before copying starts.** Before you make changes to libraries, it is important to obtain directory listings of the input and output libraries in order to visually check for duplicate data set names.

To copy files from the procedure input library (specified in the PROC DATASETS statement), use the COPY statement. Here is the syntax of the COPY statement:

**COPY** OUT=*libref* <options>;

*libref*
   specifies the libref for the SAS library to which you want to copy the files. You must specify an output library.

For example, the library PRECIP contains data sets for snowfall and rainfall amounts, and the library CLIMATE contains data sets for temperature. The following program lists the contents so that they can be visually compared before any action is taken:

```
proc datasets library=precip;
   contents data=_all_ nods;
   contents data=climate._all_ nods;
run;
```

The PROC DATASETS statement starts the procedure and specifies the procedure input library PRECIP. The first CONTENTS statement produces a directory listing of the library PRECIP. Then, the second CONTENTS statement produces a directory listing of the library CLIMATE.

The following SAS output displays the two directory listings:

*Figure 39.1* Checking the Directory PRECIP Before Copying

The SAS System

The DATASETS Procedure

| Directory | |
|---|---|
| Libref | PRECIP |
| Engine | V9 |
| Physical Name | c:\Users\userid\precip |
| Filename | c:\Users\userid\precip |

| # | Name | Member Type | File Size | Last Modified |
|---|---|---|---|---|
| 1 | RAIN | DATA | 131072 | 04/17/2013 08:26:48 |
| 2 | SNOW | DATA | 131072 | 04/17/2013 08:26:48 |

*Figure 39.2* Checking Directory CLIMATE Before Copying

The SAS System

The DATASETS Procedure

| Directory | |
|---|---|
| Libref | CLIMATE |
| Engine | V9 |
| Physical Name | c:\Users\userid\climate2 |
| Filename | c:\Users\userid\climate2 |

| # | Name | Member Type | File Size | Last Modified |
|---|---|---|---|---|
| 1 | HIGHTEMP | DATA | 131072 | 04/17/2013 08:26:48 |
| 2 | LOWTEMP | DATA | 131072 | 04/17/2013 08:26:48 |

There are no duplicate names in the directories, so the COPY statement can be issued to achieve the desired results.

```
   copy out=climate;
run;
```

The following SAS log shows the messages as the data sets in the library PRECIP are copied to the library CLIMATE. There are now two copies of the data sets RAIN and SNOW: one in the PRECIP library and one in the CLIMATE library.

*Log 39.1  Messages Sent to the SAS Log during Copying*

```
143     copy out=climate;
144  run;

NOTE: Copying PRECIP.RAIN to CLIMATE.RAIN (memtype=DATA).
NOTE: There were 5 observations read from the data set PRECIP.RAIN.
NOTE: The data set CLIMATE.RAIN has 5 observations and 4 variables.
NOTE: Copying PRECIP.SNOW to CLIMATE.SNOW (memtype=DATA).
NOTE: There were 3 observations read from the data set PRECIP.SNOW.
NOTE: The data set CLIMATE.SNOW has 3 observations and 4 variables.
```

## Copying from Other Libraries

You can copy from a library other than the procedure input library without using another PROC DATASETS statement. To do so, use the IN= option in the COPY statement to override the procedure input library. Here is the syntax for the option:

**COPY** OUT=*libref-1* IN=*libref-2*;

*libref-1*
   specifies the libref for the SAS library to which you want to copy files.

*libref-2*
   specifies the libref for the SAS library from which you want to copy files.

The IN= option is a useful tool when you want to copy more than one library into the output library. You can use one COPY statement for each input library without repeating the PROC DATASETS statement.

For example, the following statements copy the libraries PRECIP, STORM, CLIMATE, and USCLIM to the library WEATHER. The procedure input library is PRECIP, which was specified in the previous PROC DATASETS statement.

```
   copy out=weather;
   copy in=storm out=weather;
   copy in=climate out=weather;
   copy in=usclim out=weather;
run;
```

**730** Chapter 39 • Copying, Moving, and Deleting SAS Data Sets

The following SAS log shows that the data sets from these libraries have been consolidated in the library WEATHER:

**Log 39.2** Copying Four Libraries into the Library WEATHER

```
142     copy out=weather;
NOTE: Copying PRECIP.RAIN to WEATHER.RAIN (memtype=DATA).
NOTE: There were 5 observations read from the data set PRECIP.RAIN.
NOTE: The data set WEATHER.RAIN has 5 observations and 4 variables.
NOTE: Copying PRECIP.SNOW to WEATHER.SNOW (memtype=DATA).
NOTE: There were 3 observations read from the data set PRECIP.SNOW.
NOTE: The data set WEATHER.SNOW has 3 observations and 4 variables.
143     copy in=storm out=weather;
NOTE: Copying STORM.HURRICANE to WEATHER.HURRICANE (memtype=DATA).
NOTE: There were 6 observations read from the data set STORM.HURRICANE.
NOTE: The data set WEATHER.HURRICANE has 6 observations and 5 variables.
NOTE: Copying STORM.TORNADO to WEATHER.TORNADO (memtype=DATA).
NOTE: There were 5 observations read from the data set STORM.TORNADO.
NOTE: The data set WEATHER.TORNADO has 5 observations and 4 variables.
144     copy in=climate out=weather;
NOTE: Copying CLIMATE.HIGHTEMP to WEATHER.HIGHTEMP (memtype=DATA).
NOTE: There were 6 observations read from the data set CLIMATE.HIGHTEMP.
NOTE: The data set WEATHER.HIGHTEMP has 6 observations and 4 variables.
NOTE: Copying CLIMATE.LOWTEMP to WEATHER.LOWTEMP (memtype=DATA).
NOTE: There were 5 observations read from the data set CLIMATE.LOWTEMP.
NOTE: The data set WEATHER.LOWTEMP has 5 observations and 4 variables.
NOTE: Copying CLIMATE.RAIN to WEATHER.RAIN (memtype=DATA).
NOTE: There were 5 observations read from the data set CLIMATE.RAIN.
NOTE: The data set WEATHER.RAIN has 5 observations and 4 variables.
NOTE: Copying CLIMATE.SNOW to WEATHER.SNOW (memtype=DATA).
NOTE: There were 3 observations read from the data set CLIMATE.SNOW.
NOTE: The data set WEATHER.SNOW has 3 observations and 4 variables.
145     copy in=usclim out=weather;
146  run;

NOTE: Copying USCLIM.BASETEMP to WEATHER.BASETEMP (memtype=CATALOG).
NOTE: Copying USCLIM.HURRICANE to WEATHER.HURRICANE (memtype=DATA).
NOTE: There were 6 observations read from the data set USCLIM.HURRICANE.
NOTE: The data set WEATHER.HURRICANE has 6 observations and 5 variables.
NOTE: Copying USCLIM.REPORT to WEATHER.REPORT (memtype=CATALOG).
NOTE: Copying USCLIM.TEMPCHNG to WEATHER.TEMPCHNG (memtype=DATA).
NOTE: There were 5 observations read from the data set USCLIM.TEMPCHNG.
NOTE: The data set WEATHER.TEMPCHNG has 5 observations and 6 variables.
NOTE: Copying USCLIM.USHIGH to WEATHER.USHIGH (memtype=DATA).
NOTE: There were 6 observations read from the data set USCLIM.USHIGH.
NOTE: The data set WEATHER.USHIGH has 6 observations and 5 variables.
NOTE: Copying USCLIM.USLOW to WEATHER.USLOW (memtype=DATA).
NOTE: There were 7 observations read from the data set USCLIM.USLOW.
NOTE: The data set WEATHER.USLOW has 7 observations and 5 variables.
```

# Copying Specific SAS Data Sets

## Selecting Data Sets to Copy

To copy only a few data sets from a large SAS library, use the SELECT statement with the COPY statement. After the keyword SELECT, list the data set names with a blank space between the names, or use an abbreviated member list (such as YRDATA1-YRDATA5) if applicable.

For example, the following statements copy the data set HURRICANE from the library USCLIM to the library STORM. The input procedure library is PRECIP, so the COPY statement includes the IN= option in order to specify the USCLIM input library.

```
   copy in=usclim out=storm;
      select hurricane;
run;
```

The following SAS log shows that only the data set HURRICANE was copied to the library STORM:

**Log 39.3** *Copying the Data Set HURRICANE to the Library STORM*

```
147     copy in=usclim out=storm;
148        select hurricane;
149  run;

NOTE: Copying USCLIM.HURRICANE to STORM.HURRICANE (memtype=DATA).
NOTE: There were 6 observations read from the data set USCLIM.HURRICANE.
NOTE: The data set STORM.HURRICANE has 6 observations and 5 variables.
```

## Excluding Data Sets from Copying

To copy an entire library except for a few data sets, use the EXCLUDE statement with the COPY statement. After the keyword EXCLUDE, list the data set names that you want to exclude with a blank space between the names. You can also use an abbreviated member list (such as YRDATA1-YRDATA5) if applicable.

The following statements copy the files in the library PRECIP to USCLIM except for the data set SNOW. The procedure input library is PRECIP, so the IN= option is not needed.

```
   copy out=usclim;
      exclude snow;
run;
```

The following SAS log shows that the data set RAIN was copied to USCLIM and that the data set SNOW remains only in the library PRECIP:

**Log 39.4** *Excluding the Data Set SNOW from Copying to the Library USCLIM*

```
150     copy out=usclim;
151        exclude snow;
152  run;

NOTE: Copying PRECIP.RAIN to USCLIM.RAIN (memtype=DATA).
NOTE: There were 5 observations read from the data set PRECIP.RAIN.
NOTE: The data set USCLIM.RAIN has 5 observations and 4 variables.
```

# Moving SAS Libraries and SAS Data Sets

## Moving Libraries

The COPY statement provides the MOVE option to move SAS data sets from the input library (either the procedure input library or the input library named with the IN= option) to the output library (named with the OUT= option). Note that with the MOVE option, SAS first copies the files to the output library, and then deletes them from the input library.

The following statements move all the data sets in the library PRECIP to the library CLIMATE:

```
copy out=climate move;
run;
```

The following SAS log shows that the data sets in PRECIP were moved to CLIMATE:

**Log 39.5** *Moving Data Sets in the Library PRECIP to the Library CLIMATE*

```
153     copy out=climate move;
154  run;

NOTE: Moving PRECIP.RAIN to CLIMATE.RAIN (memtype=DATA).
NOTE: There were 5 observations read from the data set PRECIP.RAIN.
NOTE: The data set CLIMATE.RAIN has 5 observations and 4 variables.
NOTE: Moving PRECIP.SNOW to CLIMATE.SNOW (memtype=DATA).
NOTE: There were 3 observations read from the data set PRECIP.SNOW.
NOTE: The data set CLIMATE.SNOW has 3 observations and 4 variables.
```

After moving files with the MOVE option, a directory listing of PRECIP from the CONTENTS statement confirms that there are no members in the library. As the output from the following statements illustrates, the library PRECIP no longer contains any data sets. Therefore, the library CLIMATE contains the only copy of the data sets RAIN and SNOW.

```
contents data=_all_ nods;
run;
```

The following outputs show the SAS log, then the directory listing for the library PRECIP:

**Log 39.6** *SAS Log from the CONTENTS Statement*

```
155     contents data=_all_ nods;
156  run;

WARNING: No matching members in directory.
```

**Figure 39.3** *Directory Listing of the Library PRECIP Showing No Data Sets*

The SAS System

The DATASETS Procedure

| Directory | |
|---|---|
| Libref | PRECIP |
| Engine | V9 |
| Physical Name | c:\Users\userid\precip |
| Filename | c:\Users\userid\precip |

*Note:* The data sets are deleted from the SAS library PRECIP, but the libref is still assigned. The name that is assigned to the library in your operating environment is not removed when you move all files from one library to another.

## Moving Specific Data Sets

You can use the SELECT and EXCLUDE statements to move one or more SAS data sets. For example, the following statements move the data set HURRICANE from the library USCLIM to the library STORM:

```
copy in=usclim out=storm move;
    select hurricane;
run;
```

The following output displays the results:

**Log 39.7** *Moving the Data Set HURRICANE from the Library USCLIM to the Library STORM*

```
157    copy in=usclim out=storm move;
158        select hurricane;
159    run;

NOTE: Moving USCLIM.HURRICANE to STORM.HURRICANE (memtype=DATA).
NOTE: There were 6 observations read from the data set USCLIM.HURRICANE.
NOTE: The data set STORM.HURRICANE has 6 observations and 5 variables.
```

Similarly, the following example uses the EXCLUDE statement to move all files except the data set SNOW from the library CLIMATE to the library USCLIM:

```
copy in=climate out=usclim move;
    exclude snow;
run;
```

The following output displays the results:

*Log 39.8  Moving All Data Sets except SNOW from the Library CLIMATE to the Library USCLIM*

```
160     copy in=climate out=usclim move;
161        exclude snow;
162  run;

NOTE: Moving CLIMATE.HIGHTEMP to USCLIM.HIGHTEMP (memtype=DATA).
NOTE: There were 6 observations read from the data set CLIMATE.HIGHTEMP.
NOTE: The data set USCLIM.HIGHTEMP has 6 observations and 4 variables.
NOTE: Moving CLIMATE.LOWTEMP to USCLIM.LOWTEMP (memtype=DATA).
NOTE: There were 5 observations read from the data set CLIMATE.LOWTEMP.
NOTE: The data set USCLIM.LOWTEMP has 5 observations and 4 variables.
NOTE: Moving CLIMATE.RAIN to USCLIM.RAIN (memtype=DATA).
NOTE: There were 5 observations read from the data set CLIMATE.RAIN.
NOTE: The data set USCLIM.RAIN has 5 observations and 4 variables.
```

# Deleting SAS Data Sets

## Specifying Data Sets to Delete

Use the DELETE statement to delete one or more data sets from a SAS library. If you want to delete more than one data set, then list the names after the DELETE keyword with a blank space between the names. You can also use an abbreviated member list if applicable (such as YRDATA1-YRDATA5).

**CAUTION:**
  **SAS immediately deletes the files in a SAS library when the program statements are submitted.** You are not asked to verify the Delete operation before it begins, so make sure that you intend to delete the files before submitting the program.

For example, the following program specifies USCLIM as the procedure input library. It then deletes the data set RAIN from the library:

```
proc datasets library=usclim;
   delete rain;
run;
```

The following output shows that SAS sends messages to the SAS log when it processes the DELETE statement:

*Log 39.9  Deleting the Data Set RAIN from the Library USCLIM*

```
163  proc datasets library=usclim;
164     delete rain;
165  run;

NOTE: Deleting USCLIM.RAIN (memtype=DATA).
```

Execute the following program to list the contents of library USCLIM after the RAIN data set has been deleted:

```
proc datasets library=usclim;
```

```
run;
quit;
```

### Specifying Data Sets to Save

To delete all data sets but a few, you can use the SAVE statement to list the names of the data sets that you want to keep. List the data set names with a blank space between the names, or use an abbreviated member list (such as YRDATA1-YRDATA5) if applicable.

The following statements delete all the data sets except TEMPCHNG from the library USCLIM:

```
proc datasets library=usclim;
   save tempchng;
run;
```

The following output shows the SAS log from the Save operation. SAS sends messages to the log, verifying that it has kept the data sets that you specified in the SAVE statement and deleted all other members of the library.

*Log 39.10   Deleting All Members of the Library USCLIM except the Data Set TEMPCHNG*

```
171  proc datasets library=usclim;
172     save tempchng;
173  run;

NOTE: Saving USCLIM.TEMPCHNG (memtype=DATA).
NOTE: Deleting USCLIM.BASETEMP (memtype=CATALOG).
NOTE: Deleting USCLIM.HIGHTEMP (memtype=DATA).
NOTE: Deleting USCLIM.LOWTEMP (memtype=DATA).
NOTE: Deleting USCLIM.REPORT (memtype=CATALOG).
NOTE: Deleting USCLIM.USHIGH (memtype=DATA).
NOTE: Deleting USCLIM.USLOW (memtype=DATA).
```

## Deleting All Files in a SAS Library

To delete all files in a SAS library at one time, use the KILL option in the PROC DATASETS statement.

*CAUTION:*

**The KILL option deletes all members of the library immediately after the statement is submitted.** You are not asked to verify the Delete operation, so make sure that you intend to delete the files before submitting the program.

For example, the following program deletes all data sets in the library WEATHER and stops the DATASETS procedure:

```
proc datasets library=weather kill;
run;
quit;
```

The following output displays the SAS log:

**Log 39.11** *Deleting All Members of the Library WEATHER*

```
174  proc datasets library=weather kill;
NOTE: Deleting WEATHER.BASETEMP (memtype=CATALOG).
NOTE: Deleting WEATHER.HIGHTEMP (memtype=DATA).
NOTE: Deleting WEATHER.HURRICANE (memtype=DATA).
NOTE: Deleting WEATHER.LOWTEMP (memtype=DATA).
NOTE: Deleting WEATHER.RAIN (memtype=DATA).
NOTE: Deleting WEATHER.REPORT (memtype=CATALOG).
NOTE: Deleting WEATHER.SNOW (memtype=DATA).
NOTE: Deleting WEATHER.TEMPCHNG (memtype=DATA).
NOTE: Deleting WEATHER.TORNADO (memtype=DATA).
NOTE: Deleting WEATHER.USHIGH (memtype=DATA).
NOTE: Deleting WEATHER.USLOW (memtype=DATA).
175  run;

176  quit;
```

*Note:* All data sets and catalogs are deleted from the SAS library, but the libref is still assigned for the session. The name that is assigned to the library in your operating environment is not removed when you delete the files that are included in the library.

# Summary

## Procedures

PROC DATASETS LIBRARY=*libref* <KILL>;
    starts the procedure and specifies the procedure input library for subsequent statements. The KILL option deletes all members and member types from the library.

## DATASETS Procedure Statements

COPY OUT=*libref* <IN=*libref*> <MOVE>;
    copies files from the procedure input library that is specified in the PROC DATASETS statement to the output library that is specified in the OUT= option. The IN= option specifies a different input library. The MOVE option deletes files from the input library after copying them to the output library.

You can use the following statements with the COPY statement:

EXCLUDE *SAS-data-set*;
    specifies a SAS data set that you want to exclude from the copy process. Files that you do not list in this statement are copied to the output library.

SELECT *SAS-data-set*;
    specifies a SAS data set that you want to copy to the output library.

DELETE *SAS-data-set*;
    deletes only the SAS data set that you specify in this statement.

SAVE *SAS-data-set*;
    deletes all members of the library except those that you specify in this statement.

## Learning More

CATALOG procedure
: The CATALOG procedure manages entries in SAS catalogs. For more information, see "CATALOG" in *Base SAS Procedures Guide*.

DATASETS procedure
: The DATASETS procedure is a utility procedure that manages your SAS files. For more information, see "DATASETS" in *Base SAS Procedures Guide*.

## Part 10

# Understanding Your SAS Environment

*Chapter 40*
 **Introducing the SAS Environment** ............................... *741*

*Chapter 41*
 **Using the SAS Windowing Environment** ......................... *753*

*Chapter 42*
 **Customizing the SAS Environment** ............................. *793*

# Chapter 40
# Introducing the SAS Environment

**Introduction to the SAS Environment** .................................. 741
    Purpose .................................................................. 741
    Prerequisites ............................................................. 742
    Operating Environment Differences ....................................... 742

**Starting a SAS Session** ................................................. 742

**Selecting a SAS Processing Mode** ........................................ 743
    Processing Modes and Categories ........................................ 743
    Processing in the SAS Windowing Environment ........................... 744
    Processing Interactively in Line Mode .................................. 747
    Processing in Batch Mode ............................................... 748
    Processing Noninteractively ............................................ 748

**Summary** ............................................................... 749
    Command ................................................................. 749
    Options ................................................................. 750
    System Options .......................................................... 750
    Statements .............................................................. 750
    Commands ................................................................ 750

**Learning More** ......................................................... 751
    Operating Environment Information ...................................... 751
    Windowing Environment Commands ......................................... 751
    Documentation ........................................................... 751

## Introduction to the SAS Environment

### Purpose

In this section, you will learn about the various ways that you can run SAS programs. More importantly, it explains the different modes that SAS can run in, and which modes are best, depending on the types of jobs that you are doing.

This section introduces the SAS windowing environment, which is the default processing mode.

### Prerequisites

To understand the discussions in this section, you should be familiar with the basics of DATA step programming that are presented in "Overview of DATA Step Processing" on page 107.

### Operating Environment Differences

Even though SAS has a different appearance for each operating environment, most of the actions that are available from the menus are the same.

One of the biggest differences between operating environments is the way that you select menu items.

If your workstation is not equipped with a mouse, then here are the keyboard equivalents to mouse actions:

*Table 40.1* Mouse Actions and Keyboard Equivalents

| Mouse Action | Keyboard Equivalent |
| --- | --- |
| Double-click the item | Type an **s** or an **x** in the space next to the item, and then press Enter |
| Right-click the item | Type **?** in the space next to the item, and then press Enter |

Examples in this documentation show SAS windows as they appear in the Microsoft Windows environment. Usually, corresponding windows in other operating environments show similar results. If you do not see the drop-down menus in your operating environment, then enter the global command PMENU at a command prompt.

## Starting a SAS Session

To start a SAS session, you must invoke SAS. At the operating environment prompt, execute the SAS command. In most cases, the SAS command is

```
sas
```

*Note:* The SAS command might vary from site to site. Consult your on-site SAS support personnel for more information.

You can customize your SAS session when it starts by specifying SAS system options, which then remain in effect throughout a session. For example, you can use the LINESIZE= system option to specify a line size for the SAS log and print file. Some system options can be specified only at initialization, and other system options can be specified during a SAS session. For more information, see "Customizing SAS Sessions and Programs at Start-up" on page 795.

# Selecting a SAS Processing Mode

## Processing Modes and Categories

### Overview of Processing Modes and Categories
All four modes that you can use to run SAS belong to one of two categories:

- foreground processing
- background processing

The following figure shows the four different modes and the processing types that they belong to. As your processing requirements change, you might find it helpful to change from one processing mode to another.

**Figure 40.1** Modes of Running SAS during Foreground or Background Processing

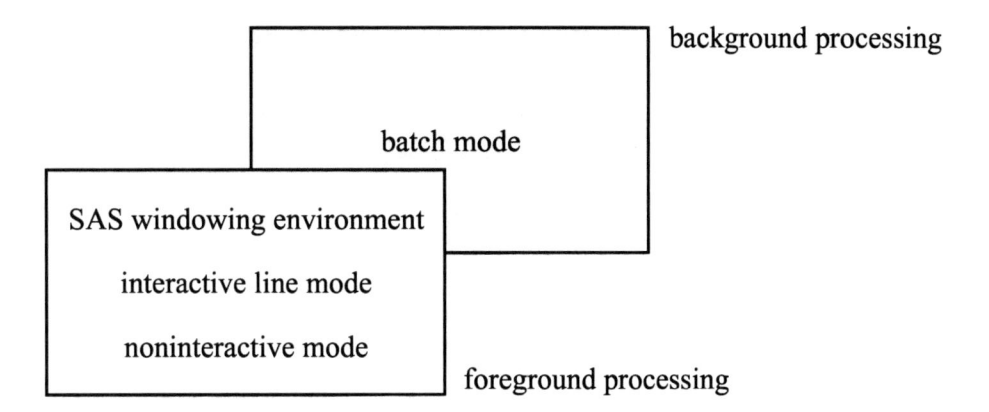

```
                    ┌──────────────────────┐
                    │                      │  background processing
                    │      batch mode      │
        ┌───────────┤                      │
        │           └──────────────────────┘
        │  SAS windowing environment       │
        │                                  │
        │  interactive line mode           │
        │                                  │
        │  noninteractive mode             │
        └──────────────────────────────────┘  foreground processing
```

### Understanding Foreground Processing
Foreground processing includes all the ways that you can run SAS in except batch mode. Foreground processing begins immediately, but as your program runs, your current workstation session is occupied, so you cannot use it to do anything else.[1] With foreground processing, you can route your output to the workstation display, to a file, to a printer, or to tape.

If you can answer yes to one or more of the following questions, then you might want to consider foreground processing:

- Are you learning SAS programming?
- Are you testing a program to see whether it works?
- Do you need fast turnaround?
- Are you processing a fairly small data file?
- Are you using an interactive application?

---

[1] In a workstation environment, you can switch to another window and continue working.

### Understanding Background Processing

Batch processing is the only way to run SAS in the background. Your operating environment coordinates all the work, so you can use your workstation session to do other work at the same time that your program runs. However, because the operating environment also schedules your program for execution and assigns it a priority, the program might have to wait in the input queue (the operating environment's list of jobs to be run) before it is executed. When your program runs to completion, you can browse, delete, or print your output.

Background processing might be required at your site. In addition, consider the following questions:

- Are you an experienced SAS user, likely to make fewer errors than a novice?
- Are you running a program that has already been tested and refined?
- Is fast turnaround less important than minimizing the use of computer resources?
- Are you processing a large data file?
- Will your program run for a long time?
- Are you using a tape?

If you answer yes to one or more of these questions, then you might want to choose background processing.

## Processing in the SAS Windowing Environment

### Overview of Processing in the SAS Windowing Environment

The SAS windowing environment is a graphical user interface (GUI) that consists of a series of windows with which you can organize files and folders, edit and execute programs, view program output, and view messages about your programs and your SAS session.

Because it is an interactive and graphical facility, you can use a single session to prepare and submit a program and, if necessary, to modify and resubmit the program after browsing the output and messages. You can move from window to window and even interrupt and return to a session at the same point that you left it.

### General Characteristics

The SAS windowing environment is the default environment for a SAS session (unless your environment is customized at your site).

*Note:* Because it is the default environment, many topics in this documentation describe tasks as you would perform them in the SAS windowing environment.

The five most commonly used windows in the SAS windowing environment are Explorer, Results, Editor, Log, and Output.

Explorer
> is a hierarchical system of folders, subfolders, and individual items. It provides a primary graphical interface to SAS from which you can do the following:
>
> - access and work with data, such as catalogs, tables, libraries, and operating environment files
> - open SAS programming windows
> - access the Output Delivery System (ODS)
> - create and define customized folders

You can use Explorer to view or set libraries and file shortcuts, view or set library members and catalog entries, or open and edit SAS files.

Note that when you start the SAS windowing environment, the Explorer might appear as a single-paned window that lists libraries that are currently available. You can add a navigational tree to the Explorer window by selecting **View** ⇨ **Show Tree** or by issuing the TREE command.

Editor or Program Editor
: provides an area to enter, edit, and submit SAS statements and to save SAS source files.

Log
: enables you to browse and scroll the SAS log. The SAS log provides messages about what is happening in your SAS session.

Output
: enables you to browse and scroll procedure output.

Results
: enables you to browse and manipulate an index of your procedure output.

Together, the Program Editor, Log, and Output windows are sometimes referred to as the programming windows.

**Figure 40.2** *SAS Windowing Environment: SAS Explorer, Log and Editor Windows, (Windows Operating System)*

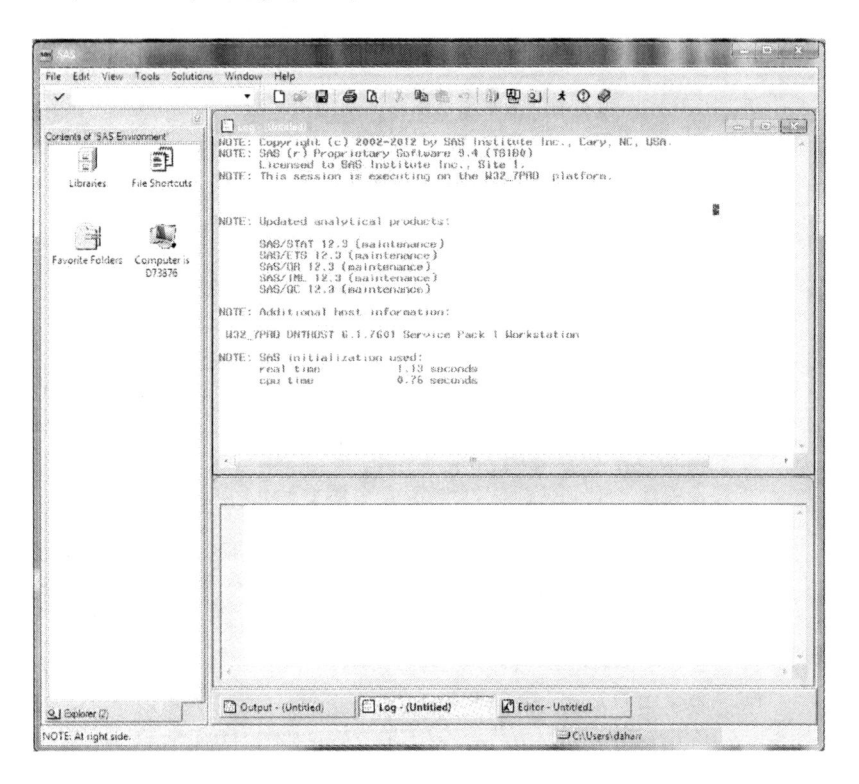

Additional windows are also available in the SAS windowing environment that enable you to do the following:

- access online Help
- view and change some SAS system options
- view and change function key settings
- create and store text information

For more information about these windows and about performing tasks in the windowing environment, see "Introduction to Using the SAS Windowing Environment" on page 754.

### *Invoking the SAS Windowing Environment*

To invoke the SAS windowing environment, execute the SAS command followed by any system options that you want to put into effect. The SAS windowing environment is set as the default method of operation for SAS, but it might not be the default setting at your work site.

If the SAS windowing environment is not the default method of operation, you can specify the DMSEXP option in the SAS command. Or, you can include the DMSEXP option in the configuration file, which contains settings for system options. For more information about the configuration file, see "Customizing SAS Sessions and Programs at Start-up" on page 795.

You specify options in the SAS command as you do any other command options on your system.

The following table shows how you would start the SAS windowing environment and specify the DMSEXP option under various operating environments:

*Table 40.2* Starting SAS with the DMSEXP Option

| Operating Environment | Command |
| --- | --- |
| z/OS | sas options ('dmsexp') |
| Windows | sas -dmsexp |
| UNIX | sas -dmsexp |

For details about how to specify command options on other systems, see the SAS documentation for your operating environment.

### *Ending a SAS Windowing Environment Session*

You can end your SAS windowing environment session with the BYE or ENDSAS command. Specify BYE or ENDSAS on the SAS command line, and then execute the command by pressing ENTER or RETURN (depending on which operating environment you use).

You can also end your session with the ENDSAS statement in the Program Editor window. Enter the following statement on a data line and submit it for execution:

```
endsas;
```

### *Interrupting a SAS Windowing Environment Session*

You might occasionally find it necessary to return to your operating environment from a SAS session. If you do not want to end your SAS session, then you can escape to the operating environment by issuing the X command. Simply execute the following command on the command line:

```
x
```

From your operating environment, you can then return to the same SAS session as you left it, by executing the appropriate operating environment command. For example, under the z/OS operating environment, the operating environment command is Return or End.

Use this form of the X command to execute a single operating environment command:

**X** *operating-environment-command*

or, if the command contains embedded blanks,

**X** *'operating-environment-command'*

For example, on many systems that you can display the current time by specifying

```
x time
```

After the command executes, you can take the appropriate action to return to your SAS session.

For information about interrupting a SAS session in other operating environments, see the SAS documentation for your operating environment.

## Processing Interactively in Line Mode

### General Characteristics

With line mode processing, you enter programming statements one line at a time; DATA and PROC steps are executed after you enter a RUN statement, or after another step boundary. Program messages and output appear on the monitor.

You can modify program statements only when you first enter them, before you press ENTER or RETURN, which means that you must enter your entries carefully.

### Invoking SAS in Line Mode

To invoke SAS in line mode, execute the SAS command followed by any system options that you want to put into effect. The NODMS system option activates an interactive line mode session. If NODMS is not the default system option at your site, you can either specify the option with the SAS command or include the NODMS specification in the configuration file, the file that contains settings for system options that are put into effect at invocation.

The following table shows you how to specify the NODMS system option with the SAS command under various operating environments.

*Table 40.3  Specifying the NODMS System Option*

| Operating environment | Command |
| --- | --- |
| z/OS | sas options ('nodms') |
| UNIX | sas -nodms |

### Using the Run Statement to Execute a Program in Line Mode

In line mode, DATA steps are executed only when a new step boundary is encountered. This occurs after you enter a RUN DATA or PROC statement. In other words, if you submit **DATA X; X=1;** in the windowing environment, then you will not see execution until the next RUN DATA or PROC statement is submitted.

At the beginning of each line, SAS prompts you with a number and a question mark to enter more statements. If you use a DATALINES statement, then a greater-than symbol (>) replaces the question mark, indicating that data lines are expected.

When you are using line mode, the log will be easier to read if you follow this programming tip: cause each DATA or PROC step to execute before you begin entering programming statements for the next step. Either an END statement or a semicolon that marks the end of data lines causes a step to execute immediately.

### Ending a Line Mode SAS Session

To end your session, type `endsas;` at the SAS prompt, and then press Enter. Your session ends, and you are returned to your operating environment.

### Interrupting a Line Mode SAS Session

In line mode, you can escape to the operating environment by executing the following statement:

x;

You can return to your SAS session by executing the appropriate operating environment command. Use this form of the X statement to execute a single operating environment command:

**X** *operating-environment-command*;

or, if the command contains embedded blanks,

**X** *'operating-environment-command'*;

For example, on many systems that you can display the current time by specifying

x time;

When you use this form of the X command, the command executes, and you are returned to your SAS session.

## Processing in Batch Mode

The first step in executing a program in batch mode is to prepare files that include the following:

- any control language statements that are required by the operating environment that you are using to manage the program
- the SAS statements necessary to execute the program

Then you submit your file to the operating environment, and your workstation session is free for other work while the operating environment executes the program. This is called background processing because you cannot view or change the program in any way until after it executes. The log and output are routed to the destination that you specify in the operating environment control language; without a specification, they are routed to the default. For examples of batch processing, see the SAS documentation for your operating environment.

## Processing Noninteractively

### General Characteristics

Noninteractive processing has some characteristics of interactive processing and some of batch processing. When you process noninteractively, you execute SAS program statements that are stored in an external file. You use a SAS command to submit the program statements to your operating environment.

*Note:* The SAS command is implemented differently under each operating environment. For example, under z/OS the command is typically a CLIST.

As in interactive processing, processing begins immediately, and your current workstation session is occupied. However, as with batch processing, you cannot interact with your program.

*Note:* For some exceptions to this, see the SAS documentation for your operating environment.

You can see the log or procedure output immediately after the program has run. Log and listing output are routed to the workstation, unlike the SAS windowing environment, where you must explicitly save output to a file. If you decide that you must correct or modify your program, then you must use an editor to make necessary changes and then resubmit your program.

### *Executing a Program in Noninteractive Mode*

When you run a program in noninteractive mode, you do not enter a SAS session as you do in interactive mode. Instead of starting a SAS session, you are executing a SAS program. The first step is to enter the SAS statements in a file, just as you would for a batch job. Then, at the system prompt, you specify the SAS command followed by the complete name of the file and any system options that you want to specify.

The following example executes the SAS statements in the member TEMP in the partitioned data set your-userid.UGWRITE.TEXT in the z/OS operating environment:

```
sas input(ugwrite.text(temp))
```

Note that the INPUT operand points to the file that contains the SAS statements for a noninteractive session.

For details about how to use noninteractive mode on other operating environments, see the SAS documentation for your operating environment. Consult your SAS Site Representative for information specific to your site.

### *Browsing the Log and Output*

Log and output information either appears in your workstation display or it is sent to a file. The default action is dependent on your operating environment. In either case, you can browse the information within your display or by opening the appropriate file.

See your operating environment documentation for more information.

# Summary

## *Command*

OPTIONS
view the option settings when you use the windowing environment.

## Options

> PROC OPTIONS *options*;
> lists the current values of all SAS system options.

## System Options

> DMS | NODMS
> at invocation, specifies whether the SAS Programming windows are to be active in a SAS session.
>
> LINESIZE=*n*
> specifies the width for SAS LISTING output.
>
> VERBOSE
> at invocation, displays a listing of all options in the configuration file and on the command line.

## Statements

> DATALINES;
> signals to SAS that the data follows immediately.
>
> ENDSAS;
> causes a SAS job or session to terminate at the end of the current DATA or PROC step.
>
> OPTIONS *option*;
> changes one or more system options from the default value set at a site.
>
> RUN;
> causes the previously entered SAS step to be executed.
>
> X '*operating-environment-command*';
> is used to issue an operating environment command from within a SAS session. Operating-environment-command specifies the command. Omitting the command puts you into the operating environment's submode.

## Commands

> BYE
> ends a SAS session.
>
> ENDSAS
> ends a SAS session.
>
> EXPLORER
> invokes the Explorer window.
>
> PMENU
> turns on drop-down menus in windows.
>
> X <'*operating-environment-command*'>
> executes the operating environment command and then prompts you to take the appropriate action to return to SAS. Omitting the command puts you into the operating environment's submode.

# Learning More

### *Operating Environment Information*

For information about specific customization options and preferences, see the documentation for your operating environment.

### *Windowing Environment Commands*

For a list of all the commands that you can use in the SAS windowing environment, see SAS online Help.

**Help ⇨ SAS Help and Documentation.**

### *Documentation*

For more examples of using the SAS windowing environment, see *Getting Started with the SAS System*.

*Chapter 41*
# Using the SAS Windowing Environment

| | |
|---|---|
| **Introduction to Using the SAS Windowing Environment** | **754** |
| Purpose | 754 |
| Prerequisites | 754 |
| Operating Environment Differences | 754 |
| **Getting Organized** | **755** |
| Overview of Data Organization | 755 |
| Exploring Libraries and Library Members | 755 |
| Assigning a Library Reference | 756 |
| Managing Library Assignment Problems | 757 |
| **Finding Online Help** | **758** |
| Accessing SAS Online Help System | 758 |
| Accessing Window Help | 758 |
| **Using SAS Windowing Environment Command Types** | **758** |
| Overview of SAS Windowing Environment Command Types | 758 |
| Using Command-Line Commands | 758 |
| Using Menus | 759 |
| Using Line Commands | 760 |
| Using Function Keys | 760 |
| **Working with SAS Windows** | **761** |
| Opening Windows | 761 |
| Managing Windows | 763 |
| Scrolling Windows | 763 |
| Example: Scrolling Windows | 764 |
| Changing Colors and Highlighting in Windows | 764 |
| Finding and Changing Text | 765 |
| Cutting, Pasting, and Storing Text | 765 |
| **Working with Text** | **766** |
| The SAS Text Editor | 766 |
| Moving and Rearranging Text | 766 |
| Displaying Columns and Line Numbers | 767 |
| Making Text Uppercase and Lowercase | 768 |
| Combining and Separating Text | 770 |
| **Working with Files** | **770** |
| Ways to Find a File | 770 |
| Issuing File-Specific Commands | 772 |
| Opening Files | 772 |
| Assigning a File Shortcut | 772 |

Modifying an Existing File Shortcut . . . . . . . . . . . . . . . . . . . . . . . . . . . . . . . . . . 774
Printing Files . . . . . . . . . . . . . . . . . . . . . . . . . . . . . . . . . . . . . . . . . . . . . . . . . . . 774

**Working with SAS Programs** . . . . . . . . . . . . . . . . . . . . . . . . . . . . . . . . . . . . . . . . . . 775
Editor Window . . . . . . . . . . . . . . . . . . . . . . . . . . . . . . . . . . . . . . . . . . . . . . . . . . 775
Output Window . . . . . . . . . . . . . . . . . . . . . . . . . . . . . . . . . . . . . . . . . . . . . . . . . 777
Log Window . . . . . . . . . . . . . . . . . . . . . . . . . . . . . . . . . . . . . . . . . . . . . . . . . . . 778
Using Other Editors . . . . . . . . . . . . . . . . . . . . . . . . . . . . . . . . . . . . . . . . . . . . . . 779
Creating and Submitting a Program . . . . . . . . . . . . . . . . . . . . . . . . . . . . . . . . . . 779
Storing a Program . . . . . . . . . . . . . . . . . . . . . . . . . . . . . . . . . . . . . . . . . . . . . . . 780
Debugging a Program . . . . . . . . . . . . . . . . . . . . . . . . . . . . . . . . . . . . . . . . . . . . 780
Opening a Program . . . . . . . . . . . . . . . . . . . . . . . . . . . . . . . . . . . . . . . . . . . . . . 780
Editing a Program . . . . . . . . . . . . . . . . . . . . . . . . . . . . . . . . . . . . . . . . . . . . . . . 781
Assigning a Program to a File Shortcut . . . . . . . . . . . . . . . . . . . . . . . . . . . . . . . 781

**Working with Output** . . . . . . . . . . . . . . . . . . . . . . . . . . . . . . . . . . . . . . . . . . . . . . . . 781
Overview of Working with Output . . . . . . . . . . . . . . . . . . . . . . . . . . . . . . . . . . . 781
Setting Output Format . . . . . . . . . . . . . . . . . . . . . . . . . . . . . . . . . . . . . . . . . . . . 782
Assigning a Default Viewer to a SAS Output Type . . . . . . . . . . . . . . . . . . . . . . 783
Working with Output in the Results Window . . . . . . . . . . . . . . . . . . . . . . . . . . 784
Working with Output Templates . . . . . . . . . . . . . . . . . . . . . . . . . . . . . . . . . . . . 787
Printing Output . . . . . . . . . . . . . . . . . . . . . . . . . . . . . . . . . . . . . . . . . . . . . . . . . 790

**Summary** . . . . . . . . . . . . . . . . . . . . . . . . . . . . . . . . . . . . . . . . . . . . . . . . . . . . . . . . . . 790
Statements . . . . . . . . . . . . . . . . . . . . . . . . . . . . . . . . . . . . . . . . . . . . . . . . . . . . . 790
Windows . . . . . . . . . . . . . . . . . . . . . . . . . . . . . . . . . . . . . . . . . . . . . . . . . . . . . . 790
Commands . . . . . . . . . . . . . . . . . . . . . . . . . . . . . . . . . . . . . . . . . . . . . . . . . . . . 791
Procedures . . . . . . . . . . . . . . . . . . . . . . . . . . . . . . . . . . . . . . . . . . . . . . . . . . . . . 792

**Learning More** . . . . . . . . . . . . . . . . . . . . . . . . . . . . . . . . . . . . . . . . . . . . . . . . . . . . . 792

# Introduction to Using the SAS Windowing Environment

## Purpose

In this section, you will learn about the SAS windowing environment, including how to get organized, how to access help, and how to find and use appropriate commands.

In addition, you will learn how to use the SAS windowing environment to work with files, SAS programs, and SAS output.

## Prerequisites

Before proceeding with this section, you should understand the concepts presented in "Introduction to the SAS Environment" on page 741.

## Operating Environment Differences

Even though SAS has a different appearance for each operating environment, most of the actions that are available from the menus are the same.

One of the biggest differences between operating environments is the way that you select menu items.

If your workstation is not equipped with a mouse, then here are the keyboard equivalents to mouse actions:

*Table 41.1   Mouse Actions and Keyboard Equivalents*

| Mouse Action | Keyboard Equivalent |
| --- | --- |
| Double-click the item | Type an **s** or an **x** in the space next to the item, and then press Enter |
| Right-click the item | Type **?** in the space next to the item, and then press Enter |

Examples in this documentation show SAS windows as they appear in the Microsoft Windows environment. Usually, corresponding windows in other operating environments will yield similar results. If you do not see the drop-down menus in your operating environment, then enter the global command PMENU at a command prompt.

# Getting Organized

## Overview of Data Organization

The SAS windowing environment helps you to organize your data, and to locate and access your files easily. In this section, you learn how to use windows to do the following:

- explore libraries and library members
- assign a library reference

## Exploring Libraries and Library Members

The SAS windowing environment opens to the Explorer window by default on many hosts. You can issue the EXPLORER command to invoke this window if it does not appear by default. You can use Explorer to view the libraries that are currently available, as well as to explore their contents.

- To list available libraries, select the Libraries folder, and then select **Open** from the menu.
- To explore the contents of a library, select a specific library, and then select `Explore from Here` from the menu.
- To explore the contents of a library member, select a specific library member, and then select **Open** from the menu.

*Note:* If the Explorer Tree view is on, then you can explore libraries and library members by expanding and collapsing tree nodes. You can expand or collapse Tree nodes by selecting their expansion icons, which look like + and - symbols. You can toggle the Explorer Tree view by selecting **View** ⇨ **Show Tree** from the Explorer window.

The following display shows an expanded Tree node:

*Figure 41.1* SAS Explorer Window with Tree View On

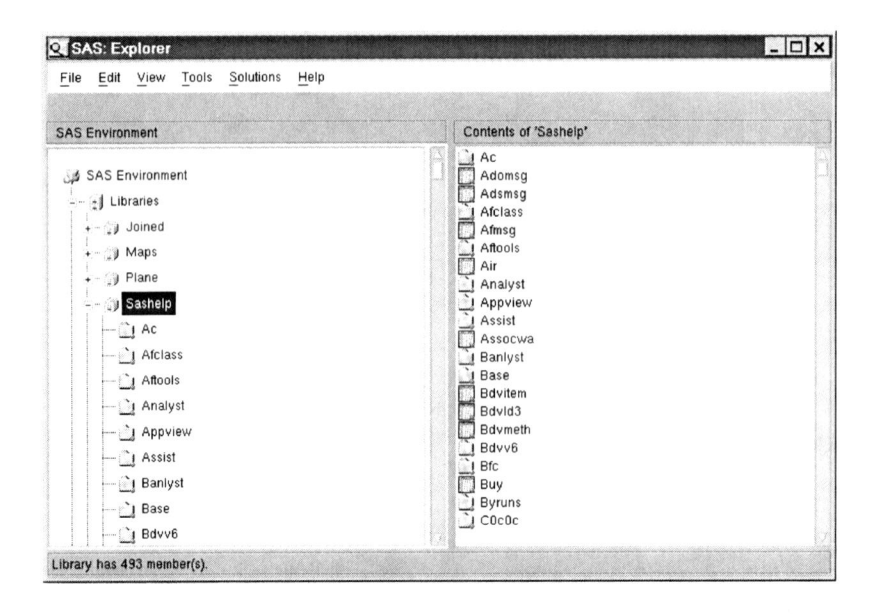

## Assigning a Library Reference

Assign a library reference before continuing your work in a SAS session, so that you can have a permanent storage location for your working SAS files:

1. From the Explorer window, select the **Libraries** folder.

2. Select **File** ⇨ **New**.

   The New Library window appears.

3. Enter a name for the library.

4. Select an engine type.

5. Enter an operating environment directory pathname or browse to select the directory.

6. Fill in any other fields as necessary for the engine, and enter any options that you want to specify.

   If you are not sure which engine to choose, then use the Default engine (which is selected automatically).

   The Default engine enables SAS to choose which engine to use for any data sets that exist at the given path of your new library. If no data sets exist, then the Base SAS engine is assigned.

7. Select **OK**. The new library will appear under the **Libraries** folder in the Explorer window.

*Note:* If you want SAS to assign the new library automatically at start-up, then select the `Enable at Startup` check box in the New Library window.

You can use the following ways to assign a library, depending on your operating environment:

Menu
   **File** ⇨ **New**

(from the Explorer window only)

Command
    DMLIBASSIGN (from any window)

Toolbar
    New Library (from any window)

## *Managing Library Assignment Problems*

If any permanent library assignment that is stored in the SAS Registry fails at start-up, then the following note appears in the SAS Log:

`NOTE: One or more library startup assignments were not restored.`

The following errors are common causes of library assignment problems:

- library dependencies are missing
- required field values for library assignment in the SAS Registry are missing
- required field values for library assignment in the SAS Registry are invalid

    For example, library names are limited to eight characters, and engine values must match actual engine names.

- encrypted password data for a library reference has changed in the SAS Registry

**CAUTION:**
  **You can correct many library assignment errors in the SAS Registry Editor.** If you are unfamiliar with library references or the SAS Registry Editor, ask for assistance. Errors can be made easily in the SAS Registry Editor, and can prevent your libraries from being assigned at start-up.

To correct a library assignment error in the SAS Registry Editor:

1. Select **Solutions** ⇨ **Accessories** ⇨ **Registry Editor** or issue the REGEDIT command.

2. Select one of the following paths, depending on your operating system, and then make modifications to keys and key values as needed:

    `CORE\OPTIONS\LIBNAMES`

    or

    `CORE\OPTIONS\LIBNAMES\CONCATENATED`

    or

    `CORE\LIBNAMES`

For example, if you determine that a key for a permanent concatenated library has been renamed to something other than a positive whole number, then you can rename that key again so that it is in compliance. Select the key, and then select `Rename` from the pop-up menu to begin the process.

## Finding Online Help

### Accessing SAS Online Help System

To access the SAS online Help, select **Help** ⇨ **SAS Help and Documentation**.

### Accessing Window Help

You can access help on an individual window in any of the following ways:

- Issue the HELP command from the command line of the window.
- Select the window's help button, if one exists.
- Select the Help icon on the toolbar.
- From the window for which you want help, select **Help** ⇨ **Using This Window**.

## Using SAS Windowing Environment Command Types

### Overview of SAS Windowing Environment Command Types

There are specific types of SAS windowing environment commands. The type of commands that you use might depend on the task that you need to complete, or on your personal preferences. These commands can be in the form of:

- command-line commands
- menu commands
- line commands (in text editing windows)
- keyboard function keys

For information about specific commands that can be issued in the SAS windowing environment, see "Working with SAS Windows" on page 761. For information about specific commands that can be used in the SAS text editor, see "Working with Text" on page 766.

### Using Command-Line Commands

Command-line commands can be entered in two places:

- on the command line (if it is turned on)
- in the Command window (if it is available)

If the command line is turned on, then you can place your cursor on the command line and type commands. You can toggle the command line on or off for a specific window by selecting **Tools** ⇨ **Options** ⇨ **Turn Command Line On** or **Tools** ⇨ **Options** ⇨ **Turn Command Line Off**.

The Command window (if it is available in your operating environment) includes a text area. You can place your cursor in this area and then issue commands.

To execute a command, type the command on the command line and then press the ENTER key, depending on which operating environment you are using. You can specify a simple one-word command, multiple commands separated by semicolons, or a command followed by an option.

For example, if you want to move from the Editor window and open both the Log and the Output windows, on the command line of the Editor window, specify

```
log; output
```

The following display shows `log; output` entered on a command line:

*Figure 41.2* Entering Commands on the Command Line

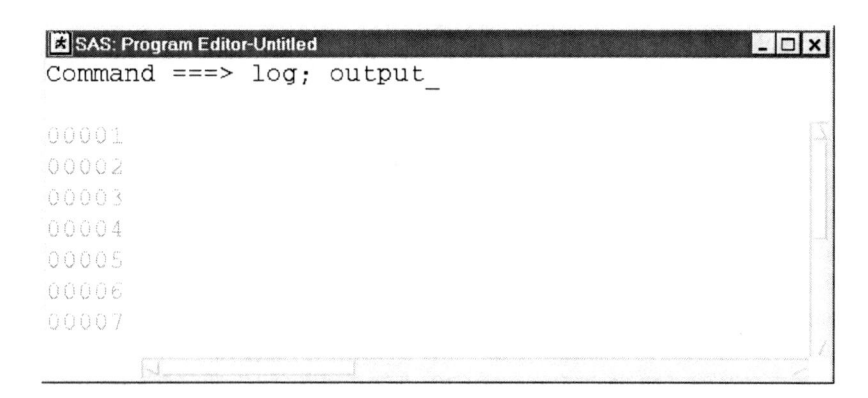

```
SAS: Program Editor-Untitled                                    _ □ x
Command ===> log; output_

00001
00002
00003
00004
00005
00006
00007
```

Next, press ENTER or RETURN to execute both commands. The Log and Output windows appear. The Output window is the active window because the command to open this window was executed last.

## Using Menus

SAS windowing environment windows can display menus instead of a command line. You can then make menu selections to do things that you would usually accomplish by entering commands.

If your operating environment does not default to using drop-down menus, then issue the PMENU command at a command line to turn on menus for all windows that support them.

You can point and click menus and menu items with a mouse to make your selections. In some operating environments, you can also make menu selections by moving your cursor over the menu items and then pressing ENTER or RETURN. Depending on the item that you select, one of three things happens:

- a command executes
- a menu appears
- a dialog box appears

In many cases, double-clicking on items and right-clicking on items will cause different menus to appear. Sometimes you might want to try one or the other when selecting an item does not give you the expected result.

In other operating environments with workstations that are not equipped with a mouse, here are the keyboard equivalents to mouse actions:

*Table 41.2* Table of Keyboard Equivalents to Mouse Actions

| Mouse Action | Keyboard Equivalent |
| --- | --- |
| Double-click | Type an **s** or an **x** in the space next to the item, and then press Enter |
| Right-click | Instead of right-clicking an item, type **?** in the space next to the item, and then press Enter |

## Using Line Commands

Line commands are one or more letters that copy, move, delete, and otherwise edit text. You can execute line commands by entering them in the numbered part of a text editing window (such as the Editor or the SAS NOTEPAD).

Although line commands are usually executed in the numbered part of the display or with function keys, they can also be executed from the command line if preceded by a colon.

*Note:* Issue the NUMBERS command to toggle line numbers on or off in text editing windows.

For more information about line commands, see "Working with Text" on page 766.

## Using Function Keys

Your keyboard includes function keys to which default values have already been assigned. You can browse or alter those values in the Keys window. To open the Keys window, select **Tools** ⇨ **Options** ⇨ **Keys** or issue the KEYS command.

To change the setting of a key in the Keys window, type the new value over the old value. The new setting takes effect immediately and is saved permanently when you execute the END command to close the Keys window.

Function keys enable you to customize your key settings to meet your needs in a particular SAS session. For example, if you might need to submit a number of programs and need to move between the Editor window and the Output window. Then each time you finish viewing your output, you must type the PGM and ZOOM commands on the command line and press ENTER or RETURN. As a shortcut, define one of your function keys to perform this action by typing the following commands over an unwanted value or where no value existed before:

```
pgm; zoom
```

Then, each time you press that function key, the commands are executed, saving you time. You can also use function keys to execute line commands. Simply precede the line command with a colon as you would if you were issuing the line command from the command line.

# Working with SAS Windows

## Opening Windows

The SAS windowing environment has numerous windows that you can use to complete tasks. You can enter commands to open windows. For more information about how to execute commands, see "Using SAS Windowing Environment Command Types" on page 758.

You can use the following commands to open a window and make it active.

*Table 41.3  Window Commands*

| Window Command | Window Name |
| --- | --- |
| AF C=library.catalog.entry.type | Build |
| DMFILEASSIGN | File Shortcut Assignment |
| DMLIBASSIGN | New Library |
| EDOP | Editor Options |
| EXPFIND | Find |
| EXPLORER | Explorer |
| FOOTNOTES | Footnotes |
| FSBROWSE | FSBrowse |
| FSEDIT | FSEdit |
| FSFORM formname | FSForm |
| FSVIEW | FSView |
| HELP | Help |
| KEYS | Keys |
| LOG | Log |
| NOTEPAD, NOTE | Notepad |
| ODSRESULTS | Results |
| ODSTEMPLATES | Templates |

| Window Command | Window Name |
|---|---|
| OPTIONS | Options |
| OUTPUT, LISTING, LIST, LST | Output |
| PROGRAM, PGM, PROG | Program Editor |
| REGEDIT | Registry Editor |
| REPOSMGR | Repository Manager |
| SASENV | Explorer (Contents Only view) |
| SETPASSWORD | Password |
| TITLES | Titles |
| VAR | Properties |

You can use window commands at any command prompt. You might find it helpful to use multiple window commands together.

For example, from the Log window, the following string of commands changes the active window, maximizes it, and changes the word paint to print:

```
pgm; zoom; change paint print
```

The following display shows that the cursor immediately moves to the Editor, which has been maximized to fill the entire display (due to the ZOOM command). The word paint has been changed to print, and the cursor rests after the last character of that text string.

*Figure 41.3* Executing a Window-Call Command in a Series

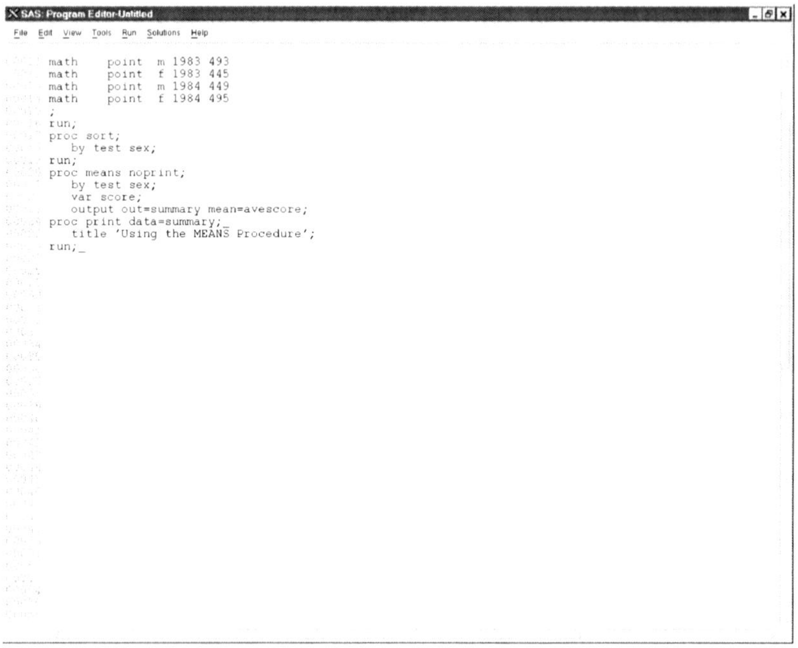

## Managing Windows

Window management commands enable you to access and use windows more efficiently. The following list includes the commands that you might use most often when managing windows:

BYE
: ends a SAS session.

CLEAR
: removes all text from an active window.

END
: closes a window. In the Editor, this command acts like the SUBMIT command.

NEXT
: moves the cursor to the next open window and makes it active.

PREVWIND
: moves the cursor to the previous open window and makes it active.

RECALL
: returns statements that are submitted from a text editor window (such as the Editor or SAS NOTEPAD) to the text editor.

ZOOM
: enlarges a window to occupy the entire display. Execute it again to return a window to its previous size. This command is not available in all operating environments.

## Scrolling Windows

Scrolling commands enable you to maneuver within text, and the command names indicate what they do. They include the following:

BACKWARD
: moves the contents of a window backward.

FORWARD
: moves the contents of a window forward.

LEFT
: moves the contents of a window to the left.

RIGHT
: moves the contents of a window to the right.

TOP
: moves the cursor to the first character of the first line in a window.

BOTTOM
: displays the last line of text.

HSCROLL, VSCROLL
: HSCROLL determines the amount that you move to the left or right when using the LEFT or RIGHT commands. VSCROLL determines the amount that you move forward or backward when using the FORWARD or BACKWARD commands.

Use the following options with the HSCROLL and VSCROLL commands as needed. HALF is the default scroll amount.

PAGE
: is the entire amount that shows in the window.

HALF
: is half the amount that shows in the window.

MAX
: is the maximum portion to the left or right or to the top or bottom that shows in the window.

n
: is n lines or columns, where n is the number that you specify.

CURSOR
: When used with HSCROLL, the cursor moves to the left or right of the display, when the LEFT or RIGHT command is executed.

    *Note:* This option is valid only in windows that allow editing.

    When used with VSCROLL, the cursor moves up and down when the FORWARD and BACKWARD command is executed.

## *Example: Scrolling Windows*

To set the automatic horizontal scrolling value to five character spaces, specify

```
hscroll 5
```

Now, when you execute the LEFT or RIGHT command, you move five character spaces in the appropriate direction. If you want to set the automatic vertical scrolling value to half a page, then specify

```
vscroll half
```

Then, when you execute the FORWARD command, half of the previous page remains on the display and half of a new page is scrolled into view.

If you need to scroll a specific number of lines forward or backward, then use the scroll amount on the FORWARD command to temporarily override the default scrolling value. You can specify scrolling values with the BACKWARD and FORWARD commands and the LEFT and RIGHT commands.

## *Changing Colors and Highlighting in Windows*

SAS gives you a simple way to customize your environment if your display supports color. You can change SAS windowing environment colors with the COLOR command. You can also change SAS code color schemes by using the SYNCONFIG command. To change windowing environment colors, simply specify the COLOR command followed by the field or window element that you want changed, and the desired color. You might also be able to change highlighting attributes, such as blinking and reverse video.

For example, to change the border of a window to red, specify

```
color border red
```

This changes the border to red.

Other available colors are blue, green, cyan, pink, yellow, white, black, magenta, gray, brown, and orange. If the color that you specify is not available, then SAS attempts to match the color to its closest counterpart.

Some color selections are valid only for certain windows.

For more information, see the online Help for the SASColor window. You can access the SASColor window with the SASCOLOR command.

You can also change the color scheme of text in the windows in which you enter code, such as the Editor window and NOTEPAD. This is useful, because you can make different elements of the SAS language appear in different colors, which makes it easier to parse code. To change the color scheme for code, use the SYNCONFIG command. The SYNCOLOR command toggles color coding off and on in these windows.

For more information about changing the color schemes for windows in which you create and edit code, see the online Help that is available when you issue the SYNCONFIG command.

## Finding and Changing Text

Often, you might want to search for a character string and change it. You can locate the character string by specifying the FIND command and then the character string. Then the cursor moves to the first occurrence of the string that you want to locate. Remember to enclose a string in quotation marks if CAPS ON is in effect.

You can change a string by specifying the CHANGE command, then a space and the current character string, and then a space and the new character string. Remember to enclose in quotation marks any string that contains an embedded blank or special characters. For both the FIND and CHANGE commands, the character string can be any length.

With both the FIND and CHANGE commands, you can specify the following options to locate or change a particular occurrence of a string:

- ALL
- FIRST
- ICASE
- LAST
- NEXT
- PREFIX
- PREV
- SUFFIX
- WORD

For details about which options you can use together, see *SAS System Options: Reference*. Note that the option ALL finds or changes all occurrences of the specified string. In the following example, all occurrences of host are changed to operating environment:

```
change host 'operating environment' all
```

To resume the search for a string that was previously specified with the FIND command, specify the RFIND command. To continue changing a string that was previously specified with the CHANGE command, specify the RCHANGE command. To find the previous occurrence of a string, specify the BFIND or FIND PREV command; you can use the PREFIX, SUFFIX, and WORD options with the BFIND command.

## Cutting, Pasting, and Storing Text

With the cut and paste facility, you can do the following:

- Identify the text that you want to manipulate.
- Store a copy of the text in a temporary storage place called a paste buffer.
- Insert text.
- List the names of all current paste buffers or delete them.

You can manipulate and store text by using the following commands:

MARK
: identifies the text that you want to cut or paste.

CUT
: removes the marked text from the display and stores it in the paste buffer.

STORE
: copies the marked text and stores it in the paste buffer.

PASTE
: inserts the text that you have stored in the paste buffer at the cursor location.

# Working with Text

## The SAS Text Editor

The SAS text editor is an editing facility that is available in the Editor and SAS NOTEPAD windows of Base SAS, SAS/FSP, and SAS/AF software. You can edit text from the command line and from any line on which code appears in an edit window.

This section provides information about commands that you can use to perform common text editing tasks by using the SAS text editor. For more information about all SAS windowing environment commands, see "Using SAS Windowing Environment Command Types" on page 758.

## Moving and Rearranging Text

Some of the basics of moving, deleting, inserting, and copying single lines of text have already been reviewed. The rules are similar for working with a block of text; simply use double letters on the beginning and ending lines that you want to edit.

For example, alphabetizing the following list requires that you move a block of text. Note the MM (move) block command on lines 5 and 6 and the B line command on line 1 of the example.

```
b 001   c signifies the line command copy
00002   d signifies the line command delete
00003   i signifies the line command insert
00004   m signifies the line command move
mm 05   a signifies the line command after
mm 06   b signifies the line command before
00007   r signifies the line command repeat
```

Press Enter to execute the changes. Here are the results:

```
00001   a signifies the line command after
00002   b signifies the line command before
00003   c signifies the line command copy
```

```
00004  d signifies the line command delete
00005  i signifies the line command insert
00006  m signifies the line command move
00007  r signifies the line command repeat
```

Mastering a few more commands greatly increases the complexity of what you can do within the text editor. Several commands enable you to justify text. Specify the JL (justify left) command to left-justify, the JR (justify right) command to right-justify, and the JC (justify center) command to center text. To justify blocks of text, use the JJL, JJR, and JJC commands. For example, if you want to center the following text,

```
00001 Study of Advertising Responses
00002 Topnotch Hotel Website
00003 Conducted by Global Information, Inc.
```

then add the JJC block command on the first and last lines and press Enter.

You can also shift text right or left the number of spaces that you choose by executing the following set of line commands:

>[n]

shifts text to the right the number of spaces that you specify; the default is one space.

<[n]

shifts text to the left the number of spaces that you specify; the default is one space.

To shift a block of text left, specify the following command on the beginning and ending line numbers of the block:

<<[n]

Specify the following command to shift a block of text to the right:

>>[n]

### Displaying Columns and Line Numbers

To display column numbers in the text editor, specify the COLS line command. This command is especially useful if you are writing an INPUT statement in column mode, as shown in the following figure:

*Figure 41.4  Executing the COLS Command*

```
Program Editor - (Untitled)
00001 data sales;
00002    input salesrep $ 1-8 sales 10-15 region $ 18-
00003    datalines;
*COLS ----|----10---|----20---|----30---|----40---|---
00005 Wilson    10498   west
00006 Lambert    9876   east
00007 Gomez      8967   west
00008 Chang     12335   north
00009 Dillon    11546   east
00010 Mohar      7945   west
00011 LaDeau    12985   north
00012 Parker     8454   south
```

To remove the COLS line command or any other pending line command, execute the RESET command on the command line. You can also execute the D (delete) line command on the line where you have specified the COLS command to achieve the same results.

The NUMBERS command numbers the data lines in the Editor and SAS NOTEPAD windows. Specify the following command to add numbers to the data lines:

```
numbers on
```

To remove the numbers, specify

```
numbers off
```

You can also use the NUMBERS command without an argument, executing the command once to turn numbers on, and again to turn them off.

## Making Text Uppercase and Lowercase

### Overview

Making text uppercase and lowercase involves two sets of commands to accomplish two types of tasks:

*Table 41.4* Changing Text Case

| Command | Action |
| --- | --- |
| CAPS | Changes the default |
| CU, CL line commands | Change the case of existing text |

### Changing the Default

To change the default case of text as you enter it, use the CAPS command. After you execute the CAPS command, the text that you enter is converted to uppercase as soon as you press ENTER or RETURN. Under some operating environments, with CAPS ON, characters that are entered or modified are translated into uppercase when you move the cursor from the line. Character strings that you specify with a FIND, RFIND, or BFIND command are interpreted as having been entered in uppercase unless you enclose the character strings in quotation marks.

For example, if you want to find the word value in the Log window, then on the command line, specify

```
find value
```

If the CAPS command has already been specified, then SAS searches for the word VALUE instead of value. You receive a message indicating that no occurrences of VALUE have been found, as shown in the following display:

*Figure 41.5* The Results of the FIND Command with CAPS ON

```
SAS: Log-Untitled                                          _ □ x
Command ===>
WARNING: No occurrences of "VALUE" found.
8    ;
9    run;
10   proc print;
11   run;

NOTE: There were 4 observations read from the datase
NOTE: PROCEDURE PRINT used:
      real time           0.26 seconds
      cpu time            0.02 seconds
```

However, specify the following command and SAS searches for the word value, and finds it:

```
find 'value'
```

Setting CAPS ON remains in effect until the end of your session or until you turn it off. You can execute the CAPS command by specifying

```
caps on
```

To discontinue the automatic uppercasing of text, specify

```
caps off
```

You can also use the CAPS command like a toggle switch, executing it once to turn the command on, and again to turn it off.

### Changing the Case of Existing Text

To uppercase or lowercase text that has already been entered, use the line commands CU and CL. Execute the CU (case upper) command to uppercase a line of text and the CL (case lower) command to lowercase a line of text.

In the following example, the CU and CL line commands each mark a line of text that will be converted to uppercase and lowercase, respectively.

```
00001 Study of Gifted Seventh Graders
cu002 Burns County Schools, North Carolina
cl003 Conducted by Educomp, Inc.
```

Press Enter to execute the commands. The lines of text are converted as follows:

```
00001 Study of Gifted Seventh Graders
00002 BURNS COUNTY SCHOOLS, NORTH CAROLINA
00003 conducted by educomp, inc.
```

For a block of text, you have two choices. First, you can execute the CCU block command to uppercase a block of text and the CCL block command to lowercase a block of text. Position the block command on both the first and last lines of text that you want to convert. Second, you can designate a number of lines that you want to uppercase or lowercase by specifying a numeric argument, as shown below:

```
cu3 1 Study of Gifted Seventh Graders
00002 Burns County Schools, North Carolina
00003 Conducted by Educomp, Inc.
```

Press Enter to execute the command. The three lines of text are converted to uppercase, as shown below:

```
00001 STUDY OF GIFTED SEVENTH GRADERS
00002 BURNS COUNTY SCHOOLS, NORTH CAROLINA
00003 CONDUCTED BY EDUCOMP, INC.
```

### Combining and Separating Text

You can combine and separate pieces of text with a number of line commands. With the TC (text connect) command, you can connect two lines of text. For example, if you want to join the following lines, then enter the TC line command as shown below. Note that the second line is deliberately started in column 2 to create a space between the last word of the first line and the first word of the second line.

```
tc001 This study was conducted by
00002  Educomp, Inc., of Annapolis, Md.
```

Press Enter to execute the command. The lines appear as shown below:

```
00001 This study was conducted by Educomp, Inc., of Annapolis, Md.
```

Conversely, the TS (text split) command shifts text after the cursor's current position to the beginning of a new line.

Remember that you can also use a function key to execute the TC line command, the TS line command, or any other line command as long as you precede it with a colon.

## Working with Files

### Ways to Find a File

#### Overview
There are a number of ways in which you can find a file or library member in the SAS windowing environment, including the following:

- using the Explorer window
- using the Find window

#### Using Explorer to Find a File
When the SAS windowing environment opens, the Explorer window also opens by default in many operating environments. You can issue the EXPLORER command to open the Explorer window if it does not open by default.

- To find a file in the Contents Only view of the Explorer window, select the **Libraries** folder or the **File Shortcuts** folder, and then select **Open** from the pop-up menu. You can continue this process with subfolders until you locate the appropriate file.

- To find a file in the Tree view of the Explorer window, use the expansion icons (+ and – icons) located in the tree until the appropriate file appears in the window.

*Note:* You might find it useful to use specific navigational tools to move through the different levels of the Explorer window:

Menu
    **View** ⇨ **Up One Level**

Command
    UPLEVEL

For more information about selecting an Explorer window view, see "Customizing the Explorer Window " on page 802.

### Using the Find Window to Find a File

The Find window enables you to search for an expression (such as a text string or a library member) that exists in a SAS library. The default search looks at everything in the library, except catalogs, but you can click the check box for the search to include the catalogs in the library as well.

The following display shows the Find window with the information for a search entered in the **Search for** and **Search in** fields:

*Figure 41.6* The Find Window

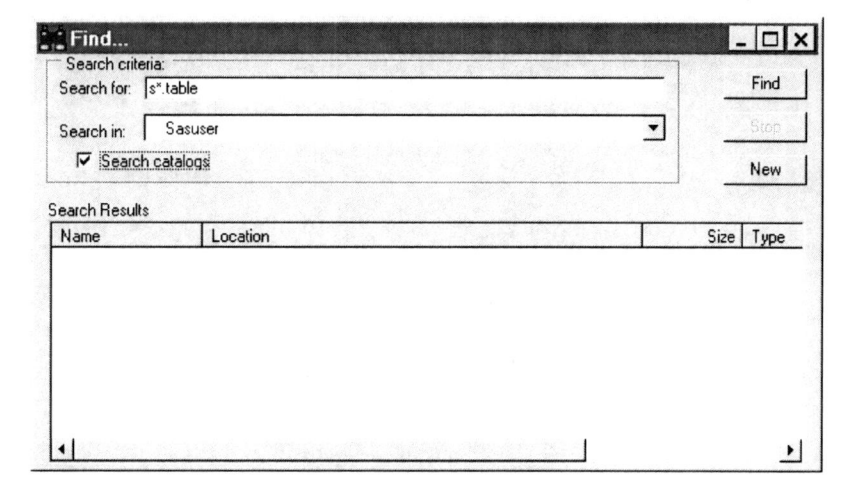

To search for a file, follow these steps:

1. Select **Tools** ⇨ **Find** from the Explorer window to open the Find window.

   Alternatively, issue the EXPFIND or EXPFIND <library-name> command. If you issue the EXPFIND command, then SASUSER is the default library. If you issue the EXPFIND WORK command, then WORK is the default library.

2. In the **Search For** field, enter the expression that you want to find. Wildcard characters are acceptable.

3. From the **Search In** drop-down list, select the library in which you want to search.

4. Click **Search Catalogs** to expand the search to include the catalogs of the library that you have selected.

   Searching catalogs can lengthen search time considerably depending on the size and number of catalogs in the library.

5. Click **Find**.

### Example: Finding Files with the Find Window

You can find TABLE files that begin with a specific letter and exist in a specific library. For a file that starts with the letter S and which exists in the SASHELP library.

1. Select **Tools** ⇨ **Find** to open the Find window.
2. Type `s*.table` in the **Search For** field.
3. Select **SASHELP** from the **Search In** drop-down list.
4. Click **Find**.

### Issuing File-Specific Commands

There are a number of commands that you can issue against a file after you find the file in the SAS windowing environment. The commands that are available are determined by the type of file with which you are working.

1. Find the file with which you want to work. For more information, see "Ways to Find a File" on page 770.
2. Select the file, and then right-click the file. A list of file-specific commands appears from which you can make a selection.

   *Operating Environment Information*
   If you are using the z/OS operating environment, then you can open a pop-up menu by entering `?` in the selection field next to an item. Alternatively, you can enter an `s` or `x` in the selection field next to an item.

### Opening Files

There are a number of ways in which you can open files in the SAS windowing environment.

To open a SAS file from Explorer:

1. Open a library and appropriate library members until you see the file that you want to open.
2. Select the file, and then select **Open** from the pop-up menu.

   Depending on the file type, you might also be able to select **Open in Editor**.

*Note:* In some cases, the pop-up menu also enables you to select **Browse in SAS Notepad**, which enables you to open a file in the SAS NOTEPAD window.

To open a file that has a file shortcut:

1. Open the `File Shortcuts` folder.
2. Select a file shortcut, and then select **Open** from the pop-up menu.

### Assigning a File Shortcut

File shortcut references provide aliases to external files (such as a .sas program file or a .dat text file). A file shortcut is the same as a file reference or fileref. In operating environments that support drag and drop functionality, you can drag file shortcuts from the Explorer window to the Editor window to display their contents.

To assign a file shortcut

1. From the Explorer window, select the `File Shortcuts` folder.
2. Select **File** ⇨ **New**.
3. In the **Name** field of the File Shortcut Assignment window, enter a name for the file shortcut.
4. Select the method or device that you want to use for the file shortcut.

   The methods or devices that are available from the **Method** drop-down list depend on your operating environment. The DISK method is the default method (if it is available for your operating environment).

5. Select the **Enable at Startup** check box if you want SAS to automatically assign the file shortcut each time SAS starts. This option is not available for all the file shortcut methods.

   If you want to stop a file shortcut from being enabled at start-up, then select the file shortcut in the SAS Explorer window, and then select **Delete** from the pop-up menu.

6. Fill in the fields of the Method Information area, including the name and location of the file for which you want to create a file shortcut. You can select **Browse** to locate the actual file. The fields that are available in this area depend on the type of method or device that you select.

   *Note:* Selecting a new method type erases any entries that you might have made in the Method Information fields.

7. Select **OK** to create the new file shortcut. The file shortcut appears in the File Shortcut folder of the SAS Explorer window.

You can use the following ways to create a file shortcut, depending on your operating environment:

Menus
> **File** ⇨ **New**
>
> while your mouse is positioned on `File Shortcuts` in the Explorer window.

Command
> DMFILEASSIGN<*file-shortuct-name*><METHOD=><AUTO=>Yes | No
>
> *file-shortcut-name*
> > specifies an existing file shortcut reference.
>
> METHOD= *method-name*
> > specifies which method to use when the File Shortcut Assignment window appears.
>
> AUTO= Yes|No
> > sets the state of the File Shortcut Assignment window's **Enable at Startup** check box when the window opens.

Pop-up
> `New File Shortcut` if you have opened the **File Shortcut** folder in the Explorer window.

Toolbar
> New (while your mouse is positioned on `File Shortcuts` in the Explorer window.)

### Modifying an Existing File Shortcut

You can modify existing file shortcut references, if needed.

From the command line:

1. Issue the following command:

   DMFILEASSIGN *file-shortcut-name*

   The File Shortcut Assignment window appears. Its fields include information that is specific to the chosen file shortcut.

2. Edit the fields of the File Shortcut Assignment window as needed.

From the SAS Explorer:

1. Right-click the **File Shortcuts** folder and select **Open**. Alternatively, you can double-click the folder to open it.

2. Right-click the file shortcut reference that you want to change, and then select **Modify**.

3. Edit the fields of the File Shortcut Assignment window as needed.

*Operating Environment Information*
   If you are using the z/OS operating environment, then you can open a pop-up menu by entering **?** in the selection field next to an item. Alternatively, you can enter an **s** or **x** in the selection field next to an item.

### Printing Files

There are a number of ways in which you can print files. Often, printing capabilities depend on the type of file with which you are working, as well as your operating environment.

Nonetheless, the following lists common ways in which you might be able to print a file.

Printing from Explorer
   Find the appropriate file in the Explorer window. Right-click over the file, and then select **Print**.

Printing from a Text Editor
   Open your file into a text editor such as the Editor or the SAS NOTEPAD. Use the text editor's printing commands.

Refer to your operating environment documentation for information about printing files.

# Working with SAS Programs

## Editor Window

### Overview
When you work with SAS programs, you typically use the SAS programming windows (the Editor, Log, and Output windows). Of these programming windows, the Editor is the window that you might use most often. It enables you to do the following:

- Enter and submit the program statements that define a SAS program.
- Edit text.
- Store your program in a file.
- Copy contents from an already-created file.
- Copy contents into another file.

*Figure 41.7* The Editor Window with Line Numbers Turned On

```
Program Editor - (Untitled)                    _ □ x
Command ===>
00001 data scores;
00002      input name $ 1-8 score 10-12;
00003 datalines;
00004
00005 Jones      85
00006 Potter     90
00007 Chang      87
00008 Gomez      89
00009 ;
00010
00011 proc print;
00012 run;
00013
00014
00015
00016
00017
00018
```

*Note:* The Program Editor window shown here includes line numbers. You might find line numbers helpful when creating or editing programs. To toggle line numbers on or off, issue the NUMBERS command.

### Command-Line Commands and the Editor
There are a number of commands that you might find useful while working on programs in the Editor. You can execute these commands from the command line:

TOP
: scrolls to the beginning of the Editor.

BOTTOM
: scrolls to the last line of text.

BACKWARD
: scrolls back toward the beginning of the text.

FORWARD
: scrolls forward toward the end of the text.

LEFT
: scrolls to the left of the window.

RIGHT
: scrolls to the right of the window.

ZOOM
: increases the size of the window. You can issue this command again to return the window to its previous size.

UNDO
: cancels the effect of the most recently submitted text editing command. Continuing to execute the UNDO command undoes previous commands, starting with the most recent and moving backward.

SUBMIT
: submits the block of statements in your current SAS windowing environment session.

RECALL
: returns to the Program Editor window the most recently submitted block of statements in your current SAS windowing environment session. Continuing to execute the RECALL command recalls previous statements, starting with the most recent and moving backward.

CLEAR
: clears a window as specified. You can clear the Editor, Log, or Output windows from another window by executing the CLEAR command with the appropriate option as shown in the following examples: `clear pgm clear log clear output`

CAPS
: converts everything that you enter to uppercase.

FIND
: searches for a specified string of characters. Enclose the string in quotation marks if it contains embedded blanks or special characters.

CHANGE
: changes a specified string of characters to another. Follow the command keyword with the first string, a space, and then the second string. The rules for embedded blanks and special characters apply. For example, you might specify `change 'operating system' platform`

This CHANGE command replaces the first occurrence of operating system with the word platform. Note that the first string must be enclosed in quotation marks because it contains an embedded blank.

*Note:* Some of the more useful command-line commands have been listed here. Almost all SAS commands are valid in the Program Editor window. For more information about other command-line commands, see "Working with SAS Windows" on page 761.

## *Line Commands and the Editor*

The left-most portion of the Program Editor window includes a numbered field. This field is where you enter line commands. These commands are denoted by one or more letters, and can move, copy, delete, justify, or insert lines.

Some common line commands include

- M — moves a line of text
- C — copies a line of text
- D — deletes a line of text
- I — inserts a line of text

When you use some line commands, you also need to specify a location. For example, if you enter an **M** in the numbered field for a line in the Editor, then you must specify where you want the line of text to be moved. You can use the **A** (after) and **B** (before) line commands to specify a location.

If you type an **A** in the numbered field for a line, then the line of text that you want to move will be placed after the line marked with an A after you press the ENTER key. If you type a **B** in the numbered field for a line, then the line of text that you want to move will be placed before the line marked with a B after you press the ENTER key.

The following examples show how to use line commands to move a line of text in the Program Editor window to a new location. To make the following lines alphabetical, place the first line after the last line. To do this, use the M and A line commands:

```
m 001 Lincoln f Wake Ligon   135
00002 Andrews f Wake Martin  140
00003 Black   m Wake Martin  149
a 004 Jones   m Wake Ligon   142
```

After pressing Enter, your Program Editor window lines appear as follows:

```
00001 Andrews f Wake Martin  140
00002 Black   m Wake Martin  149
00003 Jones   m Wake Ligon   142
00004 Lincoln f Wake Martin  135
```

There are many other line commands and combinations of line commands that you can use to edit the statements of a program in the Program Editor window. For more information, see "Working with Text" on page 766.

## Output Window

You can browse and scroll procedure output from your current SAS session with the Output window. The results of submitting a program, if it contains a PROC step that produces output, are usually displayed in the Output window.

*Figure 41.8* The Output Window Showing the Results of a Submitted Procedure

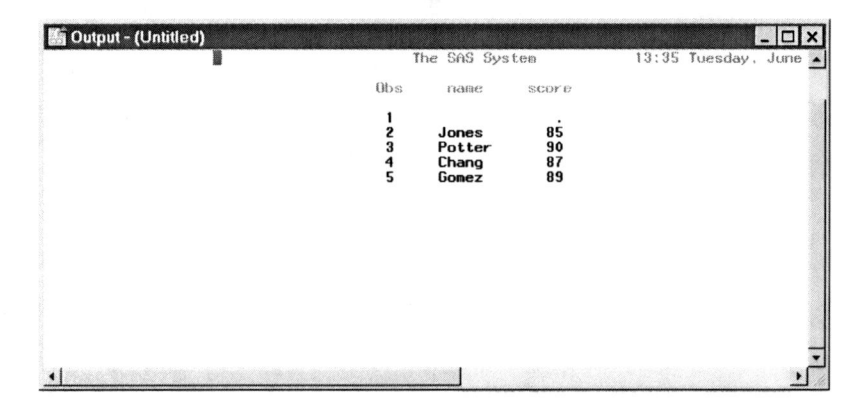

Most of the command-line commands described earlier for the Program Editor window can be used in the Output window. The CLEAR command is particularly useful in the

Output window because all output is appended to the previous output within a SAS session. If you want to avoid accumulating output, then execute the CLEAR command before you submit your next program. From any other window, you can clear the Output window by specifying

```
clear output
```

## Log Window

The Log window enables you to:

- recognize when you have made programming errors
- understand what is necessary to correct those errors
- receive feedback on the steps that you take to correct errors

*Figure 41.9* The Log Window Showing Information about a SAS Session

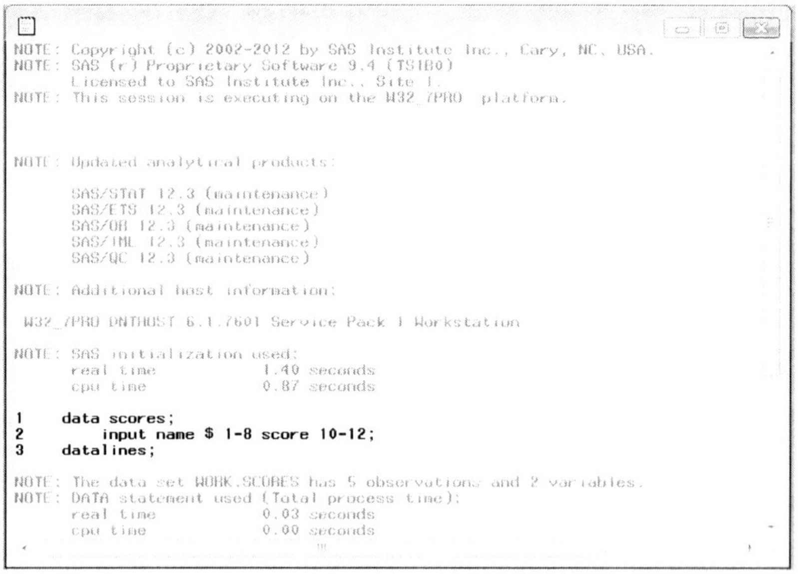

The Log window shows the SAS statements that you have submitted as well as messages from SAS concerning your program. Under most operating environments, the Log window tells you:

- when the program was executed
- the release of SAS under which the program was run
- details about the computer installation and its site number
- the number of observations and variables for a given output data set
- the computer resources that each step used

You can use command-line commands in the Log window, just as you can in the Editor and Output windows. For more information, see " Editor Window" on page 775.

## Using Other Editors

### NOTEPAD Window

Although the Editor was designed for writing SAS programs, you can also use the NOTEPAD window to create and edit SAS programs. The NOTEPAD is a text editor that you can use to create, edit, save, and submit SAS programs. You might find NOTEPAD useful as a separate place to work on code. To open NOTEPAD, issue the NOTEPAD or NOTES command.

*Figure 41.10* The SAS NOTEPAD Window with Line Numbers Turned On

```
NOTEPAD <NOTEPAD.SOURCE>
00001 data scores;
00002     input name $ 1-8 score 10-12;
00003 datalines;
00004
00005 Jones     85
00006 Potter    90
00007 Chang     87
00008 Gomez     89
00009 ;
00010
00011 proc print;
00012 run;
00013
00014
00015
```

*Note:* The NOTEPAD window shown here includes line numbers. You might find line numbers helpful when you create or edit programs. To toggle line numbers on or off in NOTEPAD, issue the NUMBERS command.

If you open multiple NOTEPADS, then you can cut, copy, and paste text between NOTEPAD windows and the Program Editor window, multiple SAS sessions, and other applications.

*Note:* To submit a program from NOTEPAD, you must either select **Run** ⇨ **Submit** or issue the NOTESUBMIT command.

*Note:* The program information that is presented in this documentation uses the Editor windows as the default editor.

## Creating and Submitting a Program

To create and submit a SAS program:

1. Enter the text of your program in the Editor.

2. Type `submit` on the command line, and then press Enter.

   You can also use the function key, menu command, or toolbar item that is assigned to submit programs in your environment.

   *Note:* If you are submitting a program from the NOTEPAD window, then you must use the NOTESUBMIT command instead of the SUBMIT command.

### Storing a Program

To store a program:

1. In the Program Editor window, create or edit a program.

2. On the command line, issue the FILE command followed by a fileref or an actual filename. If you use an actual filename, then enclose it in quotation marks.

The FILE command does not clear the contents of the Program Editor window. You can store one copy of a program and then continue working in the Program Editor window.

If you try to store a program with a fileref or filename that already exists, then SAS displays a dialog box. The dialog box enables you to choose to

- overwrite the contents of the existing file with the new file
- append the new file to the existing file
- cancel the FILE command

Often you will want to replace a file with an updated version. To suppress the dialog box, add the REPLACE option to the FILE command after the fileref or complete filename. To add the text in the Program Editor window to the end of an existing file, specify the APPEND option with the FILE command after the fileref or complete filename.

*Note:* You can also store a program as a SAS object or as a file that is specific to your operating environment. After you have created or edited a program, select **File ⇨ Save As Object** or **File ⇨ Save As**, respectively.

### Debugging a Program

You or someone in your organization might be able to help debug a program with the information that appears in the Log window after a program is submitted. If you are having problems with your program, save the contents of the Log window to an external file, if you need to study it after your SAS session has ended.

To save the contents of the Log window to an external file:

1. Open the Log window if it is not already open.

2. From the command line, execute the FILE command followed by a fileref or an actual filename. If you use a filename, then enclose the name in quotation marks.

The FILE command stores a copy of the information in the Log window without removing what is currently displayed. If you specify the name of an existing fileref or file, then a dialog box appears and offers you three choices: overwriting the contents of the existing file with the new file, appending the new file to the existing file, or canceling the command.

### Opening a Program

There is more than one way to open a SAS program. Two of the most popular methods are listed in this section.

To open a SAS program from the Program Editor window:

1. Select **File ⇨ Open**.

2. Use the Open window to locate the appropriate SAS program file.

To open a SAS program with commands:

1. Open the Program Editor window if it is not already open.

2. On the command line, specify the INCLUDE command followed by an assigned fileref or an actual filename. Remember to enclose an actual filename in single or double quotation marks.

   By default, a program is appended to the end of any existing program statements.

*Note:* If program statements already exist in the Editor, then you can determine where your program is appended by using the B (before) or A (after) line commands. For more information about line commands, see "Using Line Commands" on page 760.

If you want to replace the text that is already in the Program Editor window with the program that you open, then specify the REPLACE option with the INCLUDE command after the fileref or filename.

### Editing a Program

To edit a program:

1. Open an existing program in the Program Editor window.

2. Edit existing program statements or append new statements to the program.

   Use command-line commands and line commands as needed.

3. Store the program.

### Assigning a Program to a File Shortcut

You can assign a program to a file shortcut to make it easier to find and work with the file in the future. For more information about file shortcuts, see "Assigning a File Shortcut" on page 772.

## Working with Output

### Overview of Working with Output

You can manage your SAS procedure output with the SAS Output Delivery System (ODS). Procedures that fully support ODS can do the following:

- combine the raw data that they produce with one or more table definitions to produce one or more output objects that contain formatted results

- store a link to each output object in the Results folder in the Results window

- can generate various types of file output, such as HTML, LISTING, and in some cases, SAS/GRAPH output

- can generate output data sets from procedure output

- provide a way for you to customize the procedure output by creating table definitions that you can use whenever you run the procedure

The SAS windowing environment enables you to use many features of ODS through the Results, Templates, Preferences, and SAS Registry Editor windows. The Results window provides pointers to the procedure output that is produced by SAS. The Templates window provides a way to manage all the table, column heading, and style templates that can be associated with procedure output.

Finally, the Preferences window and the SAS Registry Editor can be used to set the types of procedure output that you want SAS to produce.

This section details only those portions of ODS that are related to the SAS windowing environment. For more information about ODS, see Chapter 24, "Directing SAS Output and the SAS Log," on page 391 and *SAS Output Delivery System: User's Guide*.

## Setting Output Format

### Overview

Depending on your operating environment, SAS output can be produced in one or more formats (or types). Listing output is the default type. Other output types include HTML, Output Data Sets, and PostScript. Pointers to procedure output appear in the Results window.

To set your output type, use either the Preferences window (if available in your operating environment), the SAS Registry Editor, or both.

### Setting Output Type with the Preferences Window

If your operating environment supports the Preferences window, you can set output type as follows:

1. Select **Tools** ⇨ **Options** ⇨ **Preferences** or issue the DLGPREF command to open the Preferences window.

2. Select the **Results** tab.

3. Select or deselect the check boxes that match the output types that you want to produce.

    If you choose to produce HTML output, then you can further define the output by selecting:

    - an HTML style

      Click the **Style** box and highlight a style. Styles among other things, define output colors and fonts.

    - the folder to which the output is saved

      Select **Use WORK folder** to save HTML output only for the duration of the current session. Your output is deleted when your current SAS session ends.

      Enter a path in the **Folder** text box to save HTML output to a folder that is not deleted when your SAS session ends.

    - the **View Results as they are Generated** check box

      If selected, then each time HTML output is produced, your browser automatically opens and loads the output.

### Setting Output Type with the SAS Registry Editor

To set output type with the SAS Registry Editor:

1. Select **Solutions** ⇨ **Accessories** ⇨ **Registry Editor** or issue the REGEDIT command to open the SAS Registry Editor.

2. From the tree on the left side, expand the ODS folder.

3. Expand the Preferences folder.

4. Select the appropriate output type.

5. On the right side, select the Value key, and then select `Modify` from the pop-up menu.

6. In the dialog box that appears, edit the Value Data field as needed.

   If this field is set to 1, then the output type is produced. If this field is set to 0, then the output type is not produced.

### Assigning a Default Viewer to a SAS Output Type

When you produce output in SAS, output pointers appear in the Results window. You can assign a default viewer for each of the types of output that you produce. After a default viewer is assigned, you can double-click an output pointer in the Results window to open output in its default viewer. For example, double-clicking on a PostScript output pointer could open Ghostview with your PostScript output loaded.

*Operating Environment Information*

   In the Windows operating environment, default viewers are established automatically with information from your Windows Registry.

To assign a default viewer to a SAS Output Type, follow these steps:

1. From the Explorer window, select **Tools** ⇨ **Options** ⇨ **Explorer**.

2. Select **Host Files** from the drop-down menu at the top of the Explorer Options window.

3. Scroll through the registered file types until you find the file type with which you want to work.

4. Select the appropriate file type, and then select **Edit**.

5. Select **Add**, and then enter an action name and action command for the file type in the Edit Action window.

   For example, add the following action name and action command to set Ghostview as the default viewer for PostScript file types:

   Action Name
      *&Edit*

   Action Command
      *x ghostview '%s' &*

6. Select **OK** from the Edit Action window.

7. Select the action that you just specified, and then select **Set Default**.

*Operating Environment Information*

   In the Windows operating environment, default viewers are established automatically with information from your Windows Registry.

### Working with Output in the Results Window

#### Overview

The Results window provides pointers to the procedure or DATA step output that SAS produces. This window might open by default when you start a SAS session. You can also open the Results window by selecting **View** ⇨ **Results** or by issuing the ODSRESULTS command.

*Figure 41.11* The Results Window in Tree View

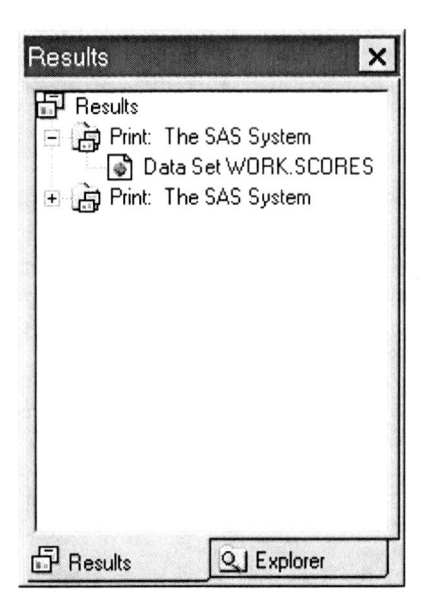

You can use the Results window to do the following:

- Navigate pointers to output.
- Delete results pointers.
- Rename results pointers.
- Save listing output to other formats.
- Quickly view the first output pointer item.
- View results properties.

#### Customizing the Results Window View

You can have the Results window display in one of three views:

- Tree
- Contents Only
- Explorer

In Tree view (the default), only a navigational tree is present. In Contents Only view, the tree is turned off, and contents appear as folders. In Explorer view, the Results window appears with two panes: one for the tree and one for the contents.

To toggle the Tree view pane, issue the TREE command from the Results window. To toggle the Contents pane, issue the CHILD command from the Results window. You can also select commands from the **View** menu of the Results window to perform the same actions, such as **Show Tree**, **Show Contents**, and others.

*Note:* By default, output pointers are listed by label rather than by name in the Tree pane. Labels are typically more descriptive than output names. You can use the following SAS system option to change this setting: LABEL.

### *Using Results Pointers to Navigate Output*

When SAS runs a procedure or a DATA step, pointers to the output are placed in the Results window. For information about using the pointers in the Results window, see "Navigating the Results Window in Tree View" on page 785, "Navigating the Results Window in Contents Only View" on page 785, or "Navigating the Results Window in Explorer View" on page 786.

### *Navigating the Results Window in Tree View*

In Tree view, output pointers appear in a procedural hierarchy. To work with your SAS output:

1. Locate the folder that matches the procedure output that you want to view.

2. Use the expansion icons (+ or – icons) next to the folder to open or hide its contents.

   You can also:

   - Double-click a folder to make it expand or collapse.

   - Select a folder, and then select **Open** from the pop-up menu.

3. When you locate the appropriate pointer, double-click the pointer or select the pointer and then select **Open** from the pop-up menu.

   The appropriate output appears.

   *Operating Environment Information*
   If you are using the z/OS operating environment, then you can open a pop-up menu by entering **?** in the selection field next to an item. Alternatively, you can enter an **s** or **x** in the selection field next to an item.

You can also use the following ways to navigate in the Tree view:

Menu
    **View** ⇨ **Up One Level**

Command
    UPLEVEL

Toolbar
    Up One Level icon

Key
    Depending on your operating environment, you might also use arrow and backspace keys to navigate.

### *Navigating the Results Window in Contents Only View*

In Contents Only view, output pointers appear in a procedural hierarchy, beginning with the top level of the hierarchy. You can drill down or roll up within the hierarchy to find the appropriate output.

When you open a folder, the current window contents are replaced with the contents of the selected folder. To work with your SAS output:

1. Locate the folder that matches the procedure output that you want to view.

2. Select the folder, and then select **Open** from the pop-up menu.

   You can also double-click a folder to open it.

3. When you locate the appropriate pointer, double-click the pointer or select the pointer, and then select **Open** from the pop-up menu.

   The appropriate output appears.

   *Operating Environment Information*
   If you are using the z/OS operating environment, then you can open a pop-up menu by entering ? in the selection field next to an item. Alternatively, you can enter an **s** or **x** in the selection field next to an item.

### Navigating the Results Window in Explorer View

In Explorer view, two window panes exist. The left pane includes a hierarchical view (the Tree view) of the procedure output that you can view. The right pane shows the contents (the Contents view) of the item that is currently in focus.

### Deleting Results Pointers

You can delete results pointers by deleting the procedure folder in which the pointers exist. When you delete a procedure folder in the Results window, any output pointer that exists in that folder is removed.

*Note:* When you delete a procedure folder that contains a listing output pointer, the actual listing output is removed from the Output window. If other output pointers exist in the folder (such as HTML), then only the pointer is removed; the actual output remains available.

To delete procedure output:

1. In the Results window, select the procedure folder that matches the procedure that you want to delete.

2. Select **Delete** from the pop-up menu.

3. Select **Yes** to confirm the deletion.

   *TIP* You can also delete output pointers by selecting the procedure folder that you want to delete, and then selecting **Edit** ⇨ **Delete**.

### Renaming Results Pointers

To rename results pointers:

1. Select the pointer that you want to rename.

2. Select **Rename** from the pop-up menu.

3. Type in a new name and/or a description, and then select **OK**.

   *TIP* You can also rename results pointers by selecting the pointer that you want to rename, and then selecting **Edit** ⇨ **Rename**.

### Saving Listing Output to Other Formats

To save listing output to a file from the Results window:

1. Expand the Results window tree until you find the appropriate listing output pointer.

2. Select the listing output pointer, and then select **Save As** from the pop-up menu.

To save listing output to a file from the Output window:

1. Access the Output window.

2. On the command line, specify the FILE command followed by a fileref or an actual filename. If you use a filename, then enclose the filename in quotation marks.

*Note:* The FILE command stores a copy of the information in the Output window without removing what is currently displayed.

To save listing output as a catalog object:

1. Expand the Results window tree until you find the appropriate listing output item.

2. Select the listing output item, and then select **Save As Object** from the pop-up menu.

### Viewing the First Output Pointer Item

To view the first output pointer item:

1. Select the appropriate results pointer.

2. Select **View** from the pop-up menu.

   The first output pointer item listed for the results pointer that you selected appears. For example, if you produced listing and HTML output for a procedure and the listing output was created first, then the listing output would appear.

### Viewing Results Properties

You can view the properties of a Results window folder, an output pointer, or an output pointer item (such as listing or HTML output).

1. In the Results window, select the appropriate folder, output pointer, or output pointer item.

2. Select **Properties** from the pop-up menu.

## Working with Output Templates

### Overview of Working with Output Templates

Templates contain descriptive information that enables the Output Delivery System (ODS) to determine the desired layout of a procedure's results.

The Templates window provides a way to manage all the templates that are currently available to SAS. Specifically, you can use the Templates window to do the following:

- Browse PROC TEMPLATE source code.

- Edit PROC TEMPLATE source code.

- View template properties.

**Figure 41.12** The Templates Window in Explorer View

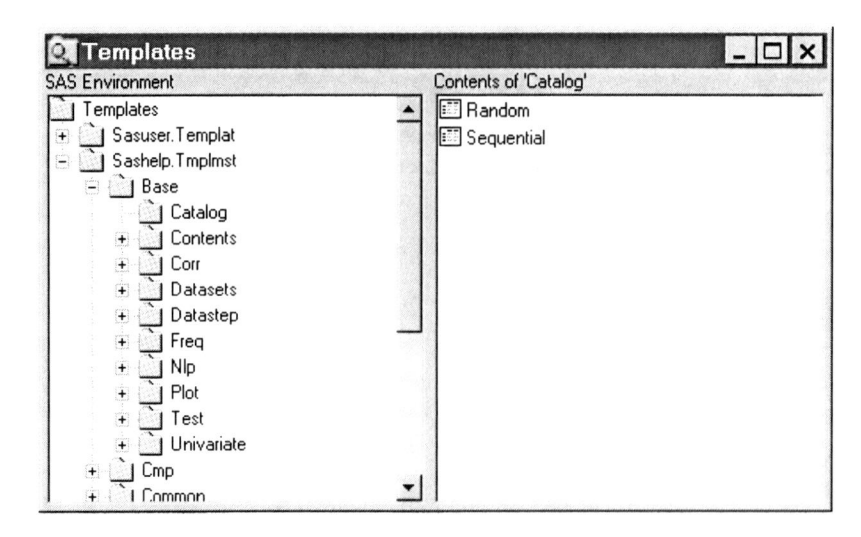

You can open the Templates window by selecting **View ⇨ Templates** from the Results window, or by issuing the ODSTEMPLATES command.

You can create or modify templates with PROC TEMPLATE.

*Note:* Templates that are supplied by SAS are stored in SASHELP. Templates that are created with PROC TEMPLATE are stored in SASUSER or whatever library you specify in the ODS PATH statement.

### Customizing the Templates Window View

The Templates window appears in one of three views:

- Explorer
- Tree
- Contents Only

In Explorer view (the default), the Templates window appears with two panes: one for the tree and one for the contents. In Tree view, only a navigational tree is present. In Contents Only view, the tree is turned off.

To toggle the Contents pane, issue the CHILD command from the Templates window. To toggle the Tree pane, issue the TREE command from the Templates window.

For more information, see "Navigating the Results Window in Tree View" on page 785, "Navigating the Results Window in Contents Only View" on page 785, or "Navigating the Results Window in Explorer View" on page 786.

### Navigating the Templates Window in Explorer View

In Explorer view, two window panes exist. The left pane includes a hierarchical view (the Tree view) of the templates that you can view. The right pane shows the contents (the Contents view) of the template currently in focus.

You can open additional template windows from the Explorer view by selecting a template, and then selecting **Explore from Here** from the pop-up menu.

### Navigating the Templates Window in Tree View

In Tree view, templates appear in a hierarchy. To work with a template:

1. Locate the folder that includes the template that you want to view.

2. Use the expansion icons (+ or – icons) next to the folder to open or hide its contents.

   You can also do the following:
   - Double-click a folder to make it expand or collapse.
   - Select a folder, and then select **Open** from the pop-up menu.

3. Double-click the template that you want to see, or select the template, and then select **Open** from the pop-up menu.

   The template code appears in a browser window.

   *Operating Environment Information*
   If you are using the z/OS operating environments, then you can open a pop-up menu by entering **?** in the selection field next to an item. Alternatively, you can double-click by entering an **s** or **x** in the selection field next to an item.

### *Navigating the Templates Window in Contents Only View*

In Contents Only view, templates appear as folders. When you open a folder, the current window contents are replaced with the contents of the selected folder. To work with your templates in this view:

1. Locate the folder that includes the template that you want to view.

2. Select the folder, and then select **Open** from the pop-up menu.

   You can also double-click on a folder to open it.

3. Double-click on the template that you want to see, or select the template, and then select **Open** from the pop-up menu.

   The template code appears in a browser window.

   *Operating Environment Information*
   If you are using the z/OS operating environments, then you can open a pop-up menu by entering **?** in the selection field next to an item. Alternatively, you can double-click by entering an **s** or **x** in the selection field next to an item.

### *Browsing PROC TEMPLATE Source Code*

To browse the PROC TEMPLATE source code:

1. Locate the appropriate template in the Templates window.

2. Select the template, and then select **Open** from the pop-up menu.

   Template code appears in a browser window.

### *Editing PROC TEMPLATE Source Code*

To edit the PROC TEMPLATE source code:

1. Locate the appropriate template in the Templates window.

2. Select the template, and then select **Edit** from the pop-up menu. Template code appears in an editor window.

3. Modify the template code as needed.

4. Select **Run** ⇨ **Submit** to submit your modified template code.

*Note:* If syntax errors occur when the code for an edited template is submitted, then the errors appear in the Log window.

*Note:* Additional information for PROC TEMPLATE is available in *SAS Output Delivery System: User's Guide*.

### Viewing Template Properties

To view template properties:

1. Locate the appropriate template in the Templates window.

2. Select the template, and then select **Properties** from the pop-up menu.

The Properties dialog box lists the type, path, size, description, and modification date for the template. You can also view this information by selecting **View** ⇨ **Details** when the Templates window is active.

## Printing Output

The method that you use to print output depends on the type of output that you produce, as well as your operating environment. SAS windowing environment windows have menus with print options that enable you to print the contents of that particular window. This feature varies from operating system to operating system, but is available in all operating environments.

If you produce HTML output, then you can open the output in a web browser, and then print the output from the web browser with the web browser's printing command.

For more information about printing, refer to your SAS operating environment companion documentation and your operating environment documentation.

# Summary

## Statements

ODS PATH location(s)
    specifies which locations to search for definitions that were created by PROC TEMPLATE, as well as the order in which to search for them.

    <libname.>item-store <READ | UPDATE | WRITE>

item-store
    identifies an item store that contains style templates, table definitions, or both.

## Windows

File Shortcut Assignment
    enables you to create or edit file shortcut references. To open this window, issue the DMFILEASSIGN command.

Find
    enables you to search for an expression that exists in a SAS library. To open this window, select **Tools** ⇨ **Find** from Explorer or issue the EXPFIND command.

Log
    enables you to review information about the programs that you have run. To open this window, select **View** ⇨ **Log** or issue the LOG command.

Output
: enables you to see listing output. To open this window, select **View** ⇨ **Output** or issue the OUTPUT command.

Editor
: enables you to enter, edit, submit, and save SAS program statements. To open this window, select **View** ⇨ **Editor** or issue the PGM command.

Results
: provides pointers to the procedure output that you produce with SAS. To open this window, select **View** ⇨ **Results** or issue the ODSRESULTS command.

SAS NOTEPAD
: enables you to enter, edit, submit, and save SAS program statements. To open this window, issue the NOTEPAD or NOTES command.

SAS Registry Editor
: enables you to edit the SAS Registry and to customize aspects of the SAS windowing environment. To access this window, issue the REGEDIT command.

Templates
: provides a way to manage the output templates that are currently available. To access this window, select **View** ⇨ **Templates** from within the Results window.

## *Commands*

AUTOEXPAND
: automatically expands the tree hierarchy when you select a tree node or when procedure output is produced.

AUTOSYNC
: enables you to automatically navigate to the first available output in the Output window by means of a single click.

CHILD
: toggles the Contents pane on and off.

CLEAR
: removes all the SAS output pointers.

DELETESELS
: removes the item currently in focus.

    *Note:* If the output pointer is associated with listing output, then the listing output is also removed.

DESELECT_ALL
: deselects any items that are selected while the Contents pane is viewable.

DETAILS
: toggles the item details about and off while the Contents pane is viewable.

DMOPTLOAD
: recalls system option settings saved by DMOPTSAVE.

DMOPTSAVE
: saves all system option settings for recall in later SAS sessions.

FIND
: searches for a match to the string that you provide.

LARGEVIEW
: displays large icons (on some operating environments) while the Contents pane is viewable.

PMENU
: turns on menus in windows.

PRINT
: writes the desired SAS listing output.

REFRESH
: refreshes the window's contents.

RENAMESELS
: enables you to rename the output pointer that currently has focus.

SELECT_ALL
: selects all items while the Contents pane is viewable.

SMALLVIEW
: displays small icons (on some operating environments) in a horizontal fashion while the Contents pane is viewable.

TREE
: toggles the Tree view (hierarchical view) on and off.

UPLEVEL
: moves focus up one level in the hierarchy.

## *Procedures*

Use PROC TEMPLATE to set template information.

# Learning More

To learn more about SAS language elements, see
*SAS Language Elements by Name, Product, and Category* on `support.sas.com`.

To learn more about printing and the SAS Output Delivery System, see
*SAS Output Delivery System: User's Guide*.

To find examples that will help you get started, see
*Getting Started with the SAS System*.

# Chapter 42
# Customizing the SAS Environment

**Introduction to Customizing the SAS Environment** .......................... 794
   Purpose .......................................................... 794
   Prerequisites ..................................................... 794
   Operating Environment Differences ................................... 794

**Customizing Your Current Session** ....................................... 795
   Ways to Customize ................................................ 795
   Customizing SAS Sessions and Programs at Start-up ..................... 795
   Customizing with SAS System Options ................................. 796

**Customizing Session-to-Session Settings** ................................. 798
   Overview of Customizing Session-to-Session Settings .................... 798
   Customizing SAS Sessions and Applications with the SAS Registry Editor ..... 798
   Customizing SAS Sessions with the Preferences Window ................. 802
   Saving System Option Settings with the DMOPTSAVE and
     DMOPTLOAD Commands ....................................... 802

**Customizing the SAS Windowing Environment** ........................... 802
   Customizing the Explorer Window .................................... 802
   Customizing an Editor ............................................. 806
   Customizing Fonts ................................................ 806
   Customizing Colors ............................................... 807
   Setting SAS Windowing Environment Preferences ...................... 807

**Summary** .......................................................... 807
   Commands ...................................................... 807
   Procedures ...................................................... 807
   Statements ...................................................... 808
   System Options .................................................. 808
   Windows ........................................................ 808

**Learning More** ..................................................... 809

# Introduction to Customizing the SAS Environment

## Purpose

In this section, you will learn how to make the following types of customizations in SAS:

- those that remain in effect for the current session only
- those that remain in effect from session to session
- those that you can apply to the SAS windowing environment, which is the default SAS environment

## Prerequisites

To use this section, you should be familiar with the SAS windowing environment. For more information about the SAS windowing environment, see "Introduction to Using the SAS Windowing Environment" on page 754.

## Operating Environment Differences

Even though SAS has a different appearance for each operating environment, most of the actions that are available from the menus are the same.

One of the biggest differences between operating environments is the way that you select menu items.

If your workstation is not equipped with a mouse, then here are the keyboard equivalents to mouse actions:

*Table 42.1* Mouse Actions and Keyboard Equivalents

| Mouse Action | Keyboard Equivalent |
| --- | --- |
| Double-click the item | Type an **s** or an **x** in the space next to the item, and then press Enter |
| Right-click the item | Type **?** in the space next to the item, and then press Enter |

Examples in this documentation show SAS windows as they appear in the Microsoft Windows environment. Usually, corresponding windows in other operating environments will yield similar results. If you do not see the drop-down menus in your operating environment, then enter the global command PMENU at a command prompt.

# Customizing Your Current Session

## Ways to Customize

As you become familiar with SAS, you will probably develop preferences for how you want SAS configured. Many options are available to you to make SAS conform to your preferred working style. Some of the things that you can change are the following:

- window color and font attributes
- library and file shortcuts
- output appearance
- file-handling capabilities
- the use of system variables

You can customize your current SAS session in the following ways:

- at the start-up of a SAS session or program
- through SAS system options
- with drop-down menu options

## Customizing SAS Sessions and Programs at Start-up

### Setting Invocation-Only Options Automatically

You can specify some system options only when you invoke SAS. These system options affect the following:

- the way SAS interacts with your operating system
- the hardware that you are using
- the way in which your session or program is configured

*Note:* There are other system options that you can specify at any time. For more information, see "Customizing with SAS System Options" on page 796.

Usually, any invocation-only options are set by default when SAS is installed at your site. However, you can specify invocation-only options on the command line each time you invoke SAS.

To avoid having to specify options that you use every time you run SAS, set the options in a configuration file. Each time you invoke SAS, SAS looks for that file and uses the customized settings that it contains. Be sure to examine the default configuration file before creating your own.

*Note:* If you specify options both in the configuration file and in the SAS command, then the options are concatenated. If you specify an option in the SAS command that also appears in the configuration file, then the setting from the SAS command overrides the setting in the configuration file.

To display the current settings for all options that are listed in the configuration file and on your command line as you invoke the system, use the VERBOSE system option in the SAS command.

### Executing SAS Statements Automatically

Just as you can set SAS system options automatically when you invoke SAS, you can also execute statements automatically when you invoke SAS by creating a special autoexec file. Each time you invoke SAS, it looks for this special file and executes any of the statements that it contains.

You can save time by using this file to execute statements that you use routinely. For example, you might add the following statements:

- OPTIONS statements that include system options that you use regularly
- FILENAME and LIBNAME statements to define the file shortcuts and libraries that you use regularly

## Customizing with SAS System Options

### Using the OPTIONS Statement and the Options Window

SAS system options determine global SAS settings. For example, the global options can affect the following:

- how your SAS output appears
- how files are handled by SAS
- how observations from SAS data sets are processed
- how system variables are used

The previous section discusses some invocation-only options that must be set at start-up. However, there are many system options that can be set at any time. These system options can be set in an OPTIONS statement as well as in the Options window.

It is important to note that system option settings remain in effect until you change them again, or until your current session ends.

There are several ways to view your system option settings. The two most common methods are the following:

- the Options window (enter OPTIONS at a command line)
- the OPTIONS procedure

To obtain a complete list of system option settings using the OPTIONS procedure, submit the following statements:

```
proc options;
run;
```

The Options window groups options by function. The left side of the window includes a tree that lists the available option groups. You can expand option groups to see subgroups.

*z/OS Specifics*

z/OS users can expand groups and subgroups by using the mouse or by typing an **s** or an **x** before the group or subgroup name. When you select a subgroup, the individual options of that subgroup appear on the right side of the window.

*Figure 42.1* SAS Options Window

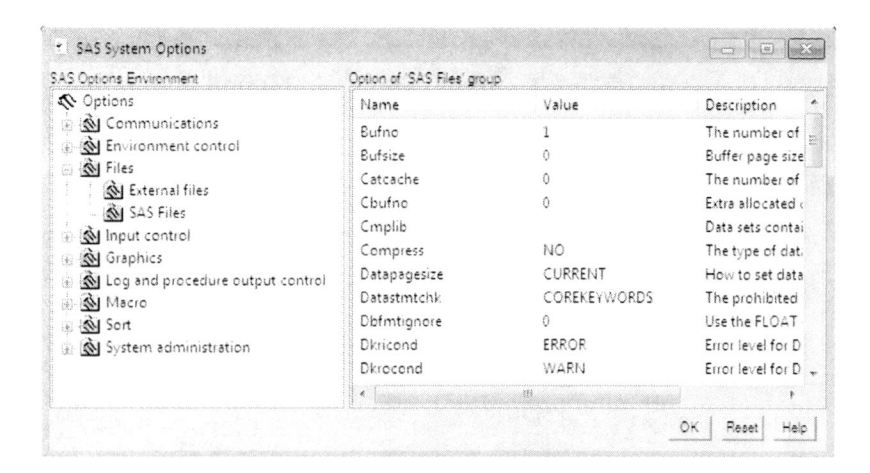

To open the Options window, do one of the following tasks:

- Issue the OPTIONS command.

- Select **Tools** ⇨ **Options** ⇨ **System**.

The options in each group or subgroup are listed alphabetically, followed by options that are specific to your operating environment (which are also listed alphabetically).

## Finding Options in the SAS Options Window

You can find options in a number of ways:

- Expand the option groups and subgroups on the left side of the window until the appropriate option appears on the right side of the window.

- Select an option group or subgroup, and then select **Find Option** from the pop-up menu. In the Find Option window, enter the name of the option that you want to locate, and then select **OK**.

## Setting Options in the SAS Options Window

1. In the Options window, find the option that you want to set.

2. Select the option from the right side of the Options window.

3. Select **Modify Value** or **Set to Default** from the pop-up menu. z/OS users can type an **s** or an **x** before the option name to access the pop-up menu.

   - If you choose **Modify Value**, then a dialog box appears that enables you to edit the option value.

   - If you choose **Set to Default**, then the option value is reset to the default SAS System value.

4. Select **OK** to save your changes. Select **Reset** to return all edited options to their previous values.

*Note:* If all the items on the pop-up menu are grayed out (that is, unavailable), then the options are invocation-only options and can be set only when a SAS session is started.

# Customizing Session-to-Session Settings

## Overview of Customizing Session-to-Session Settings

The previous section discusses making customizations that stay in effect for the duration of the current SAS session only. This section provides information about making customizations that remain from SAS session to SAS session.

You can make customizations that remain from session to session by using one of the following windows:

- SAS Registry Editor
- Preferences window
- Options window

## Customizing SAS Sessions and Applications with the SAS Registry Editor

### Understanding the SAS Registry

The SAS Registry stores information about specific SAS sessions and applications. Unlike system options, customizations to the SAS Registry remain in effect for more than one SAS session. You can make SAS Registry customizations by using either PROC REGISTRY or the SAS Registry Editor.

This section shows you how to use the SAS Registry Editor, which is a graphical alternative to PROC REGISTRY. For more information about PROC REGISTRY, see the *Base SAS Procedures Guide*.

CAUTION:
> **Changes to SAS Registry should be well planned.** In many cases, it is appropriate to have a designated person in charge of SAS Registry edits. Inappropriate SAS Registry edits can adversely affect your SAS session performance.

SAS Registry Editor values, which store data, exist in keys and subkeys. Keys and subkeys, which look like folders, appear in a tree on the left side of the SAS Registry Editor. If a key has subkeys, then you can expand or collapse it with the + and – icons that are found in the tree. If a key or subkey has values, then the values appear on the right side of the window.

*Operating Environment Information*
In the z/OS operating environment, you can select a + or – icon by positioning your cursor on it and then pressing the ENTER key.

*Figure 42.2* The SAS Registry Editor

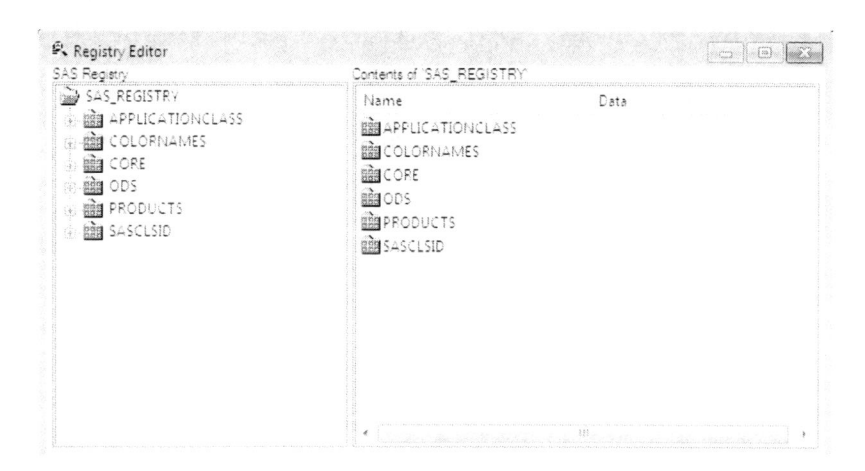

To customize SAS sessions and applications, use the SAS Registry Editor to add, modify, rename, and delete keys and key values.

You can also use the SAS Registry Editor to the following:

- import registry files (starting at any key)
- export the contents of the registry (starting at any key)
- unregister a registry file

### Opening the SAS Registry Editor

To open the SAS Registry Editor, select **Solutions** ⇨ **Accessories** ⇨ **Registry Editor**, or issue the REGEDIT command.

### Finding Information in the SAS Registry Editor

You can search for specific information in the SAS Registry Editor, including specific keys, key value names, and key value data:

1. Select the key from which you want to start a search.
2. Open the drop-down menu and select **Find**.
3. In the Registry Editor Find window, type your search string in the Find What field.
4. Check one or more of the **Keys**, **Value Name**, or **Value Data** check boxes, depending on where you want to perform your search.
5. Select **Find** to begin the search.

### Setting Keys in the SAS Registry Editor

You can add, modify, rename, or delete keys in the SAS Registry Editor. For example, you might want SAS to be able to work with a new paper type when printing output. Therefore, you might need to create a new key that represents the paper type. Also, you would have to create and set key values for this new paper type. For more information, see "Setting New Key Values in the SAS Registry Editor" on page 800.

*Note:* When you add a key, the new key becomes a subkey of the most recently selected key.

To set a key in the SAS Registry Editor:

1. Expand or collapse the keys on the left side of the SAS Registry Editor (using the + and – icons) until you find the appropriate key.

2. With a key selected, select an action from the drop-down menu (such as **New Key**, **Rename**, or **Delete**). A dialog box appears that enables you to enter additional information or confirm an action.

   *CAUTION:*
   **Delete removes all subkeys and values (if any) under the key that you are deleting.**

### Setting New Key Values in the SAS Registry Editor

If you create a new key, then you might want to add values to that key. Adding values includes assigning a value name as well as the value data.

*Note:* If your new key is similar to an existing key, then you might want to review that key's subkeys and key values. The review process might help you determine which subkeys and key values you should have for the new key.

To add a new key value, do the following:

1. Select the new key on the left side of the SAS Registry Editor.

2. Select an action from the pop-up menu (such as **New String Value**, **New Binary Value**, or **New Double Value**).

3. In the dialog box, enter a name and a value for the new key value.

4. Select **OK** to complete the process.

### Editing Existing Key Values in the SAS Registry Editor

1. Select a key on the left side of the SAS Registry Editor.

2. If the key contains subkeys, then continue to expand the key by selecting the + icon.

3. Select the key value that you want to edit on the right side of the SAS Registry Editor.

4. Select the appropriate action from the pop-up menu (such as **Modify**, **Rename**, or **Delete**). A dialog box appears that enables you to enter additional information or confirm an action.

### Importing Registry Files

You can import a registry file to populate and modify the SAS Registry quickly. Registry files are text files that you create with a text editor. For information about registry file syntax, see "REGISTRY" in *Base SAS Procedures Guide*.

1. Select **File** ⇨ **Import Registry File**.

2. Select the file that you want to import, and then select **OK**.

If errors occur during the import, then a message appears in the status bar and the errors are reported in the Log window. All registry changes can be sent to the log if you use the SAS Registry Editor option `Output full status to Log`. For more information, see "Setting Registry Editor Options" on page 801.

### Exporting Registry Files

You can export (or copy) all or a portion of the SAS Registry to a file:

1. Select the key in the existing registry from where you want to begin exporting the file. Selecting a root key exports the entire tree, beginning at the root key that you select.

2. Select **File ⇨ Export Registry File**.

3. Enter the full path to the file or browse to select the file to which you want to save the existing registry, and then select **OK**.

If errors occur during the export, then a message appears in the status bar and the errors are reported in the Log window. All registry changes can be sent to the log if you use the `Output full status to Log` SAS Registry Editor option.

### *Uninstalling an Imported Registry File*

The uninstall function reads an imported registry file and removes the keys found in the file from the registry. If any errors occur during this process, then a message appears in the status bar and errors are reported in the Log window.

*Note:* SAS is shipped with a set of ROOT keys. ROOT keys are not removed during an uninstall process.

1. Select **File ⇨ Uninstall Registry File**.

2. Select the external registry file that you want to uninstall from the SAS Registry, and then select **OK**. A message appears in the message line when the uninstall is complete.

### *Setting Registry Editor Options*

1. Open the SAS Registry Editor if it is not already open.

2. From the Registry Editor window, select **Tools ⇨ Options ⇨ Registry Editor**.

3. In the Select Registry View group box, choose a view for the Registry Editor.

   - View Overlay mode enables you to modify data anywhere in the registry. The HKEY_USER_ROOT overlays the HKEY_SYSTEM_ROOT. The parent root for Overlay View mode is shown as SAS REGISTRY.

   - In View All mode, the Registry Editor shows all the entries that are contained in the two main entry points into the registry: HKEY_SYSTEM_ROOT and HKEY_USER_ROOT. Typically, the HKEY_SYSTEM_ROOT tree is stored in the SASHELP library and the HKEY_USER_ROOT is stored in the SASUSER library.

4. Select or deselect appropriate check boxes:

   Open **HKEY_SYSTEM_ROOT** for Write access
   : enables you to open the registry for Write access if you have Write access to SASHELP.

   **Output full status to Log**
   : writes to the log all changes that were made when the registry file was imported or uninstalled. Usually, only errors appear in the Log window.

   **View unsigned integers in hexadecimal format**
   : enables you to view unsigned integers in the value list in HEX or DECIMAL format.

You can select **Reset all options** to return all Registry Editor Options window settings to the default values.

### Customizing SAS Sessions with the Preferences Window

The Preferences window includes a series of tabs that you can access to set SAS preferences. Preferences enable you to customize and control your SAS environment. For example, you might use the **General** tab to select a start-up logo, or the **Results** tab to control your output preferences, or even the **Editing** tab to set editor preferences, if, for example, your cursor inserts or overtypes text in an editor.

Preference window settings remain in effect from one SAS session to the next.

To access the Preferences window, select **Tools** ⇨ **Options** ⇨ **Preferences** or issue the DLGPREF command.

*Operating Environment Information*
The Preferences window is unavailable in some operating environments. Also, some preference settings are specific to your operating environment. Refer to the SAS documentation for your operating environment for more information about setting preferences.

### Saving System Option Settings with the DMOPTSAVE and DMOPTLOAD Commands

Perhaps the easiest way to save your system option settings from one SAS session to another is to use the global commands DMOPTSAVE and DMOPTLOAD. After you set up your system options in a way that is most appropriate for your working style, type `DMOPTSAVE` at the command line and press ENTER. This saves the current system option settings for later use. Later, when you have started another SAS session and would like to retrieve your saved settings, type `DMOPTLOAD` at the command line and press ENTER. This changes your system option settings back to the system option settings in effect when you issued the DMOPTSAVE command.

The DMOPTSAVE and DMOPTLOAD commands have other useful features:

- You can issue parameters to name different sets of system option settings and control where they are saved.

- You can view the saved system option settings by using SAS Explorer, because they are saved by default as a data set.

- You can also issue parameters to save the system option settings to a registry key.

When you issue a DMOPTSAVE command without parameters, SAS saves a data set (myopts) that contains the system option settings to the default library. The default library is usually the library where the current user profile is. In most cases, this is the SASUSER library.

See SAS online Help for more details about using these commands.

## Customizing the SAS Windowing Environment

### Customizing the Explorer Window

#### Ways to Customize the Explorer Window
You can customize the Explorer window in these ways:

- Select Contents Only view or Explorer view.
- Change how items appear in the contents view.
- Add and remove folders (including one that adds access to files in your operating environment).
- Enable member, entry, and operating environment file types to appear.
- Add a pop-up menu action.
- Hide member, entry, and operating environment file types.

### Selecting Contents Only View or Explorer View

The Explorer window can appear in either Explorer view or Contents Only view. In Explorer view, the Explorer window includes two sides: a tree view on the left that lists folders, and a contents view on the right that shows the contents of the folder that is selected in the tree view.

**Figure 42.3** *The Explorer Window with Explorer View Enabled*

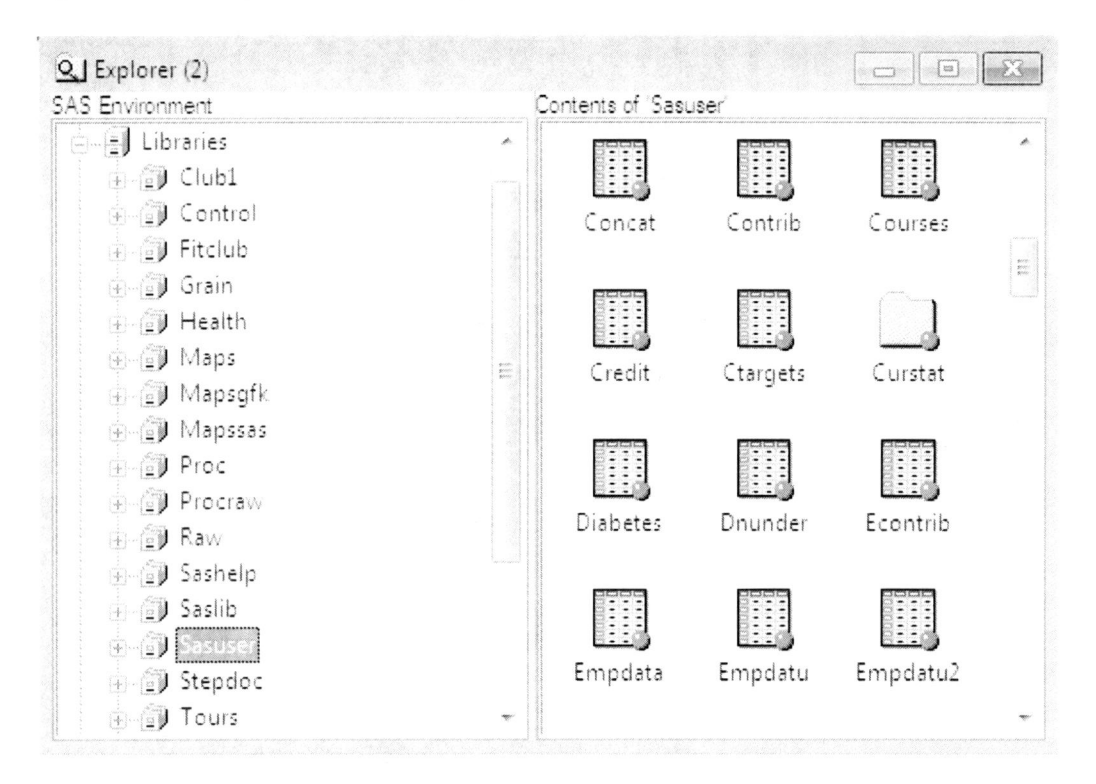

In Contents Only view, the Explorer window is a single-paned window that shows the contents of your SAS environment. As you open folders, the folder contents replace the previous contents in the same window. In Contents Only view, you navigate the Explorer window using pull-down and pop-up menu actions, and toolbar items (if a toolbar is available).

**Figure 42.4** The Explorer Window with Contents Only View Enabled

*Operating Environment Information*
> In most operating environments, the Explorer appears in Contents Only view by default.

Depending on your operating environment, you can toggle between the two views in these ways:

Menu:
> **View** ⇨ **Show Tree**

Command:
> TREE

Toolbar
> Toggle the Tree tool button

## Changing How Items Appear in the Contents View

You can make selections from the **View** menu to determine how files appear in the Contents view of the Explorer window. All possible selections follow, although not all the selections might be available in your operating environment:

Large Icons
> displays a large icon for each file.

Small Icons
> displays a small icon for each file (only available on PC hosts).

List
> displays a left-justified list of files.

Details
> lists files along with columns of descriptive information (such as file size, type, and so on).

You might also be able to use the following commands in your operating environment instead of making selections from the **View** menu:

DETAILS
> lists files along with columns of descriptive information (such as file size, type, and so on).

LARGEVIEW
> displays a large icon for each file.

SMALLVIEW
> depending on your operating environment, this command displays either a list of files or a small icon for each file.

### Adding and Removing Folders

The Explorer window shows the Libraries and File Shortcuts folders by default in many operating environments. You can turn off these folders, or turn on other folders, including Extensions, My Favorite Folders, and Results.

1. From the Explorer window, select **Tools** ⇨ **Options** ⇨ **Explorer**.

2. From the drop-down list at the top of the window, select **Initialization**.

3. Select the folder that you want to add or remove, and then select **Add** or **Remove**. The Description field changes to On or Off to reflect your change.

   *Operating Environment Information*
   The My Favorite Folders window enables you to access operating environment-specific files from the Explorer. This feature is not available in the z/OS operating environment.

### Enabling Member, Entry, and Operating Environment File Types to Appear

Commonly used members, catalog entries, and operating environment files are registered and appear in the Explorer window. Registered types must have at least an icon defined and might also have pop-up menu actions defined. Undefined types do not appear in the Explorer window and have no actions associated with them.

To add (register) an undefined type:

1. From the Explorer window, select **Tools** ⇨ **Options** ⇨ **Explorer**.

2. From the drop-down list at the top of the window, select a category (such as Members, Catalog Entries, or Host Files). The registered types are displayed in the window.

3. Select the **View Undefined Types** check box to see the undefined types for the category.

4. Select a type and then select **Edit**.

5. Select **Select Icon**.

6. In the Select Icon dialog box, choose a category from the drop-down list at the top, select an icon, and then select **OK** to close the dialog box.

7. Add actions for the type (if desired) and then select **OK**. For more information about adding actions to a type, see "Adding a Menu Action to a Member, Entry, or Operating Environment File Type" on page 805. The type is added to the Registered Types list.

### Adding a Menu Action to a Member, Entry, or Operating Environment File Type

You can add a menu action to any catalog entry, member, or operating environment file type.

1. From the Explorer window, select **Tools** ⇨ **Options** ⇨ **Explorer**.

2. From the drop-down list at the top of the window, select a category (such as Members, Catalog Entries, or Host Files). The registered types are displayed in the window.

3. Select the registered type that you want to edit.

4. Select **Edit**.

5. In the Options dialog box for that entry, select **Add**.

6. Enter a name for the action (this is the action that will appear on the pop-up menu for the item), and an action command. To see examples of action commands, look at the commands for registered types.

7. Select **OK**.

*Note:* The letter immediately after the ampersand (&) in the Action section denotes the shortcut key that can be used to perform that action.

### Hiding Member, Entry, and Host File Types

You can hide members, catalog entries, and host files so that they do not appear in the Explorer window:

1. From the Explorer window, select **Tools** ⇨ **Options** ⇨ **Explorer**.

2. From the drop-down list at the top of the window, select a category (such as Members, Catalog Entries, or Host Files). The registered types are displayed in the window.

3. Select the registered type that you want to remove from view.

4. Select **Remove**. Confirm the removal by selecting **OK** when prompted.

When you remove a registered type, it is moved to the View Undefined Types view. To add the registered type back, you must redefine its icon.

## Customizing an Editor

You can customize general and text editing options for your editor. For example, if you use line commands when you edit programs, then you might always want the Program Editor to appear with line numbers.

To customize your editor, do the following:

1. Select a SAS programming window (such as the Program Editor, Log, Output, or Notepad window).

2. Select **Tools** ⇨ **Options** ⇨ **Editor**.

3. From the drop-down list, select the category of options that you want to edit.

4. In the Options group box, select an option, and then select **Modify** from the pop-up menu.

5. In the dialog box that appears, edit the option name, value, or both.

## Customizing Fonts

You can set default font information for the SAS windowing environment with the Font window. To access the Font window, issue the DLGFONT command, or select **Tools** ⇨ **Options** ⇨ **Fonts**.

The Font window is host-specific. Refer to your host documentation for more information.

### Customizing Colors

*Note:* Changes made with the SASColor window are visible only after affected SAS windows are closed and then reopened.

You can also change the default colors in edit windows, such as the Notepad and the Program Editor by using the SYNCONFIG command. This command controls the color of SAS language and programming elements, which makes it easier to parse through a SAS program and understand how it works. SYNCONFIG opens the Edit Scheme window, which gives you several color schemes to select. You can also modify the provided color schemes.

### Setting SAS Windowing Environment Preferences

You can use the Preferences window to customize portions of the SAS windowing environment to your liking. For more information, see "Customizing SAS Sessions with the Preferences Window" on page 802.

## Summary

### Commands

DLGFONT
: opens the Font window, which is used to control the fonts in the SAS windowing environment.

DLGPREF
: opens the Preferences window, in some operating environments.

OPTIONS
: opens the SAS System Options window.

PMENU
: turns on the menu bar in the windowing environment.

REGEDIT
: opens the Registry Editor window.

SASCOLOR
: opens the SASColor window, which is used to change the color of window elements, such as backgrounds and borders.

SYNCONFIG
: opens the Edit Scheme window, which is used to edit color schemes in the Editor, NOTEPAD, or Program Editor windows.

### Procedures

PROC OPTIONS <SHORT | LONG>;
: lists the current values of all SAS system options. The SHORT and LONG options determine the format in which you want SAS system options listed.

   *Note:* You can also use the SAS System Options window to see the current values of all SAS system options.

PROC REGISTRY <options>;
: maintains the SAS Registry.

*Note:* You can also use the SAS Registry Editor to maintain the SAS Registry.

## Statements

OPTIONS *option-1* <... *option-n*>;
: changes the value of one or more SAS system options.

## System Options

VERBOSE | NOVERBOSE
: controls whether SAS writes the settings of all the system options that are specified in the configuration file to either the workstation or batch log.

## Windows

Editor Options window
: enables you to set options for specific SAS windowing environment windows, such as the Program Editor. To open the Editor Options window, go to the window that you want to change, and then select **Tools** ⇨ **Options** ⇨ **Editor** or issue the EDOPT command.

Explorer Options window
: enables you to set Explorer window options. To open this window, select **Tools** ⇨ **Options** ⇨ **Explorer Options** or issue the EXPOPTS command.

Fonts window
: enables you to select the default font that you want to use in the SAS windowing environment. To access this window, issue the DLGFONT command.

*Note:* This window is specific to your operating environment.

Preferences window
: enables you to set SAS system preferences. To access this window, issue the DLGPREF command.

*Note:* This window is specific to your operating environment.

SASColor window
: enables you to change the default colors for the different window elements in your SAS windows. To access this window, issue the SASCOLOR command.

SAS Registry Editor
: enables you to edit the SAS Registry and to customize aspects of the SAS windowing environment. To access this window, issue the REGEDIT command.

SAS System Options window
: enables you to view or change current SAS system options. To access this window, issue the OPTIONS command.

# Learning More

Customizations
: For information about operating environment-specific customization options and preferences, refer to the SAS documentation for your operating environment.

SAS Procedures
: For more information about SAS procedures, see *Base SAS Procedures Guide*.

Statements and Options
: For more information about the statements and options that are discussed in this section, see *SAS Statements: Reference* and *SAS System Options: Reference*.

# Part 11

# Appendix

*Appendix 1*
**Complete DATA Steps for Selected Examples** .................... *813*

*Appendix 2*
**DATA Step Debugger Commands** .............................. *823*

# Appendix 1
# Complete DATA Steps for Selected Examples

| | |
|---|---:|
| **Complete DATA Steps for Selected Examples** | 813 |
| **The CITY Data Set** | 814 |
| DATA Step to Create the CITY Data Set | 814 |
| **The UNIVERSITY_TEST_SCORES Data Set** | 815 |
| DATA Step to Create the UNIVERSITY_TEST_SCORES Data Set | 815 |
| **The YEAR_SALES Data Set** | 816 |
| DATA Step to Create the YEAR_SALES Data Set | 816 |
| **The HIGHLOW Data Set** | 817 |
| DATA Step to Create the HIGHLOW Data Set | 817 |
| **The GRADES Data Set** | 818 |
| DATA Step to Create the GRADES Data Set | 818 |
| **The USCLIM Data Sets** | 819 |
| DATA Step to Create the USCLIM.HIGHTEMP Data Set | 819 |
| DATA Step to Create the USCLIM.HURRICANE Data Set | 819 |
| DATA Step to Create the USCLIM.LOWTEMP Data Set | 819 |
| DATA Step to Create the USCLIM.TEMPCHNG Data Set | 820 |
| Note about the USCLIM.BASETEMP and USCLIM.REPORT Catalogs | 820 |
| **The CLIMATE, PRECIP, and STORM Data Sets** | 820 |
| DATA Step to Create the CLIMATE.HIGHTEMP Data Set | 820 |
| DATA Step to Create the CLIMATE.LOWTEMP Data Set | 820 |
| DATA Step to Create the PRECIP.RAIN Data Set | 821 |
| DATA Step to Create the PRECIP.SNOW Data Set | 821 |
| DATA Step to Create the STORM.TORNADO Data Set | 821 |

## Complete DATA Steps for Selected Examples

This documentation shows how to create the data sets that are used in each section. However, when the input data are lengthy or the actual contents of the data set are not crucial to the section, then the DATA steps or raw data that is used to create data sets are listed in this appendix instead of within the section.

Only the DATA steps that are not provided in detail in the section, are included here.

## The CITY Data Set

### DATA Step to Create the CITY Data Set

```
data city;
   input Year 4. @7 ServicesPolice comma6.
         @15 ServicesFire comma6. @22 ServicesWater_Sewer comma6.
         @30 AdminLabor comma6. @39 AdminSupplies comma6.
         @45 AdminUtilities comma6.;
   ServicesTotal=ServicesPolice+ServicesFire+ServicesWater_Sewer;
   AdminTotal=AdminLabor+AdminSupplies+AdminUtilities;
   Total=ServicesTotal+AdminTotal;
   label              Total='Total Outlays'
             ServicesTotal='Services: Total'
            ServicesPolice='Services: Police'
              ServicesFire='Services: Fire'
       ServicesWater_Sewer='Services: Water & Sewer'
                AdminTotal='Administration: Total'
                AdminLabor='Administration: Labor'
             AdminSupplies='Administration: Supplies'
            AdminUtilities='Administration: Utilities' ;
   datalines;
1980  2,819  1,120    422    391      63     98
1981  2,477  1,160    500    172      47     70
1982  2,028  1,061    510    269      29     79
1983  2,754    893    540    227      21     67
1984  2,195    963    541    214      21     59
1985  1,877    926    535    198      16     80
1986  1,727  1,111    535    213      27     70
1987  1,532  1,220    519    195      11     69
1988  1,448  1,156    577    225      12     58
1989  1,500  1,076    606    235      19     62
1990  1,934    969    646    266      11     63
1991  2,195  1,002    643    256      24     55
1992  2,204    964    692    256      28     70
1993  2,175  1,144    735    241      19     83
1994  2,556  1,341    813    238      25     97
1995  2,026  1,380    868    226      24     97
1996  2,526  1,454    946    317      13     89
1997  2,027  1,486  1,043    226       .     82
1998  2,037  1,667  1,152    244      20     88
1999  2,852  1,834  1,318    270      23     74
2000  2,787  1,701  1,317    307      26     66
;
```

# The UNIVERSITY_TEST_SCORES Data Set

## DATA Step to Create the UNIVERSITY_TEST_SCORES Data Set

The raw data in the DATA step is used to create the following data sets:

- OUT.UNIVERSITY_TEST_SCORES
- OUT.UNIVERSITY_TEST_SCORES2
- OUT.UNIVERSITY_TEST_SCORES3
- OUT.UNIVERSITY_TEST_SCORES4
- OUT.UNIVERSITY_TEST_SCORES5
- OUT_ERROR1
- OUT.ERROR2
- OUT.ERROR3

```
libname out 'your-data-library';

data out.university_test_scores;
   input Test $ Gender $ Year TestScore;
   datalines;
Verbal m 2005 504
Verbal f 2005 496
Verbal m 2006 504
Verbal f 2006 497
Verbal m 2007 501
Verbal f 2007 497
Verbal m 2008 505
Verbal f 2008 502
Verbal m 2009 507
Verbal f 2009 503
Verbal m 2010 507
Verbal f 2010 503
Verbal m 2011 509
Verbal f 2011 502
Math   m 2005 521
Math   f 2005 484
Math   m 2006 524
Math   f 2006 484
Math   m 2007 523
Math   f 2007 487
Math   m 2008 525
Math   f 2008 490
Math   m 2009 527
Math   f 2009 492
Math   m 2010 530
Math   f 2010 494
Math   m 2011 531
Math   f 2011 496
;
```

# The YEAR_SALES Data Set

## DATA Step to Create the YEAR_SALES Data Set

```
data year_sales;
   input Month $ Quarter $ SalesRep $14. Type $ Units Price @@;
   AmountSold=Units*price;
   datalines;
01 1 Hollingsworth Deluxe    260 49.50 01 1 Garcia        Standard    41 30.97
01 1 Hollingsworth Standard  330 30.97 01 1 Jensen        Standard   110 30.97
01 1 Garcia        Deluxe    715 49.50 01 1 Jensen        Standard   675 30.97
02 1 Garcia        Standard 2045 30.97 02 1 Garcia        Deluxe      10 49.50
02 1 Garcia        Standard   40 30.97 02 1 Hollingsworth Standard 1030 30.97
02 1 Jensen        Standard  153 30.97 02 1 Garcia        Standard    98 30.97
03 1 Hollingsworth Standard  125 30.97 03 1 Jensen        Standard   154 30.97
03 1 Garcia        Standard  118 30.97 03 1 Hollingsworth Standard    25 30.97
03 1 Jensen        Standard  525 30.97 03 1 Garcia        Standard   310 30.97
04 2 Garcia        Standard  150 30.97 04 2 Hollingsworth Standard   260 30.97
04 2 Hollingsworth Standard  530 30.97 04 2 Jensen        Standard  1110 30.97
04 2 Garcia        Standard 1715 30.97 04 2 Jensen        Standard   675 30.97
05 2 Jensen        Standard   45 30.97 05 2 Hollingsworth Standard 1120 30.97
05 2 Garcia        Standard   40 30.97 05 2 Hollingsworth Standard 1030 30.97
05 2 Jensen        Standard  153 30.97 05 2 Garcia        Standard    98 30.97
06 2 Jensen        Standard  154 30.97 06 2 Hollingsworth Deluxe      25 49.50
06 2 Jensen        Standard  276 30.97 06 2 Hollingsworth Standard   125 30.97
06 2 Garcia        Standard  512 30.97 06 2 Garcia        Standard  1000 30.97
07 3 Garcia        Standard  250 30.97 07 3 Hollingsworth Deluxe      60 49.50
07 3 Garcia        Standard   90 30.97 07 3 Hollingsworth Deluxe      30 49.50
07 3 Jensen        Standard  110 30.97 07 3 Garcia        Standard    90 30.97
07 3 Hollingsworth Standard  130 30.97 07 3 Jensen        Standard   110 30.97
07 3 Garcia        Standard  265 30.97 07 3 Jensen        Standard   275 30.97
07 3 Garcia        Standard 1250 30.97 07 3 Hollingsworth Deluxe      60 49.50
07 3 Garcia        Standard   90 30.97 07 3 Jensen        Standard   110 30.97
07 3 Garcia        Standard   90 30.97 07 3 Hollingsworth Standard   330 30.97
07 3 Jensen        Standard  110 30.97 07 3 Garcia        Standard   465 30.97
07 3 Jensen        Standard  675 30.97 08 3 Jensen        Standard   145 30.97
08 3 Garcia        Deluxe    110 49.50 08 3 Hollingsworth Standard   120 30.97
08 3 Hollingsworth Standard  230 30.97 08 3 Jensen        Standard   453 30.97
08 3 Garcia        Standard  240 30.97 08 3 Hollingsworth Standard   230 49.50
08 3 Jensen        Standard  453 30.97 08 3 Garcia        Standard   198 30.97
08 3 Hollingsworth Standard  290 30.97 08 3 Garcia        Standard  1198 30.97
08 3 Jensen        Deluxe     45 49.50 08 3 Jensen        Standard   145 30.97
08 3 Garcia        Deluxe    110 49.50 08 3 Hollingsworth Standard   330 30.97
08 3 Garcia        Standard  240 30.97 08 3 Hollingsworth Deluxe      50 49.50
08 3 Jensen        Standard  453 30.97 08 3 Garcia        Standard   198 30.97
08 3 Jensen        Deluxe    225 49.50 09 3 Hollingsworth Standard   125 30.97
09 3 Jensen        Standard  254 30.97 09 3 Garcia        Standard   118 30.97
09 3 Hollingsworth Standard 1000 30.97 09 3 Jensen        Standard   284 30.97
09 3 Garcia        Standard  412 30.97 09 3 Jensen        Deluxe     275 49.50
09 3 Garcia        Standard  100 30.97 09 3 Jensen        Standard   876 30.97
09 3 Hollingsworth Standard  125 30.97 09 3 Jensen        Standard   254 30.97
09 3 Garcia        Standard 1118 30.97 09 3 Hollingsworth Standard   175 30.97
```

```
09 3 Jensen         Standard  284 30.97 09 3 Garcia        Standard   412 30.97
09 3 Jensen         Deluxe    275 49.50 09 3 Garcia        Standard   100 30.97
09 3 Jensen         Standard  876 30.97 10 4 Garcia        Standard   250 30.97
10 4 Hollingsworth  Standard  530 30.97 10 4 Jensen        Standard   975 30.97
10 4 Hollingsworth  Standard  265 30.97 10 4 Jensen        Standard    55 30.97
10 4 Garcia         Standard  365 30.97 11 4 Hollingsworth Standard  1230 30.97
11 4 Jensen         Standard  453 30.97 11 4 Garcia        Standard   198 30.97
11 4 Jensen         Standard   70 30.97 11 4 Garcia        Standard   120 30.97
11 4 Hollingsworth  Deluxe    150 49.50 12 4 Garcia        Standard  1000 30.97
12 4 Jensen         Standard  876 30.97 12 4 Hollingsworth Deluxe     125 49.50
12 4 Jensen         Standard 1254 30.97 12 4 Hollingsworth Standard   175 30.97
;
```

# The HIGHLOW Data Set

## DATA Step to Create the HIGHLOW Data Set

```
data highlow;
   input Year @7 DateOfHigh:date9. DowJonesHigh @26 DateOfLow:date9. DowJonesLow;
   format LogDowHigh LogDowLow 5.2 DateOfHigh DateOfLow date9.;
   LogDowHigh=log(DowJonesHigh);
   LogDowLow=log(DowJonesLow);
datalines;
1968   03DEC1968  985.21   21MAR1968  825.13
1969   14MAY1969  968.85   17DEC1969  769.93
1970   29DEC1970  842.00   06MAY1970  631.16
1971   28APR1971  950.82   23NOV1971  797.97
1972   11DEC1972 1036.27   26JAN1972  889.15
1973   11JAN1973 1051.70   05DEC1973  788.31
1974   13MAR1974  891.66   06DEC1974  577.60
1975   15JUL1975  881.81   02JAN1975  632.04
1976   21SEP1976 1014.79   02JAN1976  858.71
1977   03JAN1977  999.75   02NOV1977  800.85
1978   08SEP1978  907.74   28FEB1978  742.12
1979   05OCT1979  897.61   07NOV1979  796.67
1980   20NOV1980 1000.17   21APR1980  759.13
1981   27APR1981 1024.05   25SEP1981  824.01
1982   27DEC1982 1070.55   12AUG1982  776.92
1983   29NOV1983 1287.20   03JAN1983 1027.04
1984   06JAN1984 1286.64   24JUL1984 1086.57
1985   16DEC1985 1553.10   04JAN1985 1184.96
1986   02DEC1986 1955.57   22JAN1986 1502.29
1987   25AUG1987 2722.42   19OCT1987 1738.74
1988   21OCT1988 2183.50   20JAN1988 1879.14
1989   09OCT1989 2791.41   03JAN1989 2144.64
1990   16JUL1990 2999.75   11OCT1990 2365.10
1991   31DEC1991 3168.83   09JAN1991 2470.30
1992   01JUN1992 3413.21   09OCT1992 3136.58
1993   29DEC1993 3794.33   20JAN1993 3241.95
1994   31JAN1994 3978.36   04APR1994 3593.35
1995   13DEC1995 5216.47   30JAN1995 3832.08
1996   27DEC1996 6560.91   10JAN1996 5032.94
```

```
                    1997   06AUG1997  8259.31   11APR1997  6391.69
                    1998   23NOV1998  9374.27   31AUG1998  7539.07
                    1999   31DEC1999 11497.12   22JAN1999  9120.67
                    2000   14JAN2000 11722.98   07MAR2000  9796.04
                    2001   21MAY2001 11337.92   21SEP2001  8235.81
                    2002   19MAR2002 10635.25   09OCT2002  7286.27
                    2003   31DEC2003 10453.92   11MAR2003  7524.06
                    2004   28DEC2004 10854.54   25OCT2004  9749.99
                    2005   04MAR2005 10940.55   20APR2005 10012.36
                    2006   27DEC2006 12510.57   20JAN2006 10667.39
                    2007   09OCT2007 14164.53   05MAR2007 12050.41
                    2008   02MAY2008 13058.20   10OCT2008  8451.19
                    ;
                    run;
```

# The GRADES Data Set

## DATA Step to Create the GRADES Data Set

```
                    data grades;
                       input Name &$14. Gender :$2. Section :$3. ExamGrade1 @@;
                    datalines;
Abdallah       F Mon 46 Anderson     M Wed 75
Aziz           F Wed 67 Bayer        M Wed 77
Bhatt          M Fri 79 Blair        F Fri 70
Bledsoe        F Mon 63 Boone        M Wed 58
Burke          F Mon 63 Chung        M Wed 85
Cohen          F Fri 89 Drew         F Mon 49
Dubos L        M Mon 41 Elliott      F Wed 85
Farmer         F Wed 58 Franklin     F Wed 59
Freeman        F Mon 79 Friedman     M Mon 58
Gabriel        M Fri 75 Garcia       M Mon 79
Harding        M Mon 49 Hazelton     M Mon 55
Hinton         M Fri 85 Hung         F Fri 98
Jacob          F Wed 64 Janeway      F Wed 51
Jones          F Mon 39 Jorgensen    M Mon 63
Judson         F Fri 89 Kuhn         F Mon 89
LeBlanc        F Fri 70 Lee          M Fri 48
Litowski       M Fri 85 Malloy       M Wed 79
Meyer          F Fri 85 Nichols      M Mon 58
Oliver         F Mon 41 Park         F Mon 77
Patel          M Wed 73 Randleman    F Wed 46
Robinson       M Fri 64 Shien        M Wed 55
Simonson       M Wed 62 Smith N      M Wed 71
Smith R        M Mon 79 Sullivan     M Fri 77
Swift          M Wed 63 Wolfson      F Fri 79
Wong           F Fri 89 Zabriski     M Fri 89
;
```

# The USCLIM Data Sets

## *DATA Step to Create the USCLIM.HIGHTEMP Data Set*

```
libname usclim 'SAS-data-library';

data usclim.hightemp;
   input State $char14. City $char14. Temp_f Date date9. Elevation;
   datalines;
Arizona        Parker         127 07jul1905 345
Kansas         Alton          121 25jul1936 1651
Nevada         Overton        122 23jun1954 1240
North Dakota   Steele         121 06jul1936 1857
Oklahoma       Tishomingo     120 26jul1943 6709
Texas          Seymour        120 12aug1936 1291
;
```

## *DATA Step to Create the USCLIM.HURRICANE Data Set*

```
libname usclim 'SAS-data-library';

data usclim.hurricane;
   input @1 State $char14. @16 Date date7. Deaths Millions Name $;
   format Date worddate18. Millions dollar6.;
   informat State $char14. Date date9.;
   label Millions='Damage';
   datalines;
Mississippi    14aug1969 256 1420 Camille
Florida        14jun1972 117 2100 Agnes
Alabama        29aug1979 5   2300 Frederick
Texas          15aug1983 21  2000 Alicia
Texas          03aug1980 28  300  Allen
North Carolina 27aug2011 6   450  Irene
;
```

## *DATA Step to Create the USCLIM.LOWTEMP Data Set*

```
libname usclim 'SAS-data-library';

data usclim.lowtemp;
   input State $char14. City $char14. Temp_f Date date9. Elevation;
   datalines;
Alaska         Prospect Creek -80 23jan1971 1100
Colorado       Maybell        -60 01jan1979 5920
Idaho          Island Prk Dam -60 18jan1943 6285
Minnesota      Pokegama Dam   -59 16feb1903 1280
North Dakota   Parshall       -60 15feb1936 1929
South Dakota   McIntosh       -58 17feb1936 2277
Wyoming        Moran          -63 09feb1933 6770
;
```

## Appendix 1 • Complete DATA Steps for Selected Examples

### DATA Step to Create the USCLIM.TEMPCHNG Data Set

```
libname usclim 'SAS-data-library';

data usclim.tempchng;
   input @1 State $char13. @15 Date date7. Start_f End_f Minutes;
   Diff=End_f-Start_f;
   informat State $char13. Date date7.;
   format Date date9.;
   datalines;
North Dakota   21feb1918 -33  50   720
South Dakota   22jan1943 -4   45   2
South Dakota   12jan1911 49  -13   120
South Dakota   22jan1943 54  -4    27
South Dakota   10jan1911 55   8    15
;
```

### Note about the USCLIM.BASETEMP and USCLIM.REPORT Catalogs

The catalogs USCLIM.BASETEMP and USCLIM.REPORT are used to show how the DATASETS procedure processes both SAS data sets and catalogs. The contents of these catalogs are not important in the context of this book. In most cases, you would use SAS/AF, SAS/FSP, or other SAS products to create catalog entries. You can test the examples in this section without having these catalogs.

## The CLIMATE, PRECIP, and STORM Data Sets

### DATA Step to Create the CLIMATE.HIGHTEMP Data Set

```
libname climate 'SAS-data-library';

data climate.hightemp;
   input Place $ 1-13 Date date9. Degree_f Degree_c;
   datalines;
South Africa  21jan2010 122 50
Israel        21jun1942 129 54
Argentina     02jan1920 120 49
Saskatchewan  05jul1937 113 45
India         18jun1905 124 51
Poland        29jul1921 104 40
;
```

### DATA Step to Create the CLIMATE.LOWTEMP Data Set

```
libname climate 'SAS-data-library';

data climate.lowtemp;
   input Place $ 1-13 Date date9. Degree_f Degree_c;
   datalines;
```

```
Antarctica   21jul83 -129 -89
Siberia      06feb33  -90 -68
Greenland    09jan54  -87 -66
Yukon        03feb47  -81 -63
Alaska       23jan71  -80 -67
;
```

## DATA Step to Create the PRECIP.RAIN Data Set

```
libname precip 'SAS-data-library';

data precip.rain;
   input Place $ 1-12 @13 Date date9. Inches Cms;
   format Date date9.;
   datalines;
La Reunion  15mar1952 74 188
Taiwan      10sep1963 49 125
Australia   04jan1979 44 114
Texas       25jul1979 43 109
Canada      06oct1964 19  49
;
```

## DATA Step to Create the PRECIP.SNOW Data Set

```
libname precip 'SAS-data-library';

data precip.snow;
   input Place $ 1-12 @13 Date date7. Inches Cms;
   format Date date9.;
   datalines;
Colorado    14apr21 76 193
Alaska      29dec55 62 158
France      05apr69 68 173
;
```

## DATA Step to Create the STORM.TORNADO Data Set

```
libname storm 'SAS-data-library';

data storm.tornado;
   input State $ 1-12 @13 Date date7. Deaths Millions;
   format Date date9. Millions dollar6.;
   label Millions='Damage in Millions';
   datalines;
Iowa         11apr65 257 200
Texas        11may70  26 135
Nebraska     06may75   3 400
Connecticut  03oct79   3 200
Georgia      31mar73   9 115
;
```

# Appendix 2
# DATA Step Debugger Commands

| Dictionary | 823 |
|---|---|
| BREAK | 823 |
| CALCULATE | 825 |
| DELETE | 826 |
| DESCRIBE | 827 |
| ENTER | 828 |
| EXAMINE | 829 |
| GO | 830 |
| HELP | 831 |
| JUMP | 831 |
| LIST | 832 |
| QUIT | 833 |
| SET | 834 |
| STEP | 834 |
| SWAP | 835 |
| TRACE | 836 |
| WATCH | 836 |

## Dictionary

### BREAK

Suspends program execution at an executable statement.

**Category:** Manipulating Debugging Requests

**Alias:** B

#### Syntax

**BREAK** *location* <AFTER *count*> <WHEN *expression*> <DO *group* >

#### Required Argument

*location*
   specifies where to set a breakpoint. *Location* must be one of these:

*label*
: a statement label. The breakpoint is set at the statement that follows the label.

*line-number*
: the number of a program line at which to set a breakpoint.

*:*
: the current line.

### Optional Arguments

**AFTER** *count*
: honors the breakpoint each time the statement has been executed *count* times. The counting is continuous. That is, when the AFTER option applies to a statement inside a DO loop, the count continues from one iteration of the loop to the next. The debugger does not reset the *count* value to 1 at the beginning of each iteration.

  If a BREAK command contains both AFTER and WHEN, AFTER is evaluated first. If the AFTER count is satisfied, the WHEN expression is evaluated.

  Tip   The AFTER option is useful in debugging DO loops.

**WHEN** *expression*
: honors a breakpoint when the expression is true.

**DO** *group*
: is one or more debugger commands enclosed by a DO and an END statement. The syntax of the DO *group* is the following:

  **DO**; *command-1*<...;*command-n*;> **END**;

  *command*
  : specifies a debugger command. Separate multiple commands by semicolons.

  A DO group can span more than one line and can contain IF-THEN/ELSE statements, as shown:

  **IF** *expression* **THEN** *command;* <ELSE *command;*>

  **IF** *expression* **THEN DO** *group;* <ELSE DO *group;*>

  IF evaluates an expression. When the condition is true, the debugger command or DO group in the THEN clause executes. An optional ELSE command gives an alternative action if the condition is not true. You can use these arguments with IF:

  *expression*
  : specifies a debugger expression. A nonzero, nonmissing result causes the expression to be true. A result of zero or missing causes the expression to be false.

  *command*
  : specifies a single debugger command.

  *DO group*
  : specifies a DO group.

### Details

The BREAK command suspends execution of the DATA step at a specified statement. Executing the BREAK command is called *setting a breakpoint*.

When the debugger detects a breakpoint, it does the following:

- checks the AFTER *count* value, if present, and suspends execution if *count* breakpoint activations have been reached
- evaluates the WHEN expression, if present, and suspends execution if the condition that is evaluated is true
- suspends execution if neither an AFTER nor a WHEN clause is present
- displays the line number at which execution is suspended
- executes any commands that are present in a DO group
- returns control to the user with a > prompt

If a breakpoint is set at a source line that contains more than one statement, the breakpoint applies to each statement on the source line. If a breakpoint is set at a line that contains a macro invocation, the debugger breaks at each statement generated by the macro.

## Example

- Set a breakpoint at line 5 in the current program:

  ```
  b 5
  ```

- Set a breakpoint at the statement after the statement label **eoflabel**:

  ```
  b eoflabel
  ```

- Set a breakpoint at line 45 that will be honored after every third execution of line 45:

  ```
  b 45 after 3
  ```

- Set a breakpoint at line 45 that will be honored after every third execution of that line only when the values of both DIVISOR and DIVIDEND are 0:

  ```
  b 45 after 3
        when (divisor=0 and dividend=0)
  ```

- Set a breakpoint at line 45 of the program and examine the values of variables NAME and AGE:

  ```
  b 45 do; ex name age; end;
  ```

- Set a breakpoint at line 15 of the program. If the value of DIVISOR is greater than 3, execute STEP. Otherwise, display the value of DIVIDEND.

  ```
  b 15 do; if divisor>3 then st;
          else ex dividend; end;
  ```

## See Also

### Commands:

- "DELETE" on page 826
- "WATCH" on page 836

# CALCULATE

Evaluates a debugger expression and displays the result.

    **Category:**    Manipulating DATA Step Variables

## Syntax

**CALC** *expression*

### Required Argument

*expression*
: specifies any debugger expression.

    Restriction   Debugger expressions cannot contain functions.

### Details

The CALCULATE command evaluates debugger expressions and displays the result. The result must be numeric.

### Example

- Add 1.1, 1.2, 3.4 and multiply the result by 0.5:

    ```
    calc (1.1+1.2+3.4)*0.5
    ```

- Calculate the sum of STARTAGE and DURATION:

    ```
    calc startage+duration
    ```

- Calculate the values of the variable SALE minus the variable DOWNPAY and then multiply the result by the value of the variable RATE. Divide that value by 12 and add 50:

    ```
    calc (((sale-downpay)*rate)/12)+50
    ```

### See Also

"Working with Expressions" on page 417

# DELETE

Deletes breakpoints or the watch status of variables in the DATA step.

**Category:** Manipulating Debugging Requests

**Alias:** D

## Syntax

**DELETE** BREAK *location*

**DELETE** WATCH *variable(s)* | _ALL_

### Required Arguments

**BREAK**
: deletes breakpoints.

Alias B

*location*
: specifies a breakpoint location to be deleted. *location* can have one of these values:

    _ALL_
    : all current breakpoints in the DATA step.

    *label*
    : the statement after a statement label.

    *line-number*
    : the number of a program line.

    \*
    : the breakpoint from the current line.

**WATCH**
: deletes watched status of variables.

    Alias W

*variable(s)*
: names one or more watched variables for which the watch status is deleted.

_ALL_
: specifies that the watch status is deleted for all watched variables.

## Example

- Delete the breakpoint at the statement label

    ```
    eoflabel
      :
    d b eoflabel
    ```

- Delete the watch status from the variable ABC in the current DATA step:

    ```
    d w abc
    ```

## See Also

### Commands:

- "BREAK" on page 823
- "WATCH" on page 836

# DESCRIBE

Displays the attributes of one or more variables.

**Category:** Manipulating DATA Step Variables
**Alias:** DESC

### Syntax

**DESCRIBE** *variable(s)* | _ALL_

### Required Arguments

*variable(s)*
: identifies one or more DATA step variables

_ALL_
: indicates all variables that are defined in the DATA step.

### Details

The DESCRIBE command displays the attributes of one or more specified variables.

DESCRIBE reports the name, type, and length of the variable, and, if present, the informat, format, or variable label.

### Example

- Display the attributes of variable ADDRESS:

  ```
  desc address
  ```

- Display the attributes of array element ARR{$i + j$}:

  ```
  desc arr{i+j}
  ```

## ENTER

Assigns one or more debugger commands to the ENTER key.

**Category:** Customizing the Debugger

### Syntax

**ENTER** *command-1* <; *command-2*; ...>

### Required Argument

*command*
: specifies a debugger command.

  Default  STEP 1

### Details

The ENTER command assigns one or more debugger commands to the ENTER key. Assigning a new command to the ENTER key replaces the existing command assignment.

If you assign more than one command, separate the commands with semicolons.

## Example

- Assign the command STEP 5 to the ENTER key:

  `enter st 5`

- Assign the commands EXAMINE and DESCRIBE, both for the variable CITY, to the ENTER key:

  `enter ex city; desc city`

# EXAMINE

Displays the value of one or more variables.

**Category:** Manipulating DATA Step Variables

**Alias:** E

## Syntax

**EXAMINE** *variable-1* <*format-1*> <*variable-2* <*format-2* ...>>

**EXAMINE** _ALL_ <*format*>

### Required Arguments

*variable*
: identifies a DATA step variable.

*_ALL_*
: identifies all variables that are defined in the current DATA step.

### Optional Argument

*format*
: identifies a SAS format or a user-created format.

## Details

The EXAMINE command displays the value of one or more specified variables. The debugger displays the value using the format currently associated with the variable, unless you specify a different format.

## Example

- Display the values of variables N and STR:

  `ex n str`

- Display the element *i* of the array TESTARR:

  `ex testarr{i}`

- Display the elements $i+1$, $j*2$, and $k-3$ of the array CRR:

  `ex crr{i+1}; ex crr{j*2}; ex crr{k-3}`

- Display the SAS date variable T_DATE with the DATE7. format:

```
ex t_date date7.
```

- Display the values of all elements in array NEWARR:

```
ex newarr{*}
```

## See Also

### Commands:

- "DESCRIBE" on page 827

# GO

Starts or resumes execution of the DATA step.

**Category:** Controlling Program Execution

**Alias:** G

## Syntax

**GO** <*line-number* | *label*>

### Without Arguments

If you omit arguments, GO resumes execution of the DATA step and executes its statements continuously until a breakpoint is encountered, until the value of a watched variable changes, or until the DATA step completes execution.

### Optional Arguments

*line-number*
: gives the number of a program line at which execution is to be suspended next.

*label*
: is a statement label. Execution is suspended at the statement following the statement label.

## Details

The GO command starts or resumes execution of the DATA step. Execution continues until all observations have been read, a breakpoint specified in the GO command is reached, or a breakpoint set earlier with a BREAK command is reached.

## Example

- Resume executing the program and execute its statements continuously:

```
g
```

- Resume program execution and then suspend execution at the statement in line 104:

```
g 104
```

## See Also

### Commands:
- "JUMP" on page 831
- "STEP" on page 834

## HELP

Displays information about debugger commands.

    **Category:**   Controlling the Windows

### Syntax

**HELP**

### Without Arguments

The HELP command displays a directory of the debugger commands. Select a command name to view information about the syntax and usage of that command. You must enter the HELP command from a window command line, from a menu, or with a function key.

## JUMP

Restarts execution of a suspended program.

    **Category:**   Controlling Program Execution

    **Alias:**   J

### Syntax

**JUMP** *line-number* | *label*

### Required Arguments

*line-number*
: indicates the number of a program line at which to restart the suspended program.

*label*
: is a statement label. Execution resumes at the statement following the label.

### Details

The JUMP command moves program execution to the specified location without executing intervening statements. After executing JUMP, you must restart execution with GO or STEP. You can jump to any executable statement in the DATA step.

*CAUTION:*
    **Do not use the JUMP command to jump to a statement inside a DO loop or to a label that is the target of a LINK-RETURN group.** In such cases, you bypass the

controls set up at the beginning of the loop or in the LINK statement, and unexpected results can appear.

JUMP is useful in two situations:

- when you want to bypass a section of code that is causing problems in order to concentrate on another section. In this case, use the JUMP command to move to a point in the DATA step after the problematic section.
- when you want to re-execute a series of statements that have caused problems. In this case, use JUMP to move to a point in the DATA step before the problematic statements and use the SET command to reset values of the relevant variables to the values that they had at that point. Then re-execute those statements with STEP or GO.

## Example

- Jump to line 5:

  j 5

## See Also

### Commands:

- "GO" on page 830
- "STEP" on page 834

## LIST

Displays all occurrences of the item that is listed in the argument.

**Category:** Manipulating Debugging Requests

**Alias:** L

### Syntax

LIST <_ALL_ | BREAK | DATASETS | FILES | INFILES | WATCH>

#### Required Arguments

**_ALL_**
   displays the values of all items.

**BREAK**
   displays breakpoints.

   Alias   B

**DATASETS**
   displays all SAS data sets used by the current DATA step.

**FILES**
   displays all external files to which the current DATA step writes.

**INFILES**
: displays all external files from which the current DATA step reads.

**WATCH**
: displays watched variables.

    Alias  W

## Example

- List all breakpoints, SAS data sets, external files, and watched variables for the current DATA step:

    l _all_

- List all breakpoints in the current DATA step:

    l b

## See Also

### Commands:

- "BREAK" on page 823
- "DELETE" on page 826
- "WATCH" on page 836

# QUIT

Terminates a debugger session.

**Category:** Terminating the Debugger

**Alias:** Q

## Syntax

**QUIT**

### Without Arguments

The QUIT command terminates a debugger session and returns control to the SAS session.

## Details

SAS creates data sets built by the DATA step that you are debugging. However, when you use QUIT to exit the debugger, SAS does not add the current observation to the data set.

You can use the QUIT command at any time during a debugger session. After you end the debugger session, you must resubmit the DATA step with the DEBUG option to begin a new debugging session; you cannot resume a session after you have ended it.

## SET

Assigns a new value to a specified variable.

    **Category:**   Manipulating DATA Step Variables

        **Alias:**   None

### Syntax

**SET** *variable=expression*

#### Required Arguments

*variable*
: specifies the name of a DATA step variable or an array reference.

*expression*
: is any debugger expression.

    Tip   *expression* can contain the variable name that is used on the left side of the equal sign. When a variable appears on both sides of the equal sign, the debugger uses the original value on the right side to evaluate the expression and stores the result in the variable on the left.

### Details

The SET command assigns a value to a specified variable. When you detect an error during program execution, you can use this command to assign new values to variables. This enables you to continue the debugging session.

### Example

- Set the variable A to the value of 3:

  ```
  set a=3
  ```

- Assign to the variable B the value **12345** concatenated with the previous value of B:

  ```
  set b='12345' || b
  ```

- Set array element ARR{1} to the result of the expression a+3:

  ```
  set arr{1}=a+3
  ```

- Set array element CRR{1,2,3} to the result of the expression crr{1,1,2} + crr{1,1,3}:

  ```
  set crr{1,2,3} = crr{1,1,2} + crr{1,1,3}
  ```

- Set the variable A to the result of the expression a+c*3:

  ```
  set a=a+c*3
  ```

## STEP

Executes statements one at a time in the active program.

**Category:** Controlling Program Execution

**Alias:** ST

## Syntax

**STEP** <*n*>

### Without Arguments
STEP executes one statement.

### Optional Argument

*n*
    specifies the number of statements to execute.

## Details

The STEP command executes statements in the DATA step, starting with the statement at which execution was suspended.

When you issue a STEP command, the debugger:

- executes the number of statements that you specify
- displays the line number
- returns control to the user and displays the > prompt.

*Note:* By default, you can execute the STEP command by pressing the ENTER key.

## See Also

### Commands:

- "GO" on page 830
- "JUMP" on page 831

# SWAP

Switches control between the SOURCE window and the LOG window.

**Category:** Controlling the Windows

**Alias:** None

## Syntax

**SWAP**

### Without Arguments
The SWAP command switches control between the LOG window and the SOURCE window when the debugger is running. When you begin a debugging session, the LOG window becomes active by default. While the DATA step is still being executed, the SWAP command enables you to switch control between the SOURCE and LOG window

## TRACE

Controls whether the debugger displays a continuous record of the DATA step execution.

**Category:** Manipulating Debugging Requests

**Alias:** T

**Default:** OFF

### Syntax

**TRACE** <ON | OFF>

#### Without Arguments

Use the TRACE command without arguments to determine whether tracing is on or off.

#### Optional Arguments

**ON**
> prepares for the debugger to display a continuous record of DATA step execution. The next statement that resumes DATA step execution (such as GO) records all actions taken during DATA step execution in the DEBUGGER LOG window.

**OFF**
> stops the display.

### Comparisons

TRACE displays the current status of the TRACE command.

### Example

- Determine whether TRACE is ON or OFF:

    ```
    trace
    ```

- Prepare to display a record of debugger execution:

    ```
    trace on
    ```

## WATCH

Suspends execution when the value of a specified variable changes.

**Category:** Manipulating Debugging Requests

**Alias:** W

## Syntax

**WATCH** *variable(s)*

### Required Argument

*variable(s)*
> specifies one or more DATA step variables.

## Details

The WATCH command specifies a variable to monitor and suspends program execution when its value changes.

Each time the value of a watched variable changes, the debugger does the following:

- suspends execution
- displays the line number where execution has been suspended
- displays the variable's old value
- displays the variable's new value
- returns control to the user and displays the > prompt.

## Example

- Monitor the variable DIVISOR for value changes:

```
w divisor
```

# Glossary

**across variable**
in the REPORT procedure, a variable whose formatted values each form a column in the report. If the variable does not have a format, then each unformatted value forms a column.

**alphanumeric character**
any of the following types of characters: alphabetic letters, numerals, and special characters or blanks. Most computer systems store strictly numeric data differently from alphanumeric or textual data.

**analysis variable**
a numeric variable that is used to calculate statistics or to display values. Usually an analysis variable contains quantitative or continuous values, but this is not required.

**arithmetic expression**
See SAS expression

**array**
in the SAS programming language, a temporary grouping of SAS variables that have the same data type, are arranged in a particular order, and are identified by an array name. The array exists only for the duration of the current DATA step.

**array name**
a name that is selected to identify a group of variables or temporary data elements. It must be a valid SAS name that is not the name of a variable in the same DATA step or SCL (SAS Component Language) program.

**array reference**
a reference to an element to be processed in an array.

**assignment statement**
a DATA step statement that evaluates an expression and stores the result in a variable.

**attribute**
See variable attribute

**autocall facility**
a feature of SAS that enables you to store the source statements that define a macro and to invoke the macro as needed, without having to include the definition in your program.

**autoexec file**
a file that contains SAS statements that are executed automatically when SAS is invoked. The autoexec file can be used to specify some of the SAS system options, as well as to assign librefs and filerefs to data sources that are used frequently.

**automatic macro variable**
a macro variable that is defined by SAS rather than by the user and that supplies information about the SAS session. For example, the SYSPROCESSID automatic macro variable contains the process ID of the current SAS process.

**background processing**
processing in which you cannot interact with the computer. Background sessions sometimes run somewhat slower than foreground sessions, because background sessions execute as processor time becomes available.

**Base SAS**
the core product that is part of SAS Foundation and is installed with every deployment of SAS software. Base SAS provides an information delivery system for accessing, managing, analyzing, and presenting data.

**batch job**
a unit of work that is submitted to an operating system for batch processing. For example, under UNIX, a batch job is a background process; under Windows, a batch job is a task; and under z/OS, a batch job is a set of JCL statements.

**batch mode**
a noninteractive method of running SAS programs by which a file (containing SAS statements along with any necessary operating system commands) is submitted to the batch queue of the operating environment for execution.

**Boolean operator**
another term for logical operator.

**break**
in the REPORT procedure, a section of the report that does one or more of the following: visually separates parts of the report; summarizes statistics and computed variables; displays text; displays values that have been calculated for a set of rows of the report; executes DATA step statements. You can create breaks when the value of a selected variable changes or at the beginning or end of a report.

**break line**
in the REPORT procedure, a line of a report that contains one or more of the following: characters that visually separate parts of the report; summaries of statistics and computed variables (called a summary line); text; values that have been calculated for a set of rows of the report.

**break variable**
in the REPORT procedure, a group variable or order variable that you select in order to specify the location of break lines. The REPORT procedure performs the actions that you specify for the break each time the value of the break variable changes.

## BY group

a group of observations or rows that have the same value for a variable that is specified in a BY statement. If more than one variable is specified in a BY statement, then the BY group is a group of observations that have a unique combination of values for those variables.

## BY group variable

See BY variable

## BY value

the value of a BY variable.

## BY variable

a variable that is named in a BY statement and whose values define groups of observations to process.

## BY-group processing

the process of using the BY statement to process observations that are ordered, grouped, or indexed according to the values of one or more variables. Many SAS procedures and the DATA step support BY-group processing. For example, you can use BY-group processing with the PRINT procedure to print separate reports for different groups of observations in a single SAS data set.

## CALL routine

a component of the SAS programming language that changes the values of variables or performs other system operations. CALL routines are similar to functions except that you cannot use CALL routines in assignment statements. All SAS CALL routines are invoked with CALL statements. That is, the name of the routine must follow the keyword CALL in the CALL statement.

## carriage-control character

a symbol that tells a printer how many lines to advance the paper, when to begin a new page, when to skip a line, and when to hold the current line for overprinting.

## catalog

See SAS catalog

## catalog directory

a part of a SAS catalog that stores and maintains information about the name, type, description, and update status of each member of the catalog.

## catalog entry

See SAS catalog entry

## category

in the TABULATE procedure, the combination of unique values of class variables. The TABULATE procedure creates a separate category for each unique combination of values that exists in the observations of the data set. Each category that PROC TABULATE creates is represented by one or more cells in the table where the pages, rows, and columns that describe the category intersect.

## character constant

a character string that is enclosed in quotation marks in a SAS statement to indicate a fixed value rather than the name of a variable. The maximum number of characters that is allowed is 32,767. Character constants are sometimes referred to as character literals.

**character format**
    a set of instructions that tell SAS to use a specific pattern for writing character data values.

**character function**
    a type of function that enables you to manipulate, compare, evaluate, or analyze character strings.

**character informat**
    a set of instructions that tell SAS to use a specific pattern for reading character data values into character variables.

**character literal**
    another term for character constant.

**character string**
    one or more consecutive alphanumeric characters, other keyboard characters, or both.

**character value**
    a value that can contain alphabetic characters, the numeric characters 0 through 9, and other special characters.

**character variable**
    a variable whose values can consist of alphabetic characters and special characters as well as numeric characters.

**chart**
    a graph in which elements, such as bars or pie slices, represent a view of the data.

**chart statistic**
    the statistical value calculated for the chart variable: frequency, cumulative frequency, percentage, cumulative percentage, sum, or mean.

**chart variable**
    a variable in the input data set whose values are categories of data represented by bars, blocks, slices, or spines.

**class variable**
    See classification variable

**classification variable**
    a variable whose values are used to classify the observations in a data set into different groups that are meaningful for analysis. A classification variable can have either character or numeric values. Classification variables include group, subgroup, category, and BY variables.

**column input**
    in the DATA step, a style of input in which column numbers are included in the INPUT statement to tell SAS which columns contain the values for each variable. This style of input is useful when the values for each variable are in the same location in all records.

**command**
    a directive to an operating system to perform a particular task.

**comment**
See comment statement

**comment statement**
information that is embedded in a SAS program and that serves as explanatory text. SAS ignores comments during processing but writes them to the SAS log. Comment syntax has several forms. For example, a comment can appear as a statement that begins with an asterisk and ends with a semicolon, as in * message ;.

**comparison operator**
in programming languages, a symbol or mnemonic code that is used in expressions to test for a particular relationship between two values or text strings. For example, the symbol < and its corresponding mnemonic, LT, are used to determine whether one value is less than another.

**compilation**
See program compilation

**composite index**
an index that locates observations in a SAS data set by examining the values of two or more key variables.

**compound expression**
an expression that contains more than one operator.

**computed variable**
a variable whose value is calculated. For example, in the REPORT procedure, the value is calculated from statements that are entered in the COMPUTE window.

**concatenate**
to join the contents of two or more elements, end to end, forming a separate element. Examples of elements are character values, tables, external files, SAS data sets, and SAS libraries.

**condition**
in a SAS program, one or more numeric or character expressions that result in a value on which some decision depends.

**configuration file**
an external file containing the SAS system options that define the environment in which to run SAS. These system options take effect each time you invoke SAS.

**configuration option**
a SAS system option that can be specified in the SAS command or in a configuration file. Configuration options affect how SAS interacts with the computer hardware and operating system.

**constant**
in SAS software, a number or a character string that indicates a fixed value.

**constant text**
the character strings that are stored as part of a macro or as a macro variable's value in open code, from which the macro processor generates text to be used as SAS statements, display manager commands, or other macro program statements. Constant text is also called model text.

**crossing**
> in the TABULATE procedure, the process that combines the effects of two or more elements.

**data error**
> a type of execution error that occurs when a SAS program analyzes data that contains invalid values. For example, a data error occurs if you specify numeric variables in the INPUT statement for character data. SAS reports these errors in the SAS log but continues to execute the program.

**data lines**
> lines of unprocessed (raw) data.

**data set**
> See SAS data set

**data set label**
> in a SAS data set, a user-defined attribute of up to 200 characters that is used for documenting the SAS data set.

**data set option**
> a SAS language element that specifies actions to apply to a particular SAS data set. For example, data set options enable you to rename variables, to select only observations for processing, to drop variables from processing, or to specify a password.

**DATA step**
> in a SAS program, a group of statements that begins with a DATA statement and that ends with either a RUN statement, another DATA statement, a PROC statement, or the end of the job. The DATA step enables you to read raw data or other SAS data sets and to create SAS data sets.

**data value**
> a unit of character, numeric, or alphanumeric information that is stored as a single item in a data record.

**data view**
> See SAS data view

**date and time format**
> instructions that tell SAS how to write numeric values as dates, times, and datetimes.

**date and time informat**
> the instructions that tell SAS how to read numeric values that are represented as dates, times, and datetimes.

**date constant**
> See SAS date constant

**date value**
> See SAS date value

**datetime constant**
> See SAS datetime constant

**datetime value**
See SAS datetime value

**declarative statement**
a statement that supplies information to SAS and that takes effect when the system compiles program statements.

**default directory**
the directory that you are working in at any given time. When you log in, your default directory is usually your home directory.

**delimiter**
a character that serves as a boundary that separates the elements of a text string.

**descriptor information**
information about the contents and attributes of a SAS data set. For example, the descriptor information includes the data types and lengths of the variables, as well as which engine was used to create the data. SAS creates and maintains descriptor information within every SAS data set.

**destination**
See ODS destination

**detail row**
in the REPORT procedure, a row of a report that contains information from a single observation in the data set or which consolidates the information for a group of observations that have a unique combination of values for all group variables.

**dimension**
See table dimension

**dimension expression**
in the TABULATE procedure, the portion of the TABLE statement that defines the content and arrangement of the rows, columns, or pages of the table.

**DO group**
a sequence of statements that starts with a simple DO statement and that ends with a corresponding END statement.

**DO loop**
a sequence of statements that starts with an iterative DO, DO WHILE, or DO UNTIL statement and that ends with a corresponding END statement. The statements are executed (usually repeatedly) according to directions that are specified in the DO statement.

**double trailing at sign**
a special symbol @@ that is used to hold a line of data in the input buffer during multiple iterations of a DATA step.

**entry type**
a characteristic of a SAS catalog entry that identifies the catalog entry's structure and attributes to SAS. When you create a SAS catalog entry, SAS automatically assigns the entry type as part of the name.

### error message
a message in the SAS log or Message window that indicates that SAS was not able to continue processing the program.

### executable statement
in the DATA step, a SAS statement that causes some action to occur while the DATA step executes rather than when SAS compiles the DATA step.

### explicit array
in the DATA step, an array that consists of a valid SAS name, a reference to the number of variables or temporary data elements, and an optional list of the array elements. When referring to an element of an explicit array, you must specify the array element's subscript.

### explicit array reference
a description of the element to be processed in an explicit array.

### exponent
in a mathematical expression, the number or expression that indicates the power to which you raise a base number or expression. For example, the exponent is 4 in the following expression: .1234 * 10 raised to the fourth power.

### external file
a file that is created and maintained by a host operating system or by another vendor's software application. An external file can read both data and stored SAS statements.

### file reference
See fileref

### file specification
the name of an external file. This name is the name by which the host operating environment recognizes the file. On directory-based systems, the file specification can be either the complete pathname or the relative pathname from the current working directory.

### fileref
a name that is temporarily assigned to an external file or to an aggregate storage location such as a directory or a folder. The fileref identifies the file or the storage location to SAS.

### FIRST. variable
a temporary variable that SAS creates to identify the first observation of each BY group. The variable is not added to the SAS data set.

### format
See SAS format

### format modifier
a special symbol that is used in the INPUT and PUT statements and which enables you to control how SAS reads input data and writes output data.

### formatted input
a style of input that uses special instructions called informats in the INPUT statement to determine how values that are entered in data fields should be interpreted.

### formatted output
a style of output that uses SAS formats in the PUT statement to specify how to write the values of variables.

### function
See SAS function

### global macro variable
a macro variable that can be referenced in either global or local scope in a SAS program, except where there is a local macro variable that has the same name. A global macro variable exists until the end of the session or program.

### header routine
a group of DATA step statements that produces page headers in print files. A header routine begins with a statement label and ends with a RETURN statement. You identify with the HEADER= option in the FILE statement.

### host
See host operating environment

### host operating environment
the operating environment (computer, operating system, and other software and hardware) that is identified by an IP address or by a domain name and that provides centralized control for software applications.

### index
a component of a SAS data set that enables SAS to access observations in the SAS data set quickly and efficiently. The purpose of SAS indexes is to optimize WHERE-clause processing and to facilitate BY-group processing.

### informat
See SAS informat

### input buffer
a temporary area of memory into which each record of data is read when the INPUT statement executes.

### inset table
a table of statistical values that is placed in or beside a plot that is produced in graphics mode. In process capability analysis, the statistics can include capability indices, specification limits, goodness-of-fit statistics, curve parameters, descriptive statistics, and quantiles.

### interactive line mode
a method of running SAS programs in which you enter one line of a SAS program at a time at the SAS session prompt. SAS processes each line immediately after you press the ENTER or RETURN key. Procedure output and informative messages are returned directly to your display device.

### interleaving
a process in which SAS combines two or more sorted SAS data sets into one sorted SAS data set based on the values of the BY variables.

### keyword
See SAS keyword

## label
descriptive text associated with a variable. By default, this text is the name of a variable or of a label previously assigned with the LABEL= option.

## LAST. variable
a temporary variable that SAS creates to identify the last observation of each BY group. This variable is not added to the SAS data set.

## library reference
See libref

## libref
a SAS name that is associated with the location of a SAS library. For example, in the name MYLIB.MYFILE, MYLIB is the libref, and MYFILE is a file in the SAS library.

## line mode
See interactive line mode

## line-hold specifier
a special symbol used in INPUT and PUT statements that enables you to hold a record in the input or output buffer for further processing. Line-hold specifiers include the trailing at sign (@) and the double trailing at sign (@@).

## list input
a style of input in which names of variables, not column locations, are specified in the INPUT statement. List input scans input records for data values that are separated by at least one blank or by some other delimiter.

## list output
a style of output in which character strings or variables are specified in a PUT statement without explicit directions that specify where SAS should place the strings or values.

## literal
a number or a character string that indicates a fixed value.

## log
See SAS log

## logical operator
an operator that is used in expressions to link sequences of comparisons. The logical operators are AND, OR, and NOT.

## macro facility
a component of Base SAS software that you can use for extending and customizing SAS programs and for reducing the amount of text that must be entered in order to perform common tasks. The macro facility consists of the macro processor and the macro programming language.

## macro invocation
another term for macro call.

## macro language
the programming language that is used to communicate with the macro processor.

**macro variable**
a variable that is part of the SAS macro programming language. The value of a macro variable is a string that remains constant until you change it. Macro variables are sometimes referred to as symbolic variables.

**macro variable reference**
the name of a macro variable, preceded by an ampersand (&name). The macro processor replaces the macro variable reference with the value of the specified macro variable.

**master data set**
in an update operation, the data set that contains the information that you want to update.

**match-merging**
a process in which SAS joins observations from two or more SAS data sets according to the values of the BY variables.

**member type**
a SAS name that identifies the type of information that is stored in a SAS file. Member types include ACCESS, AUDIT, DMBD, DATA, CATALOG, FDB, INDEX, ITEMSTOR, MDDB, PROGRAM, UTILITY, and VIEW.

**merging**
the process of combining observations from two or more SAS data sets into a single observation in a new SAS data set.

**missing value**
a type of value for a variable that contains no data for a particular row or column. By default, SAS writes a missing numeric value as a single period and a missing character value as a blank space.

**mnemonic operator**
an arithmetic or logical (Boolean) operator that consists of letters rather than symbols (for example, EQ rather than =).

**modified list input**
a style of input that uses special instructions called informats and format modifiers in the INPUT statement. Modified list input scans input records for data values that are separated by at least one blank (or by some other delimiter), or in some cases, by multiple blanks.

**multi-panel report**
output that uses sets of columns on a page to display the values of variables. For example, telephone books are usually arranged in multiple panels of names, addresses, and telephone numbers on a single page.

**named input**
a style in which equal signs appear in the INPUT statement to read data values in the form variable=data-value.

**named output**
a style in which equal signs appear in the PUT statement to write variable values in the form variable=data-value.

### noninteractive mode
a method of running SAS programs in which you prepare a file of SAS statements and submit the program to the operating system. The program runs immediately and comprises your current session.

### noninteractive processing
See noninteractive mode

### null statement
a statement that consists of a single semicolon or four semicolons. The null statement is most commonly used to designate the end of instream data in a DATA step.

### null value
a special value that indicates the absence of information. Null values are analogous to SAS missing values.

### numeric constant
a number that appears in a SAS expression.

### numeric format
a set of instructions that tell SAS to use a specific pattern for writing the values of numeric variables.

### numeric informat
a set of instructions that tell SAS to use a specific pattern for reading numeric data values.

### numeric value
a value that usually contains only numbers, which can include numbers in E-notation and hexadecimal notation. A numeric value can sometimes contain a decimal point, a plus sign, or a minus sign. Numeric values are stored in numeric variables.

### numeric variable
a variable that contains only numeric values and related symbols, such as decimal points, plus signs, and minus signs.

### observation
a row in a SAS data set. All of the data values in an observation are associated with a single entity such as a customer or a state. Each observation contains either one data value or a missing-value indicator for each variable.

### observation number
a number that indicates the relative position of an observation in a SAS data set when you read the entire data set sequentially. This number is not stored internally.

### ODS
See Output Delivery System

### ODS destination
a designation that the Output Delivery System uses to generate a specific type of output. Types of ODS destinations include but are not limited to HTML, XML, listing, PostScript, RTF, and SAS data sets.

**one-to-one matching**
the process of combining observations from two or more data sets into one observation, using two or more SET statements to read observations independently from each data set.

**one-to-one merging**
the process of using the MERGE statement (without a BY statement) to combine observations from two or more data sets based on the observations' positions in the data sets.

**output buffer**
in the DATA step, the area of memory that a PUT statement writes to before it writes to a designated file or output device.

**Output Delivery System**
a component of SAS software that can produce output in a variety of formats such as markup languages (HTML, XML), PDF, listing, RTF, PostScript, and SAS data sets. Short form: ODS.

**output object**
a programming object that contains the data that is generated by a DATA step or a PROC step and which can also contain a table definition that provides information about how to format that data.

**padding a value with blanks**
in SAS software, a process in which the software adds blanks to the end of a character value that is shorter than the length of the variable.

**PDV**
See program data vector

**permanent SAS data set**
a SAS data set that is not deleted after the current program or interactive SAS session ends. Permanent SAS data sets are available for future SAS sessions.

**permanent SAS file**
a file in a SAS library that is not deleted when the SAS session or job terminates.

**permanent SAS library**
a SAS library that is not deleted when a SAS session ends, and which is therefore available to subsequent SAS sessions.

**physical filename**
the name that an operating system uses to identify a file.

**pointer**
in the DATA step, a programming tool that SAS uses to keep track of its position in the input or output buffer.

**pointer control**
the process of instructing SAS to move the pointer before reading or writing data.

**print file**
an external file that contains carriage-control (printer-control) information.

**procedure**
See SAS procedure

**Profile catalog**
See Sasuser.Profile catalog

**program compilation**
the process of checking syntax and translating a portion of a program into a form that the computer can execute.

**program data vector**
the temporary area of computer memory in which SAS builds a SAS data set, one observation at a time. The program data vector is a logical concept and does not necessarily correspond to a single contiguous area of memory. Short form: PDV.

**programming error**
a flaw in the logic of a SAS program that can cause the program to fail or to perform differently than the programmer intended.

**propagation of missing values**
a consequence of using missing values in which a missing value in an arithmetic expression causes SAS to set the result of the expression to missing. Using that result in another expression causes the next result to be missing, and so on.

**raw data**
in statistical analysis, data (including data in SAS data sets) that has not had a particular operation, such as standardization, performed on it.

**raw data file**
an external file whose records contain data values in fields. A DATA step can read a raw data file by using the INFILE and INPUT statements.

**SAS catalog**
a SAS file that stores many different kinds of information in smaller units called catalog entries. A single SAS catalog can contain different types of catalog entries.

**SAS catalog entry**
an individual storage unit within a SAS catalog. Each entry has an entry type that identifies its purpose to SAS.

**SAS command**
a command that invokes SAS. This command can vary depending on the operating environment and site.

**SAS compilation**
the process of converting statements in the SAS language from the form in which you enter them to a form that is ready for SAS to use.

**SAS data file**
a type of SAS data set that contains data values as well as descriptor information that is associated with the data. The descriptor information includes information such as the data types and lengths of the variables, as well as the name of the engine that was used to create the data.

**SAS data set**
a file whose contents are in one of the native SAS file formats. There are two types of SAS data sets: SAS data files and SAS data views.

**SAS data set option**
an option that appears in parentheses after a SAS data set name. Data set options specify actions that apply only to the processing of that SAS data set.

**SAS data view**
a type of SAS data set that retrieves data values from other files. A SAS data view contains only descriptor information such as the data types and lengths of the variables (columns) plus other information that is required for retrieving data values from other SAS data sets or from files that are stored in other software vendors' file formats. Short form: data view.

**SAS date constant**
a string in the form 'ddMMMyy'd or 'ddMMMyyyy'd that represents a date in a SAS statement. The string is enclosed in quotation marks and is followed by the character d (for example, '6JUL01'd, '06JUL01'd, '6 JUL2001'd, or '06JUL2001'd).

**SAS date value**
an integer that represents a date in SAS software. The integer represents the number of days between January 1, 1960, and another specified date. For example, the SAS date value 366 represents the calendar date January 1, 1961.

**SAS datetime constant**
a string in the form 'ddMMMyy:hh:mm:ss'dt or 'ddMMMyyyy:hh:mm:ss'dt that represents a date and time in SAS. The string is enclosed in quotation marks and is followed by the characters dt (for example, '06JUL2001:09:53:22'dt).

**SAS datetime value**
an integer that represents a date and a time in SAS software. The integer represents the number of seconds between midnight, January 1, 1960, and another specified date and time. For example, the SAS datetime value for 9:30 a.m., June 5, 2000, is 1275816600.

**SAS expression**
a type of macro expression consisting of a sequence of operands and arithmetic operators that form a set of instructions that are evaluated to produce a numeric value, a character value, or a Boolean value. Examples of operands are constants and system functions. SAS uses arithmetic expressions in program statements to create variables, to assign values, to calculate new values, to transform variables, and to perform conditional processing.

**SAS file**
a specially structured file that is created, organized, and maintained by SAS. A SAS file can be a SAS data set, a catalog, a stored program, an access descriptor, a utility file, a multidimensional database file, a financial database file, a data mining database file, or an item store file.

**SAS format**
a type of SAS language element that applies a pattern to or executes instructions for a data value to be displayed or written as output. Types of formats correspond to the data's type: numeric, character, date, time, or timestamp. The ability to create user-defined formats is also supported. Examples of SAS formats are BINARY and DATE. Short form: format.

### SAS function

a type of SAS language element that can be used in an expression or assignment statement to process zero or more arguments and to return a value. Examples of SAS functions are MEAN and SUM. Short form: function.

### SAS informat

a type of SAS language element that applies a pattern to or executes instructions for a data value to be read as input. Types of informats correspond to the data's type: numeric, character, date, time, or timestamp. The ability to create user-defined informats is also supported. Examples of SAS informats are BINARY and DATE. Short form: informat.

### SAS initialization

the process of setting global characteristics that must be in effect in order for a SAS session to begin. SAS performs initialization by setting certain SAS system options called initialization options. SAS initialization happens automatically when you invoke SAS.

### SAS keyword

a literal that is a primary part of the SAS language. For example, SAS keywords include DATA, PROC, RUN, names of SAS language elements, names of SAS statement options, and system variables.

### SAS language

a programming language that includes procedures for data analysis and reporting, statements and functions for managing SAS files and manipulating data, options that define the SAS environment, a macro facility, Help menus, and a windowing environment for text editing and file management.

### SAS log

a file that contains a record of the SAS statements that you enter, as well as messages about the execution of your program.

### SAS name

a name that is assigned to items such as SAS variables and SAS data sets. For most SAS names, the first character must be a letter or an underscore. Subsequent characters can be letters, numbers, or underscores. Blanks and special characters (except the underscore) are not allowed. However, the VALIDVARNAME= system option determines what rules apply to SAS variable names. The maximum length of a SAS name depends on the language element that it is assigned to.

### SAS procedure

a program that provides specific functionality and that is accessed with a PROC statement. For example, SAS procedures can be used to produce reports, to manage files, or to analyze data. Many procedures are included in SAS software.

### SAS program

a group of SAS statements that guide SAS through a process or series of processes in order to read and transform input data and to generate output. The DATA step and the procedure step, used alone or in combination, form the basis of SAS programs.

### SAS session

the activity between invoking and exiting a specific SAS software product.

### SAS statement
a string of SAS keywords, SAS names, and special characters and operators that instructs SAS to perform an operation or that gives information to SAS. Each SAS statement ends with a semicolon.

### SAS system option
an option that affects the processing of an entire SAS program or interactive SAS session from the time the option is specified until it is changed. Examples of items that are controlled by SAS system options include the appearance of SAS output, the handling of some files that are used by SAS, the use of system variables, the processing of observations in SAS data sets, features of SAS initialization, and the way SAS interacts with your host operating environment.

### SAS time constant
a string in the form 'hh:mm:ss't that represents a time in a SAS statement. The string is enclosed in quotation marks and is followed by the character t (for example, '09:53:22't).

### SAS time value
an integer that represents a time in SAS software. The integer represents the number of seconds between midnight of the current day and another specified time value. For example, the SAS time value for 9:30 a.m. is 34200.

### SAS variable
a column in a SAS data set or in a SAS data view. The data values for each variable describe a single characteristic for all observations (rows).

### Sasuser library
a default, permanent SAS library that is created at the beginning of your first SAS session. The Sasuser library contains a PROFILE catalog that stores the customized features or settings that you specify for SAS.

### Sasuser.Profile catalog
a SAS catalog in which SAS stores information about attributes of the SAS windowing environment for a particular user or site. It contains function-key definitions, fonts for graphics applications, window attributes, and other information that is used by interactive SAS procedures.

### simple expression
a SAS expression that uses only one operator.

### simple index
an index that uses the values of only one variable to locate observations.

### site number
the number that SAS uses to identify the company or organization to which SAS software is licensed. The site number appears near the top of the log in every SAS session.

### split character
in some SAS procedures, a character that splits headers across multiple lines. If you use the split character in a column header, the procedure breaks the header when it reaches that character and continues the header on the next line. The split character itself is not part of the column header.

**standard data**
  data in which each digit or character occupies one byte of storage.

**statement**
  See SAS statement

**statement label**
  a SAS name followed by a colon that prefixes a statement in a DATA step so that other statements can direct execution to that statement as necessary, bypassing other statements in the step.

**statement option**
  a word that you specify in a particular SAS statement and which affects only the processing that that statement performs.

**step boundary**
  a point in a SAS program when SAS recognizes that a DATA step or PROC step is complete.

**string**
  See character string

**sum statement**
  a DATA step statement that adds the result of the expression on the right side of the plus sign to the accumulator variable on the left side of the plus sign. A sum statement has the following form: variable+expression;

**summary table**
  output that provides a concise overview of the information in a data set.

**syntax checking**
  the process by which SAS checks each SAS statement for proper usage, correct spelling, proper SAS naming conventions, and so on.

**syntax error**
  an error in the spelling or grammar of a SAS statement. SAS finds syntax errors as it compiles each SAS step before execution.

**system option**
  See SAS system option

**table definition**
  a set of instructions that describe how to format output in the Output Delivery System (ODS).

**table dimension**
  one of the basic elements of a table, such as a page, column, or row.

**temporary SAS data set**
  a data set that exists only for the duration of the current program or interactive SAS session. Temporary SAS data sets are not available for future SAS sessions.

**temporary SAS file**
  a SAS file in a SAS library (usually the Work library) that is deleted at the end of the SAS session or job.

### temporary SAS library
a library that exists only for the current SAS session or job. The most common temporary library is the Work library.

### text string
See character string

### text-editing command
a command that is used with a particular text editor.

### time constant
See SAS time constant

### time value
See SAS time value

### title
a heading that is printed at the top of each page of SAS output or of the SAS log.

### trailing at sign
a special symbol @ that is used to hold a line of input or output so that SAS can read from it or write to it in a subsequent INPUT or PUT statement.

### transaction data set
in an update operation, the data set that contains the information that is needed in order to update the master data set.

### updating
a process in which SAS replaces the values of variables in the master data set with values from observations in the transaction data set.

### user-defined format
a format that you define with the FORMAT procedure or with C, PL/I, FORTRAN, or IBM 370 assembler language using SAS/TOOLKIT software.

### user-defined informat
an informat that you define with the FORMAT procedure or with C, PL/I, FORTRAN, or IBM 370 assembler language using SAS/TOOLKIT software.

### variable
See SAS variable

### variable attribute
any of the following characteristics that are associated with a particular variable: name, label, format, informat, data type, and length.

### variable type
the classification of a variable as either numeric or character. Type is an attribute of SAS variables.

### WHERE expression
defines the criteria for selecting observations.

### WHERE processing
a method of conditionally selecting rows for processing by using a WHERE expression.

**Work library**
a temporary SAS library that is automatically defined by SAS at the beginning of each SAS session or SAS job. Unless you have specified a User library, any newly created SAS file that has a one-level name will be placed in the Work library by default and will be deleted at the end of the current SAS session or job.

# Index

**Special Characters**
_ (underscore), in SAS names  6
_DSEMTR_ code  339
_DSENMR_ code  339
_ERROR_ variable  408
_IORC_ automatic variable  339
_N_ variable  408
_SOK_ code  339
, (comma), in input data  61
; (semicolon)
  end-of-data indicator  44
  in statements  6
: (colon)
  character comparisons  162
  format modifier  64
!! (exclamation points), concatenation operator  139
/ (slash), column-pointer control
  description  69, 87
  forcing pointer to next line  79
/ (slash), splitting column headers  516
. (period)
  as missing value  30, 134
  in informat names  61
  in input data  60
' (quotation mark)
  as literal character  131
  variable indicator  131
[ ] (square brackets), in STYLE= option  493
{ } (braces), in STYLE= option  493
@ (trailing @)
  description  87
  reading raw data records  72
  releasing held output lines  601
  writing output lines  600, 610
@@ (double trailing @)
  DATA step execution and  74
  definition  73
  description  87
@n, column-pointer control  69, 87
@n, pointer control  604
  *See also* column-pointer controls
$ (dollar sign)
  defining character variables  42
  in input data  60
  in variable names  131
#n, column-pointer control  69, 87
#n, line-pointer control
  DATA step execution and  81
  skipping input variables  80
%INCLUDE statement
  description  15
  interactive line mode  13
%LET statement  471
%LIST statement  13
%PUT macro statement  387
+n, column-pointer control  69, 87
= (equal sign)
  defining summary table labels  491
  drawing lines with  522
|| (vertical bars), concatenation operator  139

**Numbers**
1969 to 2010 option
  tick mark values  538

**A**
A (after) command  776
absolute column-pointer control  63
ACCESS member type  691
ACROSS variable  503, 518, 519
adding numbers
  *See* numeric variables, calculations on
  *See* observations, calculations on
  *See* summing numbers
aliases for files  46
aliases for libraries
  *See* librefs
aligned raw data
  *See* column input
aligning values  137
ampersand
  format modifier  64
  in macro variable names  467

ANALYSIS variable 503, 510
analysis variables, specifying 477
apostrophe
    *See* quotation mark (')
APPEND procedure 279
    concatenating SAS data sets 274
    description 279
    versus SET statement 278
applications, customizing 798
arithmetic operations
    *See* numeric variables, calculations on
    *See* observations, calculations on
array processing 219
    *See also* DO groups
    defining arrays 219
    iterative DO loops 220
    selecting current variable 220
ARRAY statement
    defining arrays 219
    description 222
arrays, definition 219
ASCII collating sequence 196
assignment statements
    arithmetic operators and 119
    description 116, 127
    in DATA step 109
    numeric expressions and 121
    overview 109
attributes, variables 262
AUTOEXPAND command 791
    *See also* TREE command
AUTOSYNC command 791

## B

B (before) command 766, 776
background processing 744
BACKWARD command 763, 775
bar charts
    horizontal 558
    vertical, creating 556
    vertical, midpoint values 563
    vertical, number of midpoints 566
batch mode 12, 748
BFIND command 104
BLANKLINE option, PROC PRINT 462
BLANKLINE= option
    PROC PRINT statement 469
blanks
    as missing values 30, 134
    in SAS names 6
    leading, removing 138
    list input delimiter 53
blanks, embedded
    *See* embedded blanks
block charts 560

BLOCK statement, CHART procedure 587
    block charts 560
    three-dimensional charts 573
BODY= option 681
BOTTOM command 763, 775
BOX option
    box borders around plots 540
    PLOT statement 548
braces, in STYLE= option 493
BREAK command
    DATA step debugger 823
break lines in reports 520
BREAK statement, REPORT procedure 524
    break lines 520
BY groups
    definition 282
    totaling 204
BY statement
    computing group subtotals 449, 453
    finding first or last observation 190
    FIRST. and LAST. variables 197
    grouping observations 185, 456
    identifying group subtotals 451
    in detail reports 449
    interleaving SAS data sets 285
    match-merging SAS data sets 296
    merging SAS data sets 315
    modifying SAS data sets 347
    PRINT procedure 470
    printing values by group 606
    SORT procedure 471, 682
    UNIVARIATE procedure 683
    updating SAS data sets 333, 338
    writing to output files 610
    writing to SAS log 610
BY values 282
BY variables
    definition 282
    duplicates 343
    selecting for SAS data set update 318
BYE command
    description 750
    ending SAS sessions 746
    managing windows 763

## C

C (copy) command 776
CALCULATE command
    DATA step debugger 825
calendar dates
    *See also* date functions
    *See also* date values
    converting to SAS date values 232, 243

versus SAS date values  226
calling windows  761
CAPS command  768, 776
case sensitivity
    character comparisons  161
    character variables  131
    converting characters to uppercase  161
    SAS language  6
    sorting observations  195
    statements  6
    variable names  6
case, changing  768
    CAPS command  768, 776
    CCL (case lower) command  769
    CCU (case upper) command  769
    CL (case lower) command  769
    CU (case upper) command  769
    UPCASE function  161, 166
case, setting default for  768
catalog management  696
CATALOG member type  691
CATALOG procedure  696
CCL (case lower) command  769
CCU (case upper) command  769
cells, report  518
CENTER option  641
    column alignment  514
    DEFINE statement  525
century cutoff
    See YEARCUTOFF= system option
CFILL= option  585
CFRAME= option  585
CFREQ option
    HBAR statement  588
    horizontal bar charts  558
CGRID= option
    HISTOGRAM statement  589
    histograms  577
CHANGE command  765, 776
CHANGE statement  723
character comparisons
    case sensitivity  161
    types of  161
character groups, selecting  162, 163, 164
character strings, scanning for  137
character variables  129
    aligning values  137
    blanks, removing leading  138
    case sensitivity  131
    contents of  42
    creating  42, 137
    definition  130
    dollar sign ($), in variable names  131
    extracting portions of  137
    identifying  131
    length, default  42

length, determining  132
length, displaying  133
length, maximum  133
length, setting  133
longer than eight bytes  64
missing values, blanks as  134
missing values, checking for  135
missing values, periods as  134
missing values, setting  136
numbers as characters  142
quotation mark ('), as literal character  131
quotation mark ('), variable indicator  131
scanning for character strings  137
truncation of  132
character variables, combining
    See character variables, concatenating
character variables, concatenating  139
    adding characters  141
    exclamation points (!!), concatenation operator  139
    simple concatenation  140
    vertical bars (||), concatenation operator  139
CHART procedure  552
    BLOCK statement, block charts  560
    BLOCK statement, description  587
    BLOCK statement, three-dimensional charts  573
    charting frequencies  555
    HBAR statement, description  587
    HBAR statement, horizontal bar charts  558
    PIE statement, description  587
    PIE statement, pie charts  561
    PROC CHART statement  586
    VBAR statement, description  587
    VBAR statement, vertical bar charts  556
charts  552
    See also CHART procedure
    See also frequency charts
    See also histograms
    See also PLOT procedure
    See also plots
    See also UNIVARIATE procedure
    See also vertical bar charts
    block charts  560
    charting every value  567
    charting means  572
    discrete versus continuous values  567
    horizontal bar charts  558
    pie charts  561
    subgroups within ranges  570

tables of statistics, suppressing 558,
575
three-dimensional 573
tools for 552
charts, midpoints for
character variables, values of 569
histograms 580
numeric variables, number of 566
numeric variables, values of 563
CHILD command
description 791
toggling Contents pane on and off 784
CITY data set 92, 814
CL (case lower) command 769
CLASS statement
comparative histograms 584
specifying summary table class
variables 477
TABULATE procedure 497, 682
UNIVARIATE procedure 588, 683
class variables
missing values 478
ordering 495
specifying 477
CLASSLEV statement
TABULATE procedure 682
CLEAR command
clearing windows 395, 763, 776
description 791
CLIMATE.HIGHTEMP data set 714,
726, 820
CLIMATE.LOWTEMP data set 714, 726,
820
collating sequences 195
ASCII 196
EBCDIC 196
magnitude of letters 163
colon (:)
character comparisons 162
format modifier 64
COLOR command 764
colors
SASCOLOR command 764
SASCOLOR statement 807
SYNCOLOR command 764
windows 764
COLS command 767
column headings, reports
centering 629
customizing 515
in specific columns 631
variables as 518
column input 57
*See also* formatted input
*See also* list input
*See also* reading raw data records

creating SAS data sets 41
definition 57
embedded blanks 58
input pointers 67
mixing input styles 65
rules for 60
sample program for 57
skipping fields 59, 80
versus list input 58
COLUMN statement
customizing ODS output 672
laying out reports 69, 507
REPORT procedure 524
TEMPLATE procedure 672, 682
column-pointer controls 62
*See also* line-pointer controls
*See also* pointer controls
/ (slash), forcing pointer to next line 79
@n 69, 87
#n, description 69, 87
+n 69, 87
absolute 63
definition 62
description of 69, 87
formatted input 62, 69
relative 63
slash (/), description 69, 87
columns (raw data) 29
columns, report
layout 507
ordering 507
spacing 514
width 514
columns, SAS data sets
*See* variables
COLWIDTH= option
column width 514
description 523
combining SAS data sets
*See* SAS data sets, concatenating
*See* SAS data sets, interleaving
*See* SAS data sets, merging
*See* SAS data sets, modifying
*See* SAS data sets, updating
combining summary table elements 485,
488
command line commands 775
Command window 758
commands
command line commands 775
file-specific 772
line commands 776
operating environment, issuing from
SAS sessions 746, 748
SAS Windowing Environment 758
commas, in input data 61

Index **863**

comparison operators 154
COMPUTED variable 503
concatenating character variables
   *See* character variables, concatenating
concatenating SAS data sets
   *See* SAS data sets, concatenating
concatenating summary table elements 488
concatenation operators
   exclamation points (!!) 139
   vertical bars (||) 139
Contents pane, toggling on and off 791
CONTENTS procedure 696
CONTENTS statement 710
   describing SAS data set contents 92
   description 103, 710
   listing SAS data sets 705
CONTENTS= option 681
COPY procedure 696
COPY statement
   copying SAS data sets 727
   description 736
   moving SAS data sets 731
copying files or members 696, 697
copying from SAS data sets
   *See* SAS data sets, copying
CPERCENT option
   HBAR statement 588
   horizontal bar charts 558
cross-tabulation 474, 485, 519
crossing summary table elements 474, 485
CU (case upper) command 769
CURSOR command 764
customizing
   *See also* Explorer window, customizing
   *See also* ODS output, customizing
   *See also* output, customizing
   *See also* plots, customizing
   *See also* SAS sessions, customizing
   *See also* SAS sessions, customizing session-to-session
   applications 798
   column headers in reports 515
   detail reports 465
   frequency charts 563
   missing values output, with a procedure 640
   missing values output, with a system option 639
   reports 458
   Results window 784
   SAS Registry Editor 801
   SAS Windowing Environment 802
   SAS windows 764
   Templates window 788

CUT command 766

# D

D (delete) command 767, 776
D suffix for date values 232
data analysis utilities 6
data errors
   definition 401
   diagnosing 408
data management facility 4
DATA member type 690
data set names
   *See* SAS names
data sets
   *See* SAS data sets
DATA statement 15
   description 15, 48
   dropping/keeping variables 103
   versus SET statement 101
DATA step 5
   assignment statements 109
   compile phase 35
   compiled program files 691
   definition 5
   descriptor information 35
   duplicate BY variables 343
   example 36
   execution phase 35
   generating reports 596
   input buffers 35
   observations, changing globally 109
   observations, changing selectively 110
   output from 615
   process overview 37
   program data vectors 35
   variables, changing 111
   variables, creating 109
   variables, defining length of 114
   variables, efficient use of 112
   variables, storage space for 114
DATA step debugger 415
   assigning commands to ENTER key 828
   assigning commands to function keys 417
   assigning new variable values 834
   continuous record of DATA step execution 836
   customizing commands with macros 418
   DATA step generated by macros 418
   debugger sessions 416
   debugging DO loops 429
   deleting breakpoints 826
   deleting watch status 826

**864** Index

displaying variable attributes 827
displaying variable values 829
entering commands 417
evaluating expressions 825
examples 419
executing statements one at a time 834
expressions and 417
formats and 424
formatted variable values 430
help on commands 831
jumping to program line 831
listing items 832
macro facility with 418
macros as debugging tools 418
quitting 833
restarting suspended programs 831
resuming DATA step execution 830
starting DATA step execution 830
suspending execution 823, 836
switching window control 835
windows 417
data values 4
DATA_NULL statement
  description 610
  writing reports from DATA step 596
data, ODS 10, 646
data, raw
  *See* raw data
DATA= option 15
  creating summary tables 477
  description 15
  PROC CHART statement 587
  PROC PLOT statement 548
  PROC PRINT statement 469
  PROC REPORT statement 523
  PROC TABULATE statement 497
  PROC UNIVARIATE statement 588
database entries, output to
  *See also* ODS
  traditional output 8
DATALINES statement
  creating SAS data sets 44
  description 503
  running SAS programs in interactive line mode 747
DATASETS procedure 698
  CHANGE statement 723
  CONTENTS statement, description 710
  CONTENTS statement, listing SAS data sets 705
  COPY statement, copying SAS data sets 727
  COPY statement, description 736
  COPY statement, moving SAS data sets 731
  definition 696

DELETE statement, deleting SAS data sets 734
DELETE statement, description 736
EXCLUDE statement, copying SAS data sets 731
EXCLUDE statement, description 736
EXCLUDE statement, moving SAS data sets 733
FORMAT statement, description 724
FORMAT statement, reformatting SAS data set variable attributes 717
LABEL statement, assigning SAS data set labels 720
LABEL statement, description 724
LABEL statement, modifying SAS data set labels 720
LABEL statement, removing SAS data set labels 720
listing SAS data sets 704
managing SAS libraries 696
MODIFY statement, assigning SAS data set labels 720
MODIFY statement, description 723
MODIFY statement, modifying SAS data set labels 720
MODIFY statement, modifying SAS data set variable attributes 716
MODIFY statement, reformatting SAS data set variable attributes 717
MODIFY statement, removing SAS data set labels 720
MODIFY statement, renaming SAS data set variable attributes 716
PROC DATASETS statement, description 103, 698, 710
PROC DATASETS statement, directory listings 702
PROC DATASETS statement, KILL option 736
PROC DATASETS statement, managing SAS libraries 697
RENAME statement, description 724
RENAME statement, renaming SAS data set variable attributes 716
RENAME statement, renaming SAS data sets 714
SAVE statement, deleting SAS data sets 735
SAVE statement, description 736
SELECT statement, copying SAS data sets 731
SELECT statement, description 736
SELECT statement, moving SAS data sets 733
date functions 239
  *See also* date values

TODAY(), description 243
WEEKDAY, description 243
WEEKDAY, returning day of the week 239
DATE option 642
date values 226
  as constants 232
  as input data 60
  calculations on 237
  calendar dates, converting to SAS date values 232, 243
  calendar dates, versus SAS date values 226
  century cutoff, determining 42, 227
  creating 238
  D suffix 232
  DATE7. informat, description 228, 243
  DATE7. informat, length of year 230
  DATE9. informat, description 228, 243
  DATE9. informat, length of year 230
  displaying 233
  entering 228
  FORMAT statement, description 243
  FORMAT statement, permanent date formats 234
  formats for 233
  in reports 465, 626
  informats for 228
  MMDDYY10. informat, description 228, 243
  MMDDYY10. informat, length of year 230
  MMDDYY8. informat, description 228, 243
  MMDDYY8. informat, length of year 230
  programming practices 230
  reading 229, 230
  SAS storage format 226
  sorting 237
  two-digit years versus four-digit 42, 227, 230
  WEEKDATE29. format, description 243
  WEEKDATE29. format, displaying dates 233
  WORDDATE18. format, description 243
  WORDDATE18. format, displaying dates 233
  YEARCUTOFF= system option, description 244
  YEARCUTOFF= system option, determining century 42, 227
date values, calculations
  comparing durations 241
  day of week, finding 239
date values, formatting
  for input 228
  for output 233
  permanently 235
  temporarily 237
DATE7. informat
  description 228, 243
  length of year 230
DATE9. informat
  description 228, 243
  length of year 230
day of week, finding 239
DBMS files, creating SAS data sets 45
DEBUGGER LOG window 417
DEBUGGER SOURCE window 417
debugging 399
  See also Log window
  See also SAS log
  DATA step debugger 415
  library assignment problems 757
  programs, Log window 778
  programs, Program Editor 780
  quality control checklist 412
debugging, with SAS Supervisor 402
  _ERROR_ variable 408
  _N_ variable 408
  data errors, definition 401
  data errors, diagnosing 408
  error types 401
  execution-time errors, definition 401
  SAS error processing 400
  semantic errors 402
  syntax checking 400
  syntax errors, diagnosing 402
DEFINE statement
  column width and spacing 514
  customizing ODS output 672
  defining GROUP variables 510
  formatting report items 517
  laying out reports 69, 508, 511
  REPORT procedure 525
  TEMPLATE procedure 672, 682
DELETE command
  DATA step debugger 826
DELETE statement 115
  See also observations, subsetting
  deleting observations 115, 171
  deleting SAS data sets 734
  description 115, 180, 736
  versus IF statement 173
DELETESELS command 791
DESCENDING option
  description 525
  report layout 511
DESCRIBE command

**866** *Index*

DATA step debugger 827
descriptive statistics, calculating for summary tables 487
DESELECT_ALL command 791
detail reports 501
　*See also* printing
　*See also* reports
　creating enhanced reports 446
　creating simple reports 436
　customizing 465
　date, including automatically 465
　definition 503
　formatting 447
　group subtotals, identifying 451
　key variables, emphasizing 439
　macro facility and 465
　observations, grouping by variable values 449
　observations, selecting 444
　observations, selecting (single comparison) 444
　sorted key variables 441
　summing numeric variables 448
　time, including automatically 465
　titles 465
　unsorted key variables 440
DETAILS command
　customizing Explorer window 804
　description 791
diagnosing and avoiding errors 399
diagnosing errors
　*See* debugging
　*See* debugging, with SAS Supervisor
dimension expressions 477
directory listings
　all files 702
　by member type 704
　definition 702
　formatting contents listings 708
DISCRETE option
　BLOCK statement 587
　discrete versus continuous values 567
　HBAR statement 587
　PIE statement 587
　VBAR statement 587
DISPLAY variable 504
DLGFONT command
　description 807
　opening Fonts window 806, 808
DLGPREF command
　customizing SAS sessions 802
　description 807
　opening Preferences window 808
　setting output formats 782
DLM= option 503
DMFILEASSIGN command

　description 790
　modifying file shortcuts 774
DMOPTLOAD command
　description 791
　retrieving system options 802
DMOPTSAVE command
　description 791
　saving system options 802
DMS option 750
DMSEXP option 746
DO groups 217
　*See also* array processing
　iterative DO loops 220
DO loops 220
　*See also* array processing
　debugging 429
DO statement
　description 223
　DO groups 217
　iterative DO loops 220
DOL option
　equal sign (=), drawing lines with 522
　RBREAK statement 526
dollar sign ($)
　defining character variables 42
　in input data 60
　in variable names 131
DOUBLE option, PROC PRINT
　double spacing LISTING reports 462
double trailing @ (@@)
　DATA step execution and 74
　definition 73
　description 87
double-clicking, keyboard equivalent 754
DROP statement 103
　description 103
　dropping variables 97
DROP= option 102
　DATA statement versus SET statement 101
　description 102
　dropping selected variables 97
　efficiency 101
　versus KEEP= option 98
dropping variables
　*See* DATA statement
　*See* DROP statement
　*See* DROP= option
　*See* SET statement
DSD option
　list input 55

**E**

EBCDIC collating sequence 196
Editor Options window

description 808
opening 808
Editor window
  *See* Program Editor
editors
  *See also* NOTEPAD window
  *See also* Program Editor
  *See also* SAS text editor
  customizing 806
EDOPT command 808
ELSE statement
  description 165
  selecting observations 151
embedded blanks 58
  in column input 58
  in list input 64
embedded special characters, reading
  *See* informats
END command 763
END statement
  description 223
  DO groups 217
  iterative DO loops 220
  TEMPLATE procedure 672, 682
END= option
  description 212, 362
  determining last observation 203, 359
ENDSAS command
  description 750
  ending interactive line mode 748
  ending SAS sessions 746
ENDSAS statement
  description 750
  ending SAS sessions 746
ENTER command
  DATA step debugger 828
equal sign (=)
  defining summary table labels 491
  drawing lines with 522
error diagnosis
  *See* debugging
  *See* debugging, with SAS Supervisor
error messages, suppressing logging of 383, 384
error processing 400
  *See also* debugging
  *See also* debugging, with SAS Supervisor
error types 401
errors
  diagnosing 402
  diagnosing and avoiding 399
  processing 400
  types of 401
ERRORS= option
  description 388

suppressing error messages 383, 384
EXAMINE command
  DATA step debugger 829
exclamation points (!!), concatenation operator 139
EXCLUDE statement
  copying SAS data sets 731
  description 736
  moving SAS data sets 733
execution-time errors
  definition 401
EXOPTS command 808
EXPFIND command
  description 790
  finding files 771
EXPLORER command
  description 750
  opening Explorer window 755
Explorer Options window
  description 808
  opening 808
Explorer window 744
  *See also* SAS Windowing Environment, windows
  definition 744
  finding files 770
  opening 755
Explorer window, customizing 802
  Contents Only view versus Explorer view 803
  Contents view 804
  editing options 806
  file types, enabling display of 805
  file types, hiding 806
  folders, adding and removing 805
  fonts 806
  icon size 804
  pop-up menu actions, adding 805
expressions
  DATA step debugger and 417
external files 46
  assigning filerefs to 46
  creating SAS data sets 44, 45
  specifying as input 45, 46
external files, output to
  *See also* ODS
  traditional output 8

# F

fields (raw data) 29
FILE command 396
  storing Log window 396
  storing Output window 396
  storing Program Editor 780
file contents, listing 707

*See also* CONTENTS statement
    all files in a library 707
    CONTENTS procedure 696
    formatting contents listings 708
    one file 705
file management
    *See* SAS Windowing Environment, file management
File Shortcut Assignment window
    assigning file shortcuts 772
    description 790
file shortcuts 772
    assigning 772, 781
    modifying 774
FILE statement
    description 610
    writing reports to SAS output files 603
FILENAME statement 31
    description 48
    filerefs for external files 46
filerefs, external files 46
files
    *See also* external files
    *See also* SAS files
    *See also* SAS Windowing Environment, file management
    copying 696, 697
    finding 770
    finding, with Explorer 770
    finding, with Find window 771
    issuing file-specific commands 772
    opening 772
    overwriting 396
    printing 774
    SAS data files 690
    working with 770
files, writing to
    *See* output routing, procedures
    *See* reports, SAS output files
    *See* SAS log, routing output to
    *See* SAS log, writing to
FIND command
    description 791
    finding and changing text 765, 776
Find window
    description 790
    finding files 771
FIRST. variable
    description 197
    finding first observation 190
FIRSTOBS= option 103
    description 103
    pointing to first observation 95
FLOWOVER option 88
    description 88
    unexpected end of record 85

Fonts window 808
    customizing fonts 806
    description 808
    opening 806, 808
fonts, SAS Windowing Environment 806
FOOTNOTE statement
    description 641
    footnotes in procedure output 620, 623
    footnotes in reports 458
    PRINT procedure 470
footnotes
    procedure output 620, 623
    reports 458, 620, 623
    reports in SAS output files 603, 608
FOOTNOTES option
    FILE statement 610
    writing reports to SAS output files 603, 608
foreground processing 743
format attribute 262
FORMAT procedure 640
FORMAT statement
    formatting charts 573, 584, 590
    formatting dates 243
    formatting detail reports 447
    formatting report items 517
    formatting variables 724
    permanent date formats 234
    PRINT procedure 470
    reformatting SAS data set variable attributes 717
    reformatting variable attributes 717
FORMAT= option
    DEFINE statement 525
    formatting report items 517
    formatting summary tables 477
    histograms 582
    INSET statement 590
    PROC TABULATE statement 497
formats
    DATA step debugger and 424
formats, date values
    WEEKDATE29. 243
    WORDDATE18. 243
formatted input 60
    *See also* column input
    *See also* list input
    *See also* reading raw data records
    absolute column-pointer control 63
    column-pointer controls 62, 69
    creating SAS data sets 41
    definition 60
    input pointers 62, 67, 69
    mixing input styles 65
    pointer positioning 62, 69
    relative column-pointer control 63

rules for 63
sample program for 61
formatting report items 517
FORWARD command 763, 776
fractions, loss of precision 127
FRAME= option 681
FREQ option
   HBAR statement 588
   horizontal bar charts 558
frequency charts 555
   character variables 569
   creating 555
   customizing 563
   midpoints for numeric variables 564
   numeric variables 555
frequency counts 518
functions 123
   *See also* date functions
   *See also* date values
   combining 123
   INDEX 166
   LEFT 143
   ROUND 127
   SCAN 143
   SUM 127
   TRIM 143
   UPCASE 166

## G

GO command
   DATA step debugger 830
GOPTIONS statement
   description 590
   histograms 575
GRADES data set 553, 818
grand total 450
graphs
   *See* charts
   *See* plots
greater-than sign, with DATALINES statement 747
grid lines, histograms 577
GRID option
   HISTOGRAM statement 589
   histograms 577
GROUP variable 504, 510
GROUP= option
   BLOCK statement 587
   HBAR statement 587
   VBAR statement 587
grouping observations
   *See* observations, grouping

## H

HAXIS= option
   PLOT statement 548
   tick mark values 538
HBAR statement
   CHART procedure 587
   horizontal bar charts 558
HEADER statement, TEMPLATE procedure 672, 682
   customizing ODS output 672
HEADER= option 590
   FILE statement 641
   headings in specific columns 631
   histograms 582
   INSET statement 590
headings, reports 515
   *See also* titles, reports
   centering 629
   customizing 515
   in SAS output files 608
   in specific columns 631
   variables as 518
HEADLINE option 523
   column headers 515
   PROC REPORT statement 523
HEADSKIP option 523
   column headers 515
   PROC REPORT statement 523
HELP command 758
   DATA step debugger 831
help, SAS Windowing Environment
   *See* SAS Windowing Environment, help
hierarchical tables 485
hierarchical view
   *See* Tree view
HIGHLOW data set 532, 817
HISTOGRAM statement, UNIVARIATE procedure 589
   histograms 574
histograms 574
   changing axes of 577
   comparative histograms 584
   grid lines 577
   HISTOGRAM statement 574
   midpoints 580
   SAS/GRAPH software 575
   simple histograms 575
   summary statistics 582
   tick marks 577
HOFFSET= option
   HISTOGRAM statement 589
   histograms 580
horizontal bar charts 558
   statistics 558
HPCT= option
   multiple plots on same page 544

## I

I (insert) command 776
ID statement
  emphasizing key variables 439
  in detail reports 451
  PRINT procedure 470
IF statement 181
  *See also* observations, subsetting
  accepting observations 173
  combining observations 358
  deleting observations 171
  description 181, 361
  versus DELETE statement 173
IF-THEN statements
  description 165
  selecting observations 149
IF-THEN/ELSE statements
  changing observations selectively 110
  description 115, 361
IN= data set option
  merging data 307
IN= option 361
  COPY statement 736
  description 361
  moving SAS data sets and libraries 731
  observations from multiple SAS data sets 353
INCLUDE command 781
INDEX function
  description 166
  finding character strings 164
INFILE DATALINES statement 503
INFILE statement
  creating SAS data sets 44
  description 48, 88
  unexpected end of record 85
informat attribute 262
informats 60
  ampersand format modifier 64
  colon (:) format modifier 64
  creating long character variables 64
  naming conventions 61
  reading embedded blanks in list input 64
  reading special characters 60
informats, date values
  DATE7. 228, 243
  DATE9. 228, 230, 243

PROC PLOT statement 548
HPERCENT= option
  multiple plots on same page 544
  PROC PLOT statement 548
HSCROLL command 763
HTML output 648

MMDDYY10. 228, 243
MMDDYY8. 228, 243
input buffers, DATA step 35
input pointers 62, 67, 69
INPUT statement
  column input 41, 57
  defining variables 42
  description 48
  forcing a new record 79
  formatted input 41, 60
  holding records 73
  list input 41, 53
  mixed input styles 65
  multiple records per observation 77
  multiple statements 77
  reading date variables 228, 243
  reading records twice 72
  skipping data lines 80
input styles
  *See also* column input
  *See also* formatted input
  *See also* list input
  *See also* reading raw data records
  effects on line pointers 66
  mixing 65
INSET statement, UNIVARIATE procedure 589
  summary statistics in histograms 582
interactive line mode 13, 747
  *See also* line mode
  interrupting SAS sessions 748
interleaving SAS data sets
  *See* SAS data sets, interleaving
invoking SAS
  in line mode 747
item-store statement 790
iterative DO loops 220
  *See also* array processing
  *See also* DO groups

## J

JC (justify center) command 767
JJC (justify center) command 767
JJL (justify center) command 767
JJR (justify center) command 767
JL (justify left) command 767
JR (justify right) command 767
JUMP command
  DATA step debugger 831

## K

KEEP statement 103
  description 103
  keeping variables 96

KEEP= option 103
  DATA statement versus SET statement 101
  description 103
  efficiency 101
  keeping selected variables 96
  versus DROP= option 98
keeping variables
  *See* DATA statement
  *See* KEEP statement
  *See* KEEP= option
  *See* SET statement
KEYS command 760
Keys window 760
keys, SAS Registry
  definition 798
  deleting 799
  setting 799
  values, editing 800
  values, setting 800
KEYWORD statement
  TABULATE procedure 682
KILL option
  deleting SAS data library members 735
  PROC DATASETS statement 736

## L

l reports
  customizing 458
label attribute 262
LABEL option
  column labels in reports 460
  PROC PRINT statement 469
LABEL statement 470
  assigning SAS data set labels 720
  column headings in detail reports 470
  column headings in reports 460
  modifying SAS data set labels 720, 724
  plot axes labels 537, 548
  PLOT procedure 548
  PRINT procedure 470
  removing SAS data set labels 720
  variable labels in procedure output 621, 623, 641
labels
  subtotal and grand total 450
labels, SAS data sets
  *See* SAS data sets, labels
labels, summary table
  defining 491
  single for multiple elements 489
LARGEVIEW command
  description 792
  setting icon size 804
LAST. variable

  description 197
  finding last observation 190
LEFT command 763, 776
LEFT function
  aligning character values 137
  description 143
LEFT option 514
length attribute 262
LENGTH statement
  concatenating SAS data sets 271
  defining length of variables 114, 116, 144
  description 127
  length of character variables 133
  length of numeric variables 114
  loss of precision 127
  positioning 133
LEVELS= option
  BLOCK statement 587
  HBAR statement 587
  number of midpoints 566
  PIE statement 587
  VBAR statement 587
LGRID= option
  HISTOGRAM statement 589
  histograms 577
LIBNAME statement 31
  assigning librefs to SAS libraries 688
  description 48
library contents, listing 696, 697
library information, listing 696, 697
LIBRARY= option
  directory listings 702
  syntax 710
librefs 688
  assigning with LIBNAME statement 688
  assigning with SAS Windowing Environment 756
  USER, reserved name 692
line commands 776
line mode 747
line size, output reports 625
line-hold specifiers
  holding lines 600, 610
  reading raw data 72, 73, 87
  writing output lines 600, 610
line-pointer controls 80
  *See also* column-pointer controls
  *See also* pointer controls
  #n, and DATA step execution 81
  #n, skipping input variables 80
LINESIZE= option
  description 642, 750
  output line size 625, 742
LINESLEFT= option

FILE statement 641
page breaks 635
links to ODS, storing 647
LIST command
DATA step debugger 832
list input 53
*See also* column input
*See also* formatted input
*See also* reading raw data records
ampersand format modifier 64
blank delimiters 53
character delimiters 54
colon (:) format modifier 64
creating long character variables 63
creating SAS data sets 41
definition 53
delimiter character 503
embedded blanks 63
embedded special characters 63
input pointers 68
mixing input styles 65
modified list input 63
rules for 56
versus column input 58
LIST statement 377, 387
listings
*See* reports
log
*See* SAS log
LOG command 790
Log window 790
*See also* SAS Windowing Environment, windows
browsing 749
clearing 395
debugging programs 778
definition 745
description 790
SAS log output 395
saving contents of 396
LOG window
DATA step debugger 835
LOG= option
description 397
routing SAS log 394
logical operators 124
loops
*See* array processing
*See* DO groups
*See* iterative DO loops
lowercasing
*See* case, changing
LPI= option
pie charts 561
PROC CHART statement 587

# M

M (move) command 776
macro facility
*See also* SAS macro facility
DATA step debugger with 418
macro variables
ampersand, in names 467
automatic 465
customizing detail reports 465
referring to 467
user-defined 466
macros
as debugging tools 418
customized debugging commands with 418
debugging a DATA step generated by 418
MARK command 766
master data sets
definition 318
modifying, adding observations 339
modifying, from a transaction data set 338
update errors 342
updating 318
match-merging SAS data sets
*See* SAS data sets, merging (match-merge)
MAX command 764
means, charting 572
members
copying 696, 697
deleting 735
members, listing contents of
*See* CONTENTS procedure
*See* CONTENTS statement
*See* file contents, listing
MEMTYPE= option
directory listings, by member type 704
PROC DATASETS statement 710
menus, displaying 742
MERGE statement
creating SAS data sets 44
description 315
merging SAS data sets 290
missing values 328
multiple observations in a BY group 329
versus MODIFY and UPDATE statements 254
versus UPDATE statement 326
merging data
IN= data set option 307
merging SAS data sets
*See* SAS data sets, merging
midpoints

character variables, values of 569
histograms 580
numeric variables, number of 566
numeric variables, values of 563
MIDPOINTS= option
   BLOCK statement 587
   HBAR statement 587
   HISTOGRAM statement 589
   midpoints for character variables 569
   midpoints for numeric variables 563
   midpoints in histograms 580
   PIE statement 587
   VBAR statement 587
MISSING option
   CLASS statement 497
   missing values in summary tables 478
   PROC TABULATE statement 497
missing values
   customizing, with a procedure 640
   customizing, with a system option 639
   MERGE statement 328
   MODIFY statement 329
   numeric variables 121, 123
   output reports 638
   reading raw records 84
   SAS data sets 252
   summary tables 478
   UPDATE statement 328, 329
   updating SAS data sets 328, 329
missing values, in character variables
   blanks as 134
   checking for 135
   periods as 134
   setting 136
MISSING= option
   description 642
   missing values in output reports 638
MISSING= system option 639
MISSOVER option 88
   description 88
   unexpected end of record 85, 86
MM (move) command 766
MMDDYY10. informat
   description 243
   length of year 230
MMDDYY8. informat
   description 243
   length of year 230
MODIFY statement 348
   assigning SAS data set labels 720
   creating SAS data sets 44
   description 348, 723
   missing values 329, 345
   modifying SAS data set labels 720
   modifying SAS data set variable attributes 716

reformatting SAS data set variable attributes 717
removing SAS data set labels 720
renaming SAS data set variable attributes 716
versus MERGE and UPDATE statements 254
mouse, keyboard equivalents 754
MOVE option
   COPY statement 736
   moving SAS data sets and libraries 731

## N

N= option
   PROC PRINT statement 469
name attribute 262
names, data set
   *See* SAS names
naming conventions
   informats 61
   SAS language 6
   SAS names 6
   variables 6
negative operators 157
NEW option
   description 397
   routing SAS log 394
NEXT command 763
NOCENTER option
   centering output 625
   description 641
NODATE option
   date values 626
   description 642
NODMS option
   description 750
   running SAS programs 747
NODS option
   CONTENTS statement 710
   directory listings 707
NOFRAME option
   INSET statement 590
   suppressing frame on inset tables 585
NOLEGEND option
   PROC PLOT statement 548
   removing plot legends 540
noninteractive mode 12, 748
NONOTES option
   description 388
   suppressing system notes 383, 384
NONUMBER option
   description 641
   page numbering 625
NOOBS option
   PROC PRINT statement 469

suppressing observation columns 438
NOPRINT option
  PROC UNIVARIATE statement 588
  suppressing statistics tables 575
NOSOURCE option
  description 388
  suppressing SAS statements 382, 384
NOSTAT option
  HBAR statement 588
  horizontal bar charts 558
NOTEPAD command
  description 791
  opening NOTEPAD window 779
NOTEPAD editor 779
NOTEPAD window 791
  description 791
  opening 779
NOTES command
  description 791
  opening NOTEPAD window 779
NOTES option
  description 388
  suppressing system notes 383
notes, suppressing logging of 383, 384
NOTESUBMIT command 779
NOTITLES option
  FILE statement 610
  writing reports to SAS output files 603
NOVERBOSE option 808
NOWINDOWS option
  bypassing REPORT window 505
  description 524
NROWS= option 585
NUMBER option 641
NUMBERS command 760, 767
numbers, formatting in reports 517
numeric comparisons, abbreviating 159
numeric variables 117
  contents of 42
  definition 118
  embedded special characters 63
  fractions, loss of precision 127
  shortening 126
  storing efficiently 126
numeric variables, calculations on 119
  See also functions
  assignment statements, and arithmetic operators 119
  assignment statements, and numeric expressions 121
  comparing variables 124
  logical operators 124
  missing values 121, 123

## O

OBS option
  PROC PRINT statement 470
OBS= option 103
  description 103
  labeling observation columns 437
  pointing to last observation 96
observations 29
  See also SAS data sets
  See also variables
  assignment statements 109
  changing globally 109
  changing selectively 110
  conditional processing 349
  definition 29
  deleting conditionally 115, 171
  deleting duplicates 193
  variables, changing 111
  variables, creating 109
  variables, efficient use of 112
  variables, storage space for 114
observations, calculations on 201
  END= option, description 212
  END= option, determining last observation 203
  printing only totals 203
  RETAIN statement, description 212
  RETAIN statement, retaining values 209
  retaining values for later observations 209
  running totals 201
  sum statement, running totals 201
  totals for each BY group 204
  writing observations to separate data sets 206
  writing totals to separate data sets 207
observations, creating
  multiple from single DATA step 99
  multiple from single record 73
  single from multiple records 77
  testing raw data records 72
observations, from multiple SAS data sets
  See also IN= option
  calculations on last observation 359
  combining selected observations 358
  determining source data set 353
  example program 354, 360
observations, grouping 185
  See also observations, sorting
  See also observations, subsetting
  by multiple variables 187
  BY statement, basic groups 185
  BY statement, description 197
  BY statement, finding first or last observation 190

finding first or last observation 189
FIRST. variable, description 197
FIRST. variable, finding first observation 190
in descending order 188
LAST. variable, description 197
LAST. variable, finding last observation 190
SORT procedure, description 197
SORT procedure, grouping observations 186
observations, selecting
   *See* observations, subsetting
observations, sorting 192
   *See also* observations, grouping
   case sensitivity 195
   collating sequences, ASCII 196
   collating sequences, EBCDIC 196
   collating sequences, magnitude of letters 163
   deleting duplicates 193
   example 192
   NODUPRECS option, deleting duplicate records 193
   NODUPRECS option, description 197
   SORT procedure, description 197
   SORT procedure, sorting observations 192
observations, subsetting 169, 185
   *See also* DATA statement
   *See also* DELETE statement
   *See also* DROP statement
   *See also* DROP= option
   *See also* FIRSTOBS= option
   *See also* IF statement
   *See also* KEEP statement
   *See also* KEEP= option
   *See also* OBS= option
   *See also* observations, sorting
   *See also* SET statement
   all conditions true (AND) 156
   alternative actions 151
   character comparisons, case sensitivity 161
   character comparisons, types of 161
   character groups, selecting 162, 163, 164
   comparison operators 154
   complex comparisons 158
   construct conditions 154
   deleting conditionally 115, 171
   efficiency 101
   ELSE statement 151
   IF-THEN statement 149
   multiple comparisons 155
   mutually exclusive conditions 152

   negative operators 157
   numeric comparisons, abbreviating 159
   one condition true (OR) 156
   options versus statements 98
   pointing to first record 95, 103
   pointing to last record 96, 103
   simple conditions 150, 155, 173
   to SAS data sets 175
observations, writing to SAS data sets 175
   *See also* OUTPUT statement
   multiple times to one or more data sets 179
   to multiple data sets, common mistake 176
   to multiple data sets, example 175
   to separate data sets 206, 207
ODS 644
   data, definition 10, 646
   features of 615
   table templates 646
ODS _ALL_ CLOSE statement 681
ODS destinations
   definition 10, 646
   opening 647
ODS destinations, closing 647
   ODS _ALL_ CLOSE statement 681
   ODS HTML CLOSE statement 681
   ODS LISTING CLOSE statement 681
   ODS OUTPUT CLOSE statement 681
   ODS PRINTER CLOSE statement 681
   ODS RTF CLOSE statement 681
ODS EXCLUDE statement
   description 680
   excluding ODS output objects 663
ODS HTML CLOSE statement 681
ODS HTML statement 681
   description 641
ODS LISTING CLOSE statement 681
ODS LISTING statement 681
   description 641
ODS output 647
   definition 11, 647
   formats, list of 9
   formats, selecting 678
   HTML, for Web browsers 648
   PostScript output, high-resolution printers 656
   PowerPoint output, for Microsoft PowerPoint 658
   RTF output, for Microsoft Word 658
   SAS data sets, creating 665
   storing links to 647
ODS OUTPUT CLOSE statement 681
ODS output objects 646
   definition 10, 646
   excluding 663

identifying 661
selecting 663
ODS OUTPUT statement
　creating SAS data sets 665
　description 681
ODS output, customizing 667
　*See also* output, customizing
　at SAS job level 667
　style templates 667
　with table templates 670
ODS PATH statement 790
ODS PDF 681
ODS PDF statement 681
ODS POWERPOINT 681
ODS POWERPOINT statement 681
ODS PRINTER CLOSE statement 681
ODS RTF 681
ODS RTF CLOSE statement 681
ODS RTF statement 681
ODS SELECT statement
　description 681
　selecting ODS output objects 663
ODS table definitions
　definition 10
ODS table templates
　customizing ODS output 670
ODS TRACE statement
　description 682
　identifying ODS output objects 661
ODSRESULTS command
　description 791
　opening Results window 784
ODSTEMPLATE command 788
OL option 524
one-dimensional summary tables 480
one-level names 691, 692
one-to-one merging SAS data sets
　*See* SAS data sets, merging
　*See* SAS data sets, merging (one-to-one)
online help, SAS Windowing
　　Environment
　*See* SAS Windowing Environment, help
operating environment
　keyboard equivalents for mouse 794
OPTIONS command
　definition 749
　description 807
　opening SAS Options window 796
　opening SAS System Options window 808
OPTIONS procedure 807
　description 807
　listing SAS system options 750
　listing system options 796
OPTIONS statement
　customizing output 624

customizing SAS sessions 796
　description 641, 750
ORDER variable 504
ORDER= option
　CLASS statement 497, 588
　DEFINE statement 525
　ordering class variables 495, 585
　PROC TABULATE statement 497
　report layouts 508, 511
OUT.ERROR1 data set 402
OUT.ERROR2 data set 402
OUT.ERROR3 data set 402
OUT.SAT_SCORES3 data set 392
OUT.SAT_SCORES4 data set 392
OUT.SAT_SCORES5 data set 392
OUT= option
　COPY statement 736
　moving SAS data sets and libraries 731
output 8
　*See also* ODS
　*See also* reports
　*See also* SAS log, routing output to
　*See also* SAS log, writing to
　*See also* SAS Windowing Environment,
　　output
　browsing 749
　from DATA step 615
　navigating with pointers 785
　printing 790
　setting format 782
　setting format, with Preferences window 782
　setting format, with Registry Editor 782
OUTPUT command 791
Output Delivery System
　*See* ODS
output objects
　*See* ODS output objects
output routing, procedures 614
　default location, SAS Windowing
　　Environment 395
　default locations 617
　overview 614
　PRINT= option 393
　PRINTTO procedure 393
　suppressing output 393
　to dummy file 393
　to permanent file 393
　to SAS catalog entry 393
　to SAS Windowing Environment 395
output routing, summary tables 492
OUTPUT statement 175
　*See also* SAS data sets, writing
　　observations to
　description 175, 181
　MODIFY statement and 348

placement 177
output templates
   *See* Templates window
Output window 745
   *See also* SAS Windowing Environment, windows
   browsing 749
   clearing 777
   definition 745
   description 791
   example 14
   procedure output 395
   saving contents of 396
output, customizing 618
   *See also* ODS output, customizing
   centering output 625, 629
   column headings, centering 629
   column headings, in specific columns 631
   date values 626
   footnotes 620, 623
   line size 625
   missing values 638
   missing values, with a procedure 640
   missing values, with a system option 639
   page breaks 635
   page numbering 625, 633
   page size 625
   report headings, symbolic values in 633
   SAS system options for 624, 626, 641
   time values 626
   titles, adding 618
   titles, centering 629
   titles, in specific columns 631
   variable labels 621, 623
output, SAS output files
   *See* reports, SAS output files
output, traditional
   database entries 8
   external files 8
   reports 8
   SAS data sets 8
   SAS files 8
   SAS log, definition 8
   SAS log, example 8
OVERLAY option
   multiple sets of variables on same axes 546
   PLOT statement 548

# P

page breaks
   output reports 635
   reports 456

page numbering
   NONUMBER option 625
   NUMBER option 625
   output reports 625, 633
   PAGENO= option 625
page size, output reports 625
PAGE statement
   description 387
PAGE= option 681
PAGEBY statement
   grouping observations in detail reports 456
   PRINT procedure 470
PAGENO= option
   description 641
   page numbering 625
PAGESIZE= option
   description 641
   page size 625
PASTE command 766
PERCENT option
   HBAR statement 588
   horizontal bar charts 558
period (.)
   as missing value 30, 134
   in informat names 61
   in input data 60
permanent SAS data sets 32, 693
PGM command 791
pie charts 561
PIE statement, CHART procedure 587
   pie charts 561
PLOT procedure 531
   *See also* plots
   LABEL statement, description 548
   LABEL statement, plot axes labels 537
   PLOT statement, description 548
   PLOT statement, plotting multiple sets of variables 541
   PLOT statement, plotting one set of variables 548
   PLOT statement, plotting symbols 539
   PROC PLOT statement, description 548
   PROC PLOT statement, multiple plots on same page 544
   TITLE statement 549
PLOT statement, PLOT procedure
   description 548
   plotting multiple sets of variables 541
   plotting one set of variables 548
   plotting symbols 539
plots 531
   *See also* charts
   *See also* PLOT procedure
plots, customizing 537

axes labels, specifying 537
box borders 537
legends, removing 540
plotting symbols 539
tick mark values 538
titling 549
plots, multiple sets of variables
multiple plots on same page 544
multiple plots on separate pages 541
multiple sets on same axes 546
plots, one set of variables
example 536
PLOT statement 535
two-dimensional plots 536
PMENU command
description 750, 792
displaying menus 742, 759
pointer controls 604
*See also* column-pointer controls
*See also* line-pointer controls
@n 604
pointer positioning 62, 69
pointers
deleting Results pointers 786
navigating output with 785
renaming Results pointers 786
POSITION= option 585
histograms 582
INSET statement 590
PostScript output 656
PowerPoint output 658
PRECIP.RAIN data set 714, 726, 821
PRECIP.SNOW data set 714, 726, 821
Preferences window 808
customizing Explorer window 807
customizing SAS sessions 802
description 808
opening 808
setting output format 782
PREVWIND command 763
PRINT command 395, 792
PRINT procedure 469
*See also* reports
PRINT= option
description 397
routing SAS log 396
printing 774
*See also* output
*See also* PRINT procedure
*See also* REPORT procedure
*See also* reports
from SAS Windowing Environment 774, 790
output 790
PRINTTO procedure 397
description 397

routing procedure output 393
routing SAS log output 394
PROC CHART statement 586
PROC DATASETS statement
description 103, 698, 710
directory listings 702
KILL option 736
managing SAS libraries 697
PROC PLOT statement
description 548
multiple plots on same page 544
PROC PRINT statement 469
PROC REPORT statement
column width and spacing 514
description 523
PROC SORT statement 682
description 197, 471
SORT procedure 682
sorting detail reports 441
PROC TABULATE statement 497
ODS output 682
PROC TEMPLATE statement 682
PROC UNIVARIATE statement
description 588
ODS output 683
procedures 6
customizing missing values output 640
procedures, description and usage
APPEND 279
CATALOG 696
CHART 552
CONTENTS 696
COPY 696
DATASETS 698, 710
FORMAT 640
OPTIONS 807
PLOT 531
PRINT 469
PRINTTO 397
REGISTRY 808
REPORT 523
SORT 197, 682
TABULATE 492
TEMPLATE 682
UNIVARIATE 552
program data vectors 35
Program Editor 775
*See also* SAS Windowing Environment, windows
command line commands 775
creating programs 779
debugging programs 780
definition 745
description 791
editing programs 781
example 14

file shortcuts, assigning 781
line commands 776
opening programs 780
overview 775
storing programs 780
submitting programs 779
PROGRAM member type 691
programming language
  *See* SAS language
programming windows 745
programs, running
  *See* Program Editor
  *See* SAS programs, running
PUT statement 387
  description 387, 610
  reports to SAS output files 596

## Q

quality control checklist 412
QUIT command
  DATA step debugger 833
QUIT statement 698
quotation mark (') 131
  as literal character 131
  variable indicator 131

## R

raw data 28
  *See also* SAS data sets
  creating SAS data sets 44
  definition 28
  fields 29
  records 29
raw data, aligned
  *See* column input
raw data, reading
  *See* reading raw data records
raw data, unaligned
  *See* list input
RBREAK statement, REPORT procedure 526
  break lines 520
RCHANGE command 104
reading raw data records 71
  *See also* column input
  *See also* formatted input
  *See also* list input
  double trailing @ (@@) 73
  holding after reading 72
  line-hold specifiers 72, 73, 87
  missing values 84
  reading twice 72
  testing for conditions 72
  trailing @ (@) 72

unexpected end of record 84
variable-length records 84
RECALL command 763, 776
records, raw data 29
records, SAS data sets
  *See* observations
REFRESH command 792
REGEDIT command 791
  *See also* SAS Registry Editor
  *See also* SAS Registry, editing
  description 791
  editing the SAS Registry 757
  opening SAS Registry Editor window 808
  setting output formats 782
REGEDIT statement 807
registry files
  exporting 800
  importing 800
  uninstalling 801
REGISTRY procedure 808
  description 808
  editing SAS Registry 798
Registry, editing
  *See* SAS Registry, editing
relative column-pointer control 63
RENAME statement
  description 724
  renaming SAS data set variable attributes 716
  renaming SAS data sets 714
RENAME= option 308
RENAMESELS command 792
renaming
  MODIFY statement 716
  output pointers 792
  RENAME statement 714, 716
  RENAME= option 308
  results pointers 786
  SAS data set variable attributes 716
  SAS data sets 714
REPLACE statement 348
REPORT procedure 523
  BREAK statement 524
  COLUMN statement 524
  DEFINE statement 525
  PROC REPORT statement 523
  RBREAK statement 526
report writing tools 502
reports 501
  *See also* ODS
  *See also* output
  *See also* output, customizing
  *See also* PRINT procedure
  *See also* REPORT procedure
  *See also* summary reports

ACROSS variable 503, 518, 519
adding blank lines 462
ANALYSIS variable 503, 510
break lines 520
cells 518
column alignment 514
column headers, customizing 515
column headers, variables as 518
column labels, defining 437, 460
column labels, multi-line 461
column layout 507
column spacing 514
column width 514
columns, ordering 507
COMPUTED variable 503
constructing 502
cross-tabulation 519
DISPLAY variable 504
displaying all variables 505
footnotes 458, 459
formatting report items 517
frequency counts 518
from DATA step 596
group subtotals, computing for multiple variables 453
group subtotals, computing for single variables 449
group summaries 521
group totals, computing 455
GROUP variable 504, 510
headings, symbolic values in 633
layout, adjusting 514
layout, constructing 503
log messages, printing 395
numbers, formatting 517
observation columns, suppressing 438
observations, consolidating 510
observations, grouping by page 456
observations, selecting (multiple comparisons) 445
observations, summarizing 520
ORDER variable 504
page breaks 456
reporting selected variables 442
row layout 508, 511
rows, ordering 508, 511
showing all variables 436
titles 458, 459
traditional output 8
types of 502
using macro variables in 466
reports, SAS output files 596
calculating totals 607
character strings 597
data values 604
designing report layout 604

footnotes 608
headings 608
line size 742
numeric data values, formatting 605
printing values by group 606
PUT statement 596
releasing held lines 601
rewriting on same line 600
routing lines to 603
variable values 599
RESET command 767
Results window 791
Contents Only view 785
customizing 784
definition 745
description 791
Explorer view 786
navigating output 785
opening 784
output pointer items, viewing 787
results pointers, deleting 786
results pointers, renaming 786
saving to other formats 786
Tree view 785
window properties, viewing 787
working with output 784
RETAIN statement
description 212
retaining values 209
RFIND command 104
RIGHT command 763, 776
RIGHT option
column alignment 514
DEFINE statement 526
right-clicking, keyboard equivalent 754
ROUND function
description 127
rounding numbers 122
rounding numbers 122, 127
routing output, SAS log
See SAS log, routing output to
rows, raw data 29
rows, reports
layout 508, 511
ordering 508, 511
rows, SAS data sets
See observations
RTF output 658
RUN statement
description 15, 750
interactive line mode 13

## S

SAS libraries
definition 688

SAS catalogs
  definition 691
  SAS/ACCESS files 691
SAS command
  starting noninteractive mode 748
  starting SAS sessions 742
SAS data files 690
  See also SAS files
  definition 690
  examples 690
SAS data libraries
  See also SAS data sets
  directory listings, all files 702
  directory listings, by member type 704
  directory listings, definition 702
  exploring with SAS Windowing Environment 755
  file contents listing, all data sets 707
  file contents listing, one data set 705
  finding expressions in 790
  formatting contents listings 708
  library assignment problems 757
  WORK 32
SAS data libraries, assigning librefs with SAS Windowing Environment 756
SAS data libraries, moving 731
  selected data sets 733
  whole libraries 731
SAS data set columns
  See variables
SAS data set names
  See SAS names
SAS data set rows
  See observations
SAS data sets 91
  See also observations
  See also raw data
  See also SAS libraries
  See also variables
  bypassing 596
  data values 4
  definition 4, 690
  function of 28
  permanent 32, 693
  raw data, definition 28
  referencing in SAS libraries 691
  renaming 714
  storing in SAS libraries 690
  structure of 29
  temporary 32, 692
SAS data sets, combining
  See SAS data sets, concatenating
  See SAS data sets, interleaving
  See SAS data sets, merging
  See SAS data sets, modifying
  See SAS data sets, updating

SAS data sets, concatenating 257
  See also SAS data sets, interleaving
  See also SAS data sets, merging
  See also SAS data sets, modifying
  See also SAS data sets, updating
  definition 250
SAS data sets, concatenating with APPEND procedure
  APPEND procedure, description 274, 279
  APPEND procedure, versus SET statement 278
  variable attributes are different 277
  variables and attributes are the same 274
  variables are different 275
SAS data sets, concatenating with SET statement
  SET statement, description 258, 279
  SET statement, versus APPEND procedure 278
  variable attributes are different 262
  variable formats are different 266
  variable informats are different 266
  variable labels are different 266
  variable lengths are different 271
  variable types are different 263
  variable types, changing 264
  variables are different 261
  variables are the same 258
SAS data sets, contents information
  DATASETS procedure 704
  formatting contents listings 708
  listing all data sets 707
  listing one data set 705
SAS data sets, copying 727
  duplicate names 727
  from other libraries 729
  from procedure input library 727
  selecting data sets for 730
SAS data sets, creating
  column input 41
  data locations 44
  formatted input 41
  from DBMS files 45
  from external files 44, 45
  from other SAS data sets 44
  from raw data in the job stream 44
  input styles 41
  list input 41
  variables, defining 42
  with ODS 665
  year values, two-digit versus four-digit 42
SAS data sets, deleting 734
  confirmation of deletion 734

specific files  734
whole libraries  735
SAS data sets, interleaving  281
   *See also* SAS data sets, concatenating
   *See also* SAS data sets, merging
   *See also* SAS data sets, modifying
   *See also* SAS data sets, updating
   BY statement  285
   BY-group processing  282
   definition  250
   process overview  284
   SET statement  285
   sorting data for  282
SAS data sets, labels  720
   assigning  720
   modifying  720
   removing  720
SAS data sets, merging  289
   *See also* SAS data sets, concatenating
   *See also* SAS data sets, interleaving
   *See also* SAS data sets, modifying
   *See also* SAS data sets, updating
   definition  251
   MERGE statement  290
   versus updating and modifying  254
SAS data sets, merging (match-merge)  251
   BY statement with  296
   definition  251
   example program  294
   multiple observations in a BY group  300
   versus one-to-one merge  310
   when to use  313
   with common variables  308
   with dropped variables  305
   without common variables  308
SAS data sets, merging (one-to-one)  251
   definition  251
   different number of observations  290
   different variables  290
   example program  292
   same number of observations  290
   same variables  293
   versus match-merge  310
   when to use  312
SAS data sets, modifying  335
   *See also* MODIFY statement
   *See also* SAS data sets, concatenating
   *See also* SAS data sets, interleaving
   *See also* SAS data sets, merging
   *See also* SAS data sets, updating
   checking for program errors  339
   definition  253
   duplicate BY variables  343
   example program  339, 343

   master data sets, from transaction data sets  338
   master data sets, update errors  342
   master data sets, with network observations  339
   missing values  345
   versus updating and merging  254
SAS data sets, moving  731
   selected data sets  733
   whole libraries  731
SAS data sets, output to
   *See also* ODS
   traditional output  8
SAS data sets, specifying for input
   *See* DATA= option
SAS data sets, subsetting
   *See* observations, subsetting
SAS data sets, updating  317
   *See also* SAS data sets, concatenating
   *See also* SAS data sets, interleaving
   *See also* SAS data sets, merging
   *See also* SAS data sets, modifying
   definition  251
   example  319
   master data sets  318
   missing values  252, 328, 329
   selecting BY variables  318
   transaction data sets  318
   UPDATE statement, description  318
   versus merging  254, 326
   versus modifying  254
   with incremental values  324
SAS data sets, used in this book
   CITY  814
   CLIMATE.HIGHTEMP  820
   CLIMATE.LOWTEMP  820
   GRADES  818
   HIGHLOW  817
   PRECIP.RAIN  821
   PRECIP.SNOW  821
   SAT_SCORES  815
   STORM.TORNADO  821
   USCLIM.BASETEMP  820
   USCLIM.HIGHTEMP  819
   USCLIM.HURRICANE  819
   USCLIM.LOWTEMP  819
   USCLIM.REPORT  820
   USCLIM.TEMPCHNG  820
   YEAR_SALES  479, 816
SAS data sets, used in this documentation
   CITY  92
   CLIMATE.HIGHTEMP  714, 726
   CLIMATE.LOWTEMP  714, 726
   GRADES  553
   HIGHLOW  532
   OUT.ERROR1  402

## Index

OUT.ERROR2 402
OUT.ERROR3 402
OUT.SAT_SCORES3 392
OUT.SAT_SCORES4 392
OUT.SAT_SCORES5 392
PRECIP.RAIN 714, 726
PRECIP.SNOW 714, 726
SAT_SCORES 366
STORM.TORNADO 714, 726
USCLIM.BASETEMP 702, 714, 726
USCLIM.HIGHTEMP 702, 714, 726
USCLIM.HURRICANE 702, 714, 726
USCLIM.LOWTEMP 702, 714, 726
USCLIM.REPORT 702, 714, 726
USCLIM.TEMPCHNG 702, 714, 726
YEAR_SALES 434, 504
SAS data sets, variable attributes
  assigning 717
  modifying 716
  reformatting 717
  removing 717
  renaming 716
SAS data sets, writing observations to
  See observations, writing to SAS data sets
SAS data views 690
SAS date constants
  See date functions
  See date values
SAS date values
  See date functions
  See date values
SAS Explorer 696
SAS files 690
  definition 690
  in SAS libraries 690
  SAS data files 690
SAS files, output to
  See ODS output
  See output
  See SAS Windowing Environment, output
SAS functions
  See functions
SAS language 5
  case sensitivity 6
  elements of 5
  naming conventions 6
SAS libraries 688
  accessing 688
  catalog management 696
  copying files or members 696, 697
  file management 696
  library contents, listing 696, 697
  library information, listing 696, 697
  locating 688

  managing 695
  referencing SAS data sets 691
  SAS Explorer 696
  storing files in 690
  storing SAS data sets 690
SAS libraries, assigning librefs with LIBNAME statement 688
SAS log 366
  See also debugging
  See also debugging, with SAS Supervisor
  browsing 749
  line size 742
  locating 369
  printing 395
  printing from SAS Windowing Environment 395
  resolving errors with 368
  role of 366
SAS log, routing output to 394
  See also SAS log, writing to
  configuration file 397
  default location, batch environment 396
  default location, changing 396
  default location, restoring 395
  default location, SAS Windowing Environment 395
  LOG= option 394
  LOG= system option 396
  NEW option 394
  PRINT= system option 396
  PRINTTO procedure 394
  to alternate location 394
  to SAS Windowing Environment 395
SAS log, suppressing
  error messages 383, 384
  ERRORS= option 383, 384
  example program for 384
  NONOTES option 383, 384
  NOSOURCE option 382, 384
  NOTES option 383
  SAS statements 382, 384
  SAS system options for 382
  SOURCE option 382
  system notes 383, 384
SAS log, writing to 595
  See also ODS
  See also SAS log, routing output to
  bypassing the SAS data set 596
  LIST statement 377
  PUT statement 375, 596
  traditional output, definition 8
  traditional output, example 8
SAS macro facility 465
  definition 465
  macro variables, automatic 465

macro variables, referring to 467
macro variables, user-defined 466
SAS macro language 471
SAS names 6
　blanks in 6
　naming conventions 6
　underscore (_) in 6
SAS Options window
　customizing system options 796
　finding system options 797
　opening 796, 797
　setting system options 797
SAS output files
　*See* reports, SAS output files
SAS procedures
　*See* procedures
SAS programs, running 11
　*See also* Program Editor
　*See also* SAS Windowing Environment
　background processing 744
　batch mode 12, 748
　foreground processing 743
　from NOTEPAD 779
　in SAS Windowing Environment 744
　interactive line mode 13, 747
　noninteractive mode 12, 749
　one line at a time 13
　SAS/ASSIST software 12
　selecting an approach 11, 743
SAS Registry Editor 798
　*See also* REGEDIT command
　*See also* SAS Registry, editing
　customizing 801
　description 791, 808
　editing the SAS Registry 757
　finding information in 799
　opening 799, 808
SAS Registry, editing 798
　*See also* REGEDIT command
　*See also* SAS Registry Editor
　key values, editing 800
　key values, setting 800
　keys, definition 798
　keys, deleting 799
　keys, setting 799
　overview 798
　registry files, exporting 800
　registry files, importing 800
　registry files, uninstalling 801
　subkeys 798
SAS sessions 742
　*See also* SAS Windowing Environment, SAS sessions
　interrupting, in line mode 748
　starting in host operating environment 742

SAS sessions, customizing 795
　at startup 795
　executing SAS statements automatically 796
　invocation-only options 795
　OPTIONS statement 796
　setting system options 796
　viewing system options 796
　with Preferences window 802
　with system options 796
SAS sessions, customizing session-to-session 798
　saving/retrieving system options 802
　with SAS Registry Editor 798
SAS statements
　*See* statements
SAS Supervisor
　*See* debugging, with SAS Supervisor
SAS System
　base software components 4
　data analysis utilities 6
　data management facility 4
　definition 3
　invoking in line mode 747
SAS system options
　customizing missing values output 639
　finding 797
　listing 796
　logging 808
　retrieving 802
　saving 802
　setting 796, 797
　viewing 796, 808
SAS System Options window
　description 808
　opening 808
SAS text editor 766
　case, changing 769
　case, setting default for 768
　column numbers, displaying 767
　combining text 770
　cutting, pasting, and copying 765
　finding and changing text 765
　justifying text 767
　line numbers, displaying 767
　moving text 766
　rearranging text 766
　separating text 770
SAS Windowing Environment 11
　*See also* SAS sessions
　*See also* Templates window
　command line commands 758
　command types 758
　definition 11
　deselecting items 791

Index **885**

function keys, assigning commands to 760
icons, large 792
icons, small 792
invoking 746
item details, toggling on and off 791
keyboard equivalents to mouse actions 754
library assignment problems 757
line commands 760
line numbers, toggling on and off 760
listing output, deleting 791
moving up one level 792
output pointers, renaming 792
printing listings 792
printing log messages 395
pull-down menus 759
refreshing contents 792
running programs 744
SAS windows, customizing 764
SAS windows, managing 763
SAS windows, opening 761
SAS windows, scrolling 763
selecting items 792
system option settings, loading 791
system option settings, saving 791
Tree view, expanding 791
Tree view, toggling on and off 744, 792
SAS Windowing Environment, customizing 802
*See also* Explorer window, customizing
editors 806
fonts 806
setting preferences 807
SAS Windowing Environment, editing programs
*See* Program Editor
SAS Windowing Environment, editing SAS Registry
*See* SAS Registry Editor
SAS Windowing Environment, editing text
*See* SAS text editor
SAS Windowing Environment, file management 770
file shortcuts, assigning 772
file shortcuts, modifying 774
file-specific commands, issuing 772
finding files 770
opening files 772
printing files 774
SAS Windowing Environment, help 758
online help system 758
window help 758
SAS Windowing Environment, output 781
*See also* Log window

*See also* Results window
*See also* Templates window
default viewers, assigning 783
format, setting with Preferences window 782
format, setting with SAS Registry Editor 782
overview 781
SAS Windowing Environment, SAS sessions 746
ending 746
example 13
interrupting 746
issuing host commands from 746
starting 746
SAS Windowing Environment, windows 761
*See also* Explorer window
*See also* Fonts window
*See also* Log window
*See also* Output window
*See also* Preferences window
*See also* Program Editor
*See also* Results window
*See also* SAS Registry Editor
calling windows 761
Command window, executing commands from 758
Contents pane, toggling on and off 791
Editor Options window, description 808
Editor window 775
Explorer Options window, description 808
File Shortcut Assignment window 772, 790
Find window, description 790
Find window, finding files 771
Keys window, keyboard assignments 760
navigating to first available output 791
NOTEPAD window, description 791
NOTEPAD window, opening 779
programming windows 745
SAS Options window, customizing system options 796
SAS Options window, opening 796
SAS System Options window, description 808
SAS System Options window, opening 808
SASColor window, description 808
SASColor window, opening 808
Templates window, description 791
SAS windows 761
customizing 764
managing 763

opening 761
scrolling 775
SAS/ASSIST software 12
SASCOLOR command
  customizing windows 764
  opening SASColor window 808
SASCOLOR statement 807
SASColor window
  description 808
  opening 808
SAT_SCORES data set 366, 815
SAVE statement 736
SCAN function
  description 143
  saving storage space 139
  scanning for character strings 137
scrolling windows 775
  BACKWARD command 763, 775
  BOTTOM command 763, 775
  CURSOR command 764
  FORWARD command 763, 776
  HSCROLL command 763
  LEFT command 763, 776
  MAX command 764
  RIGHT command 763, 776
  TOP command 763, 775
  VSCROLL command 763
search and replace 776
SELECT statement
  copying SAS data sets 731
  description 736
  moving SAS data sets 733
SELECT_ALL command 792
selecting observations
  *See* observations, subsetting
semantic errors 402
semicolon (;)
  end-of-data indicator 44
  in statements 6
SET command
  DATA step debugger 834
SET statement 212
  combining observations 358
  concatenating SAS data sets 258
  creating SAS data sets 44
  description 212, 258, 361
  determining last observation 203, 359
  determining source of observations 353
  interleaving SAS data sets 285
  keeping selected variables 96
  versus APPEND procedure 278
  versus DATA statement 101
shift left command 767
shift right command 767
SHORT option
  CONTENTS statement 710

formatting contents listings 708
shortcuts
  *See* file shortcuts
SKIP option
  blank lines, inserting 522
  BREAK statement 524
  RBREAK statement 526
SKIP statement
  description 388
skipping input variables 59, 80
slash (/), column-pointer control
  description 69, 87
  forcing pointer to next line 79
slash (/), splitting column headers 516
SMALLVIEW command
  description 792
  setting icon size 804
SORT procedure 197
  BY statement 682
  description 197, 471
  grouping observations 186
  sorting detail reports 441
  sorting observations 192
sorting date values 237
sorting observations
  *See* observations, sorting
sorting sequences
  *See* collating sequences
SOURCE option
  description 388
  suppressing SAS statements 382
SOURCE window
  DATA step debugger 835
SPACING= option
  column spacing 514
  DEFINE statement 526
  PROC REPORT statement 523
special characters, reading
  *See* informats
SPLIT= option
  multi-line column labels 461
  PROC PRINT statement 470
  PROC REPORT statement 524
square brackets, in STYLE= option 493
statements 5
  case sensitivity 6
  executing automatically at startup 796
  line continuation 6
  rules for writing 5
  semicolon (;) in 6
  suppressing logging of 382, 384
statements, submitting
  *See* SAS programs, running
STEP command
  DATA step debugger 834
STOPOVER option 88

description 88
unexpected end of record 85
storage space, defining for variables 114
storage space, saving
  SCAN function 139
  treating numbers as variables 142
STORE command 766
STORM.TORNADO data set 714, 726, 821
STYLE option
  PROC PRINT statement 470
style templates 667
STYLE= option
  in dimension expressions 493
  ODS PRINTER statement 667
styles, summary table 492
SUBGROUP= option
  BLOCK statement 587
  HBAR statement 587
  subgroups within ranges 570
  VBAR statement 587
subkeys, SAS Registry 798
SUBMIT command 776
subsetting observations
  See observations, subsetting
subtotal 450
subtraction 212
SUM function
  description 127
  summing numbers 123
sum statement
  description 212
  running totals 201
SUM statement
  PRINT procedure 470
  totals in detail reports 448
SUMBY statement
  computing group totals 455
  PRINT procedure 471
SUMLABEL option
  PROC PRINT statement 470
SUMMARIZE option
  BREAK statement 524
  RBREAK statement 526
  summary lines 522
summary reports 501
  See also reports
  creating 510, 521
  definition 503
summary tables 474
  analysis variables, specifying 477
  class variables, missing values 478
  class variables, ordering 495
  class variables, specifying 477
  combining elements 485, 488
  concatenating elements 488

cross-tabulation 474, 485
crossing elements 474, 485
defining structure of 477
definition 474
descriptive statistics, calculating 487
dimension expressions 477
formatting output 487
input data sets, specifying 477
labels, defining 491
labels, single for multiple elements 489
missing values 478
output destination 492
reducing code 489
reporting on subgroups 485
styles 492
summaries for all variables 490
summary tables, creating
  hierarchical tables 485
  multiple tables per PROC TABULATE step 484
  one-dimensional 480
  three-dimensional 482
  two-dimensional 481
summing numbers 123, 127
  See also numeric variables, calculations on
  See also observations, calculations on
summing numeric variables 448
SUMVAR= option
  BLOCK statement 587
  charting means 572
  HBAR statement 587
  PIE statement 587
  VBAR statement 587
SUPPRESS option 524
SWAP command
  DATA step debugger 835
SYNCOLOR command 764
SYNCONFIG command 764
SYNCONFIG statement 807
syntax checking 400
syntax errors
  diagnosing 402
SYSDATE9 automatic macro variable
  dates in detail reports 465
  description 471
system notes, suppressing logging of 383, 384

# T

table definitions (ODS)
  See also Templates window
  definition 10
TABLE statement
  TABULATE procedure 682

TABLE statement, TABULATE procedure 498
  defining summary table structure 477
  restrictions 478
table templates (ODS) 670
  customizing ODS output 670
  definition 646
tables
  *See* summary tables
TABULATE procedure 497
  CLASS statement 477, 497
  CLASS statement, description 682
  KEYLABEL statement 682
  KEYWORD statement 682
  KEYWORD statement, description 682
  PROC TABULATE statement 497, 682
  required statements 476
  TABLE statement 477, 478, 498
  VAR statement 477, 498, 682
TC (text connect) command 770
TEMPLATE procedure 682
  COLUMN statement, customizing ODS output 672
  COLUMN statement, description 682
  DEFINE statement, customizing ODS output 672
  DEFINE statement, description 682
  DELETE statement 682
  END statement, customizing ODS output 672
  END statement, description 682
  HEADER statement, customizing ODS output 672
  HEADER statement, description 682
  PROC TEMPLATE statement 682
  setting template information 792
  source code, browsing 789
  source code, editing 789
Templates window 787
  *See also* SAS Windowing Environment, windows
  *See also* table templates (ODS)
  Contents Only view 789
  Contents pane 788
  customizing 788
  description 791
  Explorer view 788
  opening 787
  overview 787
  printing output 790
  source code, browsing 789
  source code, editing 789
  Tree view 788
  window properties, viewing 790
temporary SAS data sets 692
text 766

changing case 768
combining and separating 770
moving and rearranging 766
SAS text editor 766
search and replace 776
text editing
  *See* NOTEPAD editor
  *See* SAS text editor
three-dimensional charts 573
three-dimensional summary tables 482
tick marks, histograms 577
time values, in reports 465, 626
TITLE statement
  as global statement 496
  centering output titles 629
  description 641
  output titles 618, 623
  plot titles 549
  PRINT procedure 471
  REPORT procedure 526
  report titles 458, 465
  titling plots 549
titles, plots 549
titles, reports 458
  *See also* headings, reports
  centering 629
  creating 459, 618, 623
  in specific columns 631
  overview 458
  symbolic values in 465
TODAY() function
  description 243
TOP command 763, 775
totaling variables
  *See* observations, calculations on
  *See* summing numbers
TRACE command
  DATA step debugger 836
trailing @
  description 87
  reading raw data records 72
  releasing held output lines 601
  writing output lines 600, 610
transaction data sets 318
TREE command 792
  *See also* AUTOEXPAND command
  description 792
  Tree view, toggling on and off 744
Tree view 744, 785
  expanding 791
  toggling on and off 744, 792
TRIM function
  adding characters 141
  description 143
troubleshooting
  *See* debugging

*See* debugging, with SAS Supervisor
truncation
  *See also* FLOWOVER option
  *See also* MISSOVER option
  *See also* STOPOVER option
  *See also* TRUNCOVER option
  character variables 132
  controlling 85
TRUNCOVER option 88
  description 88
  unexpected end of record 85, 87
TS (text split) command 770
TURNVLABELS option 585
two-dimensional summary tables 481
two-level names 691, 693
type attribute 262
TYPE= option
  BLOCK statement 587
  charting means 572
  HBAR statement 587
  PIE statement 587
  VBAR statement 587

## U

unaligned raw data
  *See* list input
underscore, in SAS names 6
UNDO command 776
UNIVARIATE procedure 552
  BY statement 683
  CLASS statement, comparative histograms 584
  CLASS statement, description 588, 683
  HISTOGRAM statement, description 589
  HISTOGRAM statement, histograms 574
  INSET statement, description 589
  INSET statement, summary statistics in histograms 582
  PowerPoint output, for Microsoft PowerPoint 658
  PROC UNIVARIATE statement 588, 683
  RTF output, for Microsoft Word 658
  VAR statement 683
UPCASE function
  converting characters to uppercase 161
  description 166
UPDATE statement
  creating SAS data sets 44
  description 318, 333
  missing values 328, 329
  multiple observations in a BY group 329

  versus MERGE and MODIFY statements 254
  versus MERGE statement 326
UPDATEMODE= option
  description 348
  modifying SAS data sets 345
  updating SAS data sets 329
updating SAS data sets 251
  *See also* SAS data sets, updating
UPLEVEL command
  description 792
  navigating Explorer window 770
  navigating Results window 786
  navigating Templates window 788
uppercasing
  *See* case, changing
USCLIM.BASETEMP data set 702, 714, 726, 820
USCLIM.HIGHTEMP data set 702, 714, 726, 819
USCLIM.HURRICANE data set 702, 714, 726, 819
USCLIM.LOWTEMP data set 702, 714, 726, 819
USCLIM.REPORT data set 702, 714, 726, 820
USCLIM.TEMPCHNG data set 702, 714, 726, 820

## V

VAR statement
  PRINT procedure 471
  reporting selected variables 442
  SORT procedure 682
  specifying summary table analysis variables 477
  TABULATE procedure 498, 682
  UNIVARIATE procedure 683
variable attributes 262
variable-length records, reading 84
variables 29
  *See also* character variables
  *See also* numeric variables
  *See also* observations
  *See also* SAS data sets
  attributes 262
  changing 111
  comparing 124
  creating 109
  defining 42
  defining length of 114
  definition 29
  efficient use of 112
  naming conventions 6
  storage space for 114

VARNUM option
  CONTENTS statement 710
  formatting contents listings 708
VAXIS= option
  HISTOGRAM statement 589
  histograms 579
  PLOT statement 548
  tick mark values 538
VAXISLABEL= option 580
VBAR statement, CHART procedure 587
  vertical bar charts 556
VERBOSE option
  customizing SAS sessions 795
  description 750, 808
vertical bar charts 556
  creating 556
  midpoint values 563
  number of midpoints 566
vertical bars, concatenation operator 139
VIEW member type 690
views 690
VMINOR= option
  HISTOGRAM statement 589
  histograms 577
VPCT= option
  multiple plots on same page 544
  PROC PLOT statement 548
VPERCENT= option
  multiple plots on same page 544
  PROC PLOT statement 548
VSCALE= option
  HISTOGRAM statement 589
  histograms 579
VSCROLL command 763

# W

WATCH command
  DATA step debugger 836
WEEKDATE29. format
  description 243
  displaying dates 233
WEEKDAY function
  description 243
  returning day of the week 239
WHERE statement
  case sensitivity 444
  PRINT procedure 471
  printing reports 444
  REPORT procedure 527
  selecting report data 505
WIDTH= option
  column width 514
  DEFINE statement 526
  PROC PRINT statement 470

window help 758
WINDOWS option 524
windows, SAS
  *See* SAS windows
windows, SAS Windowing Environment
  *See* SAS Windowing Environment, windows
WORDDATE18. format
  description 243
  displaying dates 233
WORK library 32
writing
  *See* ODS
  *See* output
writing reports
  *See* PRINT procedure
  *See* REPORT procedure
  *See* reports
writing to output files
  *See* DATA step
  *See* PUT statement
  *See* reports, SAS output files
writing to SAS log
  *See* PUT statement
  *See* SAS log, writing to

# X

X command
  description 750
  interrupting interactive line mode 748
  interrupting SAS sessions 746
  issuing commands from host environment 746
X statement
  description 750
  interrupting interactive line mode 748

# Y

year values, two-digit versus four-digit 42, 227, 230
  *See also* date functions
  *See also* date values
YEAR_SALES data set 434
  creating 816
  using 479, 504
YEARCUTOFF= system option 244
  description 244
  determining century 42, 227

# Z

ZOOM command 763, 776

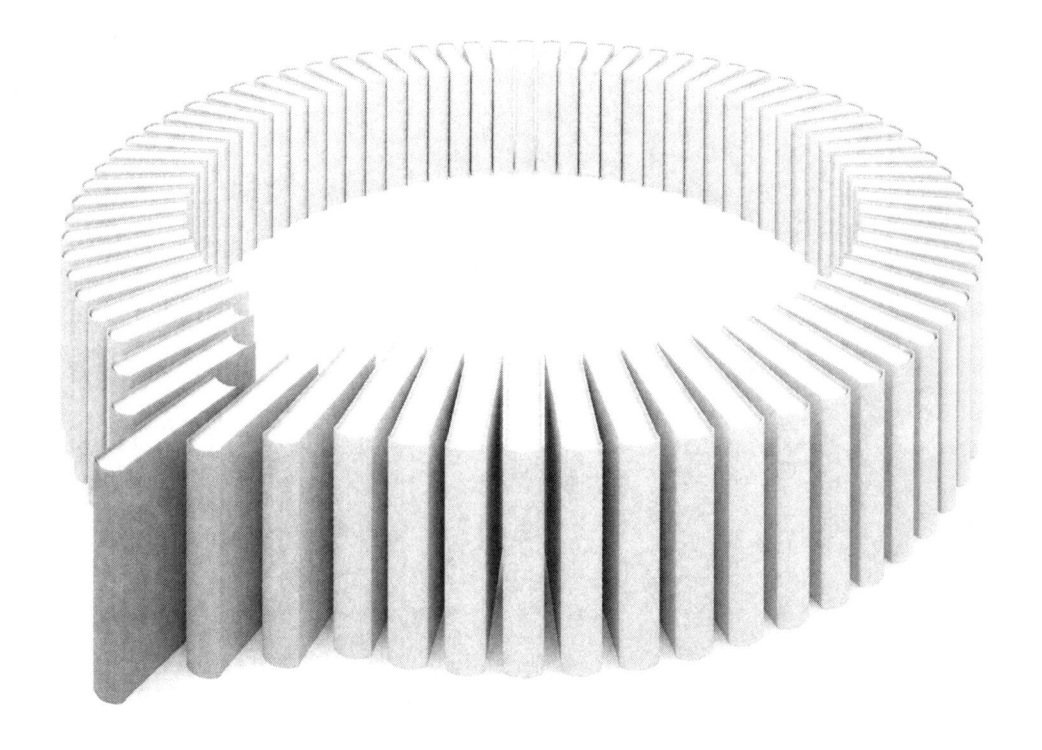

# Gain Greater Insight into Your SAS® Software with SAS Books.

Discover all that you need on your journey to knowledge and empowerment.

support.sas.com/bookstore
for additional books and resources

**§.sas**
THE POWER TO KNOW.

CPSIA information can be obtained
at www.ICGtesting.com
Printed in the USA
LVOW09s0743090518
576552LV00004B/214/P